Pure Mathematics 1

Also by the same authors:

APPLIED MATHEMATICS I
APPLIED MATHEMATICS II
PURE MATHEMATICS II
MATHEMATICS – THE CORE COURSE FOR A-LEVEL
FURTHER PURE MATHEMATICS
MATHEMATICS – MECHANICS AND PROBABILITY
FURTHER MECHANICS AND PROBABILITY

Pure
Mathematics I

L. Bostock, B.Sc.
formerly Senior Mathematics Lecturer
Southgate Technical College

S. Chandler, B.Sc.
formerly of the Godolphin and Latymer School

Stanley Thornes (Publishers) Ltd.

First published in 1978 by Stanley Thornes (Publishers) Ltd.,
Old Station Drive, Leckhampton,
CHELTENHAM GL53 0DN

Reprinted 1979
Reprinted 1980
Reprinted 1981
Reprinted 1982 (twice)
Reprinted 1983
Reprinted 1984 (twice)
Reprinted 1985 with minor corrections
Reprinted 1987

ISBN 0 85950 092 6

Typeset at the Alden Press
Oxford London and Northampton
Printed and bound in Great Britain at The Bath Press, Avon.

PREFACE

Recent developments in the Mathematics syllabuses of many boards
at advanced level have created a need for a new approach to Pure Mathematics.

This book, the first of two volumes, together with *Applied Mathematics
Volume I*, covers the work necessary for a traditional single subject
Mathematics course at advanced level. The second volume completes coverage of
an advanced level course in Pure Mathematics, and together with *Applied
Mathematics Volume II*, provides a course for Further Mathematics at advanced
level.

We have incorporated many modern approaches to mathematical
understanding whilst keeping the best of the traditional methods by making a
feature of many and varied worked examples and exercises to illustrate each
new concept at each stage of its development. The student will find stimulation
in examples to be found in the miscellaneous exercises at the end of most
chapters, some of which are suitable for the student of 'special paper' calibre.

The starting point for an advanced level text on Pure Mathematics always
presents problems, especially as there are so many varied syllabuses at ordinary
level. We have assumed knowledge only of a 'common core' syllabus, beginning
most topics from first principles.

The teaching order presents another problem. The arrangement of topics in
this book forms a logical sequence suitable for a student working on his own.
However the course teacher will find considerable flexibility.

There are many people who have helped us in the writing of this book and
we would like to thank all of them. In particular we are very grateful to
Miss J. Broughton for her helpful criticism and correction of the text and to
Mr. C. Eva for undertaking the mammoth task of working all the exercises. We
would also like to thank Miss E. Clarke for typing the manuscript.

Finally we are grateful to the following Examination boards for permission
to reproduce questions from their past examination papers (part questions are
indicated by the suffix p):

University of London   (U of L)
Joint Matriculation Board   (J M B)
University of Cambridge Local Examinations Syndicate   (C)
Oxford Delegacy of Local Examinations   (O)
The Associated Examining Board   (A E B)

L. Bostock
S. Chandler

CONTENTS

NOTES ON USE OF THE BOOK

Notation

$=$	is equal to
$\equiv$	is identical to
$\simeq$	is approximately equal to*
$>$	is greater than
$\geqslant$	is greater than or equal to
$<$	is less than
$\leqslant$	is less than or equal to
∞	infinitely large
$\rightarrow$	approaches
$\Rightarrow$	implies
$\Leftarrow$	is implied by
$\Longleftrightarrow$	implies and is implied by

A stroke through any of the above symbols negates it. i.e. $\neq$ means 'is not equal to', $\not>$ means 'is not greater than'.

Abbreviations

$\parallel$	parallel
+ve	positive
−ve	negative
w.r.t.	with respect to

*Practical problems rarely have exact answers. Where numerical answers are given they are correct to two or three decimal places depending on their context, e.g. π is 3.142 correct to 3 d.p. and although we write $\pi = 3.142$ it is understood that this is not an exact value. We reserve the symbol $\simeq$ for those cases where the approximation being made is part of the method used.

Instructions for answering Multiple Choice Exercises

These exercises are at the end of most chapters. The questions are set in groups, each group representing one of the variations that may arise in examination papers. The answering techniques are different for each type of question and are classified as follows:

TYPE I

These questions consist of a problem followed by several alternative answers, only *one* of which is correct.

Write down the letter corresponding to the correct answer.

TYPE II

In this type of question some information is given and is followed by a num-number of possible responses. *One or more* of the suggested responses follow(s) directly and necessarily from the information given.

Write down the letter(s) corresponding to the correct response(s).

e.g. PQR is a triangle

(a) $\hat{P} + \hat{Q} + \hat{R} = 180°$

(b) PQ + QR is less than PR

(c) if $\hat{P}$ is obtuse, $\hat{Q}$ and $\hat{R}$ must both be acute.

(d) $\hat{P} = 90°$, $\hat{Q} = 45°$, $\hat{R} = 45°$.

The correct responses are (a) and (c).

(b) is definitely incorrect and (d) may or may not be true of triangle PQR, i.e. it does not follow directly and necessarily from the information given. Responses of this kind should not be regarded as correct.

TYPE III

Each problem contains two independent statements (a) and (b).

1) If (a) always implies (b) but (b) does not always imply (a)	write A.
2) If (b) always implies (a) but (a) does not always imply (b)	write B.
3) If (a) always implies (b) *and* (b) always implies (a)	write C.
4) If (a) denies (b) and (b) denies (a)	write D.
5) If none of the first four relationships apply	write E.

TYPE IV

A problem is introduced and followed by a number of pieces of information. You are not required to solve the problem but to decide whether:

1) the given information is *all* needed to solve the problem. In this case write A;

2) the total amount of information is insufficient to solve the problem. If so write I;

3) the problem can be solved without using one or more of the given pieces of information. In this case write down the letter(s) corresponding to the items not needed.

TYPE V

A single statement is made. Write T if it is true and F if it is false.

CHAPTER 1

ALGEBRAIC RELATIONSHIPS

THE NATURE OF ALGEBRAIC EXPRESSIONS

Consider the expressions
$$(x + 2)^2 = 2x + 7 \qquad [1]$$
$$(x + 2)^2 = x^2 + 4x + 4 \qquad [2]$$
$$(x - 2) > 1 \qquad [3]$$
$$(x + 2)^2 \qquad [4]$$

By inspection we see that these four expressions are all different in nature and by investigating each of them we shall try to identify these differences.

(1) $(x + 2)^2 = 2x + 7$

Substituting 1 for x in both the left hand side (LHS) and right hand side (RHS) separately we find

$$\text{LHS} = (1 + 2)^2 = 3^2 = 9$$
$$\text{RHS} = 2 + 7 \quad = 9$$

i.e. LHS = RHS   when $x = 1$.

However substituting 2 for x in a similar way we find

$$\text{LHS} = (2 + 2)^2 = 16$$
$$\text{RHS} = 4 + 7 \quad = 11$$

i.e. LHS $\neq$ RHS   when $x = 2$.

Now rearranging the original expression gives

$$x^2 + 4x + 4 = 2x + 7$$
$$\Rightarrow\; x^2 + 2x - 3 = 0$$

i.e. $$(x + 3)(x - 1) = 0$$

from which we see that   LHS = RHS   if and only if

either $x + 3 = 0$ i.e. $x = -3$

or $x - 1 = 0$ i.e. $x = 1$

Thus $(x + 2)^2 = 2x + 7$ only if $x = -3$ or if $x = 1$ and the equality is not true for any other value of x.

Expressions of this type are called equations and the equality is true only for a number of distinct values of the unknown quantity (or quantities).

The process of finding these values is referred to as solving the equation.

(2) $(x + 2)^2 = x^2 + 4x + 4$

If we substitute 1 for x in both sides of this expression we find

$$\text{LHS} = (1 + 2)^2 \qquad = 9$$
$$\text{RHS} = 1^2 + 4.1 + 4 = 9$$

i.e.   LHS = RHS   when $x = 1$

If we substitute -1 for x as before we find

$$\text{LHS} = (-1 + 2)^2 = 1$$
$$\text{RHS} = (-1)^2 + (4)(-1) + 4 = 1$$

i.e.   LHS = RHS   when $x = -1$.

Whatever other numerical value we substitute for x we find that   LHS = RHS, so it appears (but is not proved) that   LHS = RHS   for all values of x.

The following geometric illustration confirms that this suspicion is correct.

Consider a square of side $x + 2$ units.

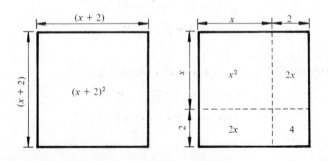

Since these two squares are identical, their areas are identical.

Hence $(x + 2)^2 = x^2 + 4x + 4$ for all values of x and we say that $(x + 2)^2$ *is identical to* $x^2 + 4x + 4$.

Using the symbol $\equiv$ for 'is identical to' the relationship is written

$$(x + 2)^2 \equiv x^2 + 4x + 4.$$

Relationships of this type are called identities and both sides are equal for any value of the unknown quantity. They are, in fact, two forms for the same expression, and we shall use the identity symbol whenever we are dealing with an identity relationship and we strongly recommend that the reader does the same.

(3) $(x - 2) > 1$

Reading from left to right the symbol $>$ means greater than (and $<$ means less than). This relationship is obviously different from the first two and it is called an inequality.

By inspection we see that for $x - 2$ to have a value greater than one then x must have a value greater than three.

i.e. if $$x - 2 > 1$$

then $$x > 3$$

Consider a line as being made up of adjacent points. We can represent all the real values that x, the variable, can have by the positions of points on a line, (known as a *number line*).

negative values of x positive values of x

The position of a point to the *left* of a second point corresponds to a value of x *less than* the value of x at the second point.

i.e. $$-99 < 1, \quad -10 < -5 \quad \dots \text{etc.}$$

The values of x given by the statement $x > 3$ can then be represented by a section of this line.

i.e.

From this we see that *not* all values of x satisfy the inequality but that there is an infinite set of values that do, i.e. the solution of an inequality is a range (or ranges) of values of the variable involved.

Note that $x = 3$ is *not* included in the range and this is indicated by an open circle at $x = 3$. For $x \geqslant 3$, which means x is greater than or equal to 3, the value $x = 3$ *is* included in the range. This is indicated on a number line by a solid circle.

i.e.

(4) $(x + 2)^2$

This expression is not related to any other expression and can take different values depending on the value given to x.

Such an expression is called a function (of x in this case). Using the symbol $f(x)$ to represent a function of x, we write

$$f(x) \equiv (x + 2)^2$$

To represent the value of this function when $x = 1$, say, we write $f(1)$, where

$$f(1) = (1 + 2)^2 \quad = \quad 9$$

similarly

$$f(3) = (3 + 2)^2 \quad = 25$$

and

$$f(-2) = (-2 + 2)^2 = \quad 0$$

Note that the values both of x and $f(x)$ are variable but whereas x can be given *any* value, the value of $f(x)$ depends on that of x, so x *is referred to as the independent variable* and $f(x)$ *as the dependent variable*.

EXERCISE 1a

1) State which of the following are equations and which are identities

(a) $x^2 - 3 = 2$ (b) $x^2 - 9 = (x - 3)(x + 3)$ (c) $\dfrac{1}{x} - \dfrac{1}{x + 1} = \dfrac{1}{x^2 + x}$

(d) $\dfrac{1}{x - 1} + \dfrac{1}{x + 1} = \dfrac{2}{x^2 - 1}$ (e) $p^2 + 2p - 3 = 3 - 2p - p^2$

(f) $(x + y)(x - y) = x^2 - y^2$ (g) $y - 1 = \dfrac{1}{y}$ (h) $\dfrac{2q}{q^2 - 1} = \dfrac{1}{q - 1} + \dfrac{1}{q + 1}$

2) Find the range of values of x for which the following inequalities are true and illustrate the range on a number line.

(a) $x - 5 > 0$ (b) $x + 1 \leqslant -1$ (c) $0 \geqslant x - 4$ (d) $3 < 4 - x$

3) Find the values of $f(0), f(1), f(-2), f(5)$ where $f(x)$ is:

(a) $x^2 - 3x$ (b) $\dfrac{1}{x - 2}$ (c) $(x - 7)(x + 2)$ (d) $\dfrac{1}{x + 1} - \dfrac{2}{2x - 3}$

POLYNOMIALS AND FRACTIONAL FUNCTIONS

If each individual term of a function is of the form ax^n, where a is a constant (i.e. has a fixed numerical value) and n is a positive integer, the function is called a *polynomial*. Thus $x^2 - 2x$, $3x^6 - 7x^4 + 6$, $(x - 4)^2$ are polynomials but $\sqrt{x}$, $\dfrac{1}{x}$, $\sqrt{x^2 - 2}$ are not.

The highest power of x that occurs in a polynomial defines *the degree or order of the polynomial*.

Thus $5x^6 - 7x^3 + 6x$ is a polynomial of degree 6.

A fractional function of the form $\dfrac{3x^2 - 7}{x^3 + 1}$ where both the numerator and denominator are polynomials, is referred to as a 'proper' fraction if the degree of the numerator is less than the degree of the denominator. However if the degree of the numerator is greater than, or equal to, the degree of the denominator the fraction is referred to as 'improper'.

An improper numerical fraction such as $\dfrac{9}{7}$ may be written as $\dfrac{7+2}{7} = 1 + \dfrac{2}{7}$

Similarly an algebraic fraction such as $\dfrac{x^2 - 1}{x^2 + 1}$ may be written as

$$\frac{x^2 + 1 - 2}{x^2 + 1} \equiv \frac{x^2 + 1}{x^2 + 1} - \frac{2}{x^2 + 1} \equiv 1 - \frac{2}{x^2 + 1}$$

PARTIAL FRACTIONS

Consider first a function such as $\quad f(x) \equiv \dfrac{2}{x + 1} + \dfrac{x}{x^2 + 1}$.

$f(x)$ may be expressed as a single fraction with a common denominator thus

$$f(x) \equiv \frac{2}{x + 1} + \frac{x}{x^2 + 1} \equiv \frac{2(x^2 + 1) + x(x + 1)}{(x + 1)(x^2 + 1)} \equiv \frac{3x^2 + x + 2}{(x + 1)(x^2 + 1)}$$

It is often useful to be able to reverse this operation, that is to take a function such as $\quad f(x) \equiv \dfrac{x - 2}{(x + 3)(x - 4)}\quad$ and express $f(x)$ as the sum of two (or in some cases more) separate fractions.

This process is called expressing $f(x)$ in partial fractions.

If the original fraction is 'proper' then the separate (or partial) fractions will also be 'proper'.

Thus $\dfrac{x + 3}{(x - 2)(x + 4)}$ can be expressed as $\dfrac{A}{x - 2} + \dfrac{B}{x + 4}$

and $\dfrac{x + 3}{(x - 2)(x^2 + 4)}$ can be expressed as $\dfrac{A}{x - 2} + \dfrac{Bx + C}{x^2 + 4}$ where A, B and C are constants to be determined. The method for evaluating these constants depends to some extent on the factors in the denominator.

EXAMPLES 1b

1) Express $\dfrac{x + 3}{(x - 2)(x + 4)}$ in partial fractions.

This example is a proper fraction with *linear* (of degree one) *factors* only and so its partial fractions are also proper. As these partial fractions have linear denominators their numerators contain only one constant.

$$\frac{x+3}{(x-2)(x+4)} \equiv \frac{A}{x-2} + \frac{B}{x+4}$$

so

$$\frac{x+3}{(x-2)(x+4)} \equiv \frac{A(x+4) + B(x-2)}{(x-2)(x+4)}$$

As the denominators are obviously identical, the numerators must also be identical,

i.e. $x+3 \equiv A(x+4) + B(x-2)$

Now   LHS = RHS   for any value of x.
Choosing to substitute 2 for x (to eliminate B) gives

$$2 + 3 = A(2+4) + B(0)$$

$\Rightarrow$ $A = \dfrac{5}{6}$

substituting -4 for x (to eliminate A) gives

$$-4 + 3 = B(-4-2)$$

$\Rightarrow$ $B = \dfrac{1}{6}$

Therefore $\dfrac{x+3}{(x-2)(x+4)} \equiv \dfrac{5}{6(x-2)} + \dfrac{1}{6(x+4)}$

2) Express $\dfrac{x^2 - 3}{(x-1)(x^2+1)}$ in partial fractions.

This example contains a *quadratic factor* in the denominator,

therefore $\dfrac{x^2 - 3}{(x-1)(x^2+1)} \equiv \dfrac{A}{x-1} + \dfrac{Bx+C}{x^2+1}$

each numerator on RHS being chosen so that each partial fraction is proper.

$\Rightarrow$ $\dfrac{x^2 - 3}{(x-1)(x^2+1)} = \dfrac{A(x^2+1) + (Bx+C)(x-1)}{(x-1)(x^2+1)}$

therefore $x^2 - 3 \equiv A(x^2+1) + (Bx+C)(x-1)$ [1]

substituting 1 for x (so eliminating B and C) gives

$$1^2 - 3 = A(1^2 + 1)$$

$\Rightarrow$ $A = -1$

There is no value which we can substitute for x to eliminate A (as there is no value of x for which $x^2 + 1 = 0$).
But substituting 0 for x will eliminate B giving

$$-3 = A(1) + C(-1)$$

As $A = -1$, $\qquad\qquad -3 = -1 - C$

$\Rightarrow \qquad\qquad\qquad\qquad C = 2$

Any other value can now be substituted for x to find B, a small value being sensible.

Substituting 2 for x gives

$$(2)^2 - 3 = A(2^2 + 1) + (2B + C)(2 - 1)$$

$$1 = 5A + 2B + C$$

But $A = -1$ and $C = 2$ so $B = 2$

Therefore $\qquad \dfrac{x^2 - 3}{(x - 1)(x^2 + 1)} \equiv \dfrac{-1}{x - 1} + \dfrac{2x + 2}{x^2 + 1}$

Alternatively the value of B may be found as follows.

Expanding the RHS of [1] gives

$$x^2 - 3 \equiv Ax^2 + A + Bx^2 - Bx + Cx - C$$

or $\qquad\qquad x^2 - 3 \equiv (A + B)x^2 + (C - B)x + A - C \qquad\qquad\qquad [2]$

As this is an identity, the coefficients (quantity) of x^2 on the LHS and RHS of [2] must be equal.

Therefore $\qquad\qquad\qquad\qquad 1 = A + B$

(this is referred to as comparing the coefficients of x^2).

As $A = -1$, $B = 2$.

In practice a mixture of these two methods will give a simple solution.

3) Express $\dfrac{x - 1}{(x + 1)(x - 2)^2}$ in partial fractions.

As $(x - 2)^2 \equiv (x - 2)(x - 2)$, it is called a *repeated factor*, but it is also quadratic so we may initially think of $\dfrac{x - 1}{(x + 1)(x - 2)^2}$ as $\dfrac{A}{x + 1} + \dfrac{Bx + C}{(x - 2)^2}$.

But this is not the simplest partial fraction form, as we shall see.

Considering just the fraction $\dfrac{Bx + C}{(x - 2)^2}$ and letting $C = -2B + D$

then $\qquad\qquad \dfrac{Bx + C}{(x - 2)^2} \equiv \dfrac{Bx - 2B + D}{(x - 2)^2}$

$$\equiv \dfrac{B(x - 2)}{(x - 2)^2} + \dfrac{D}{(x - 2)^2}$$

$$\equiv \dfrac{B}{x - 2} + \dfrac{D}{(x - 2)^2}$$

In general any repeated factor of the form $(ax + b)^2$ in a denominator will give rise to two partial fractions of the form $\dfrac{A}{ax + b}$ and $\dfrac{B}{(ax + b)^2}$.

Similarly a repeated factor $(ax + b)^3$ gives rise to three partial fractions of form $\dfrac{A}{ax + b}$, $\dfrac{B}{(ax + b)^2}$ and $\dfrac{C}{(ax + b)^3}$.

Returning to the original problem we have

$$\frac{x - 1}{(x + 1)(x - 2)^2} \equiv \frac{A}{x + 1} + \frac{B}{x - 2} + \frac{D}{(x - 2)^2}$$

Therefore $x - 1 \equiv A(x - 2)^2 + B(x + 1)(x - 2) + D(x + 1)$

substituting 2 for x gives

$$1 = 3D \quad \text{so} \quad D = \frac{1}{3}$$

substituting -1 for x gives

$$-2 = A(-3)^2 \quad \text{so} \quad A = -\tfrac{2}{9}$$

comparing the coefficients of x^2 gives

$$0 = A + B \quad \text{so} \quad B = \tfrac{2}{9}$$

therefore $\dfrac{x - 1}{(x + 1)(x - 2)^2} \equiv -\dfrac{2}{9(x + 1)} + \dfrac{2}{9(x - 2)} + \dfrac{1}{3(x - 2)^2}$

4) Express $\dfrac{x^3}{(x + 1)(x - 3)}$ in partial fractions.

This function is an *improper fraction* and it is necessary first to divide the denominator into the numerator to obtain a mixed fraction.

$$
\begin{array}{r}
x + 2 \\
x^2 - 2x - 3\,\overline{\big)\,x^3 } \\
\underline{x^3 - 2x^2 - 3x} \\
2x^2 + 3x \\
\underline{2x^2 - 4x - 6} \\
7x + 6
\end{array}
$$

It is not necessary to go any further as the remainder at this stage is of degree one (less than the degree of the divisor) so we can say

$$\frac{x^3}{(x + 1)(x - 3)} \equiv x + 2 + \frac{\text{Remainder}}{(x + 1)(x - 3)}$$

$$\equiv x + 2 + \frac{A}{x + 1} + \frac{B}{x - 3}$$

therefore $x^3 \equiv (x + 2)(x + 1)(x - 3) + A(x - 3) + B(x + 1)$

Substituting 3 for x gives $\qquad 27 = 4B \quad \Rightarrow \quad B = \frac{27}{4}$

Substituting -1 for x gives $\quad -1 = -4A \Rightarrow A = \frac{1}{4}$

Therefore

$$\frac{x^3}{(x+1)(x-3)} \equiv x + 2 + \frac{1}{4(x+1)} + \frac{27}{4(x-3)}$$

EXERCISE 1b

Express in partial fractions

1) $\dfrac{3}{(x+1)(x-1)}$

2) $\dfrac{x}{(x-4)(x-1)}$

3) $\dfrac{x-1}{(x+2)(x-2)}$

4) $\dfrac{2}{(2x-1)(x-2)}$

5) $\dfrac{x+3}{x(x+1)}$

6) $\dfrac{2x-1}{(x+1)(3x+2)}$

7) $\dfrac{3x}{(x-1)(x-2)(x-3)}$

8) $\dfrac{x^2-2x+4}{2x(x-3)(x+1)}$

9) $\dfrac{(x-2)(2x+3)}{(x-1)(x^2-9)}$

10) $\dfrac{2}{(x-1)(x^2+1)}$

11) $\dfrac{x-3}{(x+4)(x^2-2)}$

12) $\dfrac{x^2+3}{x(x^2+2)}$

13) $\dfrac{2x^2+x+1}{(x-3)(2x^2+1)}$

14) $\dfrac{x^3-1}{(x+2)(2x+1)(x^2+1)}$

15) $\dfrac{x^2+1}{x(2x^2-1)(x-1)}$

16) $\dfrac{x}{(x-1)(x-2)^2}$

17) $\dfrac{x^2-1}{x^2(2x+1)}$

18) $\dfrac{3}{x(3x-1)^2}$

19) $\dfrac{x^2+x+1}{(x^2-1)(x^2+1)}$

20) $\dfrac{x^2}{(x-1)(x+1)}$

QUADRATIC EQUATIONS

Any equation of the form

$$ax^2 + bx + c = 0$$

is called a quadratic equation.

Solution of Quadratic Equations which Factorise

Consider the equation $\qquad 2x^2 - 7x + 3 = 0$

the LHS factorises and the equation becomes

$$(2x - 1)(x - 3) = 0$$

from which we see that either

$$2x - 1 = 0 \quad \Rightarrow \quad x = \tfrac{1}{2}$$

or $\qquad\qquad\qquad x - 3 = 0 \quad \Rightarrow \quad x = 3$

Solutions to equations can sometimes be 'lost' if care is not taken with apparently straightforward equations. Consider, for example, the equation $2x^2 - 14x = 0$ and the following two solutions.

(1)

$$2x^2 - 14x = 0$$

$\div 2$

$$x^2 - 7x = 0$$

$\div x$

$$x - 7 = 0$$

therefore

$$\underline{x = 7}$$

(2)

$$2x^2 - 14x = 0$$

factorizing, $2x(x - 7) = 0$

therefore either $2x = 0$ or $x - 7 = 0$

giving $\underline{x = 0 \text{ or } x = 7}$

This shows that the first solution results in the loss of the answer $x = 0$ and this is because the equation was divided by the common factor x.

Therefore although it is correct (and desirable) to divide by constant factors, division by a common factor containing the unknown quantity will lead to a loss of some solutions. This must be remembered when solving any equation, quadratic or otherwise.

Now consider the equation $t(t - 3) = t^2 - 4$ and the following two solutions.

(1) $t(t - 3) = t^2 - 4$

$$t^2 - 3t = t^2 - 4$$

$$4 - 3t = 0$$

$$\underline{t = \tfrac{4}{3}}$$

(2) Substituting $\dfrac{1}{m}$ for t gives

$$\frac{1}{m}\left(\frac{1}{m} - 3\right) = \frac{1}{m^2} - 4$$

$$\frac{1}{m^2} - \frac{3}{m} = \frac{1}{m^2} - 4$$

$$\Rightarrow\ 1 - 3m = 1 - 4m^2$$

$$4m^2 - 3m = 0$$

$$m(4m - 3) = 0$$

$$m = 0 \quad \text{or} \quad m = \tfrac{3}{4}$$

i.e. $\dfrac{1}{t} = 0$ or $\dfrac{1}{t} = \dfrac{3}{4}$

therefore $\underline{t = \infty \text{ or } t = \tfrac{4}{3}}$

Note that the symbol ∞ means infinitely large.

This shows that the first solution resulted in the loss of the answer $t = \infty$

This is because the t^2 terms on the LHS and RHS were equal and were cancelled. In most problems an infinite solution would have no practical meaning but in some cases it would be applicable. For example if t represents $\tan\theta$, then $t = \infty$ gives the solution $\theta = 90°$.

So if, in a quadratic equation, the squared term apparently disappears, remember to consider the infinite solution.

EXERCISE 1c

Solve the following quadratic equations.

1) $x^2 + 5x - 6 = 0$ 2) $3x^2 - 7x = 0$ 3) $4x^2 - 8x + 4 = 0$

4) $x(x - 3) = x^2 - 6$ 5) $3 - x - 2x^2 = 0$ 6) $x^2 - 2ax + a^2 = 0$

7) $\dfrac{1}{x^2} - 1 = \dfrac{1}{x} - 1$ 8) $x(1 - x) = x(2x - 1)$

9) $(x - 2)(x + 3) = (x - 2)(4 - x)$ 10) $\dfrac{1}{x + 1} + \dfrac{2}{x + 2} = 1$

Solution of Quadratic Equations which do not Factorise

Consider the equation $2x^2 - 5x + 1 = 0$

Dividing by 2 gives $x^2 - \dfrac{5}{2}x + \dfrac{1}{2} = 0$

$$x^2 - \frac{5}{2}x = -\frac{1}{2}$$

Adding $\left(\dfrac{1}{2} \times \dfrac{5}{2}\right)^2$ to both sides to make the LHS a perfect square gives

$$x^2 - \frac{5}{2}x + \left(\frac{5}{4}\right)^2 = -\frac{1}{2} + \left(\frac{5}{4}\right)^2$$

i.e. $\left(x - \dfrac{5}{4}\right)^2 = \dfrac{17}{16}$

therefore $x - \dfrac{5}{4} = \dfrac{\pm\sqrt{17}}{4}$

$$x = \frac{5}{4} \pm \frac{4.123}{4}$$

therefore either $x = 2.28$ or $x = 0.22$.

This method is called 'completing the square' and when applied to the general quadratic equation, viz. $ax^2 + bx + c = 0$,   results in the familiar formula

$$x = \frac{-b \pm \sqrt{b^2 - 4ac}}{2a}$$

Nature of the Roots of a Quadratic Equation

Using the formula to solve the equation $ax^2 + bx + c = 0$ we see that

either $x = \dfrac{-b + \sqrt{b^2 - 4ac}}{2a}$ or $x = \dfrac{-b - \sqrt{b^2 - 4ac}}{2a}$

Therefore in general a quadratic equation has two solutions (called roots).

If $b^2 - 4ac$ is positive, $\sqrt{b^2 - 4ac}$ can be evaluated and the equation will have two real and distinct (i.e. different) roots.

If $b^2 - 4ac$ is zero the equation is satisfied by only one value of x $\left(\text{viz. } -\dfrac{b}{2a}\right)$ and we say that it has a repeated root or equal roots.

If $b^2 - 4ac$ is negative, $\sqrt{b^2 - 4ac}$ has no real value so the equation has no real roots.

To summarize, the equation $\quad ax^2 + bx + c = 0$

has two real distinct roots   if $b^2 - 4ac > 0$

has equal roots   if $b^2 - 4ac = 0$

has no real roots   if $b^2 - 4ac < 0$

$$b^2 - 4ac \quad \text{is called the discriminant}$$

EXAMPLES 1d

1) Determine the nature of the roots of the equations

(a) $4x^2 - 7x + 3 = 0$ (b) $x^2 + ax + a^2 = 0$ (c) $x^2 - px - q^2 = 0$

(a) $4x^2 - 7x + 3 = 0$

therefore $b^2 - 4ac = (-7)^2 - 4(4)(3) = 1$

i.e. $b^2 - 4ac > 0$

So the equation has two distinct real roots.

(b) $x^2 + ax + a^2 = 0$

therefore   'b^2 - 4ac' $= (a)^2 - 4(1)(a^2) = -3a^2$

As a^2 is positive irrespective of the value of a, '$b^2 - 4ac$' < 0

So the equation has no real roots.

(c) $x^2 - px - q^2 = 0$

$$b^2 - 4ac = (-p)^2 - 4(1)(-q^2) = p^2 + 4q^2$$

As p^2 and q^2 are both positive, $b^2 - 4ac > 0$

therefore the equation has two real distinct roots.

2) Find the value of k if $2x^2 - kx + 8 = 0$ has equal roots.

For the roots of $2x^2 - kx + 8 = 0$ to be equal,

$$b^2 - 4ac = 0$$

i.e. $(-k)^2 - 4(2)(8) = 0$

$$k^2 = 64$$

therefore $\qquad\qquad\qquad\qquad k = \pm 8$

EXERCISE 1d

Solve the following quadratic equations by completing the square.

1) $2x^2 - 6x + 4 = 0$ 2) $x^2 + 4x - 8 = 0$ 3) $2x^2 + 7x + 3 = 0$

4) $x^2 - 2x + a = 0$ 5) $x^2 - 2ax + b = 0$ 6) $ax^2 + bx + c = 0$

Determine the nature of the roots of the following equations but do not solve the equations.

7) $x^2 - 6x + 9 = 0$ 8) $x^2 - 6x + 10 = 0$ 9) $2x^2 - 5x + 3 = 0$

10) $3x^2 + 4x + 2 = 0$ 11) $4x^2 - 12x + 9 = 0$ 12) $4x^2 - 12x - 9 = 0$

13) For what values of k is $9x^2 + kx + 16$ a perfect square?

14) The roots of $3x^2 + kx + 12 = 0$ are equal. Find k.

15) Find a if $x^2 - 5x + a = 0$ has equal roots.

16) Prove that $kx^2 + 2x - (k - 2) = 0$ has real roots for any value of k.

17) Show that the roots of $ax^2 + (a + b)x + b = 0$ are real for all values of a and b.

18) Find a relationship between p and q if the roots of $px^2 + qx + 1 = 0$ are equal.

19) If x is real and $p = \dfrac{3(x^2 + 1)}{2x - 1}$,   prove that $p^2 - 3(p + 3) \geqslant 0$.

Relationships Between the Roots and Coefficients of a Quadratic Equation

Let α and β be the roots of the equation $ax^2 + bx + c = 0$

i.e. $\qquad\qquad (x - \alpha)(x - \beta) = 0 \qquad [1]$

and $\qquad\qquad ax^2 + bx + c = 0 \qquad [2]$ have the same solution

but $\qquad\qquad (x - \alpha)(x - \beta) \equiv x^2 - (\alpha + \beta)x + \alpha\beta$

and dividing $ax^2 + bx + c = 0$ by a gives $x^2 + \dfrac{b}{a}x + \dfrac{c}{a} = 0$

therefore $x^2 - (\alpha + \beta)x + \alpha\beta = 0 \qquad [3]$

and $\qquad\qquad x^2 + \dfrac{b}{a}x + \dfrac{c}{a} = 0 \qquad [4]$ have the same solution.

As the LHS of [3] and [4] have the same coefficient of x^2 it follows that the coefficients of x and the constant terms are also equal

i.e. $\qquad\qquad x^2 - (\alpha + \beta)x + \alpha\beta \equiv x^2 + \dfrac{b}{a}x + \dfrac{c}{a}$

(**Note** that the LHS of [2] and [3] are not identical unless $a = 1$.)

Therefore

$$\alpha + \beta = -\frac{b}{a}$$

$$\alpha\beta = \frac{c}{a}$$

and the equation may be written

$$x^2 - (\text{sum of roots}) \, x + (\text{product of roots}) = 0$$

So if $2x^2 - 3x + 6 = 0$ has roots α and β,

the sum of its roots $(\alpha + \beta)$ is $-\left(-\frac{3}{2}\right) = \frac{3}{2}$

and the product of its roots $(\alpha\beta)$ is $\frac{6}{2} = 3$.

Also if a quadratic equation has roots whose sum is 7 and whose product is 10 the equation can be written as $x^2 - 7x + 10 = 0$.

EXAMPLES 1e

1) The roots of the equation $2x^2 - 7x + 4 = 0$ are α and β. Find the values of $\frac{1}{\alpha} + \frac{1}{\beta}$ and $\frac{1}{\alpha\beta}$

Hence write down the equation whose roots are $\frac{1}{\alpha}$ and $\frac{1}{\beta}$.

Given $2x^2 - 7x + 4 = 0$ we see that $\alpha + \beta = -\left(-\frac{7}{2}\right) = \frac{7}{2}$

and $\alpha\beta = \frac{4}{2} = 2$

To evaluate $\frac{1}{\alpha} + \frac{1}{\beta}$ we need to express it in terms of $\alpha + \beta$ and $\alpha\beta$ whose values are known.

Expressing $\frac{1}{\alpha} + \frac{1}{\beta}$ as a single fraction gives

$$\frac{1}{\alpha} + \frac{1}{\beta} = \frac{\alpha + \beta}{\alpha\beta} = \frac{\frac{7}{2}}{2} = \frac{7}{4}$$

and $$\frac{1}{\alpha\beta} = \frac{1}{2}$$

Therefore the required equation has roots whose sum $\left(\frac{1}{\alpha} + \frac{1}{\beta}\right)$ is $\frac{7}{4}$ and whose product $\frac{1}{\alpha\beta}$ is $\frac{1}{2}$,

so the equation is $$x^2 - \tfrac{7}{4}x + \tfrac{1}{2} = 0$$

or $$4x^2 - 7x + 2 = 0$$

Alternatively, for the given equation $2x^2 - 7x + 4 = 0$, $x = \alpha, \beta$

for the required equation $X = \dfrac{1}{\alpha}, \dfrac{1}{\beta}$

therefore $X = \dfrac{1}{x} \Rightarrow x = \dfrac{1}{X}$

Substituting $\dfrac{1}{X}$ for x in the given equation we get

$$2\left(\frac{1}{X}\right)^2 - 7\left(\frac{1}{X}\right) + 4 = 0$$

i.e. $$4X^2 - 7X + 2 = 0$$

and this is the required equation.

Note. This method can be used only if each new root depends in the same way on each original root. For example, if the given equation has roots α, β and the required equation has roots α^2, β^2 it can be used, but if the required equation has roots $\alpha + \beta, \alpha - \beta$ it cannot.

2) If α and β are the roots of $x^2 + 3x - 2 = 0$ find the values of $\alpha^3 + \beta^3$ and $\alpha^3\beta^3$. Write down the equation whose roots are α^3 and β^3.

From $x^2 + 3x - 2 = 0$ we see that $\alpha + \beta = -3$

and $$\alpha\beta = -2$$

To express $\alpha^3 + \beta^3$ in terms of $\alpha + \beta$ and $\alpha\beta$ we can use

$$(\alpha + \beta)^3 \equiv \alpha^3 + 3\alpha^2\beta + 3\alpha\beta^2 + \beta^3$$

$$\equiv \alpha^3 + \beta^3 + 3\alpha\beta(\alpha + \beta)$$

therefore $$\alpha^3 + \beta^3 \equiv (\alpha + \beta)^3 - 3\alpha\beta(\alpha + \beta)$$

$$= (-3)^3 - 3(-2)(-3)$$

$$= -45$$

$$\alpha^3\beta^3 \equiv (\alpha\beta)^3 = (-2)^3 = -8$$

As the required equation has roots α^3 and β^3 the sum of its roots is $\alpha^3 + \beta^3 = -45$ and the product of its roots is $\alpha^3\beta^3 = -8$.

Therefore the required equation is $x^2 - (-45)x + (-8) = 0$

i.e. $$x^2 + 45x - 8 = 0$$

Note that although the alternative method could be used in this example, as $X = x^3$, it is not recommended as the resulting equation $(\sqrt[3]{X})^2 + 3(\sqrt[3]{X}) + 4 = 0$ is not easy to simplify.

3) Find the range of values of k for which the equation $x^2 - 2x - k = 0$ has real roots. If the roots of this equation differ by one, find the value of k.

If $x^2 - 2x - k = 0$ has real roots $b^2 - 4ac$ is greater than or equal to zero (this includes the case of equal roots)

i.e. $(-2)^2 - 4(1)(-k) \geqslant 0$

i.e. $4 + 4k \geqslant 0$

or $1 + k \geqslant 0$

$$k \geqslant -1$$

Let one root of the equation be α, then the other is $\alpha + 1$.

Sum of the roots is $2\alpha + 1 = -(-2)$

$$2\alpha = 1$$

$$\alpha = \tfrac{1}{2}$$

Product of the roots is $\alpha(\alpha + 1) = -k$

Therefore $k = -\tfrac{3}{4}$

EXERCISE 1e

1) Write down the sums and products of the roots of the following equations:

(a) $x^2 - 3x + 2 = 0$ (b) $4x^2 + 7x - 3 = 0$ (c) $x(x - 3) = x + 4$

(d) $\dfrac{x-1}{2} = \dfrac{3}{x+2}$ (e) $x^2 - kx + k^2 = 0$ (f) $ax^2 - x(a + 2) - a = 0$

2) Write down the equation, the sum and product of whose roots are:

(a) $3, 4$ (b) $-2, \tfrac{1}{2}$ (c) $\tfrac{1}{3}, -\tfrac{2}{5}$ (d) $-\tfrac{1}{4}, 0$

(e) a, a^2 (f) $-(k + 1), k^2 - 3$ (g) $\dfrac{b}{a}, \dfrac{c^2}{b}$

3) The roots of the equation $2x^2 - 4x + 5 = 0$ are α and β. Find the value of:

(a) $\dfrac{1}{\alpha} + \dfrac{1}{\beta}$ (b) $(\alpha + 1)(\beta + 1)$ (c) $\alpha^2 + \beta^2$ (d) $\alpha^2\beta + \alpha\beta^2$

(e) $(\alpha - \beta)^2$ (f) $\dfrac{\alpha}{\beta} + \dfrac{\beta}{\alpha}$ (g) $\dfrac{1}{\alpha + 1} + \dfrac{1}{\beta + 1}$ (h) $\dfrac{1}{2\alpha + \beta} + \dfrac{1}{\alpha + 2\beta}$

(i) $\dfrac{1}{\alpha^2 + 1} + \dfrac{1}{\beta^2 + 1}$

4) The roots of $x^2 - 2x + 3 = 0$ are α and β. Find the equation whose roots are:

(a) $\alpha + 2, \beta + 2$ (b) $\dfrac{1}{\alpha}, \dfrac{1}{\beta}$ (c) α^2, β^2 (d) $\dfrac{\alpha}{\beta}, \dfrac{\beta}{\alpha}$

(e) $\alpha - \beta, \beta - \alpha$

5) *Write down* and simplify the equation whose roots are the reciprocals of the roots of $3x^2 + 2x - 1 = 0$, without solving the given equation.

6) *Write down* and simplify the equation whose roots are double those of $4x^2 - 5x - 2 = 0$, without solving the given equation.

7) *Write down* and simplify the equation whose roots are one less than those of $5x^2 + 3x - 1 = 0$.

8) Write down the equation whose roots are minus those of $2x^2 - 3x - 1 = 0$.

9) Find the value of k if the roots of $3x^2 + 5x - k = 0$ differ by two.

10) Find the value of p if one root of $x^2 + px + 8 = 0$ is the square of the other.

11) If α and β are the roots of $ax^2 + bx + c = 0$, find the equation whose roots are $\dfrac{1}{\alpha}, \dfrac{1}{\beta}$.

12) Find a relationship between a and c if the roots of $ax^2 + bx + c = 0$ are the reciprocal of each other.

13) If one root of $ax^2 + bx + c = 0$ is treble the other prove that $3b^2 - 16ac = 0$.

14) For what value of k are the roots of $3x^2 + (k-1)x - 2 = 0$ equal and opposite?

SUMMARY

Partial Fractions

1. Linear factors:

$$\frac{3x - 4}{(2x - 3)(x + 5)} \equiv \frac{A}{2x - 3} + \frac{B}{x + 5}$$

2. Quadratic factors:

$$\frac{3x - 4}{(2x - 3)(x^2 + 5)} \equiv \frac{A}{2x - 3} + \frac{Bx + C}{x^2 + 5}$$

3. Repeated factors:

$$\frac{3x - 4}{(2x - 3)(x + 5)^2} \equiv \frac{A}{2x - 3} + \frac{B}{x + 5} + \frac{C}{(x + 5)^2}$$

Quadratic Equations

The equation $ax^2 + bx + c = 0$ has two roots

given by $$x = \frac{-b \pm \sqrt{b^2 - 4ac}}{2a}$$

These roots are real and distinct if $\quad b^2 - 4ac > 0$

real and equal if $\quad b^2 - 4ac = 0$

not real if $\quad b^2 - 4ac < 0$

If these roots are α and β

$$\alpha + \beta = -\frac{b}{a}$$

$$\alpha\beta = \frac{c}{a}$$

MULTIPLE CHOICE EXERCISE 1

(Instructions for answering these questions are given on page xii.)

TYPE I

1) If $\dfrac{x+p}{(x-1)(x-3)} \equiv \dfrac{q}{x-1} + \dfrac{2}{x-3}$, the values of p and q are:

(a) $p = -2, q = 1$ (b) $p = 2, q = 1$ (c) $p = 1, q = -2$
(d) $p = 1, q = 1$ (e) $p = 1, q = -1$.

2) If $x^2 + px + 6 = 0$ has equal roots and $p > 0$, p is:
(a) $\sqrt{48}$ (b) 0   (c) $\sqrt{6}$ (d) 3   (e) $\sqrt{24}$.

3) If $x^2 + 4x + p \equiv (x+q)^2 + 1$, the values of p and q are:
(a) $p = 5, q = 2$ (b) $p = 1, q = 2$ (c) $p = 2, q = 5$
(d) $p = -1, q = 5$ (e) $p = 0, q = -1$.

4) If the equation $2x^2 + 3x + 1 = 0$ has roots α, β the equation whose roots are $\dfrac{1}{\alpha}, \dfrac{1}{\beta}$ is:

(a) $3x^2 + 2x + 1 = 0$ (b) $x^2 + 3x + 2 = 0$ (c) $2x^2 + x + 3 = 0$
(d) $x^2 - 3x + 2 = 0$ (e) none of these.

5) If α and β are the roots of the equation $x^2 - px + q = 0$ the value of $\alpha^2 + \beta^2$ is:

(a) $p - q$ (b) $p^2 + 2q$ (c) p^2 (d) $p^2 - 2q$ (e) $-p^2 - 2q$.

6) $x - 3 > 2$ corresponds to:
(a) $x > 3$ (b) $x > 5$ (c) $x > 1$ (d) $x < 3$ (e) $x > 0$.

7) $f(x) \equiv x^2 + \dfrac{1}{x} + 1$ corresponds to:

(a) $f(1) = 1$ (b) $f(-1) = 3$ (c) $f(0) = 1$ (d) $f(1) = 3$
(e) $f(-1) = -1$.

8) $\dfrac{2}{(x+1)(x-1)} \equiv \dfrac{A}{x+1} + \dfrac{B}{x-1}$ corresponds to:

(a) $A = 1, B = 1$ (b) $A = -1, B = 1$ (c) $A = x, B = 1$
(d) $A = 0, B = 2$ (e) $A = x - 1, B = x + 1$.

TYPE II

9) $2x^2 - 7x + 4 = 0$ has roots α and β.
(a) $\alpha\beta = -2$.
(b) $\alpha + \beta = 3\frac{1}{2}$.
(c) α and β are real and unequal.

10) $f(x) \equiv \dfrac{2x}{(x+2)(x-2)}$.

(a) $f(x)$ is an improper fraction.
(b) $f(0) = 2$.

(c) $f(x) \equiv \dfrac{1}{x-2} + \dfrac{1}{x+2}$.

11) $\dfrac{2}{(x-1)(x^2+1)} \equiv \dfrac{A}{x-1} + \dfrac{Bx+C}{x^2+1}$.

(a) $A = 1$.
(b) $2 \equiv A(x^2+1) + (Bx+C)(x-1)$.
(c) $A + B = 0$.

12) $f(x) \equiv x^2 - 2x + 2$.
(a) $f(x) \equiv (x-1)^2 + 1$.
(b) $f(1) = 0$.
(c) $f(x) = 0$ has equal roots.

TYPE III

13) (a) $x(x-2) \equiv x^2 - 2x$.
    (b) $x = 2$.

14) (a) $x + 1 > 2$.
    (b) $x < 0$.

15) (a) $f(3) = -5$.
    (b) $f(x) \equiv x^2 - 6x + 4$.

16) (a) $x^2 - 4x + 2 = 0$ has roots α and β.

    (b) $2x^2 - 4x + 1 = 0$ has roots $\dfrac{1}{\alpha}$ and $\dfrac{1}{\beta}$.

TYPE IV

17) Solve the equation $ax^2 + bx + c = 0$.
(a) $a = 1$.
(b) One root is twice the other root.
(c) $c = 2$.

18) Write down the value of $f(2)$.
(a) $f(X)$ is a polynomial of degree 1.
(b) $f(0) = 1$.
(c) $f(1) = 2$.

19) Express $f(x)$ in partial fractions.
(a) $f(x)$ is a proper fraction.
(b) $f(0) = 1$.
(c) $f(x) \equiv \dfrac{A}{(x-1)^2(x+1)}$.

TYPE V

20) A quadratic equation always has two real solutions.

21) The relationship $x(x^2 + 4) = x^2 + 4x$ is an identity.

22) The expression $\dfrac{x-2}{(x-1)(x+1)^2}$ can be expressed as three separate fractions.

23) The inequality $x + 5 > 3$ is satisfied by a finite number of values of x.

24) The equation $ax^2 + bx + c = 0$ has two roots.

25) If $f(x) \equiv \dfrac{x}{(x^2+1)(x-1)}$ then $f(x) \equiv \dfrac{A}{x^2+1} + \dfrac{B}{x+1} + \dfrac{C}{x-1}$.

MISCELLANEOUS EXERCISE 1

1) Express $\dfrac{9x}{(2x+1)^2(1-x)}$ as a sum of partial fractions with constant
numerators.   (U of L)p

2) Resolve the expression $\dfrac{(x-2)}{(x^2+1)(x-1)^2}$ into its simplest partial fractions.
                                                                (U of L)p

3) Express the function $\dfrac{7x+4}{(x-3)(x+2)^2}$ as the sum of three partial fractions
with numerators independent of x.   (JMB)p

4) If α, β are the roots of the equation
$$ax^2 - bx + c = 0$$
form the equation whose roots are $\alpha + \dfrac{1}{\alpha}, \beta + \dfrac{1}{\beta}$.   (U of L)p

5) If the roots of $x^2 + px + q = 0$ are α and β, where α and β are non-zero, form the equation whose roots are $\dfrac{2}{\alpha}, \dfrac{2}{\beta}$.   (U of L)p

6) If α and β are the roots of the equation $ax^2 + bx + c = 0$ show that the roots of the equation $acx^2 - (b^2 - 2ac)x + ac = 0$ are $\dfrac{\alpha}{\beta}$ and $\dfrac{\beta}{\alpha}$.

(AEB)'71p

7) The roots of the quadratic equation $x^2 - px + q = 0$ are α and β. Form, in terms of p and q, the quadratic equation whose roots are $\alpha^3 - p\alpha^2$ and $\beta^3 - p\beta^2$.   (AEB)'75p

8) Form a quadratic equation with roots which exceed by 2 the roots of the quadratic equation $3x^2 - (p - 4)x - (2p + 1) = 0$.
Find the values of p for which the given equation has equal roots.

(U of L)p

CHAPTER 2

ALGEBRAIC TOPICS

INDICES

If $a^3 \equiv a \times a \times a$ then $a^2 \times a^3 \equiv (a \times a) \times (a \times a \times a) \equiv a^5$.

From similar illustrations it can be seen that

$$a^n \times a^m \equiv a^{n+m} \qquad [1]$$

and
$$a^n \div a^m \equiv a^{n-m} \qquad [2]$$

and
$$(a^n)^m \equiv a^{nm} \qquad [3]$$

These three identities form the basic laws of indices and if they are to hold for values of m and n other than positive integral values, we must attach some meaning to zero, negative and fractional indices.

From [2], $\qquad a^3 \div a^5 \equiv a^{-2}$

But $\qquad a^3 \div a^5 \equiv \dfrac{a \times a \times a}{a \times a \times a \times a \times a} \equiv \dfrac{1}{a^2}$

therefore $\qquad a^{-2}$ means $\dfrac{1}{a^2}$

In general $\qquad a^{-n} \equiv \dfrac{1}{a^n}$

Also from [2] $\qquad a^3 \div a^3 \equiv a^0$

but $$\frac{a^3}{a^3} \equiv 1$$

therefore the meaning of a^0 is that a fraction has cancelled completely to unity,

i.e. $$a^0 \equiv 1.$$

Using [1], a^1 may be written as $a^{\frac{1}{2}} \times a^{\frac{1}{2}}$

i.e. $a^1 \equiv a^{\frac{1}{2}} \times a^{\frac{1}{2}}$, so $a^{\frac{1}{2}} \equiv \sqrt{a}$

similarly $a^1 \equiv a^{\frac{1}{5}} \times a^{\frac{1}{5}} \times a^{\frac{1}{5}} \times a^{\frac{1}{5}} \times a^{\frac{1}{5}} \Rightarrow a^{\frac{1}{5}} \equiv \sqrt[5]{a}$

In general $$a^{\frac{1}{n}} \equiv \sqrt[n]{a}$$

Also $a^{\frac{3}{4}} \equiv a^{\frac{1}{4}} \times a^{\frac{1}{4}} \times a^{\frac{1}{4}}$

i.e. $a^{\frac{3}{4}}$ means the cube of the fourth root of a

In general $$a^{\frac{n}{m}} \equiv (\sqrt[m]{a})^n \text{ or } \sqrt[m]{a^n}$$

EXAMPLES 2a

1) Simplify $\left(\dfrac{125}{27}\right)^{-\frac{2}{3}}$

$\left(\dfrac{125}{27}\right)^{-\frac{2}{3}}$ means the reciprocal of $\left(\dfrac{125}{27}\right)^{\frac{2}{3}}$

i.e. $\left(\dfrac{125}{27}\right)^{-\frac{2}{3}} = \left(\dfrac{27}{125}\right)^{\frac{2}{3}} = \left(\sqrt[3]{\dfrac{27}{125}}\right)^2 = \left(\dfrac{3}{5}\right)^2 = \dfrac{9}{25}$

2) Simplify $\dfrac{x^{-\frac{1}{2}}(x-1)^{\frac{1}{2}} + x^{\frac{1}{2}}(x-1)^{-\frac{1}{2}}}{x^{\frac{1}{2}}}$

$\dfrac{x^{-\frac{1}{2}}(x-1)^{\frac{1}{2}} + x^{\frac{1}{2}}(x-1)^{-\frac{1}{2}}}{x^{\frac{1}{2}}} \times \dfrac{x^{\frac{1}{2}}(x-1)^{\frac{1}{2}}}{x^{\frac{1}{2}}(x-1)^{\frac{1}{2}}}$

$\equiv \dfrac{x^0(x-1) + x(x-1)^0}{x(x-1)^{\frac{1}{2}}}$

$\equiv \dfrac{x-1+x}{x(x-1)^{\frac{1}{2}}} \equiv \dfrac{2x-1}{x(x-1)^{\frac{1}{2}}}$

EXERCISE 2a

Evaluate

1) $\left(\dfrac{1}{2}\right)^{-2}$

2) $\dfrac{1}{2^{-1}}$

3) $27^{-\frac{1}{3}}$

4) $\left(\dfrac{16}{49}\right)^{\frac{1}{2}}$

5) $(125)^{-\frac{1}{3}}$

6) $(121)^{\frac{3}{2}}$

7) $\left(\dfrac{1}{4}\right)^{\frac{5}{2}}$

8) $\left(\dfrac{1}{9}\right)^{-\frac{3}{2}}$

9) $\left(\dfrac{100}{9}\right)^{-\frac{3}{2}}$

10) $\left(\dfrac{27}{8}\right)^{\frac{2}{3}}$

11) $\left(-\dfrac{1}{7}\right)^{-2}$

12) $(0.36)^{\frac{1}{2}}$

13) $(0.04)^{-2}$

14) $(2.56)^{-\frac{1}{2}}$

15) $(2\frac{7}{9})^{\frac{3}{2}}$

16) $\left(\dfrac{125}{27}\right)^{0}$

17) $3^{-1} . 2^2 . 4^0$

18) $12^{\frac{1}{2}} . 3^{\frac{1}{2}}$

19) $27^{\frac{1}{4}} . 3^{\frac{1}{4}}$

20) $32^{\frac{1}{2}} . 2^{-\frac{1}{2}}$

21) $\dfrac{9^{\frac{1}{2}} . 8^{\frac{1}{2}}}{2^{\frac{1}{2}}}$

22) $\dfrac{5^{\frac{1}{3}} . 5^0 . 25^{\frac{1}{3}}}{125^{\frac{1}{3}}}$

23) $\dfrac{8^{\frac{1}{3}} . 16^{\frac{1}{3}}}{32^{-\frac{1}{3}}}$

24) $\dfrac{9^{\frac{1}{3}} . 27^{-\frac{1}{2}}}{3^{-\frac{1}{6}} . 3^{-\frac{2}{3}}}$

Simplify

25) $\dfrac{y^{\frac{1}{6}} . y^{-\frac{2}{3}}}{y^{\frac{1}{4}}}$

26) $\dfrac{p^{\frac{1}{2}} . p^{-\frac{3}{4}}}{p^{-\frac{1}{4}}}$

27) $\dfrac{\sqrt{x} . \sqrt{x^3}}{x^{-3}}$

28) $\dfrac{(\sqrt{t})^3 . t^2}{\sqrt{(t^5)}}$

29) $\dfrac{(x-1)^{\frac{1}{2}} + (x-1)^{-\frac{1}{2}}}{(x-1)^{\frac{1}{2}}}$

30) $\dfrac{(x+2)^{-\frac{1}{2}} + 2(x+2)^{-\frac{3}{2}}}{(x+2)}$

31) $\dfrac{x(x+1)^{\frac{1}{2}} - (x+1)^{-\frac{1}{2}}}{x^2}$

32) $\dfrac{m(m^2+1)^{\frac{1}{2}} - 2m^2(m^2+1)^{-\frac{1}{2}}}{(m^2+1)^{-1}}$

33) $\dfrac{(t+1)^{\frac{1}{3}} - \frac{1}{3}t(t+1)^{-\frac{2}{3}}}{(t+1)^{\frac{2}{3}}}$

34) $\dfrac{(p+q)^{-\frac{1}{4}} - (p-q)^{-\frac{1}{4}}}{(p^2-q^2)^{\frac{3}{4}}}$

35) $\dfrac{(x^2-y^2)^{\frac{1}{2}} - (x-y)^{-\frac{1}{2}}}{(x+y)^{-\frac{1}{2}}}$

36) $\dfrac{x^2(1-x)^{\frac{1}{2}} - x(1-x)^{-\frac{1}{2}}}{(1-x)^{-\frac{1}{2}}}$

37) $\dfrac{x^{-\frac{1}{2}}(x^2+1)^{\frac{1}{2}} - x^{\frac{1}{2}}(x^2+1)^{-\frac{1}{2}}}{x^{\frac{1}{2}}(x^2+1)^{\frac{1}{2}}}$

SURDS

Expressions such as $\sqrt{4}$, $\sqrt{25}$ have exact numerical values, viz. $\sqrt{4} = 2$, $\sqrt{25} = 5$.

But expressions such as $\sqrt{2}, \sqrt{3}, \sqrt{5}, \ldots$ cannot be written as numerically exact quantities.

For example we might say that $\sqrt{2} = 1.4$ correct to 2 s.f.

or that $\sqrt{2} = 1.4142136$ correct to 8 s.f.

but we can never find an exact quantity equal to $\sqrt{2}$.

Such numbers are called irrational and it is often convenient to leave them in the form $\sqrt{2}$ (or $\sqrt{3} \ldots$) and they are then called surds.

(**Note**. $\sqrt{2}$ means the positive square root of 2, so although the solution to $x^2 = 4$ is $x = \pm 2$, $\sqrt{4} = 2$)

As surds occur frequently in solutions it is useful to be able to simplify them.

EXAMPLES 2b

1) Express $\sqrt{48}$ as the simplest possible surd.

$$\sqrt{48} = \sqrt{(16 \times 3)} = \sqrt{16} \cdot \sqrt{3} = 4\sqrt{3}$$

2) Expand and simplify:

(a) $(2 - 3\sqrt{3})(3 + 2\sqrt{3})$ (b) $(5 - 2\sqrt{7})(5 + 2\sqrt{7})$.

(a) $(2 - 3\sqrt{3})(3 + 2\sqrt{3}) = 6 - 9\sqrt{3} + 4\sqrt{3} - 6(\sqrt{3})^2$

$$= 6 - 5\sqrt{3} - 6 \times 3$$

$$= -12 - 5\sqrt{3}$$

(b) $(5 - 2\sqrt{7})(5 + 2\sqrt{7}) = 25 - 10\sqrt{7} + 10\sqrt{7} - 4(\sqrt{7})^2$

$$= 25 - 28$$

$$= -3$$

When the solution to a problem results in an answer containing surds it is accepted practice to leave that answer (simplified as far as possible) in surd form unless an approximation is asked for (e.g. give your answer correct to three significant figures). Simplification of a fractional answer can often be achieved by removing surds from the denominator and this process is called *rationalizing the denominator*.

3) Rationalize the denominator of $\dfrac{3}{\sqrt{2}}$.

$$\frac{3}{\sqrt{2}} = \frac{3}{\sqrt{2}} \cdot \frac{\sqrt{2}}{\sqrt{2}} = \frac{3\sqrt{2}}{2}$$

4) Simplify $\dfrac{3 - \sqrt{5}}{1 + 3\sqrt{5}}$.

Referring back to Example 2(b) we see that expanding brackets of the form $(a + b)(a - b)$ gives $a^2 - b^2$ thus eliminating any surds contained in either a or b. So to rationalize the denominator of $\dfrac{3 - \sqrt{5}}{1 + 3\sqrt{5}}$ we multiply numerator and denominator by $1 - 3\sqrt{5}$.

i.e.

$$\frac{3 - \sqrt{5}}{1 + 3\sqrt{5}} = \frac{(3 - \sqrt{5})(1 - 3\sqrt{5})}{(1 + 3\sqrt{5})(1 - 3\sqrt{5})} = \frac{18 - 10\sqrt{5}}{1^2 - (3\sqrt{5})^2}$$

$$= \frac{9 - 5\sqrt{5}}{-22} = \frac{5\sqrt{5} - 9}{22}$$

EXERCISE 2b

1) Express in terms of the simplest possible surds:
(a) $\sqrt{8}$ (b) $\sqrt{12}$ (c) $\sqrt{50}$ (d) $\sqrt{18}$ (e) $\sqrt{200}$
(f) $\sqrt{72}$ (g) $\sqrt{125}$ (h) $\sqrt{288}$ (i) $\sqrt{450}$ (j) $\sqrt{2000}$.

2) Simplify:
(a) $\sqrt{2}(3 - \sqrt{2})$ (b) $\sqrt{2}(3 - 2\sqrt{2})$ (c) $\sqrt{3}(\sqrt{27} - 1)$
(d) $(\sqrt{2} - 1)(\sqrt{2} + 1)$ (e) $(\sqrt{3} - 2)(\sqrt{3} - 1)$ (f) $(2\sqrt{2} + 1)(\sqrt{2} - 2)$
(g) $(3\sqrt{3} - 2)(3\sqrt{3} + 2)$ (h) $(2\sqrt{5} + 3)(3\sqrt{5} - 2)$ (i) $(\sqrt{3} - 1)(\sqrt{2} + 1)$
(j) $(2\sqrt{6} - 3)^2$ (k) $(\sqrt{x} - 1)(\sqrt{x} + 1)$ (l) $(2\sqrt{x} - 1)^2$

3) Rationalise the denominators and simplify:

(a) $\dfrac{1}{\sqrt{3}}$ (b) $\dfrac{1}{\sqrt{8}}$ (c) $\dfrac{2}{\sqrt{32}}$ (d) $\dfrac{1}{\sqrt{2} + 1}$

(e) $\dfrac{1}{\sqrt{3} - 1}$ (f) $\dfrac{3}{2 - \sqrt{3}}$ (g) $\dfrac{5}{2 + \sqrt{5}}$ (h) $\dfrac{2}{2\sqrt{3} - 3}$

(i) $\dfrac{1}{3 + 2\sqrt{5}}$ (j) $\dfrac{1}{\sqrt{6} - \sqrt{5}}$ (k) $\dfrac{1}{2\sqrt{3} + \sqrt{2}}$ (l) $\dfrac{1}{\sqrt{2} + 1} + \dfrac{1}{\sqrt{2} - 1}$

(m) $\dfrac{1}{\sqrt{x} + 1} - \dfrac{1}{\sqrt{x} - 2}$ (n) $\dfrac{1}{(3\sqrt{2} - 1)^2}$

LOGARITHMS

Logarithm is another word for an index or power.
Now $2^3 = 8$

i.e.   3 is the power to which the base 2 must be raised to obtain 8

or   *3 is the logarithm which, with a base 2, gives 8.*

This is written simply as $3 = \log_2 8$
Similarly $3^2 = 9$

i.e.   2 is the logarithm which, with a base 3, gives 9

or $2 = \log_3 9$

Also $(\tfrac{1}{5})^{-2} = 25$

i.e. -2 is the logarithm which, with a base $\tfrac{1}{5}$, gives 25

or $-2 = \log_{\frac{1}{5}} 25$

So we see that the base of a logarithm may be any number. The tables of common logarithms, which are usually used for calculations, have a base 10. From these tables (or from a calculator) it is found that the logarithm of 5 is 0.6990.

i.e. $10^{0.6990} = 5$ or $\log_{10} 5 = 0.6990$

However it is usual to omit the base 10 and to write simply

$$\log 5 = 0.6990 \quad \text{or} \quad \lg 5 = 0.6990$$

but any base other than 10 must be stated.

Therefore $\qquad\qquad \log 100 = 2 \qquad$ (i.e. $\quad 100 = 10^2$)

and $\qquad\qquad\quad \log_{0.1} 100 = -2 \qquad$ (i.e. $\quad 100 = 0.1^{-2}$)

In general $\qquad\qquad\qquad \log_a b = c \Longleftrightarrow b = a^c$

EXERCISE 2c

1) Express in logarithmic form:

(a) $5^3 = 125$ (b) $7^2 = 49$ (c) $4096 = 8^4$ (d) $4^{\frac{5}{2}} = 32$

(e) $1331 = (121)^{\frac{3}{2}}$ (f) $10^{-2} = 0.01$ (g) $5^0 = 1$ (h) $8^{-\frac{1}{3}} = \frac{1}{2}$

(i) $125 = (\frac{1}{5})^{-3}$ (j) $9^{-\frac{3}{2}} = \frac{1}{27}$ (k) $1 = a^0$ (l) $\pi^2 = 9.8696$

(m) $p = q^2$ (n) $a^c = b$ (o) $x^y = 2$

2) Express in index form:

(a) $\log_5 625 = 4$ (b) $\log 1000 = 3$ (c) $\log_3 27 = 3$ (d) $\log 1 = 0$

(e) $\log_{\frac{1}{2}} 4 = -2$ (f) $\log_{25} 5 = \frac{1}{2}$ (g) $\log_a 1 = 0$ (h) $\log_x y = 2$

(i) $\log_4 p = q$ (j) $\log_a 5 = b$ (k) $\log_x y = z$ (l) $p = \log_q r$

3) Evaluate

(a) $\log_4 64$ (b) $\log 10\,000$ (c) $\log_9 3$ (d) $\log_{\frac{1}{2}} 4$

(e) $\log_{0.1} 10$ (f) $\log_{125} 25$ (g) $\log_{121} 11$ (h) $\log_{81} 3$

(i) $\log 0.1$ (j) $\log_8 0.5$ (k) $\log_9 9^{\frac{1}{2}}$ (l) $\log_2 2^3$

(m) $\log 1$ (n) $\log_a 1$ (o) $\log_a a^3$ (p) $\log_a a^b$

(q) $10^{\log 10}$ (r) $5^{\log_5 25}$

The Laws of Logarithms

Let $\quad \log_a b \equiv x \quad$ and $\quad \log_a c \equiv y$

therefore $\quad a^x \equiv b \quad$ and $\qquad a^y \equiv c$

$$a^x \cdot a^y \equiv bc$$

or $$a^{x+y} \equiv bc$$

therefore $$x + y \equiv \log_a bc$$

i.e. $$\log_a b + \log_a c \equiv \log_a bc \qquad\qquad [1]$$

Similarly
$$a^x \div a^y \equiv \frac{b}{c}$$

or
$$a^{x-y} \equiv \frac{b}{c}$$

therefore
$$x - y \equiv \log_a \frac{b}{c}$$

i.e.
$$\log_a b - \log_a c \equiv \log_a \frac{b}{c} \qquad [2]$$

Let
$$\log_a b^n \equiv z$$

therefore
$$a^z \equiv b^n$$

so
$$a^{\frac{z}{n}} \equiv b$$

therefore
$$\log_a b \equiv \frac{z}{n}$$

or
$$n \log_a b \equiv z$$

i.e.
$$n \log_a b \equiv \log_a b^n \qquad [3]$$

The identities [1], [2], [3] are known as the three basic laws of logarithms, i.e.

$$\log_a x + \log_a y \equiv \log_a xy$$

$$\log_a x - \log_a y \equiv \log_a \frac{x}{y}$$

$$\log_a x^y \equiv y \log_a x$$

Using these three laws an expression such as $\log\left(\dfrac{a^2}{b^3}\right)$ may be written in terms of $\log a$ and $\log b$

i.e.
$$\log \frac{a^2}{b^3} \equiv \log a^2 - \log b^3$$

$$\equiv 2 \log a - 3 \log b$$

Conversely an expression such as $\log 100 - 2 \log 50$ may be written as the logarithm of a single number

i.e.
$$\log 100 - 2 \log 50 = \log 100 - \log 50^2$$

$$= \log 100 - \log 2500$$

$$= \log \left(\frac{100}{2500}\right)$$

$$= \log \frac{1}{25}$$

Changing the Base of a Logarithm

Tables are not readily available which list the values of expressions such as $\log_7 2$.

However if	$\log_7 2 = x$
then	$7^x = 2$
so	$x \log 7 = \log 2$
or	$x = \dfrac{\log 2}{\log 7}$

i.e. we have changed the base from 7 to 10 and can now use our ordinary tables or calculator to evaluate $\log_7 2$.

Therefore
$$\log_7 2 = \frac{\log 2}{\log 7} = \frac{0.3010}{0.8451} = 0.3562$$

A general formula for changing from base a to base b can be derived in the same way.

If	$\log_a c \equiv x$
then	$c \equiv a^x$
so	$\log_b c \equiv x \log_b a$
$\Rightarrow$	$x \equiv \dfrac{\log_b c}{\log_b a}$
or	$\log_a c \equiv \dfrac{\log_b c}{\log_b a}$

In the special case when $c = b$, this identity becomes
$$\log_a b \equiv \frac{\log_b b}{\log_b a}$$

or
$$\log_a b \equiv \frac{1}{\log_b a}$$

EXPONENTIAL EQUATIONS (i.e. where the variable is an index)

To solve an equation like $5^x = 10$ we can make use of the third law of logarithms to 'bring down' the power. Taking logs of both sides gives

$$\log 5^x = \log 10$$

$$x \log 5 = \log 10$$

$$x = \frac{\log 10}{\log 5} = \frac{1}{0.6990} = 1.43 \text{ to } 3 \text{ s.f.}$$

Now consider an equation of the type

$$2^{2x} + 3(2^x) - 4 = 0$$

Taking logs is no help this time as $2^{2x} + 3(2^x)$ cannot be combined into a single expression, i.e. $\log [2^{2x} + 3(2^x)]$ cannot be simplified. But if y is substituted for 2^x, the equation becomes quadratic in y

i.e. $$y^2 + 3y - 4 = 0$$

$$(y + 4)(y - 1) = 0$$

$$y = -4 \quad \text{or} \quad y = 1$$

i.e. $$2^x = -4 \quad \text{or} \quad 2^x = 1$$

There are no real values of x for which $2^x = -4$.
But if $2^x = 1$ then $x = 0$, which is the only solution to this equation.

EXAMPLES 2d

1) Solve for x and y the equations

$$xy = 80$$

$$\log x - 2 \log y = 1$$

$$xy = 80 \tag{1}$$

$$\log x - 2 \log y = 1 \tag{2}$$

Using the laws of logs, [2] becomes

$$\log \frac{x}{y^2} = 1 \qquad \text{i.e.} \qquad \frac{x}{y^2} = 10 \quad \text{or} \quad x = 10y^2$$

Substituting $10y^2$ for x in [1] we get

$$10y^3 = 80 \Rightarrow y^3 = 8$$

Therefore $$y = 2 \text{ and } x = 40$$

2) Solve the equation $\log_3 x - 4 \log_x 3 + 3 = 0$.

Using the identity $\log_a b \equiv \dfrac{1}{\log_b a}$, $\log_x 3$ may be replaced by $\dfrac{1}{\log_3 x}$

So the given equation becomes $\log_3 x - \dfrac{4}{\log_3 x} + 3 = 0$

$\Rightarrow$ $$(\log_3 x)^2 - 4 + 3 \log_3 x = 0$$

Substituting y for $\log_3 x$ gives

$$y^2 + 3y - 4 = 0$$

$$(y + 4)(y - 1) = 0$$

therefore $\qquad\qquad\qquad\qquad y = -4 \text{ or } 1$

i.e. either $\qquad\qquad \log_3 x = -4 \qquad \text{or} \quad \log_3 x = 1$

so $\qquad\qquad\qquad x = 3^{-4} = \frac{1}{81} \quad \text{or} \qquad x = 3$

EXERCISE 2d

1) Express in terms of $\log a$, $\log b$ and $\log c$:

(a) $\log ab$ 　　(b) $\log abc$ 　　(c) $\log \dfrac{a}{b}$ 　　(d) $\log \dfrac{ab}{c}$ 　　(e) $\log \dfrac{a}{bc}$

(f) $\log \dfrac{1}{a}$ 　　(g) $\log a^2 b$ 　　(h) $\log \sqrt{\dfrac{a}{b}}$ 　　(i) $\log \dfrac{a^2}{b}$ 　　(j) $\log \dfrac{a^3}{10}$

(k) $\log \dfrac{1}{10a}$ 　　(l) $\log \dfrac{100}{\sqrt{b}}$

2) Simplify:
(a) $\log 3 + \log 4$ 　　(b) $\log 6 - \log 2$ 　　(c) $\log 2 + \log 6 - \log 4$
(d) $2 \log 3 + \log 2$ 　　(e) $\frac{1}{2} \log 4 - \log 6$ 　　(f) $2 - 2 \log 5$

(g) $\frac{1}{2} \log 9 + 1$ 　　(h) $\frac{1}{2} \log 25 - 2 \log 3 + 2 \log 6$ 　　(i) $\dfrac{\log 9}{\log 3}$

(j) $\log (x + 1) - \log (x^2 - 1)$ 　　(k) $\dfrac{\log 125}{\log 5}$

(l) $2 \log_a 5 + \log_a 4 - 2 \log_a 10$.

3) Solve the equations:
(a) $3^x = 6$ 　　(b) $5^x = 4$ 　　(c) $2^{2x} = 5$ 　　(d) $3^{x-1} = 7$
(e) $4^{2x+1} = 3$ 　　(f) $(5^x)(5^{x-1}) = 10$.

4) Evaluate:
(a) $\log_2 10$ 　　(b) $\log_7 5$ 　　(c) $\log_{20} 2$ 　　(d) $\log_{0.3} 5$.

5) Solve the equations:
(a) $2(2^{2x}) - 5(2^x) + 2 = 0$ 　　(b) $3^{2x+1} - 26(3^x) - 9 = 0$
(c) $4^x - 6(2^x) - 16 = 0$.

6) If $\log_2 x + \log_x 2 = 2$, find x.

7) Solve the simultaneous equations $\log_x y = 2$ 　$xy = 8$.

8) Solve simultaneously $2 \log y = \log 2 + \log x$ 　and 　$2^y = 4^x$.

9) Solve the simultaneous equations $\log_3 x = y = \log_9 (2x - 1)$.

10) Find the positive value of x that satisfies the equation $\log_2 x = \log_4 (x + 6)$.

11) Show that $(\log_a b^2) \times (\log_b a^3) \equiv 6$.

12) Solve the simultaneous equations $\log(x+y) = 0$

and $2 \log x = \log(y+1)$.

THE REMAINDER THEOREM

If the remainder is required when a polynomial is divided by a linear function it can be found by long division as follows:

$$
\begin{array}{r}
x^2 - 5x - 4 \\
x - 2 \overline{)\,x^3 - 7x^2 + 6x - 2} \\
x^3 - 2x^2 \\
\hline
-5x^2 + 6x - 2 \\
-5x^2 + 10x \\
\hline
-4x - 2 \\
-4x + 8 \\
\hline
-10
\end{array}
$$

Thus when $f(x) \equiv x^3 - 7x^2 + 6x - 2$ is divided by $x - 2$ there is a remainder of -10 and a quotient $x^2 - 5x - 4$.

This result can be written in the form

$$f(x) \equiv x^3 - 7x^2 + 6x - 2 \equiv (x-2)(x^2 - 5x - 4) - 10$$

If 2 is substituted for x in this identity so that $x - 2 = 0$, the quotient is eliminated giving $f(2) = -10$.

In general if $f(x)$ is a polynomial function of x which, when divided by $x - a$, gives a quotient $Q(x)$ and a remainder R, the relationship between these expressions is

$$f(x) \equiv (x-a)Q(x) + R$$

Substituting a for x in this identity gives

$$f(a) = R$$

This result is known as the remainder theorem and can be summarized as follows.

When $f(x)$ is divided by $x - a$ the remainder is $f(a)$.

Note. This theorem gives a (simple) method for evaluating the remainder only. If the quotient is required, long division must be used.

So, for example, if we want the remainder when $x^3 - 2x^2 + 6$ is divided by $x + 3$ we write

$$f(x) \equiv x^3 - 2x^2 + 6$$

and the remainder, R, is given by

$$R = f(-3) = (-3)^3 - 2(-3)^2 + 6 = -39$$

And if we want the remainder when $6x^2 - 7x + 2$ is divided by $2x - 1$ then as

$$f(x) \equiv 6x^2 - 7x + 2 \equiv (2x - 1)Q(x) + R$$
$$R = f(\tfrac{1}{2})$$

THE FACTOR THEOREM

The remainder theorem states that when $f(x)$ is divided by $(x - a)$ the remainder, R, is $f(a)$.
Now if $x - a$ is a factor of $f(x)$ there will be no remainder,

i.e. $\qquad\qquad\qquad R = 0 \quad \text{and} \quad f(a) = 0.$

This property is known as the factor theorem and is defined as follows:

If, for a given polynomial function $f(x)$, $f(a) = 0$, then $x - a$ is a factor of $f(x)$.

This theorem is very useful when factorizing polynomials of degree greater than 2.

EXAMPLE 2e

Factorise completely $x^4 - 3x^3 + 4x^2 - 8$

Let $\qquad\qquad\qquad f(x) \equiv x^4 - 3x^3 + 4x^2 - 8$

then $\qquad\qquad\qquad f(1) = 1 - 3 + 4 - 8 \neq 0$

therefore $x - 1$ *is not* a factor of $f(x)$.

$$f(-1) = 1 + 3 + 4 - 8 = 0$$

therefore $x + 1$ *is* a factor of $f(x)$.
Now that one factor has been found it should be 'taken out', either by inspection or by long division.

By inspection $x^4 - 3x^3 + 4x^2 - 8 \equiv (x + 1)(x^3 - 4x^2 + 8x - 8)$

Now let $\qquad\qquad\qquad g(x) \equiv x^3 - 4x^2 + 8x - 8$

$$g(-1) = -1 - 4 - 8 - 8 \neq 0$$

Therefore $x + 1$ *is not* a factor of $g(x)$

$$g(2) = 8 - 16 + 16 - 8 = 0$$

Therefore $x - 2$ *is* a factor of $g(x)$.

By inspection $\quad x^3 - 4x^2 + 8x - 8 \equiv (x - 2)(x^2 - 2x + 4)$

and $x^2 - 2x + 4$ has no linear factors.

Therefore $\qquad x^4 - 3x^3 + 4x^2 - 8 \equiv (x + 1)(x - 2)(x^2 - 2x + 4)$

Note (1) The factors of 8 are $1, 2, 4, 8$ so the values we choose for a must belong to the set $\{\pm 1, \pm 2, \pm 4, \pm 8\}$.

(2) After having taken out the first factor from $f(x)$ it must be tried again as a possible factor of $g(x)$ since repeated factors occur frequently.

The Factors of $a^3 - b^3$ and $a^3 + b^3$

$$a^3 - b^3 = 0 \quad \text{when} \quad a = b$$

therefore $a - b$ is a factor of $a^3 - b^3$.

Hence $\qquad a^3 - b^3 \equiv (a - b)(a^2 + ab + b^2)$

$$a^3 + b^3 = 0 \quad \text{when} \quad a = -b$$

so $a + b$ is a factor of $a^3 + b^3$

giving $\qquad a^3 + b^3 \equiv (a + b)(a^2 - ab + b^2)$

These two identities are very useful and should be memorized.

EXERCISE 2e

1) Find the remainder when the following functions are divided by the linear factors indicated.

(a) $x^3 - 2x + 4, x - 1$ (b) $x^3 + 3x^2 - 6x + 2, x + 2$

(c) $2x^3 - x^2 + 2, x - 3$ (d) $x^4 - 3x^3 + 5x, 2x - 1$

(e) $9x^5 - 5x^2, 3x + 1$ (f) $x^3 - 2x^2 + 6, x - a$

(g) $x^2 + ax + b, x + c$

2) Determine whether the following linear functions are factors of the given polynomials.

(a) $x^3 - 7x + 6, x - 1$ (b) $2x^2 + 3x - 4, x + 1$

(c) $x^3 - 6x^2 + 6x - 2, x - 2$ (d) $x^3 - 27, x - 3$

(e) $2x^4 - x^3 - 6x^2 + 5x - 1, 2x - 1$

(f) $x^3 + ax^2 - a^2x - a^3, x + a$

3) Factorize the following functions as far as possible.

(a) $x^3 + 2x^2 - x - 2$ (b) $x^3 - x^2 - x - 2$ (c) $x^4 - 1$

(d) $x^4 + x^3 - 3x^2 - 4x - 4$ (e) $2x^3 - x^2 + 2x - 1$

(f) $27x^3 - 1$ (g) $x^3 + a^3$ (h) $x^3 - y^3$

4) If $x^2 - 7x + a$ has a remainder 1 when divided by $x + 1$, find a.

5) If $x - 2$ is a factor of $ax^2 - 12x + 4$ find a.

6) One solution of the equation $x^2 + ax + 2 = 0$ is $x = 1$, find a.

7) One root of the equation $x^2 - 3x + a = 0$ is 2. Find the other root.

BINOMIAL EXPRESSIONS

A binomial is the sum (or difference) of two terms.

e.g. $a + b$, $2x + 3y$, $p^2 - r$ are binomials.

It is often necessary to expand (i.e. multiply out) a power of a binomial, e.g.

$$(x + y)^3 \equiv (x + y)^2(x + y) \equiv (x^2 + 2xy + y^2)(x + y) \equiv x^3 + 3x^2y + 3xy^2 + y^3$$

This multiplication is a tedious process for powers of 3 and over so we describe below a far quicker way of obtaining such expansions.

Pascal's Triangle

Consider the following expansions:

$$(1 + x)^0 \equiv 1$$

$$(1 + x)^1 \equiv 1 + x \qquad [1]$$

$$(1 + x)^2 \equiv 1 + 2x + x^2 \qquad [2]$$

$$(1 + x)^3 \equiv 1 + 3x + 3x^2 + x^3 \qquad [3]$$

$$(1 + x)^4 \equiv 1 + 4x + 6x^2 + 4x^3 + x^4 \qquad [4]$$

Closer inspection of [4], i.e.

$$(1 + x)^4 \equiv (\boxed{1} + \boxed{3x} + \boxed{3x^2} + \boxed{x^3})(1 + x)$$

$$\equiv 1 + x + 3x + 3x^2 + 3x^2 + 3x^3 + x^3 + x^4$$

$$\equiv \boxed{1} + \boxed{(1 + 3)x} + \boxed{(3 + 3)x^2} + \boxed{(3 + 1)x^3} + \boxed{x^4}$$

$$\equiv 1 + 4x + 6x^2 + 4x^3 + x^4$$

shows that the coefficient of any one power of x is the sum of the coefficients of the same and preceding power of x in the previous expansion.

For the expansions [1], [2], [3] and [4], writing the coefficients only as a triangular array gives

Knowing that any one number is obtained by adding together the two numbers in the row above as shown, we can add as many further rows as we wish, e.g.

$$
\begin{array}{ccccccccc}
& & & & 1 & & 1 & & \\
& & & 1 & & 2 & & 1 & \\
& & 1 & & 3 & & 3 & & 1 \\
& 1 & & 4 & & 6 & & 4 & & 1 \\
1 & & 5 & & 10 & & 10 & & 5 & & 1 \\
\end{array}
$$

1   6   15   20   15   6   1

1   7   21   35   35   21   7   1

The numbers in row five give the coefficients in the expansion of $(1 + x)^5$, and so on for the subsequent rows.

This triangular array is known as Pascal's Triangle.

Using this triangle we may write

$$(1 + x)^5 \equiv 1 + 5x + 10x^2 + 10x^3 + 5x^4 + x^5$$

EXAMPLES 2f

1) Expand $(1 + 3y)^3$.

From Pascal's Triangle

$$(1 + x)^3 \equiv 1 + 3x + 3x^2 + x^3$$

replacing x by $3y$ we have

$$(1 + 3y)^3 \equiv 1 + 3(3y) + 3(3y)^2 + (3y)^3$$
$$\equiv 1 + 9y + 27y^2 + 27y^3$$

2) Expand $(a + b)^4$ and $(a + b)^6$.

From Pascal's Triangle

$$(1 + x)^4 \equiv 1 + 4x + 6x^2 + 4x^3 + x^4$$

Writing $\qquad (a + b)^4 \equiv a^4 \left(1 + \dfrac{b}{a}\right)^4$

and replacing x by $\dfrac{b}{a}$ we have

$$(a + b)^4 \equiv a^4 \left[1 + 4\left(\frac{b}{a}\right) + 6\left(\frac{b^2}{a^2}\right) + 4\left(\frac{b^3}{a^3}\right) + \frac{b^4}{a^4}\right]$$
$$\equiv a^4 + 4a^3b + 6a^2b^2 + 4ab^3 + b^4$$

Note that (a) the sum of the powers of a and b is always four in each term,
(b) that as the powers of a decrease, the powers of b increase,
(c) the numerical coefficients are those for the expansion of $(1 + x)^4$.

From these observations we can write out the expansion of $(a + b)^6$ directly

$$(a + b)^6 \equiv a^6 + 6a^5b + 15a^4b^2 + 20a^3b^3 + 15a^2b^4 + 6ab^5 + b^6$$

3) Expand $(2x - 3y)^3$.

Using $\quad (a + b)^3 \equiv a^3 + 3a^2b + 3ab^2 + b^3$

and replacing a by $2x$, b by $-3y$, we have:

$$(2x - 3y)^3 \equiv (2x)^3 + 3(2x)^2(-3y) + 3(2x)(-3y)^2 + (-3y)^3$$
$$\equiv 8x^3 - 36x^2y + 54xy^2 - 27y^3$$

EXERCISE 2f

Expand the following binominal expressions.

1) $(1 + 2x)^4$ 　　　 2) $(1 - x)^3$ 　　　 3) $(x - 1)^3$

4) $(1 - 3y)^4$ 　　　 5) $(1 + x)^7$ 　　　 6) $(2x - 1)^3$

7) $(2x + y)^3$ 　　　 8) $\left(x + \dfrac{1}{x}\right)^4$ 　　　 9) $(p - 2q)^3$

10) $(x^2 - y)^5$ 　　 11) $\left(x - \dfrac{1}{x}\right)^3$ 　　 12) $(a - b)^3(a + b)^3$

Use Pascal's Triangle to simplify:

13) $(1 + \sqrt{2})^3$ 　　 14) $(\sqrt{3} - \sqrt{2})^4$ 　　 15) $(1 + \sqrt{3})^3 + (1 - \sqrt{3})^3$

16) By expanding $(1 + 0.01)^3$ evaluate $(1.01)^3$ without using tables.

17) Find, without using tables, the value of $(2.1)^4$.

SUMMARY

Indices:
$$a^{-n} \equiv \frac{1}{a^n}$$

$$a^0 \equiv 1$$

$$a^{\frac{1}{n}} \equiv \sqrt[n]{a}$$

$$a^{\frac{n}{m}} \equiv \sqrt[m]{a^n}$$

Logarithms:
$$\log_a b = c \iff b = a^c$$

$$\log_a x + \log_a y \equiv \log_a xy$$

$$\log_a x - \log_a y \equiv \log_a \frac{x}{y}$$

$$\log_a x^y \equiv y \log_a x$$

$$\log_a c \equiv \frac{\log_b c}{\log_b a}$$

$$\log_a b \equiv \frac{1}{\log_b a}$$

Remainder Theorem: When $f(x)$ is divided by $x - a$ the remainder is $f(a)$.

Factor Theorem: If $f(a) = 0$, $x - a$ is a factor of $f(x)$.

$$a^3 + b^3 \equiv (a + b)(a^2 - ab + b^2)$$

$$a^3 - b^3 \equiv (a - b)(a^2 + ab + b^2)$$

MULTIPLE CHOICE EXERCISE 2

(Instructions for answering these questions are given on page xii.)

TYPE I

1) If $\log_x y = 2$:

(a) $x = 2y$ (b) $x = y^2$ (c) $x^2 = y$ (d) $y = 2x$ (e) $y = \sqrt{x}$.

2) $x^3 - 3x^2 + 6x - 2$ has remainder 2 when divided by:

(a) $x - 1$ (b) $x + 1$ (c) x (d) $x + 2$ (e) $2x - 1$.

3) $\log 5 - 2 \log 2 + \frac{3}{2} \log 16$ is equal to:

(a) $\log 80$ (b) 10   (c) 0   (d) $2 \log 12$ (e) 1.

4) $x^3 - 3x^2 + 2x - 6$ has a factor:

(a) $x - 3$ (b) $x - 2$ (c) $x - 4$ (d) $x + 3$ (e) $x + 2$.

5) If $2^{2x+1} - 6(2^x) = 0$ then x is:

(a) 1.5   (b) $\log_2 3$ (c) $\log 3$ (d) $\log_3 2$ (e) 3.

6) $\dfrac{p^{-\frac{1}{2}} \times p^{\frac{3}{4}}}{p^{-\frac{1}{4}}}$ simplifies to:

(a) 1   (b) $p^{-\frac{1}{2}}$ (c) $p^{\frac{3}{4}}$ (d) p (e) $p^{\frac{1}{2}}$.

7) In the expansion of $(a - 2b)^3$ the coefficient of b^2 is:

(a) $-2a^2$ (b) $-8a$ (c) $12a$ (d) $-4a$ (e) -12.

TYPE II

8) $f(x) \equiv x^3 - 7x^2 + 3x - 1$.
(a) $f(x)$ has a remainder -15 when divided by $x - 2$,
(b) $f(x)$ has no linear factors with integral coefficients,
(c) $f(x)$ is a polynominal of degree 3.

9) $\frac{1}{2} \log 16 - 1$:
(a) can be expressed as a single logarithm,
(b) has an exact numerical value,
(c) is equal to $\log 7$.

10) $f(x) \equiv 2x^2 + 3x - 2$.
(a) $f(x)$ can be expressed as the sum of two partial fractions,
(b) the equation $f(x) = 0$ has two real distinct roots,
(c) $x + 2$ is a factor of $f(x)$.

11) $\dfrac{2\sqrt{3} - 2}{2\sqrt{3} + 2}$:

(a) can be expressed as a fraction with a rational denominator,
(b) is an irrational number,
(c) is equal to -1.

12) $f(x) \equiv (3 - 5x)^4$.
(a) $f(x)$ has a remainder 16 when divided by $x - 1$,
(b) the expansion of $f(x)$ contains four terms,
(c) the equation $f(x) = 0$ is satisfied by only one value of x.

TYPE III

13) (a) $a = \log x$.
    (b) $a \log_x 10 = 1$.

14) (a) $f(x) \equiv x^2 + 2x + 1$.
    (b) $f(x)$ has a remainder 1 when divided by x.

15) (a) $\log_b a = c$.
    (b) $a = b^c$.

16) (a) $a = \log c + \log b$.
    (b) $a = \log (c + b)$.

TYPE V

17) If $f(x)$ is divided by $ax - 1$ the remainder is $f\left(\dfrac{1}{a}\right)$.

18) If $x - a$ is a factor of $x^2 + px + q$, the equation $x^2 + px + q = 0$ has a root equal to a.

19) $3 \log x + 1 = \log 10x^3$ is an equation.

20) In the expansion of $(1 + x)^6$ the coefficient of x is 6.

MISCELLANEOUS EXERCISE 2

1) Find the real values of x for which $\log_3 x - 2\log_x 3 = 1$. (U of L)p

2) Solve for x, correct to two significant figures, the equations:
(a) $4^{2x+1} \cdot 5^{x-2} = 6^{1-x}$
(b) $4^x - 2^{x+1} - 3 = 0$. (AEB)'71p

3) Without using tables, write down the values of:
(a) $\log_4 8\sqrt{2}$
(b) $(\log_5 49) \times (\log_7 125)$ (U of L)p

4) If a and b are both positive and unequal, and $\log_a b + \log_b a^2 = 3$ find b in terms of a. (U of L)p

5) If $x = \log_a b$, write down an expression for b in terms of a and x. Hence prove that
$$\log_s t = \frac{\log_r t}{\log_r s}.$$
Given that $\log_3 6 = m$ and $\log_6 5 = n$, express $\log_3 10$ in terms of m and n. (JMB)p

6) (i) Solve for real x the equation $4(3^{2x+1}) + 17(3^x) - 7 = 0$.
(ii) If s and t are positive numbers other than 1, prove that:
(a) $\log_s t + \log_{1/s} t = 0$,
(b) $\log_s t = \dfrac{1}{\log_t s}$. (U of L)

7) Solve the simultaneous equations $\log_y x = 2$, $5y = x + 12\log_x y$. (U of L)p

8) Solve the simultaneous equations $\log_2 x + \log_2 y = 3$, $\log_y x = 2$. (AEB)'76p

9) Solve the simultaneous equations $\log(x-2) + \log 2 = 2\log y$
$\log(x - 3y + 3) = 0$. (U of L)p

10) Show that $\log_{16}(xy) = \frac{1}{2}\log_4 x + \frac{1}{2}\log_4 y$. Hence, or otherwise, solve the simultaneous equations
$$\log_{16}(xy) = 3\tfrac{1}{2}$$
$$\frac{(\log_4 x)}{(\log_4 y)} = -8$$ (AEB)'75p

11) If $P(x) \equiv p_n x^n + p_{n-1} x^{n-1} + \ldots + p_0$ is divided by $(x - a)$, show that the remainder is $P(a)$.
If $Q(x) \equiv x^4 + hx^3 + gx^2 - 16x - 12$ has factors $(x + 1)$ and $(x - 2)$, find the constants h, g and the remaining factors. (U of L)p

12) If $f(x) \equiv x^6 - 5x^4 - 10x^2 + k$, find the value of k for which $x - 1$ is a factor of $f(x)$.

When k has this value, find another factor of $f(x)$, of the form $x + a$, where a is a constant. (C)p

13) If $f(x) \equiv ax^2 + bx + c$ leaves remainders $1, 25, 1$ on division by $x - 1$, $x + 1, x - 2$ respectively, show that $f(x)$ is a perfect square.

(U of L)p

14) Without using tables, solve each of the following equations for x, expressing your answers as simply as possible:

(a) $9 \log_x 5 = \log_5 x$,

(b) $\log_8 \dfrac{x}{2} = \dfrac{\log_8 x}{\log_8 2}$. (JMB)

15) Prove that when a polynomial $f(x)$ is divided by $ax + b$, where $a \neq 0$, the remainder is $f(-b/a)$.

Find the polynomial in x of the third degree, which vanishes when $x = -1$ and $x = 2$, has the value 8 when $x = 0$ and leaves the remainder $16/3$ when divided by $3x + 2$. (JMB)

16) Express $\log_9 xy$ in terms of $\log_3 x$ and $\log_3 y$.

Without using tables, solve for x and y the simultaneous equations

$$\log_9 xy = \frac{5}{2}$$

$$\log_3 x \log_3 y = -6$$

expressing your answers as simply as possible. (JMB)

CHAPTER 3

QUADRATIC FUNCTIONS

GRAPHICAL REPRESENTATION OF FUNCTIONS (CURVE SKETCHING)

Consider the function $f(x) \equiv x^2 - 2x + 6$.

For any arbitrarily chosen value of x, the corresponding value of $f(x)$ can be calculated, e.g.

x	-3	-2	-1	0	1	2	3	4
$f(x)$	21	14	9	6	5	6	9	14

Having arranged the values of $f(x)$ in the order of ascending values of x, we can see that there is some pattern to the values obtained for $f(x)$.

This pattern is easier to visualise if the results are displayed graphically, i.e. if the values of $f(x)$ are plotted against the corresponding values of x on a graph.

The following graph has been drawn using only eight values of x, separated by intervals of one unit. We could choose values of x from a larger range and use smaller intervals. The corresponding values of $f(x)$ so obtained would all be found to lie on the smooth curve we have already drawn.

Therefore this curve is a graphical representation of the varying values of $f(x)$ obtained from varying values of x and all possible related values of $f(x)$ and x lie on the curve.

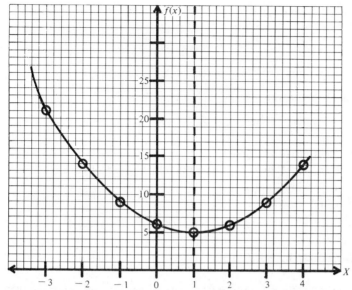

We can now use the graph to deduce some properties of the function.

(a) The lowest point on the curve is where $x = 1$ and $f(x) = 5$ so although x, as the independent variable, can be given any real value, the corresponding value of $f(x)$ is always greater than 5, or equal to 5 when $x = 1$. We say that the function has a least value of 5.

(b) The curve is symmetrical about the line $x = 1$, as two values of x equidistant from $x = 1$, $(1 + a$ and $1 - a)$, give the same value for $f(x)$.

(c) Conversely, for any one value of $f(x)$, there are two corresponding values for x of the form $1 + a$ and $1 - a$.

All these properties can be confirmed algebraically as follows:

completing the square on R.H.S gives

$$f(x) \equiv x^2 - 2x + 6 \equiv (x - 1)^2 + 5$$

(a) Whatever value we substitute for x, $(x - 1)^2$ is always positive as it is a squared quantity. Therefore the least value of $(x - 1)^2$ is zero and this occurs when $x = 1$. Hence the lowest value that $f(x)$ can have is 5, i.e. $f(x)$ *has a least value of* 5 *when* $x = 1$.

(b) Consider two values of x symmetrically placed on either side of $x = 1$, i.e. $x = 1 + a$ and $x = 1 - a$

$$f(1 + a) = a^2 + 5$$
$$f(1 - a) = a^2 + 5$$

Therefore values of x that are symmetrical about $x = 1$ give the same value for $f(x)$.

(c) Conversely, for any given value of $f(x)$, c say, the corresponding values of x are given by the solution of the equation

$$c = x^2 - 2x + 6$$

or $$x^2 - 2x + 6 - c = 0$$

From the formula, the roots of this equation are

$$x = 1 + \sqrt{c - 5} \quad \text{and} \quad x = 1 - \sqrt{c - 5}$$

Thus for x to have real values, $c \geqslant 5$, i.e. $f(x)$ has a least value of 5 and these two real values of x, corresponding to one value of $f(x)$, are symmetrical about $x = 1$.

THE QUADRATIC FUNCTION

Functions of similar form usually have properties in common, so their graphs are usually similar in shape. A knowledge of the common characteristics of such functions will allow us to draw a sketch of the graph for any one particular function without calculating several corresponding values of $f(x)$ and x. The function analysed above is a quadratic function and any function whose general form is

$$f(x) \equiv ax^2 + bx + c$$

where a, b and c are constants, is called a quadratic function.

Completing the square on the R.H.S of this expression

gives
$$f(x) \equiv a \left[x^2 + \frac{b}{a} x + \frac{c}{a} \right]$$

$$\equiv a \left[x^2 + \frac{b}{a} x + \frac{b^2}{4a^2} + \frac{c}{a} - \frac{b^2}{4a^2} \right]$$

$$\equiv a \left[\left(x + \frac{b}{2a} \right)^2 + \left(\frac{c}{a} - \frac{b^2}{4a^2} \right) \right]$$

$$\equiv a \left[\left(x + \frac{b}{2a} \right)^2 + \left(\frac{4ac - b^2}{4a^2} \right) \right]$$

$$\equiv a \left(x + \frac{b}{2a} \right)^2 + \left\{ \frac{4ac - b^2}{4a} \right\}$$

i.e.
$$f(x) \equiv \left\{ \frac{4ac - b^2}{4a} \right\} + a\left(x + \frac{b}{2a} \right)^2$$

Whatever value x takes, $\left(\dfrac{4ac - b^2}{4a} \right)$ is constant, K say,

and $\left(x + \dfrac{b}{2a} \right)^2 \geqslant 0$ as it is a squared quantity.

Therefore $f(x) \equiv K + a\,(\text{zero or} + \text{ve quantity}).$

Therefore if a is *positive* $(a > 0)$, $f(x)$ is at least equal to K

i.e. $f(x)$ has a *least* value of $\dfrac{4ac - b^2}{4a}$, occuring when $x = -\dfrac{b}{2a}$

When a is *negative*, $(a < 0)$, $f(x)$ can never be greater than K

i.e. $f(x)$ has a *greatest* value of $\dfrac{4ac - b^2}{4a}$, when $x = -\dfrac{b}{2a}$

Now $x = -\dfrac{b}{2a}$ is the value of x corresponding to the greatest or least value

of $f(x)$.

Taking two values of x that are symmetrical about $-\dfrac{b}{2a}$, i.e. $x = -\dfrac{b}{2a} \pm k$,

gives $f\left(-\dfrac{b}{2a} + k \right) = f\left(-\dfrac{b}{2a} - k \right) = ak^2 + \dfrac{4ac - b^2}{4a}$

i.e. values of x that are symmetrical about $x = -\dfrac{b}{2a}$ correspond to the same

value of $f(x)$.

From this analysis we deduce that $f(x) \equiv ax^2 + bx + c$
  has a least value if $a > 0$
  has a greatest value if $a < 0$

and is symmetrical in shape about the line $x = -\dfrac{b}{2a}$ which is called the

axis of the curve.

The diagrams below represent the two alternative graphs of a quadratic function.

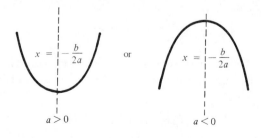

$x = -\dfrac{b}{2a}$ or $x = -\dfrac{b}{2a}$

$a > 0$ $a < 0$

Note. The expression for K found above can be used as a formula to obtain the greatest or least value of any quadratic function. However it is better to find K by completing the square in each individual problem.

The curve representing any particular quadratic function can now be sketched using this information. So, for example, to sketch the curve representing

$$f(x) \equiv 2x^2 - 7x - 4$$

we proceed as follows

$$f(x) \equiv 2x^2 - 7x - 4 \equiv 2(x^2 - \tfrac{7}{2}x - 2) \equiv 2[(x - \tfrac{7}{4})^2 - \tfrac{81}{16}]$$

Therefore $f(x)$ has a least value of $-\tfrac{81}{8}$ when $x = \tfrac{7}{4}$.

To locate the curve accurately on the axes we need one more pair of corresponding values for x and $f(x)$.
$f(0)$ is easy to find, i.e. $f(0) = -4$.

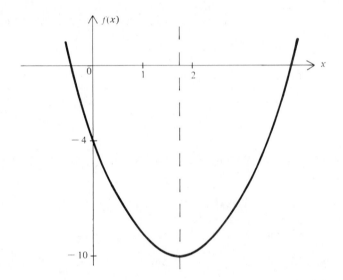

Alternatively a quick sketch of a quadratic function can be obtained as follows

$$f(x) \equiv 2x^2 - 7x - 4 \equiv (2x + 1)(x - 4)$$

The coefficient of x^2 is positive, $(a > 0)$, so $f(x)$ has a least value.
When $f(x) = 0$ the corresponding values of x are roots of the quadratic equation

$$(2x + 1)(x - 4) = 0$$

i.e. $x = -\frac{1}{2}$ and $x = 4$

The average of these values $\left(\frac{7}{4}\right)$ gives the value of x about which the curve is symmetrical.

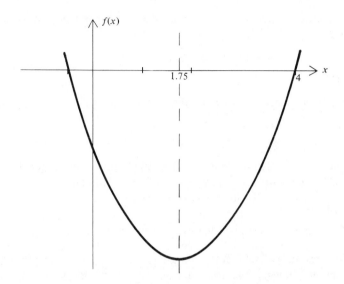

Note. This method is suitable only when the function factorises.

EXERCISE 3a

Find the greatest or least value of the following functions.

1) $x^2 - 3x + 5$　　　2) $2x^2 - 4x + 5$　　　　3) $3 - 2x - x^2$

4) $7 + x - x^2$　　　　5) $x^2 - 2$　　　　　　6) $2x - x^2$

Sketch the graphs of the following quadratic functions, showing clearly the greatest or least value of $f(x)$ and the value of x at which it occurs, where $f(x)$ is

7) $x^2 - 2x + 5$　　　　8) $x^2 + 4x - 8$　　　　9) $2x^2 - 6x + 3$

10) $4 - 7x - x^2$　　　11) $x^2 - 10$　　　　　12) $2 - 5x - 3x^2$

Draw a quick sketch of the graph of each of the following functions, showing the axis of symmetry clearly.

13) $(x - 1)(x - 3)$　　14) $(x + 2)(x - 4)$　　15) $(2x - 1)(x - 3)$

16) $(1 + x)(2 - x)$　　17) $x^2 - 9$　　　　　18) x^2

19) $4 - x^2$　　　　　20) $3 - 7x - 6x^2$

INEQUALITIES

Consider the real numbers 5 and 2.

Now $5 > 2$.

The introduction of an extra term on both sides leaves the inequality sign unchanged

e.g. $5 + 4 > 2 + 4$ and $5 - 4 > 2 - 4$

i.e. $9 > 6$ $1 > -2$

Multiplication of both sides by a positive number also leaves the inequality sign unchanged.

e.g. $5 \times 2 > 2 \times 2$ i.e. $10 > 4$

However if we multiply both sides by a negative number, the inequality is no longer true. For example, multiplying by -2, we find that the L.H.S. becomes -10 and the R.H.S. becomes -4,

i.e.   L.H.S. $<$ R.H.S.   since $-10 < -4$

This illustrates the general fact that multiplication or division of both sides by a negative number reverses the inequality sign.

To summarise, if a and b are real numbers such that

$$a > b$$

then $a + k > b + k$ for *all* real values of k

and $ak > bk$ for positive values of k

but $ak < bk$ for negative values of k

EXAMPLE 3b

Find the range of values of x satisfying the inequality $x - 3 < 2x + 5$

$$x - 3 < 2x + 5$$

$$x < 2x + 8 \quad \text{(adding 3 to both sides)}$$

$$-x < 8 \qquad \text{(subtracting } 2x \text{ from both sides)}$$

$$x > -8 \quad \text{(multiplying by } -1)$$

Therefore the range of values of x that satisfies the inequality is

$$x > -8$$

QUADRATIC INEQUALITIES

Any inequality relationship that involves a quadratic function is called a quadratic inequality. For example
$$(x - 2)(2x + 1) > 0$$

The range of values of x satisfying this inequality can be found graphically as follows:

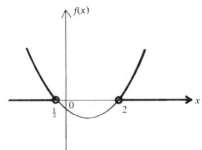

Let $f(x) \equiv (x - 2)(2x + 1)$.
The diagram on the right shows
a sketch of $f(x)$ from which
we see that $f(x) > 0$
(i.e. the curve is above the x-axis)
for values of x greater than 2
and less than $-\frac{1}{2}$.

Therefore the ranges of values of x that satisfy
$$(x - 2)(2x + 1) > 0 \quad \text{are} \quad x < -\tfrac{1}{2} \quad \text{and} \quad x > 2$$

EXERCISE 3b

Find the range (or ranges) of values of x that satisfy the following inequalities.

1) $x + 2 > 4 - x$

2) $2x - 1 < x - 4$

3) $x > 5x - 2$

4) $3x - 5 < 4$

5) $2(x - 1) > 3(x - 1)$

6) $2(x - 1) < 2(x + 1)$

7) $(x - 1)(x - 2) > 0$

8) $(x + 1)(x - 2) > 0$

9) $(x - 3)(x - 5) < 0$

10) $(2x - 1)(x + 1) < 0$

11) $x^2 - 4x > 5$

12) $4x^2 < 1$

13) $5x^2 > 3x + 2$

14) $(2 - x)(x + 4) < 0$

15) $(3 - 2x)(x + 5) > 0$

16) $3x > x^2 + 2$

17) $(x - 1)^2 > 4x^2$

18) $(x - 1)(x + 2) < x(4 - x)$

Problems involving Quadratic Inequalities

EXAMPLES 3c

1) Find the range of values of k for which the equation
$$x^2 - kx + (k + 3) = 0 \quad \text{has real roots.}$$

For $x^2 - kx + (k + 3) = 0$ to have real roots

$$b^2 - 4ac \geqslant 0$$

i.e. $$(-k)^2 - 4(k+3) \geqslant 0$$

$$k^2 - 4k - 12 \geqslant 0$$

Let $$f(k) \equiv k^2 - 4k - 12$$

$$\equiv (k-6)(k+2)$$

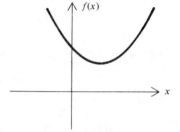

From the sketch of $f(k)$ we see that

$$(k-6)(k+2) \geqslant 0$$

for $k \leqslant -2$ and $k \geqslant 6$.

2) Find the set of values of p for which

$$f(x) \equiv x^2 + 3px + p$$

is greater than zero for all real values of x.

$f(x) \equiv x^2 + 3px + p$ is a quadratic function of x and, as the coefficient of x^2 is positive, $f(x)$ has a least value.

So if $f(x) > 0$ for all x, the least value
of $f(x)$ has to be greater than zero,
i.e. $f(x)$ has a graph of the type in the sketch.
Completing the square on the R.H.S. gives

$$f(x) \equiv \left(x + \frac{3p}{2}\right)^2 + p - \frac{9p^2}{4}$$

Therefore the least value of $f(x)$ is $p - \dfrac{9p^2}{4}$.

So for $f(x) > 0$ for all x,

$$p - \frac{9p^2}{4} > 0$$

$$4p - 9p^2 > 0$$

$$p(4 - 9p) > 0$$

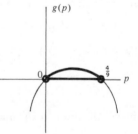

Let $g(p) \equiv p(4 - 9p)$.

From the sketch of $g(p)$ we see that

$$p(4 - 9p) > 0 \quad \text{for} \quad p > 0 \quad \text{and} \quad p < \tfrac{4}{9}$$

$$\text{or} \quad 0 < p < \tfrac{4}{9}$$

Therefore $f(x) > 0$ for all real x, for the set of values of p given by
$0 < p < \tfrac{4}{9}$.

3) Find the ranges of values of x for which

$$2x - 1 < x^2 - 4 < 12$$

There are two inequality relationships here

viz: (a) $2x - 1 < x^2 - 4$ and   (b) $x^2 - 4 < 12$

and we are looking for the ranges of values of x that satisfy *both* of these inequalities.

(a) $2x - 1 < x^2 - 4$ (b) $x^2 - \ 4 < 12$

   $-x^2 + 2x + 3 < 0$ $x^2 - 16 < 0$

   $x^2 - 2x - 3 > 0$ If $g(x) \equiv x^2 - 16$

If $f(x) \equiv x^2 - 2x - 3$ $\equiv (x - 4)(x + 4)$

   $\equiv (x - 3)(x + 1)$ the graph of $g(x)$ is

the graph of $f(x)$ is

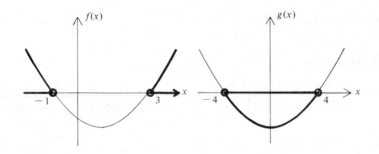

For $f(x) > 0$,   For $g(x) < 0$

   $x < -1$ and $x > 3$ $-4 < x < 4$

Illustrating these ranges on a number line

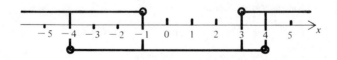

we see that the ranges of values of x that satisfy both (a) and (b), i.e. where the lines overlap, are

$$-4 < x < -1 \qquad \text{and} \qquad 3 < x < 4$$

4) Find the set of values for which

$$\frac{(x-1)^2}{(x+5)} < 1$$

$(x+5)$ may be positive or negative, so we avoid multiplying by $x+5$ and proceed as follows

$$\frac{(x-1)^2}{(x+5)} - 1 < 0$$

$$\Rightarrow \qquad \frac{x^2 - 2x + 1 - (x+5)}{x+5} < 0$$

$$\Rightarrow \qquad \frac{x^2 - 3x - 4}{x+5} < 0$$

$$\Rightarrow \qquad \frac{(x-4)(x+1)}{x+5} < 0 \quad \text{or} \quad \frac{f(x)}{g(x)} < 0$$

This fraction is negative (<0) if the numerator, $f(x)$,
and the denominator, $g(x)$, have opposite signs.

Consider $f(x) \equiv (x-4)(x+1)$

From the graph we see that

$f(x) > 0$ for $x < -1$ and $x > 4$

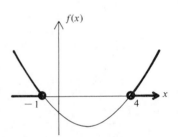

Considering $g(x) \equiv x + 5$

$g(x) > 0$ for $x > -5$

Illustrating these ranges on a number line

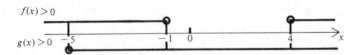

We see that $f(x)$ and $g(x)$ have opposite signs (where the positive ranges do *not* coincide) for the set of values

$$x < -5 \qquad \text{and} \qquad -1 < x < 4$$

5) If x is real, find the range of possible values of

$$\frac{x+1}{2x^2 + x + 1}$$

Let $f(x) \equiv \frac{x+1}{2x^2 + x + 1}$

and let p be a possible value of $f(x)$. Then the values of x corresponding to $f(x) = p$ can be found from the equation

$$\frac{x+1}{2x^2 + x + 1} = p$$

$$p(2x^2 + x + 1) = x + 1$$

$$2px^2 + (p-1)x + (p-1) = 0$$

As x must be real, this equation must have real roots

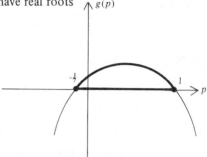

i.e. $b^2 - 4ac \geqslant 0$

or $(p-1)^2 - 8p(p-1) \geqslant 0$

$$(p-1)(p-1-8p) \geqslant 0$$

$$(1-p)(7p+1) \geqslant 0$$

Let $g(p) \equiv (1-p)(7p+1)$.

From the sketch, $g(p) \geqslant 0$ when $-\frac{1}{7} \leqslant p \leqslant 1$
Therefore the range of possible values of $f(x)$ is $-\frac{1}{7} \leqslant f(x) \leqslant 1$.

EXERCISE 3c

1) Find the range of values of x for which $\dfrac{3-x}{x-2} < 4$.

2) Find the range of values of x for which $\dfrac{x-5}{2-x} > 3$.

3) Find the ranges of values of k for which the equation

$$x^2 + (k-3)x + k = 0$$

has (a) real distinct roots, (b) roots of the same sign.   (JMB)p

4) If x is real and $x^2 + (2-k)x + 1 - 2k = 0$ show that k cannot lie between certain limits, and find these limits.   (JMB)p

5) Find the limitations required on the values of the real number c in order that the equation $x^2 + 2cx - c + 2 = 0$ shall have real roots.

       (JMB)p

6) Prove that, if $x^2 > k(x+1)$ for all real x, then $-4 < k < 0$.

       (C)p

7) If x is real, find the range of possible values of

$$\frac{4(x-2)}{4x^2 + 9}$$

       (AEB)'76p

8) Find the condition that must be satisfied by k in order that the expression

$$2x^2 + 6x + 1 + k(x^2 + 2)$$

may be positive for all real values of x.   (JMB)p

9) Find the range of values of x for which

$$(x - 4) < x(x - 4) \leqslant 5$$ (U of L)p

10) Find the range of values of x for which

$$\frac{x^2 + 56}{x} > 15$$ (U of L)p

11) If x is real, find the set of possible values of

$$\frac{(2x + 1)}{(x^2 + 2)}$$ (U of L)p

12) (a) If $x = 2$ is a root of the equation

$$\alpha^2 x^2 + 2(2\alpha - 5)x + 8 = 0$$

find the possible value (or values) of α and the corresponding value (or values) of the other root.

(b) Find the range (or ranges) of possible values of the real number α if

$$\alpha^2 x^2 + 2(2\alpha - 5)x + 8 > 0$$

for all real values of x.   (C)

13) Find the solution set of the inequality $\dfrac{12}{x - 3} < x + 1.$ (C)p

14) Determine for each of the three expressions $f(x)$, $g(x)$ and $h(x)$ the range (or ranges) of values of x for which it is positive. Give your answers correct to two places of decimals. Explain, briefly, the reasons for your answers.

(a) $f(x) \equiv x^2 + 4x - 6$ (b) $g(x) \equiv -x^2 - 8x + 2$

(c) $h(x) \equiv \dfrac{x^2 + 4x - 6}{-x^2 - 8x + 2}$ (C)

15) (a) State the range of values of x for which the following expressions are negative:

(i) $2x^2 + 5x - 12$ (ii) $\dfrac{x - 2}{(x - 1)(x - 3)}$

(b) The value of the constant a is such that the quadratic function $f(x) \equiv x^2 + 4x + a + 3$ is never negative. Determine the nature of the roots of the equation $af(x) = (x^2 + 2)(a - 1)$. Deduce the value of a for which this equation has equal roots.   (AEB)'73

16) By eliminating x and y from the equations

$$\frac{1}{x} + \frac{1}{y} = 1, \quad x + y = a, \quad \frac{y}{x} = m$$

where $a \neq 0$, obtain a relation between m and a. Given that a is real, determine the ranges of values of a for which m is real. (JMB)p

17) If $a > 0$, prove that the quadratic expression $ax^2 + bx + c$ is positive for all real values of x when $b^2 < 4ac$. Hence find the range of values of p for which the quadratic function of x

$$f(x) \equiv 4x^2 + 4px - (3p^2 + 4p - 3)$$

is positive for all real values of x.
Illustrate your result by making sketch graphs of $f(x)$ for each of the cases $p = 0$ and $p = 1$. (U of L)

18) (a) If a is a positive constant, find the set of values of x for which $a(x^2 + 2x - 8)$ is negative. Find the value of a if this function has a minimum value of -27.

(b) Find two quadratic functions of x which are zero at $x = 1$, which take the value 10 when $x = 0$ and which have a maximum value of 18. Sketch the graphs of these two functions. (U of L)

19) Find the set of values of k for which $f(x) \equiv 3x^2 - 5x - k$ is greater than unity for all real values of x.
Show that, for all k, the minimum value of $f(x)$ occurs when $x = \frac{5}{6}$.
Find k if this minimum value is zero. (U of L)p

20) Prove that $3x^2 - 4x + 2 > 0$ for all real values of x.

(U of L)p

21) Find the value of $k (\neq 1)$ such that the quadratic function of x

$$k(x + 2)^2 - (x - 1)(x - 2)$$

is equal to zero for only one value of x.
Find also (a) the range of values of k for which the function possesses a minimum value, (b) the range of values of k for which the value of the function never exceeds 12.5.
Sketch the graph of the function for $k = \frac{1}{2}$ and for $k = 2\frac{1}{2}$. (U of L)

SOME OTHER SIMPLE FUNCTIONS

We will now consider the graphical representation of some other simple functions. A full analysis of the general form of these functions is not possible at this stage, but enough information can be deduced to give a rough idea of the shapes of the curves that represent them.

Exponential Functions

An exponential function is one where the variable appears as an exponent, (i.e. an index)

e.g. 2^x, 3^{-x}, 10^{x+1}, $5^{-3x} + 2$

are all exponential functions of x.

Consider the function $f(x) \equiv 2^x$ for which the following table shows corresponding values of x and $f(x)$.

x	$\ldots$	-10	$\ldots$	-5	-4	-3	-2	-1	0	1	2	3	4	5	$\ldots$	10	$\ldots$
$f(x) \equiv 2^x$	$\ldots$	$\frac{1}{1024}$	$\ldots$	$\frac{1}{32}$	$\frac{1}{16}$	$\frac{1}{8}$	$\frac{1}{4}$	$\frac{1}{2}$	1	2	4	8	16	32	$\ldots$	1024	$\ldots$

From this table we see that:
1. $f(x) > 0$ for all real values of x,
2. as x increases $f(x)$ increases at a rapidly accelerating rate,
3. $f(x) = 1$ when $x = 0$
4. as x decreases (i.e. $x = -10, -100, \ldots$) $f(x)$ quickly becomes numerically smaller and we say that, as x approaches minus infinity, $f(x)$ approaches the value zero. This is written,

$$x \to -\infty, \quad f(x) \to 0$$

From these observations the sketch of $f(x) \equiv 2^x$ is drawn:

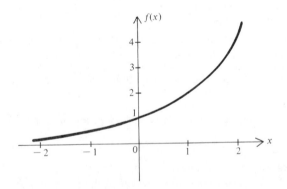

The curve approaches the negative x-axis but never actually touches it or crosses it and we say that the x-axis is an *asymptote* to the curve.

Another way of expressing the behaviour of $f(x)$ for negative values of x is based on the fact that

$f(x)$ approaches a limiting value (or limit) of zero as x approaches minus infinity.

i.e. the limit of $f(x)$ as $x \to -\infty$ is zero.

This property is written

$$\lim_{x \to -\infty} [f(x)] = 0$$

Any function of the form a^x, where $a > 1$, is represented by a curve similar to that deduced for 2^x.

Rational Functions

A function where both numerator and denominator are polynomials is called a rational function.

e.g.
$$\frac{1}{x}, \quad \frac{x}{x^2 - 1}, \quad \frac{2x^2 - 7x}{x + 1}$$

are rational functions of x.

Consider the function $f(x) \equiv \dfrac{1}{x}$ and the following table of corresponding values for x and $f(x)$

x	...	-3	-2	-1	0	1	2	3	4	...	10	...
$f(x) \equiv \dfrac{1}{x}$	...	$-\dfrac{1}{3}$	$-\dfrac{1}{2}$	-1	?	1	$\dfrac{1}{2}$	$\dfrac{1}{3}$	$\dfrac{1}{4}$	...	$\dfrac{1}{10}$	...

From this table we see that:
1. For $x > 0$, $f(x) > 0$ and as $x \to \infty$, $f(x) \to 0$.
2. For $x < 0$, $f(x) < 0$ and as $x \to -\infty$, $f(x) \to 0$
   i.e. x and $f(x)$ have the same sign.

3. For $x = 0$, $f(0) = \dfrac{1}{0}$ which has no finite value and is said to be

undefined. In such circumstances we investigate the behaviour of $f(x)$ as x approaches zero.
Now x can approach zero in two ways
i.e. decrease from positive values towards zero (approach zero from above)
or   increase from negative values towards zero (approach zero from below)

From (1) and (2) above we see that $f(x)$ and x have the same sign. As x decreases, $f(x)$ increases.
From the following table

x	-1	...	$-\dfrac{1}{10}$	...	$-\dfrac{1}{100}$	$\to$	...	0	...	$\leftarrow$	$\dfrac{1}{100}$	...	$\dfrac{1}{10}$	...	1	..
$f(x)$	-1	...	-10	...	-100	$\to$	..		..	$\leftarrow$	100	...	10	...	1	..

we see that
as x decreases to zero, $f(x) \to \infty$
and as x increases to zero, $f(x) \to -\infty$.

From these observations we can draw the following sketch representing
$f(x) \equiv \dfrac{1}{x}$.

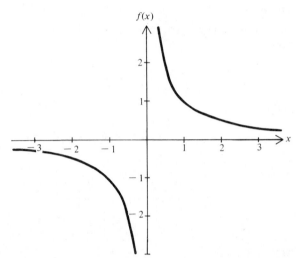

This curve has two asymptotes, the horizontal and vertical axes.

All the curves looked at so far have been unbroken or *continuous*, but this curve has a 'break' or *discontinuity* at the point where $x = 0$ and $f(0)$ is *undefined*.

In general, if $f(x)$ is any function of x and if $f(x)$ is undefined for a finite value of x, $x = a$ say, then the curve representing $f(x)$ has a discontinuity where $x = a$.

Also, we see that

$$\lim_{x \to \infty} \left[\frac{1}{x} \right] = 0 \qquad \text{and} \qquad \lim_{x \to -\infty} \left[\frac{1}{x} \right] = 0$$

but $\lim_{x \to 0} \left[\dfrac{1}{x} \right]$ does not have a unique value as $\dfrac{1}{x}$ approaches ∞ or $-\infty$ depending on whether x approaches zero from above or below. In this case we say that $\lim_{x \to 0} \left[\dfrac{1}{x} \right]$ does not exist.

In general, for $\lim_{x \to a} \left[f(x) \right]$ to exist with a value k, say,

$$\left. \begin{array}{l} \text{then as } x \to a \text{ from above} \\ \textit{and as } x \to a \text{ from below} \end{array} \right\} \; f(x) \to k$$

EXERCISE 3d

1) Write down the values of $f(x) \equiv (\tfrac{1}{2})^x$ corresponding to $x = 2, 4, 6$ and to $x = -2, -4, -6$. From these values deduce the behaviour of $f(x)$ as $x \to \infty$

and as $x \to -\infty$. Sketch the graph of $f(x) \equiv (\frac{1}{2})^x$, marking any asymptotes.

2) Draw sketch graphs of the following functions:

(a) 3^x (b) 2^{x-1} (c) 3^{-x} (d) 2^{2x} (e) $1 + 2^x$ (f) $-(2^x)$

showing clearly, in each case, any asymptotes.

3) For what value of x is $f(x) \equiv \dfrac{1}{1-x}$ undefined? Describe the behaviour of $f(x)$ as x approaches this value from above and from below.
Write down also, $\lim\limits_{x \to \infty} [f(x)]$ and $\lim\limits_{x \to -\infty} [f(x)]$.

Use this information to sketch the graph of $f(x) \equiv \dfrac{1}{1-x}$ marking the asymptotes clearly.

4) By following a procedure similar to that indicated in (3), or otherwise, draw sketch graphs of the following functions:

(a) $-\dfrac{1}{x}$ (b) $\dfrac{1}{x-2}$ (c) $\dfrac{2}{x+1}$ (d) $1 + \dfrac{1}{x}$ (e) $\dfrac{x-1}{x}$

(f) $\dfrac{x}{x+1}$

5) Write down the values of x for which $f(x) \equiv \dfrac{1}{(x-1)(x+1)}$ is undefined and investigate the behaviour of $f(x)$ in the neighbourhood of these values.
Find the ranges of values of x for which $f(x)$ is
(a) positive   (b) negative.
Find the ranges of values which $f(x)$ can take for real x.
Draw a sketch of the graph representing $f(x)$.

6) What is the meaning of $\log_2(-8)$?
For what range of values of x is the function $f(x) = \log_2 x$
(a) undefined   (b) negative   (c) positive?
Draw a sketch of the graph representing $f(x) = \log_2 x$.
Compare your sketch with that obtained for 2^x.

CHAPTER 4

COORDINATE GEOMETRY I

LOCATION OF A POINT IN A PLANE

Graphical methods lend themselves particularly well to the investigation of the geometric properties of many kinds of curves and surfaces. At this stage we will restrict ourselves to plane figures (i.e. those that can be described fully using only two dimensions). To represent any figure on a graph we need, as a start, a simple and unambiguous way of describing the position of a point.

Consider the problem of describing the location of a town, Birmingham say. There are many ways in which this can be done, but all require reference to at least one known place and known directions (called a system, or frame, of reference). Within this frame of reference, two measurements (or coordinates) will be needed to locate the town precisely.

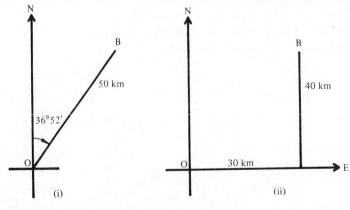

The position of B is described in two alternative ways in the diagrams above.

In (i) the system of reference is the fixed point O and the direction due N from O.

The coordinates of B are 50 km from O on a bearing $36°52'$.

In (ii) the system of reference is the directions due E and due N from a fixed point O.

The coordinates of B are 30 km E of O and 40 km N of O.

The two systems most frequently used for mathematical analysis are basically similar to the two practical systems described above.

POLAR COORDINATES

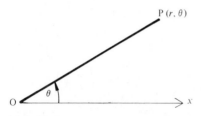

The system of reference is a fixed point O, called the pole and a fixed direction from O, the line Ox, called the initial line.

The coordinates of a point P are the distance of P from O and the angle OP makes with Ox, measured in an anticlockwise sense from Ox.

These coordinates are written as an ordered pair (r, θ), i.e. (distance, angle).

Using this system the diagram represents the position of the point whose coordinates are $(2, 30°)$

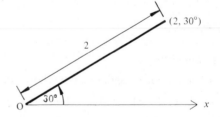

CARTESIAN COORDINATES

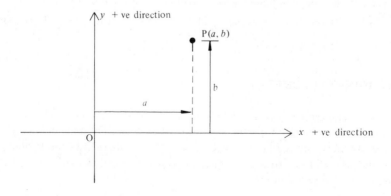

The system of reference is a fixed point O, the origin, and a pair of perpendicular lines through O.

It is usual to draw these lines as shown in the diagram. The horizontal line is called the *x*-axis, the vertical line the *y*-axis.

The coordinates of a point P are the directed distances of P from O parallel to the axes. A positive coordinate is a distance measured in the positive direction of the axis and a negative coordinate is a distance in the opposite direction.

The coordinates are given as an ordered pair (a, b) with the *x coordinate* or *abscissa* first and the *y coordinate* or *ordinate* second.

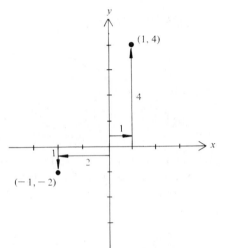

Taking the positive direction of each axis as shown, the diagram represents the points whose Cartesian coordinates are

$(1, 4)$ and $(-2, -1)$,

[referred to in future as the points $(1, 4)$ and $(-2, -1)$].

EXERCISE 4a

Represent on a diagram points whose polar coordinates are:

1) $(1, 45°)$ 2) $(3, 90°)$ 3) $(2, 60°)$ 4) $(1, 150°)$ 5) $(1, 270°)$

6) $(2, 200°)$ 7) $(3, 300°)$

Represent on a diagram points whose Cartesian coordinates are:

8) $(1, 1)$ 9) $(4, 2)$ 10) $(2, 4)$ 11) $(-1, 5)$ 12) $(3, -1)$ 13) $(0, 3)$

14) $(-2, -5)$ 15) $(0, 0)$

COORDINATE GEOMETRY

Coordinate geometry is the name given to the analysis, using graphical methods, of geometric properties. The properties of straight lines and many curves are most simply found using Cartesian coordinates and so this system of reference is used more frequently than any other. For this analysis we need to refer to three types of points:

(a) fixed points whose coordinates are known, e.g. the point $(4, 5)$,

(b) fixed points whose coordinates are not known numerically. These are referred to as the points $(x_1, y_1), (x_2, y_2) \ldots$ etc. or (a, b) etc.

(c) points which are not fixed (general points). A general point is referred to as the point (x, y).

It is conventional to use the letters $P, Q, R \ldots$ etc. for general points and the letters $A, B, C \ldots$ for fixed points.

In order to avoid distorting the shape of a curve when drawing it on a Cartesian plane, the two axes are graduated using identical scales.

The Length of a Line Joining Two Points

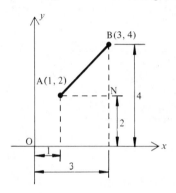

From the diagram we see that the length of the line joining $A(1, 2)$ and $B(3, 4)$ can be found using Pythagoras' Theorem where

$$AB^2 = AN^2 + BN^2$$
$$= (3 - 1)^2 + (4 - 2)^2$$
$$= 8$$

Therefore $AB = \sqrt{8} = 2\sqrt{2}$.

In general, if $A(x_1, y_1)$ and $B(x_2, y_2)$ are any two points

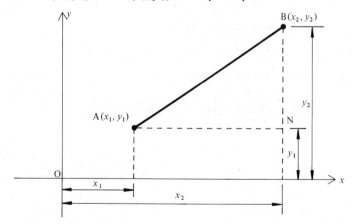

we see from the diagram, and using Pythagoras' Theorem, that

$$AB^2 = AN^2 + BN^2$$
$$= (x_2 - x_1)^2 + (y_2 - y_1)^2$$

Therefore $\qquad AB = \sqrt{(x_2 - x_1)^2 + (y_2 - y_1)^2}$

Note that this formula still holds when some, or all, of the coordinates are negative, e.g.

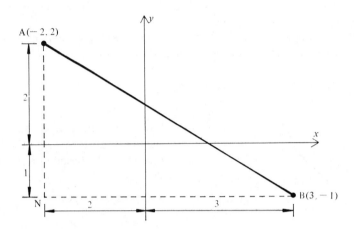

$$AN = 2 + 1 = 2 - (-1) = (y_1 - y_2)$$

Therefore $\qquad AN^2 = (y_1 - y_2)^2 = (y_2 - y_1)^2$

Similarly $\qquad NB = 2 + 3 = 3 - (-2) = (x_2 - x_1)$

Therefore $\qquad NB^2 = (x_2 - x_1)^2$

Therefore $\qquad AB^2 = (x_2 - x_1)^2 + (y_2 - y_1)^2$

Therefore *the length of the line joining $A(x_1, y_1)$ to $B(x_2, y_2)$ is given by*

$$AB = \sqrt{(x_2 - x_1)^2 + (y_2 - y_1)^2}$$

The Midpoint of the Straight Line Joining Two Given Points

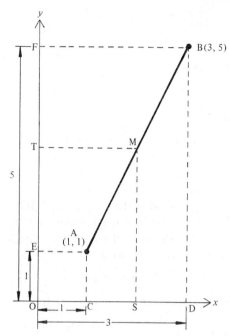

From the diagram we see that if M is the midpoint of AB, then S is the midpoint of CD, Therefore the x coordinate of M is given by OS where

$$OS = OC + \tfrac{1}{2}CD$$
$$= 1 + \tfrac{1}{2}(3 - 1)$$
$$= \tfrac{1}{2}(3 + 1) = 2$$

Similarly T is the midpoint of EF, so the y coordinate of M is given by OT, where

$$OT = OE + \tfrac{1}{2}EF$$
$$= 1 + \tfrac{1}{2}(5 - 1)$$
$$= \tfrac{1}{2}(5 + 1) = 3$$

Therefore M is the point $(2, 3)$.

In general, if M is the midpoint of the line joining $A(x_1, y_1)$ and $B(x_2, y_2)$ then we see from the diagram that

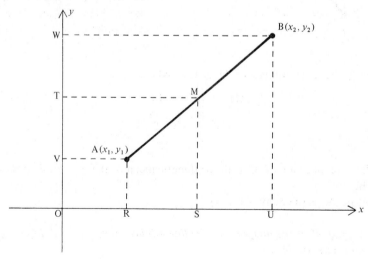

the x coordinate of M is OS, where

$$OS = OR + \tfrac{1}{2}RU$$

$$= x_1 + \tfrac{1}{2}(x_2 - x_1)$$

$$= \tfrac{1}{2}(x_1 + x_2)$$

and the y coordinate of M is OT, where

$$OT = OV + \tfrac{1}{2}VW$$

$$= y_1 + \tfrac{1}{2}(y_2 - y_1)$$

$$= \tfrac{1}{2}(y_1 + y_2)$$

Note that this formula also holds when some, or all, of the coordinates are negative:

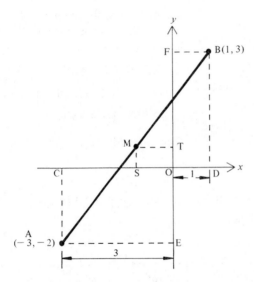

If M is the midpoint of AB, then S is the midpoint of CD. Therefore the x coordinate of M is $-OS$ where

$$OS = OC - CS$$

$$= OC - \tfrac{1}{2}CD$$

$$= 3 - \tfrac{1}{2}(3 + 1)$$

$$= \tfrac{1}{2}(3 - 1)$$

Therefore $-OS = \tfrac{1}{2}(1 - 3)$ i.e. the x coordinate of M is the arithmetic mean of the x coordinates of A and B.

Similarly the y coordinate of M is OT, where

$$OT = OF - \tfrac{1}{2}FE$$

$$= 3 - \tfrac{1}{2}(3 + 2)$$

$$= \tfrac{1}{2}(3 - 2)$$

i.e. the y coordinate of M is the arithmetic mean of the y coordinates of A and B.
Therefore M is the point $(-1, \tfrac{1}{2})$

Therefore, *if M is the midpoint of the line joining* $A(x_1, y_1)$ *and* $B(x_2, y_2)$, *the coordinates of M are*

$$\left[\tfrac{1}{2}(x_1 + x_2), \tfrac{1}{2}(y_1 + y_2) \right]$$

EXERCISE 4b

1) Find the length of the line joining the following pairs of points:
(a) $(1, 2), (4, 6)$ (b) $(3, 1), (2, 0)$ (c) $(4, 2), (2, 5)$
(d) $(-1, 4), (2, 6)$ (e) $(0, 0), (-1, -2)$ (f) $(-1, -4), (-3, -2)$.

2) Find the coordinates of the midpoints of the lines joining the pairs of points given in 1.

3) Find the length of the line from the origin to the point $(7, 4)$.

4) Show, using Pythagoras' Theorem, that the lines joining $A(1, 6)$, $B(-1, 4)$ and $C(2, 1)$ form a right angled triangle.

5) Show that $\triangle ABC$ is isosceles where A, B and C are the points $(7, 3)$, $(-4, 1)$ and $(-3, -2)$.

6) Find the midpoint of the base of $\triangle ABC$ in No. 5. Hence find the area of $\triangle ABC$.

7) Prove that the lines OA and OB are perpendicular where A, B are the points $(4, 3)$, $(3, -4)$ respectively.

8) In the triangle ABC, A, B, and C are the points $(0, 2)$, $(1, 5)$ and $(-1, 4)$. Find the coordinates of the point D such that AD is a median and find the length of this median.

9) A, B and M are three points such that M is the midpoint of AB. The coordinates of A and M are $(5, 7)$ and $(0, 2)$ respectively. Find the coordinates of B.

GRADIENT

The gradient of a straight line is a measure of its slope with respect to the *x*-axis, and is defined as *the increase in the y coordinate divided by the increase in the x coordinate between one point on the line and another point on the line.*

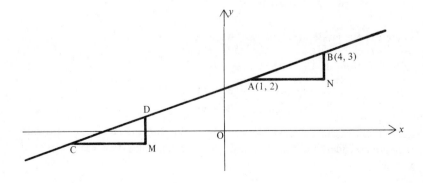

Consider the straight line passing through the points $A(1, 2)$ and $B(4, 3)$.
From A to B

> the increase in the y coordinate is 1
>
> the increase in the x coordinate is 3

Therefore the gradient of AB is $\frac{1}{3}$.

Note that NB measures the increase in the y coordinate
and AN measures the increase in the x coordinate

so the gradient of $AB = \dfrac{NB}{AN}$.

If C and D are any two other points on the same line, then as $\triangle ABN$ is similar to $\triangle CDM$,

$$\frac{NB}{AN} = \frac{MD}{CM} = \frac{1}{3}$$

i.e. the gradient of a line may be found from *any* two points on the line.

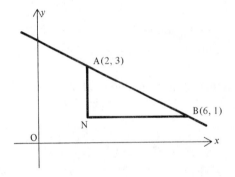

Consider the line passing through the points $A(2, 3)$ and $B(6, 1)$

From A to B,

> the y coordinate *decreases by* 2, i.e. *increases by* -2
>
> the x coordinate *increases by* 4

Therefore the gradient of the line AB is $\dfrac{-2}{4} = -\dfrac{1}{2}$.

Note that from B to A the gradient is

$$\frac{\text{increase in } y \ (B \rightarrow A)}{\text{increase in } x \ (B \rightarrow A)} = \frac{2}{-4} = -\frac{1}{2}$$

i.e. it does not matter in which order the two points are considered, provided they are considered in the *same* order when calculating the increases in both x and y.

From the two examples considered we see that the gradient of a line may be positive or negative.

From the first example we see that a *positive gradient* indicates an 'uphill' slope with respect to the positive direction of the x-axis, i.e. a line which makes an acute angle with the positive sense of the x-axis.

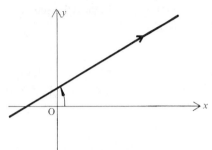

From the second example we see that a *negative gradient* indicates a 'down-hill' slope with respect to the positive direction of the x-axis, i.e. a line which makes an obtuse angle with the positive sense of the x-axis.

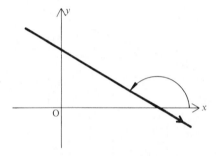

Note that in both cases:

(a)

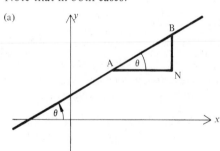

gradient of $AB = \dfrac{BN}{AN} = \tan\theta$

(b)

gradient of $AB = \dfrac{-AN}{BN} = -\tan\alpha$

$= \tan\theta$

Therefore the *gradient of a line* may also be defined as *the tangent of the angle that it makes with the positive direction of the x-axis,* where the angle is measured in an anticlockwise sense.

In general

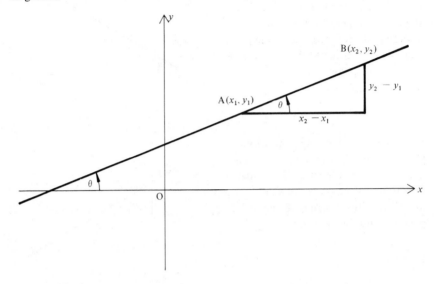

The gradient of the line passing through $A(x_1, y_1)$ and $B(x_2, y_2)$ is

$$\frac{\text{the increase in the } y \text{ coordinate}}{\text{the increase in the } x \text{ coordinate}} = \frac{y_2 - y_1}{x_2 - x_1} = \tan \theta$$

As the gradient of a straight line is the increase in y divided by the increase in x from one point to another point on the line, *gradient measures the increase in y per unit increase in x, i.e. the rate of increase of y with respect to x.*

Parallel Lines

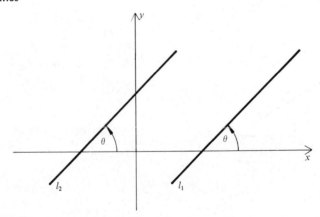

If l_1 and l_2 are parallel lines, they are equally inclined to the positive direction of the x-axis (corresponding angles), so $\tan\theta$ is the gradient of both l_1 and l_2, i.e. *parallel lines have equal gradients.*

Perpendicular Lines

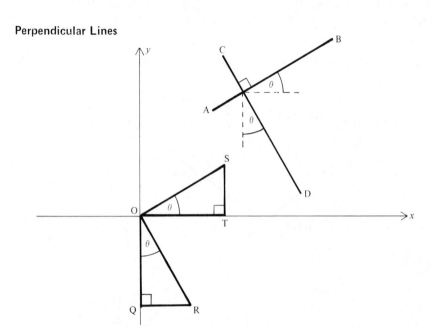

Consider the perpendicular lines AB and CD whose gradients are m_1 and m_2 respectively.

If AB makes an angle θ with the x axis
then CD makes an angle θ with the y axis.

If OS is drawn parallel to AB and OR is drawn parallel to CD then triangles OST and ORQ are similar,

therefore
$$\frac{ST}{OT} = \frac{QR}{OQ}$$

But gradient of OS $= \dfrac{ST}{OT} =$ gradient of AB $= m_1$

and gradient of OR $= -\dfrac{OQ}{QR} =$ gradient of CD $= m_2$.

Therefore $\qquad m_1 = -\dfrac{1}{m_2} \quad$ or $\quad m_1 m_2 = -1$

i.e. *the product of the gradients of perpendicular lines is -1,* or, if one line has a gradient m, the gradient of any line perpendicular to it is $-\dfrac{1}{m}$.

EXAMPLE 4c

Show that the point $(-\frac{6}{7}, 0)$ is on the median through A of triangle ABC where A, B, C are the points $(2, 4), (-2, 3), (1, -2)$. If, also, the point (a, b) is on this median, find a relationship between a and b.

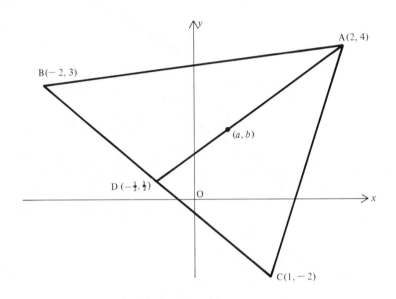

If AD is the median through A, D is the midpoint of BC. i.e. D is the point $(-\frac{1}{2}, \frac{1}{2})$.

If $E(-\frac{6}{7}, 0)$ is on AD then

gradient AD and gradient AE should be equal.

Now gradient AD $= \dfrac{4 - \frac{1}{2}}{2 + \frac{1}{2}} = \dfrac{7}{5}$

and gradient AE $= \dfrac{4 - 0}{2 + \frac{6}{7}} = \dfrac{7}{5}$

Therefore E is on the median AD.

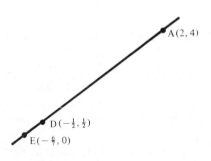

A condition that $P(a, b)$ should be on AD is
gradient AP = gradient AD.

Gradient AP $= \dfrac{4-b}{2-a}.$

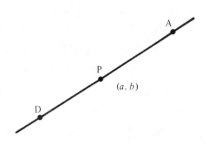

Therefore P lies on AD if

$$\dfrac{4-b}{2-a} = \dfrac{7}{5}$$

$\Rightarrow \qquad 20 - 5b = 14 - 7a$

$\Rightarrow \quad 7a - 5b + 6 = 0$

EXERCISE 4c

1) Find the gradients of the lines passing through the following pairs of points:
(a) $(0,0), (1,3)$ (b) $(1,4), (3,7)$ (c) $(5,4), (2,3)$
(d) $(-1,4), (3,7)$ (e) $(-1,-3), (-2,1)$ (f) $(-1,-6), (0,0)$
(g) $(-2,5), (1,-2)$ (h) $(3,-2), (-1,4)$ (i) $(h,k), (0,0)$

2) Write down the gradients of the lines which are inclined at the following angles to the positive direction of the x-axis.
(a) $45°$ (b) $135°$ (c) 0 (d) $90°$

3) Determine, by comparing gradients, whether the three points whose coordinates are given, are collinear (i.e. lie on the same straight line).
(a) $(0,-1)$, $(1,1)$, $(2,3)$
(b) $(0,2)$, $(2,5)$, $(3,7)$
(c) $(-1,4)$, $(2,1)$, $(-2,5)$

4) Determine whether AB is parallel or perpendicular to CD where:
(a) $A(0,-1)$, $B(1,1)$, $C(1,5)$, $D(-1,1)$
(b) $A(1,1)$, $B(3,2)$, $C(-1,1)$, $D(0,-1)$
(c) $A(3,3)$, $B(-3,1)$, $C(-1,-1)$, $D(1,-7)$
(d) $A(2,6)$, $B(-1,-9)$, $C(2,11)$, $D(0,1)$

Questions 5–13 are miscellaneous problems on coordinate geometry. A clear, reasonably accurate diagram showing all the given information will usually suggest the most direct method for answering a particular problem.

5) $A(1,3)$, $B(5,7)$, $C(4,8)$, $D(a,b)$ form a rectangle ABCD. Find a and b.

6) The points $A(1,5)$, $B(4,-1)$ and $C(-2,-4)$ form triangle ABC. Prove that the triangle is right angled, and find its area.

7) ABCD is a quadrilateral where A, B, C and D are the points $(3,-1)$, $(6,0), (7,3)$ and $(4,2)$. Prove that the diagonals bisect each other at right angles and hence find the area of ABCD.

8) The vertices of a $\triangle ABC$ are at the points $A(a, 0)$, $B(0, b)$, $C(c, d)$. If $\angle B = 90°$, find a relationship between a, b, c and d.

9) A circle, radius two units, centre the origin, cuts the x-axis at A and B and cuts the positive y-axis at C. Prove that AB subtends a right angle at C.

10) In No. 9, if $D(a, b)$ is a point on the circumference of the circle, find a relationship between a and b.

11) Prove that the point $(5, -1)$ is on the perpendicular bisector of the line joining $A(1, -3)$ to $B(3, 3)$. If $D(h, k)$ is another point on this perpendicular bisector, find a relationship between h and k.

12) The point $A(5, 0)$ lies on a circle, centre the origin. Find the radius of this circle. Prove that the points $B(4, -3)$, $C(-3, 4)$ are also on the circumference of this circle. If this circle cuts the negative x-axis at D, find the area of the quadrilateral ABDC.

13) A point $P(a, b)$ is equidistant from the y-axis and from the point $(4, 0)$. Find a relationship between a and b.

THE MEANING OF EQUATIONS

The Cartesian system of reference provides a means of defining the position of any point in a plane. This plane is called the xy plane.

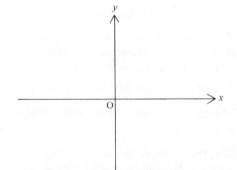

In general, x and y are independent variables (i.e. they can each take any value, independently of the value of the other) unless some restriction is placed on them.

Consider the set of points for which $x = 2$.

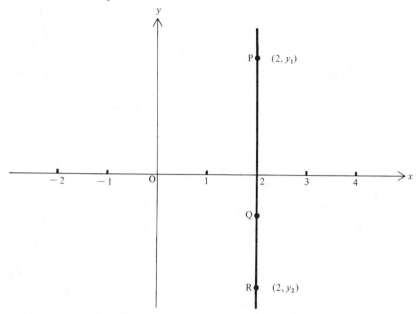

As the value of y is not restricted, these points all lie on the line parallel to the y-axis, passing through P, Q and R as shown.

So the equation $x = 2$ defines the line through P, Q, R in the xy plane.

$x = 2$ is called the equation of the line through P, Q, R which is briefly referred to as 'the line $x = 2$'.

Now consider the set of points for which $x > 2$.

All points to the right of the line $x = 2$ have an x coordinate which is greater than 2.

So the inequality $x > 2$ defines the shaded region of the xy plane, shown in Fig. 1.

Similarly the inequality $x < 2$ defines the region of the xy plane left unshaded in Fig. 1.

Note. The region defined by $x > 2$ does not include the line $x = 2$.

When a region does not include points on the boundary lines these are drawn as broken lines. When a region does include points on the boundary lines these are drawn as solid lines.

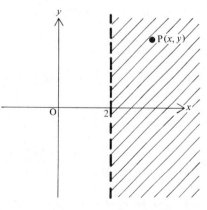

Fig. 1

Now consider the function

$$f(x) \equiv (x-3)(x+1)$$

Fig. 2 shows the curve representing
this function and Fig. 3 shows the
same curve drawn on the xy plane.

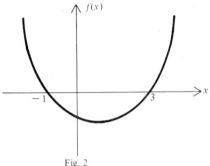

Fig. 2

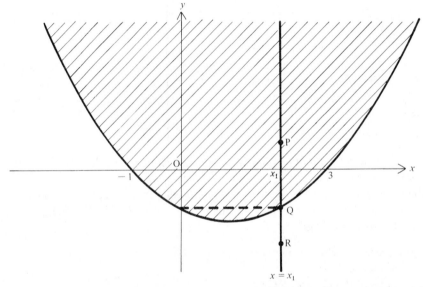

If P, Q and R are points on the line $x = x_1$ as shown, the y coordinate of
Q is $f(x_1)$,

i.e. at Q, $y = (x_1 - 3)(x_1 + 1)$

the y coordinate of P is greater than the y coordinate of Q,

i.e. at P, $y > (x_1 - 3)(x_1 + 1)$

the y coordinate of R is less than the y coordinate of Q,

i.e. at R, $y < (x_1 - 3)(x_1 + 1)$

This argument applies for all values of x.
Therefore the inequality $y > (x-3)(x+1)$ defines the set of points
contained in the shaded region of the xy plane,
and the inequality $y < (x-3)(x+1)$ defines the unshaded region of the
xy plane.

Whereas *only for points on the curve is* $y = (x - 3)(x + 1)$

> $y = (x - 3)(x + 1)$ is called the equation of the curve

which is often referred to simply as *the curve* $y = (x - 3)(x + 1)$.

Note. An equation such as $x^2 - 7x + 3 = 0$ contains only one variable and its solution comprises a finite set of values of x. Equations containing two variables, such as $y = (x - 3)(x + 1)$,   have as their solution, an infinite set of ordered pairs (x, y) and the elements of such a solution set are definable as follows.

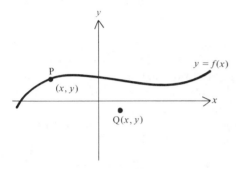

If **A** is the solution set of the equation $y = f(x)$,   the elements of **A** are the coordinates, (x, y), of all points on the curve $y = f(x)$. Conversely the coordinates (x, y) of points *not* on the curve are *not* elements of **A**.

Thus $P(x, y) \in A$,   but $Q(x, y) \notin A$.

In general if $f(x)$ is any function of x, then in the xy plane

> $y = f(x)$ defines a curve

and is called the equation of that curve.

> $y > f(x)$ and $y < f(x)$ define regions of the plane

EXAMPLES 4d

1) Determine whether the points $(5, 11)$ and $(-2, -20)$ are on the curve $y = (x - 4)(x + 6)$.

Substituting 11 for y in L.H.S. of the equation of the curve gives

$$\text{L.H.S} = 11$$

Substituting 5 for x in R.H.S. gives

$$\text{R.H.S} = (5 - 4)(5 + 6) = 11$$

Therefore   L.H.S = R.H.S   when $x = 5$ and $y = 11$.
Therefore $(5, 11)$ is a member of the solution set of $y = (x - 4)(x + 6)$ and is therefore on the curve.
Substituting -20 for y in L.H.S. gives

$$\text{L.H.S} = -20$$

Substituting -2 for x in R.H.S. gives

$$\text{R.H.S} = (-2-4)(-2+6) = -24$$

So L.H.S $>$ R.H.S when $x = -2$ and $y = -20$.
Therefore $(-2, -20)$ is not a point on the curve.

2) Draw a sketch to show the region of the xy plane defined by the following inequalities $0 \leqslant x \leqslant 2$, $y \geqslant 0$, $y \leqslant x^2$.

The relationship $0 \leqslant x \leqslant 2$ contains two inequalities, viz. $x \geqslant 0$, $x \leqslant 2$, which must be considered separately,

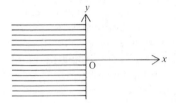 $x \geqslant 0$ indicates the line $x = 0$ (i.e. the y-axis) and the region to the right of the y-axis, $(x > 0)$. We eliminate the region to the left of the y-axis by shading it out, as it does not satisfy $x \geqslant 0$, i.e. we shade the *unwanted* region.

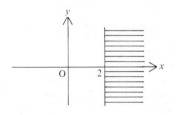 $x \leqslant 2$ indicates the line $x = 2$ and the region to the left of this line, so we shade out the region to the right of this line that does not satisfy $x \leqslant 2$.

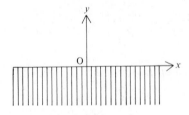 $y \geqslant 0$ is the line $y = 0$ and the region above the x-axis, so we shade out the region below the x-axis.

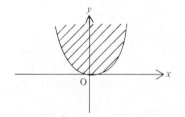 $y \leqslant x^2$ is the curve $y = x^2$ and the region below it, so we shade out the region above $y = x^2$.

Combining these four diagrams gives the diagram below

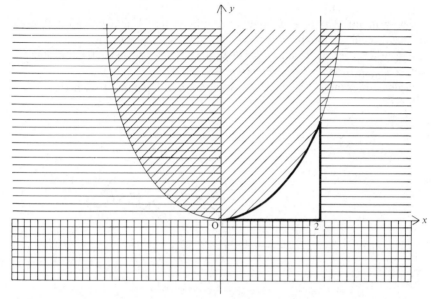

Therefore the unshaded region, including the boundary lines, is the set of points that satisfies all the given inequalities.

EXERCISE 4d

Draw a sketch of the curve whose equation is:

1) $y = x^2 - 3x + 4$ 2) $y = 1/x$ 3) $y = x(1-x)$ 4) $y = 2^{-x}$

Determine whether the given point lies on the given curve:

5) $(0, 1)$, $y = x^2 - 2$ 6) $(2, -3)$, $y = 3 - x - x^2$

7) $(-4, -0.2)$, $y = \dfrac{1}{x}$ 8) $(3, 2)$, $y^2 = x - 1$

9) $(-3, -6)$, $y + x^2 = 3$ 10) $(1, -2)$, $\dfrac{1}{x} + \dfrac{1}{y} = \dfrac{1}{2}$

Draw a sketch showing the region of the xy plane defined by the following inequalities:

11) $x > 3$ 12) $y < 2$ 13) $x < 0$ 14) $y > (x+3)(x-2)$

15) $0 < x < 2$ 16) $-1 < y < 3$

17) $y < 2$ and $y > (x-2)(x+2)$ 18) $x \geqslant 2,\ y \leqslant 4$

19) $1 \leqslant x \leqslant 3,\ y \leqslant \dfrac{1}{x}$ 20) $2 \geqslant x \geqslant 0,\ y \geqslant -7,\ y \leqslant (x-2)(x+5)$

21) $y \geqslant 0,\ 1 \leqslant x \leqslant 4,\ y \leqslant \dfrac{1}{x}$

THE EQUATION OF A PARTICULAR CURVE

So far in our work on functions and graphs we have begun with a function and have deduced from its properties the curve that represents it.

The reverse process in which we begin with a curve, given geometrically, and deduce its equation, will now be examined.

Consider, for example, the circle whose centre is at the point $(4, 2)$ and whose radius is 2.

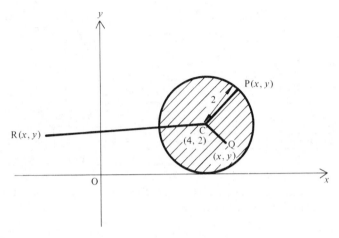

Any point P on the circumference of this circle is such that $PC = 2$.

Any point Q inside the circle satisfies the inequality $CQ < 2$.

Any point R outside the circle satisfies the inequality $CR > 2$.

The distance of any point (x, y) from C is given by

$$\sqrt{(x - 4)^2 + (y - 2)^2}$$

Therefore the coordinates (x, y) of P must satisfy the equation

$$\sqrt{(x - 4)^2 + (y - 2)^2} = 2$$

or

$$(x - 4)^2 + (y - 2)^2 = 4$$

Therefore this equation defines the set of points on the circumference of the circle and so is the equation of the circle.

Similarly the coordinates (x, y) of Q satisfy the inequality

$$(x - 4)^2 + (y - 2)^2 < 4$$

So this inequality defines the region inside the circle and the inequality

$$(x - 4)^2 + (y - 2)^2 > 4$$

defines the region outside the circle.

EQUATION OF A STRAIGHT LINE

Straight lines play an important part in any geometric analysis and we will now concentrate our attention on these. We will return to the problem of finding the equations of particular curves later in the book.

A straight line may be defined in many ways, e.g.:

1) a line which passes through the origin and has a gradient of $\frac{1}{2}$,

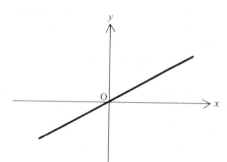

2) a line which passes through the points $(2, 1)$ and $(-4, -2)$.

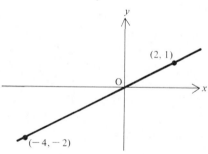

and there are many other ways of defining a straight line.

1) The equation of this line can be found as follows:

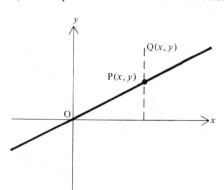

If $P(x, y)$ is any point on the line, the gradient of $OP = \frac{1}{2}$.

The gradient of OP is given by $\dfrac{y-0}{x-0} = \dfrac{y}{x}$

Therefore the coordinates of P satisfy the equation

$$\frac{y}{x} = \frac{1}{2} \quad \text{or} \quad 2y = x$$

Therefore $2y = x$ is the equation of the line.

Note. For any point $Q(x, y)$ above P, $y > \frac{1}{2}x \;\Rightarrow\; 2y > x$.
Therefore the inequality $2y > x$ defines the region above the line.
Similarly $2y < x$ defines the region below the line.

2) To obtain the equation of this line we know that
$P(x, y)$ is a point on the line $\Longleftrightarrow$ gradient of PA (or PB) = gradient of AB.

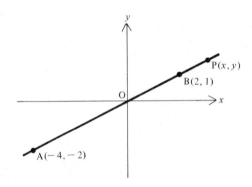

Gradient of PA $= \dfrac{y-1}{x-2}$

Gradient of AB $= \dfrac{1-(-2)}{2-(-4)} = \dfrac{1}{2}$

Therefore the coordinates of P satisfy the equation

$$\frac{y-1}{x-2} = \frac{1}{2}$$

or $2y = x$

Note. It is conventional to use integers for coefficients whenever possible.
Note also that these apparently different definitions give the same line, and
although there are other ways of defining a straight line we will concentrate on
these two, which are the commonest.

Consider the more general case of the line whose gradient is m and which passes
through the origin.

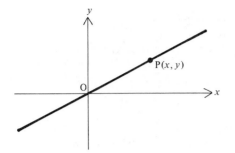

If $P(x, y)$ is any point on this line, then the gradient of OP is m.

Therefore the coordinates of P satisfy the equation

$$\frac{y}{x} = m$$

or $$y = mx$$

Generalising even further to cover any straight line, consider the line whose gradient is m and which cuts the y-axis at a directed distance c from the origin.

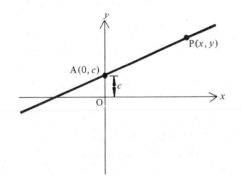

Note: c is called the *intercept* on the y-axis

$P(x, y)$ is any point on the line $\Longleftrightarrow$ gradient of AP $= m$.
Therefore the coordinates of P satisfy the equation

$$\frac{y - c}{x - 0} = m$$

or $$y = mx + c$$

The equation above is called the standard form for the equation of a straight line. It follows that

(a) an equation of the form

$$y = mx + c$$

represents a straight line with gradient m and intercept c on the y-axis.

(b) any equation involving a linear relationship between x and y is the equation of a straight line, i.e.

$$ax + by + c = 0$$

where a, b and c are constants, is the equation of a straight line.

EXAMPLES 4e

1) Write down the gradient of the line $3x - 4y + 2 = 0$ and find the equation of the line through the origin which is perpendicular to the given line.

Writing $3x - 4y + 2 = 0$ in standard form gives

$$y = \frac{3}{4}x + \frac{1}{2}$$

Comparing with $$y = mx + c$$

we see that the gradient, m, of the line is $\frac{3}{4}$
So the gradient of the perpendicular line is $-\frac{4}{3}$ $(= -1/m)$
The required line passes through the origin (i.e. has zero intercept on the y-axis).
Therefore its equation is

$$y = -\frac{4}{3}x + 0 \qquad (y = mx + c)$$

$\Rightarrow$ $$3y + 4x = 0$$

2) Sketch the line $x - 2y + 3 = 0$.

This line can be located accurately in the xy plane when we know two points on the line. As the intercepts on the axes can be found by inspection (i.e. $x = 0 \Rightarrow y = \frac{3}{2}$ and $y = 0 \Rightarrow x = -3$), we shall use these to place the line on the xy plane.

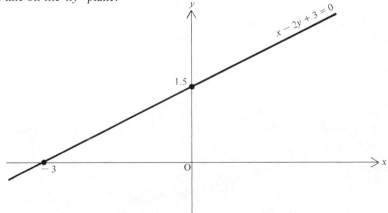

EXERCISE 4e

1) Write down the equation of the line passing through the origin and with gradient
(a) 2   (b) -1 (c) $\frac{1}{3}$ (d) $-\frac{1}{4}$ (e) 0   (f) ∞
Draw a sketch showing these lines on the same pair of axes.

2) Write down the equation of the line passing through the given point and with the given gradient.
(a) $(0, 1), \frac{1}{2}$ (b) $(0, 0), \frac{1}{2}$ (c) $(-1, -4), \frac{1}{2}$
Sketch these lines on the same pair of axes.

3) Write down the equation of the line passing through the given points.
(a) $(0, 0), (2, 1)$ (b) $(1, 4), (3, 0)$ (c) $(2, 0), (0, 4)$
(d) $(-1, 3), (-4, -3)$

4) Write down the inequality which defines the region:
(a) above the line through the origin with gradient 1,
(b) below the line through $(1, 2)$ and $(0, 4)$,
(c) above the line $x + y - 2 = 0$,
(d) below the line $2x - y + 4 = 0$.

5) Write down the equation of the line passing through the origin and perpendicular to:
(a) $3x + 2y - 4 = 0$ (b) $x - 2y + 3 = 0$

6) Write down the equation of the line passing through the given point and perpendicular to the given line.
(a) $(2, 1), 3x + y - 2 = 0$ (b) $(-1, -2), 2x - 3y + 6 = 0$

7) Write down the equation of the line passing through $(3, -2)$ and parallel to:
(a) $5x - y + 3 = 0$ (b) $x + 7y - 5 = 0$

The Equation of a Line with Gradient m passing through the Point (x_1, y_1)

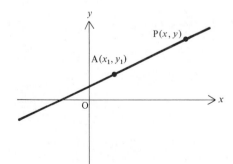

If $P(x, y)$ is any point on the line, the gradient of AP is m.

Therefore the coordinates of P satisfy the equation

$$\frac{y - y_1}{x - x_1} = m$$

$\Rightarrow$ $y - y_1 = m(x - x_1)$ [1]

EXAMPLES 4f

1) Find the equation of the line with gradient $-\frac{1}{3}$, passing through $(2, -1)$.

Substituting $-\frac{1}{3}$ for m, 2 for x_1 and -1 for y_1 in [1] gives the equation of the line as

$$y - (-1) = -\tfrac{1}{3}(x - 2)$$

or $x + 3y + 1 = 0$

Alternatively the equation can be found as follows:
As any straight line has an equation $y = mx + c$,
the equation of this line can be written

$$y = -\tfrac{1}{3}x + c$$

As the point $(2, -1)$ lies on this line, its coordinates satisfy the equation.

i.e. $-1 = -\tfrac{1}{3}(2) + c$

$\Rightarrow$ $c = -\tfrac{1}{3}$

Therefore the equation is $y = -\tfrac{1}{3}x - \tfrac{1}{3}$

or $x + 3y + 1 = 0$

Note. The worked examples in the text must necessarily contain a lot of explanation but this should not mislead the reader into thinking that his solutions should be equally long. The temptation to 'overwork' a problem should be avoided, particularly in the case of coordinate geometry problems which are basically simple. The reader will find that, with a little practice, either of the methods illustrated above will enable the required equation of a line to be written down directly.

The Equation of the Line passing through (x_1, y_1) and (x_2, y_2)

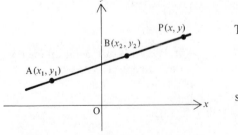

The gradient of AB is

$$\frac{y_1 - y_2}{x_1 - x_2}$$

so $y - y_1 = m(x - x_1)$

gives

$$y - y_1 = \frac{y_1 - y_2}{x_1 - x_2}(x - x_1)$$

e.g. the line through $(1, -2)$ and $(3, 5)$ has equation

$$y - 5 = \frac{5 - (-2)}{3 - 1}(x - 3) \quad \Rightarrow \quad 7x - 2y - 11 = 0$$

INTERSECTION

If two curves cut at a point A, A is called a point of intersection.

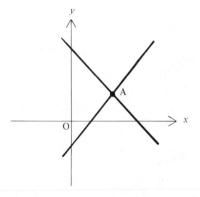

If A is the point of intersection of the lines

$$y - 3x + 1 = 0 \qquad [1]$$

and $\qquad y + x - 2 = 0 \qquad [2]$

then the coordinates of A satisfy both equations [1] and [2]. So A can be found by solving these equations simultaneously.

$$[2] - [1] \quad \Rightarrow \quad 4x - 3 = 0 \quad \Rightarrow \quad x = \tfrac{3}{4}, \; y = \tfrac{5}{4}$$

Therefore $(\tfrac{3}{4}, \tfrac{5}{4})$ is the point of intersection.

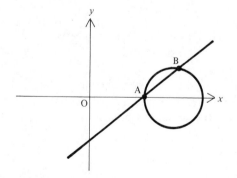

If A and B are the points of intersection of the circle

$$x^2 + y^2 - 3x + 2 = 0 \quad [1]$$

and the line $\qquad y = x - 1 \quad [2]$

then the coordinates of both A and B will satisfy both equations [1] and [2].

Solving these equations simultaneously by substituting $x - 1$ for y in [1] we have

$$x^2 + (x - 1)^2 - 3x + 2 = 0$$

$$\Rightarrow \qquad 2x^2 - 5x + 3 = 0$$

$$\Rightarrow \qquad (2x - 3)(x - 1) = 0$$

$$\Rightarrow \qquad x = \tfrac{3}{2} \quad \text{or} \quad 1.$$

Substituting $\tfrac{3}{2}$ and 1 for x in [2] gives $y = \tfrac{1}{2}$ and 0.
Therefore A and B are the points $(\tfrac{3}{2}, \tfrac{1}{2}), (1, 0)$.

In general the coordinates of the points of intersection of two curves $y = f(x)$ and $y = g(x)$ can be found from the simultaneous solution of the equations $y = f(x)$ and $y = g(x)$.

EXAMPLES 4f (contd.)

2) Find the equation of the line through $(1, 2)$ which is perpendicular to the line $3x - 7y + 2 = 0$.

Writing $3x - 7y + 2 = 0$ in standard form gives $y = \tfrac{3}{7}x + \tfrac{1}{2}$ showing that the given line has a gradient of $\tfrac{3}{7}$.
So the required line has gradient $-\tfrac{7}{3}$ and passes through $(1, 2)$.
Using $y - y_1 = m(x - x_1)$ gives its equation as

$$y - 2 = -\tfrac{7}{3}(x - 1)$$

or $\qquad\qquad\qquad\qquad 7x + 3y - 13 = 0$

Note that the line perpendicular to $\qquad\qquad 3x - 7y + 2 = 0$

has an equation $\qquad\qquad\qquad\qquad 7x + 3y - 13 = 0$

i.e. the coefficients of x and y have been transposed and the sign between the x and y terms has changed.

In fact, given the line l with equation $ax + by + c = 0$, any line perpendicular to l has equation $bx - ay + k = 0$ and this property of perpendicular lines can be used to shorten the working of problems.

3) A, B and C are the points $(0, 4), (2, 3)$ and $(-2, -1)$.
Find the circumcentre of triangle ABC.

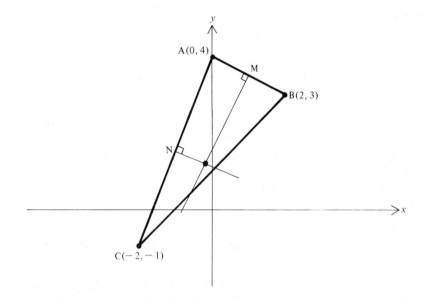

The circumcentre of a triangle is the point of intersection of the perpendicular bisectors of its sides.

AC has gradient $$\frac{4-(-1)}{0-(-2)} = \frac{5}{2}$$

and its midpoint is $$\left(\frac{0-2}{2}, \frac{4-1}{2}\right) \Rightarrow \left(-1, \frac{3}{2}\right)$$

Therefore the perpendicular bisector of AC has gradient $-\frac{2}{5}$ and passes through $(-1, \frac{3}{2})$

Hence its equation is $$y = -\tfrac{2}{5}x + \tfrac{11}{10}$$

i.e. $$4x + 10y - 11 = 0 \qquad [1]$$

Similarly the gradient of AB is $\dfrac{4-3}{0-2} = -\dfrac{1}{2}$ and its midpoint is $\left(1, \dfrac{7}{2}\right)$.

Therefore the perpendicular bisector of AB has gradient $+2$ and passes through $(1, \frac{7}{2})$

and its equation is $$y = 2x + \tfrac{3}{2}$$

i.e. $$4x - 2y + 3 = 0 \qquad [2]$$

Solving equations [1] and [2] simultaneously gives

$$12y - 14 = 0 \Rightarrow y = \tfrac{7}{6}, \ x = -\tfrac{1}{6}$$

Therefore the circumcentre of triangle ABC is the point $(-\frac{1}{6}, \frac{7}{6})$.

EXERCISE 4f

1) Find the equation of the line perpendicular to $x + 2y + 3 = 0$ and through the point $(5, 2)$.

2) Find the equation of the line through the origin which is parallel to $x - y + 2 = 0$.

3) Find the equation of the line joining the points:
(a) $(0, 1), (2, 4)$ (b) $(-1, 2), (1, 5)$ (c) $(3, -1), (3, 2)$

4) Determine which of the following pairs of lines are perpendicular:
(a) $x - 2y + 4 = 0$, $2x + y - 3 = 0$
(b) $x + 3y - 6 = 0$, $3x + y + 2 = 0$
(c) $x + 3y - 2 = 0$, $y = 3x + 2$
(d) $y + 2x + 1 = 0$, $x = 2y - 4$

5) Find the coordinates of the points of intersection of the pairs of lines given in Question 4.

6) Find the equations of the perpendicular bisectors of the lines joining:
(a) $(0, 0), (2, 4)$ (b) $(3, -1), (-5, 2)$ (c) $(5, -1), (0, 7)$

7) The line $4x - 5y + 20 = 0$ cuts the x-axis at A and the y-axis at B. Find the equation of the median through O of triangle OAB.

8) Find the equation of the altitude through O of triangle OAB defined in Question 7.

9) The sides of a triangle are the lines $y = 0$, $x - 3y + 5 = 0$ and $2x + y - 7 = 0$. Find the coordinates of the vertices of the triangle.

10) Find the equation of the perpendicular from the point $A(5, 3)$ to the line $2x - y + 4 = 0$. Hence find the distance of A from the line.

SUMMARY

If A and B are the points (x_1, y_1) and (x_2, y_2)

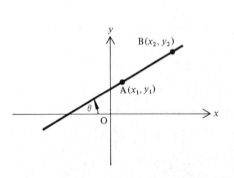

The *length* of AB is

$$\sqrt{(x_1 - x_2)^2 + (y_1 - y_2)^2}$$

The *midpoint* of AB is the point

$$\frac{x_1 + x_2}{2}, \frac{y_1 + y_2}{2}$$

The *gradient* of AB is

$$\frac{y_1 - y_2}{x_1 - x_2} = \tan \theta$$

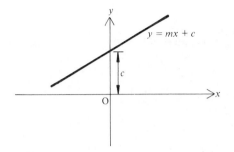

The *equation* $y = mx + c$ defines the straight line with gradient m and intercept c on the y-axis.

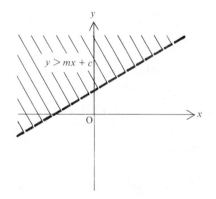

The inequality $y > mx + c$ defines the region of the xy plane above the line $y = mx + c$.

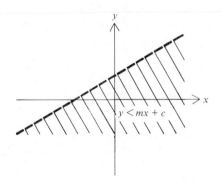

The inequality $y < mx + c$ defines the region below the line $y = mx + c$.

If lines l_1, l_2 have equations $y = m_1 x + c_1$, $y = m_2 x + c_2$ then

l_1 and l_2 are *parallel* if $m_1 = m_2$

l_1 and l_2 are *perpendicular* if $m_1 m_2 = -1$

The equation of any line perpendicular to $ax + by + c = 0$ is of the form $bx - ay + k = 0$.

MULTIPLE CHOICE EXERCISE 4

(Instructions for answering these questions are given on p. xii)

TYPE I

1) The length of the line joining $(3, -4)$ to $(-7, 2)$ is:
(a) $2\sqrt{13}$ (b) 16   (c) $2\sqrt{34}$ (d) $2\sqrt{5}$ (e) 6.

2) The midpoint of the line joining $(-1, -3)$ to $(3, -5)$ is:
(a) $(1, 1)$ (b) $(0, 0)$ (c) $(2, -8)$ (d) $(1, -4)$ (e) $(1, -1)$.

3) The gradient of the line joining $(1, 4)$ and $(-2, 5)$ is:
(a) $\frac{1}{3}$ (b) $-\frac{1}{3}$ (c) 3   (d) -3 (e) 1.3.

4) The gradient of the line perpendicular to the join of $(-1, 5)$ and $(2, -3)$ is:
(a) $\frac{3}{8}$ (b) $-2\frac{2}{3}$ (c) $\frac{1}{2}$ (d) 2   (e) $2\frac{2}{3}$.

5) The line joining $(1, 3)$ to (a, b) has unit gradient.
(a) $b - a = 2$ (b) $a - b = 2$ (c) $a + b = 2$ (d) $b - a = 4$
(e) $a - b = 4$.

6) The equation of the line through the origin and perpendicular to $3x - 2y + 4 = 0$ is:
(a) $3x + 2y = 0$ (b) $2x + 3y + 1 = 0$ (c) $2x + 3y = 0$
(d) $2x - 3y - 1 = 0$ (e) $3x - 2y = 0$.

7) The equation of the line with gradient 1 passing through the point (h, k) is:
(a) $y = x + k - h$ (b) $y = \dfrac{k}{h}x + 1$ (c) $y = x + h - k$
(d) $ky = hx - 1$ (e) $y + x = k - h$.

8) The two lines $x + y = 0$ and $2x - y + 3 = 0$ intersect at the point:
(a) $(-\frac{1}{3}, \frac{1}{3})$ (b) $(1, -1)$ (c) $(-3, 3)$
(d) $(-1, 1)$ (e) $(3, -3)$.

9) The shaded region defined by $y > 0$, $y > x(x - 1)$ is:

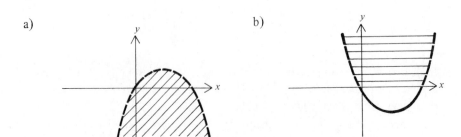

a)

b)

c)

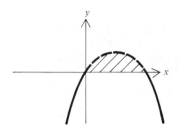

d)

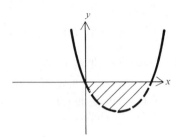

e)

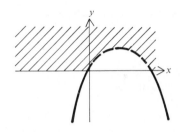

10) The curves $y = x^2$, $y = x(2-x)$ intersect at:
(a) $(0,0), (1,1)$ (b) $(2,4)$ (c) $(0,0), (2,4)$
(d) $(0,0), (-1,1)$ (e) $(1,1)$.

TYPE II

11) A and B are two points with coordinates $(3,4)$, $(-1,6)$.
(a) Gradient of AB is $-\frac{1}{2}$.
(b) Midpoint of AB is the point $(2,5)$.
(c) Length of AB is $2\sqrt{5}$.

12) A, B and C are the points $(5,0), (-5,0), (2,3)$.
(a) AB and BC are perpendicular.
(b) Area of triangle ABC is 15 square units.
(c) A, B and C are collinear.

13) A, B, C are the points $(0,13), (0,-13), (5,-12)$.
(a) A, B, C lie on the circumference of a circle, centre the origin.
(b) The equation of AC is $5x + y - 13 = 0$.
(c) The midpoint of BC is the origin.

14) The equation of a line l is $y = 2x - 1$.
(a) The line through the origin perpendicular to l is $y + 2x = 0$.
(b) The line through $(1,2)$ parallel to l is $y = 2x - 3$.
(c) l passes through $(1,1)$.

15) The equation of a line l is $7x - 2y + 4 = 0$.
(a) l has a gradient of $3\frac{1}{2}$.
(b) l is parallel to $7x + 2y - 3 = 0$.
(c) l is perpendicular to $2x + 7y - 5 = 0$.

TYPE III

16) (a) Two lines l_1 and l_2 have gradients m and $\dfrac{1}{m}$ respectively.

(b) Two lines l_1 and l_2 are perpendicular.

17) A, B and C are three points where:
(a) C is equidistant from A and B.
(b) The coordinates of A, B and C are $(1, 2)$, $(3, 6)$ and $(2, 4)$.

18) (a) A straight line has equation $x + y - 1 = 0$.
(b) A straight line has intercepts 1 and 1 on the x and y axes respectively.

TYPE IV

19) Find the equation of a line.
(a) The line is perpendicular to $x + y = 0$.
(b) The line cuts the y-axis at the point $(0, 3)$.
(c) The area of the triangle enclosed by the line and the coordinate axes is $4\frac{1}{2}$ square units.

20) Find the equation of the median through A of triangle ABC.
(a) A is the point $(5, -1)$.
(b) The equation of BC is $y = 3x - 4$.
(c) AC is parallel to $x + 2y - 3 = 0$.

21) Find the point of intersection of the two lines l_1 and l_2.
(a) l_1 passes through $(2, 5)$ and $(7, 3)$.
(b) l_2 is parallel to $y = 3x - 4$.
(c) The intercept of l_2 on the x-axis is 1.

22) Find the point of intersection of the curve $y = f(x)$ and the line l.
(a) l is perpendicular to $y = 3x - 4$.
(b) $f(x) \equiv (x - 2)(x + 5)$.
(c) l cuts $y = f(x)$ in two points.

23) Find the coordinates of the circumcentre of triangle ABC.
(a) The equation of AC is $x - y + 3 = 0$.
(b) The equation of BC is $x + 2y - 4 = 0$.
(c) The equation of AB is $y = 5$.

TYPE V

24) The line joining $(0, 0)$ and $(1, 3)$ is equal in length to the line joining $(0, 1)$ and $(3, 0)$.

25) If a line has gradient m and intercept d on the x-axis, its equation is $y = mx - md$.

26) The line $2x - y + 5 = 0$ has intercepts 5 and 2 on the y-axis and x-axis respectively.

27) The line passing through $(3, 1)$ and $(-2, 5)$ is perpendicular to the line $4y = 5x - 3$.

28) The line $3y = 7x - 2$ has an intercept of -2 on the y-axis.

MISCELLANEOUS EXERCISE 4

1) Show that the triangle whose vertices are $(1, 1)$, $(3, 2)$, $(2, -1)$ is isosceles.

2) Find the area of the triangular region defined by $y \geqslant 2x - 1$, $x \geqslant 0$, $y \leqslant 0$.

3) Find the coordinates of the vertices of the triangular region defined by $y \geqslant 0$, $y \leqslant x + 5$, $y \leqslant -\frac{1}{2}x + 3$.

4) Write down the equation of the line which goes through the point $(7, 3)$ and which is inclined at $45°$ to the positive direction of the x-axis.
Find the area enclosed by this line and the coordinate axes.

5) Find the coordinates of the points of intersection of $y = x^2 - 9$ and $y = x - 3$. Find the length of the line joining these two points.

6) Find the equation of the line through $A(5, 2)$ which is perpendicular to the the line $y = 3x - 5$. Hence find the coordinates of the foot of the perpendicular from A to the line.

7) The coordinates of a point P are $(t + 1, 2t - 1)$.
Plot the position of P when $t = -1, 0, 1, 2$. Show that these four points are collinear and find the equation of the line on which they lie.

8) Write down the equation of the perpendicular bisector of the line joining $(a, b), (2a, -3b)$.

9) A circle has a radius 4 and centre at the point $(2, 0)$. If $P(x, y)$ is any point inside the circumference of this circle, write down the condition that must be satisfied by the coordinates of P.

10) Write down the equation of the circumference of the circle defined in Question 9.

11) Find the coordinates of the points of intersection of $y = x^2$ and $y = x(4 - x)$. Draw a sketch showing the region defined by
$$x(4 - x) > y > x^2$$

12) The equation of a circle is $(x - 1)^2 + (y - 1)^2 = 4$.
Find the coordinates of A and B, the points of intersection of the line $x + y = 2$ and the circle.

13) Show that the point $C(1, 3)$ is on the circumference of the circle whose equation is given in Question 12. Find the midpoint M of AB and show that $MC = MA = MB$. What can you deduce about the line AB?

14) The equations of two adjacent sides of a rhombus are $y = 2x + 4$, $y = -\frac{1}{3}x + 4$. If $(12, 0)$ is one vertex and all vertices have positive coordinates, find the coordinates of the other three vertices.

15) A line is drawn through the point $A(1, 2)$ to cut the line $2y = 3x - 5$ in P and the line $x + y = 12$ in Q.
If $AQ = 2AP$, find the coordinates of P and Q.   (U of L)p

16) One side of a rhombus lies along the line $5x + 7y = 1$ and one of the vertices is $(3, -2)$. One diagonal of the rhombus is the line $3y = x + 1$. Find the coordinates of the other vertices and the equations of the three remaining sides.   (U of L)

CHAPTER 5

DIFFERENTIATION I

TANGENTS AND NORMALS

If A, B are two points on a curve (any curve) then

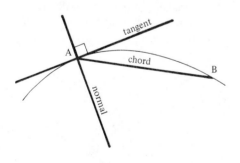

the line joining A and B is called a *chord*,

the line touching the curve at A is called the *tangent* to the curve at A,

the line perpendicular to the tangent at A is called the *normal* to the curve at A

GRADIENT OF A CURVE

The gradient of a curve, which is a measure of its slope, changes continually as one moves along it.

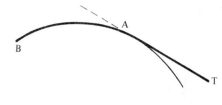

Suppose, when moving along the curve in the diagram (in the sense $B \rightarrow A$), that at the point A the gradient stops changing and remains constant. One would then move along the straight line AT, i.e. along the tangent at A.

So the gradient of the curve at the point A is the same as the gradient of the tangent at A.

The gradient of a curve at any point is defined as the gradient of the tangent to the curve at that point and measures the rate of increase of y with respect to x.

An *approximate* value for the gradient of a curve at a point can be found by plotting the curve, drawing the tangent by eye and measuring its gradient. This method has to be used for a curve when the coordinates of a finite number of points are known, but its equation is not, e.g. data from an experiment. When the equation of a curve is known, an accurate method for determining gradients is necessary so that we can further our analysis of curves and functions. Consider first the problem of finding the gradient of a curve at a given point A.

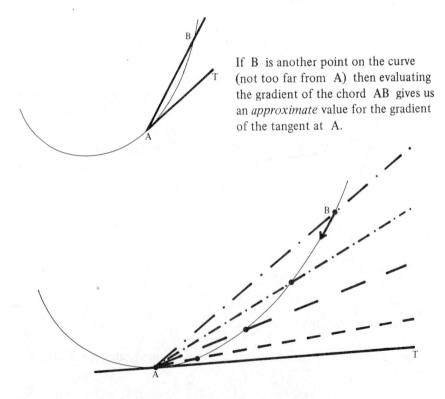

If B is another point on the curve (not too far from A) then evaluating the gradient of the chord AB gives us an *approximate* value for the gradient of the tangent at A.

The closer B is to A, the better is the approximation.

i.e.   as $B \to A$

            gradient of chord AB $\to$ gradient of tangent AT

or $\underset{\text{as } B \to A}{\text{limit}}$ {gradient of chord AB} = gradient of tangent AT

Let us now consider an example where we can use this definition to find the gradient of a curve at a particular point.
Find the gradient of the curve $y = x(2x - 1)$ at the point on the curve where $x = 1$.

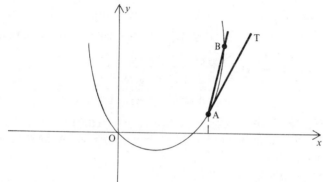

From the definition above

$$\text{Gradient of } AT = \lim_{B \to A} \{\text{gradient of chord } AB\}$$

So we must first calculate the gradient of a chord AB and observe what happens to this gradient as B progressively approaches A.
One method is to take a succession of points $B_1, B_2, B_3 \ldots$
where $x = 1.5, 1.25, 1.125 \ldots$, (i.e. halving the remaining difference between the x coordinates of A and B) and calculating the gradients of the chords $AB_1, AB_2 \ldots$ etc.

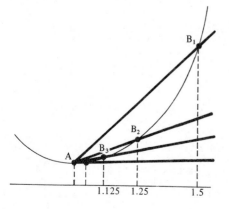

grad AB_1 = 4

grad AB_2 = 3.5

grad AB_3 = 3.25

grad AB_4 = 3.125

grad AB_5 = 3.0625

From this sequence of values we observe that

$$\text{as } B \to A, \text{ gradient of chord } AB \to 3,$$

i.e. we deduce that the gradient of the curve at A is 3.
However this method is unsatisfactory, not least because of the large amount of numerical calculation involved. So instead of placing B at particular positions we will introduce a variable quantity for the difference between the x coordinates of A and B.

The Delta Prefix

A variable quantity, prefixed by δ, means a small increase in that quantity:

i.e. δx is a small increase in x

δy is a small increase in y

δv is a small increase in v

(**Note** δ is only a prefix; it cannot be treated as a factor.)
If δx is the increase in the x coordinate in moving from A to B, then the x coordinate of B is $1 + \delta x$.
For all points on the curve, $y = x(2x - 1)$.
So at A, $x = 1$, $y = 1(2 \times 1 - 1) = 1$
   at B, $x = 1 + \delta x$, $y = (1 + \delta x)[2(1 + \delta x) - 1] = (1 + \delta x)(2\delta x + 1)$

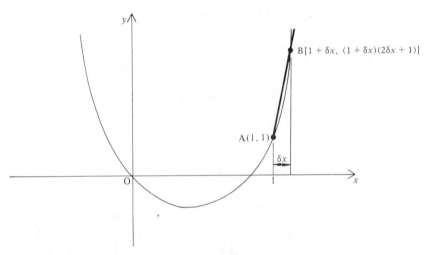

Therefore the gradient of AB which is $\dfrac{\text{increase in } y}{\text{increase in } x}$

$$= \frac{(1 + \delta x)(2\delta x + 1) - 1}{(1 + \delta x) - 1}$$

$$= \frac{2(\delta x)^2 + 3\delta x}{\delta x}$$

$$= 2\delta x + 3$$

As B approaches A, the difference between their x coordinates approaches zero, i.e. $\delta x \to 0$.

Therefore   as $B \to A$, gradient of $AB \to 3$,

or   gradient at $A = \lim_{B \to A} \{\text{gradient of chord } AB\}$

$$= \lim_{\delta x \to 0} \{2\delta x + 3\}$$

$$= 3$$

EXAMPLE 5a

Using the method described above find the gradient of the curve $y = \frac{1}{x}$ at the point where $x = 2$.

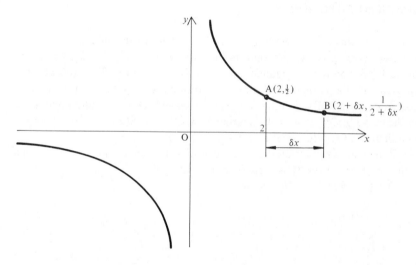

Let δx be the increase in moving from $A(2, \frac{1}{2})$ to a near point B on the curve so that the coordinates of B are $\left(2 + \delta x, \frac{1}{2 + \delta x}\right)$

$$\text{The gradient of chord } AB = \left(\frac{1}{2 + \delta x} - \frac{1}{2}\right)\Big/\delta x$$

$$= \frac{-\delta x}{2\delta x(2 + \delta x)}$$

$$= \frac{-1}{2(2 + \delta x)} = -\frac{1}{4 + 2\delta x}$$

Therefore gradient at $A = \lim_{\delta x \to 0} \left(-\frac{1}{4 + 2\delta x} \right)$

$$= -\frac{1}{4}$$

EXERCISE 5a

Use the method of Example 5a to find the gradient of the given curve at the point indicated.

1) $y = x^2 + 1$ where $x = 3$ 2) $y = (x + 1)(x - 1)$ where $x = 2$

3) $y = 2x(x - 4)$ where $x = 0$ 4) $y = (x + 2)(x - 1)$ where $x = -3$

5) $y = x^3 - 3$ where $x = 1$ 6) $y = \frac{1}{x^2}$ where $x = 1$

GRADIENT FUNCTION

   Earlier in this chapter we found that the gradient of the curve $y = x(2x - 1)$ is 3 at the point on the curve where $x = 1$.   We will now derive a function for the gradient at *any* point on the curve. Then we can find the gradient at a particular point by substitution into this derived function. Instead of taking a fixed point on the curve we will take A as any point (x, y) on the curve. Its y coordinate can then be written as $x(2x - 1)$ since both coordinates satisfy the equation $y = x(2x - 1)$.

Let B be another point on the curve such that the increase in the x coordinate in moving from A to B is δx. Then B is the point $[x + \delta x, (x + \delta x)(2x + 2\delta x - 1)]$.

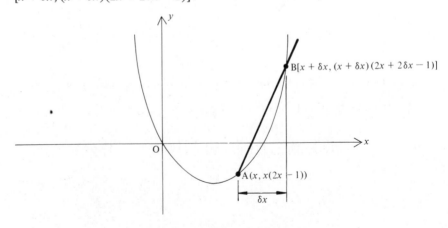

$$\text{gradient of chord AB} = \frac{(x + \delta x)(2x + 2\delta x - 1) - x(2x - 1)}{\delta x}$$

$$= \frac{2x^2 + 4x\delta x + 2(\delta x)^2 - \delta x - x - 2x^2 + x}{\delta x}$$

$$= \frac{4x\delta x - \delta x + 2(\delta x)^2}{\delta x}$$

$$= 4x - 1 + 2\delta x$$

Then the gradient at any point A on the curve

$$= \lim_{\delta x \to 0} \{4x - 1 + 2\delta x\}$$

$$= 4x - 1$$

So the function $4x - 1$ gives the gradient at any point on the curve $y = x(2x - 1)$.

We can find the gradient of the curve (i.e. the rate of increase of y with respect to x) at a particular point on $y = x(2x - 1)$ by substituting the x co-ordinate of that point into the function $4x - 1$.

Thus $4x - 1$ is called the *gradient function* of $y = x(2x - 1)$ and the process of deriving it is called *differentiation with respect to x*.

Now $4x - 1$ was derived from the function $x(2x - 1)$ so $4x - 1$ is called the *derivative*, or *derived function*, of $x(2x - 1)$.

Using $\dfrac{d}{dx}$ as a symbol to denote 'the derivative w.r.t. x of'

we may write $\dfrac{d}{dx}[x(2x - 1)] = 4x - 1$,

or, when $y = x(2x - 1)$, $\dfrac{dy}{dx} = 4x - 1$

where $\dfrac{dy}{dx}$ means the *derivative* of y w.r.t. x and is sometimes called the *differential coefficient* of y.

As an alternative notation we can use the symbol D for derivative, or derived function, of

i.e. $D[x(2x - 1)] = 4x - 1$

where D is referred to as the differential operator.

Now $4x - 1$ represents the rate of increase of y w.r.t. x

therefore $\dfrac{dy}{dx}$ represents also the rate of increase of y w.r.t. x.

Similarly $\dfrac{dv}{dt}$ means the derivative of v w.r.t. t or the rate of increase

of v w.r.t. t etc.

EXAMPLE 5b

Differentiate $y = x^3 + 3$ with respect to x.

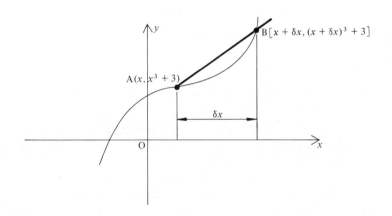

Let A be the point $(x, x^3 + 3)$ and B the point $[x + \delta x, (x + \delta x)^3 + 3]$ on $y = x^3 + 3$.

$$\text{Gradient of } AB = \frac{(x + \delta x)^3 + 3 - (x^3 + 3)}{\delta x}$$

which simplifies to $3x^2 + 3x\delta x + (\delta x)^2$.

$$\text{Gradient at } A = \lim_{\delta x \to 0} \{3x^2 + 3x\delta x + (\delta x)^2\}$$

$$= 3x^2$$

i.e. $$\frac{dy}{dx} = 3x^2$$

EXERCISE 5b

Use the method of Example 5b to differentiate the following with respect to x.

1) $y = x^2$
2) $y = x^3$
3) $y = x^4$
4) $y = x$
5) $y = 3x^2$

6) $y = 5x^3$
7) $y = \dfrac{1}{x^2}$
8) $y = x^2 + 3x$
9) $y = x^2 - 2x + 1$

GENERAL DIFFERENTIATION

Consider any curve $y = f(x)$.

Let $A[x, f(x)]$ be any point on $y = f(x)$.

Let δx be the increase in the x coordinate in moving from A to another point B on the same curve. So B is the point $[x + \delta x, f(x + \delta x)]$.

Let δy be the corresponding increase in the y coordinate in moving from A to B,

i.e. $\delta y = f(x + \delta x) - f(x)$.

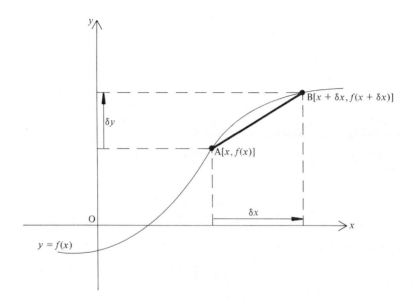

$$\text{Gradient of } AB = \frac{\delta y}{\delta x} = \frac{f(x + \delta x) - f(x)}{\delta x}$$

Therefore the gradient at A is

$$\lim_{\delta x \to 0} \left(\frac{\delta y}{\delta x} \right) = \lim_{\delta x \to 0} \left\{ \frac{f(x + \delta x) - f(x)}{\delta x} \right\}$$

i.e. $$\frac{dy}{dx} = \lim_{\delta x \to 0} \left(\frac{\delta y}{\delta x} \right) \quad \text{or} \quad \frac{d}{dx} f(x) = \lim_{\delta x \to 0} \left\{ \frac{f(x + \delta x) - f(x)}{\delta x} \right\}$$

This formal definition of differentiation can be used to differentiate any function that we have not yet met. This process is called differentiating from first principles.

Fortunately it is not always necessary to go back to first principles because certain categories of functions can be differentiated by rules, some of which we will now find.

Differentiation of a Constant

Consider the equation $y = c$. This represents a straight line parallel to the x-axis and so it has zero gradient.

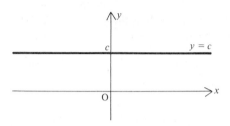

i.e.
$$\frac{dy}{dx} = 0 \qquad \text{or} \qquad \frac{d}{dx}(c) = 0.$$

Differentiation of ax where a is a Constant

Consider the equation $y = ax$. This represents a straight line with gradient a,

i.e.
$$\frac{dy}{dx} = a \qquad \text{or} \qquad \frac{d}{dx}(ax) = a$$

Differentiation of x^n

The table below summarizes the results from some of the questions in Exercise 5b.

$f(x)$	x^2	x^3	x^4	$x^{-2}\left(=\dfrac{1}{x^2}\right)$
$\dfrac{d}{dx}f(x)$	$2x$	$3x^2$	$4x^3$	$-2x^{-3}$

From this table it appears that to differentiate a power of x we multiply by that power and then subtract one from the power, i.e.

$$\frac{d}{dx}(x^n) = nx^{n-1} \qquad\qquad [1]$$

This result, deduced from a few examples, is in fact valid for all powers of x including fractional and negative powers, although proof is not possible at this stage. (This is one example of many such 'rules' whose validity, for the time being, we must take on trust.)

Accepting that rule [1] above can be used for any power of x, the following examples illustrate its application,

$$\frac{d}{dx}(x^9) = 9x^8$$

$$\frac{d}{dx}(x^{-4}) = -4x^{-4-1} = -4x^{-5}$$

$$\frac{d}{dx}(\sqrt{x}) = \frac{d}{dx}(x^{\frac{1}{2}}) = \tfrac{1}{2}x^{\frac{1}{2}-1} = \tfrac{1}{2}x^{-\frac{1}{2}}$$

EXERCISE 5c

Differentiate by rule the following functions with respect to x.

1) x^5 2) x^{10} 3) x^{-3} 4) $x^{\frac{1}{3}}$ 5) x^{-1} 6) $x^{-\frac{1}{2}}$

7) $\dfrac{1}{x^4}$ 8) $\sqrt{x^3}$ 9) $2x$ 10) $\dfrac{1}{x^5}$ 11) $\dfrac{1}{\sqrt{x}}$ 12) $\sqrt[4]{x}$

13) 5 14) $-4x$ 15) $\dfrac{1}{x^{-\frac{1}{2}}}$ 16) $\sqrt[3]{(x^2)}$

Differentiation of ax^n

Selecting some more results from Exercise 5b we see that

$$\frac{d}{dx}(3x^2) = 3(2x)$$

and

$$\frac{d}{dx}(5x^3) = 5(3x^2)$$

These results suggest that

$$\frac{d}{dx}(ax^n) = a\frac{d}{dx}(x^n) = anx^{n-1} \qquad\qquad [2]$$

where a is a constant.
Again selecting from Exercise 5b we have

$$\frac{d}{dx}(x^2 + 3x) = 2x + 3 = \frac{d}{dx}(x^2) + \frac{d}{dx}(3x)$$

and $$\frac{d}{dx}(x^2 - 2x + 1) = 2x - 2 = \frac{d}{dx}(x^2) - \frac{d}{dx}(2x) + \frac{d}{dx}(1)$$

This suggests that the operation 'differentiate' is distributive across addition and subtraction of functions: i.e.

$$\frac{d}{dx}\left[f(x) + g(x)\right] = \frac{d}{dx}f(x) + \frac{d}{dx}g(x) \qquad\qquad [3]$$

Therefore $\dfrac{d}{dx}(3x^2 - 7x) = 3(2x) - 7 = 6x - 7$

and $\dfrac{d}{dx}\left(x + \dfrac{1}{x}\right) = \dfrac{d}{dx}(x + x^{-1}) = 1 - x^{-2}$

Note that differentiation is *not* distributive across multiplication and division:

e.g. $\dfrac{d}{dx}\Big[x(x - 3)\Big] = \dfrac{d}{dx}(x^2 - 3x) = 2x - 3$

but $\left[\dfrac{d}{dx}(x)\right] \times \left[\dfrac{d}{dx}(x - 3)\right] = (1) \times (1) = 1$

i.e. $\dfrac{d}{dx}\Big[x(x - 3)\Big] \neq \dfrac{d}{dx}(x) \times \dfrac{d}{dx}(x - 3)$

In order to differentiate at this stage, therefore, any product must be expanded and any quotient must be divided out to give terms which are added or subtracted.

EXAMPLES 5d

1) Differentiate $f(x) \equiv \dfrac{4x^2 + x - 1}{2x}$.

$$f(x) \equiv \dfrac{4x^2 + x - 1}{2x} \equiv \dfrac{4x^2}{2x} + \dfrac{x}{2x} - \dfrac{1}{2x}$$

therefore $f(x) \equiv 2x + \tfrac{1}{2} - \tfrac{1}{2}x^{-1}$

therefore $\dfrac{d}{dx}f(x) = 2 + 0 - \tfrac{1}{2}(-x^{-2}) = 2 + \tfrac{1}{2}x^{-2} = 2 + \dfrac{1}{2x^2}$

2) Find the gradient of the curve $y = (x - 3)(x^2 + 2)$ at the point on the curve where $x = 1$.

$$y = (x - 3)(x^2 + 2) = x^3 - 3x^2 + 2x - 6$$

therefore $\dfrac{dy}{dx} = 3x^2 - 6x + 2.$

When $x = 1$, $\dfrac{dy}{dx} = 3(1)^2 - 6(1) + 2 = -1$

therefore the gradient of $y = (x - 3)(x^2 + 2)$ is -1 at the point where $x = 1$.

3) Find the coordinates of the point on the curve $y = \dfrac{2}{x^2}$ at which its gradient is $\tfrac{1}{2}$.

$$y = \frac{2}{x^2} = 2x^{-2}$$

therefore

$$\frac{dy}{dx} = -4x^{-3}$$

when $\frac{dy}{dx} = \frac{1}{2}$, $\qquad -4x^{-3} = \frac{1}{2}$

$\Rightarrow \qquad\qquad\qquad -\dfrac{4}{x^3} = \frac{1}{2}$

$\Rightarrow \qquad\qquad\qquad x^3 = -8$

$\Rightarrow \qquad\qquad\qquad x = -2$

At the point on $\;y = \dfrac{2}{x^2}\;$ where the gradient is $\frac{1}{2}$, the x coordinate is -2

and when $\;x = -2, \;\; y = \frac{1}{2}$

Therefore the gradient of $\;y = \dfrac{2}{x^2}\;$ is $\frac{1}{2}$ at the point $(-2, \frac{1}{2})$.

EXERCISE 5d

Differentiate with respect to x.

1) $3x^2 - 7x$ 2) $x^4 - 9x^3 + 6$ 3) $(x-3)(2x+5)$ 4) $(x-4)^2$

5) $(7x+1)(7x-1)(x-1)$ 6) $\dfrac{(x^3+2)}{x}$ 7) $x^{-2}(1+x)$

8) $\dfrac{x^2 - 7x + 4}{x^3}$ 9) $\sqrt{x}(x-1)$ 10) $\dfrac{(x-1)}{\sqrt{x}}$

11) $(2x+1)^3$ 12) $1 + \dfrac{1}{x} + \dfrac{1}{x^2} + \dfrac{1}{x^3}$

Find the gradient of the given curve at the given point on the curve.

13) $y = x^2 - 3$ where $x = 1$ 14) $y = 3x^2 - 2$ where $x = 3$

15) $y = \sqrt{x}$ where $x = 2$ 16) $y = (x-3)(x+4)$ where $x = -1$

17) $y = \dfrac{1}{x}$ where $x = 3$ 18) $y = (2x-3)(x+1)$ where $x = 0$

19) $y = 1 - \dfrac{1}{x}$ where $x = 3$ 20) $y = x^3 + 7x - 4$ where $x = -3$

21) $y = \dfrac{(3x - 1)}{x}$ where $x = \tfrac{1}{2}$ 22) $y = 2\sqrt{x}(1 - \sqrt{x})$ where $x = 4$

23) $y = x^2 + \dfrac{3}{x}$ where $x = -1$ 24) $y = \dfrac{(\sqrt{x} - 1)}{\sqrt{x}}$ where $x = 9$

Find the coordinates of the point(s) on the given curve at which its gradient has the given value.

25) $y = x^2 - x + 3$,   1   26) $y = 5 + 3x - 2x^2$, -3

27) $y = (x - 1)(x + 1)$,   2   28) $y = (x + 3)(x - 5)$,   0

29) $y = x^3 + x^2$,   1   30) $y = \tfrac{1}{3}x^3 + 3x^2 + 7x + 1$, -1

31) $y = x + \dfrac{1}{x}$,   2   32) $y = \sqrt{x}$,   2

33) $y = 1 - \dfrac{1}{x}$,   4   34) $y = \dfrac{1}{x^2}$, $\dfrac{1}{4}$

35) $y = \dfrac{1}{x^3}$, -3 36) $y = x^3$, -1

EQUATIONS OF TANGENTS AND NORMALS

Now that we know how to find the gradient of a curve at a given point on the curve, we can find the equation of the tangent or normal to the curve at that point.

EXAMPLES 5e

1) Find the equation of the tangent to the curve $y = x^2 - 3x + 2$ at the point where it cuts the y-axis.

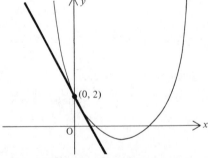

$y = x^2 - 3x + 2$ cuts the y-axis
where $x = 0$, $y = 2$.

The gradient of the tangent at $(0, 2)$ is the value of $\dfrac{dy}{dx}$ when $x = 0$.

As $y = x^2 - 3x + 2$, $\dfrac{dy}{dx} = 2x - 3$.

When $x = 0$ the gradient of the curve is -3.
Therefore the tangent has gradient -3 and passes through $(0, 2)$.
So its equation is $y = -3x + 2$,
i.e. $3x + y - 2 = 0$.

2) Find the equation of the normal to the curve $y = \dfrac{1}{x}$ at the point on the
curve where $x = 2$.   Find the coordinates of the point where this normal cuts
the curve again.

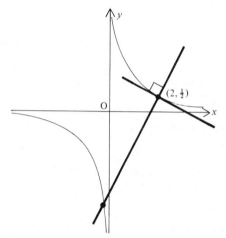

As $y = \dfrac{1}{x}, \quad \dfrac{dy}{dx} = -\dfrac{1}{x^2}$

So when $x = 2$,

$y = \dfrac{1}{2}$ and $\dfrac{dy}{dx} = -\dfrac{1}{4}$

Therefore the tangent at $(2, \frac{1}{2})$ has gradient $-\frac{1}{4}$
and the normal (i.e. the line perpendicular to the tangent) at $(2, \frac{1}{2})$ has
gradient 4.
The equation of this normal is $y - \frac{1}{2} = 4(x - 2)$ or $8x - 2y - 15 = 0$.
The points of intersection of the normal and the curve are given by solving
simultaneously

$$\begin{cases} 8x - 2y - 15 = 0 & \text{[1]} \\[2mm] y = \dfrac{1}{x} & \text{[2]} \end{cases}$$

i.e. $8x - \dfrac{2}{x} - 15 = 0$

$8x^2 - 15x - 2 = 0$ [3]

We already know that [1] and [2] cut where $x = 2$,   so $x - 2$ is a factor
of [3]

Hence $(x - 2)(8x + 1) = 0$.

Therefore [1] and [2] cut again where $x = -\frac{1}{8}$ and $y = -8$ (from [1])
i.e. at the point $(-\frac{1}{8}, -8)$.

EXERCISE 5e

Find the equation of the tangent to the given curve at the given point on the curve.

1) $y = x^2 - 2$ where $x = 1$ 2) $y = x^2 + 3x - 1$ where $x = 0$

3) $y = \dfrac{1}{x}$ where $x = -1$ 4) $y = (x - 2)(x^2 + 1)$ where $x = -1$

5) $y = x^2 - 5x + 2$ where $x = 3$ 6) $y = x^2 - 2$ where $x = 0$

7) Find the equations of the normals to the curves in Questions 1–6 at the given points.

8) Find the equation of the normal to the curve $y = x^2 + 3x - 2$ at the point where the curve cuts the y-axis.

9) Find the equation of the tangent to the curve $y = x^2 + 5x - 2$ at the point where this curve cuts the line $x = 4$.

10) Find the equations of the tangents to the curve $y = (2x - 1)(x + 1)$ at the points where the curve cuts the x-axis. Find the point of intersection of these tangents.

11) Find the equations of the normals to the curve $y = x^2 - 5x + 6$ at the points where the curve cuts the x-axis.

12) Find the equations of the tangents to the curve $y = 3x^2 + 5x - 1$ at the points of intersection of the curve and the line $y = x - 1$.

13) Find the coordinates of the point on $y = x^2$ at which the gradient is 2. Hence find the equation of the tangent to $y = x^2$ whose gradient is 2.

14) Find the coordinates of the point on $y = x^2 - 5$ at which the gradient is 3.
Hence find the value of c for which the line $y = 3x + c$ is a tangent to $y = x^2 - 5$.

15) Find the equation of the normal to $y = x^2 - 3x + 2$ which has a gradient of $\frac{1}{2}$.

16) Find the equation of the tangent to $y = 2x^2 - 3x$ which has a gradient of 1.

17) Find the value of k for which $y = 2x + k$ is a normal to $y = 2x^2 - 3$.

18) Find the equation of the tangent to $y = (x - 5)(2x + 1)$ which is parallel to the x-axis.

STATIONARY VALUES

A stationary value of a function $f(x)$ is any value of $f(x)$ at which its rate of change with respect to x is zero,

i.e. stationary values of $f(x)$ occur when $\dfrac{d}{dx}\left[f(x)\right] = 0.$

For a graphical representation of a stationary value, consider the curve whose equation is $y = f(x)$.

At a stationary value of $f(x)$, $\dfrac{dy}{dx} = 0$

i.e. the gradient of the curve is zero
i.e. the tangent to the curve is parallel to the x-axis.

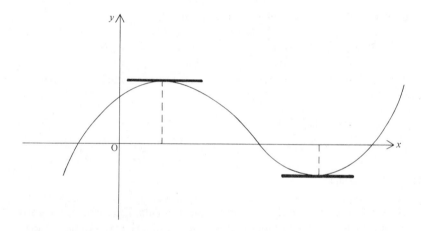

Therefore stationary values of $f(x)$ are the values of the y coordinates of points on the curve $y = f(x)$ at which the tangent is parallel to the x-axis.

EXAMPLE 5f

Find the stationary values of $x^3 - 3x^2 + 2$.

If
$$f(x) \equiv x^3 - 3x^2 + 2$$

$$\frac{d}{dx}f(x) \equiv 3x^2 - 6x$$

At stationary values of $f(x)$, $\dfrac{d}{dx}f(x) = 0$

$$\Rightarrow \qquad 3x^2 - 6x = 0$$

$$x(x - 2) = 0$$

i.e. stationary values of $x^3 - 3x^2 + 2$ occur when $x = 0$ and when $x = 2$. These stationary values are:

$$f(0) = 2 \quad \text{and} \quad f(2) = 2^3 - 3(2)^2 + 2 = -2$$

Note that for the curve $y = x^3 - 3x^2 + 2$,

$$\frac{dy}{dx} = 3x^2 - 6x = 0 \quad \text{when} \quad x = 0 \text{ or } 2$$

so the gradient of $y = f(x)$ is zero at the points $(0, 2)$ and $(2, -2)$.

EXERCISE 5f

Find the value(s) of x at which the following functions have stationary values:

1) $x^2 - 3$ 2) $x^2 - 5x + 1$ 3) $x^3 - 12x + 1$ 4) $2x^4 - 2x^3 - x^2$

Find the stationary value(s) of the following functions:

5) $x^2 - 6x + 3$ 6) $x + \dfrac{1}{x}$ 7) $3x^3 - 4x + 2$

Find the coordinates of the points on the given curves at which the gradient is zero.

8) $y = (x - 2)(x + 3)$ 9) $y = x^3 - 5x^2 + 3x - 2$ 10) $y = \dfrac{x^2 - 2}{x}$

TURNING POINTS

The gradient of a curve can be zero at several points. The shape of the curve in the immediate neighbourhood of one of these points belongs to one of the three categories shown below.

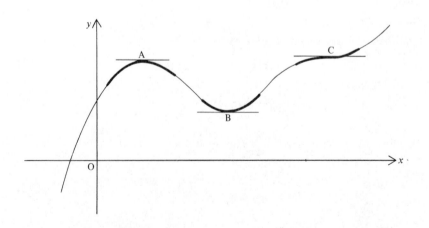

Moving along the curve in the positive direction of the x-axis:

(a) In the neighbourhood of A, the gradient changes from positive, through zero at A, to negative.
A is called a *maximum turning point*.
The y coordinate of A is called a *maximum value of y* [or of $f(x)$ where $y = f(x)$].

(b) In the neighbourhood of B, the gradient changes from negative, through zero at B, to positive.
B is called a *minimum turning point*.
The y coordinate of B is called a *minimum value of y* [or of $f(x)$].

Note that the gradient at a maximum or minimum turning point *must* be zero.
Note also that the terms maximum and minimum values are not synonymous with greatest and least values. Maxima and minima apply to the behaviour of a function in the *immediate neighbourhood only* of its stationary values.

(c) The curve does not turn at C, i.e. although the gradient is zero at C, it does not change sign when moving through C. However the sense in which the curve is turning does change (from clockwise to anticlockwise).

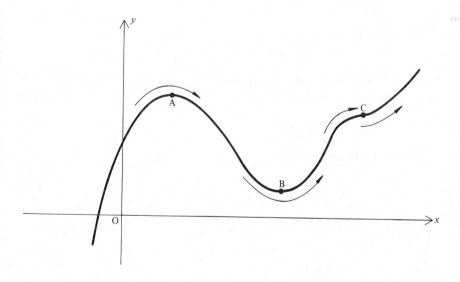

Any point on a curve at which the sense of turning changes is called a *point of inflexion*. Apart from C, there are two other points of inflexion in the diagram, one between A and B and another between B and C. Thus the gradient at a point of inflexion is not necessarily zero.

INVESTIGATING THE NATURE OF STATIONARY VALUES

We know how to find the coordinates of points on $y = f(x)$ at which $f(x)$ has stationary values, but we need to investigate further so that we can distinguish between them, and there are several ways in which this can be done.

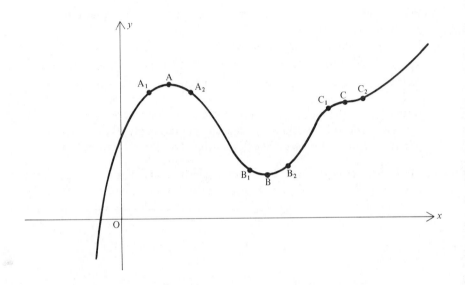

Consider the points A_1 and A_2, B_1 and B_2, C_1 and C_2 which are left and right respectively of A, B, C and *close* to them.

(1) Consider the values of y.

For A (a maximum value)

$$y \text{ at } A_1 < y \text{ at } A$$
$$y \text{ at } A_2 < y \text{ at } A$$

For B (a minimum value)

$$y \text{ at } B_1 > y \text{ at } B$$
$$y \text{ at } B_2 > y \text{ at } B$$

For C (a point of inflexion)

$$y \text{ at } C_1 < y \text{ at } C$$
$$y \text{ at } C_2 > y \text{ at } C$$

i.e.

	Maximum	Minimum	Inflexion
Values of y either side of stationary value	Both smaller	Both larger	One smaller and one larger

(2) Now consider the behaviour of the gradient $\dfrac{dy}{dx}$.

For A   at A_1, $\dfrac{dy}{dx}$ is $+$ ve

                     at A, $\dfrac{dy}{dx}$ is zero

                     at A_2, $\dfrac{dy}{dx}$ is $-$ ve

For B   at B_1, $\dfrac{dy}{dx}$ is $-$ ve

                     at B, $\dfrac{dy}{dx}$ is zero

                     at B_2, $\dfrac{dy}{dx}$ is $+$ ve

For C   at C_1, $\dfrac{dy}{dx}$ is $+$ ve

                     at C, $\dfrac{dy}{dx}$ is zero

                     at C_2, $\dfrac{dy}{dx}$ is $+$ ve

i.e.

	Maximum	Minimum	Inflexion
Sign of $\dfrac{dy}{dx}$ when moving through a stationary value	$+$ 0 $-$ ╱ ― ╲	$-$ 0 $+$ ╲ ― ╱	$+0+$ $-0-$ or ╱―╱ ╲―╲

(3) When passing through A, $\dfrac{dy}{dx}$ changes from positive to negative.

i.e. $\dfrac{dy}{dx}$ decreases as x increases

or the rate of increase w.r.t. x of $\dfrac{dy}{dx}$ is negative

i.e. $\dfrac{d}{dx}\left(\dfrac{dy}{dx}\right)$ is negative.

This clumsy notation for the rate of increase w.r.t. x of $\dfrac{dy}{dx}$ is condensed to $\dfrac{d^2y}{dx^2}$.

Similarly when passing through B, $\dfrac{dy}{dx}$ changes from negative to positive

i.e. $\dfrac{dy}{dx}$ increases as x increases

or $\dfrac{d^2y}{dx^2}$ is positive.

Points of inflexion are not so easily dealt with by this method. It is true to say that $\dfrac{d^2y}{dx^2} = 0$ at such points but $\dfrac{d^2y}{dx^2}$ can also be zero at maxima and minima.

A more detailed analysis of points of inflexion will be carried out later.

i.e.

	Maximum	Minimum
Sign of $\dfrac{d^2y}{dx^2}$	negative (or zero)	positive (or zero)

The three tables above summarize the three alternative methods for determining the nature of stationary values. The third method fails if, at a stationary value of y, $\dfrac{d^2y}{dx^2}$ is found to be zero. In this case either of the first two methods has to be used.

EXAMPLES 5g

1) Find the points on $y = x^4 + 4x^3 - 6$ at which the gradient is zero and determine the nature of these points.

If
$$y = x^4 + 4x^3 - 6 \qquad [1]$$

$$\frac{dy}{dx} = 4x^3 + 12x^2 \qquad [2]$$

and
$$\frac{d^2 y}{dx^2} = 12x^2 + 24x \qquad [3]$$

The gradient of [1] is zero when $\frac{dy}{dx} = 0$

i.e. when $4x^3 + 12x^2 = 0$

$\Rightarrow$ $4x^2(x + 3) = 0$

$\Rightarrow$ $x = -3$ and $x = 0$

So when $x = -3$, $\frac{dy}{dx} = 0$

and from [3] $\frac{d^2 y}{dx^2} = 12(-3)^2 + 24(-3) = 36 \quad (>0)$

therefore y has a minimum value here.

From [1] $y = (-3)^4 + 4(-3)^3 - 6 = -33$

therefore $(-3, -33)$ is a minimum turning point.

When $x = 0$, $\frac{dy}{dx} = 0, \ y = -6$

and $\frac{d^2 y}{dx^2} = 0$ which is inconclusive

Let us now see what the sign of $\frac{dy}{dx}$ is on either side of the point where $x = 0$.

x	-1	0	$+1$
$\frac{dy}{dx}$	$+$	0	$+$
slope	╱	—	╱

From this table we see that $(0, -6)$ is a point of inflexion.

2) Sketch the curve $y = 2x^3 + x^2 - 4x + 1$.

Finding the maximum and minimum turning points will help to give a general idea of the shape and position of this curve.

If $y = 2x^3 + x^2 - 4x + 1$

$$\frac{dy}{dx} = 6x^2 + 2x - 4$$

$$\frac{d^2 y}{dx^2} = 12x + 2$$

At turning points $\dfrac{dy}{dx} = 0$,   i.e. $6x^2 + 2x - 4 = 0$

$$(3x - 2)(x + 1) = 0$$

$$x = \tfrac{2}{3} \quad \text{and} \quad x = -1.$$

When $x = \tfrac{2}{3}$, $\dfrac{dy}{dx} = 0$

$$\frac{d^2 y}{dx^2} = 12(\tfrac{2}{3}) + 2 > 0$$

$$y = 2(\tfrac{2}{3})^3 + (\tfrac{2}{3})^2 - 4(\tfrac{2}{3}) + 1 = -\tfrac{17}{27}$$

Therefore $(\tfrac{2}{3}, -\tfrac{17}{27})$ is a minimum turning point.   [1]

When $x = -1$, $\dfrac{dy}{dx} = 0$

$$\frac{d^2 y}{dx^2} = (12)(-1) + 2 < 0$$

$$y = 2(-1)^3 + (-1)^2 - 4(-1) + 1 = 4$$

Therefore $(-1, 4)$ is a maximum turning point.   [2]

As $y = 2x^3 + x^2 - 4x + 1$,   the curve cuts the y-axis at $(0, 1)$.   [3]
From results [1], [2] and [3] we can sketch the curve.

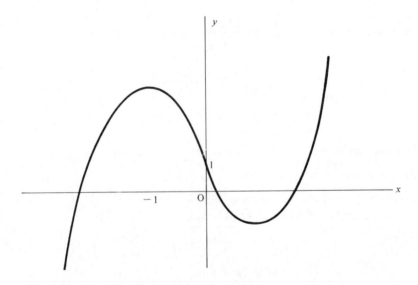

Note that in some cases, the intercepts on the x-axis can be found. It is not
always easy to solve the equation which gives these values however,

e.g. in the example above, $y = 0 \Rightarrow 2x^3 + x^2 - 4x + 1 = 0$.

3) A farmer has an adjustable electric fence that is 100 m long. He uses this fence to enclose a rectangular grazing area on three sides, the fourth side being a fixed hedge. Find the maximum area he can enclose.

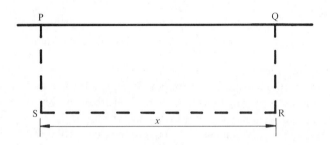

The farmer can vary the length (up to 100 m) of his enclosure and the width and area are then dependent on this chosen length.
Let x m be the length SR.

Then as   PS + SR + RQ = 100

$$PS = \tfrac{1}{2}(100 - x)$$

Therefore the area, A, of the enclosure is given by

$$A = x[\tfrac{1}{2}(100 - x)] = \tfrac{1}{2}x(100 - x)$$

Now A is a quadratic function of x.

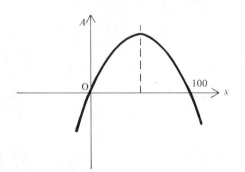

From the sketch and our knowledge of quadratic functions we can see that A is greatest when $x = 50$.

Therefore the maximum area that can be enclosed under the given conditions is

$$50 \times 25 \text{ m}^2 = 1250 \text{ m}^2.$$

Note that the maximum value of the function $\tfrac{1}{2}x(100 - x)$ can also be found as follows:

If $\qquad\qquad A = 50x - \tfrac{1}{2}x^2, \qquad \dfrac{dA}{dx} = 50 - x$

and A has a stationary value when $\dfrac{dA}{dx} = 0$, i.e. when $x = 50$.

Now $\dfrac{d^2A}{dx^2} = -1$.

So when $x = 50$, $\dfrac{dA}{dx} = 0$, $\dfrac{d^2A}{dx^2} < 0$, hence A is maximum.

The maximum value is $25 \times 50\,\text{m}^2 = 1250\,\text{m}^2$.

This second method, using differentiation, is necessary when finding maximum or minimum values of functions that are not quadratic. However the first method if preferable for quadratic functions when knowledge of their properties allows their maximum or minimum value to be found by inspection.

Note that in this case greatest value and maximum value are the same.

i.e. if $f(x) \equiv ax^2 + bx + c$

then the curve $y = ax^2 + bx + c$ crosses the x-axis where $ax^2 + bx + c = 0$.

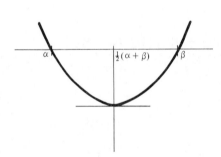

If α and β are the roots of this equation then the turning point of the curve has an x coordinate $\tfrac{1}{2}(\alpha + \beta)$.

But $\tfrac{1}{2}(\alpha + \beta) = -\dfrac{b}{2a}$.

Therefore $f(x) \equiv ax^2 + bx + c$ has a stationary value when $x = -\dfrac{b}{2a}$.

EXERCISE 5g

Find the stationary values of the following functions and investigate their nature:

1) $x^3 - 3x$

2) $x + \dfrac{4}{x}$

3) $x - \dfrac{4}{x^2}$

4) $x^2(3 - x)$

5) $(x - 3)(2x + 1)$

6) x^3

7) x^4

8) $x^2(x^2 - 2)$

9) $x^2(3 + 2x - 3x^2)$

Find the coordinates of the turning points of the following curves and sketch the curves.

10) $y = x^2 - x^3$ 11) $y = 1 - x^4$ 12) $y = x + \dfrac{1}{x}$

13) $y = x^2 - 4$ 14) $y = 3 - x + x^2$ 15) $y = 3 + 24x - 21x^2 - 4x^3$

16) $y = x^3 + 3$ 17) $y = 2x^4$ 18) $y = x^5 - 5x$

MULTIPLE CHOICE EXERCISE 5

(*Instructions for answering these questions are given on page xii*)

TYPE I

1) The function $x^3 - 12x + 5$ has a stationary value when:
(a) $x = \sqrt{6}$ (b) $x = -2$ (c) $x = 0$ (d) $x = 4$ (e) $x = 1$.

2) When $x = 1$ the function $x^3 - 3x^2 + 7$ is:
(a) stationary (b) increasing (c) maximum (d) decreasing (e) minimum.

3) The rate of increase of the function $x^2 - \dfrac{1}{x^2}$ w.r.t. x is:

(a) $2x + \dfrac{2}{x^3}$ (b) $2x - \dfrac{1}{2x}$ (c) $2x - \dfrac{2}{x^3}$ (d) $2x + \dfrac{3}{x^3}$ (e) $2x + \dfrac{1}{2x}$.

4) When $x = 0$ the function $x^3 - 2$ is:
(a) stationary (b) maximum (c) minimum (d) increasing (e) decreasing.

5) The function $\dfrac{1}{x}$ has a stationary value when:

(a) $x = 1$ (b) $x = 0$ (c) $x = -1$ (d) $x = 2$

(e) there is no real finite value of x for which $\dfrac{1}{x}$ is stationary.

6) The gradient function of $y = (x - 3)(x^2 + 2)$ is:
(a) $2x$ (b) $2x - 3$ (c) $3x^2 - 6x + 2$ (d) $-3(2x + 2)$
(e) $x^3 - 3x^2 + 2x - 6$.

7) The point $(1, 1)$ on the curve $y = x^3 - 3x^2 + 3x$ is:
(a) a maximum point   (b) a point on inflexion
(c) a minimum point   (d) none of these.

TYPE II

8) $f(x) \equiv x + \dfrac{1}{x}$.

(a) $f(x)$ is stationary when $x = -1$.

(b) $\dfrac{d}{dx} f(x) = 1 - \dfrac{1}{x^2}$.

(c) $y = f(x)$ has no turning points.

9) $y = x^3 - 4x + 5$.

(a) $\dfrac{d^2 y}{dx^2} = 9x$.

(b) The curve has two turning points.
(c) y is increasing when $x = 2$.

10) $y = x^4$.
(a) y is decreasing when $x = 1$.
(b) x^4 has only one stationary value.

(c) $\dfrac{dy}{dx} = 4x^3$.

TYPE III

11) (a) $y = f(x)$ is maximum when $x = 2$.

   (b) $\dfrac{d^2 y}{dx^2} < 0$ when $x = 2$.

12) (a) $y = f(x)$ and $\dfrac{dy}{dx} = 0$ and $\dfrac{d^2 y}{dx^2} > 0$ when $x = a$.

   (b) $y = f(x)$ and $f(a)$ is a minimum value of y.

13) When $x = 2$ (a) $\dfrac{dy}{dx} = 0$.

              (b) y is stationary.

14) When $x = a$ (a) $\dfrac{dy}{dx} > 0$.

              (b) $\dfrac{d^2 y}{dx^2} > 0$.

TYPE IV

15) Find the turning points on $y = f(x)$.

(a) $\dfrac{d^2 y}{dx^2} = 3x - 2$.

(b) $\dfrac{dy}{dx}$ is a quadratic function of x.

(c) The equation $\dfrac{dy}{dx} = 0$ has two real roots, $x = 0$, $x = 1$.

16) Find the values of x which correspond to stationary values of $f(x)$.
(a) $f(x)$ is a polynomial of degree 4.

(b) $\dfrac{d}{dx} f(x) = g(x)$.

(c) $g(x) \equiv x(x-1)(x-2)$.

17) A beer can is made from sheet metal. Find the radius of the base.
(a) The can holds 0.5 litre.
(b) The can is a cylinder.
(c) The surface area is to be as small as possible.

18) Sketch the curve $y = f(x)$.
(a) The curve has two stationary values.
(b) There are no discontinuities.
(c) The curve passes through $(2, 0)$ and $(0, 1)$.

MISCELLANEOUS EXERCISE 5

1) If $x^3 y = 2$, find the value of $\dfrac{d^2 y}{dx^2}$ when $x = 1$.

2) If $vt = 7$, find the value of $\dfrac{dv}{dt}$ when $v = 3$.

3) Find the derivatives of the following functions:
(a) $t^2 - 5t$ (b) $(y-1)^2$ (c) $(v - 2v^3)/v$.

4) Differentiate the function $f(x) \equiv x + \dfrac{1}{x}$ from first principles.

5) Plot the graph of $y = \sqrt{(1+x)}$ for integral values of x between -1 and 4.
Use your graph to estimate the value of $\dfrac{dy}{dx}$ when $x = 2$.

6) Find the coordinates of the point on $y = x^2 - 7x + 3$ at which the gradient is 2. Hence find the equation of the normal to $y = x^2 - 7x + 3$ which is parallel to $x + 2y - 1 = 0$.

7) Find the equations of the tangents to $y = x^3 + 3x$ which are parallel to the line $y = 15x + 2$.

8) Find the equation of the normal to $y = x^2 - 4$ which is parallel to $x + 3y - 1 = 0$.

9) Find the turning points on $y = 3x^4 + 4x^3 - 12x^2$. Give a rough sketch of the curve.

10) Find the stationary values of $f(x) \equiv x + \dfrac{1}{x}$ and use these to sketch the curve $y = x + \dfrac{1}{x}$.

11) Sketch the curve $y = (x - 3)^3$.

12) A closed cylindrical can has height h and base radius r. The volume is $0.01\,\mathrm{m}^3$. Show that $h = \dfrac{1}{100\pi r^2}$.

Show further that S, the surface area, is given by $S = 2\pi r^2 + \dfrac{1}{50r}$.

Hence find the value of r for which S is minimum.

13) The gradient function of $y = ax^2 + bx + c$ is $4x + 2$. The function has a minimum value of 1. Find the values of a, b and c.

14) An open rectangular box is made from a square sheet of cardboard by removing a square from each corner and joining the cut edges. If the cardboard is of edge $0.5\,\mathrm{m}$, find the maximum volume of the box.

15) A cylinder is cut from a solid sphere of radius $5\,\mathrm{cm}$. If the height of the cylinder is $2h$, show that the volume of the cylinder is $2\pi h(25 - h^2)$, assuming that the curved edges of the cylinder reach the surface of the sphere. Find the maximum volume of such a cylinder.

16) If a piece of string of fixed length is made to enclose a rectangle, show that the enclosed area is greatest when the rectangle is a square.

17) A point P moves along a straight line such that, after a time t seconds, its displacement s metres from a fixed point on the line is given by

$$s = 5t^2 + 1$$

What does $\dfrac{ds}{dt}$ represent?

If velocity is the rate of increase of displacement w.r.t. time, find the velocity of P when $t = 20$.

18) Find the maximum displacement of a particle from a point O, if its displacement s metres from O after time t seconds is given by

$$s = 2 + 3t - t^2$$

19) The table below shows the results of an experiment in which some hot liquid was left to cool and its temperature measured at intervals of one minute.

Time (minutes)	0	1	2	3	4	5
Temperature ($^\circ$C)	100°	93°	87°	83°	80°	78°

By drawing a graph of these results, estimate the rate of decrease of the temperature after three minutes.

20) In an experiment a small ball bearing was catapulted into a tank of viscous liquid. The table below records the penetration of the ball at second intervals.

Time (seconds)	0	1	2	3	4	5	6
Penetration (cm)	0	20	30	35	37	38	38.2

Plot these results on a graph and use your graph to estimate the rate of penetration after four seconds.

21) Show that the point $(p, 4p^2)$ lies on the curve $y = 4x^2$ for all real values of p. Find the equation of the tangent to $y = 4x^2$ at $(p, 4p^2)$.

22) Show that $\left(t, \dfrac{1}{t}\right)$ lies on the curve $y = \dfrac{1}{x}$ for all values of t. Find the equation of the tangent at $\left(t, \dfrac{1}{t}\right)$ to $y = \dfrac{1}{x}$. Find the area of the triangle enclosed by this tangent and the coordinate axes.

CHAPTER 6

TRIGONOMETRIC FUNCTIONS

CIRCULAR FUNCTIONS

Measurement of Rotation

When a line OP is pivoted at O and rotates from its initial position OP_0 to a new position OP_1, the angle P_0OP_1 is a measure of the rotation of OP.

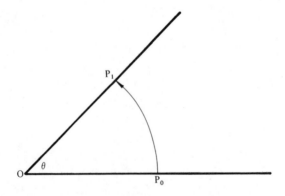

The angle θ is usually measured in one of two units.

The Degree

The ancient Babylonian Mathematicians, because they thought that the solar year was 360 days long, divided one complete revolution into 360 equal parts,

128

each part now being known as one degree $(1°)$. Using the degree as the unit of rotation, half a revolution corresponds to $180°$ and quarter of a revolution, i.e. a right angle, corresponds to $90°$.

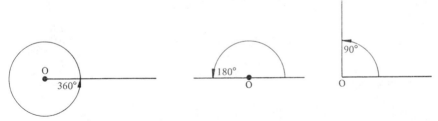

Angles smaller than a degree are usually given as decimal parts, e.g. half a degree is $0.5°$, but in some fields (e.g. navigation) a degree may be divided into 60 minutes $(60')$ and each minute into 60 seconds $(60'')$.

The Radian

If an arc P_0P_1 of a circle is drawn so that it is equal in length to the radius of the circle, then the angle P_0OP_1 is called one *radian* (1^c)

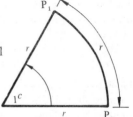

The number of radians in one complete revolution is therefore given by the ratio
$$\frac{\text{circumference}}{\text{radius}} .$$

But the circumference of a circle of radius r is of length $2\pi r$.

Thus there are $\dfrac{2\pi r}{r} = 2\pi$ radians in one revolution.

i.e. $\qquad\qquad\boxed{2\pi \text{ radians } = \ 360 \text{ degrees}}$

Further, half a revolution is π radians $= \ 180$ degrees

and one right angle is $\dfrac{\pi}{2}$ radians $= \ 90$ degrees

When an angle is quoted in terms of π it is normal to omit the radian symbol. Thus we would write $180° = \pi$ (not π^c)

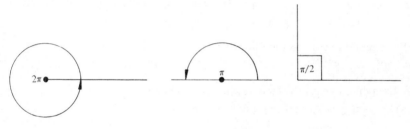

Other angles which are simple fractions of $180°$ can easily be expressed in radians in terms of π, using the relationship $180° = \pi$.

e.g.

$$60° = 60 \times \frac{\pi}{180} = \frac{\pi}{3}$$

$$135° = 135 \times \frac{\pi}{180} = \frac{3\pi}{4}$$

and conversely

$$\frac{7\pi}{6} = \frac{7\pi}{6} \times \frac{180°}{\pi} = 210°$$

$$\frac{5\pi}{3} = \frac{5\pi}{3} \times \frac{180°}{\pi} = 300°$$

Unit conversion carried out in this way is not convenient for angles which are not simple fractions of a revolution. It would not, for instance, be easy to express $47° \, 34'$ as a multiple of π. We would first have to express $34'$ as a decimal part of a degree (i.e. $34' = 0.567°$) and then use

$$47.567° = 47.567 \times \frac{\pi}{180} = 0.830^c$$

Calculations such as this are tedious and can be avoided either by using conversion tables or by using an electronic calculator which offers this facility. In order to visualise the size of an angle of 1 radian, it helps to remember that

$$\pi \text{ radians} = 180°$$

and

$$\pi = 3.14 \text{ (to 2 d.p.)}$$

so

$$1 \text{ radian} = \frac{180°}{3.14} = 57.3° \text{ (to 3 s.f.)}$$

Thus one radian is a little less than $60°$.

EXERCISE 6a

1) Without using tables, express the following angles in radians, giving your answer in terms of π:

$30°$; $270°$; $120°$; $45°$; $240°$; $150°$; $20°$; $300°$; $22.5°$; $80°$

2) Without using tables, express the following angles in degrees:

$\dfrac{3\pi}{4}$; $\dfrac{5\pi}{6}$; $\dfrac{\pi}{10}$; $\dfrac{3\pi}{2}$; $\dfrac{11\pi}{6}$; $\dfrac{\pi}{3}$; $\dfrac{4\pi}{3}$; $\dfrac{\pi}{12}$; $\dfrac{\pi}{8}$; $\dfrac{4\pi}{9}$

3) Use tables to express the following angles in radians:

$35° \, 30'$; $78° \, 12'$; $54° \, 45'$; $70°$; $16°$

4) Use tables to express the following angles in degrees:

2.86^c; 4.21^c; 1^c; 3.47^c; 3^c

MENSURATION OF A CIRCLE

The reader will already be familiar with the formulae for the area and the circumference of a circle of radius r.

$$\text{Circumference} = 2\pi r \qquad \text{Area} = \pi r^2$$

These formulae can be used to derive further results.

Length of an Arc

Consider an arc which subtends an angle θ at the centre of the circle, *where θ is measured in radians*.

From the definition of a radian, the arc which subtends 1 radian at the centre is of length r. Hence an arc which subtends θ radians at the centre is of length $r\theta$.

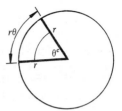

Area of a Sector

The area of a sector containing an angle of θ radians at the centre can be found by considering the sector as a fraction of a circle. The ratio of the area of the sector to the area of the circle is equal to the ratio of the angle θ contained in the sector to the angle 2π contained in the circle.

i.e.
$$\frac{\text{area of sector}}{\pi r^2} = \frac{\theta}{2\pi}$$

$\Rightarrow$
$$\text{area of sector} = \tfrac{1}{2}r^2\theta$$

Hence if an arc AB subtends an angle θ radians at the centre, O, of a circle of radius r

Length of arc $\quad$ AB $= r\theta$

Area of sector $\quad$ AOB $= \tfrac{1}{2}r^2\theta$

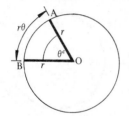

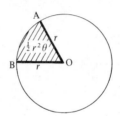

EXAMPLES 6b (π is taken as 3.142)

1) A railway line changes direction by $20°$ when passing round a circular arc of length $500\,\text{m}$. What is the radius of the arc?

Length of arc $= r\theta$
(where θ is in radians)

and $\quad 20° = \dfrac{20}{180} \times \pi = \dfrac{\pi}{9}$ radians

Hence $\qquad 500 = r\left(\dfrac{\pi}{9}\right)$

i.e. $\qquad\qquad r = \dfrac{4500}{\pi}$

The radius of the track is therefore $1432\,\text{m}$.

2) A chord AB divides a circle of radius $2\,\text{m}$ into two segments. If AB subtends an angle of $60°$ at the centre of the circle, find the area of the minor segment.

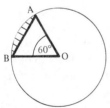

$$60° = \frac{\pi}{3}\text{ radians}$$

Area of sector $AOB = \tfrac{1}{2}r^2\theta = \tfrac{1}{2}\times 2^2 \times \dfrac{\pi}{3} = \dfrac{2\pi}{3} = 2.094\,\text{m}^2$

Area of triangle $AOB = \tfrac{1}{2}r^2 \sin\theta = \tfrac{1}{2}\times 2^2 \sin 60° = 1.732\,\text{m}^2$

Area of minor segment (shaded in the diagram) $= (2.094 - 1.732)\,\text{m}^2$

$\qquad\qquad\qquad\qquad\qquad\qquad\qquad\qquad\qquad\qquad = 0.362\,\text{m}^2$

3) Two discs of radii $3\,\text{cm}$ and $4\,\text{cm}$ are laid on a table with their centres $5\,\text{cm}$ apart. Find the perimeter of the 'figure-eight' shape so formed.

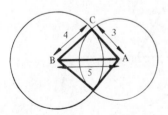 

Let the discs intersect at C and have their centres at A and B.
Triangle ABC is right angled at C $(5^2 = 3^2 + 4^2)$

Thus $\qquad\qquad\qquad \tan \alpha = \frac{4}{3}$ and $\alpha = 53.13° = 0.927^c$

Hence $\qquad\qquad\qquad \beta = 90° - \alpha = 36.87° \qquad = 0.644^c$

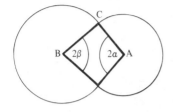 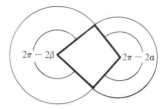

The perimeter is made up of
an arc subtending $(2\pi - 2\alpha)^c$ in the circle of radius 3 cm and
an arc subtending $(2\pi - 2\beta)^c$ in the circle of radius 4 cm.
Hence, using arc length $= r\theta$, we have

$$\text{perimeter} = 3(6.284 - 1.854) + 4(6.284 - 1.288)\,\text{cm}$$

$$= 33.3\,\text{cm}.$$

EXERCISE 6b

1) If d is the length of an arc of a circle of radius r and θ is the angle
subtended by the arc at the centre of the circle, complete the following table:

d (cm)		9	2		5	
r (cm)	3		4	2		10
θ	30°	$\dfrac{2\pi}{3}$		$x°$	1.3^c	$47° \, 18'$

2) The moon subtends an angle of $31'$ at the earth and its distance from the
earth is 382 100 km. Find, in kilometres, the diameter of the moon.

3) Calculate, in degrees and minutes, the angle subtended at the centre of a
circle of radius 3.12 cm by an arc of length 6.14 cm.

4) If a is the area of a sector of a circle of radius r and θ is the angle
contained in the sector at the centre of the circle, complete the following table:

a (m²)		16	8		6	
r (m)	4		3	7		5
θ	30°	$\dfrac{3\pi}{4}$		$x°$	2.13^c	$64° \, 28'$

5) Calculate in degrees and minutes the angle at the centre of a circle of radius 5.29 cm contained in a sector of area $4.13\,\text{cm}^2$.

6) An arc AB of length 5 cm is marked on a circle of radius 3 cm. Find the area of the sector bounded by this arc and the radii from A and B.

7) A chord AB of length 4 cm divides a circle of radius 3.3 cm into two segments. Find the area of each segment.

8) A chord AB of length 5.2 cm subtends an angle of $120°$ at the centre of a circle. Calculate:
(a) the length of the arc AB,
(b) the area of the sector containing the angle $120°$,
(c) the area of the minor segment cut off by AB.

9) Two discs, each of radius 11.6 cm, are laid on a table with their centres 14 cm apart. Find:
(a) the length of their common chord,
(b) the area common to the two discs.

10) The area of a sector of a circle, diameter 10.23 cm, is $22.86\,\text{cm}^2$. What is the length of the arc of the sector?

CIRCULAR FUNCTIONS

For any acute angle θ there are six trigonometric ratios, each of which is defined by referring to a right angled triangle containing θ.

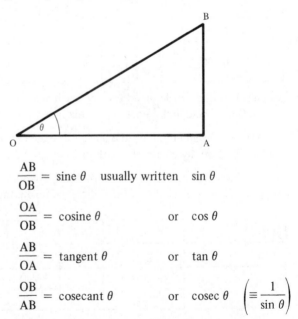

$$\frac{AB}{OB} = \text{sine } \theta \quad \text{usually written} \quad \sin \theta$$

$$\frac{OA}{OB} = \text{cosine } \theta \qquad \text{or} \quad \cos \theta$$

$$\frac{AB}{OA} = \text{tangent } \theta \qquad \text{or} \quad \tan \theta$$

$$\frac{OB}{AB} = \text{cosecant } \theta \qquad \text{or} \quad \text{cosec } \theta \quad \left(\equiv \frac{1}{\sin \theta} \right)$$

$$\frac{OB}{OA} = \text{secant } \theta \qquad \text{or} \quad \sec \theta \qquad \left(\equiv \frac{1}{\cos \theta}\right)$$

$$\frac{OA}{AB} = \text{cotangent } \theta \qquad \text{or} \quad \cot \theta \qquad \left(\equiv \frac{1}{\tan \theta}\right)$$

Each of the above ratios has a unique value for any one acute angle and these values are available in tables or from electronic calculators.

Since we are now regarding an angle as the measure of rotation from a given position of a straight line about a fixed point, it is clear that the size of an angle is unlimited, as the line can keep on rotating indefinitely. The meaning of the six trigonometric ratios is, as yet, restricted to acute angles, since the definition used so far for each ratio refers to an angle in a right angled triangle.

If we wish to extend the application of trigonometric ratios to angles of any size, they must be defined in a more general way.

TRIGONOMETRIC RATIOS FOR A GENERAL ANGLE

The system of reference in which a general angle is measured is very similar to that used for polar coordinates (see, p. 61). The point about which the line OP rotates is the pole or origin, O, and the position from which the angle is measured is the initial line or x axis. An angle formed when the line rotates anticlockwise is taken as positive, while clockwise rotation corresponds to negative angles.

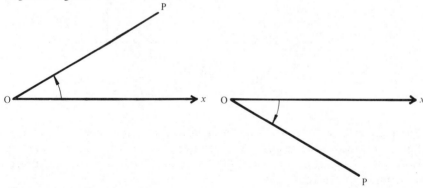

The pair of Cartesian axes divides the plane of rotation into four quadrants numbered 1, 2, 3, 4 as shown.

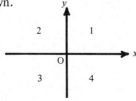

As the line OP rotates, P moves round the first quadrant and in this region its coordinates are both positive. As OP moves into the second quadrant its x coordinate becomes negative. In the third quadrant both the x and y coordinates of P are negative. Finally in the fourth quadrant we have a positive x coordinate and a negative y coordinate. Further rotation of OP causes the path of P to be retraced through the four quadrants in the same order. The length, r, of OP (the radius vector) is taken always to be positive.

From any position of P, a vertical line drawn from P to meet the x-axis at Q, forms a right angled triangle OPQ.

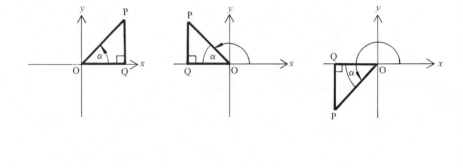

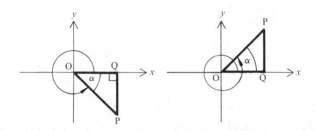

The angle POQ so formed is always acute regardless of the value of θ, and is called the associated acute angle, α. The value of α for a particular value of θ is the difference between θ and π (180°) or 2π (360°) or further multiples of π for larger angles.

e.g. when $\theta = 160°$ $\alpha = 180° - 160° = 20°$

when $\quad \theta = \dfrac{7\pi}{6} \qquad \alpha = \dfrac{7\pi}{6} - \pi = \dfrac{\pi}{6}$

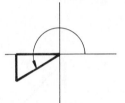

when $\quad \theta = 275° \qquad \alpha = 360° - 275° = 85°$

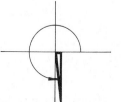

when $\quad \theta = 517° \qquad \alpha = 540° - 517° = 23°$

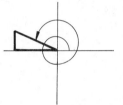

The associated angle *and* the signs of the coordinates of P in each quadrant, form the basis for defining the trigonometric ratios of *any* angle θ.

1st Quadrant

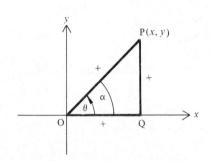

All six ratios are positive and, since θ is acute, their numerical values can be obtained directly from tables.

If OP has rotated through more than a complete revolution, $\alpha = \theta - 360°$

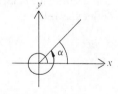

2nd Quadrant

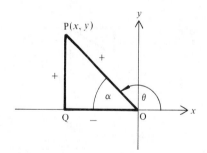

The associated acute angle, α is $180° - \theta$.

Now in triangle OPQ, using the signs of the coordinates as shown, we see that:

the sine ratio, $\dfrac{QP}{PO}$, is positive,

the cosine ratio, $\dfrac{OQ}{OP}$, is negative,

the tangent ratio, $\dfrac{QP}{OQ}$, is negative.

The trigonometric ratios of θ when P is in the second quadrant are therefore defined as follows:

$$\left. \begin{array}{l} \sin \theta = + \sin \alpha \\ \cos \theta = - \cos \alpha \\ \tan \theta = - \tan \alpha \end{array} \right\} \quad \text{where} \quad \alpha = 180° - \theta$$

3rd Quadrant

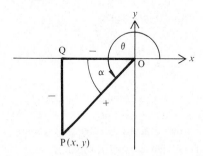

Here $\alpha = \theta - 180°$.

In triangle OPQ,

the tangent ratio, $\dfrac{QP}{OQ}$, is positive

while the sine ratio, $\dfrac{QP}{OP}$,

and the cosine ratio, $\dfrac{OQ}{OP}$,

are both negative.

The trigonometric ratios of θ when P is in the third quadrant are therefore defined as follows:

$$\left. \begin{array}{l} \sin \theta = - \sin \alpha \\ \cos \theta = - \cos \alpha \\ \tan \theta = + \tan \alpha \end{array} \right\} \quad \text{where} \quad \alpha = \theta - 180°$$

4th Quadrant

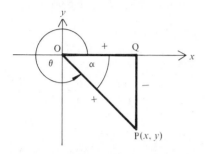

In this quadrant $\alpha = 360° - \theta$ and we see that, in triangle OPQ, the cosine ratio is positive but the tangent and sine ratios are negative.

The trigonometric ratios of θ when P is in the fourth quadrant are therefore defined as follows:

$$\left.\begin{array}{l} \sin \theta = -\sin \alpha \\ \cos \theta = +\cos \alpha \\ \tan \theta = -\tan \alpha \end{array}\right\} \text{ when } \alpha = 360° - \theta$$

These results can be summarised in a quadrant diagram as shown.

$s+$	$s+$
$c-$	$c+$
$t-$	$t+$
$s-$	$s-$
$c-$	$c+$
$t+$	$t-$

(i)

or

s	
s	c
	t
t	c

(ii)

In diagram (ii) the ratios shown are those which are positive in each quadrant. The quadrant rule together with the value of α, the associated acute angle, enable us to find the value of any trig ratio of any angle.

Negative Angles

If OP rotates clockwise so that P moves through the quadrants in reverse order, i.e. 4th, 3rd, 2nd, 1st, θ is taken to be negative,

e.g.

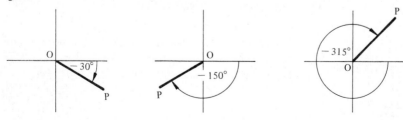

Every position of OP could be reached either by anticlockwise or by clockwise rotation and therefore corresponds to two different values of θ, one positive and one negative.

e.g.

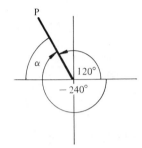

$\theta = + 120°$ or $\theta = -240°$
In both cases $\alpha = 60°$, hence the angle $+120°$ and the angle $-240°$ have the same trig ratios.

P is in the second quadrant where only the sine ratio is positive.

Hence
$$\sin 120° = \sin(-240°) = +\sin 60°$$
$$\cos 120° = \cos(-240°) = -\cos 60°$$
$$\tan 120° = \tan(-240°) = -\tan 60°$$

EXAMPLES 6c

1) Find the sine, cosine and tangent of $243°$.

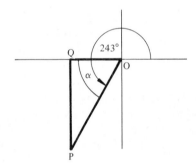

$\theta = 243°$

$\alpha = \theta - 180° = 63°$

P is in the third quadrant where only the tangent ratio is positive.

Hence
$$\sin 243° = -\sin 63° = -0.8910$$
$$\cos 243° = -\cos 63° = -0.4540$$
$$\tan 243° = +\tan 63° = 1.9626$$

2) If $\cos \theta = 0.866$ and $\tan \theta$ is negative, find $\sin \theta$.

θ is in a quadrant where the cosine ratio is positive and the tangent ratio is negative, i.e. in the fourth quadrant where $\alpha = 360° - \theta$.

Now $\qquad\qquad\qquad\qquad \cos \alpha = 0.866$

$\Rightarrow \qquad\qquad\qquad\qquad\qquad \alpha = 30°$

Hence $\qquad\qquad\qquad\qquad\quad \theta = 330°$

In the fourth quadrant the sine ratio is negative.

Therefore $\qquad\qquad\qquad \sin \theta = -\sin \alpha = -\sin 30°$

i.e. $\qquad\qquad\qquad\qquad \sin \theta = -0.5$

3) Given that $\tan \theta = 1$ and $-2\pi < \theta < 2\pi$, give four possible values for θ.

The tangent ratio is positive in the first and third quadrants.

When $\quad \tan \alpha = 1, \quad \alpha = \dfrac{\pi}{4}$ (or $45°$).

In this problem the range of values of θ is specified in radians so the solution should also be given in radians. Thus we use $\quad \alpha = \dfrac{\pi}{4}$.

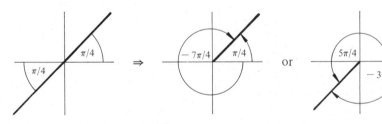

Hence $\quad \theta = \dfrac{\pi}{4}$ or $-\dfrac{7\pi}{4}, \quad \dfrac{5\pi}{4}$ or $-\dfrac{3\pi}{4}$.

In this example we are solving a simple trig equation.

4) An angle θ has an associated acute angle of $53°$. Find $\sin \theta$.

Only α is given in this case, so θ could be in any of the four quadrants.

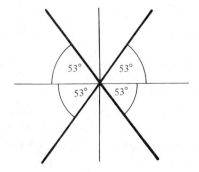

In quadrants 1 and 2,
$\sin \theta = \sin 53° = 0.7986$.

In quadrants 3 and 4,
$\sin \theta = -\sin 53° = -0.7986$.

Therefore $\sin \theta = \pm 0.7986$.

EXERCISE 6c

1) Write down the associated acute angle when θ is:

$93°$; $\dfrac{3\pi}{4}$; $308°$; $\dfrac{8\pi}{3}$; $22°$; $864°$; $\dfrac{7\pi}{6}$; $\dfrac{11\pi}{4}$; $-\dfrac{5\pi}{4}$; $-176°$

2) Complete the following table

	sin θ	cos θ	tan θ	α	θ
(a)	0.3907		-0.4245		
(b)		-0.7071	1.0000		
(c)					$\dfrac{5\pi}{6}$
(d)					$300°$
(e)					$170°$
(f)	0.5000	-0.8660			
(g)		0.5000			
(h)				$45°$	

3) Within the range $-360° \leqslant \theta \leqslant 360°$, give all values of θ for which:
(a) $\sin \theta = 0.4$ (b) $\cos \theta = -0.5$ (c) $\tan \theta = 1.2$
(d) $\operatorname{cosec} \theta = -1.5$ (e) $\sec \theta = 2.5$ (f) $\cot \theta = -2$

4) Find the smallest (positive or negative) angle for which:
(a) $\cos \theta = 0.8$ and $\sin \theta$ is positive,
(b) $\sin \theta = -0.6$ and $\tan \theta$ is negative,
(c) $\cos \theta$ is positive and $\tan \theta = \sin 30°$.

THE GRAPHS OF THE CIRCULAR FUNCTIONS

Since there is a value for each trigonometric ratio of every angle, we are now able to plot a graph showing the behaviour of each trigonometric function as θ varies. The graphs of the three major circular functions are particularly important and these will now be examined.

1) $f(\theta) \equiv \sin \theta$.

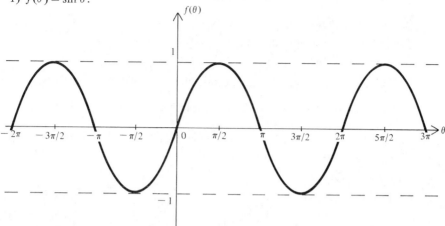

The above graph of the sine function has the following characteristics:

(a) It is continuous (i.e. it has no sudden changes in position).

(b) It lies entirely within the range $-1 \leqslant \sin \theta \leqslant +1$.

(c) The shape of the graph from $\theta = 0$ to $\theta = 2\pi$ is repeated for each
further complete revolution. Such a function is said to be *periodic* or *cyclic*.
The width of the repeating pattern, as measured on the horizontal axis, is
called the *period*.

Thus $f(\theta) \equiv \sin \theta$ is a periodic function with a period of 2π, a maximum
value of 1 and a minimum value of -1. A graph of this shape, for obvious
reasons, is known as a *sine wave*.

The greatest value of $|\sin \theta|$ is called the *amplitude* of the sine wave and its
value is 1.

2) $f(\theta) \equiv \cos \theta$.

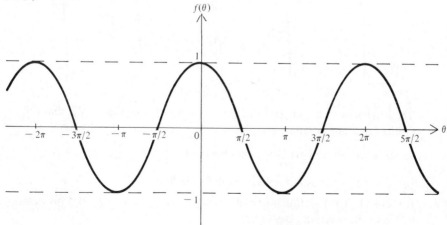

The characteristics of the graph of the cosine function are as follows:

(a) It is continuous.

(b) It lies entirely within the range $-1 \leqslant \cos \theta \leqslant 1$.

(c) It is periodic with a period of 2π.

(d) It has the same shape as the sine graph but is displaced a distance $\dfrac{\pi}{2}$ to the left on the horizontal axis. Such a displacement is known as a phase difference or phase shift.

Thus $f(\theta) \equiv \cos \theta$ is a cyclic function with period 2π, a maximum value of 1 and a minimum value of -1.

3) $f(\theta) \equiv \tan \theta$.

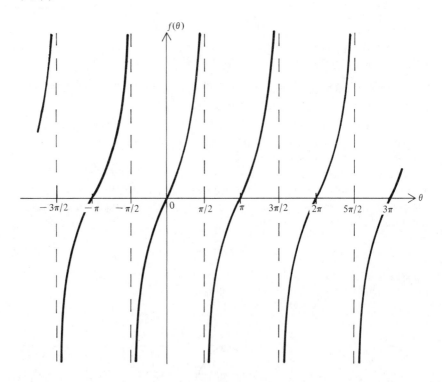

The behaviour of the tangent function is different from that of the sine and cosine functions in several respects.

(a) It is not continuous, being *undefined* when $\theta = -\dfrac{\pi}{2}, \dfrac{\pi}{2}, \dfrac{3\pi}{2},$ etc.

(b) The range of possible values of $\tan \theta$ is unlimited.

(c) The tangent function is periodic but the period in this case is π (not 2π as in the previous cases).

Special Values

It is useful to note the angles whose trig ratios have the values $0, \pm 1$ and (for $\tan \theta$ only) $\pm \infty$.

Reference to the graph of $f(\theta) \equiv \sin \theta$ shows that

$$\sin \theta = 0 \qquad \text{when} \quad \theta = \ldots -2\pi, -\pi, 0, \pi, 2\pi, 3\pi \ldots$$

$$\text{i.e. when} \quad \theta = \text{any whole multiple of } \pi$$

Thus $\sin \theta = 0$ when $\theta = n\pi$ where n is an integer

$$\sin \theta = 1 \qquad \text{when} \quad \theta = \ldots -\frac{3\pi}{2}, \frac{\pi}{2}, \frac{5\pi}{2} \ldots$$

$$= \ldots \left(-2\pi + \frac{\pi}{2}\right), \frac{\pi}{2}, \left(2\pi + \frac{\pi}{2}\right) \ldots$$

$$= (\text{an even multiple of } \pi) + \frac{\pi}{2}$$

Thus $\sin \theta = 1$ when $\theta = 2n\pi + \dfrac{\pi}{2}$

$$\sin \theta = -1 \qquad \text{when} \quad \theta = \ldots -\frac{\pi}{2}, \frac{3\pi}{2}, \frac{7\pi}{2} \ldots$$

$$= (\text{an even multiple of } \pi) - \frac{\pi}{2}$$

Thus $\sin \theta = -1$ when $\theta = 2n\pi - \dfrac{\pi}{2}$

Similar examination of the graph $f(\theta) \equiv \cos \theta$ shows that

$$\cos \theta = 0 \qquad \text{when} \quad \theta = \ldots -\frac{3\pi}{2}, -\frac{\pi}{2}, \frac{\pi}{2}, \frac{3\pi}{2}, \frac{5\pi}{2} \ldots$$

$$\text{i.e. when} \quad \theta = \text{an odd multiple of } \frac{\pi}{2}$$

Thus $\cos \theta = 0$ when $\theta = (2n + 1)\dfrac{\pi}{2}$

$$\cos \theta = 1 \qquad \text{when} \quad \theta = \ldots -2\pi, 0, 2\pi, 4\pi \ldots$$

$$= \text{an even multiple of } \pi$$

Thus $\cos \theta = 1$ when $\theta = 2n\pi$

$$\cos \theta = -1 \qquad \text{when} \quad \theta = \ldots -\pi, \pi, 3\pi, 5\pi \ldots$$

$$= \text{an odd multiple of } \pi$$

Thus $\cos \theta = -1$ when $\theta = (2n + 1)\pi$

Now considering the graph of $f(\theta) \equiv \tan \theta$

$$\tan \theta = 0 \qquad \text{when} \quad \theta = \ldots - \pi, 0, \pi, 2\pi \ldots$$

Thus $\tan \theta = 0$ when $\theta = n\pi$

$$\tan \theta = \pm \infty \quad \text{when} \quad \theta = \ldots \frac{\pi}{2}, \frac{3\pi}{2}, \frac{5\pi}{2} \ldots$$

Thus $\tan \theta = \pm \infty$ when $\theta = (2n + 1)\dfrac{\pi}{2}$

Note: Frequent use will be made, throughout any study of mathematics, of the expressions used above for odd and even numbers. In general we use $2n$ to represent all even numbers and $(2n + 1)$ to represent all odd numbers, provided that (as in the previous analysis) n is an integer.

THE RECIPROCAL (MINOR) TRIGONOMETRIC RATIOS

The three major trig ratios, $\sin \theta$, $\cos \theta$ and $\tan \theta$, are used much more frequently than the three minor trig ratios, $\operatorname{cosec} \theta$, $\sec \theta$ and $\cot \theta$. Consequently our analysis will tend to concentrate on the major ratios. But the reciprocal ratios must not be overlooked completely so at this stage we will examine the graphs and properties of these functions.

The graph of the function $f(\theta) \equiv \operatorname{cosec} \theta$ can be drawn without any reference to tables of values, simply by observing the graph of $f(\theta) \equiv \sin \theta$ and using the following properties of *any* reciprocal functions.

1) (a) The reciprocal of zero is $\pm \infty$.
   (b) The reciprocal of $\pm \infty$ is zero.

2) (a) The reciprocal of 1   is 1.
   (b) The reciprocal of -1 is -1.

3) If one expression has a maximum value its reciprocal has a minimum value (and conversely).

4) If one expression is increasing, its reciprocal is decreasing (and conversely).

5) An expression and its reciprocal have the same sign.

(**Note**: The above properties are logical enough to accept without detailed analysis at this stage. A fuller treatment of reciprocal curves is covered in Chapter 12).

Considering the range of values $-\pi \leqslant \theta \leqslant 2\pi$ for $f(\theta) \equiv \sin \theta$ and $f(\theta) \equiv \operatorname{cosec} \theta$ we have:

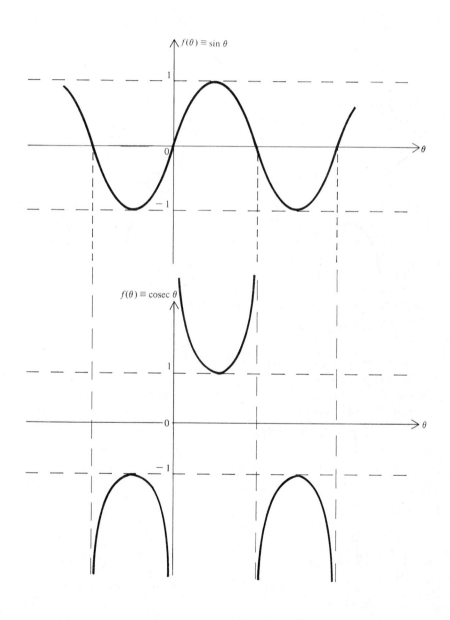

Similarly we can deduce the graphs of $f(\theta) \equiv \sec \theta$ and $f(\theta) \equiv \cot \theta$. These curves are shown below.

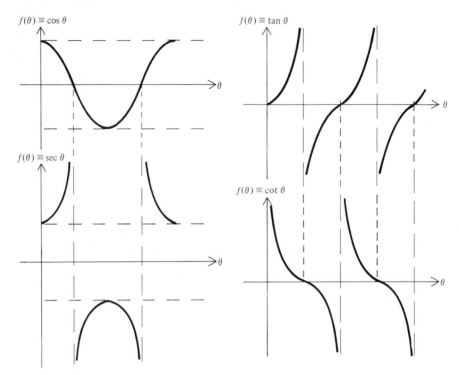

$f(\theta) \equiv \cos \theta$

$f(\theta) \equiv \tan \theta$

$f(\theta) \equiv \sec \theta$

$f(\theta) \equiv \cot \theta$

Common Trigonometric Ratios

The angles $30°$, $45°$ and $60°$ (and other angles for which these are the associated acute angles, e.g. $150°, 225°, 300° \ldots$) are used frequently in problems so it is well worth noting the values of these trig ratios.
Consider first an equilateral triangle ABC which is bisected by the line AD.

Then,

   if AB $= 2$ units

     BD $= 1$ unit

and AD $= \sqrt{3}$ units   (Pythagoras)

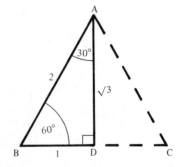

In $\triangle$ BAD

$$\angle B = 60° \left(\frac{\pi}{3}\right), \quad \text{since } \triangle \text{ ABC is equilateral}$$

and $\quad \angle A = 30° \left(\frac{\pi}{6}\right), \quad$ since angle BAC is bisected

Therefore

$$\sin 30° = \sin\frac{\pi}{6} = \frac{1}{2} \qquad \sin 60° = \sin\frac{\pi}{3} = \frac{\sqrt{3}}{2}$$

$$\cos 30° = \cos\frac{\pi}{6} = \frac{\sqrt{3}}{2} \qquad \cos 60° = \cos\frac{\pi}{3} = \frac{1}{2}$$

$$\tan 30° = \tan\frac{\pi}{6} = \frac{1}{\sqrt{3}} \qquad \tan 60° = \tan\frac{\pi}{3} = \sqrt{3}$$

Now consider a triangle ABC in which AB = BC and $\angle$B is a right angle.
If AB = BC = 1 unit, then AC = $\sqrt{2}$ units (Pythagoras).

Also since AB = BC,

$$\angle A = \angle C = 45° \left(\frac{\pi}{4}\right)$$

Therefore

$$\sin 45° = \sin\frac{\pi}{4} = \frac{1}{\sqrt{2}} = \frac{\sqrt{2}}{2}$$

$$\cos 45° = \cos\frac{\pi}{4} = \frac{1}{\sqrt{2}} = \frac{\sqrt{2}}{2}$$

$$\tan 45° = \tan\frac{\pi}{4} = 1$$

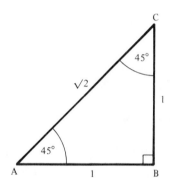

Complementary Angles

If the sum of two acute angles is $90° \left(\frac{\pi}{2}\right)$ they are said to be complementary
and each is the complement of the other.
Consider a right angled triangle ABC containing the angles α and β as shown.

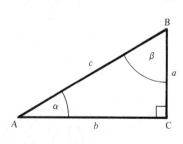

$$\sin\alpha = \frac{a}{c} = \cos\beta$$

$$\cos\alpha = \frac{b}{c} = \sin\beta$$

$$\tan\alpha = \frac{a}{b} = \cot\beta$$

$$\cot\alpha = \frac{b}{a} = \tan\beta$$

But α and β are complementary. Therefore we have shown that
the sine of an angle is the cosine of its complement,
the tangent of an angle is the cotangent of its complement.
Because of this property, the sine and cosine of an angle are called
complementary ratios.
Similarly the tangent and cotangent ratios are also said to be complementary.

Thorough familiarity with these special values and relationships and with the
shapes and properties of the graphs of the trigonometric functions is essential in
all further trigonometric analysis.

THE SOLUTION OF TRIGONOMETRIC EQUATIONS

If at least one term in an equation contains a trig ratio it is called a trig
equation. Its solution involves finding the angle or angles for which it is valid.
Consider the simple equation $\sin \theta = 0$

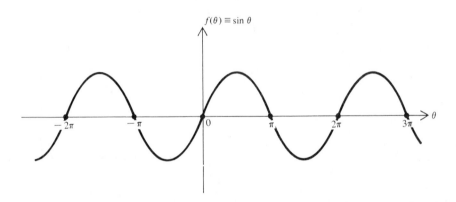

Referring to the graph of the sine function we see that
$\sin \theta = 0$ when θ is any multiple of π,

i.e. $\theta = n\pi$ where n is any integer.

The full, or general, solution to this simple equation is therefore the infinite set
of angles $\theta = n\pi$ (or $\theta = 180n°$).
Sometimes it is necessary to extract certain specific values of θ from the infinite
set. Consider, for instance, a modified form of the above problem, viz:
Solve the equation $\sin \theta = 0$ for $-\pi \leqslant \theta \leqslant \pi$.
This time the finite solution set is $\theta = -\pi, 0, \pi$.
There are two basic approaches to finding the solution of a trig equation. One of
them was used above and involves reference to the graph of the appropriate
circular function. This approach is usually best for dealing with trig ratios of
± 1 (for sine and cosine), $\pm \infty$ (for tangent) and zero.

Alternatively, the position of a rotating line OP in the appropriate quadrants can lead to a clear solution.

But in all cases, the first step in solving a trig equation is to find the *principal solution* which is the *principal value* of θ.

Principal Values

1) $f(\theta) \equiv \sin\theta$.

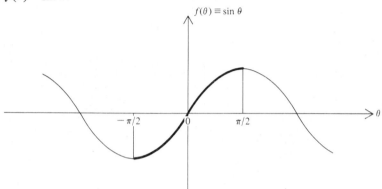

From the graph of the sine function it can be seen that,

within the range $-\dfrac{\pi}{2} \leqslant \theta \leqslant \dfrac{\pi}{2}$ every possible value of $\sin\theta$ occurs once and only once.

Thus any equation

$$\sin\theta = s \quad (-1 \leqslant s \leqslant 1)$$

has one and only one solution in this range and this is the *principal value* of θ (PV).

The position of the principal solution is therefore in either the first or the fourth quadrant.

e.g.   if $\sin\theta = \tfrac{1}{2}$ the principal solution is $\theta = \dfrac{\pi}{6}$

          if $\sin\theta = -\tfrac{1}{2}$ the principal solution is $\theta = -\dfrac{\pi}{6}$

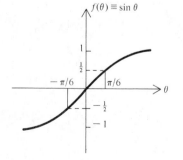
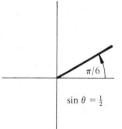
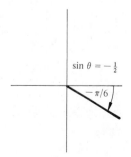

2) $f(\theta) \equiv \cos\theta$.

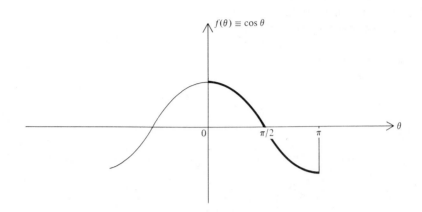

Every possible value of $\cos\theta$ occurs once and only once within
the range $0 \leqslant \theta \leqslant \pi$ so there is one and only one solution of the equation

$$\cos\theta = c \quad (-1 \leqslant c \leqslant 1)$$

within this range. This is the principal value of θ and the
principal solution is therefore in either the first or the second quadrant,

e.g.   if $\cos\theta = \frac{1}{2}$ the principal solution is $\theta = \dfrac{\pi}{3}$

         if $\cos\theta = -\frac{1}{2}$ the principal solution is $\theta = \dfrac{2\pi}{3}$

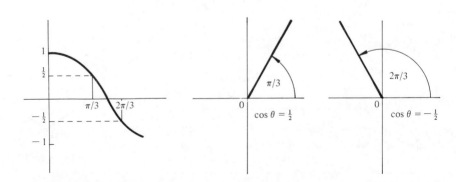

3) $f(\theta) \equiv \tan \theta$.

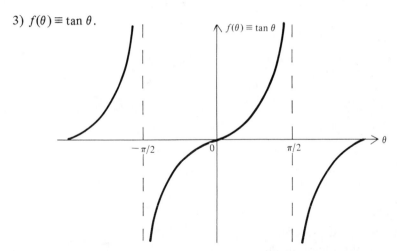

Every possible value of $\tan \theta$ occurs once and only once for angles in the range $-\dfrac{\pi}{2} \leqslant \theta \leqslant \dfrac{\pi}{2}$. One and only one solution of the equation

$$\tan \theta = t$$

is in this range and it is the principal value of θ. The principal solution is therefore in the first or the fourth quadrant,

e.g.   if $\tan \theta = 1$ the principal solution is $\theta = \dfrac{\pi}{4}$

          if $\tan \theta = -1$ the principal solution is $\theta = -\dfrac{\pi}{4}$

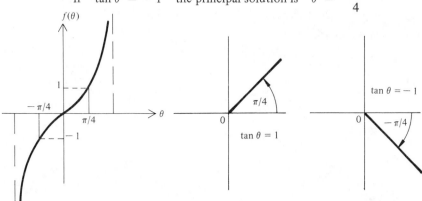

Secondary Values

Having determined the quadrant in which the principal value of a trig equation lies, it is usually found that a second angle with the same trig ratio occurs in the range $-\pi \leqslant \theta \leqslant \pi$.

This solution lies in a different quadrant and is called the *secondary value* of θ (SV), or the *secondary solution* of the equation,

e.g.

If $\sin\theta = \frac{1}{2}$ the secondary
solution is in the second
quadrant where the sine ratio is also
positive.

The secondary value is $\theta = \dfrac{5\pi}{6}$.

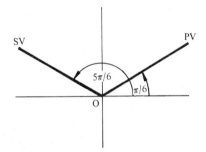

If $\sin\theta = -\frac{1}{2}$ the secondary
solution is in the third quadrant
where the sine ratio is also
negative.

The secondary value is $\theta = -\dfrac{5\pi}{6}$.

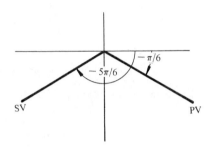

If $\cos\theta = \frac{1}{2}$ the secondary
value is in the fourth quadrant

and is $\theta = -\dfrac{\pi}{3}$.

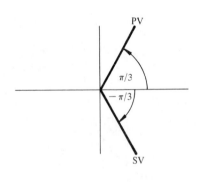

If $\cos\theta = -\frac{1}{2}$ the secondary
value is in the third quadrant

and is $\theta = -\dfrac{2\pi}{3}$.

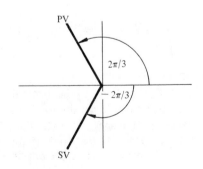

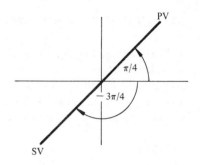

If $\tan \theta = 1$ the secondary
solution is in the third quadrant,
and is $\theta = \dfrac{-3\pi}{4}$.

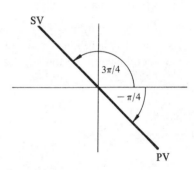

If $\tan \theta = -1$ the secondary
value is in the second quadrant
and is $\theta = \dfrac{3\pi}{4}$.

EXERCISE 6d

1) Determine the principal solutions of the following equations. In each case
indicate your solution on the graph of the appropriate circular function.

(a) $\sin \theta = \dfrac{\sqrt{3}}{2}$ (b) $\cos \theta = -\dfrac{\sqrt{2}}{2}$ (c) $\tan \theta = \sqrt{3}$

(d) $\cos \theta = \dfrac{\sqrt{3}}{2}$ (e) $\sin \theta = -1$ (f) $\tan \theta = 0$

(g) $\sin \theta = -\dfrac{\sqrt{2}}{2}$ (h) $\cos \theta = 0$ (i) $\sin \theta = 0$

(j) $\cos \theta = -1$ (k) $\tan \theta = -\dfrac{\sqrt{3}}{3}$ (l) $\sin \theta = 1$

(m) $\sin \theta = 0.63$ (n) $\cos \theta = 0.44$ (p) $\tan \theta = 2.84$

2) Find the principal and secondary solutions of the following equations. In
each case draw a quadrant diagram showing your solutions:

(a) $\sin \theta = \dfrac{1}{\sqrt{2}}$ (b) $\cos \theta = -\dfrac{\sqrt{3}}{2}$ (c) $\tan \theta = -\sqrt{3}$

(d) $\sin \theta = -0.54$ (e) $\cos \theta = 0.63$ (f) $\tan \theta = 1.5$

3) By referring to the graph of the appropriate circular function, explain why
the following equations have no secondary solution:
(a) $\sin \theta = -1$ (b) $\cos \theta = 1$
Write down two more equations which have no secondary solution.

SOLUTION OF TRIGONOMETRIC EQUATIONS IN A SPECIFIED RANGE

In attempting to solve a trig equation we find first the principal angle and the secondary angle (except in those cases where there is no secondary angle). At this stage a quadrant diagram can be drawn showing the principal and secondary solution positions. Then any angle measured from the positive x-axis to either of the solution positions is a solution to the given equation.

EXAMPLES 6e

1) Solve the equation $\sin \theta = 0.4$ within the range $-360° \leqslant \theta \leqslant 360°$.

The principal solution is $\theta = 23.58°$.
The secondary solution is in the second quadrant (sine ratio is positive in first and second quadrants) and is $\theta = 156.42°$.

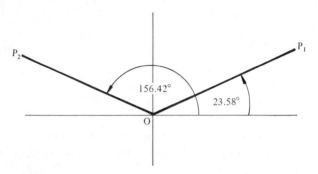

Therefore the following angles are the solutions within the specified range.

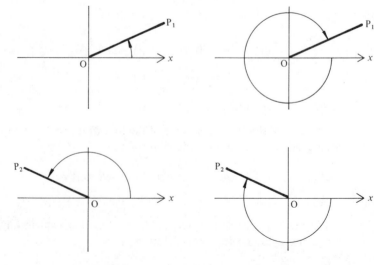

i.e. $\theta = 23.58°, \ -336.42°, \ 156.42°, \ -203.58°$

2) Solve the equation $\tan \theta = -\dfrac{1}{\sqrt{3}}$ in the range $0 \leqslant \theta \leqslant 2\pi$.

If $\tan \theta = -\dfrac{1}{\sqrt{3}}$, the principal solution is in the fourth quadrant and the secondary solution is in the second quadrant.

i.e.   PV is $-\dfrac{\pi}{6}$

and   SV is $\dfrac{5\pi}{6}$

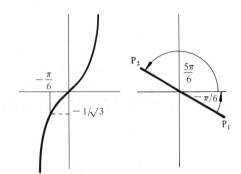

Within the specified range, the solution set is

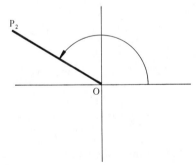

i.e. $\theta = \dfrac{5\pi}{6}, \dfrac{11\pi}{6}$.

(**Note.** The principal value is not always included in the solution set.)

3) Find the angles in the range $-360° \leqslant \theta \leqslant 0$ which satisfy the equation

$$\cos \theta = 0.7$$

Since $\cos \theta$ is positive we have the principal solution in the first quadrant and the secondary solution in the fourth quadrant.

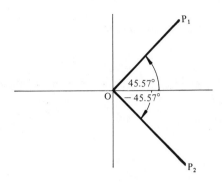

i.e.   PV is $45.57°$

SV is $-45.57°$

In the range $-360° \leqslant \theta \leqslant 0$,   the solution set is

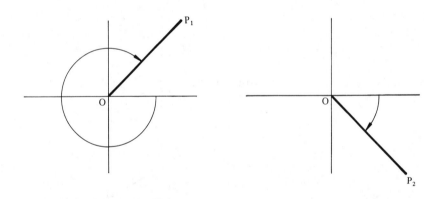

i.e. $\theta = -314.43°, -45.57°$.

4) Solve, within the range $0 \leqslant \theta \leqslant 360°$ the equation

$$\sin \theta + 3 \sin \theta \cos \theta = 0.$$

First the equation must be reduced to a simpler form

$$\sin \theta + 3 \sin \theta \cos \theta = 0$$

$\Rightarrow$ $$\sin \theta \, (1 + 3 \cos \theta) = 0$$

Therefore either $\sin \theta = 0$ [1]

or $1 + 3 \cos \theta = 0 \Rightarrow \cos \theta = -\dfrac{1}{3}$ [2]

Now considering [1] by referring to the sine graph

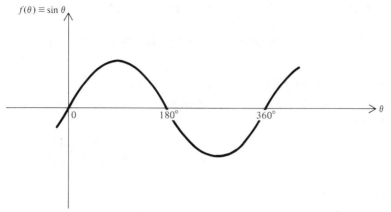

we see that $\quad \theta = 0, 180°, 360°$.

For [2] the principal solution is in the second quadrant
and the secondary solution is in the third quadrant

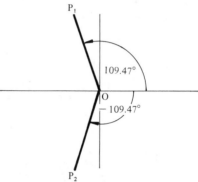

i.e. $\qquad$ PV is $109.47°$

$\qquad\quad$ SV is $-109.47°$

Within the specified range the solutions are

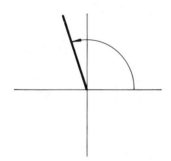 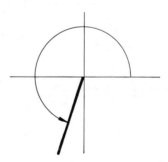

i.e. $\quad \theta = 109.47°, 250.53°$.

So the complete solution set from 0 to $360°$ is

$$\theta = 0, 109.47°, 180°, 250.53°, 360°.$$

(**Note** that, although the solution to [1] can conveniently be expressed in radians, degrees are used because the required range is specified in degrees. In any case, units must not be mixed in any one example.)

GENERAL SOLUTION OF TRIGONOMETRIC EQUATONS

In looking for a general solution to a given trig equation, use can be made of:

1) the circular function graphs,
2) the period of each circular function,
3) the principal solution and, except when the tangent ratio is involved, the secondary solution.

The general solution is an expression which represents all angles which satisfy the given equation, i.e. the general solution is an infinite set of angles.

EXAMPLES 6e (contd.)

5) Find the general solution of the equation $\tan \theta = 1$.

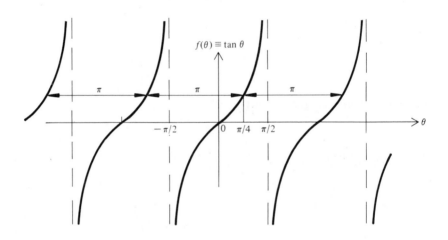

The range $-\dfrac{\pi}{2} \leqslant \theta \leqslant \dfrac{\pi}{2}$,   in which the principal value lies, is also the period π of the tangent function. So, by adding any multiple of π to the principal value, or subtracting any multiple of π from it, we get an angle with the same tangent.

Now the principal solution is $\theta = \dfrac{\pi}{4}$.

So the general solution is $\dfrac{\pi}{4} + n\pi$, where n is any integer, positive or negative. The above method can be used to find the general solution of all equations of the form $\tan \theta = t$. The result, which is quotable, is

$$\theta = PV + n\pi \qquad (\text{or } PV + 180n°)$$

e.g. if $\tan \theta = -\sqrt{3}$, the principal solution is $\theta = -\dfrac{\pi}{3}$ and the general solution is therefore

$$\theta = -\frac{\pi}{3} + n\pi$$

6) Find the general solution of the equation $\cos \theta = \dfrac{1}{\sqrt{2}}$

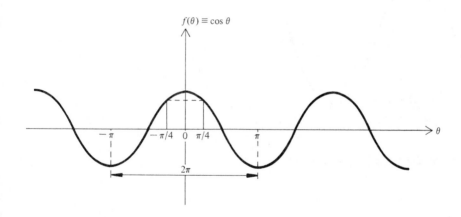

$f(\theta) \equiv \cos \theta$

The period of the cosine function is 2π, a range which includes both the principal and the secondary value of θ. So, starting with each of these values and adding or subtracting multiples of 2π (or $360°$) we get all the other angles with the same cosine.

The principal solution is $\theta = \dfrac{\pi}{4}$ and the secondary solution, in the fourth quadrant, is $\theta = -\dfrac{\pi}{4}$.

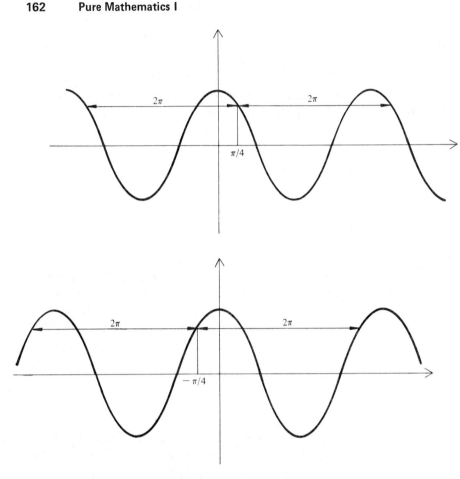

So the general solution includes $\quad \theta = \dfrac{\pi}{4} + 2n\pi$

and $\qquad\qquad\qquad\qquad\qquad \theta = -\dfrac{\pi}{4} + 2n\pi$

Now these two expressions can be combined to give a single expression

$$\theta = \pm\dfrac{\pi}{4} + 2n\pi$$

The result in all cases is of this form and can be quoted as

$$\theta = \pm\,\mathrm{PV} + 2n\pi \quad (\text{or } \pm\,\mathrm{PV} + 360n°)$$

e.g. if $\cos\theta = -\frac{1}{2}$, the principal solution is $\theta = \dfrac{2\pi}{3}$ so the general solution is

$$\theta = \pm\dfrac{2\pi}{3} + 2n\pi$$

7) Find the general solution set of the equation $\sin \theta = \frac{1}{2}$.

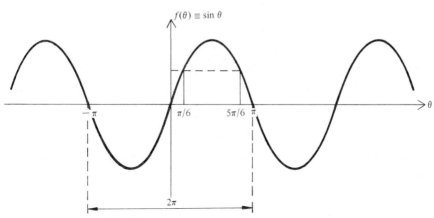

As we found in Example 2 for the cosine function, the period of the sine function is 2π, a range which includes both the principal and the secondary value. So again the general solution includes each of these values, together with multiples of 2π.

The principal value when $\sin \theta = \dfrac{1}{2}$ is $\theta = \dfrac{\pi}{6}$ and the secondary value, in

the second quadrant, is $\theta = \dfrac{5\pi}{6}$.

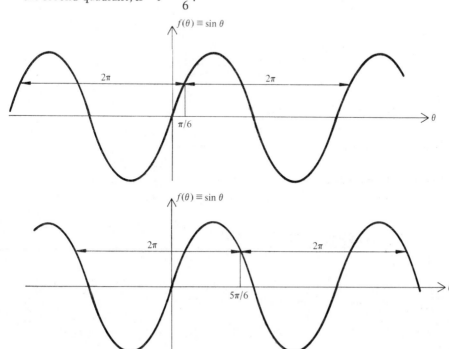

So the general solution includes $\theta = \dfrac{\pi}{6} + 2n\pi$

and $\theta = \dfrac{5\pi}{6} + 2n\pi$

There is no obvious way of combining these two expressions so the general solution in all cases of this type should include

$$\theta = PV + 2n\pi$$
$$and \quad \theta = SV + 2n\pi$$
or
$$\begin{cases} PV + 360n° \\ SV + 360n° \end{cases}$$

Although not obvious, there is a way of combining the two parts of the general solution giving:

When $\sin \theta = \frac{1}{2}$ $\theta = (-1)^n \dfrac{\pi}{6} + n\pi$

When $\sin \theta = s$ $\theta = (-1)^n PV + n\pi$

Students who prefer this quotable form may use it instead of the two-part formula.

8) Find the general solution of the equation

$$4 \sin \theta \,(2 \tan \theta + 3) + 6 \tan \theta + 9 = 0.$$

First we must simplify the equation.

$$4 \sin \theta (2 \tan \theta + 3) + 3(2 \tan \theta + 3) = 0$$
$$\Rightarrow \qquad (4 \sin \theta + 3)(2 \tan \theta + 3) = 0$$

So   either $4 \sin \theta + 3 = 0$ or $2 \tan \theta + 3 = 0$

Therefore either $\sin \theta = -\frac{3}{4}$ [1]

or $\tan \theta = -\frac{3}{2}$ [2]

For [1] the principal solution is $\theta = -48.59°$
and the secondary solution, in the third quadrant, is $\theta = -131.41°$.

Hence the general solution is $\begin{cases} \theta = -48.59° + 360n° \\ \theta = -131.41° + 360n° \end{cases}$

For [2] the principal solution is the only one we need and it is $\theta = -56.31°$.
The general solution then is $\theta = -56.31° + 180n°$

Combining the results of [1] and [2], the general solution set of the given equation is

$$\theta \; = \; \begin{cases} -48.59° + 360n° \\ -131.41° + 360n° \\ -56.31° + 180n° \end{cases}$$

EXERCISE 6e

1) Solve the following equations for angles in the range
$0 \leqslant \theta \leqslant 2\pi$ (or $0 \leqslant \theta \leqslant 360°$)

(a) $\cos \theta = \dfrac{1}{2}$ (b) $\tan \theta = -\dfrac{1}{\sqrt{3}}$ (c) $\sin \theta = \dfrac{\sqrt{2}}{2}$

(d) $\sec \theta = -3$ (e) $\cot \theta = 2$ (f) $\operatorname{cosec} \theta = 4$

(g) $\cos \theta = 0.84$ (h) $\operatorname{cosec} \theta = -2.5$ (i) $\tan \theta = 0.75$

(j) $\sqrt{3} \tan \theta = 2 \sin \theta$ $\left(Hint\colon\ \tan \theta \equiv \dfrac{\sin \theta}{\cos \theta} \right)$

(k) $2 \sin \theta \cos \theta + \sin \theta = 0$ (l) $4 \cos \theta = \cos \theta \operatorname{cosec} \theta$

2) Find the general solution of the following equations, illustrating your results by reference to the graphs of the circular functions and/or quadrant diagrams.

(a) $\sin \theta = \dfrac{\sqrt{3}}{2}$ (b) $\cos \theta = 0$ (c) $\tan \theta = -\sqrt{3}$

(d) $\sin \theta = -\frac{1}{4}$ (e) $\cos \theta = 0.371$ (f) $\cot \theta = \frac{1}{2}$

(g) $\operatorname{cosec} \theta = 3$ (h) $\sec \theta = 1$ (i) $(\sin \theta)^2 = \frac{1}{4}$

MULTIPLE ANGLES

Equations are frequently met in which the angle involved is a multiple of θ,

e.g. $\cos 2\theta = \frac{1}{2}$, $\tan 3\theta = -2$

Such equations are solved by determining first the necessary values of the multiple angle and then, by division, the corresponding values of θ.

EXAMPLES 6f

1) Find the general solution of the equation $\cos 2\theta = \frac{1}{2}$.

Let $2\theta \equiv \phi$ so that $\cos \phi = \frac{1}{2}$.

The principal value of ϕ is $\dfrac{\pi}{3}$ so the general solution for ϕ is

$$\phi \; = \; \pm\frac{\pi}{3} + 2n\pi$$

i.e.

$$2\theta = \pm\frac{\pi}{3} + 2n\pi$$

Hence

$$\theta = \pm\frac{\pi}{6} + n\pi$$

2) Find the angles within the range $-180° \leqslant \theta \leqslant 180°$ which satisfy the equation $\tan 3\theta = -2$.

Let $3\theta \equiv \phi$ so that $\tan \phi = -2$.
The principal value is in the fourth quadrant and the secondary value is in the second quadrant. All angles which satisfy the equation therefore correspond to the solution positions shown

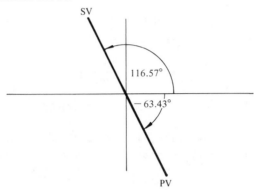

Values of θ are required in the range $-180° \leqslant \theta \leqslant 180°$. But $\phi \equiv 3\theta$ so we need to consider values of ϕ in the range $3(-180°) \leqslant \phi \leqslant 3(180°)$.
Hence, within the range $-540° \leqslant \phi \leqslant 540°$, we have

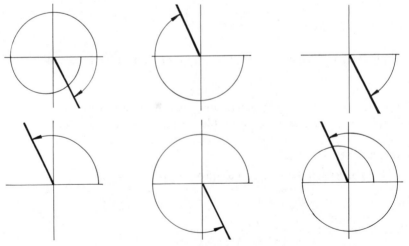

i.e. $\phi = -423.43°, -243.43°, -63.43°, 116.57°, 296.57°, 476.57°$

Therefore $\theta = -141.14°, -81.14°, -21.14°, 38.86°, 98.86°, 158.86°$.

(Alternatively, quoting the general solution for ϕ gives

$$\phi = -63.43° + 180n° \Rightarrow \theta = -21.14° + 60n°$$

Giving n the values which cover the required range for θ
i.e. $n = -2, -1, 0, 1, 2, 3,$ the correct values of θ can be found.)

3) Find the solution of the equation $\sin \dfrac{\theta}{2} = 0.6$ giving values of θ between 0 and 360°.

Let $\dfrac{\theta}{2} \equiv \phi$ so that $\sin \phi = 0.6$

The principal value of ϕ is $36.87°$ and the secondary value, in the second quadrant, is $143.13°$.
So the solution positions for ϕ are

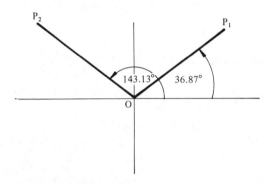

Now the required range of values of θ is from 0 to $360°$. Hence the range of values of ϕ $\left(\equiv \dfrac{\theta}{2}\right)$ is from 0 to $180°$,

i.e.

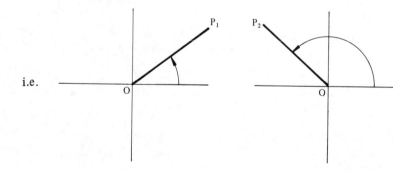

Hence $\phi = 36.87°, 143.13°$.
Therefore $\theta = 73.74°, 286.26°$.

EXERCISE 6f

1) Within the range $0 \leqslant \theta \leqslant 360°$, solve the following equations:
(a) $\tan 2\theta = 1$ (b) $\sin 3\theta = 0.7$ (c) $\cos \frac{1}{2}\theta = 0.85$ (d) $\sec 5\theta = 2$

2) Find the general solution of the following equations:

(a) $\operatorname{cosec} \dfrac{\theta}{3} = 1.5$ (b) $\cot 4\theta = 3$ (c) $\cos 2\theta = 0.63$

(d) $\sin \dfrac{\theta}{3} = 0.5$

3) Sketch the graph of $f(\theta) \equiv \cos \theta$ from -3π to $+3\pi$. *Use your graph* to derive a general solution of the equation $\cos \theta = -1$.

4) Sketch the graph of $f(\theta) \equiv \tan \theta$ for the range $-2\pi \leqslant \theta \leqslant 2\pi$. Mark on the graph the points where $\tan \theta = \sqrt{3}$. Explain why these points are equidistant and derive the general solution of the equation $\tan \theta = \sqrt{3}$.

THE EQUATION   cos A = cos B

This type of equation can be solved using a factor formula but we will now look at an alternative, very neat, method of solution.
Let $\cos A = \cos B = c$ say $(-1 \leqslant c \leqslant 1)$.
In general, for $\cos B = c$ there are two solution positions, OP_1 and OP_2.

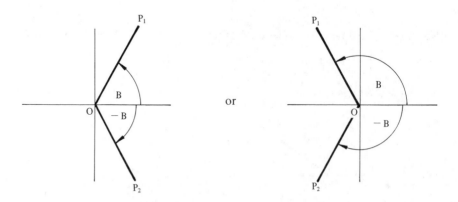

The set of angles represented by OP_1 and OP_2 in either case is $2n\pi \pm B$. But we also know that $\cos A = c$ so OP_1 and OP_2 together represent all possible values of A.

Thus $A = 2n\pi \pm B$

i.e. $A = $ general solution set for B

The same conclusion is reached when considering the equations

$$\tan A = \tan B \quad \Rightarrow \quad A = n\pi + B$$

and
$$\sin A = \sin B \quad \Rightarrow \quad A = \begin{cases} 2n\pi + B \\ (2n+1)\pi - B \end{cases}$$

EXAMPLES 6g

1) Solve the equation $\cos 4\theta = \cos \theta$.

Using only the conclusion to the argument above, we can write

$$4\theta = 2n\pi \pm \theta$$

Hence
$$5\theta = 2n\pi \quad \text{or} \quad 3\theta = 2n\pi$$

$\Rightarrow$
$$\theta = \frac{2n\pi}{5}, \frac{2n\pi}{3}$$

2) Find the values in the range $0 \leqslant \theta \leqslant 360°$ which satisfy the equation $\tan (3\theta - 40°) = \tan \theta$.

The general solution is
$$3\theta - 40° = 180n° + \theta$$

$\Rightarrow$
$$2\theta = 180n° + 40°$$

$\Rightarrow$
$$\theta = 90n° + 20°$$

For $0 \leqslant \theta \leqslant 360°$, let $n = 0, 1, 2, 3$

giving
$$\theta = 20°, 110°, 200°, 290°$$

This very neat method can be used only for equations containing two terms involving the same trig ratio. This situation can sometimes be arranged in an apparently unsuitable case as in Example 3.

3) Find the general solution of $\cos 3\theta = \sin \theta$.

We know that $\sin \theta \equiv \cos \left(\dfrac{\pi}{2} - \theta \right)$ (complementary angles)

So the equation can be written

$$\cos 3\theta = \cos \left(\frac{\pi}{2} - \theta \right)$$

giving
$$3\theta = 2n\pi \pm \left(\frac{\pi}{2} - \theta \right)$$

Therefore either $4\theta = 2n\pi + \dfrac{\pi}{2}$ or $2\theta = 2n\pi - \dfrac{\pi}{2}$

$\Rightarrow$ $\theta = \dfrac{n\pi}{2} + \dfrac{\pi}{8}, \; n\pi - \dfrac{\pi}{4}$

EXERCISE 6g

Find the general solution of the following equations:

1) $\cos 4\theta = \cos 3\theta$ 2) $\tan 7\theta = \tan 2\theta$ 3) $\cos 3\theta = \sin 2\theta$

4) $\cot 4\theta = \tan 5\theta$ 5) $\sin 4\theta = \sin 3\theta$ 6) $\cos 5\theta \sec \theta = 1$

7) $\cos\left(\theta - \dfrac{\pi}{4}\right) = \cos\left(4\theta + \dfrac{\pi}{4}\right)$ 8) $\tan 2\theta = \cot \theta$

Find the solutions, from 0 to π inclusive, of the following equations:

9) $\cos 3\theta = \cos 7\theta$ 10) $\tan 3\theta = \cot 2\theta$ 11) $\sin 7\theta = \sin 2\theta$

12) $\sec 6\theta = \sec 5\theta$

Solve the following equations giving values from $-180°$ to $+180°$:

13) $\cos(2\theta + 60°) = \cos \theta$ 14) $\tan 3\theta = \tan(\theta - 50°)$

15) $\sin \theta = \cos(2\theta + 60°)$

THE GRAPHS OF CIRCULAR FUNCTIONS OF COMPOUND ANGLES

Consider $f(\theta) \equiv \sin 2\theta$.
The following table gives pairs of corresponding values of θ and $f(\theta)$.

θ	0	$\dfrac{\pi}{4}$	$\dfrac{\pi}{2}$	$\dfrac{3\pi}{4}$	π	$\dfrac{5\pi}{4}$	$\dfrac{3\pi}{2}$	$\dfrac{7\pi}{4}$	2π
2θ	0	$\dfrac{\pi}{2}$	π	$\dfrac{3\pi}{2}$	2π	$\dfrac{5\pi}{2}$	3π	$\dfrac{7\pi}{2}$	4π
$f(\theta)$	0	1	0	-1	0	1	0	-1	0

Thus

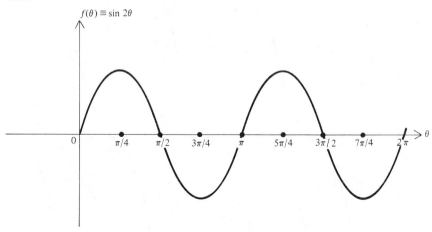

$f(\theta) \equiv \sin 2\theta$

The following characteristics can be observed.

(a) The function $\sin 2\theta$ is cyclic and its period is π $\left(\text{i.e. } \dfrac{2\pi}{2}\right)$

(b) The greatest and least values are 1 and -1.

(c) The shape is a sine wave.

(d) Within the range $0 \leqslant \theta \leqslant 2\pi$, there are two complete curve patterns compared with only one for the basic function $f(\theta) \equiv \sin\theta$, i.e. the complete *cycle* appears with twice the frequency.

When the same investigation is carried out on $f(\theta) \equiv \sin 3\theta$ we find that the function is cyclic with a period $\dfrac{2\pi}{3}$, so that three complete cycles occur between 0 and 2π.

It seems likely then (although it has not been generally proved) that the graph of the function $f(\theta) \equiv \sin k\theta$ is a sine wave with a period of $\dfrac{2\pi}{k}$ and a frequency k times that of $f(\theta) \equiv \sin\theta$.

These properties are, in fact, valid for all values of k and similar properties can be deduced for $f(\theta) \equiv \cos k\theta$ and $f(\theta) \equiv \tan k\theta$,

e.g.

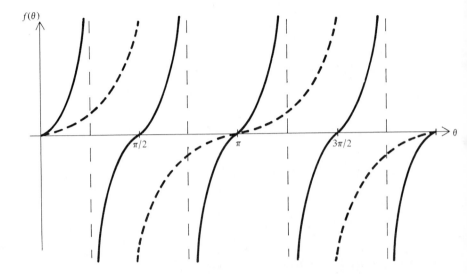

$f(\theta) \equiv \tan 2\theta$ —————

$f(\theta) \equiv \tan \theta$ — — — — —

Frequency of $\tan 2\theta$ is 2

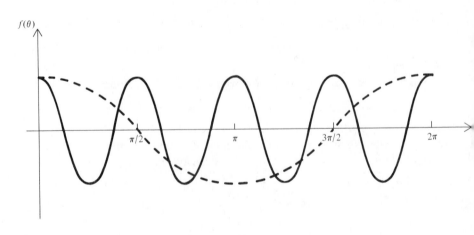

$f(\theta) \equiv \cos 4\theta$ —————

$f(\theta) \equiv \cos \theta$ — — — — —

Frequency of $\cos 4\theta$ is 4

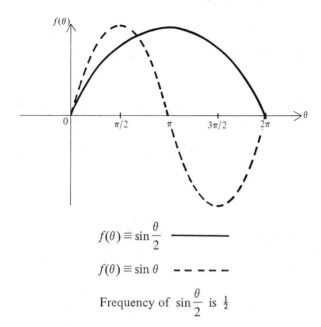

$$f(\theta) \equiv \sin\frac{\theta}{2} \quad \underline{\hspace{2cm}}$$

$$f(\theta) \equiv \sin\theta \quad \text{-----}$$

Frequency of $\sin\dfrac{\theta}{2}$ is $\tfrac{1}{2}$

Now consider the function $f(\theta) \equiv \cos(\theta - \alpha)$.
This is clearly a cosine function but

(a) $f(\theta) = 0$ when $(\theta - \alpha) = \dfrac{\pi}{2}, \dfrac{3\pi}{2}$, etc.

i.e. when $\theta = \dfrac{\pi}{2} + \alpha, \dfrac{3\pi}{2} + \alpha$ etc.

(b) $f(\theta) = 1$ when $(\theta - \alpha) = 0, 2\pi, 4\pi$, etc.

i.e. when $\theta = \alpha, 2\pi + \alpha, 4\pi + \alpha$, etc.

(c) $f(\theta) = -1$ when $(\theta - \alpha) = \pi, 3\pi, 5\pi$, etc.

i.e. when $\theta = \pi + \alpha, 3\pi + \alpha, 5\pi + \alpha$, etc.

e.g. if $\alpha = \dfrac{\pi}{4}$, we have

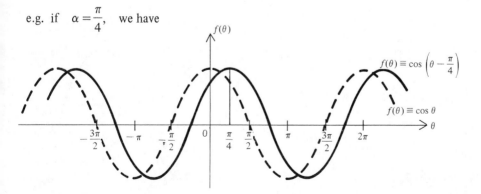

$$f(\theta) \equiv \cos\left(\theta - \frac{\pi}{4}\right)$$

$$f(\theta) \equiv \cos\theta$$

From the sketch we see that the graph of $f(\theta) \equiv \cos\left(\theta - \dfrac{\pi}{4}\right)$ is identical in shape to the graph of $f(\theta) \equiv \cos\theta$ but is in a position given by moving the standard cosine curve a horizontal distance $\dfrac{\pi}{4}$ to the *right*.

Similarly it can be shown that the graph of the function $f(\theta) \equiv \cos(\theta + \alpha)$ is given by moving the standard cosine curve a horizontal distance α to the *left*.

e.g. the graph of $f(\theta) \equiv \cos\left(\theta + \dfrac{\pi}{3}\right)$ is

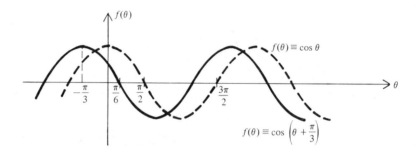

Now consider the function $f(\theta) \equiv \sin(2\theta + \alpha)$.
We know that *adding* a constant angle causes a curve to move *leftward*.

In this case $\sin(2\theta + \alpha) = 0$ when $\theta = -\dfrac{\alpha}{2}$ so the graph of this function is obtained by moving the graph of $\sin 2\theta$ a distance $\dfrac{\alpha}{2}$ to the left

e.g. if $f(\theta) \equiv \sin\left(2\theta + \dfrac{\pi}{2}\right)$, the graph is

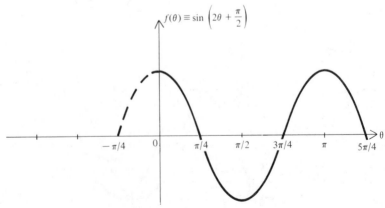

Similarly if $f(\theta) \equiv \cos\left(3\theta - \dfrac{\pi}{2}\right)$, by putting $3\theta - \dfrac{\pi}{2} = 0$, we see that the required graph is obtained by moving the graph of $f(\theta) \equiv \cos 3\theta$ a distance $\dfrac{\pi}{6}$ to the *right*

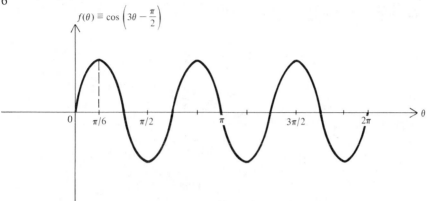

and if $f(\theta) \equiv \tan\left(\dfrac{\theta}{2} + \dfrac{\pi}{6}\right)$ we have a tangent curve with a frequency of $\dfrac{1}{2}$ which is moved a distance $\dfrac{\pi}{3}$ to the left $\left(\dfrac{\theta}{2} + \dfrac{\pi}{6} = 0 \Rightarrow \theta = -\dfrac{\pi}{3}\right)$

i.e.

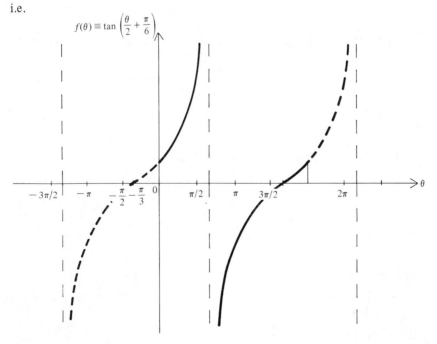

Note. An equation containing a compound angle can be solved in the following way.

If
$$\cos\left(2\theta - \frac{\pi}{6}\right) = \tfrac{1}{2}$$

the principal value of $\left(2\theta - \frac{\pi}{6}\right)$ is $\frac{\pi}{3}$

i.e.
$$2\theta - \frac{\pi}{6} = \pm\frac{\pi}{3} + 2n\pi$$

$\Rightarrow$
$$2\theta = \pm\frac{\pi}{3} + 2n\pi + \frac{\pi}{6}$$

$\Rightarrow$
$$\theta = \pm\frac{\pi}{6} + n\pi + \frac{\pi}{12}$$

i.e.
$$\theta = \begin{cases} n\pi + \dfrac{\pi}{4} \\[2mm] n\pi - \dfrac{\pi}{12} \end{cases}$$

EXERCISE 6h

Sketch the graphs of the following functions in the range $0 \leqslant \theta \leqslant 2\pi$, in each case state the period of the function and its frequency.

1) $\sin 4\theta$ 2) $\sec 2\theta$ 3) $\tan\dfrac{\theta}{4}$ 4) $\sin 2\theta$ 5) $\cot 2\theta$

6) $\sin\left(\theta + \dfrac{\pi}{3}\right)$ 7) $\cos\left(3\theta - \dfrac{\pi}{4}\right)$ 8) $\operatorname{cosec} 3\theta$ 9) $-\sin\dfrac{\theta}{4}$

10) $\cos\left(2\theta + \dfrac{\pi}{2}\right)$ 11) $\tan\left(\theta - \dfrac{\pi}{6}\right)$

Find the general solutions of the following equations:

12) $\cos\left(\theta + \dfrac{\pi}{4}\right) = \dfrac{1}{2}$ 13) $\tan\left(2\theta - \dfrac{\pi}{3}\right) = -1$

14) $\cos\left(3\theta - \dfrac{\pi}{3}\right) = -\dfrac{\sqrt{3}}{2}$ 15) $\sin\left(2\theta + \dfrac{\pi}{6}\right) = \dfrac{1}{2}$

SUMMARY

An angle of 1 radian is subtended at the centre of a circle of radius r by an arc of length r.

π radians $= 180°$.

For a circle of radius r
the length of an arc subtending θ radians at the centre is $r\theta$.

For a circle of radius r
the area of a sector containing θ radians at the centre is $\frac{1}{2}r^2\theta$.

For any angle θ, the associated acute angle α is the difference between θ and the nearest multiple of $180°$ (π).

If the principal solution of $\quad \cos\theta = c \quad$ is $\quad \theta = \theta_1$

the general solution is $\qquad\qquad \theta = 2n\pi \pm \theta_1$

If the principal solution of $\quad \tan\theta = t \quad$ is $\quad \theta = \theta_1$

the general solution is $\qquad\qquad \theta = n\pi + \theta_1$

If the principal solution of $\quad \sin\theta = s \quad$ is $\quad \theta = \theta_1$

the general solution is $\qquad\qquad \theta = \begin{cases} 2n\pi + \theta_1 \\ (2n+1)\pi - \theta_1 \end{cases}$

MULTIPLE CHOICE EXERCISE 6

(The instructions for answering these questions are on p. xii)

TYPE I

1) An angle of 1 radian is equivalent to:
(a) $90°$ (b) $60°$ (c) $67° 18'$ (d) $57° 18'$ (e) $45°$.

2) An arc PQ subtends an angle of $60°$ at the centre of a circle of radius 1 cm. The length of PQ is:

(a) 60 cm   (b) 30 cm   (c) $\frac{\pi}{6}$ cm   (d) $\frac{\pi}{3}$ cm   (e) $\frac{\pi^2}{18}$ cm.

3) If $\quad \theta = \dfrac{13\pi}{6}, \quad \cos\theta$ is:

(a) $\dfrac{1}{2}$ (b) $-\dfrac{1}{2}$ (c) $\dfrac{\sqrt{3}}{2}$ (d) $-\dfrac{\sqrt{3}}{2}$ (e) $\dfrac{\sqrt{2}}{2}$.

4) The associated acute angle for $280°$ is:
(a) $100°$ (b) $10°$ (c) $80°$ (d) $-80°$ (e) $190°$.

5) If $\cos\theta = \frac{1}{2}$, the general solution is:

(a) $\theta = 2n\pi \pm \dfrac{\pi}{6}$ (b) $\theta = n\pi + \dfrac{\pi}{3}$ (c) $\theta = 2n\pi + \dfrac{\pi}{3}$

(d) $\theta = 2n\pi \pm \dfrac{\pi}{3}$ (e) $\theta = n\pi \pm \dfrac{\pi}{6}$.

6) There is a solution of the equation $4 \sin \theta + 1 = 0$ in quadrants:
(a) 1 and 2   (b) 1 and 3   (c) 3 and 4   (d) 2 and 3.

7) The graph of the function $f(\theta) \equiv \cos \left(2\theta - \dfrac{\pi}{2} \right)$ has a period

(a) 2π (b) π (c) $\dfrac{\pi}{2}$ (d) $-\dfrac{\pi}{2}$ (e) none of these.

TYPE II

8) $\theta = 60°$.

(a) $\sin \theta = \dfrac{1}{2}$.   (b) $\tan \theta = \cot 30°$.   (c) $\theta = \dfrac{\pi}{3}$.   (d) $\sec \theta = 2$.

9)   An angle θ is such that $\tan \theta = 1$ and $\cos \theta$ is negative.

(a) $\sin \theta$ is positive.   (b) $\cos \theta = -\dfrac{\sqrt{2}}{2}$.   (c) $\cot \theta = -1$.
(d) $\sec \theta$ is negative.

10) $f(\theta) \equiv \cos \theta$.

(a) For $-\dfrac{\pi}{2} < \theta < \dfrac{\pi}{2}$, $f(\theta) > 0$.

(b) $f(\theta)$ is undefined when $\theta = (2n + 1)\dfrac{\pi}{2}$.

(c) $-1 \leqslant f(\theta) \leqslant 1$.

(d) $f(\theta)$ is periodic with a period of π.

11) π can represent:
(a) the ratio of the circumference to the radius of a circle,
(b) half a revolution,
(c) the period of the function $f(\theta) \equiv \tan \theta$,
(d) an angle whose cosine is zero.

12) Given that $\tan \theta = 1$
(a) θ lies in quadrants 1 and 2,

(b) the principal solution is $\theta = \dfrac{\pi}{4}$,

(c) $\cos \theta = \sqrt{2}$,

(d) the general solution is $\theta = n\pi \pm \dfrac{\pi}{4}$.

13)

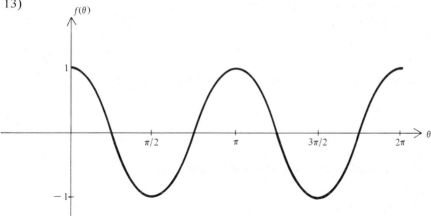

In the sketch above $f(\theta)$ could be:

(a) $\cos 2\theta$ (b) $\cos \dfrac{\theta}{2}$ (c) $\sin \left(2\theta - \dfrac{\pi}{4}\right)$ (d) $\sin \left(2\theta + \dfrac{\pi}{2}\right)$

(e) $\sin \left(2\theta + \dfrac{\pi}{4}\right)$.

14) The graph of $f(\theta) \equiv \cos \theta$ compared with the graph of $f(\theta) \equiv \sin \theta$ is:
(a) inverted   (b) 90° to the left
(c) 90° to the right   (d) of equal period.

TYPE III

15) (a) $\sin \theta = 1$

    (b) $\theta = \dfrac{\pi}{2}$.

16) (a) The graph of a trigonometric function $f(\theta)$ is continuous.
    (b) $f(\theta) \equiv \cot \theta$.

17) In a circle of radius r an arc PQ subtends an angle θ at the centre 0.
    (a) The length of PQ is $4r$.
    (b) The area of the sector POQ is $2r^2$.

18) $f(\theta)$ is a circular function.
    (a) $f(\theta) \equiv \sec \theta$.

    (b) $f(\theta)$ is undefined at $\theta = \dfrac{\pi}{2}$.

19) (a) $\cos \theta = a$.

    (b) $\sin \left(\dfrac{\pi}{2} - \theta\right) = a$.

20) (a) A circular function $f(\theta)$ has a period of π.
    (b) $f(\theta) \equiv \cos 2\theta$.

TYPE IV

21) Find the length of an arc PQ of a circle if:

(a) the angle POQ is $\dfrac{\pi}{2}$,

(b) the area of the circle is $9\pi\,\mathrm{cm}^2$,
(c) O is the centre of the circle,
(d) the radius of the circle is measured in cm.

22) Evaluate the angle θ,
(a) $\tan\theta$ is given,
(b) $\cos\theta$ is positive,
(c) $0 \leqslant \theta \leqslant 2\pi$,
(d) there is only one value of θ.

23) Sketch the graph of $f(\theta)$ given that:
(a) $f(\theta)$ is a major circular function,
(b) $f(\theta)$ is continuous,
(c) $f(\theta)$ is zero when $\theta = n\pi$,
(d) $f(\theta)$ is periodic.

TYPE V

24) π radians $= 360°$.

25) The function $f(\theta) \equiv \cos\theta$ is such that $f(\theta) \leqslant 1$.

26) $\sin\theta = 0$ when $\theta = n\pi$.

27) The function $f(\theta) \equiv \sec\theta$ is undefined when $\theta = (2n+1)\dfrac{\pi}{2}$.

28) If a function $f(x)$ is positive and decreasing for $1 < x < 2$ then the function $\dfrac{1}{f(x)}$ is negative and increasing for $1 < x < 2$.

MISCELLANEOUS EXERCISE 6

1) A chord of a circle subtends an angle of θ radians at the centre of the circle. If the area of the minor segment cut off by the chord is one sixth of the area of the circle prove that $\sin\theta = \theta - \dfrac{\pi}{3}$.

2) Three circular discs each of radius a lie on a table touching each other. Find, in terms of a, the area enclosed between them.

3) A chord AB of a circle of radius $5a$ is of length $3a$. The tangents to the circle at A and B meet at T. Find the area enclosed by TA, TB and the *major* arc AB.

4) Three cylinders are placed in contact with each other with their axes parallel. The radii of the cylinders are 3 cm, 4 cm, 5 cm. An elastic band is stretched round the three cylinders so that the plane of the elastic band is perpendicular to the axes of the cylinders. Calculate the length of the part of the band in contact with the largest cylinder.   (U of L)p

5) P and Q are points on a circle of radius r, and the chord PQ subtends an angle 2θ radians at its centre O. If A is the area enclosed by the minor arc PQ and the chord PQ, and if B is the area enclosed by the arc PQ and the tangents to the circle at P and Q, prove that

$$A - B \equiv r^2(2\theta - \tan\theta - \sin\theta\cos\theta)$$

6) Find the general solutions of the equations:

(a) $\sin\left(\theta + \dfrac{\pi}{4}\right) = \dfrac{1}{2}$,

(b) $\cos\left(\theta - \dfrac{\pi}{4}\right) = \dfrac{1}{2}$.

By drawing sketches of $f(\theta) \equiv \sin\left(\theta + \dfrac{\pi}{4}\right)$ and $g(\theta) \equiv \cos\left(\theta - \dfrac{\pi}{4}\right)$, explain your answers to (a) and (b).

7) Sketch the function $f(\theta) \equiv 2\sin\left(2\theta + \dfrac{\pi}{2}\right)$ for values of θ in the range $-\pi \leqslant \theta \leqslant \pi$.

Solve the equation $f(\theta) = 1$, indicating your solutions on the sketch.

Find the general solutions of the equations:

8) $\tan\left(3\theta - \dfrac{\pi}{2}\right) = \sqrt{3}$.

9) $\sec\left(2\theta + \dfrac{\pi}{2}\right) = 2$.

10) $\tan 5\theta = \cot 2\theta$.

11) Explain briefly what is meant by:
(a) a periodic function,
(b) the amplitude of a sine wave,
(c) the frequency of a circular function,
(d) the general solution of a trigonometric equation,
(e) a radian.

12) What is the period of the function $f(\theta) \equiv \cos k\theta$? Draw sketches to illustrate your answer when $k = 2$ and $k = \frac{1}{2}$. In each of these cases, write down the general solution of the equations $f(\theta) = 0$, $f(\theta) = 1$, $f(\theta) = -1$.

13) Without the use of tables or calculator find, for each of the following equations, all the solutions in the interval $0° \leqslant x \leqslant 180°$.
(a) $\cos (x + 30°) = \cos (60° - 3x)$
(b) $\sin (x + 20°) = \cos 3x$.   (JMB)

14) Without using tables or calculator, show that $x = \dfrac{\pi}{14}$ is a solution of

the equation
$$\sin 3\theta = \cos 4\theta$$

Find the general solution.

15) (a) Find the solutions of the equation $\cos 3\theta = \cos 2\theta$ for which $0 < \theta \leqslant 2\pi$.

(b) Write down the general solution of the equation
$$\sin k\theta = \sin \theta$$

Hence find the value of k for which this equation and the equation $\cos 3\theta = \cos 2\theta$ have, within the range $0 < \theta \leqslant 2\pi$,
(i) five common solutions,
(ii) one and only one common solution.

CHAPTER 7

TRIGONOMETRIC IDENTITIES

In the previous chapter we saw that any one angle is associated with six different trigonometric ratios and that a particular value of one trigonometric ratio applies to an infinite set of angles. It is not surprising, therefore, to find that relationships exist between the various circular functions. Some of these are very useful in the development of trigonometry. This chapter deals with their derivation and applications.

Consider, for instance, the simple relationship between the sine, cosine and tangent of any angle.

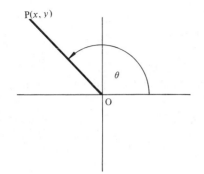

For any angle θ we have:

$$\sin \theta = \frac{y}{OP}, \quad \cos \theta = \frac{x}{OP},$$

$$\tan \theta = \frac{y}{x}$$

But $\dfrac{\sin \theta}{\cos \theta} = \dfrac{y}{OP} \Big/ \dfrac{x}{OP} = \dfrac{y}{x}$

Thus for all angles

$$\tan \theta \equiv \frac{\sin \theta}{\cos \theta}.$$

THE PYTHAGOREAN OR 'SQUARED RATIO' GROUP

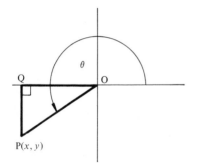

For any position of OP a right angled triangle OPQ can be drawn for which

$$x^2 + y^2 = OP^2$$

Dividing throughout, in turn, by OP^2, x^2 and y^2 gives

$$\left(\frac{x}{OP}\right)^2 + \left(\frac{y}{OP}\right)^2 = 1 \tag{1}$$

$$1 + \left(\frac{y}{x}\right)^2 = \left(\frac{OP}{x}\right)^2 \tag{2}$$

$$\left(\frac{x}{y}\right)^2 + 1 = \left(\frac{OP}{y}\right)^2 \tag{3}$$

[1] becomes $\qquad (\cos\theta)^2 + (\sin\theta)^2 = 1$

[2] becomes $\qquad 1 + (\tan\theta)^2 = (\sec\theta)^2$

[3] becomes $\qquad (\cot\theta)^2 + 1 = (\text{cosec}\,\theta)^2$

The use of brackets when raising a trigonometric ratio to a power can be avoided by writing $\cos^2\theta$ for $(\cos\theta)^2$ etc.
The above relationships are valid for any position of OP, i.e. for any angle θ, so for all angles:

$$\cos^2\theta + \sin^2\theta \equiv 1$$

$$1 + \tan^2\theta \equiv \sec^2\theta$$

$$\cot^2\theta + 1 \equiv \text{cosec}^2\theta$$

These identities are very useful in the solution of certain trig equations.

EXAMPLES 7a

1) Solve the equation $\quad 2\cos^2\theta - \sin\theta = 1 \quad$ for values of θ between 0 and 2π.

Using $\quad \cos^2\theta + \sin^2\theta \equiv 1 \quad$ gives

$$2(1 - \sin^2\theta) - \sin\theta = 1$$

or $\qquad\qquad\qquad 2\sin^2\theta + \sin\theta - 1 = 0$

This is now a quadratic equation (of the form $2x^2 + x - 1 = 0$)

Hence $(2 \sin \theta - 1)(\sin \theta + 1) = 0$

$\Rightarrow$ $\sin \theta = \frac{1}{2}$ or -1

If $\sin \theta = \frac{1}{2}$, $\theta = \dfrac{\pi}{6}, \dfrac{5\pi}{6}$

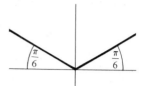

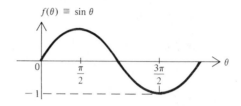

If $\sin \theta = -1$, $\theta = \dfrac{3\pi}{2}$

Therefore the solution of the equation is

$$\theta = \frac{\pi}{6}, \frac{3\pi}{2}, \frac{5\pi}{6}$$

2) If $3 \sec^2\theta - 5 \tan \theta - 4 = 0$ find the general solution.

Using $1 + \tan^2\theta \equiv \sec^2\theta$ we have

$$3(1 + \tan^2\theta) - 5 \tan \theta - 4 = 0$$

i.e. $3 \tan^2\theta - 5 \tan \theta - 1 = 0$

Again we have a quadratic equation but because it has no simple factors we solve it by formula

$\Rightarrow$ $\tan \theta = \dfrac{5 \pm \sqrt{25 + 12}}{6}$

$\Rightarrow$ $\tan \theta = 1.8471$ or -0.1805

If $\tan \theta = 1.8471$,   the principal solution is $\theta = 61.57°$.

If $\tan \theta = -0.1805$ the principal solution is $\theta = -10.23°$.

The complete general solution is therefore

$$\theta = \begin{cases} 180n° + 61.57° \\ 180n° - 10.23° \end{cases}$$

Other applications of the standard identities include:
derivation of a variety of further trigonometric relationships,
elimination of trigonometric terms from pairs of equations,
calculation of the remaining trigonometric ratios of any angle for which only
one trigonometric ratio is known.

3) Prove that $(1 - \cos A)(1 + \sec A) \equiv \sin A \ \tan A$.

Because this relationship has yet to be proved, we must not assume its truth by
using the complete identity in our working. The left and right hand sides must
be isolated throughout the proof.
Consider the L.H.S.

$$(1 - \cos A)(1 + \sec A) \equiv 1 + \sec A - \cos A - \cos A \sec A$$

$$\equiv 1 + \sec A - \cos A - \cos A\left(\frac{1}{\cos A}\right)$$

$$\equiv \sec A - \cos A$$

$$\equiv \frac{1 - \cos^2 A}{\cos A}$$

$$\equiv \frac{\sin^2 A}{\cos A} \qquad\qquad (\text{since } \cos^2 A + \sin^2 A \equiv 1)$$

$$\equiv \sin A\left(\frac{\sin A}{\cos A}\right)$$

$$\equiv \sin A \tan A \quad \text{which is identical to the R.H.S.}$$

4) Prove that $(\operatorname{cosec} A - \sin A)(\sec A - \cos A) \equiv \dfrac{1}{\tan A + \cot A}$.

Considering the L.H.S:

$$(\operatorname{cosec} A - \sin A)(\sec A - \cos A) \equiv \left(\frac{1}{\sin A} - \sin A\right)\left(\frac{1}{\cos A} - \cos A\right)$$

$$\equiv \left(\frac{1 - \sin^2 A}{\sin A}\right)\left(\frac{1 - \cos^2 A}{\cos A}\right)$$

$$\equiv \left(\frac{\cos^2 A}{\sin A}\right)\left(\frac{\sin^2 A}{\cos A}\right)$$

$$\equiv \cos A \sin A$$

Now this is already a very simple form but is not obviously identical to the
given R.H.S. So this time we begin working independently on the R.H.S.

$$\frac{1}{\tan A + \cot A} \equiv 1 \div \left(\frac{\sin A}{\cos A} + \frac{\cos A}{\sin A}\right)$$

$$\equiv 1 \div \left(\frac{\sin^2 A + \cos^2 A}{\cos A \sin A}\right)$$

$$\equiv \cos A \, \sin A$$

Since both L.H.S and R.H.S reduce to cos A sin A they are identical.

5) Eliminate θ from the equations $x = 2 \cos \theta$ and $y = 3 \sin \theta$.

Now $\cos \theta = \dfrac{x}{2}$ and $\sin \theta = \dfrac{y}{3}$

Using $\cos^2\theta + \sin^2\theta \equiv 1$

gives $\left(\dfrac{x}{2}\right)^2 + \left(\dfrac{y}{3}\right)^2 = 1$

$\Rightarrow$ $9x^2 + 4y^2 = 36$

Note. In this problem we see that both x and y initially depend on θ, a variable angle. Used in this way, θ is called a *parameter*, a type of variable which plays an important part in the analysis of curves and functions.

6) If $\sin A = \frac{1}{3}$ and A is obtuse, find cos A and cot A without using tables or calculator.

Using $\cos^2 A + \sin^2 A \equiv 1$

$\Rightarrow$ $\cos^2 A + \frac{1}{9} = 1$

Therefore $\cos A = \pm\sqrt{\dfrac{8}{9}} = \pm\dfrac{2\sqrt{2}}{3}$

But A is obtuse so cos A is negative,

i.e. $\cos A = -\dfrac{2\sqrt{2}}{3}$

Also $\cot A \equiv \dfrac{\cos A}{\sin A}$

Hence $\cot A = -\dfrac{2\sqrt{2}}{3} \Big/ \dfrac{1}{3} = -2\sqrt{2}$

Note: This type of problem can often be done more directly by drawing the appropriate right-angled triangle and using Pythagoras' Theorem.

e.g.

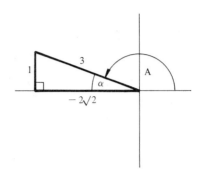

EXERCISE 7a

Solve the following equations for angles in the range $-180° \leqslant \theta \leqslant 180°$.

1) $\sec^2\theta + \tan^2\theta = 6$ 2) $4\cos^2\theta + 5\sin\theta = 3$

3) $\cot^2\theta = \operatorname{cosec}\theta$ 4) $\tan\theta + \cot\theta = 2$

5) $\tan\theta + 3\cot\theta = 5\sec\theta$ 6) $\sec\theta = 1 - 2\tan^2\theta$

Find the general solution of the following equations.

7) $5\cos\theta - 4\sin^2\theta = 2$ 8) $4\cot^2\theta + 12\operatorname{cosec}\theta + 1 = 0$

9) $4\sec^2\theta - 3\tan\theta = 5$ 10) $2\cos\theta - 4\sin^2\theta + 2 = 0$

Prove the following identities:

11) $\cot\theta + \tan\theta \equiv \sec\theta \operatorname{cosec}\theta$

12) $\dfrac{\cos A}{1 - \tan A} + \dfrac{\sin A}{1 - \cot A} \equiv \sin A + \cos A$

13) $\tan^2\theta + \cot^2\theta \equiv \sec^2\theta + \operatorname{cosec}^2\theta - 2$

14) $\dfrac{\sin A}{1 + \cos A} \equiv \dfrac{1 - \cos A}{\sin A}$ (*Hint*: Multiply L.H.S by $(1 - \cos A)$ in both numerator and denominator)

15) $(\sec^2\theta + \tan^2\theta)(\operatorname{cosec}^2\theta + \cot^2\theta) \equiv 1 + 2\sec^2\theta \operatorname{cosec}^2\theta$

16) $\dfrac{\sin A}{1 + \cos A} + \dfrac{1 + \cos A}{\sin A} \equiv \dfrac{2}{\sin A}$

17) $\sec^2 A \equiv \dfrac{\operatorname{cosec} A}{\operatorname{cosec} A - \sin A}$

18) $(1 + \sin\theta + \cos\theta)^2 \equiv 2(1 + \sin\theta)(1 + \cos\theta)$

19) $\dfrac{\tan^2 A + \cos^2 A}{\sin A + \sec A} \equiv \sec A - \sin A$

20) Eliminate θ from the following pairs of equations:

(a) $x = 4 \sec \theta$
$\quad y = 5 \tan \theta$

(b) $x = a \csc \theta$
$\quad y = b \cot \theta$

(c) $x = 2 \tan \theta$
$\quad y = 3 \cos \theta$

(d) $x = 1 - \sin \theta$
$\quad y = 1 + \cos \theta$

(e) $x = 2 + \tan \theta$
$\quad y = 2 \cos \theta$

(f) $x = a \sec \theta$
$\quad y = b \sin \theta$

21) Simplify the following expressions:

(a) $\dfrac{1 - \sec^2 A}{1 - \csc^2 A}$

(b) $\dfrac{\sin \theta}{\sqrt{(1 - \cos^2 \theta)}}$

(c) $\dfrac{\sin \theta}{\cos \theta} + \dfrac{\cos \theta}{\sin \theta}$

(d) $\dfrac{\sqrt{(1 + \tan^2 \theta)}}{\sqrt{(1 - \sin^2 \theta)}}$

(e) $\dfrac{1}{\cos \theta \sqrt{(1 + \cot^2 \theta)}}$

22) Without reference to tables or calculator, complete the following table:

	$\sin \theta$	$\cos \theta$	$\tan \theta$	type of angle
a)		$-\dfrac{5}{13}$		reflex
b)	$\dfrac{3}{5}$			obtuse
c)			$\dfrac{7}{24}$	acute
d)		-1		

COMPOUND ANGLE IDENTITIES

It is often useful to be able to express the trig ratios of angles such as $A + B$ or $A - B$ in terms of the trig ratios of A and of B.
At first sight it is dangerously easy to think, for instance, that
$\sin (A + B)$ is $\sin A + \sin B$.
That this is false can be seen by considering $\sin (45° + 45°) = \sin 90° = 1$

whereas $\quad \sin 45° + \sin 45° = \dfrac{\sqrt{2}}{2} + \dfrac{\sqrt{2}}{2} = \sqrt{2} \neq 1.$

So we see that the sine function is *not distributive* (and similarly for the other trig ratios).
The correct expression is $\quad \sin (A + B) \equiv \sin A \cos B + \cos A \sin B.$
This formula can be proved geometrically when A and B are both acute, by using the diagram overleaf.

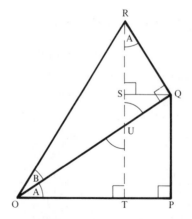

The right angled triangles OPQ and OQR contain angles A and B as shown.
The dotted lines are construction lines and the angle URQ is equal to A.

$$\sin (A + B) \equiv \frac{TR}{OR} \equiv \frac{TS + SR}{OR} \equiv \frac{PQ + SR}{OR}$$

$$\equiv \frac{PQ}{OQ} \times \frac{OQ}{OR} + \frac{SR}{QR} \times \frac{QR}{OR}$$

$$\equiv \sin A \cos B + \cos A \sin B$$

However a neater proof which is valid for all angles is given in Volume II.
Accepting at this stage the validity of this formula for all angles, it can be
adapted to give the full set of compound angle identities. The reader is given the
opportunity in the following exercise to derive these for himself.

EXERCISE 7b

1) In the identity $\sin (A + B) \equiv \sin A \cos B + \cos A \sin B$,   replace
B by $-B$ to show that $\sin (A - B) \equiv \sin A \cos B - \cos A \sin B$.

2) In the identity derived in (1), replace A by $\left(\frac{\pi}{2} - A\right)$ to show that
$\cos (A + B) \equiv \cos A \cos B - \sin A \sin B$.

3) In the identity derived in (2), replace B by $-B$ to show that
$\cos (A - B) \equiv \cos A \cos B + \sin A \sin B$.

4) Use $\dfrac{\sin (A + B)}{\cos (A + B)}$ to show that $\tan (A + B) \equiv \dfrac{\tan A + \tan B}{1 - \tan A \tan B}$.

5) In the identity derived in (4), replace B by $-B$ to show that
$\tan (A - B) \equiv \dfrac{\tan A - \tan B}{1 + \tan A \tan B}$

Collating these results we have:

$$\sin (A + B) \equiv \sin A \cos B + \cos A \sin B$$
$$\sin (A - B) \equiv \sin A \cos B - \cos A \sin B$$

$$\cos (A + B) \equiv \cos A \cos B - \sin A \sin B$$
$$\cos (A - B) \equiv \cos A \cos B + \sin A \sin B$$

$$\tan (A + B) \equiv \frac{\tan A + \tan B}{1 - \tan A \tan B}$$

$$\tan (A - B) \equiv \frac{\tan A - \tan B}{1 + \tan A \tan B}$$

The similarity between the pairs of identities for $A + B$ and $A - B$ makes it clear that care must be taken with signs when using these formulae.

EXAMPLES 7c

1) Without using tables or calculator, evaluate:

(a) $\sin 75°$ (b) $\cos 105°$ (c) $\tan (- 15°)$

(a) $\sin 75° = \sin (45° + 30°) = \sin 45° \cos 30° + \cos 45° \sin 30°$

$$= \left(\frac{\sqrt{2}}{2}\right)\left(\frac{\sqrt{3}}{2}\right) + \left(\frac{\sqrt{2}}{2}\right)\left(\frac{1}{2}\right)$$

$$= (\sqrt{3} + 1)\frac{\sqrt{2}}{4}$$

(b) $\cos 105° = \cos (60° + 45°) = \cos 60° \cos 45° - \sin 60° \sin 45°$

$$= \left(\frac{1}{2}\right)\left(\frac{\sqrt{2}}{2}\right) - \left(\frac{\sqrt{3}}{2}\right)\left(\frac{\sqrt{2}}{2}\right)$$

$$= \frac{(1 - \sqrt{3})\sqrt{2}}{4}$$

Note that this result is negative which is consistent with the cosine of an angle in the second quadrant.

(c) $\tan (- 15°) = \tan (45° - 60°) = \dfrac{\tan 45° - \tan 60°}{1 + \tan 45° \tan 60°}$

$$= \frac{1 - \sqrt{3}}{1 + (1)(\sqrt{3})}$$

$$= \sqrt{3} - 2$$
(rationalising the denominator)

Note: In each part of this example there are alternative compound angles which could be used,

e.g. $75° = 120° - 45°; \quad 105° = 150° - 45°; \quad -15° = 30° - 45°.$

2) A is obtuse and $\sin A = \frac{3}{5}$, B is acute and $\sin B = \frac{12}{13}$. Without finding the values of A and B, evaluate:
(a) $\cos(A + B)$,   (b) $\tan(A - B)$

In order to use the relevant compound angle formulae we need values for $\cos A$, $\cos B$, $\tan A$ and $\tan B$. These are most simply obtained by using Pythagoras in the appropriate right angled triangles.

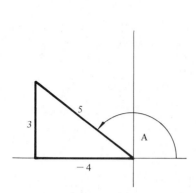
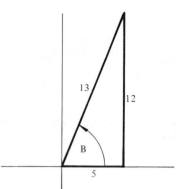

(a) $\cos(A + B)$

$$\equiv \cos A \cos B - \sin A \sin B$$

$$= \left(-\frac{4}{5}\right)\left(\frac{5}{13}\right) - \left(\frac{3}{5}\right)\left(\frac{12}{13}\right)$$

$$= -\frac{56}{65}$$

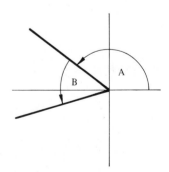

(b) $\tan(A - B)$

$$\equiv \frac{\tan A - \tan B}{1 + \tan A \tan B}$$

$$= \frac{\left(-\frac{3}{4}\right) - \left(\frac{12}{5}\right)}{1 + \left(-\frac{3}{4}\right)\left(\frac{12}{5}\right)}$$

$$= \frac{63}{16}$$

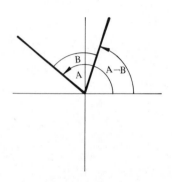

3) Prove that $\dfrac{\sin (A - B)}{\cos A \cos B} + \dfrac{\sin (B - C)}{\cos B \cos C} + \dfrac{\sin (C - A)}{\cos C \cos A} \equiv 0.$

L.H.S. becomes:

$$\dfrac{\sin A \cos B - \cos A \sin B}{\cos A \cos B} + \dfrac{\sin B \cos C - \cos B \sin C}{\cos B \cos C} + \dfrac{\sin C \cos A - \cos C \sin A}{\cos C \cos A}$$

$$\equiv \dfrac{\sin A \cos B}{\cos A \cos B} - \dfrac{\cos A \sin B}{\cos A \cos B} + \dfrac{\sin B \cos C}{\cos B \cos C} - \dfrac{\cos B \sin C}{\cos B \cos C} + \dfrac{\sin C \cos A}{\cos C \cos A} - \dfrac{\cos C \sin A}{\cos C \cos A}$$

$$\equiv \tan A - \tan B + \tan B - \tan C + \tan C - \tan A$$

$$\equiv 0$$

4) Solve the equation $\quad 2 \cos \theta = \sin (\theta + 30°)\quad$ giving the general values of θ.

It is very important to appreciate that this is *not an identity*. Only certain distinct values of θ satisfy this *equation*.

$$2 \cos \theta = \sin (\theta + 30°)$$

$$= \sin \theta \cos 30° + \cos \theta \sin 30°$$

$$= \frac{\sqrt{3}}{2} \sin \theta + \frac{1}{2} \cos \theta$$

Therefore $\qquad \dfrac{3}{2} \cos \theta = \dfrac{\sqrt{3}}{2} \sin \theta$

$\Rightarrow \qquad\qquad \dfrac{3}{\sqrt{3}} = \dfrac{\sin \theta}{\cos \theta}$

$\Rightarrow \qquad\qquad \tan \theta = \sqrt{3}$

The principal solution is $\quad \theta = \dfrac{\pi}{3}$

So the general solution is $\quad \theta = n\pi + \dfrac{\pi}{3}$

EXERCISE 7c

(Do not use tables or a calculator in Questions 1–3.)

1) Evaluate:
(a) $\cos 80° \cos 20° + \sin 80° \sin 20°$
(b) $\sin 37° \cos 7° - \cos 37° \sin 7°$
(c) $\cos 15°$ (d) $\sin 165°$ (e) $\cos 75°$ (f) $\tan 75°$

2) Complete the following table:

	(a)	(b)	(c)	(d)	(e)	(f)
A	$60°$	$\dfrac{3\pi}{4}$	Acute	Obtuse	Acute	
sin A			$\frac{7}{25}$	$\frac{3}{5}$		$-\frac{5}{13}$
cos A						$\frac{12}{13}$
tan A					$\frac{7}{24}$	
B	$150°$	$\dfrac{7\pi}{6}$	Obtuse	Obtuse	Acute	
sin B			$\frac{4}{5}$			
cos B					$\frac{24}{25}$	$\frac{4}{5}$
tan B				$-\frac{12}{5}$		
sin (A + B)						$\frac{16}{65}$
cos (A + B)						
tan (A − B)						

Prove the following identities:

3) $\cot (A + B) \equiv \dfrac{\cot A \cot B - 1}{\cot A + \cot B}$

4) $\sin \left(\dfrac{\pi}{4} + A\right) + \sin \left(\dfrac{\pi}{4} - A\right) \equiv \sqrt{2} \cos A$

5) $(\sin A + \cos A)(\sin B + \cos B) \equiv \sin (A + B) + \cos (A - B)$

6) $\dfrac{\sin (A + B)}{\cos A \cos B} \equiv \tan A + \tan B$

7) $\sin (\theta + 60°) \equiv \sin (120° - \theta)$

8) $\tan (x + y) - \tan x \equiv \dfrac{\sin y}{\cos x \cos (x + y)}$

Solve the following equations, giving angles from $0°$ to $360°$:

9) $\cos (45° - \theta) = \sin (30° + \theta)$

10) $3 \sin x = \cos (x + 60°)$

11) $\tan (A - \theta) = \frac{2}{3}$ and $\tan A = 3$

12) $\sin (x + 60°) = \cos x$

THE DOUBLE ANGLE IDENTITIES

The compound angle formulae deal with any two angles A and B and can therefore be used for two equal angles   (B = A).

Replacing B by A in the compound angle formulae for (A + B) gives

$$\sin 2A \equiv 2 \sin A \cos A$$

$$\cos 2A \equiv \cos^2 A - \sin^2 A$$

$$\tan 2A \equiv \frac{2 \tan A}{1 - \tan^2 A}$$

The second of this group can be expressed in several forms because

$$\cos^2 A - \sin^2 A \equiv (1 - \sin^2 A) - \sin^2 A \equiv 1 - 2 \sin^2 A$$

$$\cos^2 A - \sin^2 A \equiv \cos^2 A - (1 - \cos^2 A) \equiv 2 \cos^2 A - 1$$

Thus
$$\cos 2A \equiv \begin{cases} \cos^2 A - \sin^2 A \\ 1 - 2 \sin^2 A \\ 2 \cos^2 A - 1 \end{cases}$$

These alternative expressions for cos 2A can themselves be rearranged to give

$$2 \sin^2 A \equiv 1 - \cos 2A$$

$$2 \cos^2 A \equiv 1 + \cos 2A$$

Complete familiarity with all the double angle formulae, including *all* the alternative forms involving cos 2A, is essential. These are probably the most useful of all the trig identities, being a powerful tool for simplifying trig functions.

EXAMPLES 7d

1) Find the general solution of the equation $\cos 2x + 3 \sin x = 2$.

Using $\cos 2x \equiv 1 - 2 \sin^2 x$ gives

$$1 - 2 \sin^2 x + 3 \sin x = 2$$

i.e.
$$2 \sin^2 x - 3 \sin x + 1 = 0$$
$$(2 \sin x - 1)(\sin x - 1) = 0$$

$\Rightarrow$ $\qquad\qquad\qquad\qquad\qquad\qquad \sin x = \frac{1}{2}$ or 1

For $\sin x = \frac{1}{2}$ the principal solution is $\dfrac{\pi}{6}$

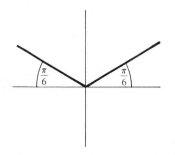

General solution is $\quad x = \begin{cases} 2n\pi + \dfrac{\pi}{6} \\ \\ (2n + 1)\pi - \dfrac{\pi}{6} \end{cases}$

For $\sin x = 1$ the principal solution is $\quad x = \dfrac{\pi}{2}$

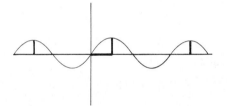

General solution is $\quad x = 2n\pi + \dfrac{\pi}{2}$

The full solution is therefore
$$x = 2n\pi + \frac{\pi}{6}, \quad (2n + 1)\pi - \frac{\pi}{6}, \quad 2n\pi + \frac{\pi}{2}$$

2) If $\tan \theta = \frac{3}{4}$ and θ is acute, find the values of $\tan 2\theta$, $\tan 4\theta$ and $\tan \dfrac{\theta}{2}$.

In this problem we use $\quad \tan 2A \equiv \dfrac{2 \tan A}{1 - \tan^2 A}$,

for three cases: $\quad A = \theta, \quad A = 2\theta, \quad A = \dfrac{\theta}{2}$

If $A = \theta$, $\qquad \tan 2\theta \equiv \dfrac{2 \tan \theta}{1 - \tan^2 \theta} = \dfrac{\frac{3}{2}}{1 - \frac{9}{16}} = \dfrac{24}{7}$

If $A = 2\theta$, $\qquad \tan 4\theta \equiv \dfrac{2 \tan 2\theta}{1 - \tan^2 2\theta} = \dfrac{2(\frac{24}{7})}{1 - (\frac{24}{7})^2} = -\dfrac{336}{527}$

If $A = \dfrac{\theta}{2}$, $\tan\theta \equiv \dfrac{2\tan\dfrac{\theta}{2}}{1-\tan^2\dfrac{\theta}{2}} = \dfrac{3}{4}$

or $\dfrac{2t}{1-t^2} = \dfrac{3}{4}$ where $t \equiv \tan\dfrac{\theta}{2}$

$\Rightarrow$ $8t = 3 - 3t^2$

$\Rightarrow$ $3t^2 + 8t - 3 = 0$

$\Rightarrow$ $(3t-1)(t+3) = 0$

$\Rightarrow$ $t = \dfrac{1}{3}$ or -3

i.e. $\tan\dfrac{\theta}{2} = \dfrac{1}{3}$ or -3

But θ is acute, so $\dfrac{\theta}{2}$ is acute and $\tan\dfrac{\theta}{2} \neq -3$.

Therefore $\tan\dfrac{\theta}{2} = \dfrac{1}{3}$

3) Prove that $\sin 3A \equiv 3\sin A - 4\sin^3 A$.

$\begin{aligned}
\sin 3A &\equiv \sin(2A + A)\\
&\equiv \sin 2A \cos A + \cos 2A \sin A\\
&\equiv (2\sin A \cos A)\cos A + (1 - 2\sin^2 A)\sin A\\
&\equiv 2\sin A \cos^2 A + \sin A - 2\sin^3 A\\
&\equiv 2\sin A(1 - \sin^2 A) + \sin A - 2\sin^3 A\\
&\equiv 3\sin A - 4\sin^3 A
\end{aligned}$

Note: This is quite a useful identity and is worth remembering.

4) Eliminate θ from the equations $x = \cos 2\theta$, $y = \sec\theta$.

Using $\cos 2\theta \equiv 2\cos^2\theta - 1$

$$x = 2\cos^2\theta - 1 \quad \text{and} \quad y = \dfrac{1}{\cos\theta}$$

Hence $x = 2\left(\dfrac{1}{y}\right)^2 - 1$

$\Rightarrow$ $(x+1)y^2 = 2$

Note: This is a *Cartesian equation* which we have obtained by eliminating the *parameter* θ from a *pair of parametric equations*.

EXERCISE 7d

(Do not use tables or calculator in Questions 1–3.)

1) Express as a single trig ratio:

(a) $2 \sin 14° \cos 14°$ (b) $\dfrac{2 \tan 35°}{1 - \tan^2 35°}$

(c) $1 - 2 \sin^2 4\theta$ (d) $\dfrac{2 \tan 3\theta}{1 - \tan^2 3\theta}$

(e) $\sqrt{1 + \cos 6\theta}$ (f) $\cos^2 26° - \sin^2 26°$

(g) $\sin \theta \cos \theta$ (h) $2 \cos^2 34° - 1$

(i) $\dfrac{1 + \tan x}{1 - \tan x}$ (*Hint*: $\tan 45° = 1$)

2) Find the values of $\sin 2\theta$ and $\cos 2\theta$ given:

(a) $\cos \theta = \frac{3}{5}$ (b) $\sin \theta = \frac{7}{25}$ (c) $\tan \theta = \frac{12}{5}$

assuming that   (i) in all cases θ is acute,

(ii) in all cases θ is not acute.

3) If $\tan \theta = -\frac{7}{24}$ and θ is obtuse, find the value of $\tan \dfrac{\theta}{2}$ and hence find $\sin \dfrac{\theta}{2}$ and $\cos \dfrac{\theta}{2}$. Use these results to evaluate $\sin \theta$ and $\cos \theta$ and check that they are consistent with the given value of $\tan \theta$.

4) By eliminating θ from the following pairs of parametric equations, find the corresponding Cartesian equation:

(a) $x = \tan 2\theta,\ y = \tan \theta$ (b) $x = \cos 2\theta,\ y = \cos \theta$

(c) $x = \cos 2\theta,\ y = \operatorname{cosec} \theta$ (d) $x = \sin 2\theta,\ y = \sec 4\theta$

Prove the following identities:

5) $\dfrac{1 - \cos 2A}{\sin 2A} \equiv \tan A$

6) $\tan \theta + \cot \theta \equiv 2 \operatorname{cosec} 2\theta$

7) $\sec 2A + \tan 2A \equiv \dfrac{\cos A + \sin A}{\cos A - \sin A}$

8) $\dfrac{1 - \cos 2A + \sin 2A}{1 + \cos 2A + \sin 2A} \equiv \tan A$

9) $\cos 4A \equiv 8 \cos^4 A - 8 \cos^2 A + 1$

10) $\sin 2\theta \equiv \dfrac{2 \tan \theta}{1 + \tan^2 \theta}$

11) $\cos 2\theta \equiv \dfrac{1 - \tan^2\theta}{1 + \tan^2\theta}$

12) $\cos 3\theta \equiv 4 \cos^3\theta - 3 \cos \theta$

Solve the following equations giving angles within the range $0°$ to $360°$. Also in each case state the general solution.

13) $\cos 2x = \sin x$

14) $\sin 2x + \cos x = 0$

15) $4 - 5 \cos \theta = 2 \sin^2\theta$

16) $\tan \theta \tan 2\theta = 2$

17) $\sin 2\theta - 1 = \cos 2\theta$

18) $5 \cos x \sin 2x + 4 \sin^2 x = 4$

THE HALF ANGLE IDENTITIES

We already know that $\tan 2A \equiv \dfrac{2 \tan A}{1 - \tan^2A}$ and, from the previous exercise

(Nos. 10 and 11), that $\sin 2A \equiv \dfrac{2 \tan A}{1 + \tan^2A}$ and $\cos 2A \equiv \dfrac{1 - \tan^2A}{1 + \tan^2A}$.

If we replace $2A$ by θ and use t to denote $\tan \dfrac{\theta}{2}$ we have

$$\tan \theta \equiv \frac{2t}{1 - t^2}$$

$$\sin \theta \equiv \frac{2t}{1 + t^2}$$

$$\cos \theta \equiv \frac{1 - t^2}{1 + t^2}$$

These three identities allow *all* the trig ratios of any one angle to be expressed in terms of a common variable t. In problems where it is not possible to apply any of the identities used previously, this group can be helpful.

EXAMPLE

Solve the equation $\sin \theta + 2 \cos \theta = 1$ for angles between $0°$ and $360°$.

$$\sin \theta + 2 \cos \theta = 1$$

Therefore $\qquad \dfrac{2t}{1 + t^2} + 2\left(\dfrac{1 - t^2}{1 + t^2}\right) = 1 \quad$ where $\quad t = \tan \dfrac{\theta}{2}$

$\Rightarrow \qquad\qquad 2t + 2 - 2t^2 = 1 + t^2$

$\Rightarrow \qquad\qquad 3t^2 - 2t - 1 = 0$

$\Rightarrow \qquad\qquad (3t + 1)(t - 1) = 0$

Therefore either $3t + 1 = 0$ or $t - 1 = 0$

i.e. $\tan \dfrac{\theta}{2} = -\dfrac{1}{3}$ or 1

The range of values specified for θ is $0°$ to $360°$,

so the range of values required for $\dfrac{\theta}{2}$ is $0°$ to $180°$.

Within this range

$$\tan \frac{\theta}{2} = -\frac{1}{3} \qquad \text{gives} \qquad \frac{\theta}{2} = 161.57°$$

$$\tan \frac{\theta}{2} = 1 \qquad \text{gives} \qquad \frac{\theta}{2} = 45°$$

Thus $\theta = 323.14°, 90°$

If we now look at a very similar equation we will see that extra care is sometimes needed in using this method. Consider the equation

$$\sin \theta - \cos \theta = 1$$

Using $t = \tan \dfrac{\theta}{2}$ $\qquad \dfrac{2t}{1 + t^2} - \dfrac{1 - t^2}{1 + t^2} = 1$

$\Rightarrow$ $2t - 1 + t^2 = 1 + t^2$

Now it was seen on page 10 that if we simply cancel out the two t^2 terms, we lose the solution $t = \pm \infty,$ so the solution proceeds

$\Rightarrow$ $2t = 2$ or $t = \pm \infty$

Hence $\tan \dfrac{\theta}{2} = 1$ or $\pm \infty$

Giving $\dfrac{\theta}{2} = n\pi + \dfrac{\pi}{4}, \quad n\pi + \dfrac{\pi}{2}$

Note: When using the half angle identities, t does not always represent $\tan \dfrac{\theta}{2}$.

For instance, in solving the equation $\sin 4\theta + \tan 2\theta = 0$ we would use $t \equiv \tan 2\theta$.

THE EXPRESSION $a \cos \theta + b \sin \theta$

It is often useful to reduce $a \cos \theta + b \sin \theta$
to a single term such as $r \cos (\theta - \alpha)$.
This is possible, provided that we can find values of r and α for which

$$r[\cos \theta \cos \alpha + \sin \theta \sin \alpha] \equiv a \cos \theta + b \sin \theta$$

Comparing the coefficients of $\cos \theta$ and $\sin \theta$

we have $r \cos \alpha = a$ [1]

and $r \sin \alpha = b$ [2]

[2] ÷ [1] gives $\tan \alpha = \dfrac{b}{a}$ $\Rightarrow$

From the triangle $\cos \alpha = \dfrac{a}{\sqrt{(a^2 + b^2)}}$

Hence from [1] $r = \sqrt{a^2 + b^2}$

i.e. r is equal to the length of the hypotenuse of the triangle containing α.

Thus $a \cos \theta + b \sin \theta \equiv r \cos (\theta - \alpha)$

where $r = \sqrt{a^2 + b^2}$ and $\alpha = \arctan \dfrac{b}{a}$

Note: arctan means 'the angle with a tangent of'
                       arcsin means 'the angle with a sine of'
                       arccos means 'the angle with a cosine of'

e.g. $60° = \arctan\sqrt{3} = \arcsin \dfrac{\sqrt{3}}{2} = \arccos \dfrac{1}{2}$

EXAMPLE

Express $3 \cos \theta + 4 \sin \theta$ in the form $r \cos (\theta - \alpha)$ giving values for r and α.

Let $r[\cos \theta \cos \alpha + \sin \theta \sin \alpha] \equiv 3 \cos \theta + 4 \sin \theta$

So that $\begin{cases} r \cos \alpha = 3 \\ r \sin \alpha = 4 \end{cases}$

Thus $\qquad\qquad\qquad \tan \alpha = \frac{4}{3}$

and $\qquad\qquad\qquad\qquad r = 5$

i.e. $\qquad 3 \cos \theta + 4 \sin \theta \equiv 5 \cos (\theta - \alpha)$

where $\qquad\qquad\qquad \alpha = \arctan \frac{4}{3}$

It is sometimes more convenient to begin by comparing $a \cos \theta + b \sin \theta$ with $r \sin (\theta + \alpha)$ so that

$$r[\sin \theta \cos \alpha + \cos \theta \sin \alpha] \equiv a \cos \theta + b \sin \theta$$

Then, since $\qquad\qquad \begin{cases} r \sin \alpha = a \\ r \cos \alpha = b \end{cases}$

we get $\qquad\qquad\qquad \tan \alpha = \dfrac{a}{b}$

and $\qquad\qquad\qquad\qquad r = \sqrt{a^2 + b^2}$

Note that the value of α is not the same as it was when we used $r \cos (\theta - \alpha)$. Further variations that could be used are $r \sin (\theta - \alpha)$ and $r \cos (\theta + \alpha)$. When using this method, it is better to work from the basic comparison, as we did in the examples above, rather than to quote values of r and α. Before going any further, the reader is recommended to carry out the following transformations:

$$5 \cos \theta + 12 \sin \theta \text{ to the forms} \quad r \cos (\theta - \alpha), \ r \sin (\theta + \alpha)$$
$$7 \cos \theta - 24 \sin \theta \text{ to the forms} \quad r \cos (\theta + \alpha), \ r \sin (\theta - \alpha)$$
$$3 \sin \theta + 4 \cos \theta \quad \text{to the forms} \quad r \sin (\theta + \alpha), \ r \cos (\theta - \alpha)$$

THE GRAPH OF THE FUNCTION $a \cos \theta + b \sin \theta$

First consider the function $f(\theta) \equiv k \cos \theta \ (k > 0)$ which has the following characteristics:

(a) $f(\theta) = \quad 0 \quad$ when $\quad \theta = \dfrac{\pi}{2}, \dfrac{3\pi}{2}, \dfrac{5\pi}{2}, \text{etc.}$

(b) $f(\theta) = \quad k \quad$ when $\quad \theta = 0, 2\pi, 4\pi, \text{etc.}$

(c) $f(\theta) = -k \quad$ when $\quad \theta = \pi, 3\pi, 5\pi, \text{etc.}$

So we see that the graph of this function is very similar to a standard cosine curve but has maximum and minimum values $\pm k$. (We say that the curve has an amplitude k),

e.g. if $k = 5$, we have

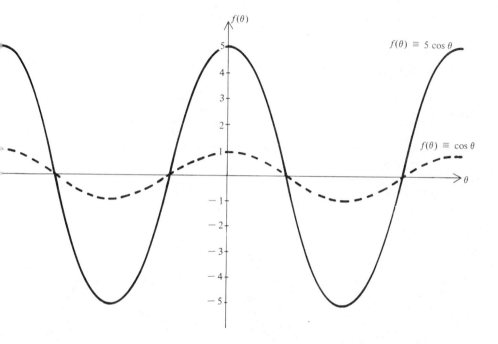

Also we saw in Chapter Six that the graph of $\cos(\theta - \alpha)$ is given by moving the graph of $\cos\theta$ a distance α to the right.

Now combining these two modifications to a standard cosine curve, the graph of the function $f(\theta) \equiv k \cos(\theta - \alpha)$ can be sketched.

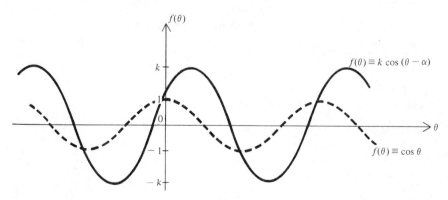

Consider now the function $a \cos \theta + b \sin \theta$.   At first sight its graph is not easy to visualise, but using $a \cos \theta + b \sin \theta \equiv r \cos (\theta - \alpha)$,   we see that the graph of $f(\theta) \equiv a \cos \theta + b \sin \theta$ is a cosine curve modified as follows:

(a) its maximum and minimum values are $\pm r$,
      i.e. its *amplitude* is r;

(b) its position is a distance α to the right of the standard curve.

EXAMPLE

Sketch the graph of the function $3 \cos \theta + 4 \sin \theta$ from $-180°$ to $+180°$.

Let $3 \cos \theta + 4 \sin \theta \equiv r \cos (\theta - \alpha)$

so that $r = 5$ and $\tan \alpha = \dfrac{4}{3} \Rightarrow \alpha = 53.13°$

Hence the graph is a cosine curve with an amplitude of 5 and a rightward phase shift of $53.13°$,

i.e.

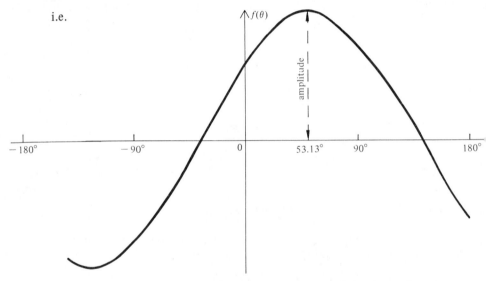

It is interesting to see how the same graph is produced if the alternative transformation $3 \cos \theta + 4 \sin \theta = r \sin (\theta + \alpha')$ is used. With this approach we have

$$r = 5 \quad \text{and} \quad \tan \alpha' = \frac{3}{4}$$

and $3 \cos \theta + 4 \sin \theta \equiv 5 \sin (\theta + \alpha')$

The R.H.S gives a sine curve with an amplitude of 5 and a displacement of α' to the left. But since $\tan \alpha' = \cot \alpha$, α' and α are complementary angles,

i.e. $\alpha' + \alpha = \dfrac{\pi}{2}$

We also know that a cosine curve is the same as a sine curve displaced $\frac{\pi}{2}$ to the left.

Thus a cosine curve moved a distance α to the right, coincides with a sine curve moved a distance α' to the left.

So any correct transformation of $a \cos \theta + b \sin \theta$ into a compound angle form provides a quick method of sketching the graph of that function and, in particular, of evaluating its maximum and minimum values (which, for sine and cosine functions, are also the greatest and least values).

THE EQUATION $a \cos \theta + b \sin \theta = c$

One way of solving an equation of this type has already been used. It depends on using the half angle formulae. An alternative method is now available using a compound angle form such as $r \cos (\theta - \alpha)$.

This approach, applied to one of the examples already solved using the 'little t' formulae, (see page 199) gives

$$\sin \theta + 2 \cos \theta = 1$$

But $2 \cos \theta + \sin \theta \equiv r[\cos \theta \cos \alpha + \sin \theta \sin \alpha] \equiv r \cos (\theta - \alpha)$

where $\begin{cases} r \cos \alpha = 2 \\ r \sin \alpha = 1 \end{cases}$

i.e. where $\tan \alpha = \frac{1}{2}$ and $r = \sqrt{5}$

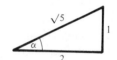

Hence $\sqrt{5} \cos (\theta - \alpha) = 1$

$$\cos (\theta - \alpha) = \frac{1}{\sqrt{5}}$$

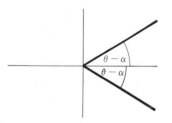

$\Rightarrow$ $\theta - \alpha = 360n° \pm 63.43°$

from which $\theta = 360n° \pm 63.43° + \alpha$

But $\alpha = \arctan \frac{1}{2} = 26.57°$

Hence $\theta = 360n° + 90°$ or $360n° - 36.86°$

Using values of n which give values of θ between $0°$ and $360°$ $(n = 0, 1)$

we have $\theta = 90°, \ 323.14°$

EXAMPLES 7e

1) Express $\sqrt{\dfrac{1-\sin 2\theta}{1+\sin 2\theta}}$ in terms of $\tan\theta$.

Using $\sin 2\theta \equiv \dfrac{2t}{1+t^2}$ where $t \equiv \tan\theta$ gives

$$1-\sin 2\theta \equiv 1-\frac{2t}{1+t^2} \equiv \frac{1+t^2-2t}{1+t^2} \equiv \frac{(1-t)^2}{1+t^2}$$

and $\qquad 1+\sin 2\theta \equiv 1+\dfrac{2t}{1+t^2} \equiv \dfrac{1+t^2+2t}{1+t^2} \equiv \dfrac{(1+t)^2}{1+t^2}$

Hence $\qquad \dfrac{1-\sin 2\theta}{1+\sin 2\theta} \equiv \dfrac{(1-t)^2}{1+t^2}\Big/\dfrac{(1+t)^2}{1+t^2} \equiv \left(\dfrac{1-t}{1+t}\right)^2$

$\Rightarrow \qquad\qquad\qquad \sqrt{\dfrac{1-\sin 2\theta}{1+\sin 2\theta}} \equiv \dfrac{1-\tan\theta}{1+\tan\theta}$

2) Find the general solution of the equation $\cos\theta - \sqrt{3}\sin\theta = 1$,
(a) by using half angle formulae,
(b) by using a compound angle transformation.

(a) $\qquad\qquad\qquad \cos\theta - \sqrt{3}\sin\theta = 1$

Therefore $\qquad \dfrac{1-t^2}{1+t^2} - \dfrac{2\sqrt{3}t}{1+t^2} = 1$ where $t \equiv \tan\dfrac{\theta}{2}$.

$\Rightarrow \qquad\qquad 1-t^2 - 2\sqrt{3}t = 1+t^2$

$\Rightarrow \qquad\qquad\qquad 2t^2 + 2\sqrt{3}t = 0$

$\Rightarrow \qquad\qquad\qquad 2t(t+\sqrt{3}) = 0$

Hence, either $\qquad\qquad\qquad t = 0 \quad\text{or}\quad t+\sqrt{3} = 0$

$\Rightarrow \qquad\qquad\qquad \tan\dfrac{\theta}{2} = 0 \quad\text{or}\quad -\sqrt{3}$

Principal values of $\dfrac{\theta}{2}$ are 0 and $-\dfrac{\pi}{3}$

So the general solution is $\dfrac{\theta}{2} = n\pi, \quad n\pi - \dfrac{\pi}{3}$

$\Rightarrow \qquad\qquad\qquad\qquad \theta = 2n\pi, \quad 2n\pi - \dfrac{2\pi}{3}$

(b) Let $\cos\theta - \sqrt{3}\sin\theta \equiv r(\cos\theta\cos\alpha - \sin\theta\sin\alpha) \equiv r\cos(\theta + \alpha)$

where $\left.\begin{array}{l} r\cos\alpha = 1 \\ r\sin\alpha = \sqrt{3} \end{array}\right\} \Rightarrow \quad \tan\alpha = \sqrt{3}$

Hence $\qquad\qquad\qquad \alpha = \dfrac{\pi}{3}$ and $r = 2$

The equation can now be written as

$$2\cos\left(\theta + \dfrac{\pi}{3}\right) = 1$$

$\Rightarrow \qquad\qquad\qquad \cos\left(\theta + \dfrac{\pi}{3}\right) = \dfrac{1}{2}$

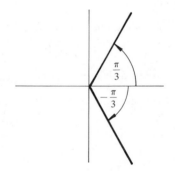

The principal value of $\theta + \dfrac{\pi}{3}$ is $\dfrac{\pi}{3}$

Therefore $\qquad\qquad\qquad \theta + \dfrac{\pi}{3} = 2n\pi \pm \dfrac{\pi}{3}$

$\Rightarrow \qquad\qquad\qquad\qquad \theta = 2n\pi \pm \dfrac{\pi}{3} - \dfrac{\pi}{3}$

$\Rightarrow \qquad\qquad\qquad\qquad \theta = \begin{cases} 2n\pi \\ 2n\pi - \dfrac{2\pi}{3} \end{cases}$

3) Express $5\sin\theta + 12\cos\theta$ in the form $r\sin(\theta + \alpha)$ giving the values of r and α.
Show that $5\sin\theta + 12\cos\theta + 7 \leqslant 20$ and find the minimum value of

$5 \sin \theta + 12 \cos \theta + 7$. Sketch the graph of the function $\dfrac{1}{5 \sin \theta + 12 \cos \theta}$ for $0 \leqslant \theta \leqslant 2\pi$.

Let $5 \sin \theta + 12 \cos \theta \equiv r[\sin \theta \cos \alpha + \cos \theta \sin \alpha] \equiv r \sin (\theta + \alpha)$

so that
$$\left. \begin{array}{r} r \cos \alpha = 5 \\[2mm] r \sin \alpha = 12 \end{array} \right\} \Rightarrow$$

giving $\qquad\qquad r = 13 \quad$ and $\quad \tan \alpha = \frac{12}{5} \Rightarrow \alpha = 67.38°$

Hence $\qquad 5 \sin \theta + 12 \cos \theta \equiv 13 \sin (\theta + \alpha)$

But $\qquad -1 \leqslant \sin (\theta + \alpha) \leqslant 1$

so $\qquad -13 \leqslant 13 \sin (\theta + \alpha) \leqslant 13$

i.e. $\qquad -13 \leqslant 5 \sin \theta + 12 \cos \theta \leqslant 13$

Adding 7 throughout gives

$$-6 \leqslant 5 \sin \theta + 12 \cos \theta + 7 \leqslant 20$$

This shows that $5 \sin \theta + 12 \cos \theta + 7 \leqslant 20$ and that the minimum value of $5 \sin \theta + 12 \cos \theta + 7$ is -6.

Now $5 \sin \theta + 12 \cos \theta \equiv 13 \sin (\theta + \alpha)$

Therefore the graph of the function $\dfrac{1}{5 \sin \theta + 12 \cos \theta}$ is also the graph of the function $\dfrac{1}{13 \sin (\theta + \alpha)}$ or $\dfrac{1}{13} \operatorname{cosec} (\theta + \alpha)$

The general shape of the graph is a typical cosec curve (see page 147) except that the maximum and minimum values are $-\frac{1}{13}$ and $+\frac{1}{13}$ and its position is a distance α to the left of the standard curve.

Thus the graph of the function $\dfrac{1}{5 \sin \theta + 12 \cos \theta}$ is deduced to be

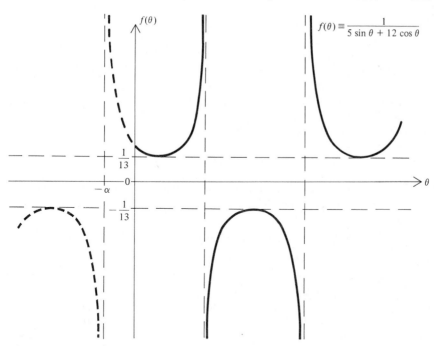

$$f(\theta) \equiv \frac{1}{5 \sin \theta + 12 \cos \theta}$$

EXERCISE 7e

1) If $\tan \theta = \dfrac{4}{3}$ and θ is acute, find the values of:

(a) $\sin 2\theta$ (b) $\tan \dfrac{\theta}{2}$ (c) $\cot 2\theta$

2) If $t \equiv \tan \dfrac{\theta}{2}$ express in terms of t:

(a) $\dfrac{1 - \cos \theta}{1 + \cos \theta}$ (b) $\dfrac{\sin \theta}{1 - \cos \theta}$ (c) $\cot \theta \cot \dfrac{\theta}{2}$

(d) $\dfrac{\cos^2 \dfrac{\theta}{2}}{3 \sin \theta + 4 \cos \theta - 1}$ (e) $\dfrac{1 - 2 \sin \theta}{2 \cos \theta + 1}$

3) Prove that $\operatorname{cosec} A + \cot A \equiv \cot \dfrac{A}{2}$

4) If $\sec \theta - \tan \theta = x$ prove that $\tan \dfrac{\theta}{2} = \dfrac{1 - x}{1 + x}$

5) Using $t \equiv \tan \dfrac{\theta}{2}$,   solve the following equations giving values of θ from $-180°$ to $180°$:

(a) $3 \cos \theta + 2 \sin \theta = 3$ (b) $5 \cos \theta - \sin \theta + 4 = 0$
(c) $\cos \theta + 7 \sin \theta = 5$ (d) $2 \cos \theta - \sin \theta = 1$

6) Transform each of the following expressions into the compound angle form suggested.

(a) $\sqrt{3} \cos \theta - \sin \theta$ $r \cos (\theta + \alpha)$
(b) $\cos \theta + 3 \sin \theta$ $r \cos (\theta - \alpha)$
(c) $4 \sin \theta - 3 \cos \theta$ $r \sin (\theta - \alpha)$
(d) $\cos 2\theta - \sin 2\theta$ $r \cos (2\theta + \alpha)$
(e) $2 \cos 3\theta + 5 \sin 3\theta$ $r \sin (3\theta + \alpha)$

7) Find the maximum and minimum values of the following functions, stating in each case the values (from $0°$ to $360°$) of θ at which the turning points occur:

(a) $\cos \theta - \sqrt{3} \sin \theta$ (b) $7 \cos \theta - 24 \sin \theta + 3$

(c) $\dfrac{1}{\cos 2\theta + \sin 2\theta}$ (d) $\dfrac{\sqrt{2}}{\cos \theta - \sqrt{2} \sin \theta}$

(e) $(3 \cos \theta + 4 \sin \theta)^2$

8) Use a compound angle transformation to find the general solution of the following equations:

(a) $\cos x + \sin x = \sqrt{2}$ (b) $7 \cos x + 6 \sin x = 2$
(c) $\cos x - 3 \sin x = 1$ (d) $2 \cos x - \sin x = 2$

9) Sketch the graphs of the functions in Question 6.

THE FACTOR FORMULAE

To factorise is to express in the form of a product. The set of identities called the factor formulae (which, the reader may be relieved to learn, is the last set in this work) converts expressions such as $\sin A + \sin B$ into a product, so *factorising* the expression.

To derive these identities, we use the compound angle group.

$$\sin A \cos B + \cos A \sin B \equiv \sin (A + B)$$

$$\sin A \cos B - \cos A \sin B \equiv \sin (A - B)$$

Adding: $2 \sin A \cos B \equiv \sin (A + B) + \sin (A - B)$ [1]

Subtracting: $2 \cos A \sin B \equiv \sin (A + B) - \sin (A - B)$ [2]

Similar treatment of $\cos (A + B)$ and $\cos (A - B)$ gives:

$$2 \cos A \cos B \equiv \cos (A + B) + \cos (A - B)$$ [3]

$$-2 \sin A \sin B \equiv \cos (A + B) - \cos (A - B)$$ [4]

The R.H.S of each of these formulae can be simplified by putting

$$\begin{cases} A + B = P \\ A - B = Q \end{cases} \Rightarrow \begin{cases} A = \frac{1}{2}(P + Q) \\ B = \frac{1}{2}(P - Q) \end{cases}$$

Then

$$\sin P + \sin Q \equiv 2 \sin \frac{P + Q}{2} \cos \frac{P - Q}{2} \qquad [5]$$

$$\sin P - \sin Q \equiv 2 \cos \frac{P + Q}{2} \sin \frac{P - Q}{2} \qquad [6]$$

$$\cos P + \cos Q \equiv 2 \cos \frac{P + Q}{2} \cos \frac{P - Q}{2} \qquad [7]$$

$$\cos P - \cos Q \equiv -2 \sin \frac{P + Q}{2} \sin \frac{P - Q}{2} \qquad [8]$$

Identities [5]–[8] are best used when a sum or difference is to be expressed as a product, while identities [1]–[4] should be used when a given product is to be changed to a sum or difference,

e.g. to express $\sin 6\theta - \sin 4\theta$ as a product we would use [6] to give

$$2 \cos \frac{6\theta + 4\theta}{2} \sin \frac{6\theta - 4\theta}{2} \equiv 2 \cos 5\theta \sin \theta$$

But to express $2 \cos 7\theta \cos 2\theta$ as a sum we would use [3] to give

$$\cos(7\theta + 2\theta) + \cos(7\theta - 2\theta) \equiv \cos 9\theta + \cos 5\theta$$

When these identities are being used regularly, it is not too difficult to remember them. Most people find it best to memorise them in words rather than as symbols,
e.g. [5] would be remembered in the form:

sum of sines ≡ twice sin(semi-sum) cos(semi difference)

And [1] would be:

twice sin cos ≡ sin (sum) + sin (difference)

Note: Numbers [4] and [8] require special care because of the minus sign. This group of identities will prove to be particularly useful when integrating certain trig functions later on.

EXAMPLES 7f

1) Prove that $\dfrac{\sin A + \sin B}{\cos A + \cos B} \equiv \tan \dfrac{A + B}{2}$

If A, B and C are the angles of a triangle, deduce that

$$\frac{\sin A + \sin B}{\cos A + \cos B} = \cot \frac{C}{2}$$

Considering the L.H.S

$$\frac{\sin A + \sin B}{\cos A + \cos B} \equiv \frac{2 \sin \dfrac{A+B}{2} \cos \dfrac{A-B}{2}}{2 \cos \dfrac{A+B}{2} \cos \dfrac{A-B}{2}} \qquad \begin{bmatrix} [5] \\ \\ [7] \end{bmatrix}$$

$$\equiv \tan \frac{A+B}{2}$$

If A, B and C are angles in a triangle

$$A + B + C = 180°$$

$$\Rightarrow \qquad \frac{A+B}{2} + \frac{C}{2} = 90°$$

$$\Rightarrow \qquad \left(\frac{A+B}{2}\right) \text{ and } \frac{C}{2} \text{ are complementary}$$

therefore

$$\tan \frac{A+B}{2} = \cot \frac{C}{2}$$

$$\Rightarrow \qquad \frac{\sin A + \sin B}{\cos A + \cos B} = \cot \frac{C}{2}$$

2) Solve the equation $\sin 5x - \sin 3x = 0$ giving the general solution.

$$\sin 5x - \sin 3x = 0$$

Therefore

$$2 \cos \frac{5x + 3x}{2} \sin \frac{5x - 3x}{2} = 0$$

Therefore   either $\cos 4x = 0$ or $\sin x = 0$

If $\cos 4x = 0$,   let $\theta = 4x$ so that $\cos \theta = 0$

then

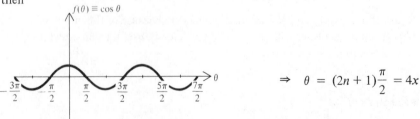

$$\Rightarrow \quad \theta = (2n + 1)\frac{\pi}{2} = 4x$$

If $\sin x = 0$

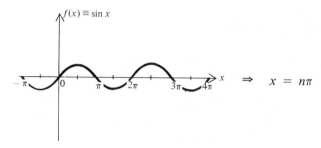

$$\Rightarrow \quad x = n\pi$$

The general solution is therefore

$$x = (2n + 1)\frac{\pi}{8}, \; n\pi$$

3) Factorise $\cos\theta - \cos 3\theta - \cos 5\theta + \cos 7\theta$

Grouping in pairs we have

$$(\cos 7\theta + \cos\theta) - (\cos 5\theta + \cos 3\theta) \equiv f(\theta)$$

But
$$\cos 7\theta + \cos\theta \equiv 2 \cos\frac{7\theta + \theta}{2} \cos\frac{7\theta - \theta}{2}$$

$$\equiv 2 \cos 4\theta \cos 3\theta$$

and
$$\cos 5\theta + \cos 3\theta \equiv 2 \cos\frac{5\theta + 3\theta}{2} \cos\frac{5\theta - 3\theta}{2}$$

$$\equiv 2 \cos 4\theta \cos\theta$$

So
$$f(\theta) \equiv 2 \cos 4\theta\,(\cos 3\theta - \cos\theta)$$

$$\equiv 2 \cos 4\theta \left(-2 \sin\frac{3\theta + \theta}{2} \sin\frac{3\theta - \theta}{2}\right)$$

$$\Rightarrow \quad \cos\theta - \cos 3\theta - \cos 5\theta + \cos 7\theta \equiv -4\cos 4\theta \sin 2\theta \sin\theta$$

Note. Other groupings of the four given terms can be used, but arranging them so that pairs of cosines are *added* makes the factorising simplest.

4) If A, B, C are the angles of a triangle show that

$$\sin A + \sin B + \sin C = 4 \cos\frac{A}{2} \cos\frac{B}{2} \cos\frac{C}{2}$$

Considering the L.H.S

$$\sin A + \sin B + \sin C = (\sin A + \sin B) + \sin C$$

$$= 2 \sin \frac{A+B}{2} \cos \frac{A-B}{2} + 2 \sin \frac{C}{2} \cos \frac{C}{2}$$

Now $A + B + C = 180° \Rightarrow \dfrac{A+B}{2} + \dfrac{C}{2} = 90°$

Therefore $\sin \dfrac{A+B}{2} = \cos \dfrac{C}{2}$ and $\sin \dfrac{C}{2} = \cos \dfrac{A+B}{2}$

So $\sin A + \sin B + \sin C = 2 \cos \dfrac{C}{2} \cos \dfrac{A-B}{2} + 2 \cos \dfrac{A+B}{2} \cos \dfrac{C}{2}$

$$= 2 \cos \frac{C}{2} \left[\cos \frac{A-B}{2} + \cos \frac{A+B}{2} \right]$$

But $\cos \dfrac{A-B}{2} + \cos \dfrac{A+B}{2} = 2 \cos \dfrac{1}{2} \left(\dfrac{A-B}{2} + \dfrac{A+B}{2} \right) \cos \dfrac{1}{2} \left(\dfrac{A-B}{2} - \dfrac{A+B}{2} \right)$

$$= 2 \cos \frac{A}{2} \cos \left(-\frac{B}{2} \right)$$

$$= 2 \cos \frac{A}{2} \cos \frac{B}{2}$$

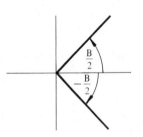

So $\qquad \sin A + \sin B + \sin C = 4 \cos \dfrac{A}{2} \cos \dfrac{B}{2} \cos \dfrac{C}{2}$

EXERCISE 7f

1) Factorise:
(a) $\sin 3A + \sin A$
(b) $\cos 5A + \cos 3A$
(c) $\sin 4A - \sin 2A$
(d) $\cos 7A - \cos A$
(e) $\sin 3A - \sin 5A$
(f) $\cos A - \cos 5A$
(g) $\sin 30° + \sin 60°$
(h) $\cos 70° + \cos 50°$
(i) $\sin 2A + 1$ (*Hint*: $\sin 90° = 1$)
(j) $1 + \cos 4A$

2) Express as a sum or difference:

(a) $2 \sin 2\theta \cos \theta$ (b) $2 \cos 3\theta \cos 2\theta$ (c) $2 \cos \theta \sin 4\theta$

(d) $-2 \sin 3\theta \sin \theta$ (e) $2 \sin 4\theta \sin 2\theta$ (f) $\cos \theta \cos 4\theta$

(g) $2 \sin 30° \cos 60°$ (h) $2 \cos 20° \cos 40°$

3) Prove the following identities:

(a) $\dfrac{\sin 2A + \sin 2B}{\sin 2A - \sin 2B} \equiv \dfrac{\tan (A + B)}{\tan (A - B)}$

(b) $\dfrac{\cos 2A + \cos 2B}{\cos 2B - \cos 2A} \equiv \cot (A + B) \cot (A - B)$

(c) $\dfrac{\sin A \sin 2A + \sin 3A \sin 6A}{\sin A \cos 2A + \sin 3A \cos 6A} \equiv \tan 5A$

(d) $\dfrac{\sin 3x + \sin 5x}{\sin 4x + \sin 6x} \equiv \dfrac{\sin 4x}{\sin 5x}$

(e) $\dfrac{\sin A + \sin 3A + \sin 5A}{\cos A + \cos 3A + \cos 5A} \equiv \tan 3A$

(f) $\dfrac{\sin A - \sin B}{\sin A + \sin B} \equiv \cot \dfrac{A + B}{2} \tan \dfrac{A - B}{2}$

(g) $\sin \theta + \sin 2\theta + \sin 3\theta \equiv \sin 2\theta (1 + 2 \cos \theta)$

(h) $1 + 2 \cos 2A + \cos 4A \equiv 4 \cos^2 A \cos 2A$

(i) $\cos 2\theta + \cos 4\theta + \cos 6\theta + \cos 12\theta \equiv 4 \cos 3\theta \cos 4\theta \cos 5\theta$

(j) $\dfrac{\cos A - \cos B}{\sin A + \sin B} \equiv \tan \dfrac{B - A}{2}$

4) Simplify:

(a) $\cos (\theta - 60°) + \cos (\theta + 60°)$

(b) $\sqrt{3} \cos x - \sin (x + 60°) - \sin (x + 120°)$

5) If A, B and C are the angles of a triangle, prove that:

(a) $\cos (B + C) = -\cos A$ (b) $\sin C = \sin (A + B)$

(c) $\sin \dfrac{A + B}{2} = \cos \dfrac{C}{2}$ (d) $\sin \dfrac{B}{2} = \cos \dfrac{A + C}{2}$

(e) $\sin B + \sin (A - C) = 2 \sin A \cos C$

(f) $\cos (A - B) - \cos C = 2 \cos A \cos B$

(g) $\sin (A + B) + \sin (B + C) = 2 \cos \dfrac{B}{2} \cos \dfrac{A - C}{2}$

(h) $\cos A + \cos B + \cos C = 1 + 4 \sin \dfrac{A}{2} \sin \dfrac{B}{2} \sin \dfrac{C}{2}$

(i) $\sin 2A + \sin 2B + \sin 2C = 4 \sin A \sin B \sin C$

(j) $\sin \dfrac{A}{2} - \cos \dfrac{B-C}{2} = -2 \sin \dfrac{B}{2} \sin \dfrac{C}{2}$

(k) $1 + \cos 2C - \cos 2A - \cos 2B = 4 \sin A \sin B \cos C$

Solve the following equations, giving values from $0°$ to $360°$:

6) $\cos 2x + \cos 4x = 0$

7) $\sin 3x - \sin x = 0$

8) $\sin 4\theta + \sin 2\theta = 0$

9) $\cos x = \cos 2x + \cos 4x$

10) $\cos x + \cos 3x = \sin x + \sin 3x$

11) $\sin 3\theta + \sin 6\theta + \sin 9\theta = 0$

12) $\sin 3\theta - \sin \theta = \cos 2\theta$

13) $\cos 5\theta - \cos \theta = \sin 3\theta$

14) $\cos 2x = \cos (30° - x)$

SUMMARY

$$\tan \theta \equiv \frac{\sin \theta}{\cos \theta}$$

$$\begin{cases} \cos^2\theta + \sin^2\theta \equiv 1 \\ \tan^2\theta + 1 \equiv \sec^2\theta \\ \cot^2\theta + 1 \equiv \operatorname{cosec}^2\theta \end{cases}$$

$$\begin{cases} \sin (A \pm B) \equiv \sin A \cos B \pm \cos A \sin B \\ \cos (A \pm B) \equiv \cos A \cos B \mp \sin A \sin B \\ \tan (A \pm B) \equiv \dfrac{\tan A \pm \tan B}{1 \mp \tan A \tan B} \end{cases}$$

$$\begin{cases} \sin 2A \equiv 2 \sin A \cos A \\ \cos 2A \equiv \cos^2A - \sin^2A \equiv 2 \cos^2A - 1 \equiv 1 - 2 \sin^2A \\ \tan 2A \equiv \dfrac{2 \tan A}{1 - \tan^2A} \end{cases}$$

$$\begin{cases} \sin^2\theta \equiv \tfrac{1}{2}(1 - \cos 2\theta) \\ \cos^2\theta \equiv \tfrac{1}{2}(1 + \cos 2\theta) \end{cases}$$

$$\begin{cases} \sin P + \sin Q \equiv 2 \sin \dfrac{P+Q}{2} \cos \dfrac{P-Q}{2} \\[2mm] \sin P - \sin Q \equiv 2 \cos \dfrac{P+Q}{2} \sin \dfrac{P-Q}{2} \\[2mm] \cos P + \cos Q \equiv 2 \cos \dfrac{P+Q}{2} \cos \dfrac{P-Q}{2} \\[2mm] \cos P - \cos Q \equiv -2 \sin \dfrac{P+Q}{2} \sin \dfrac{P-Q}{2} \end{cases}$$

$$\begin{cases} 2 \sin A \cos B \equiv \sin (A + B) + \sin (A - B) \\ 2 \cos A \sin B \equiv \sin (A + B) - \sin (A - B) \\ 2 \cos A \cos B \equiv \cos (A + B) + \cos (A - B) \\ -2 \sin A \sin B \equiv \cos (A + B) - \cos (A - B) \end{cases}$$

$$\begin{cases} \sin \theta \equiv \dfrac{2t}{1 + t^2} \\[2mm] \cos \theta \equiv \dfrac{1 - t^2}{1 + t^2} \quad \text{where} \quad t \equiv \tan \dfrac{\theta}{2} \\[2mm] \tan \theta \equiv \dfrac{2t}{1 - t^2} \end{cases}$$

$$\sin 3\theta \equiv 3 \sin \theta - 4 \sin^3 \theta$$

$$\cos 3\theta \equiv 4 \cos^3 \theta - 3 \cos \theta$$

MULTIPLE CHOICE EXERCISE 7

(*Instructions for answering these questions are given on p. xii.*)

TYPE I

1) $\cos (A + B) + \cos (A - B) \equiv$

(a) $2 \cos A \sin B$ (b) $-2 \sin A \cos B$ (c) $2 \cos A \cos B$ (d) $-2 \sin A \sin B$.

2) Using $t \equiv \tan \dfrac{\theta}{2}$ converts the equation $2 \cos \theta + 3 \sin \theta + 4 = 0$ into:

(a) $2t^2 + 3t + 6 = 0$ (b) $3 + 3t + t^2 = 0$

(c) $6 + 6t - 2t^2 = 0$ (d) $t^2 + 6t + 5 = 0$.

3) One of the following expressions is not identical to any of the others. Which one is it?

(a) $\dfrac{2 \tan \theta}{1 + \tan^2 \theta}$ (b) $2 \cos^2 \dfrac{\theta}{2}$ (c) $1 - \sin^2 \theta$

(d) $\dfrac{1}{1 + \tan^2 \theta}$ (e) $\sin 2\theta$.

4)

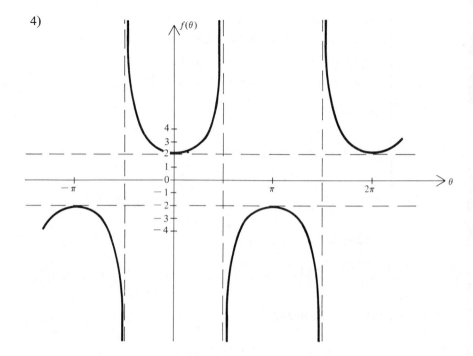

$f(\theta)$ could be:
(a) cosec 2θ (b) 2 sec θ (c) sec 2θ (d) 2 cosec θ (e) 2 cot θ

5) The greatest value of $5 \cos \theta - 4 \sin \theta$ is:
(a) 3   (b) 1   (c) $\sqrt{41}$ (d) ± 5 (e) $\pm\sqrt{41}$.

TYPE II

6) $3 \cos \theta - 4 \sin \theta \equiv$
(a) $5 \cos (\theta + \alpha)$ where $\tan \alpha = \frac{3}{4}$ (b) $5 \sin (\alpha - \theta)$ where $\tan \alpha = \frac{3}{4}$
(c) $5 \cos (\theta + \alpha)$ where $\tan \alpha = \frac{4}{3}$ (d) $-5 \cos (\theta - \alpha)$ where $\tan \alpha = \frac{4}{3}$.

7) The general solution of the equation $\cos 2\theta = \frac{1}{2}$ is the same as the solution of the equation

(a) $\sin 2\theta = \dfrac{\sqrt{3}}{2}$ (b) $\tan 2\theta = \sqrt{3}$

(c) $\cos \theta = \frac{1}{4}$ (d) $\cos (-2\theta) = \frac{1}{2}$ (e) $4 \cos^2 \theta = 3$.

8) If $x = 1 - \tan \theta$ and $y = \sec \theta$ the Cartesian equation given by eliminating θ is:
(a) $x^2 + y^2 = 2x$ (b) $x^2 - y^2 = 2x$ (c) $x^2 - y^2 + 2 = 2x$
(d) $(1 - x)^2 = (y - 1)(y + 1)$ (e) $(x - 1)^2 = (1 - y)(1 + y)$.

9) $\sin\left(\theta - \dfrac{\pi}{2}\right) \equiv$

(a) $\cos\left(\theta + \dfrac{\pi}{2}\right)$ (b) $\sin\left(\dfrac{\pi}{2} - \theta\right)$ (c) $\cos\theta$ (d) $\sin\left(\theta + \dfrac{3\pi}{2}\right)$

10) If $f(\theta) \equiv \cos\theta - \sin\theta$

(a) $-1 \leqslant f(\theta) \leqslant 1$ (b) $f(\theta)$ is cyclic

(c) $[f(\theta)]^2 \equiv 1 - \sin 2\theta$ (d) The amplitude of $f(\theta)$ is $\sqrt{2}$.

11)

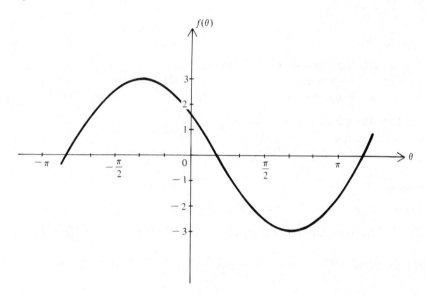

In the sketch $f(\theta)$ could be:

(a) $3 \sin\left(\theta - \dfrac{\pi}{6}\right)$ (b) $3 \cos\left(\theta - \dfrac{\pi}{6}\right)$ (c) $3 \sin\left(\theta - \dfrac{5\pi}{6}\right)$

(d) $3 \cos\left(\theta + \dfrac{\pi}{6}\right)$ (e) $3 \cos\left(\theta + \dfrac{\pi}{3}\right)$.

TYPE III

12) (a) $\theta = n\pi$

(b) $\cos\theta = 0$

13) (a) $\cos A = \sin B$

(b) $A + B = 90°$

14) (a) $f(\theta) \equiv \cos \theta$

(b) $-1 \leqslant f(\theta) \leqslant 1$

15) (a) $f(\theta) \equiv \sqrt{\dfrac{1 - \cos \theta}{1 + \cos \theta}}$

(b) $f(\theta) \equiv \tan \dfrac{\theta}{2}$

16) (a) $\sin 3\theta = x$

(b) $3 \sin \theta = x$

TYPE IV

17) Identify the function $f(\theta)$ given that:

(a) $f(\theta)$ is a trig function   (b) $f(\theta)$ is cyclic

(c) $f(\theta)$ is continuous   (d) $f(\theta) = 1$ when $\theta = 0$.

18) Find the value of θ if:

(a) $\cos 2\theta$ is given   (b) $0 < \theta < 180°$

(c) $\cos \theta$ is positive   (d) $\cos 2\theta \equiv 2 \cos^2 \theta - 1$

TYPE V

19) The graph of $\tan \theta$ is continuous because it is cyclic.

20) The general solution of the equation $\cos \theta = -1$ is $\theta = (2n + 1)\pi$.

21) $\sin 2\theta \equiv \dfrac{2t}{1 + t^2}$ where $t \equiv \tan \dfrac{\theta}{2}$.

22) $\cos 4\theta = \sin 3\theta \;\Rightarrow\; \cos 4\theta = \sin\left(3\theta - \dfrac{\pi}{2}\right)$.

MISCELLANEOUS EXERCISE 7

1) (a) If $\sin \alpha = \tfrac{2}{3}$ and $\cos \beta = -\tfrac{2}{7}$,   find the possible values of $\cos (\alpha + \beta)$.

(b) Find the values of θ between $-180°$ and $180°$ which satisfy the equation $3 \cos \theta - 5 \sin \theta = 2$.   (C)

2) (a) Solve the equations:

(i) $\cos \dfrac{3x}{4} = \tan 163°$ (ii) $7 \cos x - 24 \sin x = 12.5$

giving in each case all the solutions between $0°$ and $360°$.

(b) Without using tables find the numerical value of

$$\sin^2\frac{\pi}{8} - \cos^4\frac{3\pi}{8}$$ (U of L)

3) (a) If A is the acute angle such that $\sin A = \frac{3}{5}$ and B is the obtuse angle such that $\sin B = \frac{5}{13}$,   find without using tables the values of $\cos(A + B)$ and $\tan(A - B)$.

(b) Find the solutions of the equation $\tan\theta + 3\cot\theta = 5\sec\theta$ for which $0 < \theta < 2\pi$.   (U of L)

4) Calculate the values of θ in the range $0 \leqslant \theta \leqslant 180°$ which satisfy the equations:

(a) $2\sin\theta + \cos\theta = 1$ (b) $2\sin\theta + \cos 2\theta = 1$.   (AEB)'72p

5) (a) Find the values of θ between $0°$ and $180°$ for which $\tan^2\theta = 5 - \sec\theta$.

(b) Find the values of θ between $0°$ and $360°$ which satisfy the equation

$$6\cos\theta + 7\sin\theta = 4$$ (C)

6) (a) Find the values of x between $0°$ and $360°$ which satisfy the equation

$$\sin 2x + 2\cos 2x = 1$$

(b) Find the general solution of the equation $\cos 3x + \cos x = \sin 2x$.   (U of L)p

7) (a) If $\sin(\theta - \alpha) = k\sin(\theta + \alpha)$ find $\tan\theta$ in terms of $\tan\alpha$ and k and so determine the possible values of θ between 0 and $360°$ when $k = \frac{1}{2}$ and $\alpha = 150°$.

(b) Show without the use of tables or calculator, that $x = \dfrac{\pi}{10}$ satisfies the equation $\cos 3x = \sin 2x$.   By expressing this equation in terms of $\sin x$ and $\cos x$ show that $\sin\dfrac{\pi}{10}$ is a root of the equation

$$4s^2 + 2s - 1 = 0$$ (C)

8) (a) Find the values of θ between 0 and 2π for which $\sin 2\theta = \sin\dfrac{\pi}{6}$.

(b) Show that $(2\cos\phi + 3\sin\phi)^2 \leqslant 13$ for all values of ϕ.   (U of L)p

9) Find all the values of θ in the range $0 \leqslant \theta \leqslant 2\pi$ for which

$$\sin\theta + \sin 3\theta = \cos\theta + \cos 3\theta$$ (JMB)

10) Find, to the nearest minute, the acute angle α for which

$$4\cos\theta - 3\sin\theta \equiv 5\cos(\theta + \alpha)$$

Calculate the values of θ in the interval $-180° \leqslant \theta \leqslant 180°$ for which the function $f(\theta) \equiv 4 \cos \theta - 3 \sin \theta - 4$ attains its greatest value, its least value and the value zero.   (JMB)

11) (a) Prove that $(\sin 2\theta - \sin \theta)(1 + 2 \cos \theta) \equiv \sin 3\theta$.
    (b) Find the values of x between $0°$ and $360°$ which satisfy the equation
$$3 \cos x + 1 = 2 \sin x \qquad \text{(C)}$$

12) Find all the solutions of the following equations for which $-180° < \theta \leqslant 180°$.
(a) $3 \sin \theta + 4 \cos \theta = 2$ (b) $7 \tan 2\theta + 4 \sin \theta = 0$ (JMB)p

13) Prove that $\sec x + \tan x = \tan \left(\dfrac{\pi}{4} + \dfrac{x}{2} \right)$ and deduce a similar expression for $\sec x - \tan x$.

Hence find in surd form the values of $\tan \dfrac{7\pi}{12}$ and $\tan \dfrac{\pi}{12}$.   (AEB)'75p

14) (a) Prove that $\cos 3\theta - \sin 3\theta \equiv (\cos \theta + \sin \theta)(1 - 4 \cos \theta \sin \theta)$.
    (b) Prove that if $\sec A = \cos B + \sin B$

      (i) $\tan^2 A = \sin 2B$ (ii) $\cos 2A = \tan^2 \left(\dfrac{\pi}{4} - B \right)$ (C)

15) (a) Find, in radians, the general solution of the equation
$$\sin x + \sin 2x = \sin 3x$$

    (b) By expressing $\sec 2x$ and $\tan 2x$ in terms of $\tan x$, or otherwise, solve the equation $2 \tan x + \sec 2x = 2 \tan 2x$, giving all solutions between $-180°$ and $+180°$.   (U of L)

16) Express $\sqrt{3} \sin \theta - \cos \theta$ in the form $R \sin (\theta - \alpha)$ where R is positive. Find all values of θ in the range $0° \leqslant \theta \leqslant 360°$ which satisfy the equation
$$4 \sin \theta \cos \theta = \sqrt{3} \sin \theta - \cos \theta \qquad \text{(JMB)}$$

17) (a) Prove that $(\cot \theta + \operatorname{cosec} \theta)^2 \equiv \dfrac{1 + \cos \theta}{1 - \cos \theta}$ and hence, or otherwise,

    solve the equation $(\cot 2\theta + \operatorname{cosec} 2\theta)^2 = \sec 2\theta$ for values of θ between $0°$ and $180°$.
    (b) Find the general solution of the equation $\sin 2x + \sin 3x + \sin 5x = 0$.
                                      (AEB)'73

18) By using the formulae expressing $\sin \theta$ and $\cos \theta$ in terms of $t \left(\equiv \tan \dfrac{\theta}{2} \right)$ or otherwise, show that $\dfrac{1 + \sin \theta}{5 + 4 \cos \theta} \equiv \dfrac{(1 + t)^2}{9 + t^2}$.

Deduce that $0 \leqslant \dfrac{1 + \sin \theta}{5 + 4 \cos \theta} \leqslant \dfrac{10}{9}$ for all values of θ.   (C)

19) (a) Show that $\cos^6 x + \sin^6 x \equiv 1 - \frac{3}{4}\sin^2 2x$.

   (b) Solve, for $\quad 0° \leqslant x \leqslant 180°,\quad$ the equation $\quad \sin x + \sin 5x = \sin 3x$.

   (c) Find the general solution of the equation $\quad 3\cos x + 4\sin x = 2$.

(U of L)

20) Prove that $\quad \csc\theta + \cot\theta \equiv \cot\dfrac{\theta}{2}$.

Hence (a) deduce the values, in surd form, of $\cot\dfrac{\pi}{8}$ and $\cot\dfrac{\pi}{12}$

   (b) express $\csc\theta + \csc 2\theta + \csc 4\theta$ as the difference of two cotangents.

   (c) prove, without using tables (or calculator), that

$$\csc\frac{4\pi}{15} + \csc\frac{8\pi}{15} + \csc\frac{16\pi}{15} + \csc\frac{32\pi}{15} = 0 \qquad\text{(C)}$$

21) (Tables should not be used for this question.)

Prove that $\quad \tan 3\theta \equiv \dfrac{3t - t^3}{1 - 3t^2},\quad$ where $\quad t \equiv \tan\theta$.

Hence, or otherwise, show that $\quad \tan\frac{1}{12}\pi = 2 - \sqrt{3}$.

Give the angle θ, between 0 and $\frac{1}{2}\pi$, for which $\tan\theta = 2 + \sqrt{3}$.   (O)

22) (a) Express $\quad 7\sin x - 24\cos x\quad$ in the form $\quad R\sin(x - \alpha),\quad$ where R is positive and α is an acute angle.

      Hence or otherwise solve the equation

$$7\sin x - 24\cos x = 15, \quad \text{for}\quad 0° < x < 360°$$

   (b) Solve the simultaneous equations

$$\cos x + \cos y = 1, \quad \sec x + \sec y = 4$$

for $\quad 0° < x < 180°,\quad 0° < y < 180°$.   (AEB)'76

23) Express $\quad \cos 2x - \sin 2x\quad$ in the form $\quad R\cos(2x + \alpha),\quad$ giving values of R and α.

Hence find the general solution of each of the following equations:

(a) $\cos 2x - \sin 2x = 1$ (b) $\cos 2x - \sin 2x = \sqrt{2}\cos 4x$.   (U of L)p

24) (a) Find in the range $\quad -180° < x < 180°\quad$ the solutions of the equation $\quad \cos 5x = \cos x$.

   (b) Prove that $\quad \dfrac{1 + \cos\theta + \sin\theta}{1 - \cos\theta + \sin\theta} \equiv \dfrac{1 + \cos\theta}{\sin\theta}$ (JMB)p

25) (a) Find, in radians, the general solution of the equation $\quad 4\sin\theta = \sec\theta$.

   (b) If $\quad \sin\theta + \sin 2\theta + \sin 3\theta + \sin 4\theta = 0,\quad$ show that θ is either a multiple of $\frac{1}{2}\pi$ or a multiple of $\frac{2}{5}\pi$.   (U of L)

26) (a) Find the values of x, for angles between $0°$ and $360°$ inclusive, for which $\quad 3\sin 2x = 2\tan x$.

(b) Find the values of x, for angles between $0°$ and $360°$ inclusive, for which $4 \cos x - 6 \sin x = 5$.   (C)

27) (a) Find the general solution of the equation

$$10 \sin \frac{\pi x}{3} + 24 \cos \frac{\pi x}{3} = 13$$

(b) Solve the equation $2 \cos \theta \cos 2\theta + \sin 2\theta = 2(3 \cos^3\theta - \cos \theta)$ for values of θ within the range $0 < \theta < 2\pi$.   (AEB)'67

28) (a) Find, in radians, the general solution of the equation $2 \sin \theta = \sqrt{3} \tan \theta$.

(b) Express $4 \sin \theta - 3 \cos \theta$ in the form $R \sin (\theta - \alpha)$, where α is an acute angle.

   (i) Solve the equation $4 \sin \theta - 3 \cos \theta = 3$, giving all solutions between $0°$ and $360°$.

   (ii) Find the greatest and least values of $\dfrac{1}{4 \sin \theta - 3 \cos \theta + 6}$ (U of L)

29) (a) Given $x = 2 \sin \left(nt + \dfrac{\pi}{3}\right)$ and $y = 4 \sin \left(nt + \dfrac{\pi}{6}\right)$, express x and y in terms of $\sin nt$ and $\cos nt$. Find the Cartesian equation of the locus of the point (x, y) as t varies.

(b) Solve the equation $3 \cos^2\theta + 5 \sin \theta - 1 = 0$ for $0° < \theta < 360°$.   (AEB)'75

30) (a) Prove that $\sin^2 2\theta (\cot^2\theta - \tan^2\theta) = 4 \cos 2\theta$.

(b) Solve the equation $\sec \theta \tan \theta = 2$, giving solutions for $0° \leqslant \theta < 360°$.   (C)

31) Write down the expansions of $\cos (A + B)$ and $\cos (A - B)$ in terms of cosines and sines of A and B.

(a) Find angles x and y, each between $0°$ and $90°$, which satisfy the simultaneous equations $\cos x \cos y = 0.6$, $\sin x \sin y = 0.2$.

(b) Prove that $\cos 3x \equiv 4 \cos^3x - 3 \cos x$. Hence find all the solutions, in the range $-180° < x \leqslant + 180°$, of the equation $2 \cos 3x + \cos 2x + 1 = 0$.   (JMB)

32) (a) Solve the equations for values of x between $0°$ and $360°$:

   (i) $\cos 2x° + \sin x° = 0$

   (ii) $\sin x° - \sin 2x° + \sin 3x° = 0$.

(b) If $A = 36°$, show that $\sin 3A = \sin 2A$, and deduce that

$$\cos 36° = \frac{\sqrt{5} + 1}{4}$$

(c) If $A + B = \dfrac{\pi}{4}$ and $\tan A = \dfrac{n}{n + 1}$, find $\tan B$ and $\tan (A - B)$.

33) (a) Solve the equation $\sin 4x = \cos x$ for values of x between $0°$ and $180°$.

 (b) If $2 \sin 2x + \cos 2x = k$,   show that
 $(1 + k) \tan^2 x - 4 \tan x - 1 + k = 0$.

 Hence, or otherwise, show that if $\tan x_1$ and $\tan x_2$ are the roots of this quadratic equation in $\tan x$, then $\tan (x_1 + x_2) = 2$.   (AEB)'72

34) (a) Solve the equation $1 - 3 \cos^2\theta = 5 \sin \theta$ for $0° \leqslant \theta \leqslant 360°$.

 (b) State the formula for $\tan (x + y)$ in terms of $\tan x$ and $\tan y$.

 If $2x + y = \dfrac{\pi}{4}$,   show that $\tan y = \dfrac{1 - 2 \tan x - \tan^2 x}{1 + 2 \tan x - \tan^2 x}$

 Deduce that $\tan \dfrac{\pi}{8}$ is a root of the equation $t^2 + 2t - 1 = 0$ and that its value is $\sqrt{2} - 1$.   (AEB)'71

CHAPTER 8

FURTHER TRIGONOMETRY

EQUALITY OF FUNCTIONS

Consider two functions $f(x) \equiv \sin x$ and $g(x) \equiv \dfrac{x}{2}$. When plotted on Cartesian axes, the curve with equation $y = f(x)$ is a sine wave, while the graph of $y = g(x)$ is a straight line through the origin.

Let us examine the two graphs when they are sketched on the same axes as shown

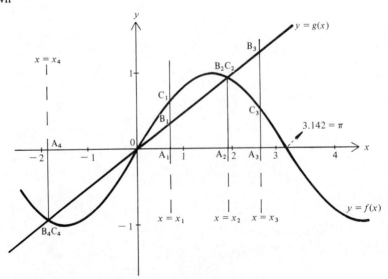

It can be seen that:

when $x = x_1$ the value of $f(x)$ is represented by A_1C_1
and the value of $g(x)$ is represented by A_1B_1

thus $f(x_1) \neq g(x_1)$

when $x = x_3$ the value of $f(x)$ is represented by A_3C_3
and the value of $g(x)$ is represented by A_3B_3

thus $f(x_3) \neq g(x_3)$

when $x = x_2$ the value of $f(x)$ is represented by A_2B_2
and the value of $g(x)$ is represented by A_2C_2

But $A_2B_2 = A_2C_2$

thus $f(x_2) = g(x_2)$.

A similar argument shows that when $x = x_4$ also, $f(x_4) = g(x_4)$,
i.e. whenever two functions $f(x)$ and $g(x)$ plotted on the same axes, intersect
at a point where $x = a$ say, then $f(a) = g(a)$. For any value of x for
which the curves do not intersect, $x = b$ say, $f(b) \neq g(b)$. Thus the set
of values of x for which the two graphs intersect is the set of values of x
satisfying the equation $f(x) = g(x)$.

The equation $\sin x = \dfrac{x}{2}$ can therefore be solved by finding the x coordinates

of the points of intersection of the graphs $y = \sin x$ and $y = \dfrac{x}{2}$.

There are three such values of x and no more, as can be seen by *sketching* the
two graphs over a wide range.

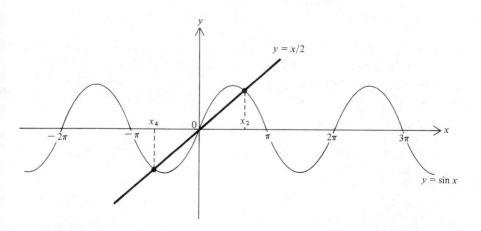

Therefore the equation $\sin x = \dfrac{x}{2}$ has three solutions (roots)

$$x = x_2, \quad x = 0 \quad \text{and} \quad x = x_4$$

However the actual values of x_2 and x_4 cannot be found from a *sketch*.
By *plotting* the curves between $-\pi$ and π (the sketch shows that all the roots are within this range), approximate values for x_2 and x_4 can be found. Great accuracy at this stage is not necessary as it is used for a first approximation; a more accurate stage comes later. The following table is quite adequate.

x (radians)	0	± 1	± 2	± 3
$\sin x$	0	± 0.84	± 0.91	± 0.14
$\dfrac{x}{2}$	0	$\pm \frac{1}{2}$	no more points are needed for a line	

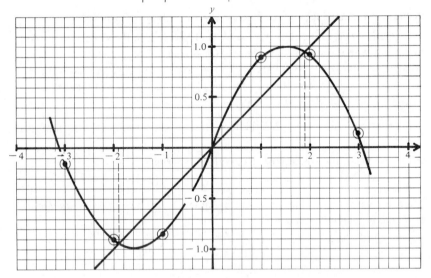

Now it is clear that the roots x_2 and x_4 are approximately ± 1.9. An accurate plot of the graphs in the region of each of these values then gives a very good approximation to the value of each root, e.g. for the positive root we could work within the range $1.7 \leqslant x \leqslant 2.1$, calculating values of $\sin x$ at intervals of 0.05^c (the graph of $y = \dfrac{x}{2}$ requires only two points in any case).

So the graphical method of solution of an equation $f(x) = g(x)$ usually proceeds as follows:

1) A *sketch* of the two graphs on the same axes indicates the number and rough positions of the roots of the equation.

2) Choosing a range indicated by the sketch, a *plot* of the two graphs on the same axes gives a fair approximation for the value of each root within that range.

3) Using a larger scale and a smaller range near to each required root, the abscissa of a point of intersection of the two graphs now gives a very good approximation for that root.

Note: It is sometimes possible to combine two of the steps listed above, but in all cases a sketch should be made first. This is because deductions can be made from a sketch, whose range is unlimited, which cannot be observed from a plot of limited range. This graphical solution of an equation is invaluable when no analytical method exists. (For $\sin x = \dfrac{x}{2}$ there are no trig identities which provide a solution.) Roots found graphically are approximations however, the degree of accuracy depending upon the quality of the graph plotting and drawing.

EXAMPLES 8a

1) Show by a suitable sketch that the equation $2^x = \cos x$ has no positive roots.
Determine an approximate value for the negative root whose value is nearest to zero.

A knowledge of the shape of the functions $f(x) \equiv 2^x$ and $g(x) \equiv \cos x$ is used to sketch the graphs $y = 2^x$ and $y = \cos x$ on the same axes as shown

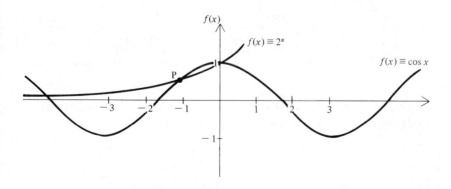

Now for $x > 0$, $\cos x \not> 1$

and $2^x > 1$

Therefore there are no points of intersection for positive values of x; i.e. no positive roots of the equation $2^x = \cos x$.

For $x < 0$, $2^x \to 0$ as $x \to -\infty$

and $\cos x$ is periodic.

Thus the graph $y = 2^x$ cuts the graph $y = \cos x$ twice in each cycle (both points of intersection are above the x axis).

So we see that there is an infinite set of negative roots and a root $x = 0$,   but no positive roots of the given equation.

The point of intersection nearest to the origin is the point P on the sketch and it lies between $x = -2$ and $x = 0$.   This, then, is the range for our first plotted graph.

x	-2	-1.5	-1	-0.5	0
2^x	0.25	0.35	0.50	0.71	1
$\cos x$	-0.42	0.07	0.54	0.88	1

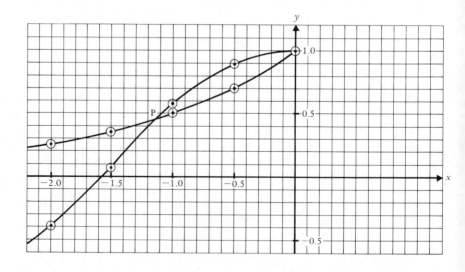

The position of P is now clearly close to $x = -1.1$ and enlarged graphs of this region can be plotted using, say $-1.3 \leqslant x \leqslant -0.9$

x	-1.3	-1.2	-1.1	-1.0	-0.9
2^x	0.406	0.435	0.467	0.5	0.536
$\cos x$	0.267	0.362	0.454	0.540	0.622

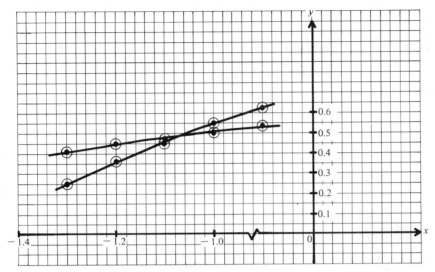

The point of intersection can now be read off as -1.08 (the answer should, at this stage, be correct to three significant figures).

Thus the negative solution of the equation $2^x = \cos x$ which is nearest to zero is $x \simeq -1.08$.

2) By sketching suitable graphs find the number of roots of the equation $x \tan x = 1$ that lie between -2π and 2π.

First we rearrange the given equation so that it is made up of two familiar functions, i.e. $\tan x = \dfrac{1}{x}$

Now we sketch $y = \tan x$ and $y = \dfrac{1}{x}$ on the same axes.

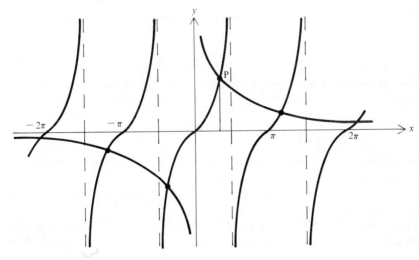

The sketch shows that the equation $\tan x = \dfrac{1}{x}$ has an infinite set of negative roots and an infinite set of positive roots, and four of these roots are between -2π and 2π.

3) A chord AB of a circle subtends an angle θ at the centre. AB divides the area of the circle into two parts in the ratio $1:5$. Find the value of θ.

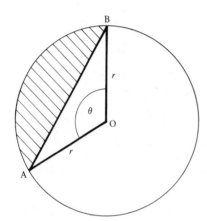

Let the radius of the circle be r.

Area of sector AOB $= \frac{1}{2}r^2\theta$

Area of triangle AOB $= \frac{1}{2}r^2 \sin\theta$

Hence the area of the minor segment (shaded) is $\frac{1}{2}r^2(\theta - \sin\theta)$.

If AB divides the area in the ratio $1:5$ then the area of the minor segment is $\frac{1}{6}$ of the area of the circle

i.e.
$$\tfrac{1}{2}r^2(\theta - \sin\theta) = \tfrac{1}{6}\pi r^2$$

$\Rightarrow$
$$\theta - \sin\theta = \frac{\pi}{3}$$

This equation cannot be solved by using standard identities so it is rearranged giving

$$\sin\theta = \theta - \frac{\pi}{3} = \theta - 1.047^c$$

Now we sketch $y = \sin\theta$ and $y = \theta - 1.047$ to find their point(s) of intersection.

Clearly $0 < \theta < \pi$ (θ is an angle in a minor segment) so only this range need be considered.

The line $y = \theta - 1.047$ is easily drawn using the points of intersection with the axes, i.e. $(0, -1.047)$ and $(1.047, 0)$.

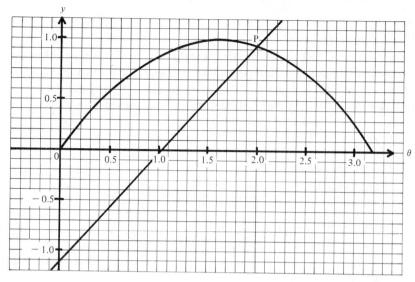

At P, $\theta \simeq 1.9^c$

A more accurate graph is now drawn using $1.7 \leqslant \theta \leqslant 2.1$.

θ	1.7	1.8	1.9	2.0	2.1
$\sin \theta$	0.992	0.974	0.946	0.909	0.863
$\theta - 1.047$	0.653				1.053

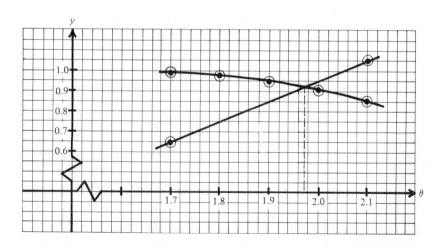

The graphs intersect where $\theta \simeq 1.97^c$ so this is the angle subtended by AB at the centre of the circle.

When one term in an equation contains an angle and another term involves a trig ratio of that angle, it is understood that the angle is measured in radians unless it is specifically indicated that degrees are to be used (a rare situation). Normally, therefore, the graphical solution of such equations requires a *horizontal axis graduated in radians.*

Some Notes on Drawing Graphs

For those readers who have not had much practice in plotting and using graphs, the following points may be helpful.

1) Use a sharp, but not too hard, pencil.

2) Mark points either with a dot, ringed for visibility ⊙, or with a 'straight' cross (+), rather than a 'diagonal' cross (×) whose centre point can easily be misplaced.

3) Choose a scale for each axis which spreads the available points over as much of the length of the axis as possible. This avoids 'crowding' the points.

e.g.

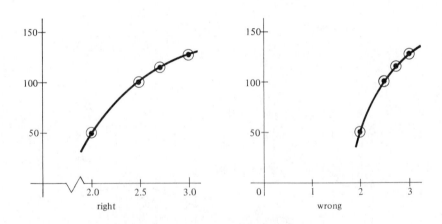

It is not necessary to use the same scales on the two axes, nor is it necessary to begin each scale from zero.

Avoid scales which are inherently difficult to use, e.g. making one square represent three units.

4) When hand drawing a curve, always make sure that you draw from the concave side,

e.g.

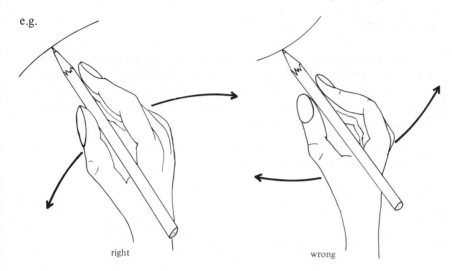

right   wrong

EXERCISE 8a

1) Sketch the pairs of graphs you could use to solve the following equations (do not carry out the solution).

(a) $\sin x = \dfrac{1}{x}$ (b) $\cos x = x^2 - 1$ (c) $2^x = \tan x$

(d) $2^x \sin x = 1$ (e) $(x^2 - 4)\cos x = 1$ (f) $x 2^x = 1$

2) Use the sketches in Question (1) to estimate, in each case, the number of roots of the given equation (some may have an infinite set of solutions).

3) Find graphically the specified roots of the following equations, giving your answer to three significant figures.

(a) $\tan \theta = 2\theta$;   the root between 0 and $\dfrac{\pi}{2}$.

(b) $x - 2 = \sin x$;   the smallest positive root.

(c) $2^{-x} = \sin x$;   the smallest positive root.

4) Show by a suitable sketch that there are no positive roots of the equation $3^{-x} = \sec x$. How many negative roots are there?

5) Show by means of a sketch that the equation $\theta = 1 + \cos \theta$ has only one root. Find this root graphically.

6) Find the length of an arc AB of a circle of radius $0.5\,\text{m}$, if the chord AB divides the area of the circle in the ratio $3 : 1$.

SMALL ANGLES

A glance at the values of $\sin\theta$ and $\tan\theta$ when θ is a very small positive angle shows that these two trig ratios are almost equal. Further, if the small angle is measured in radians it is found that $\sin\theta \simeq \tan\theta \simeq \theta$.
These relationships are important in the development of analytical trigonometry and can be defined precisely as follows.

Consider a small angle θ subtended by an arc AB at the centre, O, of a circle of radius r. The area of the sector OAB is $\frac{1}{2}r^2\theta$.

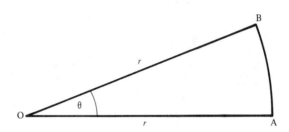

Now if AC is drawn perpendicular to OA, to cut OB produced at C, OAC is a right angled triangle with base r and height $r\tan\theta$.
Its area is therefore $\frac{1}{2}r^2\tan\theta$.
Further, when the chord AB is drawn, an isosceles triangle OAB is formed with area $\frac{1}{2}r^2\sin\theta$.

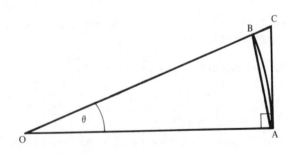

Now   area $\triangle$OAB $<$ area sector OAB $<$ area $\triangle$OAC

i.e. $\frac{1}{2}r^2\sin\theta < \frac{1}{2}r^2\theta < \frac{1}{2}r^2\tan\theta$

Dividing throughout by $\frac{1}{2}r^2$ (which is positive) we have

$$\sin \theta < \theta < \tan \theta$$

But $\sin \theta$, θ and $\tan \theta$ are all positive since θ is a small positive angle. Therefore we can divide throughout by any of these terms.

Thus $$\frac{\sin \theta}{\sin \theta} < \frac{\theta}{\sin \theta} < \frac{\tan \theta}{\sin \theta}$$

$\Rightarrow$ $$1 < \frac{\theta}{\sin \theta} < \sec \theta$$

For small positive values of θ, $\sec \theta > 1$ and $\sec \theta \to 1$ as $\theta \to 0$.

Hence, as $\theta \to 0$, $\dfrac{\theta}{\sin \theta}$ lies between 1 and a number which approaches 1 from above and we can say

$$\text{as } \theta \to 0, \quad \frac{\theta}{\sin \theta} \to 1 \quad \text{from above}$$

Alternatively, dividing the first inequalities by $\tan \theta$ gives

$$\frac{\sin \theta}{\tan \theta} < \frac{\theta}{\tan \theta} < \frac{\tan \theta}{\tan \theta}$$

i.e. $$\cos \theta < \frac{\theta}{\tan \theta} < 1$$

But for small positive values of θ, $\cos \theta < 1$ and $\cos \theta \to 1$ as $\theta \to 0$.

Hence, as $\theta \to 0$, $\dfrac{\theta}{\tan \theta}$ lies between 1 and a number which approaches 1 from below and we can say

$$\text{as } \theta \to 0, \quad \frac{\theta}{\tan \theta} \to 1 \quad \text{from below}$$

These limiting values verify that, for small positive values of θ,

$$\sin \theta \simeq \theta$$

and $$\tan \theta \simeq \theta$$

So far we have not found an approximate value for $\cos \theta$ when θ is small. To do this we use the double angle identity

$$\cos \theta \equiv 1 - 2 \sin^2 \frac{\theta}{2}$$

So if $\dfrac{\theta}{2}$ is small $\quad \sin \dfrac{\theta}{2} \simeq \dfrac{\theta}{2}$

Hence
$$\cos\theta \simeq 1 - 2\left(\frac{\theta}{2}\right)^2$$

i.e. if θ is small $\cos\theta \simeq 1 - \dfrac{\theta^2}{2}$

Now let us consider the case of a small *negative* angle, ϕ, so that $\phi = -\theta$.
Then $\sin\phi = \sin(-\theta) = -\sin\theta$,
and $\tan\phi = \tan(-\theta) = -\tan\theta$
We know that $\sin\theta < \theta < \tan\theta$.
Multiplying this inequality throughout by -1, which reverses the inequality
signs, we have

$$-\sin\theta > -\theta > -\tan\theta$$

i.e.
$$\sin\phi > \phi > \tan\phi$$

It must be remembered that all three terms in this last inequality are *negative
numbers*.
Hence, when we divide throughout by $\sin\phi$, the inequality signs are again
reversed giving

$$1 < \frac{\phi}{\sin\phi} < \sec\phi$$

For small negative angles, $\sec\phi \to 1$ so again we find that, as $\phi \to 0$, $\dfrac{\phi}{\sin\phi} \to 1$
from above.
Similarly we can show that, as $\phi \to 0$, $\dfrac{\phi}{\tan\phi} \to 1$ from below.

Thus, for any small angle θ, measured in radians,

$$\lim_{\theta \to 0}\left(\frac{\theta}{\sin\theta}\right) = 1$$

and
$$\lim_{\theta \to 0}\left(\frac{\theta}{\tan\theta}\right) = 1$$

Further,
$$\cos\phi \equiv 1 - 2\sin^2\frac{\phi}{2}$$

$\Rightarrow$
$$\cos\phi \simeq 1 - 2\left(\frac{\phi}{2}\right)^2 \quad \text{when } \phi \text{ is small.}$$

This is the same approximation as was found for a small positive angle θ. Hence
for all small angles, positive or negative, we have

$$\sin \theta \simeq \theta$$

$$\tan \theta \simeq \theta$$

$$\cos \theta \simeq 1 - \frac{\theta^2}{2}$$

These approximations are correct to 3 s.f. for angles in the range

$$-0.105^c < \theta < 0.105^c$$

i.e. $$-6° < \theta < 6°$$

EXAMPLES 8b

1) Find an approximation for the expression $\dfrac{\sin 3\theta}{1 + \cos 2\theta}$ when θ is small.

When θ is small, 3θ is small, so $\sin 3\theta \simeq 3\theta$

also 2θ is small, so $$\cos 2\theta \simeq 1 - \frac{(2\theta)^2}{2}$$

so when θ is small, $$\frac{\sin 3\theta}{1 + \cos 2\theta} \simeq \frac{3\theta}{2(1 - \theta^2)}$$

2) Without using tables find an approximate value for $\tan 61°$ given $\sqrt{3} = 1.732$ and $1° = 0.017^c$ giving your answer to 3 decimal places.

$$\tan(60° + 1°) = \frac{\tan 60° + \tan 1°}{1 - \tan 60° \tan 1°}$$

$$= \frac{\sqrt{3} + \tan 1°}{1 - \sqrt{3} \tan 1°}$$

Now $\tan \theta \simeq \theta$ when θ is small and measured in radians
Therefore $\tan 1° = \tan 0.017^c \simeq 0.017$

Hence $$\tan 61° \simeq \frac{1.732 + 0.017}{1 - (1.732)(0.017)}$$

$\Rightarrow$ $$\tan 61° \simeq 1.802$$

EXERCISE 8b

1) If θ is small enough to neglect θ^3, find approximations for the following expressions:

(a) $\dfrac{2\theta}{\sin 4\theta}$ (b) $\dfrac{\theta \sin \theta}{\cos 2\theta}$ (c) $\sin \dfrac{\theta}{2} \sec \theta$

(d) $\dfrac{\theta \tan \theta}{1 - \cos \theta}$ (e) $\sin\left(\alpha + \dfrac{\theta}{2}\right) \sin \dfrac{\theta}{2}$ (f) $\dfrac{2 \sin \dfrac{\theta}{2}}{\theta}$

(g) $\dfrac{\sin\theta\,\tan\theta}{\theta^2}$ (h) $\dfrac{\cos\left(\alpha+\dfrac{\theta}{2}\right)-\cos\alpha}{\theta}$

2) If θ is small enough to neglect θ^2 show that:

(a) $2\cos\left(\dfrac{\pi}{3}+\theta\right)\simeq 1-\sqrt{3}\theta$ (b) $\tan\left(\dfrac{\pi}{4}+\theta\right)\simeq\dfrac{1+\theta}{1-\theta}$

(c) $4\sin\left(\dfrac{\pi}{4}-\theta\right)\simeq 2\sqrt{2}(1-\theta)$

3) Using only the values $\sqrt{3}=1.732\,05$, $5°=0.087\,27^c$, $\sin 2^c=0.909\,30$, $\cos 2^c=-0.416\,15$, $\tan 2^c=-2.185\,04$, $\sin 3^c=0.141\,12$, $\cos 3^c=-0.989\,99$, $\tan 3^c=-0.142\,55$, find, to three decimal places, approximate values for:

(a) $\tan 2.01^c$ (b) $\sin 55°$ (c) $\cos 115°$ (d) $\sin 3.005^c$

(e) $\tan 175°$ (f) $\cos 1.98^c$

INVERSE CIRCULAR FUNCTIONS

If $\sin\theta=x$ then θ is an angle with a sine of value x, i.e. $\theta=\arcsin x$. In general, if $f(x)\equiv\arcsin x$,   then $f(x)$ is called an *inverse circular function* or *inverse trig function* of x.

Principal values of the angle θ are indicated by $\arcsin x$ (using a *small initial letter*) whereas *general* values of θ are denoted by $\text{Arcsin}\,x$ (*capital* A). The other inverse circular functions are:

$\arccos x$, $\arctan x$, $\text{arccosec}\,x$, $\text{arcsec}\,x$ and $\text{arccot}\,x$.

Graphs of the Inverse Circular Functions

The two statements

$$y = \text{Arcsin}\,x$$

and $$x = \sin y$$

express, in different ways, the same relationship between x and y. Consequently the sketch of the curve with equation $y=\text{Arcsin}\,x$ is the same shape as the curve with equation $x=\sin y$

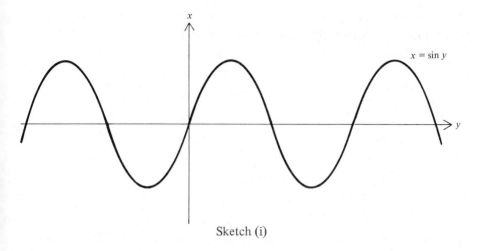

Sketch (i)

Without moving this sketch relative to the axes, a suitable reorientation produces the sketch of $y = \text{Arcsin } x$.

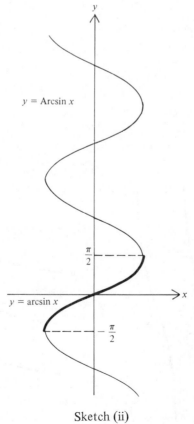

Sketch (ii)

(Sketch (i) is transformed into Sketch (ii) by rotating the axis plane complete with sketch, through $90°$ anticlockwise and then reflecting the plane in the y axis.)

Similarly considering the sketches of $\begin{cases} y = \text{Arccos } x \\ x = \cos y \end{cases}$ we have

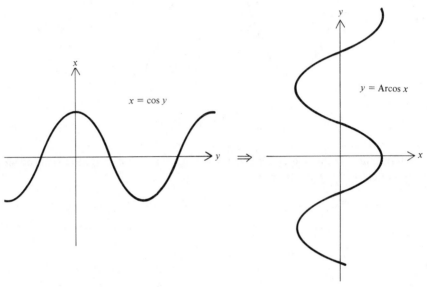

and for $\begin{cases} y = \text{Arctan } x, \\ x = \tan y \end{cases}$ we have

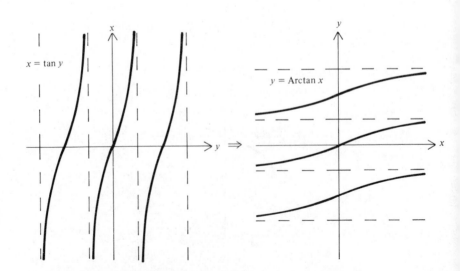

The graphs of the minor inverse trig functions can be deduced in the same way. (**Note:** The notation $\sin^{-1} x$ is occasionally used in place of arcsin x, etc.)

When an equation involves an inverse trig function, careful note must be made of whether a general solution or a principal value solution is required.

For example the equation $\qquad x = \arctan\sqrt{3}$

has only one solution $\qquad\qquad x = \dfrac{\pi}{3}$

But the equation $\qquad\qquad x = \text{Arctan}\sqrt{3}$

has an infinite set of solutions $\quad x = \dfrac{\pi}{3} + n\pi$

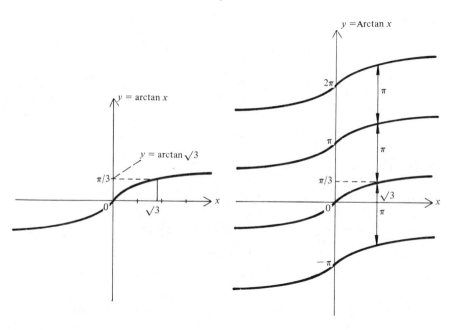

EXAMPLES 8c

1) Find x if $\quad \arcsin x + \arccos\dfrac{x}{2} = \dfrac{5\pi}{6}$

Let $\quad \arcsin x \equiv \theta \quad \Rightarrow \quad \sin\theta \equiv x \quad \Rightarrow$

and $\arccos \dfrac{x}{2} \equiv \phi \;\Rightarrow\; \cos \phi \equiv \dfrac{x}{2} \;\Rightarrow$

The given equation then becomes

$$\theta + \phi = \frac{5\pi}{6}$$

Hence $\qquad\qquad\qquad\qquad \sin(\theta + \phi) = \tfrac{1}{2}$

$\Rightarrow \qquad\qquad\qquad \sin\theta\,\cos\phi + \cos\theta\,\sin\phi = \tfrac{1}{2}$

$\Rightarrow \qquad\qquad (x)\left(\dfrac{x}{2}\right) + \sqrt{(1-x^2)}\left(\dfrac{\sqrt{4-x^2}}{2}\right) = \tfrac{1}{2}$

$\Rightarrow \qquad\qquad\qquad \sqrt{(1-x^2)}\sqrt{(4-x^2)} = 1 - x^2$

Squaring both sides (*not* cancelling $\sqrt{1-x^2}$)

gives $\qquad\qquad\qquad 4 - 5x^2 + x^4 = 1 - 2x^2 + x^4$

Therefore, either $\quad x^4 = \infty \quad$ (impossible since x is a sine ratio)

or $\qquad\qquad\qquad\qquad\qquad 3x^2 = 3$

$\Rightarrow \qquad\qquad\qquad\qquad\qquad x = \pm 1$

But the principal value of $\arcsin(-1)$ is $-\dfrac{\pi}{2}$ and the principal value of

$\arccos(-\tfrac{1}{2})$ is $\dfrac{2\pi}{3}$. So $\;x = -1\;$ does not satisfy the given equation and the only solution is $\;x = 1$.

2) Prove that $\quad \arctan 3 + 2\arctan 2 = \pi + \text{arccot } 3$

Let $\qquad\qquad \arctan 3 = \theta \;\Rightarrow\; \tan\theta = 3 \;\Rightarrow\; \dfrac{\pi}{4} < \theta < \dfrac{\pi}{2}$

and $\qquad\qquad \arctan 2 = \phi \;\Rightarrow\; \tan\phi = 2 \;\Rightarrow\; \dfrac{\pi}{4} < \phi < \dfrac{\pi}{2}$

The L.H.S then becomes $\qquad\qquad \theta + 2\phi$

Now $\qquad \tan(\theta + 2\phi) \equiv \dfrac{\tan\theta + \tan 2\phi}{1 - \tan\theta \tan 2\phi}$

Also $\qquad \tan 2\phi \equiv \dfrac{2\tan\phi}{1 - \tan^2\phi} = \dfrac{2 \times 2}{1 - 4} = -\dfrac{4}{3}$

Hence $\qquad \tan(\theta + 2\phi) = \dfrac{3 - \frac{4}{3}}{1 - 3(-\frac{4}{3})} = \dfrac{\frac{5}{3}}{5} = \dfrac{1}{3}$

But $\quad \dfrac{\pi}{4} < \theta < \dfrac{\pi}{2}$ and $\dfrac{\pi}{4} < \phi < \dfrac{\pi}{2}$ $\Rightarrow$ $\dfrac{3}{4}\pi < \theta + 2\phi < \dfrac{3}{2}\pi$

Therefore $\qquad \theta + 2\phi = \pi + \arctan\frac{1}{3} = \pi + \text{arccot }3$

i.e. $\qquad \arctan 3 + 2\arctan 2 = \pi + \text{arccot }3$

3) Simplify $\quad \arctan x + \arctan\left(\dfrac{1-x}{1+x}\right)$

Let $\qquad \alpha \equiv \arctan x \quad \Rightarrow \quad \tan\alpha \equiv x$

and $\qquad \beta \equiv \arctan\left(\dfrac{1-x}{1+x}\right) \quad \Rightarrow \quad \tan\beta \equiv \dfrac{1-x}{1+x}$

So we have to simplify $\alpha + \beta$.

Using $\qquad \tan(\alpha + \beta) \equiv \dfrac{\tan\alpha + \tan\beta}{1 - \tan\alpha\tan\beta}$

gives $\qquad \tan(\alpha + \beta) \equiv \dfrac{x + \dfrac{1-x}{1+x}}{1 - x\left(\dfrac{1-x}{1+x}\right)}$

$$\equiv \dfrac{x^2 + 1}{1 + x^2} \equiv 1$$

Hence $\qquad \alpha + \beta \equiv \arctan 1 \equiv \dfrac{\pi}{4}$

Thus $\qquad \arctan x + \arctan\dfrac{1-x}{1+x} \equiv \dfrac{\pi}{4}$

EXERCISE 8c

1) Evaluate, using tables where necessary:

$\arctan 3$; $\arccos\dfrac{1}{2}$; $\arcsin\dfrac{1}{\sqrt{2}}$; $\arccos\dfrac{1}{3}$;

$\text{arcsec }4$; $\arctan\dfrac{1}{\sqrt{3}}$; $\text{arccosec }2$; $\arcsin(-1)$

Prove the following relationships:

2) $\arcsin x + \arccos x \equiv \dfrac{\pi}{2}$

3) $\arctan \dfrac{1}{3} + \arctan \dfrac{1}{2} = \dfrac{\pi}{4}$

4) $2 \arctan \dfrac{1}{2} = \arccos \dfrac{3}{5}$

5) $2 \arctan \dfrac{3}{4} = \arccos \dfrac{7}{25}$

6) $4 \operatorname{arccot} 2 + \arctan \dfrac{24}{7} = \pi$

Simplify:

7) $\sin (2 \arctan x)$

8) $\arctan x + \arctan \dfrac{1}{x}$

9) $\tan (\arctan \tfrac{1}{3} + \arctan \tfrac{1}{4})$

Solve the following equations:

10) $\arctan 2 = \arctan 4 - \arctan x$

11) $\arctan (1 + x) + \arctan (1 - x) = \arctan 2$

12) $\arcsin \left(\dfrac{x}{x-1} \right) + 2 \arctan \left(\dfrac{1}{x+1} \right) = \dfrac{\pi}{2}$

MULTIPLE CHOICE EXERCISE 8

(Instructions for answering these questions are given on p. xii)

Unless otherwise stated, angles are measured in radians.

TYPE II

1) The equation $\sin x = \dfrac{1}{x}$ has:

(a) 2 roots   (b) 4 roots   (c) an infinite set of negative roots
(d) an infinite set of positive roots   (e) an infinite set of roots.

2) Given that θ is a small positive angle:

(a) $\dfrac{\sin \theta}{\theta}$ is very slightly greater than 1.

(b) $\cos \theta$ is very slightly less than 1.

(c) $\dfrac{\tan \theta}{\theta}$ is very slightly greater than 1.

(d) $\dfrac{\sin \theta}{\cos \theta}$ is very slightly greater than 1.

(e) $\dfrac{\sin \theta}{\tan \theta}$ is very slightly greater than 1.

3) To solve graphically the equation $x \cot x = 2,$ we can draw on the same axes the graphs:

(a) $y = \cot x$ and $y = 2x$.

(b) $y = \dfrac{2}{x}$ and $y = \tan x$.

(c) $y = \dfrac{x}{2}$ and $y = \cot x$.

(d) $y = 2 \tan x$ and $y = x$.

(e) $y = \cot x$ and $y = \dfrac{2}{x}$.

4) The approximate value, when θ is small, of the expression $\dfrac{2\theta - \sin \theta}{\sin 2\theta - \theta}$ is:

(a) 1   (b) 2   (c) -1 (d) -2 (e) none of these.

5)

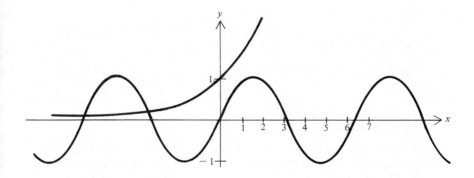

This graph might be drawn in order to solve the equation:

(a) $\sin x = 2^x$ (b) $2^x \sin x = 1$ (c) $\sin x = 3^x$

(d) $3^x \sin x = 1$ (e) none of these.

6) Given that α is a very small angle:

(a) $\sin (2\pi + \alpha) = \sin \alpha$ (b) $\sin \alpha \simeq \alpha$

(c) $\sin (2\pi + \alpha) \simeq 2\pi + \alpha$ (d) $\cos \alpha \simeq \alpha$ (e) $\tan \alpha \simeq \alpha$.

7) $f(x) \equiv \arccos x$:

(a) $-1 \leqslant f(x) \leqslant 1$ (b) $f(x) \equiv \dfrac{1}{\cos x}$

(c) $f(0) = 1$ (d) $f(x) \equiv \arcsec \dfrac{1}{x}$

TYPE III

8) (a) $\sin \theta \simeq \tan \theta$.

   (b) θ is very small.

9) (a) The graphs $y = f(x)$ and $y = g(x)$ cut at a point where $x = a$.

   (b) $f(a) = g(a)$.

10) (a) $\theta = 0.5°$.

    (b) $\sin \theta \simeq 0.5$

11) (a) The equation $f(x) = g(x)$ has no positive roots.

    (b) The equation $f(x) = g(x)$ has an infinite set of negative roots.

12) (a) $\tan \theta \to 0$

    (b) $\theta \to 0$.

13) (a) $\theta = \arctan 1$.

    (b) $\theta = \dfrac{\pi}{4} + n\pi$.

14) (a) $f(\theta) \equiv \arccos x$.

    (b) $f(\theta) \equiv \arcsin \left(\dfrac{\pi}{2} - x \right)$.

MISCELLANEOUS EXERCISE 8

1) Plot on the same axes the graphs whose equations are $y = 2 \sin x$ and $y = \cos 2x$ for values of x from 0 to π. Hence give approximate values of x which satisfy the equation $2 \sin x = \cos 2x$. Check your answers by solving the equation exactly.

2) Find approximations, when θ is very small, for the following expressions.

(a) $\cos \theta + \sin \theta$

(b) $\dfrac{2 \tan \theta - \theta}{\sin 2\theta}$

(c) $\cot \theta \, (1 - \cos \theta)$

(d) $\dfrac{\sqrt{2} - \sin \theta}{\cos \theta}$

3) State the approximate relationships between $\sin \theta$, $\tan \theta$ and θ (radians) when θ is a very small angle. Use these relationships to find approximate values for:

(a) $\sin 2\theta$ when $\cos \theta = 0.998$

(b) $\tan 59°$

(c) $\sin 1.05^c$

(Take $\pi = 3.1416$ and $1° = 0.0175^c$.)

4) Using a graphical method, find good approximations to the roots of the following equations, within the range $0 \leqslant x \leqslant \pi$.

(a) $x = \cos x$

(b) $\tan x = \cos x$

(c) $1 + \cos x = 2 \sin x$

(d) $\dfrac{1}{x} = 2 \tan x$

(e) $\cos \left(x - \dfrac{\pi}{4} \right) = x - 1$

5) By drawing suitable sketches, state the number of (i) positive, (ii) negative roots of the following equations:

(a) $\sin x = x^2$

(b) $x + 1 = \cos x$

(c) $2^x = \dfrac{1}{x}$

(d) $3^x = \tan x$

(e) $3^x = \tan 3x$

(f) $\sin x + x^2 = 1$

6) Without using tables, find in radians the smallest positive value of θ correct to three decimal places when $\sin \theta = 0.842\,01$, given that to five decimal places the sine and cosine of one radian are $0.841\,47$ and $0.540\,30$ respectively.

(U of L)p

7) Two equal circles of radius a intersect and the common chord subtends an angle of 2θ radians at either centre. Find an expression for the area of the region common to the two circles.

If this area is equal to half the area of either circle show that $\sin 2\theta = 2\theta - \tfrac{1}{2}\pi$.

Using a graphical method, or otherwise, estimate the value of θ which satisfies this equation.

(C)

8) By using sketch graphs, or otherwise, show that the equation

$$x = 3 \sin x$$

where the angle is measured in radians, has only one positive root.

Verify, with the use of tables, that this root lies between 2.2 and 2.3.

Determine the root correct to two places of decimals.

(C)

9) The chord AB of a circle divides the circle into two portions whose areas are in the ratio $3:1$. If AB makes an angle θ with the diameter passing through A show that θ satisfies the equation

$$\sin 2\theta = \frac{\pi}{2} - 2\theta$$

Solve this equation graphically.   (AEB)'66

10) A circle has centre O and radius r. Two parallel chords AB and CD are on the same side of O: the angle AOB is $\frac{1}{3}\pi$ and the angle COD is $(\frac{1}{3}\pi + 2\theta)$. Show that the area of the part of the circle between AB and CD is

$$\tfrac{1}{4}r^2 [4\theta + \sqrt{3} - 2 \sin (\tfrac{1}{3}\pi + 2\theta)]$$

If θ is small deduce an approximation for this area in the form $a + b\theta + c\theta^2$ and state the values of the constants a, b, c.   (JMB)

11) By using a graphical method, find approximate solutions to the equation $2 \sin \theta + \cos 3\theta = 1$ for values of θ in the range $0 \leqslant \theta \leqslant 90°$.   (AEB)'72 p

12) By means of the substitution $\tan \theta \equiv t$,   or otherwise, find the values of θ in the range $0 \leqslant \theta \leqslant \dfrac{\pi}{2}$ such that

$$(2 - \tan \theta)(1 + \sin 2\theta) - 2 = 0$$

Show that, when θ is small

$$(2 - \tan \theta)(1 + \sin 2\theta) - 2 \simeq 3\theta$$ (JMB)

13) Prove that $\arctan x + \arctan y \equiv \arctan \dfrac{x + y}{1 - xy}$

Use this relationship to show that, if $\arctan x + \arctan y + \arctan z = \dfrac{\pi}{2}$ then $xy + yz + zx = 1$.

14) Find, without using tables, the value of x when

$$\arctan \tfrac{1}{2} - \arctan \tfrac{1}{3} = \arcsin x.$$ (AEB)'75 p

15) Solve the equation $\arctan \left(\dfrac{1 - x}{1 + x}\right) = \dfrac{1}{2} \arctan x.$ (U of L)p

CHAPTER 9

DIFFERENTIATION II

DIFFERENTIATION OF TRIGONOMETRIC FUNCTIONS

For any function $f(x)$, $\quad \dfrac{d}{dx} f(x) = \lim\limits_{\delta x \to 0} \left[\dfrac{f(x + \delta x) - f(x)}{\delta x} \right]$

If $f(x) \equiv \sin x$, $\quad \dfrac{d}{dx} \sin x = \lim\limits_{\delta x \to 0} \left[\dfrac{\sin (x + \delta x) - \sin x}{\delta x} \right]$

Now $\sin (x + \delta x) - \sin x \equiv 2 \cos \left(x + \dfrac{\delta x}{2} \right) \sin \dfrac{\delta x}{2}$ (factor formulae)

Therefore $\quad \dfrac{d}{dx} \sin x = \lim\limits_{\delta x \to 0} \left[\cos \left(x + \dfrac{\delta x}{2} \right) \dfrac{\sin \delta x/2}{\delta x/2} \right]$

Provided that x is measured in radians

$$\lim\limits_{\delta x \to 0} \left[\cos \left(x + \dfrac{\delta x}{2} \right) \right] = \cos x \quad \text{and} \quad \lim\limits_{\delta x \to 0} \left[\dfrac{\sin \delta x/2}{\delta x/2} \right] = 1$$

Hence $\dfrac{d}{dx} \sin x = \cos x$.

Similarly it is found that $\quad \dfrac{d}{dx} (\cos x) = - \sin x$.

Hence, provided that x is measured in radians,

$$\dfrac{d}{dx} (\sin x) = \cos x \quad \text{and} \quad \dfrac{d}{dx} (\cos x) = - \sin x$$

251

The derivatives of the remaining trig functions are dealt with later in this chapter.

EXAMPLE 9a

Find the smallest positive value of θ for which the curve $y = 2\theta - 3 \sin \theta$ has a gradient of $\frac{1}{2}$.

$$y = 2\theta - 3 \sin \theta \quad \text{gives} \quad \frac{dy}{d\theta} = 2 - 3 \cos \theta$$

when
$$\frac{dy}{d\theta} = \frac{1}{2} \qquad 2 - 3 \cos \theta = \frac{1}{2}$$
$$3 \cos \theta = \frac{3}{2}$$
$$\cos \theta = \frac{1}{2}$$

The smallest positive value of θ for which $\cos \theta = \frac{1}{2}$, is $\dfrac{\pi}{3}$.

Note that the answer *must* be given in radians as the differentiation is valid only if θ is measured in radians.

EXERCISE 9a

1) Find $\dfrac{d}{dx}(\cos x)$ from first principles.

2) Write down the derivatives of the following functions:
(a) $\cos x + \sin x$ (b) $3 - \cos x$ (c) $2 \sin \theta$
(d) $4 \cos \theta$ (e) $2 \sin \theta - 3 \cos \theta$ (f) $4 \sin t - 6$

3) Expand and simplify the following functions and write down their derivatives:

(a) $\sin\left(x + \dfrac{\pi}{2}\right)$ (b) $\cos\left(x - \dfrac{\pi}{4}\right)$ (c) $\sin\left(\dfrac{\pi}{3} - x\right)$

4) Find the gradient of the following curves at the point whose x coordinate is given:

(a) $y = \sin x, \quad \pi$ (b) $y = \cos x + \sin x, \quad \dfrac{\pi}{2}$

(c) $y = 3 \sin x - 2 \cos x, \quad \dfrac{\pi}{4}$ (d) $y = x^2 - \sin x, \quad \dfrac{\pi}{3}$

5) Find the smallest positive value of x for which the gradient of each of the following curves has the value indicated:
(a) $y = x + \sin x, \quad \frac{1}{2}$ (b) $y = \cos x - 2 \sin x, \quad 1$
(c) $y = \sin x - 3 \cos x, \quad 0$

6) Find the values of θ for which the following functions have (a) maximum values, (b) minimum values:

$\cos \theta - 3$, $2 \sin \theta + \cos \theta$, $3 \cos \theta - 4 \sin \theta$

7) Find the equation of the tangent to the following curves at the point indicated.

(a) $y = 3 \cos x - 2 \sin x$ where $x = \dfrac{\pi}{4}$

(b) $y = \theta^2 - 3 \sin \theta$ where $\theta = \dfrac{\pi}{3}$

INTRODUCING THE EXPONENTIAL FUNCTION

We have already met exponential functions such as 2^x (Chapter 3) and the diagram below shows a few members of the family of curves $y = a^x$ $(a \geqslant 0)$

viz: $y = 1^x (= 1)$, $y = 2^x$, $y = 3^x$, $y = 4^x$.

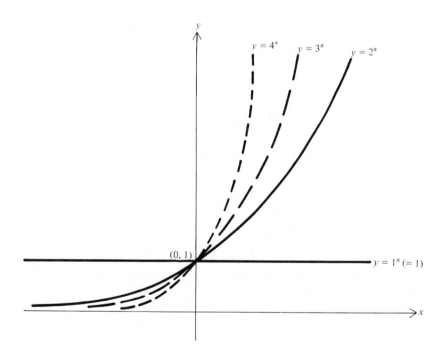

Note that all the curves pass through $(0, 1)$.

In fact any member of the family of curves $y = a^x$ passes through this point (when $x = 0$, $y = a^0 = 1$).

Note also that the higher the value of the base (a), the greater is the gradient of the curve at the point where it cuts the y-axis.

By plotting these curves accurately we find the following approximate values for the gradient of each curve as it cuts the y-axis:

	gradient at $(0, 1)$
$y = 1^x$	0
$y = 2^x$	0.7
$y = 3^x$	1.1
$y = 4^x$	1.4

From this table we deduce that there is a number between 2 and 3, which we call e, for which the gradient of $y = e^x$ is 1 at $(0, 1)$.
Now consider the general form $y = a^x$:

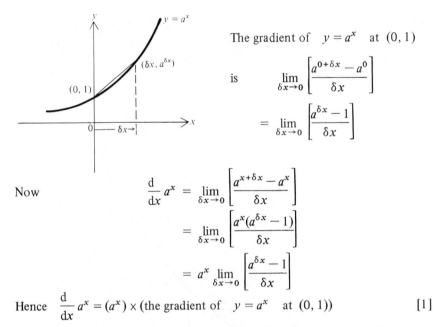

The gradient of $y = a^x$ at $(0, 1)$

is $\displaystyle\lim_{\delta x \to 0}\left[\frac{a^{0+\delta x} - a^0}{\delta x}\right]$

$\displaystyle= \lim_{\delta x \to 0}\left[\frac{a^{\delta x} - 1}{\delta x}\right]$

Now $\displaystyle\frac{d}{dx}a^x = \lim_{\delta x \to 0}\left[\frac{a^{x+\delta x} - a^x}{\delta x}\right]$

$\displaystyle= \lim_{\delta x \to 0}\left[\frac{a^x(a^{\delta x} - 1)}{\delta x}\right]$

$\displaystyle= a^x \lim_{\delta x \to 0}\left[\frac{a^{\delta x} - 1}{\delta x}\right]$

Hence $\displaystyle\frac{d}{dx}a^x = (a^x) \times$ (the gradient of $y = a^x$ at $(0, 1)$)   [1]

If we now consider our earlier observation that there is a number e, whose value lies between 2 and 3, for which $y = e^x$ has a gradient of 1 at $(0, 1)$ we have a special case of [1],

$$\frac{d}{dx}e^x = (e^x) \times (1) = e^x$$

The function e^x is clearly important as it is the only function that remains unaltered when differentiated.

The number e came to the notice of many mathematicians at about the same time in the early 18th century. It appeared as a result of several different lines of investigation and became known as a natural number.

The number e is irrational, i.e. it cannot be given an exact numerical value (like $\pi, \sqrt{2}$, etc.), but we may say

$$e = 2.718 \quad \text{to } 4 \text{ s.f.}$$

Working with the function e^x involves evaluating powers of e, such as e^3, e^{-4}, $e^{2.72}$ ... etc. These values are obtainable from books of 4 figure tables or from a suitable calculator.

To summarize

$a^x \ (a > 0)$ is an exponential function

$e^x \ (e \simeq 2.718)$ is *the* exponential function

and if $y = e^x$ then

$$\frac{d}{dx} e^x = e^x \quad \text{and} \quad \frac{dy}{dx} = y$$

The diagrams below show sketches of the graphs of e^x and some simple variations.

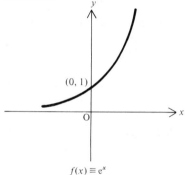

$f(x) \equiv e^x$

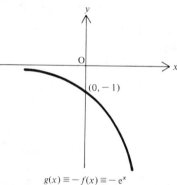

$g(x) \equiv -f(x) \equiv -e^x$

In general if y is replaced by $-y$ in the equation $y = f(x)$ to give $y = -f(x)$ then the curve $y = -f(x)$ is the reflection in the x-axis of the curve $y = f(x)$.

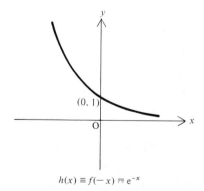

$h(x) \equiv f(-x) \equiv e^{-x}$

In general if x is replaced by $-x$ in the equation $y = f(x)$ to give $y = f(-x)$ then the curve $y = f(-x)$ is the reflection in the y-axis of the curve $y = f(x)$.

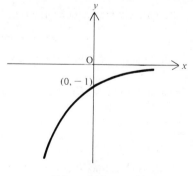

$j(x) \equiv -h(x) \equiv -e^{-x}$

EXAMPLE 9b

Find the coordinates of the point on the curve

$$y = x - e^x$$

at which y has a stationary value and sketch the curve.

For $y = x - e^x$, $\dfrac{dy}{dx} = 1 - e^x$

y is stationary when $1 - e^x = 0$

$\Rightarrow$ $e^x = 1$

$\Rightarrow$ $x = 0$ and $y = 0 - e^0 = -1$

Therefore y has a stationary value at $(0, -1)$.

The diagram shows the line $y = x$ and the curve $y = -e^x$.

We may obtain the curve $y = x - e^x$ by adding the ordinates of $y = x$ and $y = -e^x$.

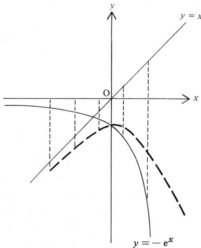

Alternatively we may observe that

for large + ve values of x, $x - e^x \simeq -e^x$

for large $-$ ve values of x, $x - e^x \simeq x$

We also know that $(0, -1)$ is a stationary point and that $\dfrac{d^2 y}{dx^2} = -e^x$,

$\Rightarrow$ when $x = 0$, $\dfrac{d^2 y}{dx^2} = -1$

so $(0, -1)$ is maximum turning point.

From this information the sketch of $y = x - e^x$ can be drawn.

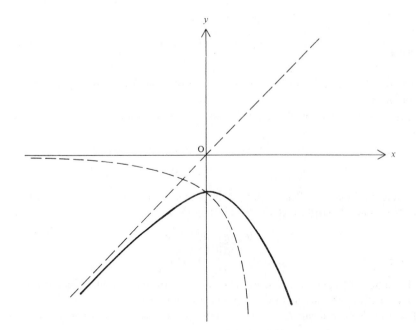

EXERCISE 9b

1) Evaluate (to 3 s.f.) e^3, e^{-2}, $e^{1.7}$, $e^{-0.2}$.

2) Solve the following equations for x by drawing a graph to obtain an approximate solution and then redrawing a part of the graph to obtain a solution correct to three significant figures:
(a) $e^x = 2$ (b) $e^{-x} = 3$.

3) Write down the differential coefficients of:
(a) $2e^x$ (b) $x^2 - e^x$ (c) $e^x + \cos x$.

4) Find the values of x for which the following functions have stationary values:
(a) $e^x - x$ (b) $2e^x - x - 1$ (c) $e^x + 1$.

5) Draw a sketch of the curves whose equations are:
(a) $y = e^x - x$ (b) $y = e^x + 1$ (c) $y = 1 - e^x$
(d) $y = x^2 + e^x$ (e) $y = 1 + x - e^x$ (f) $y = 1 - e^{-x}$

Naperian Logarithms

Consider the equation $e^x = 0.78$.
We can solve this equation by taking logs of both sides,

i.e. $$x \log e = \log 0.78$$

giving $$x = \frac{\log 0.78}{\log e}$$

Equations, such as this one, arise frequently when working with exponential functions.

Introducing logarithms with base e makes the calculation simpler. We can then write

$$x \log_e e = \log_e 0.78 \ (\log_e e = 1)$$

$\Rightarrow$ $$x = \log_e 0.78$$

Such logarithms are called natural, or Naperian, logarithms. To avoid confusion with common logarithms (base 10) we denote

$$\log_e a \quad \text{by} \quad \ln a$$

where

$$\ln a = b \iff a = e^b$$

The values of Naperian logarithms can be found from a suitable calculator and they are also given in most books of four figure tables.

The three laws of logarithms (which apply to any base) are useful when working with Naperian logarithms and, as a reminder, they are listed below.

$$\log_c a + \log_c b \equiv \log_c ab$$

$$\log_c a - \log_c b \equiv \log_c \frac{a}{b}$$

$$\log_c a^n \equiv n \log_c a$$

EXAMPLE 9c

Express $\ln (\tan x)$ as a sum or difference of logarithms.

$$\ln (\tan x) \equiv \ln \left(\frac{\sin x}{\cos x}\right) \equiv \ln \sin x - \ln \cos x$$

EXERCISE 9c

1) Evaluate: $\ln 7.821$, $\ln 0.483$, $\ln 21.5$.

2) Express as a sum or difference of logarithms:

(a) $\ln \left(\dfrac{x}{x+1}\right)$ (b) $\ln (3x^2)$ (c) $\ln (x^2 - 4)$

(d) $\ln \sqrt{\left(\dfrac{x-1}{x+1}\right)}$ (e) $\ln \cot x$ (f) $\ln (4 \cos^2 x)$

3) Express as a single logarithm:
(a) $\ln x - 2 \ln (1 - x)$ (b) $1 - \ln x$
(c) $\ln \sin x + \ln \cos x$ (d) $3 \ln x - \frac{1}{2} \ln (x - 1)$

4) Solve the following equations for x:

(a) $e^x = 8.2$ (b) $e^{2x} + e^x - 2 = 0$ (*Hint*: see Ch. 2, Ex. 2d)

(c) $e^{2x-1} = 3$ (d) $e^{4x} - e^x = 0$

THE LOGARITHMIC FUNCTION

Consider the function $f(x) \equiv \ln x$ and the curve whose equation is $y = \ln x$.

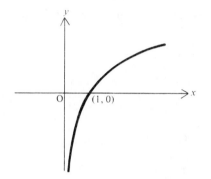

If $y = \ln x$ then $x = e^y$ so the curve $y = \ln x$ has the same shape as $y = e^x$ with the x and y axes interchanged (this is achieved by rotating $90°$ clockwise in the xy plane and reflecting in the horizontal axis).

Note that there is no part of the curve in the 2nd and 3rd quadrants.

If $\log_a b = c$, i.e. $b = a^c$, there is no real value of c for which b is negative.

So $\ln x$ does not exist for negative values of x

If $y = f(x)$ where $f(x)$ is any function of x, then

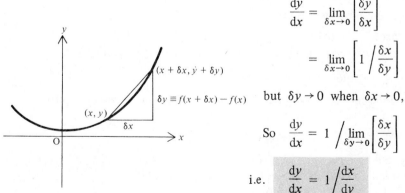

$$\frac{dy}{dx} = \lim_{\delta x \to 0} \left[\frac{\delta y}{\delta x} \right]$$

$$= \lim_{\delta x \to 0} \left[1 \Big/ \frac{\delta x}{\delta y} \right]$$

but $\delta y \to 0$ when $\delta x \to 0$,

So $\dfrac{dy}{dx} = 1 \Big/ \lim_{\delta y \to 0} \left[\dfrac{\delta x}{\delta y} \right]$

i.e. $\dfrac{dy}{dx} = 1 \Big/ \dfrac{dx}{dy}$

$(x + \delta x, y + \delta y)$

$\delta y \equiv f(x + \delta x) - f(x)$

(x, y)

δx

When $y = \ln x$ we may write $x = e^y$.

Differentiating e^y w.r.t. y we have

$$\frac{dx}{dy} = e^y = x$$

so

$$\frac{dy}{dx} = 1 \bigg/ \frac{dx}{dy} = \frac{1}{x}$$

i.e.

$$\frac{d}{dx} \ln x = \frac{1}{x}$$

This result can be used to differentiate compound logarithmic functions if they are first simplified using the identities on page 258.

EXAMPLE 9d

Write down the derivative of:

(a) $\ln (2x^3)$ (b) $\ln (1/\sqrt{x})$

(a) $\ln (2x^3) = \ln 2 + \ln x^3 = \ln 2 + 3 \ln x$

Therefore $\dfrac{d}{dx} \ln 2x^3 = \dfrac{d}{dx} [\ln 2] + \dfrac{d}{dx} [3 \ln x]$

$$= 0 + 3 \left(\frac{1}{x} \right) = \frac{3}{x}$$

(b) $\ln (1/\sqrt{x}) = \ln 1 - \ln x^{\frac{1}{2}} = 0 - \frac{1}{2} \ln x$

Therefore $\dfrac{d}{dx} [\ln (1/\sqrt{x})] = \dfrac{d}{dx} [-\frac{1}{2} \ln x] = -\dfrac{1}{2x}$

EXERCISE 9d

1) Write down the derivatives of :

(a) $2 \ln x$ (b) $\ln x^5$ (c) $\ln (3x^2)$ (d) $\ln (x^{-\frac{3}{2}})$

(e) $\ln (3/\sqrt{x})$ (f) $\ln (3x/\sqrt{x})$ (g) $\ln \sqrt{(2/x^5)}$ (h) $\ln \sqrt{(\frac{1}{3}x^{-\frac{1}{2}})}$

(i) $\log x$ (*Hint*: change the base to e.)

2) Find the coordinates of points at which the following curves have zero gradient.

(a) $y = x - \ln x$ (b) $y = \ln (x^2) - x^2$

3) Sketch the curves whose equations are:

(a) $y = \ln (-x)$ (b) $y = -\ln x$ (c) $y = x - \ln x$

(d) $y = 1 + \ln x$ (e) $y = \log x$

COMPOUND FUNCTIONS

We have shown that differentiation is distributive across addition and subtraction,

i.e. $\dfrac{d}{dx}[f(x) + g(x)] = \dfrac{d}{dx}[f(x)] + \dfrac{d}{dx}[g(x)]$

but that differentiation is *not* distributive across multiplication or division of functions, and so we have not attempted to differentiate functions such as xe^x. Nor have we attempted to differentiate compound functions of the form e^{x^2-1}, $\sin(2x-4)\dots$ etc.

Before we derive rules to differentiate such compound functions it is important to recognise into which category a particular function can be placed.

1) *Function of a Function*

Functions such as e^{x^2-1}, $\sin(3x-2)$, $(x^2+1)^{\frac{1}{2}}$ are all of the form $g[f(x)]$, sometimes written $gf(x)$,

i.e. each is 'a function of (a function of x)' and by substituting u for $f(x)$ such functions may be written as $g(u)$,

e.g. $e^{x^2-1} \equiv e^u$ where $u \equiv x^2 - 1$.

2) *Products*

Functions such as $e^x \sin x$, $x \ln x \dots$ are of the form $[f(x)] \times [g(x)]$, i.e. each is a product of two functions, and by the substitutions $u \equiv f(x)$ and $v \equiv g(x)$ such a function may be written as uv,

e.g. $e^x \sin x \equiv uv$ where $u \equiv e^x$ and $v \equiv \sin x$.

3) *Quotients*

Functions such as $\dfrac{e^x}{\sin x}$, $\dfrac{\sin x}{\cos x}$, $\dots$ are of the form $\dfrac{f(x)}{g(x)}$, i.e. each is a quotient of two functions, and such a function may be written as $\dfrac{u}{v}$,

e.g. $\dfrac{e^x}{\sin x} \equiv \dfrac{u}{v}$ where $u \equiv e^x$ and $v \equiv \sin x$.

Some functions may fall into more than one of these categories, for example

$$e^x \sqrt{1-x} \text{ is a product of } e^x \text{ and } \sqrt{1-x}$$

where $\sqrt{1-x}$ is a function of a function.

Substituting u for e^x and v for $1-x$

$$e^x \sqrt{1-x} \text{ may be written as } u\sqrt{v}$$

EXERCISE 9e

Write the following functions in terms of u and or v, stating clearly the substitutions that you have made.

1) $e^x(x^2-1)$ 2) e^{3x^2} 3) $\sin(x^2-2)$ 4) $x^3 \cos x$

5) $\sqrt{\dfrac{x-1}{x+1}}$ 6) $(x+1)^4$ 7) $\ln(x^2-1)$ 8) $(x^2-1)(x-2)^5$

9) $\tan x^2$ 10) $\ln\left(\dfrac{x+1}{x-1}\right)$ 11) $\cos^2 x$ 12) $\sqrt{\ln x}$

DIFFERENTIATION OF A FUNCTION OF A FUNCTION

Consider $y = g(u)$ where $u = f(x)$.

If δx is a small increase in x let δu and δy be the corresponding increases in u and y.

Then as $\delta x \to 0$, δu and δy also tend to zero.

Hence $\dfrac{dy}{dx} = \lim\limits_{\delta x \to 0} \left[\dfrac{\delta y}{\delta x}\right] = \lim\limits_{\delta x \to 0} \left[\dfrac{\delta y}{\delta u} \times \dfrac{\delta u}{\delta x}\right]$

$= \lim\limits_{\delta u \to 0} \left[\dfrac{\delta y}{\delta u}\right] \times \lim\limits_{\delta x \to 0} \left[\dfrac{\delta u}{\delta x}\right]$

i.e. $\boxed{\dfrac{dy}{dx} = \dfrac{dy}{du} \times \dfrac{du}{dx}}$

This rule may now be used to differentiate a function of a function.

EXAMPLES 9f

1) Differentiate $\sqrt{(x^2-1)}$ w.r.t. x.

$$y = \sqrt{(x^2-1)} = (x^2-1)^{\frac{1}{2}}$$

Let $u \equiv x^2 - 1$ $\Rightarrow$ $y = u^{\frac{1}{2}}$

Therefore $\dfrac{du}{dx} = 2x$ and $\dfrac{dy}{du} = \tfrac{1}{2} u^{-\frac{1}{2}}$

Now $\dfrac{dy}{dx} = \dfrac{dy}{du} \times \dfrac{du}{dx}$

$= (\tfrac{1}{2} u^{-\frac{1}{2}})(2x)$

$= \dfrac{x}{u^{\frac{1}{2}}}$

$= \dfrac{x}{\sqrt{(x^2-1)}}$

Therefore $\dfrac{d}{dx}\sqrt{(x^2-1)} = \dfrac{x}{\sqrt{(x^2-1)}}$

2) Differentiate $\sin\left(2\theta - \dfrac{\pi}{4}\right)$ w.r.t. θ.

$$y = \sin\left(2\theta - \dfrac{\pi}{4}\right)$$

Let $\qquad\qquad u \equiv 2\theta - \dfrac{\pi}{4} \qquad\Rightarrow\qquad y = \sin u$

$$\dfrac{du}{d\theta} = 2 \qquad \text{and} \qquad \dfrac{dy}{du} = \cos u$$

Then $\qquad\qquad \dfrac{dy}{d\theta} = \dfrac{dy}{du} \times \dfrac{du}{d\theta}$

$$= 2 \times \cos u$$

$$= 2 \cos\left(2\theta - \dfrac{\pi}{4}\right)$$

Therefore $\quad \dfrac{d}{d\theta}\left[\sin\left(2\theta - \dfrac{\pi}{4}\right)\right] = 2 \cos\left(2\theta - \dfrac{\pi}{4}\right)$

The reader will find that, with a little practice, the substitution $u \equiv f(x)$ can often be carried out mentally so that $\dfrac{d}{dx} gf(x)$ can be written down directly as $\left(\dfrac{d}{du} g(u)\right)\left(\dfrac{d}{dx} f(x)\right)$ and it is important that this facility should be acquired.

3) Find the derivative of $2 \ln (x\sqrt{x^2-1})$.

$$2 \ln (x\sqrt{x^2 - 1}) \equiv 2 \ln x + 2 \ln (x^2 - 1)^{\frac{1}{2}} \equiv 2 \ln x + \ln (x^2 - 1)$$

Therefore $\quad \dfrac{d}{dx}[2 \ln (x\sqrt{x^2 - 1})] = \dfrac{d}{dx}[2 \ln x] + \dfrac{d}{dx}[\ln (x^2 - 1)]$

$$= \dfrac{2}{x} + (2x)\left(\dfrac{1}{x^2 - 1}\right)$$

$$= \dfrac{2(2x^2 - 1)}{x(x^2 - 1)}$$

In the example above, transforming the given log function into a sum at the start made differentiating it simpler. *Any* function should be simplified, whenever possible, *before* differentiation.

4) Differentiate $\cos^3 x$ with respect to x.
Note that $\cos^3 x$ means $(\cos x)^3$

Let $\qquad\qquad y = (\cos x)^3 \quad$ and $\quad u \equiv \cos x \quad$ so $\quad y = u^3$

then $\qquad\qquad \dfrac{dy}{dx} = 3u^2(-\sin x)$

$$= -3 \cos^2 x \sin x$$

EXERCISE 9f

Differentiate the following functions:

1) e^{3x} $\qquad\qquad$ 2) $\ln (x - 1)^2$ $\quad$ 3) $(x + 1)^5$ $\qquad$ 4) $\cos\left(3\theta - \dfrac{\pi}{4}\right)$

5) $(x^2 + 1)^5$ 6) e^{x^2} 7) $4 \ln (x^2 + 1)$ 8) $(3x^2 + 4)^{\frac{1}{2}}$

9) $5e^{2x}$ 10) $\sin (\theta^2)$ 11) $\sin^2\theta$ 12) $(e^x)^3$

13) $(x + 1)^{-1}$ 14) $\dfrac{1}{x^2 + 1}$ 15) $\sqrt{\dfrac{1}{x-1}}$

Write down directly the derivatives of the following functions:

16) $(x + 1)^6$ 17) $(2x - 4)^3$ 18) $e^{(x^2+2)}$ 19) $\sin \left(3\theta + \dfrac{\pi}{4}\right)$

20) $\cos^2\theta$ 21) $\ln (x^2 + 2)$ 22) $(2x + 3)^{-2}$ 23) $\sqrt{\dfrac{1}{2x-1}}$

24) $e^x - e^{-x}$ 25) $\dfrac{e^x - 1}{e^{2x}}$ 26) $3 \sin \left(2\theta - \dfrac{\pi}{4}\right)$ 27) $\dfrac{1}{\sin \theta}$

28) $\dfrac{1}{\cos \theta}$ 29) $2e^{-3x} + e^{4x}$ 30) $\ln (\sin x)$ 31) e^{-kt}

32) $\ln (x - 1)^3$ 33) $e^{\sin x}$ 34) $\dfrac{1}{\sin^2 x}$ 35) $\sec^2 t$

36) $2e^{\sin 3\theta}$ 37) $4 \ln \sqrt{x-1}$ 38) $2 \sin (\theta^2 + 4)^{\frac{1}{2}}$ 39) e^{e^x}

40) $2 \cos^2(x^2 + 1)$ 41) $3 \sin (e^x)$ 42) $\sqrt{(3 - \cos^2\theta)}$ 43) $\ln (\sin x \cos x)$

44) $e^{\sqrt{1-x}}$ 45) $\sin^2(x^2 + 1)$ 46) $2e^{\frac{1}{x}}$

47) $\ln \dfrac{\sin x}{1 - \cos x}$ 48) $(e^x - e^{-x})^{-1}$

DIFFERENTIATION OF PRODUCTS

Consider $y = uv$ where $u \equiv f(x)$ and $v \equiv g(x)$.
If δx is a small increase in x, and $\delta y, \delta u, \delta v$ are the corresponding increases
in y, u, v

then $y + \delta y = (u + \delta u)(v + \delta v) = uv + u\delta v + v\delta u + \delta u\delta v$

As $y = uv$, $\delta y = u\delta v + v\delta u + \delta u\delta v$

Therefore $\dfrac{\delta y}{\delta x} = u \dfrac{\delta v}{\delta x} + v \dfrac{\delta u}{\delta x} + \delta u \dfrac{\delta v}{\delta x}$

When $\delta x \to 0$: $\dfrac{\delta y}{\delta x} \to \dfrac{dy}{dx}, \quad \dfrac{\delta u}{\delta x} \to \dfrac{du}{dx}, \quad \dfrac{\delta v}{\delta x} \to \dfrac{dv}{dx}, \quad \delta u \to 0.$

Therefore $\dfrac{dy}{dx} = \lim_{\delta x \to 0} \left[\dfrac{\delta y}{\delta x}\right]$

$= u \dfrac{dv}{dx} + v \dfrac{du}{dx} + 0$

i.e.
$$\frac{d}{dx}[uv] = v\frac{du}{dx} + u\frac{dv}{dx}$$

Thus if
$$y = e^x \sin 2x$$

and
$$u \equiv e^x, \qquad v \equiv \sin 2x$$

$$\frac{du}{dx} = e^x, \qquad \frac{dv}{dx} = 2\cos 2x$$

$\Rightarrow \qquad \dfrac{dy}{dx} = 2e^x \cos 2x + (\sin 2x)\,e^x = e^x(2\cos 2x + \sin 2x)$

As with functions of a function, after some practice in the use of the rule for differentiating a product, such derivatives can be written down directly,

e.g.
$$\frac{d}{dx}[x \sin 3x] = x(3\cos 3x) + (\sin 3x)(1)$$
$$= 3x\cos 3x + \sin 3x$$

DIFFERENTIATION OF QUOTIENTS

Consider $y = \dfrac{u}{v}$ where $u \equiv f(x)$ and $v \equiv g(x)$ as above

then
$$y + \delta y = \frac{u + \delta u}{v + \delta v}$$

as $\;y = \dfrac{u}{v}$,
$$\delta y = \frac{u + \delta u}{v + \delta v} - \frac{u}{v}$$
$$= \frac{v\delta u - u\delta v}{v^2 + v\delta v}$$

Therefore
$$\frac{\delta y}{\delta x} = \frac{v\dfrac{\delta u}{\delta x} - u\dfrac{\delta v}{\delta x}}{v^2 + v\delta v}$$

and
$$\frac{dy}{dx} = \lim_{\delta x \to 0}\left[\frac{\delta y}{\delta x}\right] = \frac{v\dfrac{du}{dx} - u\dfrac{dv}{dx}}{v^2}$$

or
$$\frac{d}{dx}\left[\frac{u}{v}\right] = \frac{v\dfrac{du}{dx} - u\dfrac{dv}{dx}}{v^2}$$

Thus if $\;y = \dfrac{e^x}{\sin x}\;$ where $\;u \equiv e^x\;$ and $\;v \equiv \sin x$

$$\frac{du}{dx} = e^x \qquad \frac{dv}{dx} = \cos x$$

$$\frac{dy}{dx} = \frac{(\sin x)(e^x) - (e^x)(\cos x)}{\sin^2 x}$$

$$= \frac{e^x(\sin x - \cos x)}{\sin^2 x}$$

Use of Partial Fractions

Rational functions with two, or more, factors in the denominator may be differentiated by first expressing the function as partial fractions (see Chapter 1).

For example, $f(x) \equiv \dfrac{x}{(x-2)(x-3)}$

$$\equiv \frac{-2}{x-2} + \frac{3}{x-3}$$

Hence $\dfrac{d}{dx} f(x) = \dfrac{d}{dx}[-2(x-2)^{-1}] + \dfrac{d}{dx}[3(x-3)^{-1}]$

$\Rightarrow$ $\dfrac{d}{dx} f(x) = 2(x-2)^{-2} - 3(x-3)^{-2} = \dfrac{2}{(x-2)^2} - \dfrac{3}{(x-3)^2}$

The 'Cover Up' Method for Expressing a Function in Partial Fractions

Consider again the function

$$f(x) \equiv \frac{x}{(x-2)(x-3)} \equiv \frac{A}{x-2} + \frac{B}{x-3} \equiv \frac{A(x-3) + B(x-2)}{(x-2)(x-3)}$$

so that $x \equiv A(x-3) + B(x-2)$

when $x = 2$, we have $A = \dfrac{2}{(2-3)} = -2$

i.e. A is the value of $\dfrac{x}{(x-3)}$ when $x = 2$

i.e. $A = f(2)$ with the factor $(x-2)$ omitted or 'covered up'

similarly $B = f(3)$ with the factor $(x-3)$ covered up

$$= \frac{3}{(3-2)} = 3$$

This method gives a quick way of expressing a function in partial fractions when linear factors only are in the denominator, e.g.

$$f(x) \equiv \frac{1}{(2x-1)(x+3)}$$

$$\equiv \frac{f(\tfrac{1}{2}) \text{ with } (2x-1) \text{ covered up}}{2x-1} + \frac{f(-3) \text{ with } (x+3) \text{ covered up}}{x+3}$$

$$\equiv \frac{\tfrac{2}{7}}{(2x-1)} - \frac{\tfrac{1}{7}}{(x+3)}$$

The cover up method can also be used to find the numerator of a partial fraction with a linear denominator, but for quadratic denominators, the methods described in Chapter 1 must be used,

e.g. $$f(x) \equiv \frac{x}{(x-1)(x^2+1)} \equiv \frac{f(1) \text{ with } (x-1) \text{ covered up}}{x-1} + \frac{Ax+B}{x^2+1}$$

$$\equiv \frac{\tfrac{1}{2}}{(x-1)} + \frac{Ax+B}{x^2+1}$$

thus $$x \equiv \tfrac{1}{2}(x^2+1) + (Ax+B)(x-1)$$

$$\Rightarrow \qquad A = -\tfrac{1}{2}, \quad B = \tfrac{1}{2} \quad \text{by comparing coefficients.}$$

EXERCISE 9g

Differentiate the following functions w.r.t. x.

1) $x \ln x$

2) $x^2(x-1)^{\frac{1}{2}}$

3) $\dfrac{x}{\ln x}$

4) $3 \sin x \cos 2x$

5) $\tan x$

6) $\dfrac{e^x}{x-1}$

7) $(x-1) \ln (x-1)$

8) $\sin x \ln x$

9) $x \sec x$

10) $\operatorname{cosec} x \cos x$

11) $\dfrac{x-1}{x+1}$

12) $e^x \sin x$

13) $\dfrac{e^x}{\sin x}$

14) $x \log x$

15) $\dfrac{e^x}{e^x - e^{-x}}$

16) $\dfrac{x}{x^2+1}$

17) $\dfrac{x^2+1}{x}$

18) $\dfrac{\sqrt{x-1}}{x}$

19) $\dfrac{\sin x}{\sqrt{x}}$

20) $\dfrac{\ln x}{\ln (x-1)}$

21) $\cot x$

Use the cover up method to express the following functions in partial fractions and hence differentiate them.

22) $\dfrac{1}{(x-1)(x+1)}$

23) $\dfrac{x}{(x-2)(x+1)}$

24) $\dfrac{2x}{(2x-1)(x-3)}$

25) $\dfrac{3x-1}{(x+2)(2x-1)}$

26) $\dfrac{5x}{(x-1)(x-2)(x-3)}$

27) $\dfrac{x^2+1}{(2x-1)(x-1)(x+1)}$

28) $\dfrac{x^2}{(x+3)(x^2+1)}$

29) $\dfrac{1}{(x-1)^2(x+1)}$

30) $\dfrac{3x}{(x+3)(x^2+1)}$

By taking results from the last two exercises we may complete the list of derivatives of trig functions:

$$\frac{d}{dx}(\tan x) = \sec^2 x$$

$$\frac{d}{dx}(\cot x) = -\operatorname{cosec}^2 x$$

$$\frac{d}{dx}(\sec x) = \sec x \tan x$$

$$\frac{d}{dx}(\operatorname{cosec} x) = -\operatorname{cosec} x \cot x$$

Also from the 'function of a function' rule we obtain the following general results

$$\frac{d}{dx}\sin f(x) = f'(x)\cos f(x) \qquad \text{where} \quad f'(x) \equiv \frac{d}{dx}[f(x)]$$

$$\frac{d}{dx}\cos f(x) = -f'(x)\sin f(x)$$

$$\frac{d}{dx}e^{f(x)} = f'(x)e^{f(x)}$$

$$\frac{d}{dx}\ln f(x) = \frac{f'(x)}{f(x)}$$

Any of these results may now be used to differentiate a particular function.

EXERCISE 9h

Find the derivatives of the following functions:

1) $3e^{\sin x}$ 2) $e^{-x}\sin x$ 3) $\dfrac{1}{1+x+x^2}$

4) $\cos x \sin^3 x$ 5) $\ln (\sec x)$ 6) $2 \sin 3t \cos 4t$

7) $e^{x^2} \ln 2x^2$ 8) $\tan^2 2x$ 9) $\ln \left(\dfrac{1 + \cos x}{1 + \sin x} \right)$

10) Find the value of $\dfrac{d^2 y}{dx^2}$ when $x = 0$ if $y = e^x \sin 2x$.

11) If $y = e^x \sin x$ show that $\dfrac{d^2 y}{dx^2} - 2 \dfrac{dy}{dx} + 2y = 0$.

12) If $y = \dfrac{1 + x}{1 - x}$ find $\dfrac{d^2 y}{dx^2}$

13) Find the gradient of the curve $y = \ln \sqrt{(1 + \sin 2x)}$ at the point where $x = \dfrac{\pi}{2}$.

14) If $y = \sqrt{(x - 1)} \, e^x \ln x$, find $\dfrac{dy}{dx}$

15) Using the identity $\dfrac{dy}{dx} \equiv 1 \Big/ \dfrac{dx}{dy}$ find $\dfrac{dy}{dx}$ when $\sin y = x$,

giving your answer as a function of x.

16) Find the gradient of the curve $y = \dfrac{1}{\sqrt{(x - 1)} \sin x}$ when $x = \dfrac{\pi}{2}$.

17) If $y = \sin^2 ax$ find a if $\dfrac{dy}{dx} = 1$ when $x = \dfrac{\pi}{4}$, given that a is an integer.

18) Find the values of k for which $\dfrac{x}{(x + 1)^2 (x - k)}$ has one stationary value.

19) Find the value of x in the range $0 \leqslant x \leqslant 2\pi$ for which $e^x \sin x$ has a minimum value. Sketch the curve $y = e^x \sin x$ for $0 \leqslant x \leqslant 2\pi$.

20) Show that $x \ln x$ has only one stationary value and find it.

IMPLICIT FUNCTIONS

All the differentiation carried out so far has involved equations of the form $y = f(x)$.
Now consider the curve whose equation is $y + xy + y^2 = 2$.
This equation is not easily transposed to the form $y = f(x)$ and we say that $y = f(x)$ is implied by the equation $y + xy + y^2 = 2$,
i.e. $f(x)$ is an implicit function.

Differentiation of Implicit Functions

Consider again the equation $y + xy + y^2 = 2$.
The equation may be rewritten as

$$f(x) + xf(x) + [f(x)]^2 = 2$$

where $y = f(x)$ and $\dfrac{dy}{dx} = f'(x)$.

Differentiating term by term we have

(i) $\dfrac{d}{dx} f(x) = f'(x) = \dfrac{dy}{dx}$

(ii) $xf(x)$ is a product

therefore $\dfrac{d}{dx} [xf(x)] = (1)f(x) + (x)f'(x) = y + x\dfrac{dy}{dx}$

i.e. $\dfrac{d}{dx}(xy) = (1)y + (x)\dfrac{dy}{dx}$

(iii) $[f(x)]^2$ is a function of a function.

Therefore $\dfrac{d}{dx} [f(x)]^2 = [2f(x)] f'(x) = 2y\dfrac{dy}{dx}$

i.e. $\dfrac{d}{dx}(y^2) = 2y\dfrac{dy}{dx}$

(iv) and finally $\dfrac{d}{dx}(2) = 0$

Therefore differentiating $y + xy + y^2 = 2$ w.r.t. x

we get $\dfrac{dy}{dx} + \left(y + x\dfrac{dy}{dx}\right) + 2y\dfrac{dy}{dx} = 0$

or $(1 + x + 2y)\dfrac{dy}{dx} + y = 0$

Note that *every* term in the equation is differentiated w.r.t. x.
Note that if $g(y)$ is any function of y where $y = f(x)$ then $g(y)$ is a function of a function of x.
Thus the derivative of $g(y)$ w.r.t. x is

$$\frac{d}{dx} g(y) = g'(y)\frac{dy}{dx}$$

where $g'(y)$ is the derivative of $g(y)$ w.r.t. y

For example
$$\frac{d}{dx} \sin y = (\cos y) \frac{dy}{dx}$$

EXAMPLES 9i

1) Differentiate the following w.r.t. x:

(a) $x^2 + xy^2 + y^3 = 2$ (b) $x = ye^x$

(a) If $x^2 + xy^2 + y^3 = 2$ then

$$\frac{d}{dx}(x^2) + \frac{d}{dx}(xy^2) + \frac{d}{dx}(y^3) = \frac{d}{dx}(2)$$

i.e.
$$2x + \left(x(2y)\frac{dy}{dx} + y^2\right) + 3y^2 \frac{dy}{dx} = 0$$

$\Rightarrow$
$$\frac{dy}{dx}(2xy + 3y^2) + 2x + y^2 = 0$$

(b) $\dfrac{dy}{dx}$ can be found as a function of x by rewriting $x = ye^x$ as $y = xe^{-x}$

giving
$$\frac{dy}{dx} = -xe^{-x} + e^{-x}$$

or alternatively, using the method for implicit functions, we have

$$\frac{d}{dx}(x) = \frac{d}{dx}(ye^x) \quad \Rightarrow \quad 1 = ye^x + \frac{dy}{dx}e^x$$

2) If $e^x y = \sin x$ show that $\dfrac{d^2 y}{dx^2} + 2\dfrac{dy}{dx} + 2y = 0$.

The temptation with a problem of this type is to use $e^x y = \sin x$ in the form $y = e^{-x} \sin x$, find $\dfrac{dy}{dx}$ and $\dfrac{d^2 y}{dx^2}$ as functions of x and then show that they satisfy the given equation (called a differential equation).
However a more direct method is to differentiate the given implicit equation.

Thus if
$$e^x y = \sin x$$

then
$$e^x y + e^x \frac{dy}{dx} = \cos x$$

and differentiating again w.r.t. x gives

$$e^x y + e^x \frac{dy}{dx} + e^x \frac{dy}{dx} + e^x \frac{d^2 y}{dx^2} = -\sin x = -e^x y$$

hence
$$e^x \frac{d^2 y}{dx^2} + 2e^x \frac{dy}{dx} + 2e^x y = 0$$

There is no finite value of x for which $e^x = 0$ so we may divide the equation by e^x, giving

$$\frac{d^2y}{dx^2} + 2\frac{dy}{dx} + 2y = 0$$

The Equation of a Tangent

Consider the curve whose equation is $3x^2 - 7y^2 + 4xy - 8x = 0$.
For brevity we may write this as $f(x, y) = 0$ where
$f(x, y) \equiv 3x^2 - 7y^2 + 4xy - 8x$.
The equation of a tangent, or normal, to $f(x, y) = 0$ can be found, *without* transposing the equation to the form $y = g(x)$, as follows.
Differentiating w.r.t. x gives

$$6x - 14y\frac{dy}{dx} + 4\left(y + x\frac{dy}{dx}\right) - 8 = 0$$

$\Rightarrow$
$$\frac{dy}{dx} = \frac{4 - 2y - 3x}{2x - 7y}$$

i.e. $\dfrac{4 - 2y - 3x}{2x - 7y}$ is the gradient function of $f(x, y) = 0$.

At the point $(-1, 1)$ on the curve

$$\frac{dy}{dx} = \frac{4 - 2 + 3}{-2 - 7} = -\frac{5}{9}$$

So the equation of the tangent to the curve at this point is

$$y - 1 = -\tfrac{5}{9}(x + 1)$$

$\Rightarrow$
$$5x + 9y - 4 = 0$$

EXAMPLES 9i (continued)

3) Find the equation of the tangent to $3x^2 - 4y^2 = 9$ at the point (x_1, y_1) on the curve.

Differentiating the equation w.r.t. x gives

$$6x - 8y\frac{dy}{dx} = 0$$

$$\frac{dy}{dx} = \frac{3x}{4y}$$

At the point (x_1, y_1), the gradient is $\dfrac{3x_1}{4y_1}$ and the equation of the tangent is

$$(y - y_1) = \frac{3x_1}{4y_1}(x - x_1)$$

$\Rightarrow$ $3xx_1 - 4yy_1 = 3x_1{}^2 - 4y_1{}^2$

but as (x_1, y_1) is on the curve, $3x_1{}^2 - 4y_1{}^2 = 9$,   so the equation of the tangent can be written

$$3xx_1 - 4yy_1 = 9$$

Note that the equation of the tangent is the same as the equation of the curve except that x^2 is replaced by xx_1 and y^2 is replaced by yy_1.
In general, if $f(x, y)$ is of degree two, the equation of the tangent at (x_1, y_1) to the curve $f(x, y) = 0$ can be written down directly by replacing, in the equation of the curve

x^2, y^2 by xx_1, yy_1 respectively

xy by $\frac{1}{2}(xy_1 + yx_1)$

x, y by $\frac{1}{2}(x + x_1), \frac{1}{2}(y + y_1)$ respectively

It is left to the reader to prove this in Exercise 9i.
Using this result, the equation of the tangent at (x_1, y_1) to

$$3x^2 - 7y^2 + 4xy - 8x = 0$$

can be written down as

$$3xx_1 - 7yy_1 + 2(xy_1 + yx_1) - 4(x + x_1) = 0$$
$\Rightarrow$ $(3x_1 + 2y_1 - 4)x + (2x_1 - 7y_1)y - 4x_1 = 0$

and at the point $(-1, 1)$ this equation becomes

$$-5x - 9y + 4 = 0 \quad \text{or} \quad 5x + 9y - 4 = 0$$

which is consistent with the result on page 272.
Note that this result gives a quick method for *writing down* the equation of a tangent. It must not be used when the derivation of the equation is required.

The Inverse Trigonometric Functions

The inverse trigonometric functions, viz. arcsin, arccos, arctan, are introduced in Chapter 8, where 'arcsin' means 'the principal angle whose sine is' and is called *the inverse sine function* i.e. $\qquad\qquad y = \text{Arcsin } x \iff \sin y = x$

similarly $\qquad\quad y = \text{Arccos } x \iff \cos y = x$

and $\qquad\qquad y = \text{Arctan } x \iff \tan y = x$

Differentiation of   arcsin x

We have already shown that $\dfrac{dy}{dx} \equiv 1 \bigg/ \dfrac{dx}{dy}$

and used it to find $\dfrac{d}{dx} \ln x$ (page 259).

This relationship is also used to find $\dfrac{d}{dx} \text{arcsin } x$.

Let $y = \text{arcsin } x$
so that $x = \sin y$
differentiating w.r.t. y gives

$$\frac{dx}{dy} = \cos y$$

therefore $\qquad\qquad\qquad \dfrac{dy}{dx} = \dfrac{1}{\cos y}$

but $\cos^2 y \equiv 1 - \sin^2 y = 1 - x^2$

Hence $\qquad\qquad\qquad \dfrac{dy}{dx} = \dfrac{1}{\sqrt{(1-x^2)}}$

i.e. $\qquad\qquad \dfrac{d}{dx}(\text{arcsin } x) = \dfrac{1}{\sqrt{(1-x^2)}}$ $\qquad$ [1]

Note that $\quad y = \text{arcsin } x \Rightarrow -\dfrac{\pi}{2} \leqslant y \leqslant \dfrac{\pi}{2},\quad$ because y is a principal value

angle.
For this range of values of y, $\cos \theta \geqslant 0$.
i.e. $\cos y = \sqrt{(1-x^2)}$ $[\text{not} -\sqrt{(1-x^2)}]$.

(Note also that, when asked to find $\dfrac{d}{dx}[f(x)]$, the answer should be given as a
function of x.)

Similarly it can be shown that

$$\frac{d}{dx}(\text{arccos } x) = -\frac{1}{\sqrt{(1-x^2)}} \qquad [2]$$

and $\qquad\qquad \dfrac{d}{dx}(\text{arctan } x) = \dfrac{1}{1+x^2}$ $\qquad\qquad$ [3]

Note. If $\quad y = f(x)$,   the inverse function of x, $f^{-1}(x)$, is implied by $f(y) = x$

or $\qquad\qquad\qquad g(y) = x \iff y = g^{-1}(x)$

Differentiation of a^x

It was shown earlier in this chapter that

$$\frac{d}{dx} a^x = (a^x) \times [\text{the gradient of } y = a^x \text{ at } (0, 1)]$$

The number e was introduced such that $y = e^x$ has a gradient of unity at $(0, 1)$, but we did not at that stage find $\frac{d}{dx} a^x$.

If $y = a^x$ then $\ln y = x \ln a$

differentiating w.r.t. y gives $\quad \dfrac{1}{y} = \dfrac{dx}{dy} \ln a$

$$\Rightarrow \quad\quad\quad\quad \frac{dx}{dy} = \frac{1}{y \ln a}$$

Therefore $\quad\quad\quad\quad \dfrac{dy}{dx} = y \ln a = a^x \ln a$

i.e. $\quad\quad\quad\quad \dfrac{d}{dx} a^x = a^x \ln a$ $\quad\quad\quad\quad\quad\quad$ [4]

The results [1], [2], [3] and [4] are standard results and may be quoted *unless* more explanation is required.

Logarithmic differentiation

We differentiated the function a^x by first taking logs of the equation $y = a^x$. This method is known as logarithmic differentiation and is necessary whenever an exponent contains a variable (unless the base is e). Logarithmic differentiation is also a useful way of simplifying the differentiation of some of the more complicated compound functions.

EXAMPLES 9i (continued)

4) Differentiate x^x w.r.t. x.

Let $\quad\quad\quad\quad\quad\quad\quad\quad y = x^x$

so $\quad\quad\quad\quad\quad\quad\quad\quad \ln y = x \ln x$

thus $\quad\quad\quad\quad\quad\quad\quad \dfrac{1}{y} \dfrac{dy}{dx} = x \dfrac{1}{x} + \ln x$

$\Rightarrow \quad\quad\quad\quad\quad\quad\quad \dfrac{dy}{dx} = y(1 + \ln x)$

Therefore $\quad\quad\quad\quad \dfrac{d}{dx} (x^x) = x^x(1 + \ln x)$

5) Differentiate $\dfrac{x-1}{x\sqrt{(x^2+1)}}$

Let $y = \dfrac{x-1}{x\sqrt{(x^2+1)}}$ so that $\ln y = \ln(x-1) - \ln x - \frac{1}{2}\ln(x^2+1)$.

Therefore

$$\frac{1}{y}\frac{dy}{dx} = \frac{1}{x-1} - \frac{1}{x} - \frac{x}{x^2+1}$$

$$\frac{dy}{dx} = y\left[\frac{-x^3 + 2x^2 + 1}{x(x-1)(x^2+1)}\right]$$

$$= \frac{(x-1)(-x^3 + 2x^2 + 1)}{x(x^2+1)^{\frac{1}{2}}x(x-1)(x^2+1)}$$

$$= \frac{1 + 2x^2 - x^3}{x^2(x^2+1)^{3/2}}$$

6) Find $\dfrac{d}{dx}[\arcsin(ax+b)]$.

Let $y = \arcsin(ax+b)$ and $u \equiv ax+b$

so that $y = \arcsin u$

Therefore $\dfrac{dy}{du} = \dfrac{1}{\sqrt{(1-u^2)}}$ and $\dfrac{du}{dx} = a$

so that $\dfrac{dy}{dx} = \dfrac{1}{\sqrt{(1-u^2)}} \times a = \dfrac{a}{\sqrt{[1-(ax+b)^2]}}$

i.e. $\dfrac{d}{dx}\arcsin(ax+b) = \dfrac{a}{\sqrt{[1-(ax+b)^2]}}$

EXERCISE 9i

Differentiate the following equations w.r.t. x.

1) $x^2 + y^2 = 4$

2) $x^2 + xy + y^2 = 0$

3) $x(x+y) = y^2$

4) $\dfrac{1}{x} + \dfrac{1}{y} = e^y$

5) $\dfrac{1}{x^2} + \dfrac{1}{y^2} = \dfrac{1}{4}$

6) $\dfrac{x^2}{4} - \dfrac{y^2}{9} = 1$

7) $\sin x + \sin y = 1$ 8) $\sin x \cos y = 2$

9) $xe^y = x + 1$ 10) $\sqrt{(1+y)(1+x)} = x$

11) Find $\dfrac{dy}{dx}$ as a function of x if $y^2 = 2x + 1$.

12) Show that $\dfrac{d}{dx}(\arctan x) = \dfrac{1}{1+x^2}$

13) Show that $\dfrac{d}{dx}(\arccos x) = -\dfrac{1}{\sqrt{(1-x^2)}}$

14) Find $\dfrac{d^2y}{dx^2}$ as a function of x if $\sin y + \cos y = x$.

15) Find the gradient of $x^2 + y^2 = 9$ at the point where $x = 1$.

16) If $y \cos x = e^x$ show that $\dfrac{d^2y}{dx^2} - 2\tan x \dfrac{dy}{dx} - 2y = 0$.

17) Find the differential equation given by differentiating $y2^x = 1$ w.r.t. x and find the gradient at the point where $x = 2$.

18) If $y = 3^x$ find $\dfrac{d^2y}{dx^2}$ when $x = -1$.

19) Find the gradient of $y = \arctan x$ when $x = 1$.

20) Sketch the curves:
(a) $y = \arccos x$ (b) $y = \arctan x$.

21) If $\sin y = 2\sin x$ show that:
(a) $\left(\dfrac{dy}{dx}\right)^2 = 1 + 3\sec^2 y$
(b) by differentiating (a) w.r.t. x show that $\dfrac{d^2y}{dx^2} = 3\sec^2 y \tan y$
and hence that $\cot y \dfrac{d^2y}{dx^2} - \left(\dfrac{dy}{dx}\right)^2 + 1 = 0$

22) Differentiate the following functions w.r.t. x:
(a) $a^x + b^x$ (b) $x^{\sin x}$ (c) $(\sin x)^x$
(d) $(x + x^2)^x$ (e) $\dfrac{x}{(x-1)(x+3)^2(x^2-1)}$ (f) $\sqrt{\dfrac{1}{(x^2-1)(3x-4)^4}}$

23) Differentiate the following equations w.r.t. x

(a) $y^x = x$ (b) $x^y = \sin x$

(c) $(1+x)y = \sin^{\frac{1}{2}}x$ (d) $y^{\sin x} = \sqrt{x}$

24) Write down the equation of the tangent to:

(a) $x^2 - 3y^2 = 4y$ (b) $x^2 + xy + y^2 = 3$

at the point (x_1, y_1).

25) Show that the equation of the tangent to $x^2 + xy + y = 0$ at the point (x_1, y_1) is

$$x(2x_1 + y_1) + y(x_1 + 1) + y_1 = 0$$

26) Write down the equation of the tangent at $(1, \frac{1}{3})$ to the curve whose equation is $2x^2 + 3y^2 - 3x + 2y = 0$.

27) Show that the equation of the tangent at (x_1, y_1) to the curve $ax^2 + by^2 + cxy + dx = 0$ is

$$axx_1 + byy_1 + \tfrac{1}{2}c(xy_1 + yx_1) + \tfrac{1}{2}d(x + x_1) = 0$$

PARAMETRIC EQUATIONS

Some relationships between x and y are so complicated that it is easier to express x and y each in terms of a third variable, called a *parameter*.

For example:
$$\begin{cases} x = t^3 \\ y = t^2 - t \end{cases}$$

[1]

[2]

The equations [1] and [2] are called the parametric equations of the curve. By eliminating t from [1] and [2] it is possible to get a direct relationship between x and y, which is the Cartesian equation of the curve. In this case we get $y = x^{\frac{2}{3}} - x^{\frac{1}{3}}$ which is clearly awkward to analyse and it is much simpler to use the parametric equations.

To get an idea of the shape of the curve whose parametric equations are

$$\begin{cases} x = t^3 \\ y = t^2 - t \end{cases}$$

we can find some points on the curve by assigning various values to t, viz:

t	-2	-1	0	1	2
x	-8	-1	0	1	8
y	6	2	0	0	2

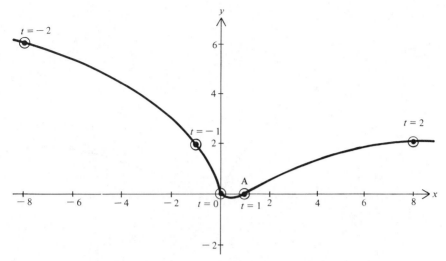

There is no finite value of t for which either x or y is infinite, so it is reasonable to assume that the curve is continuous.

We have, however, assumed that there is one turning point (between O and A). We can use differentiation to confirm this.

Differentiation of a Curve whose Equation is given Parametrically

By using $\dfrac{dy}{dx} = \dfrac{dy}{dt}\dfrac{dt}{dx}$ and $\dfrac{dt}{dx} = 1 \bigg/ \dfrac{dx}{dt}$

we can find $\dfrac{dy}{dx}$ in terms of the parameter t, as follows,

$$y = t^2 - t \quad \text{and} \quad x = t^3$$

$$\frac{dy}{dt} = 2t - 1 \quad \text{and} \quad \frac{dx}{dt} = 3t^2 \quad \Rightarrow \quad \frac{dt}{dx} = \frac{1}{3t^2}$$

$$\frac{dy}{dx} = \frac{dy}{dt}\frac{dt}{dx} = \frac{2t-1}{3t^2}$$

From this we note that:

1) $\dfrac{dy}{dx} = 0$ when $t = \tfrac{1}{2}$

i.e. there *is* a turning point between O and A.

2) $\dfrac{dy}{dx}$ is infinite when $t = 0$.

i.e. the tangent to the curve at O is parallel to the y-axis.

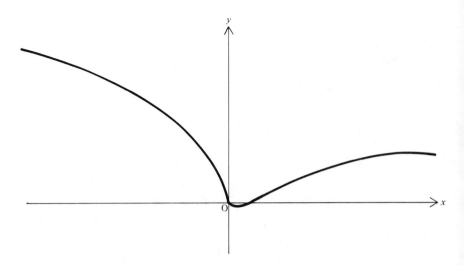

The Second Derivative of Parametric Equations

Consider the equations $\begin{cases} x = \sin\theta \\ y = \cos 2\theta \end{cases}$

$$\frac{dx}{d\theta} = \cos\theta \quad \text{and} \quad \frac{dy}{d\theta} = -2\sin 2\theta$$

Therefore $\dfrac{dy}{dx} = \dfrac{dy}{d\theta} \bigg/ \dfrac{dx}{d\theta} = \dfrac{-2\sin 2\theta}{\cos\theta} = -4\sin\theta$

now $\dfrac{d^2y}{dx^2} = \dfrac{d}{dx}\left(\dfrac{dy}{dx}\right) = \dfrac{d}{dx}(-4\sin\theta) = \dfrac{d}{d\theta}(-4\sin\theta)\dfrac{d\theta}{dx}$

$$= (-4\cos\theta)\left(\frac{1}{\cos\theta}\right) = -4$$

Note that if $x = f(t)$ and $y = g(t)$

then $\dfrac{d^2y}{dx^2} = \dfrac{d}{dx}\left[\dfrac{dy}{dx}\right] = \dfrac{d}{dx}\left[\dfrac{dy}{dt}\dfrac{dt}{dx}\right]$

$$= \frac{d}{dx}\text{ [a product]}$$

so that $\dfrac{d^2y}{dx^2}$ is *not* equal to $\dfrac{d^2y}{dt^2} \times \dfrac{d^2t}{dx^2}$, *nor* to $1 \bigg/ \dfrac{d^2x}{dy^2}$.

Tangents and Normals

If a curve has an equation expressed in the form $y = f(x)$, we may use $[x, f(x)]$ as the coordinates of a general point on that curve. Similarly, for a curve expressed parametrically

i.e.
$$\begin{cases} x = f(t) \\ y = g(t) \end{cases}$$

we may use $[f(t), g(t)]$ as the coordinates of a general point on the curve.

Thus for
$$\begin{cases} x = t^3 \\ y = t^2 - t \end{cases}$$

$(t^3, t^2 - t)$ are the coordinates of any point on the given curve.
Using the coordinates of a general point enables us to find an equation for a tangent to the curve at any point.

For example, for the curve
$$\begin{cases} x = t^2 - 1 \\ y = 3t \end{cases}$$

we find that
$$\frac{dy}{dx} = \frac{3}{2t}$$

This is the gradient function of the curve, i.e. the gradient at a general point on the curve.
Therefore, using $y - y_1 = m(x - x_1)$, the general equation of the tangent at $(t^2 - 1, 3t)$ is

$$y - 3t = \frac{3}{2t}[x - (t^2 - 1)]$$

$\Rightarrow \qquad\qquad 3x - 2ty + (3t^2 + 3) = 0$

The equation of the tangent at a particular point can be found by substituting the value of t at this point into the general equation.

EXAMPLES 9j

1) Find the turning points on the curve
$$\begin{cases} y = t + \dfrac{1}{t} \\ x = t^2 \end{cases}$$

and distinguish between them. Sketch the curve.

$$\left.\begin{array}{l} y = t + \dfrac{1}{t}, \quad \dfrac{dy}{dt} = 1 - \dfrac{1}{t^2} \\[2mm] x = t^2, \quad \dfrac{dx}{dt} = 2t \end{array}\right\} \quad\Rightarrow\quad \dfrac{dy}{dx} = \dfrac{dy}{dt} \Big/ \dfrac{dx}{dt} = \dfrac{t^2 - 1}{2t^3}$$

At turning points, $\qquad \dfrac{dy}{dx} = 0, \quad$ i.e. $\quad \dfrac{t^2 - 1}{2t^3} = 0$

$\Rightarrow$ $$t^2 - 1 = 0$$

$\Rightarrow$ $$t = \pm 1$$

Value of t	$-0.75, -1, -1.25$	$0.75,\ 1,\ 1.25$
Sign of $\dfrac{dy}{dx}$	+, 0, — ╱ — ╲	—, 0, + ╲ — ╱

When $t = -1$, $x = 1$ and $y = -2$
When $t = 1$, $x = 1$ and $y = 2$.
From the table above we see that $(1, -2)$ is a maximum point and that $(1, 2)$ is a minimum point.
To sketch the curve we use our knowledge of the turning points together with the following observations:

$x = t^2 \quad \Rightarrow \quad \begin{cases} x \geqslant 0 \quad \text{for all values of } t & [1] \\ \\ x = 0 \quad \text{when} \quad t = 0 & [2] \end{cases}$

$y = t + \dfrac{1}{t} \quad \Rightarrow \quad \begin{cases} y \text{ is infinite when } t \text{ is zero} & [3] \\ \\ \text{since} \quad y = \dfrac{t^2 + 1}{t}, \quad y \text{ cannot be zero} & [4] \end{cases}$

From [2] and [3] we see that the y-axis is an asymptote.

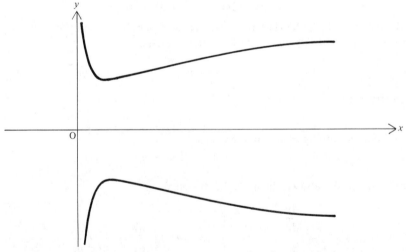

2) The parametric equations of a curve are: $x = \sin^2\theta$, $y = 2\sin\theta$.
Show that the tangent to the curve at the point P $(\sin^2\theta,\ 2\sin\theta)$ has an equation $x - y\sin\theta + \sin^2\theta = 0$ and find the points on the curve at which the tangent is parallel to the y axis.

$$\frac{dy}{d\theta} = 2 \cos \theta \quad \text{and} \quad \frac{dx}{d\theta} = 2 \sin \theta \cos \theta$$

Therefore $\quad \dfrac{dy}{dx} = \dfrac{dy}{d\theta} \Big/ \dfrac{dx}{d\theta} = \dfrac{2 \cos \theta}{2 \sin \theta \cos \theta} = \dfrac{1}{\sin \theta}$

Using $\;y - y_1 = m(x - x_1),\;$ the general tangent to the curve has equation

$$y - 2 \sin \theta = \frac{1}{\sin \theta}(x - \sin^2 \theta)$$

$$y \sin \theta - 2 \sin^2 \theta = x - \sin^2 \theta$$

$$x - y \sin \theta + \sin^2 \theta = 0$$

The tangents which are parallel to the y-axis have infinite gradient.

i.e. $\qquad \dfrac{1}{\sin \theta} = \infty \;\Rightarrow\; \sin \theta = 0 \;\Rightarrow\; x = 0 \;\text{ and }\; y = 0$

So the point where the tangent is parallel to the y-axis is the origin.

EXERCISE 9j

1) Find the Cartesian equation of the curves whose parametric equations are:

(a) $x = t^2$ (b) $x = \cos \theta$ (c) $x = 2t$

    $y = 2t$ $y = \sin \theta$ $y = \dfrac{1}{t}$

2) By taking values of t from -2 to $+2$ at intervals of one unit for (a) and (c), and values of θ from $-\pi$ to $+\pi$ at intervals of $\dfrac{\pi}{4}$ for (b), draw sketches of the curves in (1).

3) Find the gradient function of each curve in (1) as a function of the parameter.

4) Find $\dfrac{d^2 y}{dx^2}$ for each curve in (1) as a function of the parameter.

5) Find the turning points of the curve whose parametric equations are $x = t, \;\; y = t^3 - t,$ and distinguish between them.

6) A curve has parametric equations $\;x = \theta - \cos \theta, \;\; y = \sin \theta.\;$ Find the coordinates of the points at which the gradient of this curve is zero.

7) The parametric equations of a curve are $\;x = e^t, \;\; y = \sin t.$

Find $\dfrac{dy}{dx}$ and $\dfrac{d^2 y}{dx^2}$ as functions of $t.$

Hence show that $\;x^2 \dfrac{d^2 y}{dx^2} + x \dfrac{dy}{dx} + y = 0.$

8) The parametric equations of a curve are $x = t$, $y = \dfrac{1}{t}$. Find the equation of the general tangent to this curve [i.e. the tangent at the point $(t, 1/t)$].

Find in terms of t the coordinates of the points at which the tangent cuts the coordinate axes. Hence show that the area enclosed by this tangent and the coordinate axes is constant.

9) A curve has parametric equations $x = t^2$, $y = 4t$.

Find the equation of the normal to this curve at $(t^2, 4t)$.

Find the coordinates of the points where the normal cuts the coordinate axes. Hence find, in terms of t, the area of the triangle enclosed by the normal and the axes.

SUMMARY

Standard Results

$f(x)$	$\dfrac{d}{dx} f(x)$
$\sin x$	$\cos x$
$\cos x$	$-\sin x$
$\tan x$	$\sec^2 x$
$\sec x$	$\sec x \tan x$
$\operatorname{cosec} x$	$-\operatorname{cosec} x \cot x$
$\cot x$	$-\operatorname{cosec}^2 x$
e^x	e^x
$\ln x$	$\dfrac{1}{x}$
a^x	$a^x \ln a$
$\arcsin x$	$1/\sqrt{1-x^2}$
$\arccos x$	$-1/\sqrt{1-x^2}$
$\arctan x$	$1/(1+x^2)$

General Results

$$\frac{dy}{dx} = 1 \Big/ \frac{dx}{dy}$$

$$\frac{dy}{dx} = \frac{dy}{du}\frac{du}{dx} = \frac{dy}{du} \Big/ \frac{dx}{du}$$

$$\frac{d}{dx}(uv) = u\frac{dv}{dx} + v\frac{du}{dx}$$

$$\frac{d}{dx}\left(\frac{u}{v}\right) = \frac{v\dfrac{du}{dx} - u\dfrac{dv}{dx}}{v^2}$$

Any of these results may be quoted, unless their derivation is asked for.

MULTIPLE CHOICE EXERCISE 9

(Instructions for answering these questions are given on page xii.)

TYPE I

1) $\dfrac{d}{dx}(e^{x^2+1})$ is:

(a) $2x$ (b) $2x\,e^{x^2+1}$ (c) $2x\,e^{2x}$ (d) $(x^2+1)\,e^{x^2}$
(e) $\ln(x^2+1)\,e^{x^2+1}$.

2) If $x^2+y^2=4$ then $\dfrac{dy}{dx}$ is:

(a) $2x+2y$ (b) $4-x^2$ (c) $-\dfrac{x}{y}$ (d) $\dfrac{y}{x}$ (e) $\dfrac{4-x}{y}$.

3) If $y=\cos x+\sin x$, $\dfrac{d^2y}{dx^2}$ is:

(a) $\cos x-\sin x$ (b) $-y$ (c) $\cos 2x$ (d) y^2 (e) $\cos x+\sin x$.

4) $\dfrac{d}{dx}\left(\dfrac{1}{1+x}\right)$ is:

(a) $\dfrac{-1}{(1+x)^2}$ (b) $\dfrac{1}{1-x}$ (c) $\ln(1+x)$ (d) $\dfrac{-1}{1+x^2}$ (e) 1.

5) $\dfrac{d}{dx}\ln\left(\dfrac{x+1}{2x}\right)$ is:

(a) $\dfrac{1}{2}$ (b) $\dfrac{1}{x+1}-\dfrac{1}{2x}$ (c) $\dfrac{2x}{x+1}$ (d) $\dfrac{1}{x+1}+\dfrac{1}{x}$ (e) $\dfrac{1}{x+1}-\dfrac{1}{x}$.

6) $\dfrac{d}{dx}a^x$ is:

(a) xa^{x-1} (b) a^x (c) $x\ln a$ (d) $a^x\ln a$ (e) none of these.

7) If $x=\cos\theta$ and $y=\cos\theta+\sin\theta$, $\dfrac{dy}{dx}$ is:

(a) $1-\cot\theta$ (b) $1-\tan\theta$ (c) $\cot\theta-1$ (d) $\cot\theta+1$
(e) $\dfrac{1}{1-\cot\theta}$.

8) If $y = x^2 - 4$, the graph of the curve $y^2 - 4 = x$ is:

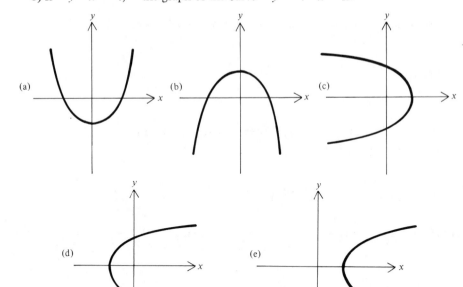

(a)

(b)

(c)

(d)

(e)

9) If $e^x = 7.3$, the value of x, to 2 s.f. is:
(a) 2.0   (b) 1.0   (c) -2.0 (d) 0   (e) 0.2.

10) $\dfrac{d}{dx}(\ln 2x^{\frac{1}{3}})$ is:

(a) $2x^{-\frac{1}{3}}$ (b) $\dfrac{1}{3x}$ (c) $\dfrac{2}{3x}$ (d) $\dfrac{x}{3}$ (e) $\dfrac{2}{3x^{2/3}}$.

TYPE II

11) $f(x) \equiv \dfrac{u}{v}$ where u and v are both functions of x.

(a) $\ln [f(x)] \equiv \ln u - \ln v$.

(b) $\dfrac{d}{dx}[f(x)] = \dfrac{1}{v}\dfrac{du}{dx} - \dfrac{u}{v^2}\dfrac{dv}{dx}$ (c) $e^{f(x)} \equiv e^u - e^v$.

12) $x = f(y)$.

(a) $\dfrac{dy}{dx} = f'(y)$ (b) $\dfrac{d}{dx}[f(y)] = f'(y)\dfrac{dy}{dx}$ (c) $\dfrac{d^2y}{dx^2} = \left(\dfrac{d^2}{dy^2}f(y)\right)\left(\dfrac{dy}{dx}\right)$

13) $xy = e^x$.

(a) $\dfrac{d^2y}{dx^2} + 2\dfrac{dy}{dx} - y = 0$.

(b) $\dfrac{dy}{dx} = \dfrac{1}{x}(e^x - y)$.   (c) $\ln y = x - \ln x$.

14) The parametric equations of a curve are $x = t^2$, $y = t^3$.

(a) $\dfrac{dy}{dx} = 6t^3$.

(b) The curve has only one stationary value.

(c) The curve is symmetrical about the x-axis.

15) $y = g(u)$ where u is a function of x.

(a) $\dfrac{dy}{dx} = g'(u)\dfrac{du}{dx}$ (b) $\dfrac{d^2y}{dx^2} = \dfrac{d^2u}{dx^2}$.

(c) y is equal to a product of functions.

TYPE III

16) (a) $f(x) \equiv e^x \cos x$

   (b) $f(-x) \equiv \dfrac{\cos x}{e^x}$

17) (a) $f(x) \equiv \dfrac{3x}{(x-2)(x+1)}$

   (b) $f(x) \equiv \dfrac{2}{x-2} + \dfrac{1}{x+1}$

18) (a) $y = \ln 3x$

   (b) $\dfrac{dy}{dx} = \dfrac{1}{x}$

19) (a) $y = \sin \theta$, $x = \sin 2\theta$.
   (b) $x^2 + y^2 = 1$.

20) (a) y has a maximum value when $t = 2$.

   (b) $\dfrac{dy}{dx} = 4 - t^2$.

TYPE IV

21) If $x = a \cos \theta$ and $y = b \sin \theta$,   find a and b.

(a) $x = 2$ when $y = 3$ (b) $\dfrac{dy}{dx} = 1$ when $\theta = \dfrac{\pi}{4}$

(c) $-5 \leqslant x \leqslant 5$ for all values of θ.

22) Does $y = ae^x$ satisfy the differential equation $\dfrac{d^2y}{dx^2} + k\dfrac{dy}{dx} + cy = 0$?

(a) $k = -2$ (b) $c = 1$ (c) $a = 2$.

23) Find the maximum value of $y = f(x)$.

(a) $\dfrac{d^2y}{dx^2} < 0$ for all values of x (b) $\dfrac{dy}{dx} = ax^3$

(c) the curve $y = f(x)$ passes through the origin.

24) Sketch the curve whose parametric equations are $x = f(t), \; y = g(t)$.

(a) $f(t)$ is a quadratic function of t.

(b) $g(t)$ is a linear function of t.

(c) the curve passes through the points $(1, 0), (0, 1), (2, 4), (3, 5)$.

TYPE V

25) $\dfrac{d}{dx} (s \times t) = \dfrac{ds}{dx} \times \dfrac{dt}{dx}$.

26) $\dfrac{d}{dy} [f(x)] = f'(x) \Big/ \dfrac{dy}{dx}$.

27) $\dfrac{d}{dx} \left[\ln \dfrac{x}{1+x} \right] = \dfrac{d}{dx} [\ln x] - \dfrac{d}{dx} [\ln (1+x)]$.

28) $y = e^{x^2+1}$ has intercept 1 on the y-axis.

29) $f(\theta) \equiv \sin 2\theta$ has a maximum value of 2.

30) If $f(x) \equiv 1 + x, \; f^{-1}(x) \equiv \dfrac{1}{1+x}$.

MISCELLANEOUS EXERCISE 9

1) (a) If $y = e^{ax} \sin bx$, express $\dfrac{d^2y}{dx^2}$ in the form

$$e^{ax} (A \sin bx + B \cos bx)$$

giving A and B in terms of a and b.

(b) If $y = \dfrac{(1 + 2x)}{(1 - 2x)}$, find $\dfrac{d^2y}{dx^2}$ in its simplest form.

(c) If $x^2 - y^2 = a^2$, find $\dfrac{dy}{dx}$ and $\dfrac{d^2y}{dx^2}$ in terms of x and y.   (U of L)

2) Show that $\dfrac{d}{dx} \left(\dfrac{x}{1+x} \right) = \dfrac{1}{(1+x)^2}$.

A curve is described by the equation

$$\dfrac{y}{1+y} + \dfrac{x}{1+x} - x^2 y^3 = 0$$

Find the equation of the tangent to the curve at the point $(1, 1)$.   (JMB)

3) (a) Differentiate the following functions with respect to x simplifying
your answers where possible:

(i) $\dfrac{1}{x^2}\sqrt{(1 + x^3)}$ (ii) $\ln\left(\dfrac{2 + \cos x}{3 - \sin x}\right)$

(b) If $y = e^{3x} \sin 4x$ show that $\dfrac{d^2y}{dx^2} - 6\dfrac{dy}{dx} + 25y = 0.$ (U of L)

4) (a) Find $\dfrac{dy}{dx}$ when:

(i) $y = \ln(\sec 2x + \tan 2x)$ (ii) $y = \dfrac{(1 + 2x^2)}{(1 + x^2)}$

and simplify your answers.

(b) If $y = \cos\left(e^x + \dfrac{\pi}{4}\right)$,   show that $\dfrac{d^2y}{dx^2} = \dfrac{dy}{dx} - e^{2x}y.$

Find the least positive value of x for which y is a minimum.   (AEB)'76

5) Differentiate with respect to x:

(a) x^x (b) $\arctan\left(\dfrac{1 - x}{1 + x}\right)$

simplifying the results where possible.   (U of L)p

6) (a) For values of $x > 0$,   the equation of a curve is $y = x \ln x$.
Find the coordinates of the turning point on this curve, and determine
whether it is a maximum or a minimum.
Sketch the graph. (You may assume that $y \to 0$ as $x \to 0$.)

(b) A curve is given parametrically by

$$x = t - \frac{1}{t}, \quad y = t + \frac{1}{t}, \quad \text{where} \quad t \neq 0$$

Find the coordinates of the points on the curve where the gradient is zero,
and find the equation of the tangent at the point where $t = 2.$ (C)

7) Show that the graph of $y = \dfrac{ax + b}{cx + d}$ has, in general, no turning points and
that

$$2\left(\frac{dy}{dx}\right)\left(\frac{d^3y}{dx^3}\right) = 3\left(\frac{d^2y}{dx^2}\right)^2$$ (U of L)

8) Given that $x = \sec\theta + \tan\theta$ and $y = \csc\theta + \cot\theta$, show that
$x + \dfrac{1}{x} = 2\sec\theta$ and $y + \dfrac{1}{y} = 2\csc\theta$. Find $\dfrac{dx}{d\theta}$ and $\dfrac{dy}{d\theta}$ in terms of θ,

and hence show that $\dfrac{dy}{dx} = -\dfrac{1 + y^2}{1 + x^2}.$ (JMB)

9) (a) Find $\dfrac{dy}{dx}$ when $2y\,e^{3x} + \dfrac{1}{x^2}\sin 2x = 0$.

(b) If $x = \dfrac{1+t}{1-2t}$ and $y = \dfrac{1+2t}{1-t}$, where t is variable find the value

of $\dfrac{dy}{dx}$ when $t = 0$.

(c) Differentiate with respect to x, $f(x) \equiv 3x + \sin x - 8\sin\frac{1}{2}x$ and deduce that $f(x)$ is positive for $x > 0$.   (AEB)'72

10) Prove that $\dfrac{d}{dx}\arcsin x = \dfrac{1}{(1-x^2)^{\frac{1}{2}}}$.

Given that the variables x and y satisfy the equation

$$\arcsin 2x + \arcsin y + \arcsin (xy) = 0$$

find $\dfrac{dy}{dx}$ when $x = y = 0$.   (JMB)

11) Sketch the curves with the equations $y = e^x$ and $y = x^2 - 1$ on the same diagram.
Using the information gained from your sketches, redraw part of the curves more accurately to find the negative real root of the equation $e^x = x^2 - 1$ to two decimal places.   (U of L)

12) (a) If $y = \sqrt{(5x^2 + 3)}$, show that $y\dfrac{d^2y}{dx^2} + \left(\dfrac{dy}{dx}\right)^2 = 5$.

(b) The parametric equations of a curve are

$$x = 3(2\theta - \sin 2\theta)$$
$$y = 3(1 - \cos 2\theta)$$

The tangent and the normal to the curve at the point P where $\theta = \dfrac{\pi}{4}$

meet the y-axis at L and M respectively. Show that the area of triangle

PLM is $\dfrac{9}{4}(\pi - 2)^2$.   (AEB)'74

13) Express the function $y = \dfrac{2x^2}{(2x-1)(x+1)}$ in partial fractions. Hence, or

otherwise, find the value of $\dfrac{d^2y}{dx^2}$ when $x = 1$.   (JMB)

14) (a) Differentiate $\ln(k\sec x) + a^x$ with respect to x, a and k being constants.

(b) If $x = \sin t$ and $y = \cos 2t$, prove that $\dfrac{d^2y}{dx^2} + 4 = 0$.

(c) Find the maximum and minimum values of $\dfrac{x-3}{x^2-x-2}$ and distinguish between them.   (U of L)

15) (a) If $x^2 y = a \cos nx$, show that

$$x^2 \frac{d^2 y}{dx^2} + 4x \frac{dy}{dx} + (n^2 x^2 + 2)y = 0.$$

(b) A curve is given by the parametric equations
$$x = t^2 + 3, \quad y = t(t^2 + 3).$$

(i) Show that the curve is symmetrical about the x-axis.

(ii) Show that there is no part of the curve for which $x < 3$.

(iii) Find $\dfrac{dy}{dx}$ in terms of t, and show that $\left(\dfrac{dy}{dx}\right)^2 \geqslant 9$.

Using the results (i), (ii), and (iii), sketch the curve.   (C)

16) (a) Differentiate with respect to x:

(i) $\ln [x + \sqrt{(x^2 + 1)}]$

(ii) $\sec^2 2x$

(iii) 10^{3x}

simplifying your answer to (a).

(b) If $x^2 + y^2 = 2y$, find $\dfrac{dy}{dx}$ in terms of x and y without first

finding y in terms of x. Prove that $\dfrac{d^2 y}{dx^2} = \dfrac{1}{(1-y)^3}$.   (U of L)

17) The point P moves in such a way that at time t its Cartesian coordinates with respect to an origin O are $x = e^{-t}, \quad y = 2t\,e^{-t}$.
The distance OP is denoted by r and the angle between OP and the x-axis by θ.
Find in terms of t:

(a) the rate of change of r^2 with respect to t,

(b) the rate of change of θ with respect to t.   (JMB)

18) If $y = x \arctan x$, show that:

(a) $x(1 + x^2) \dfrac{dy}{dx} = x^2 + (1 + x^2)y$

(b) $(1 + x^2) \dfrac{d^2 y}{dx^2} + 2x \dfrac{dy}{dx} - 2y = 2.$

Draw a rough sketch of the curve.   (U of L)p

19) If $\tan y = x$ find the value of $\dfrac{d^2 y}{dx^2}$ when $y = \dfrac{\pi}{4}$.   (U of L)p

20) (a) Differentiate with respect to x:

(i) $x^2 \sin 3x$

(ii) $e^{-2/x}$

(iii) $\left\{\dfrac{x-1}{2-x}\right\}^2$

(b) Given that $y = \ln (1 + \sin x)$, prove that $\dfrac{d^2 y}{dx^2} + e^{-y} = 0$. What

can be deduced from the equation about all the stationary values of y?

(AEB)'74

21) Given that u and v are functions of x, prove from first principles that

$$\frac{d}{dx}\left(\frac{u}{v}\right) = \frac{v \, du/dx - u \, dv/dx}{v^2}$$

Find, in a simplified form, the derivative of the function

$$\frac{2 + \ln (1 + x)^2}{2 - \ln (1 - x)^2} \qquad \text{(JMB)}$$

22) If $x = a(\theta - \sin \theta)$, $y = a(1 - \cos \theta)$, show that $\dfrac{dy}{dx} = \cot \tfrac{1}{2}\theta$. As θ

varies, the point $P(x, y)$ traces out a curve. When $\theta = \tfrac{1}{2}\pi$, P is at the point A
and when $\theta = \tfrac{3}{2}\pi$, P is at the point B. Find the coordinates of the points A
and B and the equations of the tangents to the curve at these two points.

(U of L)

23) Given that $-1 < x < 1$, that $0 < \arccos x < \pi$ and that $(1 - x^2)^{\frac{1}{2}}$
denotes the positive square root of $1 - x^2$, find the derivative of the function

$$f(x) \equiv \arccos x - x(1 - x^2)^{\frac{1}{2}},$$

expressing your answer as simply as possible.
Prove that, as x increases in the interval $-1 < x < 1$, $f(x)$ decreases, and
sketch the graph of $f(x)$ in this interval. (JMB)

CHAPTER 10

INTEGRATION

INDEFINITE INTEGRATION

When x^2 is differentiated with respect to x the derived function is $2x$. Conversely, given that an unknown function has a derived function of $2x$, it is clear that the unknown function could be x^2. This process of finding a function from its derived function is called *integration* and it reverses the operation of differentiation.

The Constant of Integration

Now consider the functions $x^2 + 1$ and $x^2 - 7$.

We have already noted that $\dfrac{d}{dx}(x^2) = 2x$

But we also see that

$$\frac{d}{dx}(x^2 + 1) = 2x \quad \text{and} \quad \frac{d}{dx}(x^2 - 7) = 2x$$

Clearly $2x$ is the derivative not only of x^2, but also of x^2 *plus any constant*. Thus the result of integrating $2x$, which is called the *integral* of $2x$, is not a unique function but is of the form $x^2 + K$ where K is called *the constant of integration*.

This is written

$$\int 2x \, dx = x^2 + K$$

where $\int \ldots dx$ means 'the integral of $\ldots$ w.r.t. x'.

As integration reverses the process of differentiation, for a function $f(x)$ we have

$$\int \frac{d}{dx} f(x) \, dx = f(x) + K$$

Similarly, since differentiating x^3 w.r.t. x gives $3x^2$, we can say

$$\int 3x^2 \, dx = x^3 + K$$

or

$$\int x^2 \, dx = \tfrac{1}{3}x^3 + K$$

(It is not necessary to write $K/3$ in the second form as K represents *any* constant in either expression.)

In general, since differentiating x^{n+1} w.r.t. x gives $(n+1)x^n$, we have

$$\int x^n \, dx = \frac{1}{(n+1)} x^{n+1} + K$$

i.e. to integrate a power of x, *increase* the power by 1 and divide by the *new* power.

This rule can be used to integrate x^n for any value of n *except* -1 (which will be considered later),

e.g.

$$\int x^7 \, dx = \tfrac{1}{8}x^8 + K$$

$$\int x^{-4} \, dx = -\tfrac{1}{3}x^{-3} + K$$

$$\int x^{\frac{1}{2}} \, dx = \tfrac{2}{3}x^{\frac{3}{2}} + K$$

Integrating a Constant, c

The result of differentiating cx is c.

Thus

$$\int c \, dx = cx + K$$

Integrating cx^n

Differentiating cx^{n+1} gives $(n+1)(c)x^n$.

Hence

$$\int (n+1)(c)x^n \, dx = cx^{n+1} + K$$

or

$$\int cx^n \, dx = \frac{c}{(n+1)}x^{n+1} + K$$

Integrating a Sum or Difference of Functions

It was shown in Chapter 5 that differentiation is a distributive process across addition or subtraction. So, since integration reverses the differential operation, integration also is distributive in this respect,

e.g.

$$\int \left(x^3 + \frac{1}{x^2} + \sqrt{x} \right) dx = \int (x^3 + x^{-2} + x^{\frac{1}{2}}) \, dx$$

$$= \int x^3 \, dx + \int x^{-2} \, dx + \int x^{\frac{1}{2}} \, dx$$

$$= \tfrac{1}{4}x^4 - x^{-1} + \tfrac{2}{3}x^{\frac{3}{2}} + K$$

EXERCISE 10a

Integrate the following functions with respect to x.

1) x^6; $x^{\frac{1}{3}}$; x^{-4}; $x^{-\frac{3}{2}}$; $\sqrt[4]{x}$; $\dfrac{1}{x^7}$

2) $2x^2 - \dfrac{1}{x^2} + x$ 3) $\sqrt{x} + \dfrac{1}{\sqrt[3]{x}}$ 4) $3x^3 + x^{-3} + 3$

5) $\dfrac{4}{x^3} - \dfrac{1}{x^2} - x^2$ 6) $2x^{\frac{5}{2}} - x^{-\frac{2}{5}}$ 7) $5x^4 - 3x^2 + 7$

8) $4x^{-3} + x^{-4} + 1$ 9) $3x^{-\frac{1}{2}} - x^{-\frac{3}{2}}$ 10) $\tfrac{1}{2}x - \dfrac{2}{\sqrt{x}} - 1$

11) $\dfrac{1}{x^4} + \dfrac{1}{\sqrt[4]{x}} - 4$ 12) $6\sqrt{x} - 3x^3 + x^{-2} + 2$

Integrating $(ax + b)^n$

First consider the function $f(x) \equiv (2x + 3)^4$.   The derivative of $f(x)$ is found by using the substitution $u \equiv 2x + 3$ so that $f(x) \equiv u^4$, giving

$$\frac{d}{dx}(2x+3)^4 = (4)(2)(2x+3)^3$$

Conversely

$$\int (4)(2)(2x+3)^3 \, dx = (2x+3)^4 + K$$

or

$$\int (2x+3)^3 \, dx = \frac{1}{(4)(2)}(2x+3)^4 + K$$

Similarly considering the general case where $f(x) \equiv (ax + b)^{n+1}$ we find that

$$\frac{d}{dx}(ax + b)^{n+1} = (n + 1)(a)(ax + b)^n$$

Hence
$$\int (ax + b)^n \, dx = \frac{1}{(n + 1)(a)}(ax + b)^{n+1} + K$$

e.g.
$$\int (3x - 8)^6 \, dx = \frac{1}{(7)(3)}(3x - 8)^7 + K$$

$$\int (2 - 5x)^9 \, dx = \frac{1}{(10)(-5)}(2 - 5x)^{10} + K$$

$$\int (\tfrac{1}{2} - \tfrac{1}{3}x)^7 \, dx = \frac{1}{(8)(-\tfrac{1}{3})}(\tfrac{1}{2} - \tfrac{1}{3}x)^8 + K$$

EXERCISE 10b

Integrate the following functions with respect to x.

1) 3; $3x^5$; $(3x)^5$; $3(x + 1)^5$

2) $4x^3 - 5x + 6$ 3) $(2x - 1)^2$ 4) $(2 + 7x)^3$

5) $4(1 - x)^{\frac{1}{2}}$ 6) $3\sqrt{2 - 5x}$ 7) $\dfrac{1}{(4x + 5)^3}$

8) $\dfrac{1}{\sqrt{(1 - 2x)}}$ 9) $\dfrac{1}{\sqrt[3]{(3 - 7x)}}$ 10) $\dfrac{3}{(2x + 1)^3} + \sqrt{1 - 2x}$

11) $4\sqrt{x} + \sqrt{4x + 1} - 4(1 - 3x)^3$ 12) $(px + q)^r$

13) $\dfrac{1}{\sqrt[4]{(3 + 5x)}}$ 14) $\dfrac{3}{\sqrt[3]{(4 - 5x)}}$ 15) $\sqrt{(1 - x)} + \dfrac{1}{\sqrt{(1 - x)}} - \dfrac{1}{(1 - x)^2}$

INTEGRATION OF TRIGONOMETRIC FUNCTIONS

Whenever a function $f'(x)$ is *recognised* as the derivative of a function $f(x)$, then

$$\frac{d}{dx}f(x) = f'(x) \Leftrightarrow \int f'(x) \, dx = f(x) + K$$

Thus, referring to the derivatives of trig functions derived in Chapter 9, we see that

$$\frac{d}{dx}(\sin x) = \cos x \Leftrightarrow \int \cos x \, dx = \sin x + K$$

$$\frac{d}{dx}(\cos x) = -\sin x \Leftrightarrow \int \sin x \, dx = -\cos x + K$$

$$\frac{d}{dx}(\tan x) = \sec^2 x \Leftrightarrow \int \sec^2 x \, dx = \tan x + K$$

$$\frac{d}{dx}(\sec x) = \sec x \tan x \Leftrightarrow \int \sec x \tan x \, dx = \sec x + K$$

$$\frac{d}{dx}(\operatorname{cosec} x) = -\operatorname{cosec} x \cot x \Leftrightarrow \int \operatorname{cosec} x \cot x \, dx = -\operatorname{cosec} x + K$$

$$\frac{d}{dx}(\cot x) = -\operatorname{cosec}^2 x \Leftrightarrow \int \operatorname{cosec}^2 x \, dx = -\cot x + K$$

(The reader should not regard these integrals as six more facts to be memorised. Knowledge of the standard derivatives is clearly sufficient.)

It can be shown, by considering the corresponding differentiation, that

$$\int c \cos x \, dx = c \sin x + K$$

and

$$\int \cos (ax + b) \, dx = \frac{1}{a} \sin (ax + b) + K$$

with similar results for the remaining trig integrals.

e.g.

$$\int 3 \sec^2 x \, dx = 3 \tan x + K$$

$$\int \sin 4\theta \, d\theta = -\tfrac{1}{4} \cos 4\theta + K$$

$$\int \operatorname{cosec}^2 (2x + 3\pi/4) \, dx = -\tfrac{1}{2} \cot (2x + 3\pi/4) + K$$

$$\int \operatorname{cosec} 5\theta \cot 5\theta \, d\theta = -\tfrac{1}{5} \operatorname{cosec} 5\theta + K$$

$$\int \sec (\pi/2 - 6x) \tan (\pi/2 - 6x) \, dx = -\tfrac{1}{6} \sec (\pi/2 - 6x) + K$$

EXERCISE 10c

Integrate the following expressions with respect to x.

1) $\sin 2x$

2) $3 \cos (4x - \pi/2)$

3) $\sec^2 (\pi/3 - 2x)$

4) $\operatorname{cosec}^2 (3x + \pi/6)$

5) $5 \sec (\pi/4 - x) \tan (\pi/4 - x)$

6) $2 \operatorname{cosec} 3x \cot 3x$

7) $2 \sin (3x + \alpha)$

8) $5 \cos (\alpha - x/2)$

9) $\cos 3x + 3 \sin x$

10) $\sec^2 2x - \operatorname{cosec}^2 4x$

INTEGRATION OF EXPONENTIAL FUNCTIONS

It is already known that $\dfrac{d}{dx} e^x = e^x,$

hence
$$\int e^x \, dx = e^x + K$$

Further, we have
$$\frac{d}{dx} (ce^x) = ce^x$$

and
$$\frac{d}{dx} e^{(ax+b)} = ae^{(ax+b)}$$

Hence
$$\int ce^x \, dx = ce^x + K$$

and
$$\int e^{ax+b} \, dx = \frac{1}{a} e^{(ax+b)} + K$$

e.g.
$$\int e^{3x} \, dx = \tfrac{1}{3} e^{3x} + K$$

$$\int 2e^{-5x} \, dx = (2)(-\tfrac{1}{5}) e^{-5x} + K$$

To integrate an exponential function in which the given base is not e but is a, say, the base must first be changed to e, as follows.

Let
$$a^x \equiv e^z$$

Taking logs to the base e gives

$$x \ln a \equiv z$$

Hence
$$a^x \equiv e^{x \ln a}$$

Then
$$\int a^x \, dx \equiv \int e^{x \ln a} \, dx$$

$$= \frac{1}{\ln a} e^{x \ln a} + K$$

i.e.
$$\int a^x \, dx = \frac{1}{\ln a} a^x + K$$

Alternatively, remembering that
$$\frac{d}{dx} (a^x) = (\ln a)a^x$$

it follows that

$$\frac{1}{\ln a} \frac{d}{dx} (a^x) = a^x \Leftrightarrow \int a^x \, dx = \frac{a^x}{\ln a} + K$$

EXERCISE 10d

Integrate the following expressions with respect to x.

1) e^{2x} 2) $3e^{-x}$ 3) e^{4x+1} 4) $4e^{5-3x}$

5) 2^x 6) 3^{2x} 7) 3^{1-x} 8) $\sqrt{e^x}$

9) $e^{2x} + \dfrac{1}{e^{2x}}$ 10) $3e^{-3x} - \frac{1}{2}e^{2x}$

Integration of $\dfrac{1}{x}$

(a) $x > 0$

At first sight it may appear that, since $1/x \equiv x^{-1}$, integrating $1/x$ requires use of the rule $\displaystyle\int x^n \, dx = \dfrac{1}{n+1}x^{n+1} + K$. But when $n = -1$, this method fails.

Taking a second look at $1/x$ it can be *recognised* as the derived function of $\ln x$. But $\ln x$ is defined only when $x > 0$. Hence, provided $x > 0$ we have

$$\frac{d}{dx} \ln x = \frac{1}{x} \Leftrightarrow \int \frac{1}{x} dx = \ln x + K$$

(b) $x < 0$

If x is negative the statement that $\displaystyle\int \frac{1}{x} dx = \ln x + K$ is not valid because the logarithm of a negative number does not exist. This problem is dealt with as follows

$$\int \frac{1}{x} dx = \int \frac{-1}{(-x)} dx \qquad \{-x > 0\}$$

$$= \ln(-x) + K$$

Thus

when $x > 0$ $\displaystyle\int \frac{1}{x} dx = \ln x + K$

when $x < 0$ $\displaystyle\int \frac{1}{x} dx = \ln(-x) + K$

Combining these results, for all values of x we have

$$\int \frac{1}{x} dx = \ln |x| + K$$

where $|x|$ denotes the positive magnitude of x,

e.g. $\qquad\qquad\qquad |-3| = 3; \qquad |-1| = 1$

This result, being fully comprehensive, will be used in all future work of this nature in this book.

Note. If the constant K is replaced by $\ln A$, $(A > 0)$, which also represents any real number, we can say

$$\int \frac{1}{x}\,dx = \ln|x| + \ln A = \ln A|x|$$

Further, $\qquad\qquad \dfrac{d}{dx}[\ln x^c] = \dfrac{d}{dx}[c \ln x] = \dfrac{c}{x}$

$\Leftrightarrow \qquad\qquad\qquad \int \dfrac{c}{x}\,dx = c \ln|x| + K$

Now consider $\qquad \dfrac{d}{dx}\ln(ax + b) = \dfrac{a}{ax + b}$

Hence $\qquad\qquad \int \dfrac{1}{ax + b}\,dx = \dfrac{1}{a}\ln|ax + b| + K$

e.g. $\qquad\qquad\qquad \int \dfrac{4}{x}\,dx = 4 \ln|x| + K$

$$\int \frac{1}{2x + 5}\,dx = \tfrac{1}{2}\ln|2x + 5| + K$$

$$\int \frac{3}{4 - 2x}\,dx = -\tfrac{3}{2}\ln|4 - 2x| + K$$

EXERCISE 10e

Integrate the following expressions with respect to x.

1) $\dfrac{1}{2x}$ $\qquad$ 2) $\dfrac{2}{x}$ $\qquad$ 3) $\dfrac{1}{3x + 1}$ $\qquad\qquad$ 4) $\dfrac{1}{1 - 3x}$

5) $\dfrac{4}{1 + 2x}$ $\qquad$ 6) $\dfrac{3}{4 - 2x}$ $\qquad$ 7) $\dfrac{3x}{(x - 1)(x - 2)}$

(*Hint.* Use partial fractions.)

Questions 8–20 are miscellaneous examples of the types considered so far in this chapter.

8) $\dfrac{1}{(2 - 3x)^3}$ $\qquad\qquad$ 9) $\dfrac{1}{\sqrt{2 - 3x}}$ $\qquad$ 10) $\dfrac{1}{2 - 3x}$

11) $\sin(\pi/2 - 3x)$ $\qquad$ 12) $(4x + 1)^2$ $\qquad$ 13) x^2 $\qquad\qquad$ 14) 4^x

15) e^{4-5x} 16) $\sec^2 4x$ 17) $\dfrac{4}{1-x}$ 18) $\sqrt{2x+3}$

19) e^{6x} 20) $\dfrac{5}{6-7x}$

THE RECOGNITION ASPECT OF INTEGRATION

The ability to integrate a given function often depends primarily on *recognising it as a derived function*. Such identification applied to certain groups of functions leads to simple integration rules, as we have seen in the preceding paragraphs.

It is equally important to use the recognition process to avoid serious errors in integration. Consider, for instance, the derived function of the product $\dfrac{x^2}{2}\cos x$.

Using

$$\frac{d}{dx}(uv) = v\frac{du}{dx} + u\frac{dv}{dx}$$

gives

$$\frac{d}{dx}\left(\frac{x^2}{2}\cos x\right) = x\cos x - \frac{x^2}{2}\sin x$$

Clearly the derived function is not a simple product.

Conversely the integral of a simple product is not itself a product,

i.e. integration is not distributive when applied to a product.

INTEGRATING PRODUCTS

Consider the function e^u where u is a function of x. Differentiating this function of a function gives

$$\frac{d}{dx}(e^u) = \frac{du}{dx}e^u$$

In this case the derived function *is* a product.

Thus any product of the form $\left(\dfrac{du}{dx}\right)e^u$ can be integrated by recognition, since

$$\int\left(\frac{du}{dx}\right)e^u\,dx = e^u + K$$

e.g.

$$\int 2x\,e^{x^2}\,dx = e^{x^2} + K \qquad (u \equiv x^2)$$

$$\int \cos x\,e^{\sin x}\,dx = e^{\sin x} + K \quad (u \equiv \sin x)$$

$$\int x^2\,e^{x^3}\,dx = \tfrac{1}{3}\int 3x^2\,e^{x^3}\,dx = \tfrac{1}{3}e^{x^3} + K \quad (u \equiv x^3)$$

In these simple examples, the method of integration uses a *mental* change of variable (from x to u). A similar approach can be made to the integration of similar, but slightly less simple, functions.

Changing the Variable

Let us consider a more general function $f(u)$ where u is a function of x, so that

$$\frac{d}{dx} f(u) = \frac{du}{dx} f'(u) \qquad \text{where } f'(u) \text{ denotes } \frac{d}{du} f(u).$$

From this we see that any product of the form $\left(\frac{du}{dx}\right) f'(u)$ can be integrated using

$$\int \left(\frac{du}{dx}\right) f'(u) \, dx = f(u) + K \qquad\qquad [1]$$

But

$$f'(u) \equiv \frac{d}{du} f(u)$$

so

$$\int f'(u) \, du = f(u) + K \qquad\qquad [2]$$

Comparing [1] and [2] gives

$$\int \frac{du}{dx} f'(u) \, dx = \int f'(u) \, du$$

or

$$\int f'(u) \frac{du}{dx} \, dx = \int f'(u) \, du$$

i.e.

$$\int \ldots \frac{du}{dx} \, dx = \int \ldots du$$

Thus $\frac{du}{dx} dx$ and du are equivalent operators (i.e. give the same results when applied to the same function).

In practice such an equivalence can be found more simply.
Suppose we have to integrate $2x(x^2 + 3)^7$ w.r.t. x.
Making the substitution $u \equiv x^2 + 3$ we have

$$\frac{du}{dx} = 2x$$

But

$$\ldots \frac{du}{dx} \, dx \equiv \ldots du$$

So

$$\ldots 2x \, dx \equiv \ldots du$$

This pair of equivalent operators can be obtained directly from

$$\frac{du}{dx} = 2x$$

if we *separate the variables*, i.e. treat $\dfrac{du}{dx}$ as though it were a fraction, giving

$$\dots du \equiv \dots 2x\, dx$$

Note that this is not an equation or an identity; it is an *equivalence of operators*. Certain functions can be integrated in this way by choosing a suitable substitution (i.e. change of variable). Products in which one factor is basically the derivative of the function in the other factor, respond to this approach.

EXAMPLES 10f

1) Integrate $x^2\sqrt{x^3 + 5}$ w.r.t. x.

Let $\qquad\qquad u \equiv x^3 + 5 \;\Rightarrow\; \dfrac{du}{dx} = 3x^2$

or $\qquad\qquad\qquad\qquad \dots du \equiv \dots 3x^2\, dx$

Thus $\qquad\displaystyle\int x^2\sqrt{x^3 + 5}\; dx \equiv \tfrac{1}{3}\int (x^3 + 5)^{\frac{1}{2}} 3x^2\, dx$

$$\equiv \tfrac{1}{3}\int u^{\frac{1}{2}}\, du$$

$$= (\tfrac{1}{3})(\tfrac{2}{3})u^{\frac{3}{2}} + K$$

i.e. $\qquad\displaystyle\int x^2\sqrt{x^3 + 5}\; dx = \tfrac{2}{9}(x^3 + 5)^{\frac{3}{2}} + K$

2) Find $\displaystyle\int \cos x \sin^3 x\, dx$.

Writing the given integral in the form

$$\int \cos x\, (\sin x)^3\, dx,$$

we see that a suitable substitution is

$$u \equiv \sin x \;\Rightarrow\; \dots du \equiv \dots \cos x\, dx$$

Thus $\qquad\displaystyle\int \cos x \sin^3 x\, dx \equiv \int (\sin x)^3 \cos x\, dx$

$$\equiv \int u^3\, du$$

$$= \frac{u^4}{4} + K$$

i.e. $$\int \cos x \sin^3 x \, dx = \tfrac{1}{4} \sin^4 x + K$$

Note. The method used above also shows that, in general

$$\int \cos \theta \sin^n \theta \, d\theta = \frac{1}{n+1} \sin^{n+1} \theta + K$$

Similarly we can show that

$$\int \sin \theta \cos^n \theta \, d\theta = \frac{-1}{n+1} \cos^{n+1} \theta + K$$

3) Find $\displaystyle\int \frac{\ln x}{x} \, dx$

As $$\int \frac{\ln x}{x} \, dx \equiv \int \left(\frac{1}{x}\right) \ln x \, dx$$

we see that a suitable change of variable is

$$u \equiv \ln x \quad \Rightarrow \quad \dots du \equiv \dots \frac{1}{x} \, dx$$

Thus $$\int \frac{1}{x} \ln x \, dx \equiv \int u \, du$$

$$= \frac{u^2}{2} + K$$

i.e. $$\int \frac{\ln x}{x} \, dx = \tfrac{1}{2}(\ln x)^2 + K$$

Note. $(\ln x)^2$ is *not* the same as $\ln x^2$.
The reader may well find that, after some practice, certain problems where a change of variable is appropriate can be integrated at sight.

EXERCISE 10f

Integrate the following expressions with respect to x.

1) $4x^3 \, e^{x^4}$ 2) $\sin x \, e^{\cos x}$ 3) $\sec^2 x \, e^{\tan x}$

4) $(2x+1) \, e^{(x^2+x)}$ 5) $\operatorname{cosec}^2 x \, e^{(1-\cot x)}$

Find the following integrals by making the substitution suggested.

6) $\displaystyle\int x(x^2 - 3)^4 \, dx$ $u \equiv x^2 - 3$

7) $\displaystyle\int x\sqrt{1 - x^2} \, dx$ $u \equiv 1 - x^2$

8) $\displaystyle\int \cos 2x (\sin 2x + 3)^2 \, dx$ $u \equiv \sin 2x + 3$

9) $\int x^2(1-x^3)\,dx$ $\qquad\qquad\qquad u \equiv 1-x^3$

10) $\int e^x\sqrt{1+e^x}\,dx$ $\qquad\qquad u \equiv 1+e^x$

11) $\int \cos x\,\sin^4 x\,dx$ $\qquad\qquad u \equiv \sin x$

12) $\int \sec^2 x\,\tan^3 x\,dx$ $\qquad\qquad u \equiv \tan x$

13) $\int x^n(1+x^{n+1})^2\,dx$ $\qquad\qquad u \equiv 1+x^{n+1}$

14) $\int \mathrm{cosec}^2 x\,\cot^2 x\,dx$ $\qquad\qquad u \equiv \cot x$

15) $\int \sqrt{x}\,\sqrt{(1+x^{\frac{3}{2}})}\,dx$ $\qquad\qquad u \equiv 1+x^{\frac{3}{2}}$

By using a suitable substitution, or by integrating at sight, find

16) $\int x^3(x^4+4)^2\,dx$ $\qquad\qquad$ 17) $\int e^x(1-e^x)^3\,dx$

18) $\int \sin\theta\sqrt{1-\cos\theta}\,d\theta$ $\qquad\qquad$ 19) $\int (x+1)\sqrt{x^2+2x+3}\,dx$

20) $\int x\,e^{x^2+1}\,dx$

FURTHER INTEGRATION OF PRODUCTS

Integration by Parts

Many products cannot be expressed in the form $\dfrac{du}{dx}f'(u)$ and so cannot be integrated by the previous method.

A different approach being needed, we look again at the differentiation of a product uv where u and v are both functions of x,

i.e. $\qquad\qquad\qquad \dfrac{d}{dx}(uv) = v\dfrac{du}{dx} + u\dfrac{dv}{dx}$

Isolating one of the products on the R.H.S. gives

$$v\dfrac{du}{dx} = \dfrac{d}{dx}(uv) - u\dfrac{dv}{dx}$$

Now $v\dfrac{du}{dx}$ can be taken to represent a product which is to be integrated w.r.t. x

Thus
$$\int v \frac{du}{dx} dx = \int \frac{d}{dx}(uv)\, dx - \int u \frac{dv}{dx} dx$$

i.e.
$$\int v \frac{du}{dx} dx = uv - \int u \frac{dv}{dx} dx$$

Integrating a product by using this formula is called integrating by parts. Care must be exercised in the choice of the factor to be replaced by v. The aim must be to ensure that $u \frac{dv}{dx}$ is simpler to integrate than $v \frac{du}{dx}$.

EXAMPLES 10g

1) Integrate $x\,e^x$ w.r.t. x.

Let
$$\begin{cases} v = x \\[2mm] \dfrac{du}{dx} = e^x \end{cases} \Rightarrow \begin{cases} \dfrac{dv}{dx} = 1 \\[2mm] u = e^x \end{cases}$$

Using
$$\int v \frac{du}{dx} dx = uv - \int u \frac{dv}{dx} dx$$

gives
$$\int x\,e^x\,dx = (e^x)(x) - \int (e^x)(1)\,dx$$
$$= x\,e^x - e^x + K$$

Hence
$$\int x\,e^x\,dx = e^x(x - 1) + K$$

2) Find $\displaystyle\int x^2 \sin x\,dx$.

Let
$$\begin{cases} v = x^2 \\[2mm] \dfrac{du}{dx} = \sin x \end{cases} \Rightarrow \begin{cases} \dfrac{dv}{dx} = 2x \\[2mm] u = -\cos x \end{cases}$$

Then
$$\int v \frac{du}{dx} dx = uv - \int u \frac{dv}{dx} dx$$

gives
$$\int x^2 \sin x\,dx = (-\cos x)(x^2) - \int (-\cos x)(2x)\,dx$$
$$= -x^2 \cos x + 2 \int x \cos x\,dx$$

At this stage the integral on the R.H.S. cannot be found without *repeating* the process of integrating by parts.

Thus, for $\int x \cos x \, dx$,

let

$$\begin{cases} v = x \\ \dfrac{du}{dx} = \cos x \end{cases} \Rightarrow \begin{cases} \dfrac{dv}{dx} = 1 \\ u = \sin x \end{cases}$$

giving

$$\int x \cos x \, dx = (\sin x)(x) - \int (\sin x)(1) \, dx$$

$$= x \sin x + \cos x + K$$

Hence

$$\int x^2 \sin x \, dx = -x^2 \cos x + 2x \sin x + 2 \cos x + K$$

3) Find $\int x^4 \ln x \, dx$.

Because $\ln x$ can be differentiated but *not* integrated, we are obliged to take $v = \ln x$.

Thus, let

$$\begin{cases} v = \ln x \\ \dfrac{du}{dx} = x^4 \end{cases} \Rightarrow \begin{cases} \dfrac{dv}{dx} = \dfrac{1}{x} \\ u = \tfrac{1}{5}x^5 \end{cases}$$

The formula for integrating by parts then gives

$$\int x^4 \ln x \, dx = (\tfrac{1}{5}x^5)(\ln x) - \int (\tfrac{1}{5}x^5)(1/x) \, dx$$

$$= \tfrac{1}{5}x^5 \ln x - \tfrac{1}{5} \int x^4 \, dx$$

$$\Rightarrow \qquad \int x^4 \ln x \, dx = \tfrac{1}{5}x^5 \ln x - \tfrac{1}{25}x^5 + K$$

Special Cases of Integration by Parts

An interesting situation arises when an attempt is made to integrate $e^x \cos x$.

Integrating by parts, let

$$\begin{cases} v = e^x \\ \dfrac{du}{dx} = \cos x \end{cases} \Rightarrow \begin{cases} \dfrac{dv}{dx} = e^x \\ u = \sin x \end{cases}$$

Hence

$$\int e^x \cos x \, dx = e^x \sin x - \int e^x \sin x \, dx \qquad\qquad [1]$$

But $\int e^x \sin x \, dx$ is very similar to $\int e^x \cos x \, dx$, so apparently we have made no progress.

However if we now apply integration by parts to $\int e^x \sin x \, dx$

we have
$$\begin{cases} v = e^x \\ \dfrac{du}{dx} = \sin x \end{cases} \Rightarrow \begin{cases} \dfrac{dv}{dx} = e^x \\ u = -\cos x \end{cases}$$

so that
$$\int e^x \sin x = -e^x \cos x + \int e^x \cos x \, dx$$

or
$$\int e^x \cos x \, dx = e^x \cos x + \int e^x \sin x \, dx \qquad [2]$$

Adding [1] and [2] gives

$$2 \int e^x \cos x \, dx = e^x (\sin x + \cos x) + K$$

Clearly the same two equations also give

$$2 \int e^x \sin x \, dx = e^x (\sin x - \cos x) + K$$

(**Note.** Neither of the equations [1] and [2] contains a completed integration process, so the constant of integration is introduced when these two equations are combined.)

Integration of ln x

So far we have found no means of integrating $\ln x$. But now, by regarding $\ln x$ as a product $(1)(\ln x)$, the method of integration by parts can be applied as follows:

Let
$$\begin{cases} v = \ln x \\ \dfrac{du}{dx} = 1 \end{cases} \Rightarrow \begin{cases} \dfrac{dv}{dx} = \dfrac{1}{x} \\ u = x \end{cases}$$

Then
$$\int v \frac{du}{dx} \, dx = uv - \int u \frac{dv}{dx} \, dx$$

becomes
$$\int \ln x \, dx = x \ln x - \int x \left(\frac{1}{x} \right) dx$$

$$= x \ln x - x + K$$

i.e.
$$\int \ln x \, dx = x (\ln x - 1) + K$$

This 'trick' of multiplying a function by 1 to convert it into a product can also be used to integrate arcsin x and other inverse trig functions. (See Exercise 10h, Questions 25 to 27.)

EXERCISE 10g

Integrate the following functions w.r.t. x

1) $x \cos x$

2) $x^2 e^x$

3) $x^3 \ln 3x$

4) $x e^{-x}$

5) $3x \sin x$

6) $e^x \sin 2x$

7) $e^{2x} \cos x$

8) $x^2 e^{4x}$

9) $e^{-x} \sin x$

10) $\ln 2x$

11) $e^x(x + 1)$

12) $x(1 + x)^7$

13) $x \sin \left(x + \dfrac{\pi}{6}\right)$

14) $x \cos nx$

15) $x^n \ln x$

16) $3x \cos 2x$

17) $2e^x \sin x \cos x$

18) $x^2 \sin x$

19) $e^{ax} \sin bx$

20) By writing $\cos^3\theta$ as $(\cos^2\theta)(\cos\theta)$ use integration by parts to find $\displaystyle\int \cos^3 \theta \, d\theta$.

Each of the following products can be integrated either:

(a) by immediate recognition, or

(b) by a suitable change of variable, or

(c) by parts.

Choose the best method in each case and hence integrate each function.

21) $(x - 1) e^{x^2 - 2x + 4}$

22) $(x + 1)^2 e^x$

23) $\sin x(4 + \cos x)^3$

24) $\cos x \, e^{\sin x}$

25) $x^4\sqrt{1 + x^5}$

26) $e^x(e^x + 2)^4$

27) $x \cos \left(\dfrac{\pi}{4} - x\right)$

28) $x \, e^{2x-1}$

29) $x(1 - x^2)^9$

30) $\cos x \sin^5 x$

INTEGRATING FRACTIONS

Type I

Consider first a function $\ln u$ where u is a function of x. Differentiating with respect to x gives

$$\frac{d}{dx} \ln u = \left(\frac{1}{u}\right)\left(\frac{du}{dx}\right) \quad \text{or} \quad \frac{du/dx}{u}$$

So

$$\int \frac{du/dx}{u} \, dx = \ln |u| + K$$

or, writing $f(x)$ and $f'(x)$ for u and $\dfrac{du}{dx}$ respectively

$$\int \frac{f'(x)}{f(x)}\,dx = \ln|f(x)| + K$$

Thus all fractions of the form $f'(x)/f(x)$ can be integrated immediately by recognition

e.g. $\displaystyle\int \frac{\cos x}{1 + \sin x}\,dx = \ln|1 + \sin x| + K$ $\left(\dfrac{d}{dx}(1 + \sin x) = \cos x\right)$

$\displaystyle\int \frac{x^2}{1 + x^3}\,dx \equiv \frac{1}{3}\int \frac{3x^2}{1 + x^3}\,dx = \frac{1}{3}\ln|1 + x^3| + K$

$\left(\dfrac{d}{dx}(1 + x^3) = 3x^2\right)$

$\displaystyle\int \frac{e^x}{e^x + 4}\,dx = \ln|e^x + 4| + K$

But $\displaystyle\int \frac{x}{\sqrt{1 + x}}\,dx \neq \ln|\sqrt{1 + x}| + K$ $\left(\dfrac{d}{dx}\sqrt{1 + x} \neq x\right)$

i.e. integrals are of Type I *only* when the numerator is basically the derivative of the *complete denominator*.

An integral whose numerator is the derivative, not of the complete denominator, but of a function *within* the denominator, belongs to the following group.

Type II (Changing the Variable)

An example of this category is $\displaystyle\int \frac{2x}{\sqrt{(x^2 + 1)}}\,dx$.

Noting that $2x$ is the derivative of $x^2 + 1$ we use a change of variable based on $x^2 + 1$.

Let $\qquad\qquad u \equiv x^2 + 1 \Rightarrow \ldots du \equiv \ldots 2x\,dx$

Thus $\qquad\qquad \displaystyle\int \frac{2x}{\sqrt{(x^2 + 1)}}\,dx \equiv \int \frac{du}{\sqrt{u}}$

But $\qquad\qquad \displaystyle\int u^{-\frac{1}{2}}\,du = 2u^{\frac{1}{2}} + K$

Hence $\qquad\qquad \displaystyle\int \frac{2x}{\sqrt{(x^2 + 1)}}\,dx = 2\sqrt{x^2 + 1} + K$

Similarly for $\displaystyle\int \frac{\cos x}{\sin^4 x}\,dx$ we note that $\cos x$ is the derivative of $\sin x$ (which is a function within the denominator) and proceed as follows:

Let $\qquad\qquad u \equiv \sin x \Rightarrow \ldots du \equiv \ldots \cos x\,dx$

Then $$\int \frac{\cos x}{\sin^4 x} \, dx \equiv \int \frac{1}{u^4} \, du$$

$$= -\tfrac{1}{3} u^{-3} + K$$

$\Rightarrow$ $$\int \frac{\cos x}{\sin^4 x} \, dx = K - \frac{1}{3 \sin^3 x}$$

Type III

A fraction which has not already fallen into one of the earlier categories, may be suitable for conversion to partial fractions before integration is attempted,

e.g. $$\frac{1}{(x+1)(x+2)} \equiv \frac{1}{x+1} - \frac{1}{x+2}$$

so $$\int \frac{1}{(x+1)(x+2)} \, dx \equiv \int \frac{1}{x+1} \, dx - \int \frac{1}{x+2} \, dx$$

$$= \ln |x+1| - \ln |x+2| + K$$

$$= \ln \left| \frac{x+1}{x+2} \right| + K$$

Note. Only proper fractions can be converted directly into partial fractions. *An improper fraction must first be divided out until the remaining fraction is proper.*

e.g. to find $\int \dfrac{x^2+1}{(x+1)(x-1)} \, dx$, we first convert $\dfrac{x^2+1}{(x+1)(x-1)}$ as follows:

$$\frac{x^2+1}{(x+1)(x-1)} \equiv \frac{x^2+1}{x^2-1} \equiv 1 + \frac{2}{x^2-1} \equiv 1 + \frac{2}{(x-1)(x+1)}$$

$$\equiv 1 + \frac{1}{x-1} - \frac{1}{x+1}$$

Hence $$\int \frac{x^2+1}{(x+1)(x-1)} \, dx \equiv \int \left(1 + \frac{1}{x-1} - \frac{1}{x+1} \right) dx$$

$$= x + \ln |x-1| - \ln |x+1| + K$$

Some very simple improper fractions do not require transformation into partial fractions, but it is still essential to reduce them to proper fraction form.

e.g. $$\frac{x+2}{x+1} \equiv \frac{x+1+1}{x+1} \equiv 1 + \frac{1}{x+1}$$

Hence $$\int \frac{x+2}{x+1} \, dx \equiv \int \left(1 + \frac{1}{x+1} \right) dx$$

$$= x + \ln |x+1| + K$$

It is very important to identify the correct category when integrating a given fraction. Otherwise lengthy or fruitless attempts are likely to be made. Correct identification requires really careful scrutiny as fractions requiring different integration techniques often *look* very similar. This is demonstrated by the following example

EXAMPLES 10h

1) Integrate the following expressions with respect to x:

(a) $\dfrac{x+1}{x^2+2x-8}$ (b) $\dfrac{x+1}{(x^2+2x-8)^4}$ (c) $\dfrac{x+2}{x^2+2x-8}$

When integrating any fraction, the check-points used to identify the type of integral are, in order:

(i) is the numerator the basic derivative of the complete denominator

i.e. is the integral of the form $\displaystyle\int \dfrac{f'(x)}{f(x)}\, dx$?

(ii) is the numerator the derivative of a function within the denominator? (use a change of variable)

(iii) are partial fractions possible?

So we apply these checks to the given problem.

(a) As the derivative of x^2+2x-8 is $2x+2$ or $2(x+1)$ we identify this integral as belonging to the group $\displaystyle\int \dfrac{f'(x)}{f(x)}\, dx$.

Thus $\displaystyle\int \dfrac{\frac{1}{2}(2x+2)}{x^2+2x-8}\, dx \;=\; \tfrac{1}{2}\ln|x^2+2x-8| + K$

(b) This time the numerator is basically the derivative of the function x^2+2x-8 *within* the denominator so we use

$$u \equiv x^2+2x-8 \;\Rightarrow\; \ldots\, du \equiv \ldots (2x+2)\, dx$$

Thus $\displaystyle\frac{1}{2}\int \dfrac{2x+2}{(x^2+2x-8)^4}\, dx \;\equiv\; \frac{1}{2}\int \dfrac{1}{u^4}\, du$

$$=\; \frac{1}{2}\left(\dfrac{u^{-3}}{-3}\right) + K$$

i.e. $\displaystyle\int \dfrac{x+1}{(x^2+2x-8)^4}\, dx \;=\; K - \dfrac{1}{6(x^2+2x-8)^3}$

(c) In $\displaystyle\int \dfrac{x+2}{x^2+2x-8}\, dx$ the numerator is not related to the derivative of the denominator so we try partial fractions (using the cover-up method).

$$\dfrac{x+2}{x^2+2x-8} \;\equiv\; \dfrac{x+2}{(x+4)(x-2)} \;\equiv\; \dfrac{1/3}{x+4} + \dfrac{2/3}{x-2}$$

Thus $\int \dfrac{x+2}{x^2+2x-8}\,dx \equiv \dfrac{1}{3}\int \dfrac{1}{x+4}\,dx + \dfrac{2}{3}\int \dfrac{1}{x-2}\,dx$

$$= \tfrac{1}{3}\ln|x+4| + \tfrac{2}{3}\ln|x-2| + K$$

or $\ln A\,|(x+4)^{\frac{1}{3}}(x-2)^{\frac{2}{3}}|$

2) Find $\int \tan x\,dx$ by writing $\tan x$ as $\dfrac{\sin x}{\cos x}$

$$\int \dfrac{\sin x}{\cos x}\,dx \equiv -\int \dfrac{-\sin x}{\cos x}\,dx$$

$$\equiv -\int \dfrac{f'(x)}{f(x)}\,dx \quad \text{where} \quad f(x) \equiv \cos x$$

Thus $\int \tan x\,dx = -\ln|\cos x| + K$

$$= K - \ln|\cos x|$$

or $K + \ln|\sec x|$

Note. This result is important and quotable.

EXERCISE 10h

Integrate the following functions w.r.t. x

1) $\dfrac{\cos x}{3+\sin x}$ 2) $\dfrac{e^x}{1-e^x}$ 3) $\dfrac{2x}{1+x^2}$ 4) $\dfrac{\sec^2 x}{1-3\tan x}$

5) $\dfrac{1}{1+e^x}$ (*Hint*: multiply throughout by e^{-x}) 6) $\dfrac{x^3}{1+x^4}$

7) $\dfrac{\cos 2x}{4\sin x\cos x + 1}$ 8) $\dfrac{\tan x}{1+\cos x}$ (*Hint*: multiply throughout by $\sec x$)

9) $\dfrac{1}{x\ln x}$ $\left(\text{i.e. } \dfrac{1/x}{\ln x}\right)$ 10) $\dfrac{2x+3}{x^2+3x+4}$

Integrate the following trig functions by writing each as a fraction as indicated.

11) $\cot x$ $\left(\dfrac{\cos x}{\sin x}\right)$ 12) $\sec x$ $\left(\dfrac{\sec x\,(\sec x + \tan x)}{(\tan x + \sec x)}\right)$

13) $\csc x$ $\left(\dfrac{\csc x\,(\csc x + \cot x)}{(\cot x + \csc x)}\right)$

Integrate the following w.r.t. x

14) $\dfrac{x}{\sqrt{(x^2 + 1)}}$

15) $\dfrac{\cos x}{\sin^6 x}$

16) $\dfrac{\cos x}{\sqrt{(1 + \sin x)}}$

17) $\dfrac{e^x}{(e^x + 4)^2}$

18) $\dfrac{\sec^2 x}{\tan^3 x}$

19) $\dfrac{\sin x}{\cos^n x}$

20) $\dfrac{\cos x}{\sin^n x}$

21) $\dfrac{e^x}{\sqrt{(1 + e^x)}}$

22) $\dfrac{\sec x \tan x}{3 - \sec x}$

23) $\dfrac{\csc^2 x}{(2 + \cot x)^4}$

24) $\dfrac{x - 1}{3x^2 - 6x + 1}$

25) $\arcsin x$

26) $\arctan x$

27) $\arccos x$

Use partial fractions to find the following integrals.

28) $\displaystyle\int \dfrac{x}{x + 1}\,dx$

29) $\displaystyle\int \dfrac{x^2 - 2}{x^2 - 1}\,dx$

30) $\displaystyle\int \dfrac{x^2}{(x + 1)(x + 2)}\,dx$

31) $\displaystyle\int \dfrac{x + 4}{x}\,dx$

32) $\displaystyle\int \dfrac{x + 4}{x + 1}\,dx$

33) $\displaystyle\int \dfrac{2x}{(x - 2)(x + 2)}\,dx$

34) $\displaystyle\int \dfrac{3u + 4}{u(u + 1)}\,du$

35) $\displaystyle\int \dfrac{x^2 + x + 5}{x(x + 1)}\,dx$

36) $\displaystyle\int \dfrac{3 - y}{(y - 1)(y - 2)}\,dy$

37) $\displaystyle\int \dfrac{2z - 5}{z^2 - 5z + 6}\,dz$

38) $\displaystyle\int \dfrac{12x}{(2 - x)(3 - x)(4 - x)}\,dx$

39) $\displaystyle\int \dfrac{x^2 + 2x + 4}{(2x - 1)(x^2 - 1)}\,dx$

40) $\displaystyle\int \dfrac{4u^2 + 3u - 2}{(u + 1)(2u + 3)}\,du$

Standard Integrals

Some of the results obtained in this exercise, together with the integral of $\tan x$ [Example 10h (2)] are useful and can be quoted.

$$\int \tan x \, dx = \ln |\sec x| + K$$

$$\int \cot x \, dx = \ln |\sin x| + K$$

$$\int \sec x \, dx = \ln |\sec x + \tan x| + K$$

$$\int \csc x \, dx = -\ln |\csc x + \cot x| + K$$

INTEGRATION OF SOME TRIGONOMETRIC EXPRESSIONS

Even Powers of sin θ or cos θ

The double angle trig identities are useful here,

e.g. to find $\int \cos^4\theta \, d\theta$ we use

$$\cos^2\theta \equiv \tfrac{1}{2}(1 + \cos 2\theta)$$

$\Rightarrow$ $$\cos^4\theta \equiv \tfrac{1}{4}(1 + 2\cos 2\theta + \cos^2 2\theta)$$

$$\equiv \tfrac{1}{4}[1 + 2\cos 2\theta + \tfrac{1}{2}(1 + \cos 4\theta)]$$

Thus $$\int \cos^4\theta \, d\theta \equiv \int (\tfrac{3}{8} + \tfrac{1}{2}\cos 2\theta + \tfrac{1}{8}\cos 4\theta) \, d\theta$$

$$= \tfrac{3}{8}\theta + \tfrac{1}{4}\sin 2\theta + \tfrac{1}{32}\sin 4\theta + K$$

Odd Powers of sin θ or cos θ

In this case we can use the identity $\cos^2\theta + \sin^2\theta \equiv 1$

e.g. to find $\int \sin^5\theta \, d\theta$ we use

$$\sin^5\theta \equiv (\sin\theta)(\sin^2\theta)^2$$

$$\equiv (\sin\theta)(1 - \cos^2\theta)^2$$

Hence $$\int \sin^5\theta \, d\theta \equiv \int (\sin\theta - 2\sin\theta\cos^2\theta + \sin\theta\cos^4\theta) \, d\theta$$

Now we saw in Example 10f (2) that

$$\int \sin\theta \cos^n\theta \, d\theta = -\frac{1}{n+1}\cos^{n+1}\theta + K$$

so $$\int \sin^5\theta \, d\theta = -\cos\theta + \tfrac{2}{3}\cos^3\theta - \tfrac{1}{5}\cos^5\theta + K$$

Powers of tan θ

The identity $\tan^2\theta + 1 \equiv \sec^2\theta$ is helpful in integrating any power of $\tan\theta$,

e.g.
$$\int \tan^3\theta \; d\theta \equiv \int (\tan\theta)(\tan^2\theta) \; d\theta$$

$$\equiv \int \tan\theta \; (\sec^2\theta - 1) \; d\theta$$

$$\equiv \int \tan\theta \; \sec^2\theta \; d\theta - \int \tan\theta \; d\theta$$

Now consider $\int \tan\theta \; \sec^2\theta \; d\theta$ in which the substitution $u \equiv \tan\theta$ is appropriate $\left(\text{since} \quad \sec^2\theta \equiv \dfrac{du}{d\theta}\right)$; thus

$$u \equiv \tan\theta \Rightarrow \ldots du \equiv \ldots \sec^2\theta \; d\theta$$

Hence
$$\int \tan\theta \; \sec^2\theta \; d\theta \equiv \int u \; du$$

$$= \frac{u^2}{2} + K$$

i.e.
$$\int \tan\theta \; \sec^2\theta \; d\theta = \tfrac{1}{2}\tan^2\theta + K$$

Further we know that $\int \tan\theta \; d\theta = \ln|\sec\theta| + K$ so, finally

$$\int \tan^3\theta \; d\theta \equiv \int \tan\theta \; (\sec^2\theta - 1) \; d\theta$$

$$= \tfrac{1}{2}\tan^2\theta - \ln|\sec\theta| + K$$

Multiple Angles

When integrating products such as $\sin 5\theta \cos 3\theta$, one of the factor formulae should be used

e.g.
$$\int \sin 5\theta \cos 3\theta \; d\theta \equiv \tfrac{1}{2} \int (\sin 8\theta + \sin 2\theta) \; d\theta$$

$$= \tfrac{1}{2}[-\tfrac{1}{8}\cos 8\theta - \tfrac{1}{2}\cos 2\theta] + K$$

$$= K - \tfrac{1}{16}(\cos 8\theta + 4\cos 2\theta)$$

The basic ideas used in the examples above can be applied to the integration of a variety of trig functions. The aims when dealing with trig integrals are usually:

(a) to convert the integral to the form $\int \dfrac{du}{dx} f'(u)\, dx$,

(b) to reduce the trig expression to a number of single trig ratios.

For instance, to find $\int \sin^2\theta \cos^3\theta\, d\theta$

we use $\cos^3\theta \equiv \cos\theta \cos^2\theta \equiv \cos\theta(1 - \sin^2\theta)$

Hence $\int \sin^2\theta \cos^3\theta\, d\theta \equiv \int (\sin^2\theta \cos\theta - \sin^4\theta \cos\theta)\, d\theta$

$$= \tfrac{1}{3}\sin^3\theta - \tfrac{1}{5}\sin^5\theta + K$$

But to find $\int \sin 3\theta \cos^2\theta\, d\theta$ it is better to transform $\cos^2\theta$ into $\tfrac{1}{2}(1 + \cos 2\theta)$ so that

$$\int \sin 3\theta \cos^2\theta\, d\theta \equiv \tfrac{1}{2}\int (\sin 3\theta + \sin 3\theta \cos 2\theta)\, d\theta$$

$$\equiv \tfrac{1}{2}\int [\sin 3\theta + \tfrac{1}{2}(\sin 5\theta + \sin\theta)]\, d\theta$$

$$= -\tfrac{1}{6}\cos 3\theta - \tfrac{1}{20}\cos 5\theta - \tfrac{1}{4}\cos\theta + K$$

EXERCISE 10i

1) Find:

(a) $\int \sin^4\theta\, d\theta$ (b) $\int \cos^3\theta\, d\theta$ (c) $\int \tan^4\theta\, d\theta$

(d) $\int \sin^3\theta\, d\theta$ (e) $\int \cos^4\theta\, d\theta$ (f) $\int \tan^5\theta\, d\theta$

2) Find:

(a) $\int 2\sin 4\theta \cos 3\theta\, d\theta$ (b) $\int 2\cos 2\theta \cos 5\theta\, d\theta$ (c) $\int \sin 2\theta \cos 6\theta\, d\theta$

(d) $\int \sin\theta \sin 3\theta\, d\theta$ (e) $\int 2\sin nx \cos mx\, dx$ (f) $\int 2\cos \dfrac{u}{2} \cos \dfrac{u}{3}\, du$

(g) $\int \cos nx \cos mx\, dx$

3) Integrate w.r.t. x:

(a) $\sin^2 x \cos^3 x$ (b) $\sin^{10} x \cos^3 x$ (c) $\sin^2 x \cos^2 x$ (use double angle formulae)

(d) $\tan^2 x \sec^4 x$ (e) $\sin^3 2x \cos^2 2x$ (f) $\sin^n x \cos^3 x$

(g) $\dfrac{\cos^2 x}{\operatorname{cosec}^3 x}$ (h) $\dfrac{\tan^3 x}{\cos^2 x}$

SYSTEMATIC INTEGRATION

At this stage it is possible to classify most of the integrals which are likely to arise.

Once correctly classified, the given expression can be integrated using the method best suited to its category.

The simplest category comprises the standard integrals listed below.

Function	Integral
x^n	$\dfrac{1}{n+1} x^{n+1} \quad (n \neq -1)$
e^x	e^x
$\dfrac{1}{x}$	$\ln |x|$
$\cos x$	$\sin x$
$\sin x$	$-\cos x$
$\sec^2 x$	$\tan x$
$-\operatorname{cosec}^2 x$	$\cot x$
$\tan x$	$\ln |\sec x|$
$\sec x$	$\ln |\sec x + \tan x|$
$\dfrac{1}{1+x^2}$	$\arctan x$
$\dfrac{1}{\sqrt{(1-x^2)}}$	$\arcsin x$

Each of these results should also be recognised when x is replaced by $ax + b$, as in the following shortened list.

Function	Integral
$(ax + b)^n$	$\dfrac{1}{a(n+1)} (ax + b)^{n+1} \quad (n \neq -1)$
e^{ax+b}	$\dfrac{1}{a} e^{ax+b}$

$$\frac{1}{ax + b} \qquad\qquad \frac{1}{a}\ln |ax + b|$$

$$\cos (ax + b) \qquad\qquad \frac{1}{a}\sin (ax + b)$$

When attempting to classify a given integral, and so determine the best method of integration, the following points should be considered.

(a) Is the integral a standard form?

(b) If it is a product, is it of the form $\dfrac{du}{dx}f'(u)$? If so, integrate at sight or change the variable. If it is not of this form, try integrating by parts.

(c) If it is a quotient, is it of the form:

(i) $\dfrac{f'(x)}{f(x)}$ (integrate at sight giving $\ln |f(x)|$) or,

(ii) $\dfrac{du/dx}{f(u)}$ (change the variable) or,

(iii) will partial fractions help?

Note. Be prepared for fractions whose *numerators* can be separated, thus producing two distinct integrals which may be of completely different types,

e.g. $\displaystyle\int \frac{x + 1}{\sqrt{(1 - x^2)}}\,dx \equiv \int \frac{x}{\sqrt{(1 - x^2)}}\,dx + \int \frac{1}{\sqrt{(1 - x^2)}}\,dx$

(d) If it is a trig function, and it has not already been classified, try to use, or adapt, one of the methods suggested on pages 315 to 317.

Although this systematic approach deals successfully with many integrals, inevitably the reader will encounter some integrals for which no method is obvious. Many expressions other than products and quotients can be integrated if an appropriate substitution is made. Because at this stage the reader cannot always be expected to 'spot' a suitable change of variable, a substitution will be suggested in most cases.

The following examples demonstrate a variety of integrals in which this technique can be applied.

EXAMPLES 10j

1) Find $\displaystyle\int x(2 - 3x)^{11}\,dx$.

Let $\qquad\qquad u \equiv 2 - 3x \Rightarrow \ldots du \equiv \ldots - 3\,dx$

Hence $\qquad \displaystyle\int x(2 - 3x)^{11}\,dx \equiv \int \left(\frac{2 - u}{3}\right)(u^{11})\left(-\frac{du}{3}\right)$

$$\equiv \frac{1}{9}\int (u^{12} - 2u^{11})\,du$$

$$= \frac{1}{9}\left[\frac{u^{13}}{13} - \frac{u^{12}}{6}\right] + K$$

$$= \left(\frac{u^{12}}{9}\right)\frac{(6u - 13)}{78} + K$$

i.e. $\int x(2 - 3x)^{11}\, dx = -\frac{(2 - 3x)^{12}(1 + 18x)}{702} + K$

2) Integrate $\dfrac{3x}{\sqrt{(4 - x)}}$ w.r.t. x.

Let $u \equiv \sqrt{4 - x} \Rightarrow u^2 \equiv 4 - x \Rightarrow \ldots 2u\, du \equiv \ldots - dx$

Hence $\int \dfrac{3x}{\sqrt{(4 - x)}}\, dx \equiv \int \dfrac{3(4 - u^2)(- 2u\, du)}{u}$

$$\equiv -6 \int (4 - u^2)\, du$$

$$= -6\left(4u - \frac{u^3}{3}\right) + K$$

$$= -2u(12 - u^2) + K$$

i.e. $\int \dfrac{3x}{\sqrt{(4 - x)}}\, dx = K - 2(8 + x)\sqrt{(4 - x)}$

3) Integrate $\sqrt{1 - x^2}$ w.r.t. x by using the substitution $x \equiv \sin\theta$.

Let $x \equiv \sin\theta \Rightarrow \ldots dx \equiv \ldots \cos\theta\, d\theta$

Hence $\int \sqrt{1 - x^2}\, dx \equiv \int \sqrt{(1 - \sin^2\theta)} \cos\theta\, d\theta$

$$\equiv \int \cos^2\theta\, d\theta$$

$$\equiv \tfrac{1}{2} \int (1 + \cos 2\theta)\, d\theta$$

$$= \frac{\theta}{2} + \frac{1}{4} \sin 2\theta + K$$

$$= \frac{\theta}{2} + \frac{1}{2} \sin\theta \cos\theta + K$$

$$= \frac{\theta}{2} + \frac{1}{2} \sin\theta \sqrt{1 - \sin^2\theta} + K$$

$\Rightarrow$ $$\int \sqrt{1-x^2}\, dx = \tfrac{1}{2}(\arcsin x + x\sqrt{1-x^2}) + K$$

EXERCISE 10j

Find the following integrals using the suggested substitution.

1) $\displaystyle\int (x+1)(x+3)^5\, dx$; $x + 3 \equiv u$

2) $\displaystyle\int \frac{1}{4+x^2}\, dx$; $x \equiv 2\tan\theta$

3) $\displaystyle\int \frac{x}{\sqrt{(3-x)}}\, dx$; $3 - x \equiv u^2$

4) $\displaystyle\int x\sqrt{x+1}\, dx$; $x + 1 \equiv u^2$

5) $\displaystyle\int \frac{2x+1}{(x-3)^6}\, dx$; $x - 3 \equiv u$

6) $\displaystyle\int \frac{1}{\sqrt{(1+x^2)}}\, dx$; $x \equiv \tan\theta$

7) $\displaystyle\int 2x\sqrt{3x-4}\, dx$; $3x - 4 \equiv u^2$

Devise a suitable substitution and hence find:

8) $\displaystyle\int 2x(1-x)^7\, dx$ 9) $\displaystyle\int \frac{1}{\sqrt{(9-x^2)}}\, dx$ 10) $\displaystyle\int \frac{x+3}{(4-x)^5}\, dx$

Classify each of the following integrals. Hence perform each integration using an appropriate method.

11) $\displaystyle\int e^{2x+3}\, dx$ 12) $\displaystyle\int x\sqrt{2x^2-5}\, dx$ 13) $\displaystyle\int \sin^2 3x\, dx$

14) $\displaystyle\int x\,e^{-x^2}\, dx$ 15) $\displaystyle\int \sin 3\theta \cos\theta\, d\theta$ 16) $\displaystyle\int u(u+7)^9\, du$

17) $\displaystyle\int \frac{x^2}{(x^3+9)^5}\, dx$ 18) $\displaystyle\int \frac{\sin 2y}{1-\cos 2y}\, dy$ 19) $\displaystyle\int \frac{1}{2x+7}\, dx$

20) $\displaystyle\int \frac{1}{\sqrt{(1-u^2)}}\, du$ 21) $\displaystyle\int \sin 3x\sqrt{1+\cos 3x}\, dx$

22) $\displaystyle\int x\sin 4x\, dx$ 23) $\displaystyle\int \frac{x+2}{x^2+4x-5}\, dx$ 24) $\displaystyle\int \frac{x+1}{x^2+4x-5}\, dx$

25) $\int \dfrac{x+2}{(x^2+4x-5)^3}\,dx$ 26) $\int 3y\sqrt{9-y^2}\,dy$ 27) $\int e^{2x}\cos 3x\,dx$

28) $\int \ln 5x\,dx$ 29) $\int \cos^3 2x\,dx$ 30) $\int \csc^2 x\,e^{\cot x}\,dx$

31) $\int \dfrac{\sin y}{\sqrt{(7+\cos y)}}\,dy$ 32) $\int x^2\,e^x\,dx$ 33) $\int \dfrac{x}{x^2-4}\,dx$

34) $\int \dfrac{x^2}{x^2-4}\,dx$ 35) $\int \dfrac{1}{x^2-4}\,dx$ 36) $\int \cos 4x\cos x\,dx$

37) $\int \sin^5 2\theta\,d\theta$ 38) $\int \cos^2 u\,\sin^3 u\,du$ 39) $\int \tan^4 \theta\,d\theta$

40) $\int \tan^5 \theta\,d\theta$ 41) $\int \dfrac{1-2x}{\sqrt{(1-x^2)}}\,dx$ 42) $\int \dfrac{1}{u\ln u}\,du$

43) $\int y^2\cos 3y\,dy$ 44) $\int \dfrac{\sec^2 x}{1-\tan x}\,dx$ 45) $\int x\sqrt{(7+x^2)}\,dx$

46) $\int \sin(5\theta-\pi/4)\,d\theta$ 47) $\int \cos\theta\,\ln\sin\theta\,d\theta$ 48) $\int \sec^2 u\,e^{\tan u}\,du$

49) $\int \dfrac{x}{(3-x)^7}\,dx$ 50) $\int \tan^2 x\,\sec^2 x\,dx$

CALCULATION OF THE CONSTANT OF INTEGRATION

When an expression is integrated the result includes a constant of unknown value.

In order to determine its value, further information is needed. For example, if a curve has a gradient function $(x+4)$ and also passes through the point $(2,5)$, the equation of the curve can be found as follows:

If the equation of the curve is $y=f(x)$ the gradient function is $\dfrac{dy}{dx}$ where

$$\frac{dy}{dx}=x+4 \quad\Rightarrow\quad y=\int(x+4)\,dx$$

$$\Rightarrow \qquad\qquad y=\frac{x^2}{2}+4x+K$$

But we also know that when $x=2,\;\; y=5$

$$\Rightarrow \qquad\qquad 5=\frac{2^2}{2}+8+K \quad\Rightarrow\quad K=-5$$

Thus the equation of the curve is

$$y = \frac{x^2}{2} + 4x - 5$$

EXERCISE 10k

In the following problems the gradient function of a curve is given, together with the coordinates of a point on the curve. Find the equation of the curve in each case.

1) $3x - 4$; $(1, 2)$

2) $3x^2 - 5x + 1$; $(0, 3)$

3) $6e^{2x}$; $(0, 2)$

4) $(7 - 5x)^2$; $(1, \frac{7}{15})$

5) $\cos 3x$; $(\pi/2, 1)$

6) A curve that passes through the origin has a gradient function $2x - 1$. Find its equation and sketch the curve.

7) If $\dfrac{dy}{dx} = e^{3x}$ and $y = 2$ when $x = 1$ find the coordinates of the point where the curve crosses the y axis.

MOTION OF A PARTICLE IN A STRAIGHT LINE

If O is a fixed point on a straight line Ox, the *directed* distance from O of a particle P moving on this line is called its *displacement* and is represented by s. The *velocity* of P, which is the rate at which the displacement increases with respect to time $\left(\dfrac{ds}{dt}\right)$ is represented by v, and a represents the *acceleration* of P, which is the rate at which the velocity increases with respect to time $\left(\dfrac{dv}{dt}\right)$,

i.e.
$$v = \frac{ds}{dt} \quad \text{and} \quad a = \frac{dv}{dt}$$

Conversely
$$s = \int v \, dt \quad \text{and} \quad v = \int a \, dt$$

Hence if the particle moves so that either its acceleration or its velocity or its displacement is a function of time, its motion can be analysed using calculus methods.

Consider, for example, a particle P moving along a straight line Ox with an acceleration $3t$ where t is the time in seconds. When $t = 2$, P has a displacement of 4 metres from O and a velocity of 7 metres per second. From this information we can find the velocity and position of P as functions of t.

Using a, v and s for acceleration, velocity and displacement we have

$$a = 3t$$

$$v = \int a \, dt = \int 3t \, dt = \frac{3t^2}{2} + K_1$$

But $v = 7$ when $t = 2$ $\Rightarrow$ $K_1 = 1$

i.e. $v = \frac{3t^2}{2} + 1$

Then $s = \int v \, dt = \int \left(\frac{3t^2}{2} + 1 \right) dt$

$$= \frac{t^3}{2} + t + K_2$$

But $s = 4$ when $t = 2$ $\Rightarrow$ $K_2 = -2$

i.e. $s = \frac{t^3}{2} + t - 2$

Thus when $t = 4$,   say, we have

$$v = 3 \times \frac{4^2}{2} + 1 = 25$$

and $s = \frac{4^3}{2} + 4 - 2 = 34$

i.e. when $t = 4$,   P is 34 metres from O along Ox and travelling with a velocity 25 metres per second.

Note. A negative value for s indicates that the particle has moved to the opposite side of O.

Similarly if v is negative the particle is moving in the direction xO (not Ox).

A negative value for a indicates that the velocity is decreasing.

EXAMPLE 10I

A particle P starts from a point O with velocity 5 metres per second and moves along a straight line Ox with an acceleration of $-2t^2$ at a time t seconds after leaving O.

Describe its motion after one second, after three seconds and after four seconds.

$$a = \frac{dv}{dt} = -2t^2$$

Hence $v = -\tfrac{2}{3}t^3 + K_1$

But $v = 5$ when $t = 0$ so $K_1 = 5$

Thus $v = \dfrac{ds}{dt} = 5 - \tfrac{2}{3}t^3$

Hence $s = 5t - \tfrac{1}{6}t^4 + K_2$

But $s = 0$ when $t = 0$ so $K_2 = 0$

Thus after t seconds:

$$a = -2t^2, \quad v = 5 - \tfrac{2}{3}t^3, \quad s = 5t - \tfrac{1}{6}t^4$$

After 1 second $a = -2$, $v = 4\tfrac{1}{3}$, $s = 4\tfrac{5}{6}$.

So the particle is $4\tfrac{5}{6}$ metres from O along Ox moving in the direction Ox with velocity $4\tfrac{1}{3}$ metres per second.

After three seconds $a = -18$, $v = -13$, $s = 1\tfrac{1}{2}$ so the particle is $1\tfrac{1}{2}$ metres along Ox, moving in the direction xO.

After four seconds $a = -32$, $v = -37\tfrac{2}{3}$, $s = -22\tfrac{2}{3}$.   So, although the particle is still travelling in the direction xO, it is now $22\tfrac{2}{3}$ metres along xO.

EXERCISE 10l

In the following problems a, v and s represent the acceleration, velocity and displacement (i.e. directed distance) of a particle P from a fixed point O along a straight line Ox. In each case find v and s after time t.

1) $a = 3t$; $s = 0$ and $v = 1$ when $t = 0$

2) $a = \sin 2t$; $s = 3$ and $v = \frac{1}{2}$ when $t = 0$

3) $a = \dfrac{1}{(t + 1)^2}$; $s = 2$ and $v = 0$ when $t = 0$

4) The acceleration of a particle P is $(2 - 6t)$, t seconds after leaving the point O.
If its velocity at O is 2 units, find the time(s) at which, after leaving O:

(a) the particle first comes to rest,

(b) the particle returns to O,

(c) the particle is at a distance 2 units from O along Ox.

DIFFERENTIAL EQUATIONS

An equation in which at least one term contains $\dfrac{dy}{dx}$, $\dfrac{d^2y}{dx^2}$ etc, is called a *differential equation*. For instance

$$x + 3\frac{dy}{dx} = 4y \quad \text{is a } \textit{first order} \text{ differential equation because it}$$

contains only *a first differential coefficient* $\left(\dfrac{dy}{dx}\right)$,

$$\frac{d^2y}{dx^2} + 3\frac{dy}{dx} + 4y = x^2 \quad \text{is a } \textit{second order} \text{ differential equation}$$

because it contains a *second differential coefficient* $\left(\dfrac{d^2y}{dx^2}\right)$.

Each of these examples is a *linear* differential equation because none of the differential coefficients is raised to a power other than 1.

Conversely $x^2 + 2\left(\dfrac{dy}{dx}\right)^2 + 3y = 0$ is *not* a linear differential equation.

Any differential equation represents a relationship between two variables, x and y, say.
The same relationship can often be expressed in a form that does not contain a differential coefficient,

e.g. $y = x^2 + K$ and $\dfrac{dy}{dx} = 2x$ express the same relationship between x

and y but $\dfrac{dy}{dx} = 2x$ is a differential equation while $y = x^2 + K$ is not.

Converting a differential relationship into a direct one is called *solving a differential equation* and this clearly involves some form of integration. There are many different types of differential equation, each requiring a particular approach and technique for its solution. At this stage however, we are going to solve only one simple type belonging to the first order, linear group.

Differential Equations with Variables Separable

Consider the differential equation

$$3y \frac{dy}{dx} = 5x^2 \qquad\qquad [1]$$

If we integrate both sides of this equation w.r.t. x we have

$$\int 3y \frac{dy}{dx}\, dx = \int 5x^2\, dx$$

But we saw earlier in this chapter (p. 302) that

$$\int \ldots \frac{dy}{dx}\, dx \equiv \int \ldots dy$$

so

$$\int 3y\, dy = \int 5x^2\, dx \qquad\qquad [2]$$

Temporarily removing the integral signs from equation [2] leaves

$$3y\, dy = 5x^2\, dx \qquad\qquad [3]$$

This form can be obtained direct from equation [1] by *separating the variables*, i.e. by separating dy from dx and collecting on one side all terms involving y together with dy, while all the x terms together with dx are collected separately.

It is important to appreciate that what we have written in [3] above does not, in itself, have any meaning. It nevertheless provides a quick method of converting the given equation [1] into the integral form [2].

Thus, the solution of the given equation is carried out as follows

$$3y \frac{dy}{dx} = 5x^2$$

Separating the variables gives $3y\, dy = 5x^2\, dx.$

Hence
$$\int 3y \, dy = \int 5x^2 \, dx$$

$\Rightarrow$
$$\frac{3y^2}{2} = \frac{5x^3}{3} + A$$

The constant of integration is usually represented by A, B, etc. rather than K when solving differential equations and is called an *arbitrary constant*.

The solution of a differential equation including the arbitrary constant is called the *general solution* (or sometimes the *complete primitive*).

Further information is required if A is to be evaluated, providing the *particular solution*.

EXAMPLES 10m

1) A curve is such that at any point the gradient multiplied by the x coordinate is equal to three times the y coordinate of that point. If the curve passes through the point $(1, 4)$ find its equation.

The given condition can be written

$$x \frac{dy}{dx} = 3y$$

$$\left[\text{Separating the variables gives} \quad \frac{dy}{y} = 3 \frac{dx}{x} \right]$$

Thus
$$\int \frac{1}{y} \, dy = \int \frac{3}{x} \, dx$$

$\Rightarrow$
$$\ln |y| = 3 \ln |x| + A$$

But $y = 4$ when $x = 1$, so $A = \ln 4$

Thus
$$\ln |y| = 3 \ln |x| + \ln 4$$

$\Rightarrow$
$$\ln |y| = \ln |4x^3|$$

$\Rightarrow$
$$y = 4x^3$$

2) Find the general solution of the differential equation

$$2(x^2 + 1) \frac{dy}{dx} = x(4 - y^2)$$

If $y = 1$ when $x = 0$, express y as a function of x

$$2(x^2 + 1) \frac{dy}{dx} = x(4 - y^2)$$

$\Rightarrow$
$$\frac{2}{4 - y^2} \frac{dy}{dx} = \frac{x}{x^2 + 1}$$

So, after separating the variables, we have,

$$\int \frac{2}{4-y^2} \, dy = \int \frac{x}{x^2+1} \, dx$$

i.e.

$$\int \frac{\frac{1}{2}}{2-y} + \frac{\frac{1}{2}}{2+y} \, dy = \int \frac{x}{x^2+1} \, dx$$

Hence $-\frac{1}{2} \ln (2-y) + \frac{1}{2} \ln (2+y) = \frac{1}{2} \ln (x^2+1) + \ln A$

$\Rightarrow$

$$\ln \sqrt{\left(\frac{2+y}{2-y}\right)} = \ln A \sqrt{(x^2+1)}$$

$\Rightarrow$

$$2+y = A^2(2-y)(x^2+1)$$

Now if $y = 1$ when $x = 0$, $A^2 = 3$, so

$$2+y = 3(2-y)(x^2+1)$$

$\Rightarrow$

$$y + 3y(x^2+1) = 6(x^2+1) - 2$$

$\Rightarrow$

$$y = \frac{6x^2+4}{3x^2+4}$$

Natural Occurrence of Differential Equations

Differential equations often arise when a physical situation is interpreted mathematically (i.e. when a mathematical model is made of the physical situation).

For example:

(a) Suppose that a body falls from rest in a medium which causes the velocity to decrease at a rate proportional to the velocity.

Using v for velocity and t for time, the rate of *decrease* of velocity can be written as $-\dfrac{dv}{dt}$.

Thus the motion of the body satisfies the differential equation

$$-\frac{dv}{dt} = kv$$

(b) During the initial stages of the growth of yeast cells in a culture, the number of cells present increases in proportion to the number already formed.

Thus n, the number of cells at a particular time t, can be found from the differential equation

$$\frac{dn}{dt} = kn$$

(c) Suppose that a chemical mixture contains two substances A and B whose weights are W_A and W_B and whose combined weight remains constant. B is converted into A at a rate which is inversely proportional to the weight of B and proportional to the square of the weight of A in the mixture at any time t. The weight of B present at time t can be found using

$$\frac{d}{dt}(W_B) = \frac{k}{W_B} \times (W_A)^2$$

But $W_A + W_B$ is constant, W say

Hence

$$\frac{d}{dt}(W_B) = \frac{k(W - W_B)^2}{W_B}$$

This differential equation now relates W_B and t.

Note. In forming (and subsequently solving) differential equations from naturally occurring data, it is not necessary to understand the background of the situation or experiment.

EXERCISE 10m

Find the general solutions of the following differential equations.

1) $y \dfrac{dy}{dx} = \sin x$

2) $\dfrac{dy}{dx} = y^2$

3) $\dfrac{1}{x}\dfrac{dy}{dx} = \dfrac{1}{x^2 + 1}$

4) $(x - 3)\dfrac{dy}{dx} = y$

5) $\tan y \dfrac{dx}{dy} = 4$

6) $u \dfrac{du}{dv} = v + 2$

7) $\dfrac{y^2}{x^3}\dfrac{dy}{dx} = \ln x$

8) $e^x \dfrac{dy}{dx} = \dfrac{x}{y}$

9) $\sec x \dfrac{dy}{dx} = e^y$

10) $r \dfrac{dr}{d\theta} = \sin^2 \theta$

11) $\dfrac{dv}{du} = \dfrac{v + 1}{u + 2}$

12) $xy \dfrac{dy}{dx} = \ln x$

13) $y(x + 1) = (x^2 + 2x)\dfrac{dy}{dx}$

14) $v^2 \dfrac{dv}{dt} = (2 + t)^3$

15) $x \dfrac{dy}{dx} = \dfrac{1}{y} + y$

16) $e^x \dfrac{dy}{dx} = e^{y-1}$

17) $\tan x \dfrac{dy}{dx} = 2y^2 \sec^2 x$

18) $r \dfrac{d\theta}{dr} = \cos^2 \theta$

19) $y \sin^3 x \dfrac{dy}{dx} = \cos x$

20) $\dfrac{uv}{u - 1} = \dfrac{du}{dv}$

Find the particular solutions of the following differential equations:

21) $\dfrac{y}{x}\dfrac{dy}{dx} = \dfrac{y^2 + 1}{x^2 + 1}$; $\quad y = 0$ when $x = 1$.

22) $e^t \dfrac{ds}{dt} = \sqrt{s}$; $\quad s = 4$ when $t = 0$.

23) $y^2 \dfrac{dy}{dx} = x^2 + 1; \quad y = 1 \quad \text{when} \quad x = 2.$

Find the equation of each of the following curves:

24) A curve passes through the points $(1, 2)$ and $(\frac{1}{4}, -10)$ and has a gradient which is inversely proportional to x^2.

25) The gradient function of a curve is $\dfrac{y+1}{x^2-1}$ and the curve passes through $(-3, 1)$.

26) A curve passes through the point $(0, -1)$ and $e^{-x} \dfrac{dy}{dx} = 1.$

Form, *but do not solve*, the differential equations representing the following data:

27) A body moves with a velocity v which is inversely proportional to its displacement s from a fixed point.

28) The rate at which the height h of a certain plant increases is proportional to the natural logarithm of the difference between its present height and its final height H.

29) The manufacturers of a certain brand of soap powder are concerned that the number, n, of people buying their product at any time t has remained constant for some months. They launch a major advertising programme which results in the number of customers increasing at a rate proportional to the square root of n.
Express as differential equations the progress of sales:
(a) before advertising   (b) after advertising.

30) In an isolated community, the number, n, of people suffering from an infectious disease is N_1 at a particular time. The disease then becomes epidemic and spreads so that the number of sick people increases at a rate proportional to n, until the total number of sufferers is N_2. The rate of increase then becomes inversely proportional to n until N_3 people have the disease. After this, the total number of sick people decreases at a constant rate. Write down the differential equation governing the incidence of the disease:

(a) for $N_1 \leqslant n < N_2$,

(b) for $N_2 \leqslant n < N_3$,

(c) for $n \geqslant N_3$.

INTEGRATION AS A PROCESS OF SUMMATION

Consider the area bounded by the x axis, the lines $x = a$ and $x = b$ and the curve $y = f(x)$ that is continuous for $a \leqslant x \leqslant b$.

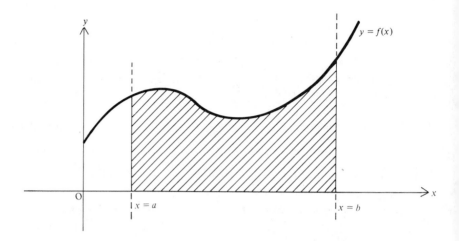

There are several ways in which this area can be estimated (e.g. counting squares on graph paper). The method we are going to use this time is to split the area into thin vertical strips and treat each strip as being approximately rectangular.

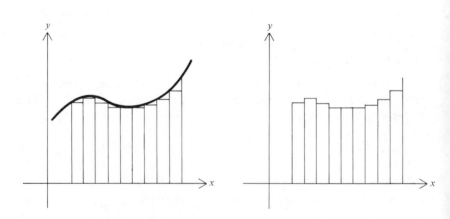

The sum of the areas of the rectangular strips then gives an approximate value for the required area. The thinner the strips are, the better is the approximation.

Note that every strip has one end on the x axis, one end on the curve and two vertical sides, i.e., they all have the same type of boundaries.

Now consider one typical strip (or element) PP′Q′Q where P is the point (x, y). The width of the strip, P′Q′, is a small increment in x and can be called δx.

Also, if A represents the area of all the strips up to PP′ then adding the area of the strip PP′Q′Q causes a small increase in A.

Thus the area of the strip can be called δA.

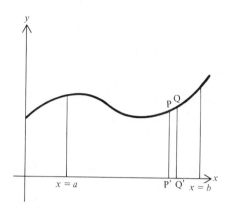

The area of the strip is approximately that of a rectangle of width δx and length y.

Thus for every strip

$$\delta A \simeq y\,\delta x \tag{1}$$

Now if the areas of all the strips are summed from $x = a$ to $x = b$,

$$\left(\text{which we denote by } \sum_{x=a}^{b} \delta A\right), \text{ the total area is obtained}$$

i.e.
$$\text{total area} = \sum_{x=a}^{x=b} \delta A$$

$\Rightarrow$
$$\text{total area} \simeq \sum_{x=a}^{x=b} y\,\delta x$$

As δx gets smaller the accuracy of the results increases until, in the limit,

$$\text{total area} = \lim_{\delta x \to 0} \sum_{x=a}^{x=b} y\,\delta x$$

Alternatively we can return to equation [1] above and consider it in the form

$$\frac{\delta A}{\delta x} \simeq y$$

This form too becomes more accurate as δx gets smaller giving, in the limiting case,

$$\lim_{\delta x \to 0} \frac{\delta A}{\delta x} = y$$

But
$$\lim_{\delta x \to 0} \frac{\delta A}{\delta x} \quad \text{is} \quad \frac{dA}{dx}$$

so
$$\frac{\mathrm{d}A}{\mathrm{d}x} = y$$

i.e.
$$\frac{\mathrm{d}A}{\mathrm{d}x} = y \implies A = \int y \, \mathrm{d}x$$

The boundary values of x defining the total area are $x = a$ and $x = b$ and we indicate this by writing

$$\text{total area} = \int_a^b y \, \mathrm{d}x$$

The total area can therefore be found in two ways, either by a process of summation or by integration

i.e.
$$\lim_{\delta x \to 0} \sum_{x=a}^{x=b} y \, \delta x = \int_a^b y \, \mathrm{d}x$$

and we conclude that integration is a process of summation.

The application of integration to problems involving summation continues in Chapter 18. In this chapter we will use integration only to find areas bounded by straight lines and a curve, but first we must investigate the meaning of $\int_a^b y \, \mathrm{d}x$.

DEFINITE INTEGRATION

Suppose that we wish to find the area bounded by the x axis, the lines $x = a$ and $x = b$ and the curve $y = 3x^2$.

Using the method above we find that $A = \int 3x^2 \, \mathrm{d}x$

i.e. $A = x^3 + K$

From this area function we can find the value of A corresponding to a particular value of x.

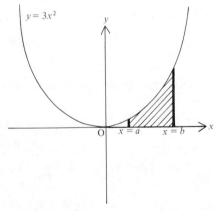

Hence, using $x = a$ gives

$$A_a = a^3 + K$$

Similarly using $x = b$ gives

$$A_b = b^3 + K$$

Then the area between $x = a$ and $x = b$ is given by $A_b - A_a$ where

$$A_b - A_a = (b^3 + K) - (a^3 + K)$$
$$= b^3 - a^3$$

Now $A_b - A_a$ is referred to as *the definite integral from a to b of $3x^2$* and is denoted by $\int_a^b 3x^2 \, dx$

i.e.

$$\int_a^b 3x^2 \, dx = (x^3)_{x=b} - (x^3)_{x=a}$$

The R.H.S. of this equation is usually written in the form $\left[x^3\right]_a^b$ where a and b are called the boundary values or limits of the integration (b is the upper limit and a the lower limit).

Whenever a definite integral is calculated, the constant of integration disappears.

Note. A definite integral can be found in this way only if the function to be integrated is defined for every value of x from a to b,

e.g. $\int_{-1}^1 \frac{1}{x} \, dx$ cannot, as yet, be found because $\frac{1}{x}$ is undefined when $x = 0$.

A method for calculating some definite integrals which are undefined for some value(s) between a and b is discussed in Volume 2.

EXAMPLES 10n

1) Evaluate $\int_1^4 \frac{1}{(x+3)^2} \, dx$.

$$\int_1^4 \frac{1}{(x+3)^2} \, dx \equiv \int_1^4 (x+3)^{-2} \, dx$$

$$= \left[-(x+3)^{-1}\right]_1^4$$

$$= \{-(4+3)^{-1}\} - \{-(1+3)^{-1}\}$$

$$= -\tfrac{1}{7} + \tfrac{1}{4} = \tfrac{3}{28}$$

2) Evaluate $\int_0^{\frac{\pi}{2}} \cos x \, dx$.

$$\int_0^{\frac{\pi}{2}} \cos x \, dx = \left[\sin x\right]_0^{\frac{\pi}{2}}$$

$$= \sin\frac{\pi}{2} - \sin 0$$

$$= 1 - 0 = 1$$

Definite Integration with Change of Variable

A definite integral can be evaluated only after the appropriate integration has been performed. In some problems, as we have already seen, this may require a change of variable, e.g. from x to u. In such cases the limits of integration, originally values of x, can be transformed into values of u allowing direct calculation of the required definite integral.

EXAMPLES 10n (continued)

3) By using the substitution $u \equiv x^3 + 1$, evaluate $\int_0^1 x^2\sqrt{x^3 + 1}\, dx$.

If $u \equiv x^3 + 1$ then $\quad \ldots du \equiv \ldots 3x^2\, dx$

and $\qquad \begin{cases} x = 0 & \Rightarrow \quad u = 1 \\ x = 1 & \Rightarrow \quad u = 2 \end{cases}$

Hence $\qquad \int_0^1 x^2\sqrt{x^3 + 1}\, dx \equiv \int_1^2 \sqrt{u}\, \dfrac{du}{3}$

$$= \left[\frac{2}{3}\frac{u^{\frac{3}{2}}}{3}\right]_1^2$$

$$= \tfrac{2}{9}\{2\sqrt{2} - 1\}$$

4) Evaluate $\displaystyle\int_0^1 \frac{1}{(4 - x^2)^{\frac{3}{2}}}\, dx$ by using the substitution $x \equiv 2\sin\theta$.

If $x \equiv 2\sin\theta$ then $\quad \ldots dx \equiv \ldots 2\cos\theta\, d\theta$

and $\qquad \begin{cases} x = 0 & \Rightarrow \quad \sin\theta = 0 \quad \Rightarrow \quad \theta = 0 \\ x = 1 & \Rightarrow \quad \sin\theta = \tfrac{1}{2} \quad \Rightarrow \quad \theta = \dfrac{\pi}{6} \end{cases}$

Hence $\qquad \displaystyle\int_0^1 \frac{1}{(4 - x^2)^{\frac{3}{2}}}\, dx \equiv \int_0^{\frac{\pi}{6}} \frac{1}{8\cos^3\theta}\, 2\cos\theta\, d\theta$

$$\equiv \frac{1}{4}\int_0^{\frac{\pi}{6}} \sec^2\theta\, d\theta$$

$$= \frac{1}{4}\Big[\tan\theta\Big]_0^{\frac{\pi}{6}}$$

$$= \frac{1}{4}\left\{\frac{1}{\sqrt{3}} - 0\right\} = \frac{\sqrt{3}}{12}$$

Definite Integration by Parts

When using the formula

$$\int v\frac{du}{dx}\, dx = uv - \int u\frac{dv}{dx}\, dx$$

it must be appreciated that the term uv on the R.H.S. is fully integrated. Consequently in a definite integration, uv must be *evaluated between the appropriate boundaries*

i.e.
$$\int_a^b v \frac{du}{dx} dx = [uv]_a^b - \int_a^b u \frac{dv}{dx} dx$$

EXAMPLES 10n (continued)

5) Evaluate $\int_0^1 x e^x dx$

$$\int x e^x dx \equiv \int v \frac{du}{dx} dx$$

where
$$v = x \quad \text{and} \quad \frac{du}{dx} = e^x$$

Hence
$$\int_0^1 x e^x dx = \left[x e^x \right]_0^1 - \int_0^1 e^x dx$$

$$= \left[x e^x \right]_0^1 - \left[e^x \right]_0^1$$

$$= (e^1 - 0) - (e^1 - e^0)$$

$$= e - e + 1$$

i.e.
$$\int_0^1 x e^x dx = 1$$

EXERCISE 10n

Evaluate the following definite integrals

1) $\int_2^3 (x^2 + 2x - 1) dx$

2) $\int_{\frac{\pi}{6}}^{\frac{\pi}{3}} \cos 2\theta \, d\theta$

3) $\int_0^1 e^{4x} dx$

4) $\int_3^4 \frac{1}{x+1} dx$

5) $\int_2^7 \sqrt{(x+2)} dx$

6) $\int_0^{\frac{\pi}{4}} \sec^2 3\theta \, d\theta$

7) $\int_0^{\frac{\pi}{6}} \sin^2 \theta \, d\theta$

8) $\int_1^2 x e^{x^2} dx$

9) $\int_{\frac{\pi}{6}}^{\frac{\pi}{2}} \sin^4 x \cos x \, dx$

10) $\int_{\frac{\pi}{3}}^{\frac{\pi}{2}} \frac{\sin \theta}{1 - \cos \theta} d\theta$

11) $\int_0^{\frac{\pi}{2}} x \sin x \, dx$

12) $\int_0^1 \frac{x+1}{(x+2)(x+3)} dx$

13) $\int_1^2 \frac{x}{x^2+1} dx$

14) $\int_0^{\frac{\pi}{4}} \cos 3\theta \cos \theta \, d\theta$

15) $\int_0^1 x^2 (x^3 + 1)^4 dx$

16) $\int_1^2 x^3 \ln x \, dx$

17) $\int_0^{\frac{\pi}{2}} \cos x \sqrt{\sin x} \, dx$

18) $\displaystyle\int_1^3 \left(x^2 - \frac{1}{x^2}\right) dx$ 19) $\displaystyle\int_1^2 \frac{(e^u + e^{-u})^2}{e^u} du$ 20) $\displaystyle\int_0^{\frac{\pi}{2}} e^x \sin x \, dx$

Evaluate the following, using the suggested change of variable, or otherwise.

21) $\displaystyle\int_0^{\sqrt{3}} \frac{x}{\sqrt{(x^2 + 1)}} dx$; $u^2 \equiv x^2 + 1$ 22) $\displaystyle\int_3^4 x(x-3)^7 \, dx$; $u \equiv x - 3$

23) $\displaystyle\int_0^{\frac{\pi}{6}} \cos\theta\sqrt{1 - 2\sin\theta} \, d\theta$; $x \equiv 1 - 2\sin\theta$

24) $\displaystyle\int_0^{\frac{\pi}{6}} \sec^3\theta \tan\theta \, d\theta$; $u \equiv \sec\theta$ 25) $\displaystyle\int_{1.5}^3 \frac{1}{\sqrt{(9 - x^2)}} dx$; $x \equiv 3\sin\theta$

26) $\displaystyle\int_{\frac{1}{4}}^{\frac{\sqrt{3}}{4}} \frac{1}{1 + 16u^2} du$; $4u \equiv \tan x$ 27) $\displaystyle\int_0^{\pi} \frac{\sin x}{\cos^4 x} dx$; $\cos x \equiv u$

28) $\displaystyle\int_0^2 (x + 1)\sqrt{(x^2 + 2x)} \, dx$; $x^2 + 2x \equiv 0$

29) $\displaystyle\int_0^2 \frac{x + 1}{\sqrt{(x^2 + 2x + 8)}} dx$; $u \equiv x^2 + 2x + 8$

30) $\displaystyle\int_{\pi/4}^{\arctan 2} \frac{1}{3\sin 2\theta - 4\cos 2\theta} d\theta$; $\tan\theta \equiv t$

$\left(\text{use also}\quad \sin 2\theta \equiv \dfrac{2t}{1 + t^2}, \quad \cos 2\theta \equiv \dfrac{1 - t^2}{1 + t^2} \quad \text{and} \quad \tan^2\theta + 1 \equiv \sec^2\theta\right)$

AREA BOUNDED PARTLY BY A CURVE

If the area, A, between two specified values of x, $(x = a$ and $x = b)$ is bounded by the x axis and a curve $y = f(x)$, we have seen that A can be found from $\displaystyle\int_a^b f(x) \, dx$

The reader is now in a position to evaluate this definite integral and so can use this method for determining certain areas.

It is recommended that the required area is first treated as a summation of areas of elements, a typical element being indicated on a diagram.

EXAMPLES 10p

1) Find the area bounded by the x axis, the y axis, the curve $y = e^x$ and the line $x = 2$.

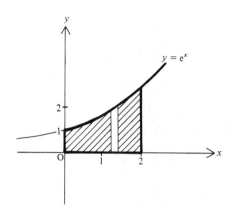

This area can be divided into vertical strips of approximate area $y\,\delta x$.

Thus the required area

$$= \lim_{\delta x \to 0} \sum_{x=0}^{x=2} y\,\delta x$$

$$= \int_0^2 y\,dx$$

But
$$\int_0^2 y\,dx = \int_0^2 e^x\,dx$$

$$= \left[e^x\right]_0^2 = e^2 - e^0$$

So the required area is $(e^2 - 1)$.

2) Find the area between the curve $x = 4 - y^2$ and the y axis.

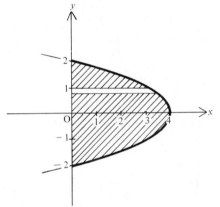

A vertical strip in this case would have *both ends on the same curve*. Its length therefore cannot be defined easily so we use a horizontal strip instead. Its approximate area is $x\,\delta y$.

The curve crosses the y axis at $y = -2$ and $y = 2$, so our summation is bounded by these values,

i.e.   required area $= \lim_{\delta y \to 0} \sum_{y=-2}^{y=2} x\,\delta y$

$$= \int_{-2}^2 x\,dy$$

But
$$\int_{-2}^2 x\,dy = \int_{-2}^2 (4 - y^2)\,dy = \left[4y - \frac{y^3}{3}\right]_{-2}^2$$

$$= (8 - \tfrac{8}{3}) - (-8 + \tfrac{8}{3})$$

Thus the required area is $10\tfrac{2}{3}$ square units.

The Meaning of a Negative Result

Consider the area bounded by $y = 4x^3$ and the x axis if the other boundaries are the lines

(a) $x = -2$ and $x = -1$ (b) $x = 1$ and $x = 2$

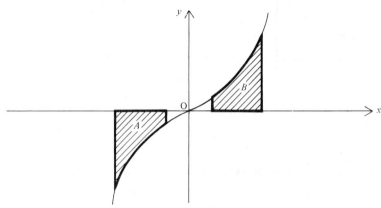

This curve is symmetrical about the origin so the two shaded areas are equal.

(a) Considering A

$$\lim_{\delta x \to 0} \sum_{x=-2}^{-1} y\,\delta x = \int_{-2}^{-1} y\,dx$$

$$= \int_{-2}^{-1} 4x^3\,dx$$

$$= \left[x^4 \right]_{-2}^{-1}$$

$$= 1 - 16 = -15$$

(b) Considering B

$$\lim_{\delta x \to 0} \sum_{x=1}^{2} y\,\delta x = \int_{1}^{2} y\,dx$$

$$= \int_{1}^{2} 4x^3\,dx$$

$$= \left[x^4 \right]_{1}^{2}$$

$$= 16 - 1 = 15$$

So we see that, while the magnitudes of the two areas are equal, the result for the area of A which is below the x axis, is negative. This is explained by the fact that the length of a strip in A was taken as y, which is negative for the part of the curve bounding A.

Note. Care must be taken with problems involving a curve that crosses the x axis between the boundary values.

EXAMPLES 10p (continued)

3) Find the area enclosed between the curve $y = x(x-1)(x-2)$ and the x axis.

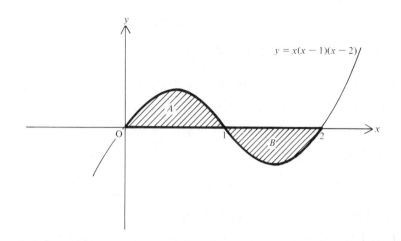

The area enclosed between the curve and the x axis is the sum of the areas A and B.

For A we use

$$\int_0^1 y \, dx = \int_0^1 (x^3 - 3x^2 + 2x) \, dx$$

$$= \left[\frac{x^4}{4} - x^3 + x^2 \right]_0^1$$

$$= \tfrac{1}{4}$$

For B we use

$$\int_1^2 (x^3 - 3x^2 + 2x) \, dx = \left[\frac{x^4}{4} - x^3 + x^2 \right]_1^2$$

$$= (4 - 8 + 4) - (\tfrac{1}{4} - 1 + 1)$$

$$= -\tfrac{1}{4}$$

The minus sign refers only to the position of area B relative to the x axis. The actual area is $\tfrac{1}{4}$ square unit.
So the total shaded area is $\tfrac{1}{4} + \tfrac{1}{4} = \tfrac{1}{2}$ square unit.

EXERCISE 10p

Find the areas bounded by the specified lines and curves in Questions 1–10.

1) The x and y axes, the line $x = 3$ and the curve $y = x^2 + 1$.

2) The x axis, the lines $x = 1$ and $x = 4$ and the curve $xy = 2$.

3) The y axis, the lines $y = 2$ and $y = 4$ and the curve $y = e^{-x}$.

4) The x axis and the curve $y = 1 - x^2$.

5) The y axis and the curve $x = 9 - y^2$.

6) The curve $y = \sin x$, the x axis and the line $x = \dfrac{\pi}{2}$.

7) The x axis, the line $x = \dfrac{\pi}{4}$ and the curve $y = \tan x$.

8) The curve $y = x^2 - 1$, the positive x axis and the negative y axis.

9) The curve $y = \ln x$, the y axis, the x axis and the line $y = 1$.

10) The curve $y = 9 - x^2$ and the x axis.

11) Evaluate:

(a) $\displaystyle\int_0^2 (x - 2)\, dx$ (b) $\displaystyle\int_2^4 (x - 2)\, dx$ (c) $\displaystyle\int_0^4 (x - 2)\, dx$

Interpret your results with the help of a graph.

12) Find the area enclosed between the curve $y = \cos x$ and the x axis bounded by the lines $x = 0$ and $x = \pi$.

13) If $y = x^2$, show by means of sketch graphs and *not* by evaluating the integrals, that

$$\int_0^1 y\, dx = 1 - \int_0^1 x\, dy$$

SUMMARY

1)

Function	Integral
x^n	$\dfrac{1}{n+1} x^{n+1}$ $(n \neq -1)$
e^x	e^x
$\dfrac{1}{x}$	$\ln |x|$
$\cos x$	$\sin x$
$\sin x$	$-\cos x$
$\sec^2 x$	$\tan x$

Function	Integral
$\csc^2 x$	$-\cot x$
$\tan x$	$\ln|\sec x|$
$\cot x$	$\ln|\sin x|$
$\dfrac{1}{\sqrt{(1-x^2)}}$	$\arcsin x$
$\dfrac{1}{1+x^2}$	$\arctan x$
$\sin^p x \cos x$	$\dfrac{1}{p+1}\sin^{p+1}x \quad (p\neq-1)$
$\cos^p x \sin x$	$-\dfrac{1}{p+1}\cos^{p+1}x \quad (p\neq-1)$
$\tan^p x \sec^2 x$	$\dfrac{1}{p+1}\tan^{p+1}x \quad (p\neq-1)$

2) $\displaystyle\int f'(x)\,e^{f(x)}\,dx = e^{f(x)} + K$

3) $\displaystyle\int \frac{f'(x)}{f(x)}\,dx = \ln|f(x)| + K$

4) $\displaystyle\int v\frac{du}{dx}\,dx = uv - \int u\frac{dv}{dx}\,dx$

5) $\displaystyle\lim_{\delta x \to 0}\sum_{x=a}^{x=b} f(x)\,\delta x = \int_a^b f(x)\,dx$

MULTIPLE CHOICE EXERCISE 10

(Instructions for answering these questions are given on p. xii.)

TYPE I

1) $\displaystyle\int\left(x^2 - \frac{1}{x^2} + \sin x\right)dx$ is:

(a) $\dfrac{x^3}{3} + \dfrac{1}{x} + \cos x + K$ (b) $\dfrac{x^3}{3} - \dfrac{2}{x} - \cos x + K$

(c) $2x - \dfrac{3}{x^2} + \cos x + K$ (d) $\dfrac{x^3}{3} + \dfrac{1}{x} - \cos x + K$

2) The shaded area in the diagram
is given by:

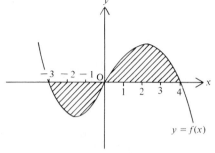

(a) $\int_{-3}^{4} f(x)\,dx$

(b) $\int_{-3}^{0} f(x)\,dx + \int_{0}^{4} f(x)\,dx$

(c) $\int_{-3}^{1} f(x)\,dx + \int_{1}^{4} f(x)\,dx$

(d) none of these.

3) e^{x^2} could be the integral w.r.t. x of:

(a) e^{2x} (b) $2x\,e^{x^2}$ (c) $\dfrac{e^{x^2}}{2x}$ (d) $x^2\,e^{x^2} - 1$ (e) none of these.

4) $x - \ln x^2 + K$ is the result of integrating w.r.t. x:

(a) $\dfrac{1}{1 - x^2}$ (b) $\dfrac{1 - 2x}{x^2}$ (c) $\dfrac{x - 2}{x}$

(d) $1 - \dfrac{2}{x^2}$ (e) none of these.

5) If $\displaystyle\int_{1}^{5} \dfrac{dx}{2x - 1} = \ln K$,   the value of K is:

(a) 9   (b) 3   (c) undefined   (d) 81   (e) 8.

6) $I = \displaystyle\int_{1}^{2} x\sqrt{x^2 - 1}\,dx$ is found as follows. Where does an error first occur?

(a) Let $u \equiv x^2 - 1$ (b) $\ldots du \equiv \ldots 2x\,dx$

(c) $I = \dfrac{1}{2}\displaystyle\int_{1}^{2} u^{\frac{1}{2}}\,du$ (d) $I = \dfrac{3}{4}\left[u^{\frac{3}{2}}\right]_{1}^{2}$

7) $\displaystyle\int_{0}^{\frac{\pi}{6}} \sin^n x \cos x \, dx = \tfrac{1}{64}$; n is:

(a) 6   (b) 5   (c) 4   (d) 3   (e) none of these.

8) The differential equation of a curve is $\dfrac{x}{y}\dfrac{dy}{dx} = 1$.   The sketch of the curve
could be:

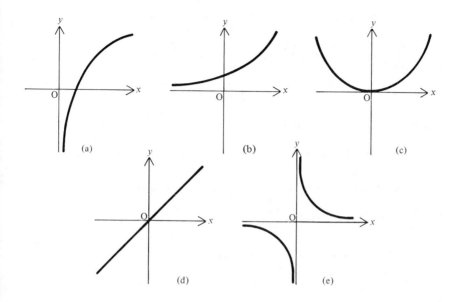

(a)   (b)   (c)

(d)   (e)

9) The value of $\int_0^2 2e^{2x}\, dx$ is:

(a) e^4 (b) $e^4 - 1$ (c) ∞ (d) $4e^4$ (e) $\tfrac{1}{2}e^4$.

TYPE II

10) $\int_0^n \tan x\, dx$ can be evaluated if:

(a) $n = \dfrac{\pi}{4}$ (b) $n = -\dfrac{\pi}{3}$ (c) $n = \dfrac{\pi}{2}$ (d) $n = -\dfrac{\pi}{2}$.

11) Which of the following differential equations can be solved by separating the variables:

(a) $x\dfrac{dy}{dx} = y + x$ (b) $xy\dfrac{dy}{dx} = x + 1$

(c) $e^{x+y} = y\dfrac{dy}{dx}$ (d) $x + \dfrac{dy}{dx} = \ln y$

12) $\int_1^2 x\, e^x\, dx$:

(a) is a definite integral   (b) is equal to $x\, e^x - e^x$

(c) is equal to $\left[\tfrac{1}{2}e^{x^2}\right]_1^2$ (d) can be integrated by parts.

13) Using $x \equiv \sin \theta$ transforms $\displaystyle\int \frac{x^2}{\sqrt{(1-x^2)}} \, dx$ into:

(a) $\displaystyle\int \frac{\sin^2 \theta}{\cos \theta} \, d\theta$ (b) $\frac{1}{2} \displaystyle\int (1 - \cos 2\theta) \, d\theta$

(c) $-\displaystyle\int \sin^2 \theta \, d\theta$ (d) $\frac{1}{2} \displaystyle\int (1 + \cos 2\theta) \, d\theta$.

14) Integration by parts can be used to find:

(a) $\displaystyle\int x^2 \, e^x \, dx$ (b) $\displaystyle\int e^x \ln x \, dx$ (c) $\displaystyle\int \ln x \, dx$ (d) $\displaystyle\int (\ln x)(\sin x) \, dx$.

15) Which of the following definite integrals can be evaluated:

(a) $\displaystyle\int_0^1 \frac{1}{x-1} \, dx$ (b) $\displaystyle\int_0^{\frac{\pi}{2}} \sin x \, dx$

(c) $\displaystyle\int_1^2 \sqrt{1-x^2} \, dx$ (d) $\displaystyle\int_{-2}^1 \ln x \, dx$

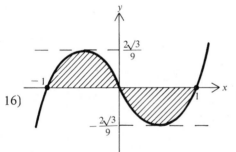

16)   The shaded area is equal to:

(a) $\displaystyle\int_{-1}^1 y \, dx$ (b) zero   (c) $2 \displaystyle\int_0^{(2\sqrt{3})/9} x \, dy$

(d) $\displaystyle\int_{-1}^0 y \, dx + \int_0^1 y \, dx$ (e) $\displaystyle\int_{-1}^0 y \, dx - \int_0^1 y \, dx$.

TYPE III

17) (a) $\dfrac{dy}{dx} = \cos x \, e^{\sin x}$

    (b) $y = e^{\sin x}$

18) (a) An area can be found by evaluating $\displaystyle\int_0^p f(x) \, dx$.

(b) An area is bounded by the curve $y = f(x)$, the x and y axes and the line $y = p$.

19) (a) $y = \int x \cos x \, dx$.

(b) $y = K + \cos x + x \sin x$.

20) (a) x and y are related by a linear differential equation.

(b) y is a quadratic function of x.

TYPE IV

21) Find the shaded area.

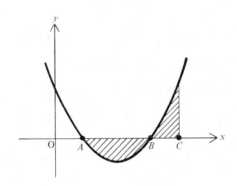

(a) The equation of the curve is $y = x^2 + ax + b$.
(b) At A, $x = 1$.
(c) The curve crosses the y axis at the point $(0, 3)$.
(d) At C, $x = 4$.

22) Find the equation of a curve.
(a) Its gradient is proportional to x.
(b) It passes through the origin.
(c) y is a quadratic function of x.
(d) It passes through the point $(1, 2)$.

23) Evaluate $\int_p^q f(x) \, dx$.

(a) The constant of integration is zero.
(b) $f(x)$ is fully defined for $p \leqslant x \leqslant q$.
(c) p is given.
(d) q is given.

TYPE V

24) A differential equation must contain $\dfrac{dy}{dx}$.

25) $\int \tan x \, dx = \sec^2 x + K.$

26) $\int_0^a f(y) \, dy = \lim_{\delta y \to 0} \sum_{y=0}^{a} f(y) \, \delta y.$

27) $\left[f(x) \right]_0^a \equiv f(a) - 0.$

MISCELLANEOUS EXERCISE 10

1) Evaluate:

(a) $\int_1^3 \dfrac{x+2}{x+1} \, dx$ (b) $\int_0^{\frac{\pi}{2}} \sin^4 x \cos x \, dx.$ (U of L)p

2) (a) Evaluate $\int_0^1 \dfrac{x^2 \, dx}{(1+x^3)^2}$ (b) Evaluate $\int_5^6 \dfrac{dx}{(x-2)(x-4)}$

giving the answer to three decimal places.   (JMB)p

3) Evaluate the integrals:

(a) $\int x(1+x^2)^5 \, dx$ (b) $\int x \, e^{-2x} \, dx$

(c) $\int_2^3 \dfrac{dx}{x(x^2-1)}$ (d) $\int_0^{\frac{\pi}{2}} \sin^2 x \cos^3 x \, dx$ (O)

4) (a) Find: (i) $\int (e^{2x} - 1)^2 \, dx$ (ii) $\int \dfrac{\cos \theta \, d\theta}{\sqrt{\sin \theta}}.$

     (b) Evaluate: (i) $\int_0^{\frac{\pi}{2}} 2 \cos 2\theta \cos \theta \, d\theta$ (ii) $\int_3^4 \dfrac{5 \, dx}{x^2 + x - 6}.$ (AEB)'74

5) Evaluate:

(a) $\int_0^{\frac{\pi}{6}} \tan 2\theta \, d\theta$ (b) $\int_0^1 x(1-x)^{\frac{1}{2}} \, dx.$ (JMB)

6) Evaluate the integrals:

(a) $\int_0^2 x(x^2+1)^3 \, dx$ (b) $\int_2^3 \dfrac{dx}{(4-x)(x-1)}$ (c) $\int_0^{\frac{\pi}{2}} \tan \tfrac{1}{2} x \, dx.$ (O)

7) (a) Evaluate:

$$\int_{\frac{\pi}{6}}^{\frac{\pi}{4}} 2 \sin 3x \cos 2x \, dx$$

giving your answer correct to two significant figures.

(b) Using the substitution $t = \tan x,$ or otherwise, find:

$$\int \dfrac{dx}{4 \cos^2 x - 9 \sin^2 x}.$$ (C)

8) Evaluate the definite integrals:

(a) $\int_1^2 \frac{(x^2 - 1)^2}{x^3} dx$ (b) $\int_{\frac{\pi}{2}}^{\pi} (\sin 3x + \cos \frac{1}{2}x) dx$ (c) $\int_1^2 \frac{x + 2}{x(x + 4)} dx.$

(U of L)

9) (a) By use of partial fractions, or otherwise, find:

$$\int \frac{x}{x^2 - 2x - 3} dx.$$

(b) Using the substitution $z = 1 - x$,   or otherwise, evaluate:

$$\int_0^1 x^2 (1 - x)^{\frac{1}{2}} dx.$$

(c) Solve the differential equation:

$$\frac{dy}{dx} = 1 - y$$

given that $y < 1$ and that $y = 0$ when $x = 0$.   (C)

10) Evaluate, correct to three decimal places:

(a) $\int_0^1 \frac{2x^2 dx}{2x + 1}$ (b) $\int_0^{\frac{\pi}{3}} \sin^2 x \cos^2 x \, dx$ (c) $\int_e^{e^2} \frac{dx}{x \ln x}.$ (U of L)

11) (a) Evaluate $\int_4^5 \frac{2x \, dx}{x^2 - 4x + 3}.$

(b) By using the substitution $x \equiv \sec^2 y$,   or otherwise, evaluate

$$\int_2^5 \frac{dx}{x^2 \sqrt{(x - 1)}}.$$ (U of L)p

12) (a) Find $\int \frac{2x - 1}{(x + 1)^2} dx$ and $\int \left(e^x - \frac{1}{e^x} \right)^2 dx.$

(b) Evaluate $\int_0^{\frac{\pi}{4}} \tan^2 x \, dx.$ (C)

13) Evaluate the following integrals:

(a) $\int_0^1 \frac{8}{3 + 4x} dx$ (b) $\int_0^1 \frac{8}{\sqrt{(3 + 4x)}} dx$ (c) $\int_0^1 \frac{8x}{3 + 4x} dx.$

(U of L)p

14) (a) If $e^x = \tan 2y$ prove that $\frac{d^2 y}{dx^2} = \frac{e^x - e^{3x}}{2(1 + e^{2x})^2}.$

(b) Find (i) $\int \frac{dx}{\sqrt{x} - x}$ (ii) $\int (1 - 3 \cos^2 x)^{\frac{1}{2}} \sin 2x \, dx.$ (AEB)'69

15) (a) Evaluate $\int_1^2 \frac{(x - 1)^2}{x^3} dx$ giving the answer correct to three decimal

places.

(b) Find $\int_{\alpha}^{7\alpha} \sin 2(x + \alpha)\,dx$ where $\alpha = \dfrac{\pi}{24}$.

(c) $\int_{0}^{3} (x + 1)^{\frac{3}{2}}\,dx$, and hence find $\int_{0}^{3} x\sqrt{(x + 1)}\,dx$.   (U of L)

16) (a) Find $\int 2 \cos 3x \sin x\,dx$ and $\int \dfrac{x - 2}{\sqrt{(x - 1)}}\,dx$.

   (b) Using the substitution $t = \tan x$, or otherwise, evaluate:

$$\int_{0}^{\frac{\pi}{3}} \frac{dx}{9 - 8 \sin^2 x}.$$ (C)

17) Find:

(a) $\int \dfrac{x + 1}{x(2x + 1)}\,dx$ and   (b) $\int \dfrac{x(2x + 1)}{x + 1}\,dx$.   (JMB)

18) (a) If $a > 1$ and $\int_{1}^{a} \dfrac{x^4 - 1}{x^3}\,dx = \dfrac{9}{8}$, find a.

   (b) If n is a positive integer, find in terms of n the three possible values

   of $\int_{\frac{\pi}{2}}^{\pi} \cos nx\,dx$.

   (c) Evaluate $\int_{0}^{\frac{\pi}{4}} \dfrac{2 \cos x - \sin x}{2 \sin x + \cos x}\,dx$, correct to three decimal places.

     (U of L)

19) Express $\dfrac{x - 2}{2x^2 - x - 3}$ in partial fractions and hence evaluate:

$$\int_{2}^{3} \frac{(x - 2)\,dx}{2x^2 - x - 3}.$$ (AEB)'73

20) (a) Find $\int \dfrac{x - 7}{2x^2 - 3x - 2}\,dx$.

   (b) Evaluate $\int_{0}^{\frac{\pi}{3}} \sin^3 x\,dx$.

   (c) Using the substitution $x = 4 \sin^2 \theta$, or otherwise, show that:

$$\int_{0}^{2} \sqrt{x(4 - x)}\,dx = \pi.$$ (C)

21) Evaluate:

(a) $\int_{0}^{\frac{\pi}{4}} \sin 5x \cos 3x\,dx$ (b) $\int_{0}^{1} xe^{-3x}\,dx$.   (JMB)p

22) Evaluate the integrals:

(a) $\int_0^{\frac{\pi}{2}} \sin 2x \cos 3x \, dx$ (b) $\int_0^{\frac{\pi}{2}} \sin^2 x \cos^2 x \, dx$

(c) $\int_0^3 \frac{x \, dx}{\sqrt{(25 - x^2)}}$ (d) $\int_0^1 \frac{x \, dx}{6 - 5x + x^2}$.   (O)

23) (a) Evaluate $\int_0^{\frac{\pi}{4}} \sin 3\theta \sin \theta \, d\theta$.

(b) Find $\int \sin^2 \theta \cos^3 \theta \, d\theta$.

(c) Find $\int \frac{x}{\sqrt{(3 + x)}} \, dx$.   (C)

24) Evaluate:

(a) $\int_1^{15} \frac{x + 2}{(x + 1)(x + 3)} \, dx$ (b) $\int_0^{\frac{\pi}{3}} x \sin 3x \, dx$ (c) $\int_{-1}^2 x^2 \sqrt{(x^3 + 1)} \, dx$.

(U of L)

25) Find y in terms of x given that:

$$x \frac{dy}{dx} = (1 - 2x^2)y, \quad (x > 0)$$

and that $y = 1$ when $x = 1$.   (U of L)p

26) (a) Evaluate $\int_0^1 \frac{1 - 4x}{3 + x - 2x^2} \, dx$.

(b) Find $\int xe^{2x} \, dx$.

(c) Solve the differential equation $\frac{dy}{dx} = xy$ given that $y = 2$ when $x = 0$.   (C)

27) Solve the differential equation $(1 + \cos 2x) \frac{dy}{dx} - (1 + e^y) \sin 2x = 0$

given that $y = 0$ when $x = \frac{\pi}{4}$.   (AEB)'73

28) Solve the differential equation $(3x + 5)^2 \frac{dy}{dx} = \frac{1 + 4y^2}{1 + y}$, given that $y = 0$ when $x = 0$.   (AEB)'76

29) (a) Evaluate:

(i) $\int_0^1 \frac{1 + x}{1 + 2x} \, dx$ (ii) $\int_0^{\frac{\pi}{2}} \sin x \cos^2 x \, dx$.

(b) Sketch the arc of the curve $y = 2x - x^2$ for which y is positive. Find the area of the region which lies between this arc and the x-axis.

(U of L)p

30) Solve the differential equation $(1 + x^2)\dfrac{dy}{dx} - y(y + 1)x = 0$, given that $y = 1$ when $x = 0$.

(AEB)'76

31) P is a point on a curve. The normal at P meets the x-axis at a point Q, and the curve is such that PQ is always equal to OP where O is the origin. Show that

$$y\frac{dy}{dx} = x$$

Hence find the equation of the curve.

32) Find the solution of the differential equation

$$\frac{dy}{dx} = \frac{x(y^2 - 1)}{y(x^2 + 1)}$$

for which $y = 3$ when $x = 1$.

(O)p

33) Solve the differential equation

$$y - x\frac{dy}{dx} = 2\left(y^2 + \frac{dy}{dx}\right).$$

(AEB)'72

34) If $\dfrac{dy}{dx} = \dfrac{y(y + 1)}{x(x + 1)}$, and $y = 2$ when $x = 1$, find y in terms of x.

(O)p

35) (a) Find:

(i) $\displaystyle\int \frac{x^2}{(x + 2)}\,dx$ (ii) $\displaystyle\int \sin 3x \cos 2x\,dx$.

(b) Find the coordinates of P, the point of intersection of the curves

$$y = e^x, \quad y = 2 + 3e^{-x}$$

If these curves cut the y-axis at the points A and B, calculate the area bounded by AB and the arcs AP and BP.

(U of L)

36) The curve $y = 2\cos x - 1$ cuts the axis of x at two points P and Q whose abscissae lie in the range $-\frac{1}{2}\pi < x < \frac{1}{2}\pi$. Find the area enclosed by the straight line PQ and the arc of the curve between P and Q.

(C)p

37) Calculate the area bounded by the lines $x = 0$, $x = 1$, $y = 1$, and the part of the graph of

$$y = \frac{x^2}{x^2 + 1}$$

between $x = 0$ and $x = 1$.

(JMB)p

38) The equations of the tangents to the curve
$$y = x(x-a)(x-b),$$
where a and b are constants, at the points $(0,0)$ and $(a,0)$ are $y=4x$ and $y=3-3x$ respectively. Calculate the values of a and b and sketch the curve.
Find the area in the right half plane $x>0$ which is bounded by the x-axis and the curve, and which lies below the x-axis.   (U of L)p

39) The curves $y=3\sin x$, $y=4\cos x$ $(0 \leqslant x \leqslant \frac{1}{2}\pi)$ intersect at the point A, and meet the axis of x at the origin O and the point $B(\frac{1}{2}\pi, 0)$ respectively. Prove that the area enclosed by the arcs OA, AB and the line OB is 2 square units.   (C)p

40) (a) Find y as a function of x when
$$\frac{1}{y}\frac{dy}{dx} - x = xy$$
and $y=1$ when $x=0$.

(b) If $x-y=z$, express $\frac{dy}{dx}$ in terms of $\frac{dz}{dx}$. Using the substitution $x-y=z$, or otherwise, solve the differential equation
$$\frac{dy}{dx} = x-y$$
given that $y=0$ when $x=0$.   (U of L)

41) In a chemical reaction in which a compound X is formed from a compound Y and other substances, the masses of X and Y present at time t are x and y respectively. The sum of the two masses is constant and at any time the rate at which x is increasing is proportional to the product of the two masses at that time. Show that the equation governing the reaction is of the form
$$\frac{dx}{dt} = kx(a-x)$$
and interpret the constant a. If $x = \dfrac{a}{10}$ at time $t=0$, find in terms of k and a the time at which $y = \dfrac{a}{10}$.   (U of L)

CHAPTER 11

COORDINATE GEOMETRY II

ANALYTICAL GEOMETRY IN TWO DIMENSIONS

Any line, straight or curved, is made up of an infinite set of points. Any point, P, can be located by a pair of coordinates in a suitable frame of reference. If there is an equation whose solution set is made up of the coordinates of every point P on the line, but excludes the coordinates of any point not on the line, we say that this is the equation of the line. (The form of the equation depends upon the shape of the line.)

Conversely, an equation which relates the coordinates of a point P can be displayed graphically by plotting corresponding coordinates on a suitable graph. (The shape of the line depends upon the form of the equation.)

Hence:

(a) the geometric properties of a line can be expressed in the form of an equation,

(b) the geometric properties of a line can be found by analysing its equation.

There have been references, in earlier chapters, to coordinates of three types; Cartesian, parametric and polar. Thus lines can have Cartesian, parametric or polar equations.

The techniques used in the analysis of straight lines and curves depend, to some extent, on the frame of reference being used. In this chapter we will examine a variety of the methods adopted when using Cartesian coordinates.

THE STRAIGHT LINE

Any geometric work makes frequent use of straight lines. Their analysis, which began in Chapter 4, will now be developed. Before proceeding, the reader should remind himself of the results already obtained. (See Summary, p. 90.)

Division of a Straight Line in a Given Ratio

Suppose that a point $P(X, Y)$ divides the line joining $A(x_1, y_1)$ to $B(x_2, y_2)$ in the ratio $\lambda : \mu$.

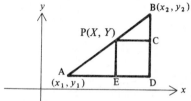

In the diagram
triangles $\begin{cases} \text{APE} \\ \text{PBC} \end{cases}$ are similar.

Therefore $\dfrac{AP}{PB} = \dfrac{AE}{PC}$.

But $AE = X - x_1$, $PC = x_2 - X$ and $\dfrac{AP}{PB} = \dfrac{\lambda}{\mu}$

Therefore

$$\frac{X - x_1}{x_2 - X} = \frac{\lambda}{\mu}$$

$\Rightarrow$

$$X = \frac{\lambda x_2 + \mu x_1}{\lambda + \mu}$$

Similarly

$$Y = \frac{\lambda y_2 + \mu y_1}{\lambda + \mu}$$

These formulae, which are quotable, apply to both internal and external division. Their use is not always necessary however. When the coordinates of A and B are known numbers a diagram, together with simple mental arithmetic, is often adequate,

e.g. to divide the line joining $A(-2, 5)$ to $B(4, 2)$ internally and externally in the ratio $2:1$ the following diagrams are effective.

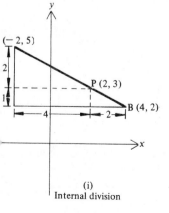

(i)
Internal division

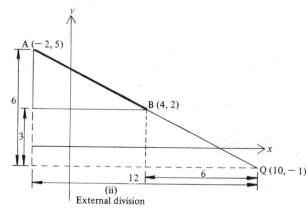

(ii)
External division

Alternatively the division formulae can be used as follows:

(a) For the internal ratio $2:1$, $\lambda = 2$ and $\mu = 1$.

Thus, at P, $x = \dfrac{2(4) + 1(-2)}{2 + 1} = 2$

and $y = \dfrac{2(2) + 1(5)}{2 + 1} = 3$

(b) For the external ratio $2:1$, $\lambda = 2$ and $\mu = -1$

Thus, at Q, $x = \dfrac{2(4) - 1(-2)}{2 - 1} = 10$

and $y = \dfrac{2(2) - 1(5)}{2 - 1} = -1$

(**Note:** In the external division the sign of μ is opposite to the sign of λ because the direction of the line segment QB is opposite to the direction of the line segment AQ.

For the same reason, external division is sometimes denoted by a negative ratio or fraction, e.g. $2:-1$ or $-\frac{2}{1}$.)

The Centroid of a Triangle

The centroid, G, of a triangle is the point of intersection of the medians and G divides each median in the ratio $2:1$.

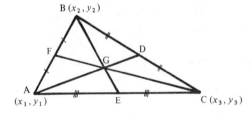

i.e. $\dfrac{AG}{GD} = \dfrac{BG}{GE} = \dfrac{CG}{GF} = \dfrac{2}{1}$

If the coordinates of A, B and C are (x_1, y_1), (x_2, y_2) and (x_3, y_3) respectively then

the coordinates of D, the midpoint of BC, are $\left(\dfrac{x_2 + x_3}{2}, \dfrac{y_2 + y_3}{2}\right)$

But $\dfrac{AG}{GD} = \dfrac{2}{1}$

Hence, at G, $x = \left[2\left(\dfrac{x_2 + x_3}{2}\right) + x_1\right] \div 3$

and
$$y = \left[2\left(\frac{y_2 + y_3}{2}\right) + y_1\right] \div 3$$

This result shows that the coordinates of the centroid of a triangle are the average of the coordinates of the vertices,

i.e.
$$\left(\tfrac{1}{3}(x_1 + x_2 + x_3),\ \tfrac{1}{3}(y_1 + y_2 + y_3)\right)$$

The Angle between Two Straight Lines

Consider two lines with gradients $m_1 \ (= \tan \theta_1)$ and $m_2 \ (= \tan \theta_2)$. The angle, α, between the lines is given by

$$\alpha = \theta_1 - \theta_2$$

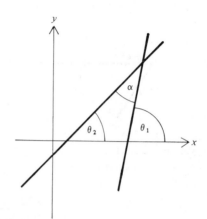

Therefore

$$\tan \alpha = \tan(\theta_1 - \theta_2)$$
$$= \frac{\tan \theta_1 - \tan \theta_2}{1 + \tan \theta_1 \tan \theta_2}$$

$\Rightarrow$
$$\tan \alpha = \frac{m_1 - m_2}{1 + m_1 m_2}$$

This result is quotable:
e.g. if two straight lines l_1 and l_2 have equations $3x - 2y = 5$ and $4x + 5y = 1$ then

$$\text{gradient of } l_1 = m_1 = \tfrac{3}{2}$$
$$\text{gradient of } l_2 = m_2 = -\tfrac{4}{5}$$

Thus if α is the angle between l_1 and l_2,

$$\tan \alpha = \frac{\tfrac{3}{2} - (-\tfrac{4}{5})}{1 + (\tfrac{3}{2})(-\tfrac{4}{5})} = \frac{23}{10} \Bigg/ \left(\frac{-2}{10}\right) = -\frac{23}{2}$$

Note: If, as in this example, $\tan \alpha$ is negative, α is the obtuse angle between the lines. The acute angle between them is $\arctan \tfrac{23}{2}$.

The Distance of a Point from a Straight Line

It is understood that the distance from a point to a line is the perpendicular distance.
Consider the line with equation $y = mx + c$ i.e. $mx - y + c = 0$

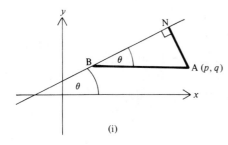

(i)

A is distant d from the line where

$$d = AN = AB \sin \theta$$

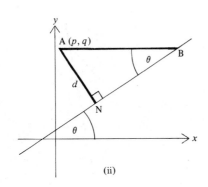

(ii)

AB is horizontal therefore

at B $y = q$

and $x = \dfrac{q-c}{m}$

Hence, in diagram (i) $AB = p - \left(\dfrac{q-c}{m}\right)$

and in diagram (ii) $AB = \left(\dfrac{q-c}{m}\right) - p$

i.e. for any point A $AB = \pm\left[\dfrac{mp - q + c}{m}\right]$

The gradient of the line is $m = \tan \theta$

Hence $\sin \theta = \dfrac{m}{\sqrt{(m^2 + 1)}}$

Therefore $AN = \pm\left[\dfrac{mp - q + c}{m}\right]\left[\dfrac{m}{\sqrt{(m^2 + 1)}}\right]$

i.e. $AN = \pm\dfrac{mp - q + c}{\sqrt{(m^2 + 1)}}$

Note that this result is equally valid for lines with negative gradient since
$AN = AB \sin (\pi - \theta) = AB \sin \theta$.

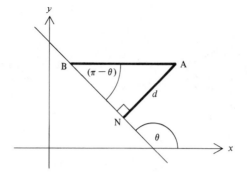

The two signs ($\pm$) in the formula arise from points which are on opposite sides of the line. If we are interested only in the length d of AN, we take the positive value of the formula. This is written

$$d = \left| \frac{mp - q + c}{\sqrt{(m^2 + 1)}} \right|$$

When the equation of the line is given in the form $ax + by + c = 0$, the argument used above leads to the result that

$$d = \left| \frac{ap + bq + c}{\sqrt{(a^2 + b^2)}} \right|$$

where d is the length of AN.

Note: If the formula is used in the form $AN = \dfrac{ap + bq + c}{\sqrt{(a^2 + b^2)}}$ then for two points on opposite sides of the line $ax + by + c = 0$, the two values given for AN are opposite in sign.

EXAMPLES 11a

1) The lines with equations $y = 2x$, $2y + x = 5$ and $4y = 3x - 5$ form a triangle. Find the coordinates of each vertex. By finding the length of one altitude and one side calculate the area of the triangle.

Let the lines $y = 2x$ and $2y + x = 5$ meet at A,

then at A, $\qquad\qquad 2(2x) + x = 5 \implies x = 1, y = 2$

If the lines $2y + x = 5$ and $4y = 3x - 5$ meet at B

then at B $\qquad\qquad 3x - 5 = 10 - 2x \implies x = 3, y = 1$

If the lines $y = 2x$ and $4y = 3x - 5$ meet at C,

then at C, $\qquad\qquad 4(2x) = 3x - 5 \implies x = -1, y = -2$

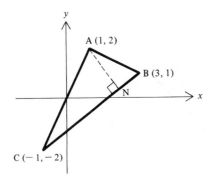

AN is an altitude whose length is the distance from $A(1, 2)$ to the line CB with equation $4y = 3x - 5$.

Re-arranging this equation in the form $3x - 4y - 5 = 0$ and using the formula

$$d = \left| \frac{ap + bq + c}{\sqrt{[a^2 + b^2]}} \right|$$

we have length of $AN = \left| \frac{3(1) - 4(2) - 5}{\sqrt{[3^2 + (-4)^2]}} \right| = \left| -\frac{10}{5} \right| = 2$

The area of $\triangle ABC = \frac{1}{2}(CB)(AN)$

and $CB = \sqrt{[1 - (-2)]^2 + [3 - (-1)]^2} = 5$

Therefore the area is $\frac{1}{2}(5)(2) = 5$ square units.

2) Show that the point $(-1, 5)$ is the reflection (image) of the point $(3, -3)$ in the line $2y = x + 1$.

A point A is the reflection of a point B in a given line if, and only if,

(a) the given line bisects AB, *and*

(b) AB is perpendicular to the given line.

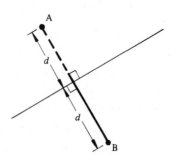

(a) The midpoint of the line joining $A(-1, 5)$ and $B(3, -3)$ is

$$\left(\frac{-1 + 3}{2}, \frac{5 - 3}{2} \right) \quad \text{i.e.} \quad (1, 1)$$

$x = 1$, $y = 1$ is a solution of the equation $2y = x + 1$.

Therefore the midpoint of AB is on the given line.

(b) The gradient of $AB = \dfrac{5-(-3)}{-1-3} = -2 = m_1$.

The gradient of the given line, $y = \frac{1}{2}(x+1)$, is $\frac{1}{2} = m_2$,

i.e. $\qquad\qquad\qquad\qquad m_1 m_2 = -1$,

showing that AB is perpendicular to the given line.
Therefore $(-1, 5)$ *is* the reflection of $(3, -3)$ in the line $2y = x + 1$.

3) Determine, without drawing a diagram, whether the points $A(-3, 4)$ and $B(-2, 3)$ are on the same side of the line $y + 3x + 4 = 0$. Find the acute angle between this line and the line AB.

Using the formula $\dfrac{ap + bq + c}{\sqrt{[a^2 + b^2]}}$ to find the directed distances of A and B from the line $3x + y + 4 = 0$ gives

for A, $\qquad\qquad\qquad \dfrac{3(-3) + 4 + 4}{\sqrt{[3^2 + 1^2]}} < 0$

for B, $\qquad\qquad\qquad \dfrac{3(-2) + 3 + 4}{\sqrt{[3^2 + 1^2]}} > 0$

The signs are opposite so A and B are on opposite sides of the line.
The gradient m_1 of the given line is -3,

the gradient m_2 of AB is $\dfrac{3-4}{-2-(-3)} = -1$.

If an angle between the given line and AB is α

then $\qquad\qquad \tan \alpha = \dfrac{m_1 - m_2}{1 + m_1 m_2} = \dfrac{-3+1}{1+3} = -\dfrac{1}{2}$

Thus the acute angle between the lines is $\arctan \frac{1}{2}$.

EXERCISE 11a

1) Find the coordinates of the point that divides AB in the given ratio in each of the following cases:
(a) $A(2, 4)$, $B(-3, 9)$ $\qquad 1:4$ internally
(b) $A(-3, -4)$, $B(3, 5)$ $\qquad 3:1$ externally
(c) $A(1, 5)$, $B(8, -2)$ $\qquad 4:3$
(d) $A(-1, 6)$, $B(3, -2)$ $\qquad 3:-2$

2) A is the midpoint of BC. If A is (X, Y) and B is (x_1, y_1) show that C has coordinates $(2X - x_1, 2Y - y_1)$.

3) Find the distance from A to the given line in each of the following cases:

(a) A(3, 4); $2x - y = 3$ (b) A($-1, -2$); $3y = 4x - 1$
(c) A(a, b); $y = mx + c$ (d) A(4, -1); $x + y = 6$
(e) A(x, y); $ax + by + c = 0$ (f) A(0, 0); $ax + by + c = 0$

4) Determine whether A and B are on the same or opposite sides of the given line in each of the following cases:
(a) A(1, 2), B(4, -3); $3x + y = 7$
(b) A(0, 3), B(7, 6); $x - 4y + 1 = 0$
(c) A($-5, 1$), B($-2, 3$); $7x + y - 6 = 0$

5) Find the tangents of the acute angles between the following pairs of lines:
(a) $2x + 3y = 7$, $x - 6y = 5$
(b) $x + 4y - 1 = 0$, $3x + 7y = 2$
(c) $a_1x + b_1y + c_1 = 0$, $a_2x + b_2y + c_2 = 0$

6) A(0, 1), B(3, 7) and C($-4, -4$) are the vertices of a triangle. Find the tangent of each of the three vertex angles and the length of each altitude.

7) Find the image of the point (5, 6) in the line:
(a) $3x - y + 1 = 0$ (b) $y = 4x + 20$ (c) $2x + 5y + 18 = 0$

8) Find the equations of the two lines through the origin which are inclined at $45°$ to the line $2x + 3y - 4 = 0$.

9) A point P(X, Y) is equidistant from the line $x + 2y = 3$ and from the point (2, 0). Find an equation relating X and Y.

10) Show that A(4, 1) and B(2, -3) are equidistant from the line $2x + 5y = 1$.   Is A the reflection of B in this line?

11) Write down the distance of the point P(X, Y) from each of the lines $5x - 12y + 3 = 0$ and $3x + 4y - 6 = 0$.   By equating these distances find the equations of two lines that bisect the angles between the two given lines. [i.e. the equations of the set of points P(X, Y).]

12) A(4, 4) and B(7, 0) are two vertices of a triangle OAB. Find the equation of the line that bisects the angle OBA. If this line meets OA at C show that C divides OA in the ratio OB : BA.

LOCI

When the possible positions of a point P are restricted to a straight line or curve, the set of such points is called the *locus of P*.
Further, if P is the point (x, y), the relationship between x and y which applies only to the set of points P is called the Cartesian equation of the locus of P.

EXAMPLES 11b

1) A point P moves so that it is equidistant from the points A(1, 2) and B($-2, -1$). Find the Cartesian equation of the locus of P.

Consider one of the possible
positions of $P(x, y)$.
The given condition can be written

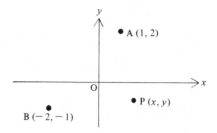

$$PA = PB$$

or $$PA^2 = PB^2$$

i.e. $$(x - 1)^2 + (y - 2)^2 = (x + 2)^2 + (y + 1)^2$$

$$\Rightarrow \qquad 0 = 6x + 6y$$

This equation is satisfied by every point on the locus and by no other point.
Thus $x + y = 0$ is the equation of the locus of the specified set of points.
Note The line $x + y = 0$ is the perpendicular bisector of **AB**.

2) A point $P(x, y)$ is twice as far from the point $(3, 0)$ as it is from the line
$x = 5$. Find the Cartesian equation of the set of points P.

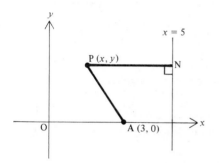

In the diagram, A is the point
$(3, 0)$.
PN is the distance of P from the
line $x = 5$.
For the set of point P,

$$PA = 2PN$$

or $$PA^2 = 4PN^2$$

Thus P satisfies the given condition $\iff y^2 + (x - 3)^2 = 4(5 - x)^2$

i.e. the equation of the set of points P is

$$y^2 - 3x^2 + 34x = 91$$

EXERCISE 11b

Find the Cartesian equation of the locus of the set of points P in each of the
following cases.

1) P is equidistant from the point $(4, 1)$ and the line $x = -2$.

2) P is equidistant from $(3, 5)$ and $(-1, 1)$.

3) P is three times as far from the line $x = 8$ as from the point $(2, 0)$.

4) P is equidistant from the lines $3x + 4y + 5 = 0$ and
$12x - 5y + 13 = 0$.

5) P is at a constant distance of two units from the point $(3, 5)$.

6) P is at a constant distance of five units from the line $4x - 3y = 1$.

7) A is the point $(-1, 0)$, B is the point $(1, 0)$ and angle APB is a right angle.

THE CIRCLE

By definition, a circle is the locus of the set of points P which are at a constant distance r from a fixed point C. This geometric property can be used to derive the Cartesian equation of a circle.

If C is the point (p, q) and P(x, y) is any point satisfying the definition,

then

$$CP = r$$

or $CP^2 = r^2$

But

$$CP^2 = (x - p)^2 + (y - q)^2$$

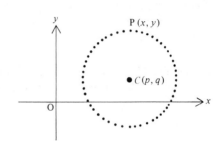

Therefore the coordinates of P must satisfy the equation

$$(x - p)^2 + (y - q)^2 = r^2 \qquad [1]$$

and this is the equation of the circle which is the locus of P.
e.g. the circle with centre $(-2, 3)$ and radius 1 has an equation

$$[x - (-2)]^2 + [y - 3]^2 = 1$$

$$\Rightarrow \qquad x^2 + y^2 + 4x - 6y + 12 = 0$$

Conversely, an equation of the form

$$x^2 + y^2 + 2gx + 2fy + c = 0 \qquad [2]$$

where g, f and c are constants, can be rearranged as follows

$$x^2 + 2gx + g^2 + y^2 + 2fy + f^2 + c = g^2 + f^2$$

i.e. $\qquad (x + g)^2 + (y + f)^2 = g^2 + f^2 - c \qquad [3]$

Comparing [1] and [3] we see that equation [3], and hence equation [2], is the equation of a circle with centre $(-g, -f)$ and radius $\sqrt{g^2 + f^2 - c}$ (provided that $g^2 + f^2 > c$).

Thus $x^2 + y^2 + 2gx + 2fy + c = 0$ is called the *general equation of a circle*.

(**Note** that the coefficients of x^2 and y^2 are equal and that no xy term is present.)

e.g. $x^2 + y^2 + 8x - 2y + 13 = 0$ is the equation of a circle whose centre and radius can be found in one of the following ways:

(a) by forming perfect squares as illustrated in the general case

i.e. $x^2 + 8x + 16 + y^2 - 2y + 1 = 16 + 1 - 13$

$\Rightarrow$ $(x + 4)^2 + (y - 1)^2 = 4$

comparing with $(x - p)^2 + (y - q)^2 = r^2$ we see that the centre is $(-4, 1)$ and the radius is 2.

(b) by comparing the given equation with the general equation and noting that $2g = 8$, $2f = -2$ and $c = 13$, from which the centre $(-g, -f)$, is $(-4, 1)$ and the radius, $\sqrt{g^2 + f^2 - c}$, is 2.

Note. A similar equation, e.g. $2x^2 + 2y^2 - 6x + 10y = 1$ also represents a circle but the centre and radius can be found *only after dividing the whole equation by 2* as follows

$$2x^2 + 2y^2 - 6x + 10y - 1 = 0$$

$\Rightarrow$ $$x^2 + y^2 - 3x + 5y - \tfrac{1}{2} = 0$$

Comparing this form with the general equation we see that

$$2g = -3, \quad 2f = 5, \quad c = -\tfrac{1}{2} \Rightarrow g = -\tfrac{3}{2}, \quad f = \tfrac{5}{2}, \quad \sqrt{g^2 + f^2 - c} = 3$$

So the centre of the circle is the point $(\tfrac{3}{2}, -\tfrac{5}{2})$ and its radius is 3.

The analytical geometry of circles makes use of some of the simple geometric properties of the circle, such as:

(a) the tangent at a point P on the circle is perpendicular to the radius PC,

(b) the distance of the centre, C, from any tangent is equal to the radius,

(c) the angle subtended at the circumference by a diameter is $90°$, i.e. an angle in a semi-circle is a right angle.

The Equation of a Tangent at a Given Point

Suppose that $A(x_1, y_1)$ is a point on a circle with centre $C(p, q)$. The gradient of AC is

$$m = \frac{y_1 - q}{x_1 - p}$$

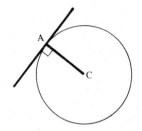

The tangent at A is perpendicular to AC, so its gradient is $-\dfrac{1}{m} = -\dfrac{x_1 - p}{y_1 - q}$

It is therefore unnecessary to use $\dfrac{dy}{dx}$ to find the gradient of the tangent at a point on a *circle*.

EXAMPLES 11c

1) Find the equation of the tangent at the point $(3, 1)$ on the circle
$$x^2 + y^2 - 4x + 10y - 8 = 0.$$

The centre of the circle is
$$C(2, -5)$$
Hence the gradient of AC is
$$\frac{1-(-5)}{3-2} = 6$$

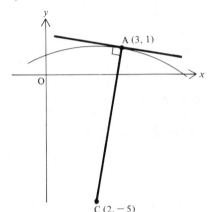

Therefore the tangent at A has a gradient of $-\frac{1}{6}$, so its equation is
$$y - 1 = -\tfrac{1}{6}(x - 3)$$
i.e. $6y + x = 9$

Condition for a Line to be Tangential to a Circle

A line is a tangent to a circle if and only if the distance of that line from the centre of the circle is equal to the radius.

EXAMPLES 11c (continued)

2) Determine whether the lines $5y = 12x - 33$ and $3x + 4y = 9$,   are tangents to the circle $x^2 + y^2 + 2x - 8y = 8$.

The equation of the circle can be written
$$x^2 + 2x + 1 + y^2 - 8y + 16 = 1 + 16 + 8$$
$$\Rightarrow \qquad (x + 1)^2 + (y - 4)^2 = 25$$

Therefore $C(-1, 4)$ is the centre of the circle and the radius is 5 units.
For the line $5y = 12x - 33$,   i.e. $12x - 5y - 33 = 0$,   the distance d_1 from the centre $(-1, 4)$ of the circle is given by
$$d_1 = \left| \frac{12(-1) - 5(4) - 33}{\sqrt{[12^2 + (-5)^2]}} \right| = \left| -\frac{65}{13} \right|$$
i.e. $d_1 = 5 = r$

Thus $12x - 5y - 33 = 0$ *is* a tangent.

For the line $3x + 4y = 9$,   i.e. $3x + 4y - 9 = 0$,

the distance d_2 from $(-1, 4)$ is given by

$$d_2 = \left| \frac{3(-1) + 4(4) - 9}{\sqrt{[3^2 + 4^2]}} \right| = \frac{4}{5}$$

i.e. $d_2 = \frac{4}{5} \neq 5$

Thus $3x + 4y - 9 = 0$ *is not* a tangent.

3) Find the equations of the tangents from the origin to the circle
$x^2 + y^2 - 5x - 5y + 10 = 0$.

The given circle has centre $(\frac{5}{2}, \frac{5}{2})$

and radius $\sqrt{(\frac{5}{2})^2 + (\frac{5}{2})^2 - 10} = \dfrac{\sqrt{10}}{2}$

Any line through the origin has equation $y = mx$

i.e. $mx - y = 0$

If the line is a tangent to the circle, the distance from the centre, $(\frac{5}{2}, \frac{5}{2})$, to the
line is equal to the radius

i.e. $\left| \dfrac{m(\frac{5}{2}) - (\frac{5}{2})}{\sqrt{[(m)^2 + (-1)^2]}} \right| = \dfrac{\sqrt{10}}{2}$

$\Rightarrow$ $(\frac{5}{2}m - \frac{5}{2})^2 = \frac{10}{4}(m^2 + 1)$

$\Rightarrow$ $3m^2 - 10m + 3 = 0$

$\Rightarrow$ $(3m - 1)(m - 3) = 0$

$\Rightarrow$ $m = \frac{1}{3}$ or 3

So the two tangents from the origin to the given circle are

$$y = 3x \quad \text{and} \quad 3y = x.$$

Orthogonal Circles

Two circles which intersect at right angles cut *orthogonally* and are called
orthogonal circles.
Consider two circles, with centres A and B, which cut orthogonally at P
and Q.

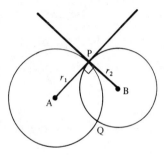

The tangent at P to the circle with
centre B is perpendicular to PB.
Similarly the tangent at P to the
circle with centre A is perpendicular to PA.

But the circles are orthogonal so the tangents at P are perpendicular, hence PA is perpendicular to PB. Thus each tangent at P passes through the centre of the other circle.

Also $\qquad\qquad\qquad AP^2 + BP^2 = AB^2$

i.e. $\qquad\qquad\qquad r_1{}^2 + r_2{}^2 = AB^2$

Thus, if two circles are orthogonal, the square of the distance between their centres is equal to the sum of the squares of the two radii.

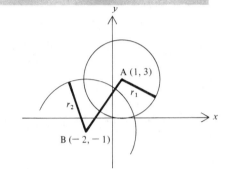

e.g. $(x-1)^2 + (y-3)^2 = 9$

and $(x+2)^2 + (y+1)^2 = 16$

are orthogonal because

$\qquad\qquad AB^2 = 3^2 + 4^2$

and $r_1{}^2 + r_2{}^2 = 9 + 16$

i.e. $\qquad AB^2 = r_1{}^2 + r_2{}^2$

Circles that Touch

Consider two circles with centres A and B, and radii r_1 and r_2.

If the circles touch externally, then

$\qquad\qquad AB = r_1 + r_2$

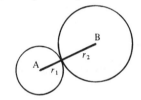

and if they touch internally, then

$\qquad\qquad AB = r_1 \sim r_2$

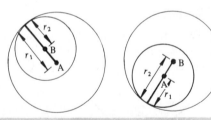

Thus two circles touch if the distance between their centres is the sum or the difference of their radii.

EXAMPLES 11c (continued)

4) Show that the circles $x^2 + y^2 - 16x - 12y + 75 = 0$ and $5x^2 + 5y^2 - 32x - 24y + 75 = 0$ touch each other and find the equation of the common tangent at their point of contact.

The circle $x^2 + y^2 - 16x - 12y + 75 = 0$ has centre $A(8, 6)$ and radius

$$r_1 = \sqrt{8^2 + 6^2 - 75} = 5$$

The circle $5x^2 + 5y^2 - 32x - 24y + 75 = 0$ can be written as

$$x^2 + y^2 - \tfrac{32}{5}x - \tfrac{24}{5}y + 15 = 0$$

and has centre $B(\tfrac{16}{5}, \tfrac{12}{5})$ and radius

$$r_2 = \sqrt{(\tfrac{16}{5})^2 + (\tfrac{12}{5})^2 - 15} = 1$$

The distance between the centres is given by

$$AB^2 = (8 - \tfrac{16}{5})^2 + (6 - \tfrac{12}{5})^2 = (\tfrac{24}{5})^2 + (\tfrac{18}{5})^2 = 36$$

Thus $AB = 6$

and $r_1 + r_2 = 5 + 1 = 6$

Therefore the circles touch externally.

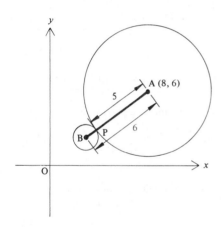

The point of contact, P, divides BA in the ratio $1:5$. Therefore the coordinates of P are

$$\begin{cases} x = \dfrac{1(8) + 5(\tfrac{16}{5})}{1 + 5} \\[2mm] y = \dfrac{1(6) + 5(\tfrac{12}{5})}{1 + 5} \end{cases}$$

i.e. $x = 4, \quad y = 3$

The common tangent at P is perpendicular to BA.

The gradient of BA is $\dfrac{6 - 12/5}{8 - 16/5} = \dfrac{3}{4}$.

Thus the gradient of the common tangent is $-\tfrac{4}{3}$ and its equation is

$$y - 3 = -\tfrac{4}{3}(x - 4)$$

$\Rightarrow$ $3y + 4x = 25$

5) A circle, whose centre is in the first quadrant, touches the x and y axes and the line $3x - 4y = 12$.   Find its equation.

In this problem there is no information which leads directly to the centre and radius of the circle. So we represent the radius by r. Then, since the centre of the circle is in the first quadrant, its coordinates are both positive. The circle also touches both axes so its centre is

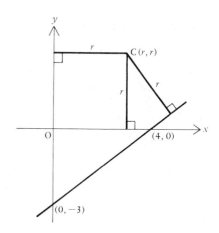

$$(r, r).$$

But the circle also touches the line $3x - 4y - 12 = 0$, so the distance CN is equal to r.

i.e.
$$r = \pm \frac{3(r) - 4(r) - 12}{\sqrt{(3^2 + 4^2)}}$$

$\Rightarrow$
$$\pm(-r - 12) = 5r$$

Thus either $6r = -12 \Rightarrow r = -2$

or $4r = 12 \Rightarrow r = 3$

But C is in the first quadrant therefore its coordinates are positive, i.e. r is positive.

Thus $r = 3$, the centre of the circle is $(3, 3)$ and its equation is

$$(x - 3)^2 + (y - 3)^2 = 3^2$$

$\Rightarrow$
$$x^2 + y^2 - 6x - 6y + 9 = 0$$

6) Find the equation of a circle with centre on the y-axis, which cuts orthogonally each of the circles $x^2 + y^2 + 6x + 2y - 9 = 0$ and $x^2 + y^2 - 2x - 2y + 1 = 0$.

The required circle has its centre on the y axis, so let its centre be the point $(0, q)$ and let its radius be r.
If it is orthogonal to the first circle

$$r^2 + 19 = [0 - (-3)]^2 + [q - (-1)]^2$$

i.e.
$$r^2 = q^2 + 2q - 9 \qquad [1]$$

and if it is orthogonal to the second circle

$$r^2 + 1^2 = [0 - 1]^2 + [q - 1]^2$$

i.e.
$$r^2 = q^2 - 2q + 1 \qquad [2]$$

Solving [1] and [2] gives $q = \frac{5}{2}$, $r^2 = \frac{9}{4}$.

Thus the equation of the circle which is orthogonal to both the given circles is

$$[x - 0]^2 + [y - \tfrac{5}{2}]^2 = \tfrac{9}{4}$$

i.e.

$$x^2 + y^2 - 5y + 4 = 0$$

There are many different ways in which a circle can be specified. The examples that follow illustrate some of varied approaches which can lead to the equation of a circle.

7) Find the equation of a circle given that it passes through the points $(0, 1)$, $(4, -1)$, $(4, 7)$.

The centre of a circle lies on the perpendicular bisector of any chord.

Using chords AB and BC we have

AB has midpoint $(2, 0)$ and gradient $-\frac{1}{2}$, so its perpendicular bisector is

$$(y - 0) = 2(x - 2)$$
$$\Rightarrow \quad y = 2x - 4 \qquad [1]$$

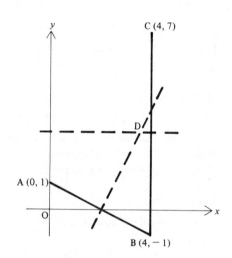

BC has midpoint $(4, 3)$ and it is vertical, so its perpendicular bisector is horizontal and its equation is

$$y = 3 \qquad\qquad\qquad [2]$$

Solving [1] and [2] gives $y = 3$, $x = \frac{7}{2}$ and these are the coordinates of the centre, D, of the circle.

The radius, r, is the length of DA (or DC, or DB)

i.e.

$$r^2 = (3 - 1)^2 + (\tfrac{7}{2} - 0)^2 = \tfrac{65}{4}$$

Therefore the equation of the circle is

$$(x - \tfrac{7}{2})^2 + (y - 3)^2 = \tfrac{65}{4}$$

$$\Rightarrow \qquad x^2 + y^2 - 7x - 6y + 5 = 0$$

8) The points $A(x_1, y_1)$ and $B(x_2, y_2)$ are the ends of a diameter of a circle. Find the equation of the circle.

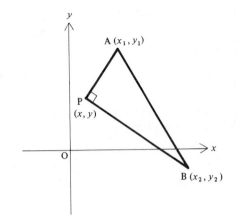

Let $P(x, y)$ be any point on the required circle. Since AB is a diameter, angle APB is $90°$.

Gradient of PA is $\dfrac{y - y_1}{x - x_1} = m_1$

Gradient of PB is $\dfrac{y - y_2}{x - x_2} = m_2$

But PA and PB are perpendicular, so $m_1 m_2 = -1$

i.e.
$$\left(\frac{y - y_1}{x - x_1}\right)\left(\frac{y - y_2}{x - x_2}\right) = -1$$

$\Rightarrow \qquad (y - y_1)(y - y_2) + (x - x_1)(x - x_2) = 0$

This is the equation of the circle on AB as diameter.
(**Note** that this method is neater than finding the midpoint and half the length of AB to determine the centre and radius of the circle.)

EXERCISE 11c

1) Write down the equation of the circle with:
(a) centre $(1, 2)$, radius 3
(b) centre $(0, 4)$, radius 1
(c) centre $(-3, -7)$, radius 2
(d) centre $(4, 5)$, radius 3

2) Find the centre and radius of the circle whose equation is:
(a) $x^2 + y^2 + 8x - 2y - 8 = 0$
(b) $x^2 + y^2 + x + 3y - 2 = 0$
(c) $x^2 + y^2 + 6x - 5 = 0$
(d) $2x^2 + 2y^2 - 3x + 2y + 1 = 0$
(e) $x^2 + y^2 = 4$
(f) $(x - 2)^2 + (y + 3)^2 = 9$
(g) $2x + 6y - x^2 - y^2 = 1$
(h) $3x^2 + 3y^2 + 6x - 3y - 2 = 0$

3) Determine whether the given line is a tangent to the given circle in each of the following cases:
(a) $3x - 4y + 14 = 0$; $x^2 + y^2 + 4x + 6y - 3 = 0$
(b) $5x + 12y = 4$; $x^2 + y^2 - 2x - 2y + 1 = 0$
(c) $x + 2y + 6 = 0$; $x^2 + y^2 - 6x - 4y + 8 = 0$
(d) $x + 2y + 6 = 0$; $x^2 + y^2 - 6x + 4y + 8 = 0$

4) Write down the equation of the tangent to the given circle at the given point:
(a) $x^2 + y^2 - 2x + 4y - 20 = 0$; $(5, 1)$

(b) $x^2 + y^2 - 10x - 22y + 129 = 0$;   (6, 7)

(c) $x^2 + y^2 - 8y + 3 = 0$; $(-2, 7)$

5) Some of the following pairs of circles touch and some pairs are orthogonal. Determine whether each of the pairs:
(i) touch,   (ii) cut orthogonally,   (iii) do neither of these.

(a) $x^2 + y^2 + 2x - 4y + 1 = 0$; $x^2 + y^2 - 6x - 10y + 25 = 0$

(b) $x^2 + y^2 + 8x + 2y - 8 = 0$; $x^2 + y^2 - 16x - 8y = 64$

(c) $x^2 + y^2 + 6x = 0$; $x^2 + y^2 + 6x - 4y + 12 = 0$

(d) $x^2 + y^2 + 2x - 8y + 1 = 0$; $x^2 + y^2 - 6y = 0$

(e) $x^2 + y^2 + 2x = 3$; $x^2 + y^2 - 6x - 3 = 0$

6) If $y = 2x + c$ is a tangent to the circle $x^2 + y^2 + 4x - 10y - 7 = 0$ find the value(s) of c.

7) Find the condition that m and c satisfy if the line $y = mx + c$ touches the circle $x^2 + y^2 - 2ax = 0$.

Find the equations of the following circles (in some cases more than one circle is possible).

8) A circle touches the negative x and y axes and also the line $7x + 24y + 12 = 0$.

9) A circle passes through the points $(1, 4)$, $(7, 5)$ and $(1, 8)$.

10) A circle has its centre on the line $x + y = 1$ and passes through the origin and the point $(4, 2)$.

11) The line joining (a, b) to $(3a, 5b)$ is a diameter of a circle.

12) A circle whose centre is in the first quadrant touches the y axis at the point $(0, 3)$ and is orthogonal to the circle $x^2 + y^2 - 8x + 4y - 5 = 0$.

13) A circle with centre $(2, 7)$ passes through the point $(-3, -5)$.

14) A circle intersects the y axis at the origin and at the point $(0, 6)$ and also touches the x axis.

Find the equations of the tangents specified in questions 15–18.

15) Tangents *from* the origin to the circle $x^2 + y^2 - 10x - 6y + 25 = 0$.

16) The tangent *at* the origin to the circle $x^2 + y^2 + 2x + 4y = 0$.

17) Tangents to the circle $x^2 + y^2 - 4x + 6y - 7 = 0$ which are parallel to the line $2x + y = 3$.

18) Show that if the point P is twice as far from the point $(4, -2)$ as it is from the origin then P lies on a circle. Find the centre and radius of this circle.

THE PARABOLA

If a point P is always equidistant from a fixed point and a fixed straight line, the locus of the set of points P is called a parabola. The general shape of a parabola can be seen by plotting some of the possible positions of P,

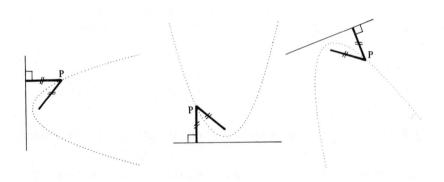

Although the shape varies slightly according to the relative positions of the fixed point (called the *focus*) and the fixed line (called the *directrix*), all parabolas have similar properties.

For instance, every parabola is symmetrical about the line through the focus which is perpendicular to the directrix. This line is called the axis of the parabola.

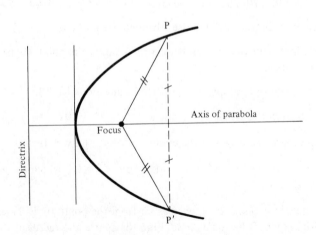

The point where a parabola crosses its axis is the *vertex*.
The distance between the vertex and the focus is the *focal length*.

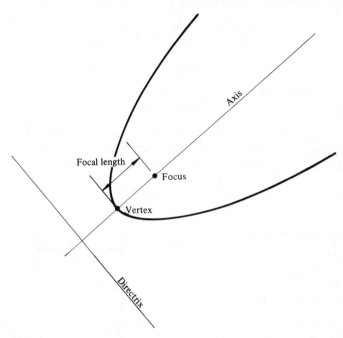

Equation of a Parabola

Consider a parabola whose focus is the point $(3, 4)$ and whose directrix is the line $x + y = 1$.

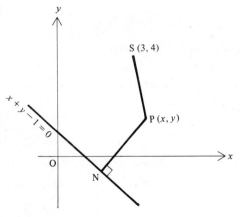

If S is the point $(3, 4)$, P is any point (x, y) on the parabola and PN is the perpendicular from P to the directrix, then P must satisfy the definition

$$PS = PN$$

or

$$PS^2 = PN^2$$

Now PN is the distance from (x, y) to $x + y - 1 = 0$

i.e.

$$PN = \left| \frac{x + y - 1}{\sqrt{(1^2 + 1^2)}} \right|$$

and $$PS^2 = (x-3)^2 + (y-4)^2$$

Thus, for all points on the parabola

$$(x-3)^2 + (y-4)^2 = \left(\frac{x+y-1}{\sqrt{2}}\right)^2$$

$\Rightarrow \qquad x^2 + y^2 - 10x - 14y - 2xy + 49 = 0$

and this is the equation of the parabola.

(**Note.** This cannot be the equation of a circle as it contains the term $2xy$.)

Rather than analyse a parabola with such an awkward equation, it is usual to study one with a simpler equation. This can be achieved by choosing the point $S(a, 0)$ for the focus and the line $x = -a$ for the directrix.

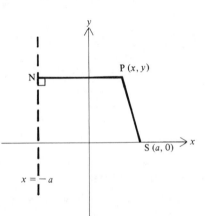

Thus, for any point P on this parabola

$$PS = PN$$

or $$PS^2 = PN^2$$

i.e.

$$(x-a)^2 + y^2 = [x-(-a)]^2$$

$\Rightarrow \qquad y^2 = 4ax$

So, for a parabola with focus $(a, 0)$ and directrix $x = -a$

$$P(x, y) \text{ is on the parabola} \iff y^2 = 4ax$$

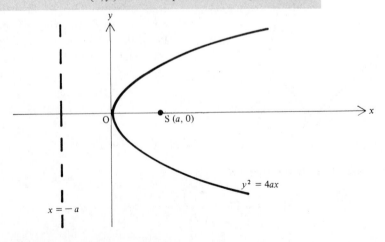

Note that when $y^2 = 4ax$:

(i)   the focal length is a,

(ii) the vertex is the origin,

(iii) the axis of the parabola is the x axis,

(iv) x is a quadratic function of y.

The vertex, focal length, focus and directrix of any parabola with a horizontal or vertical axis can be identified by comparing its equation with that of the standard parabola $Y^2 = 4aX$

whose vertex is
$$\begin{cases} X = 0, \\ Y = 0 \end{cases}$$
whose focus is
$$\begin{cases} X = a \\ Y = 0 \end{cases}$$

and whose directrix is $X = -a$.

For example:

1) If $y^2 = 12x$, comparing with $Y^2 = 4aX$ gives

$a = 3$, $Y = y$, $X = x$

Thus

the vertex $(X = 0, Y = 0)$

is $(0, 0)$

the focus $(X = a, Y = 0)$

is $(3, 0)$

the directrix $(X = -a)$

is $x = -3$.

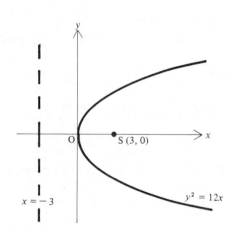

2) If $y^2 = 4(1 - x)$,   comparing with $Y^2 = 4aX$ gives

$a = 1$, $Y = y$, $X = 1 - x$

Thus the vertex is $(1, 0)$

the focus is $(0, 0)$

the directrix is $x = 2$.

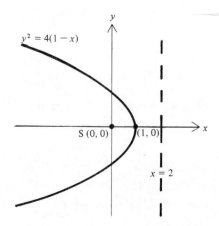

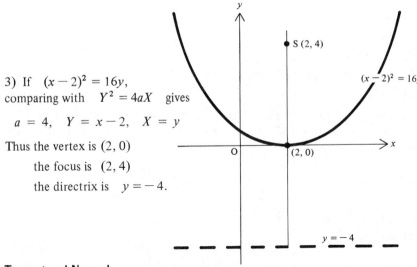

3) If $(x-2)^2 = 16y$,
comparing with $Y^2 = 4aX$ gives

$a = 4$, $Y = x-2$, $X = y$

Thus the vertex is $(2, 0)$

the focus is $(2, 4)$

the directrix is $y = -4$.

Tangent and Normal

The equation $y^2 = 4ax$ can be differentiated as an implicit function,

i.e. $$2y \frac{dy}{dx} = 4a$$

Thus, at any point on the parabola

$$\frac{dy}{dx} = \frac{2a}{y}$$

Hence the gradient at a particular point can be determined.
For instance, to find the equations of the tangent and the normal to the parabola
$y^2 = 8x$ at the point $(2, 4)$ we say

$$y^2 = 8x \implies 2y \frac{dy}{dx} = 8$$

$$\implies \qquad \frac{dy}{dx} = \frac{4}{y}$$

At the point $(2, 4)$ $\qquad \dfrac{dy}{dx} = \dfrac{4}{4} = 1$

Thus the equation of the tangent is $y - 4 = (1)(x-2)$ i.e. $y = x+2$
and the equation of the normal is $y - 4 = (-1)(x-2)$ i.e. $y + x = 6$.

EXERCISE 11d

1) Find the Cartesian equation of the locus of the set of points P if P is
equidistant from the given point and the given line. In each case sketch the locus
of P.

(a) $(1, 1)$; $x + y + 3 = 0$
(b) $(3, 4)$; $y = 0$
(c) $(-a, 0)$; $x = a$

2) Find the equations of the tangent and the normal to the given parabola at the given point in each of the following cases:

(a) $y^2 = 4x$; $(1, 2)$ (b) $(y - 1)^2 = 6x$; $(0, 1)$
(c) $y = 8x^2$; $(\frac{1}{2}, 2)$ (d) $y^2 + 4x = 0$; $(-4, -4)$

3) Sketch each of the following parabolas, marking the focus, the directrix, the vertex, the axis of the parabola and the focal length.

(a) $y^2 = 12x$ (b) $y^2 = 4x$ (c) $y^2 = -4x$
(d) $4y = x^2$ (e) $y^2 = 4(x - 1)$

THE ELLIPSE

If a point P moves so that its distances from a fixed point and a fixed straight line are in a constant ratio (< 1), the locus of the set of points P is called an ellipse. Its general shape is shown below.

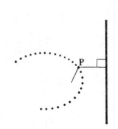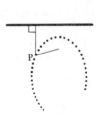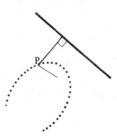

The position of the locus depends upon the positions of the fixed point and line (again called the *focus* and the *directrix*) and the shape of the locus depends upon the value of the constant ratio (called the *eccentricity* of the ellipse). (**Note** that the definition of a parabola is the same as the definition of an ellipse with an eccentricity of 1.)

Although there is an infinite set of ellipses, our analysis concentrates on an ellipse with a conveniently simple Cartesian equation. This is achieved by choosing the point $S(ae, 0)$ and the line $x = \dfrac{a}{e}$ as focus and directrix where e is the eccentricity $(e < 1)$.

Then, if $P(x, y)$ is any point on this ellipse, the given condition can be expressed as

$$PS = ePN$$

or $PS^2 = e^2 PN^2$

But

$$PS^2 = (y - 0)^2 + (x - ae)^2$$

and

$$PN = \frac{a}{e} - x$$

Thus

$$y^2 + (x - ae)^2 = e^2 \left(\frac{a}{e} - x \right)^2$$

i.e.

$$x^2(1 - e^2) + y^2 = a^2(1 - e^2)$$

Replacing $a^2(1 - e^2)$ by b^2 we have

$$\frac{x^2}{a^2} + \frac{y^2}{b^2} = 1$$

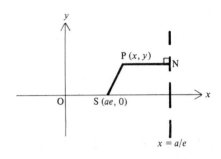

This equation is valid if and only if $P(x, y)$ satisfies the given condition. It is therefore the equation of the locus of P, which is an ellipse.

The shape of the locus can be deduced by rearranging the equation as follows:

1)
$$x^2 = \frac{a^2}{b^2}(b^2 - y^2)$$

$$x^2 \geqslant 0 \quad \text{so} \quad b^2 - y^2 \geqslant 0$$

$$\Rightarrow \quad (b - y)(b + y) \geqslant 0$$

Hence $-b \leqslant y \leqslant b$.

Also, since $x = \pm \frac{a}{b} \sqrt{b^2 - y^2}$, the curve is symmetrical about the y axis.

2)
$$y^2 = \frac{b^2}{a^2}(a^2 - x^2)$$

$$y^2 \geqslant 0 \quad \text{so} \quad a^2 - x^2 \geqslant 0$$

$$\Rightarrow \quad (a - x)(a + x) \geqslant 0$$

Hence $-a \leqslant x \leqslant a$

Also, since $y = \pm \frac{b}{a} \sqrt{a^2 - x^2}$, the curve is symmetrical about the x axis.

3) When
$$\begin{cases} x = 0, & y = \pm b \\ y = 0, & x = \pm a \end{cases}$$

4) $\dfrac{dy}{dx} = -\dfrac{b^2 x}{a^2 y}$ (implicit differentiation).

Thus
$$\begin{cases} \text{when} \quad x = 0 & \dfrac{dy}{dx} = 0 \\[2ex] \text{when} \quad y = 0 & \dfrac{dy}{dx} = \infty \end{cases}$$

It is now possible to sketch the ellipse.

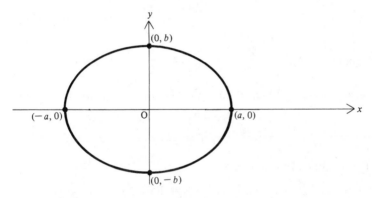

The ellipse is symmetrical about both the x axis and the y axis and so has two symmetrical foci and directrices.

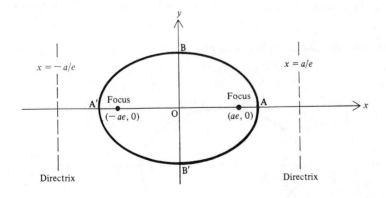

The line $A'A$ is called the *major axis* of the ellipse and is of length $2a$ and $B'B$ is called the *minor axis* of the ellipse and is of length $2b$.

The midpoint of AA′ (or BB′) is called the *centre* of the ellipse. (In this case the centre of the ellipse is the origin.)

If a chord of an ellipse passes through the centre it is called a *diameter*.

Tangent and Normal

The gradient of an ellipse can be found by differentiating implicitly its equation.

For example the equations of the tangent and normal at the point $(1, \frac{3}{2})$ to the ellipse $\dfrac{x^2}{4} + \dfrac{y^2}{3} = 1$ can be found as follows.

Differentiating we have
$$\frac{2x}{4} + \frac{2y}{3}\frac{dy}{dx} = 0$$

or
$$\frac{dy}{dx} = -\frac{3x}{4y}$$

At the point $(1, \frac{3}{2})$,
$$\frac{dy}{dx} = -\frac{3}{4(\frac{3}{2})} = -\frac{1}{2}$$

Hence, at this point, the equation of the tangent is
$$(y - \tfrac{3}{2}) = -\tfrac{1}{2}(x - 1)$$

$\Rightarrow$
$$2y + x = 4$$

and the equation of the normal is
$$(y - \tfrac{3}{2}) = 2(x - 1)$$

$\Rightarrow$
$$2y = 4x - 1$$

Note. Comparing the given equation with the standard form $\dfrac{x^2}{a^2} + \dfrac{y^2}{b^2} = 1$ we see that $a^2 = 4$ and $b^2 = 3$.

Thus the given ellipse can be sketched.

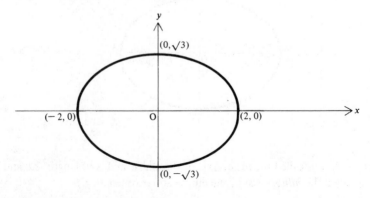

It is interesting to observe that, when $a = b$, the equation of the ellipse becomes

$$x^2 + y^2 = a^2$$

i.e. the ellipse becomes a circle.

EXERCISE 11e

Write down the values of a and b and hence sketch the ellipse in each of the following cases.

1) $\dfrac{x^2}{9} + \dfrac{y^2}{4} = 1$

2) $\dfrac{x^2}{16} + \dfrac{y^2}{9} = 1$

3) $x^2 + 4y^2 = 4$

4) $4x^2 + 9y^2 = 36$

5) $x^2 + 25y^2 = 1$ $\left(Hint \quad 25 = \dfrac{1}{1/25}\right)$

Find the equation of the tangent and the normal to the following ellipses at the points specified.

6) $\dfrac{x^2}{9} + \dfrac{y^2}{4} = 1$; $\left(2, \dfrac{2\sqrt{5}}{3}\right)$

7) $\dfrac{x^2}{16} + \dfrac{y^2}{9} = 1$; $\left(\sqrt{7}, \dfrac{9}{4}\right)$

8) $x^2 + 4y^2 = 4$; $\left(\sqrt{3}, \dfrac{1}{2}\right)$

CARTESIAN ANALYSIS OF A GENERAL CURVE

The Cartesian equation of any two dimensional curve expresses a relationship between x and y and can be given in the form

$$f(xy) = 0$$

The gradient function is given by the equation

$$\frac{d}{dx} f(xy) = 0$$

INTERSECTION

The points of intersection of any two lines or curves whose equations are $f(xy) = 0$ and $g(xy) = 0$ are found by solving simultaneously the two equations.
Each real solution gives a point of intersection.

Coincident Points of Intersection

The equation obtained by eliminating x (or y) from the equations $f(xy) = 0$ and $g(xy) = 0$ in order to solve them simultaneously may have two or more equal roots.

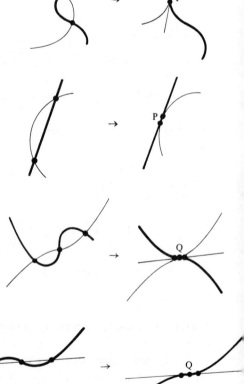

If there are two equal roots the curves meet twice at the same point P, i.e. they touch at P and have a common tangent at P.
In particular, when a line and a curve meet twice at the same point P, the line is a tangent to the curve at P.
If there are three equal roots the curves meet three times at the same point Q. The curves have a common tangent at Q but this time each curve crosses, at Q, to the opposite side of the common tangent. In particular, when a line and a curve meet at three coincident points Q, the line is a tangent to the curve at Q and the curve has a point of inflexion at Q.

Taking this argument further it becomes clear that,

(a) when the number of coincident points of intersection of a line and a curve is even, the curve touches the line and remains on the same side of the line.

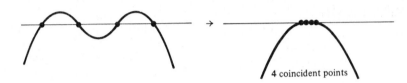

4 coincident points

(b) when the number of coincident points of intersection is odd, the curve touches the line and crosses it. Thus the curve has a point of inflexion.

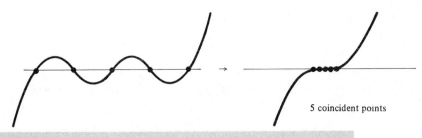

5 coincident points

so if the solution of $\begin{cases} y = mx + c \\ f(x, y) = 0 \end{cases}$

has $\begin{cases} \text{distinct roots, the curve crosses the line at distinct points.} \\ \text{repeated roots, the line touches the curve.} \end{cases}$

The Curves $y = kx^n$ where n is a Positive Integer

If $y = kx^n$,

the points where the curve meets the x axis are given by $x^n = 0$

i.e. by n equal zero values of x.

Thus if n is even, the curve touches the x axis at the origin at a minimum point $(k > 0)$ or a maximum point $(k < 0)$

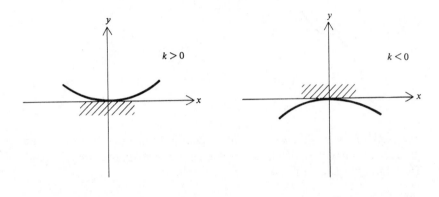

But if n is odd the curve has a point of inflexion at the origin.

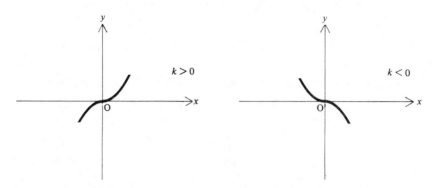

Further, since $\dfrac{\mathrm{d}y}{\mathrm{d}x} = knx^{n-1}$, the only stationary point is where $x = 0$.

Thus the graph of $y = kx^n$ is

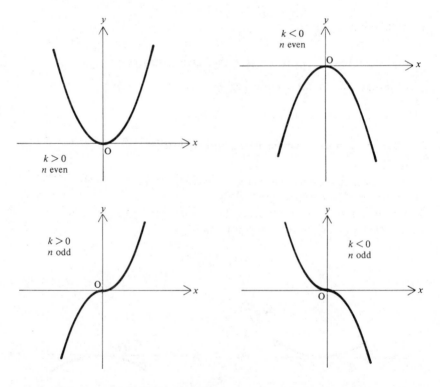

The graph of $y = k(x - a)^n$ is very similar to the graphs shown above. But the intersection with the x axis occurs when $x = a$ in this case.

Thus, for $y = k(x - a)^n$ we have

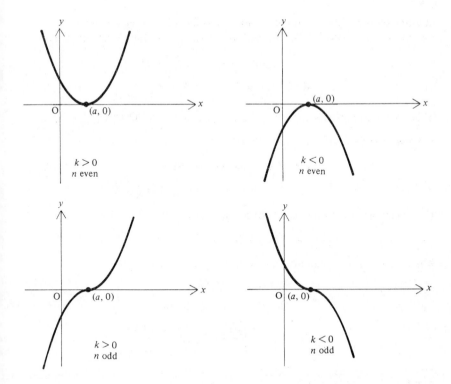

$k > 0$
n even

$(a, 0)$

$k < 0$
n even

$(a, 0)$

$k > 0$
n odd

$k < 0$
n odd

EXAMPLES 11f

1) Show that the line $y = mx + c$ touches the parabola $y^2 = 4ax$ if and only if $mc = a$.

The line $y = mx + c$ intersects the parabola $y^2 = 4ax$ at points whose y coordinates are the roots of the equation

$$y^2 = 4a\left(\frac{y - c}{m}\right)$$

i.e. $\qquad\qquad my^2 - 4ay + 4ac = 0$ \hfill [1]

This equation is quadratic so it cannot have more than two roots. Therefore the line and the parabola cannot have more than two points of intersection. The line touches the parabola if the two points of intersection coincide, i.e. if and only if equation [1] has equal roots.

$\Rightarrow\qquad\qquad (-4a)^2 = 4(m)(4ac)$

$\Rightarrow\qquad\qquad a = mc$

2) Find the equations of the tangents with gradient 2 to the ellipse $4x^2 + 9y^2 = 36$ and find their points of contact.

Any line with gradient 2 has equation $y = 2x + c$. This line meets the ellipse $4x^2 + 9y^2 = 36$ at points whose x coordinates are given by

$$4x^2 + 9(2x + c)^2 = 36$$

i.e. $\qquad\qquad\qquad 40x^2 + 36cx + (9c^2 - 36) = 0 \qquad\qquad\qquad [1]$

If the line is a tangent, equation [1] has equal roots $\left(\text{in this case} \quad x = -\frac{``\;b\;"}{2a}\right.$

and gives the point of contact$\Big)$

i.e. $\qquad\qquad\qquad (36c)^2 - 4(40)(9c^2 - 36) = 0$

$\Rightarrow \qquad\qquad\qquad\qquad\qquad\qquad c = \pm 2\sqrt{10}$

Thus there are two tangents, whose equations are

$$y = 2x \pm 2\sqrt{10}$$

When $c = 2\sqrt{10}$, equation [1] is satisfied only by

$$x = \frac{-36c}{2(40)} = -\frac{9\sqrt{10}}{10}$$

and $\qquad\qquad y = 2\left(\frac{-9\sqrt{10}}{10}\right) + 2\sqrt{10} = \frac{\sqrt{10}}{5}$

When $c = -2\sqrt{10}$, $x = \frac{9\sqrt{10}}{10}$ and $y = \frac{19\sqrt{10}}{5}$

So the points of contact are $\left(\frac{-9\sqrt{10}}{10}, \frac{\sqrt{10}}{5}\right)$ and $\left(\frac{9\sqrt{10}}{10}, \frac{19\sqrt{10}}{5}\right)$

3) Find the point(s) of intersection of the curves $xy = 1$ and $x^2 + y^2 + 2x + 2y - 6 = 0$.
Use sketches to illustrate your solution.

Points of intersection satisfy both equations simultaneously.
Eliminating y from the two equations we have

$$x^2 + \left(\frac{1}{x}\right)^2 + 2x + 2\left(\frac{1}{x}\right) - 6 = 0$$

i.e. $\qquad\qquad x^4 + 2x^3 - 6x^2 + 2x + 1 = 0 \qquad\qquad [1]$

The roots of this equation are the x coordinates of the points of intersection of the two curves.
Using the factor theorem to factorize the L.H.S. we find that [1] becomes

$$(x - 1)(x - 1)(x^2 + 4x + 1) = 0$$

Hence $x = 1, 1, -2 \pm \sqrt{3}$,
i.e. there are two real equal roots and two other real distinct roots.

Therefore, calculating the corresponding y-coordinates from the equation $xy = 1$, we see that

the two curves *touch* at the point $A(1, 1)$ and *cut* at the points

$$B\left(-2 + \sqrt{3}, \frac{1}{-2 + \sqrt{3}}\right) \quad \text{and} \quad C\left(-2 - \sqrt{3}, \frac{1}{-2 - \sqrt{3}}\right).$$

The curve with equation

$$xy = 1, \quad \text{or} \quad y = \frac{1}{x},$$

is called a rectangular hyperbola.
(Any curve with equation $xy = K$
is a rectangular hyperbola.)

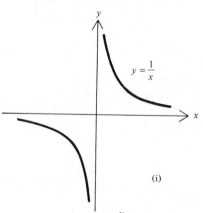

(i)

The equation
$$x^2 + y^2 + 2x + 2y - 6 = 0,$$
or $(x + 1)^2 + (y + 1)^2 = 8$, can
be recognised as a circle.
The points common to the two
curves are shown in diagram (iii)

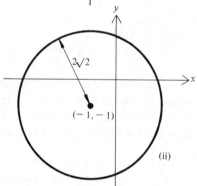

(ii)

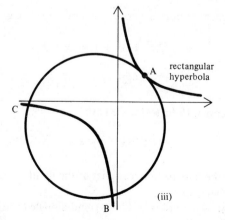

(iii)

4) Find the midpoint of the chord cut off by the curve $y^2 - 3y = 5x$ on the line $3x + y = 1$.

Points common to the line and the curve have y coordinates given by the equation

$$y^2 - 3y = 5\left(\frac{1-y}{3}\right)$$

i.e. $3y^2 - 4y - 5 = 0$

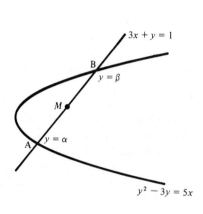

If the roots are α and β
the line cuts the curve at A and B
where $y = \alpha$ and $y = \beta$.
Thus at M, the midpoint of AB

$$y = \tfrac{1}{2}(\alpha + \beta)$$

But $\alpha + \beta = -\dfrac{b}{a} = \dfrac{4}{3}$

Therefore, at M,
$y = \tfrac{2}{3}$ and $x = \tfrac{1}{9}$ (from $3x + y = 1$). Thus M is the point $(\tfrac{1}{9}, \tfrac{2}{3})$.

Note: In this problem there is the alternative method of actually solving the equation $3y^2 - 4y - 5 = 0$ to determine α and β. This is practical only when the coefficients are numbers. The method shown above however is equally suitable when the line and curve have general equations.

5) Find the points in which the line $y = x + 2$ meets the curve $y = x^4 - 2x^3 + 3x + 1$ showing that one of them is an inflexion on the curve.

The line and curve meet at points whose x coordinates are given by

$$x^4 - 2x^3 + 3x + 1 = x + 2$$

i.e.

$$x^4 - 2x^3 + 2x - 1 = 0$$

Using the factor theorem to factorize we have

$$(x - 1)^3(x + 1) = 0$$

Thus the curve and the line meet three times at the point where $x = 1$ and once at the point where $x = -1$.
Thus there is a point of inflexion at $P(1, 3)$ and a single crossing at $Q(-1, 1)$.

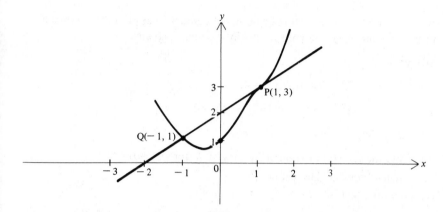

EXERCISE 11f

Investigate the possible intersection of the following lines and curves giving the coordinates of all common points. State clearly those cases where the line touches the curve.

1) $y = x + 1$; $y^2 = 4x$

2) $2y + x = 3$; $x^2 - y^2 - 3y + 3 = 0$

3) $y = x - 5$; $x^2 + 2y^2 = 7$

4) $2y - x = 4$; $x^2 + y^2 - 4x = 4$

5) $y = 0$; $y = x^2 - 3x + 2$

6) $y = 0$; $y = x^3 + 5x^2 + 6x$

7) $y = 0$; $y = (x - 1)^2 (x - 2)^2$

8) $y = 0$; $y = (x + 3)^3 (x + 2)$

9) $x = 0$; $x = y^4$

Find the value of k such that the given line shall touch the given curve.

10) $y = x + 2$; $y^2 = kx$

11) $y = kx + 3$; $xy + 9 = 0$

12) $y = 3x - k$; $x^2 + 2y^2 = 8$

Without finding the coordinates of the points of intersection of the following lines and curves, find the coordinates of the midpoint of the chord cut off by the curve on the line.

13) $y = 2x + 3$; $y^2 = 4x + 8$

14) $y = x$; $2x^2 + 5y^2 = 10$

15) $3y + x + 7 = 0$; $xy = 1$

Find the points of intersection or points of contact (if any) of the following pairs of curves. Illustrate your results by drawing diagrams.

16) $y^2 = 8x;\quad xy = 1$

17) $x^2 + y^2 + 2x - 7 = 0;\quad y^2 = 4x$

18) $xy = 2;\quad 2x^2 + 2y^2 - 6x + 3y - 10 = 0$

19) $9x^2 = 2y;\quad y^2 = 6x$

20) Find the value(s) or ranges of values of λ for which the line $y = 2x + \lambda$:
(a) touches   (b) cuts in real points
(c) does not meet, the curve $y^2 + 2x^2 = 4$.

21) Sketch the curves $y = 3x^4,\quad y = 4(2-x)^5,\quad y = 2(x+3)^7,$
$y = -5x^6$.

22) Find the equation(s) of the tangent(s):
  (i) from the point $(1, 0)$,
  (ii) with gradient $-\frac{1}{2}$,
  to each of the following curves,
  (a) $y^2 + 4x = 0$ (b) $xy = 9$ (c) $x^2 = 6y$

THE CUBIC CURVE

Observation of the equation of the general cubic curve

$$y = ax^3 + bx^2 + cx + d$$

shows that:
(i) if x is very large, $y \simeq ax^3$, thus

as	$x \to \infty$	$y \to \infty$	if $a > 0$
and as	$x \to -\infty$	$y \to -\infty$	
while as	$x \to \infty$	$y \to -\infty$	if $a < 0$
and as	$x \to -\infty$	$y \to \infty$	

(ii) there are no finite values of x for which $y \to \infty$ so the curve is continuous.

(iii) $\dfrac{dy}{dx} = 3ax^2 + 2bx + c$,   therefore solutions of the equation

$$3ax^2 + 2bx + c = 0$$

are the values of x for which y is stationary.
This quadratic equation may have:

(a) two real distinct roots, in which case the curve has two distinct turning points which must be a maximum and a minimum (as the curve is continuous),

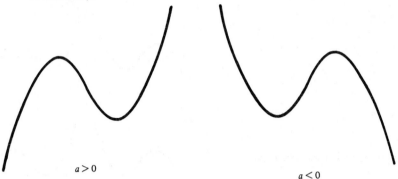

$a > 0$

$a < 0$

(b) two equal real roots, in which case the maximum and minimum coincide and the curve therefore has a point of inflexion.

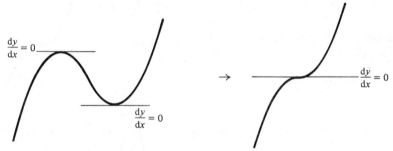

$\frac{dy}{dx} = 0$

$\frac{dy}{dx} = 0$

$\rightarrow$

$\frac{dy}{dx} = 0$

(c) no real roots, indicating that the curve has no real stationary points.

Combining these observations we deduce that a cubic curve has one of the following shapes.

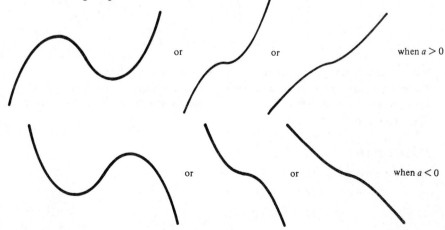

or

or

when $a > 0$

or

or

when $a < 0$

Depending upon the position of the curve relative to the x-axis we also see that a cubic curve can

cross the x-axis three times

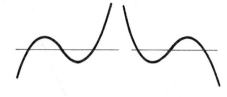

or cross and touch the x-axis

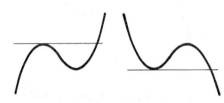

or touch the x-axis at a point of inflexion

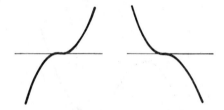

or cross the x-axis once only

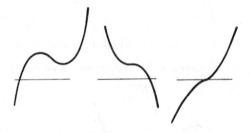

Note: There is always *at least one point of intersection* with the x-axis. Therefore, since a value of x where the curve meets the x-axis is a solution of the equation $ax^3 + bx^2 + cx + d = 0$, we see that *a cubic equation has at least one real root.*

EXAMPLES 11g

Sketch the curves

(1) $y = x(x-1)(x-2)$

(2) $y = x(x-1)^2$

(3) $y = (1-x)^3$

1) If $y = x(x-1)(x-2) = x^3 - 3x^2 + 2x$

   $y = 0$ when $x = 0, 1, 2$

Also $a = 1$ i.e. $a > 0$

Thus

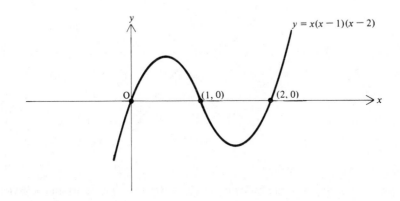

$$y = x(x-1)(x-2)$$
$(1, 0)$ $(2, 0)$

2) If $y = x(x-1)^2$

   $y = 0$ when $x = 0, 1, 1$

Hence the curve *touches* the x-axis when $x = 1$.

Again $a > 0$

Thus

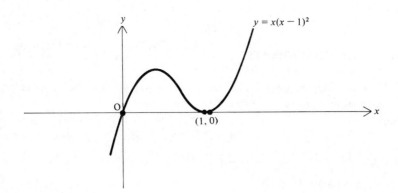

$$y = x(x-1)^2$$
$(1, 0)$

3) If $y = (1-x)^3 = 1 - 3x + 3x^2 - x^3$

   $y = 0$ when $x = 1, 1, 1$

The triple solution $x = 1$ indicates a point of inflexion.

In this case $a = -1$ i.e. $a < 0$

so the graph is

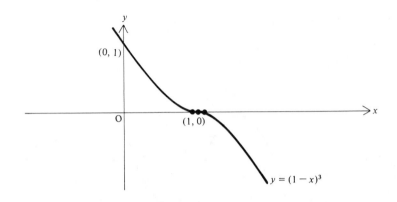

EXERCISE 11g

Sketch the graphs of the following cubic curves showing clearly the behaviour of the curve at the points where it meets the x-axis and at stationary points.

1) $y = (x - 1)(x - 2)(x - 3)$ 2) $y = 1 - x - x^2 + x^3$

3) $y = x^2(1 - x)$ 4) $y = (x - 2)^3$

5) $y = (x + 1)(x - 3)^2$ 6) $y = x^3 - 1$

7) $y = a + x^2 - x - x^3$ 8) $y = (1 + x)(1 - x)(2 - x)$

9) $y = x^3 + 3x^2 + 3x + 1$ 10) $y + x^3 = 0$

PARAMETRIC COORDINATES

The relationship between the x and y coordinates of a point can often be expressed more simply in the form of two equations

$$\begin{cases} x = f(p) \\ y = g(p) \end{cases}$$

where p is a parameter.

The use of parametric equations to find the gradient function of a curve, and hence the equations of the tangent and normal at a particular point, was demonstrated in Chapter 9. There are many other ways in which the parametric approach helps with problems on curve analysis, and we will now explore some of them.

Relation between Cartesian and Parametric Equations

When the parametric equations of a curve are given, the Cartesian equation can be found by eliminating the parameter.

It is not always so easy to convert a Cartesian equation into a suitable pair of parametric equations. In some cases, however, the form of a Cartesian equation suggests clearly what parameter should be used,

e.g. if $x^2 + y^2 = 4$, the similarity to the trig identity $\cos^2\theta + \sin^2\theta \equiv 1$

indicates that $\begin{cases} x = 2\cos\theta \\ y = 2\sin\theta \end{cases}$ are suitable parametric equations.

In this case the parameter, θ, has graphical significance, as can be seen in the diagram below.

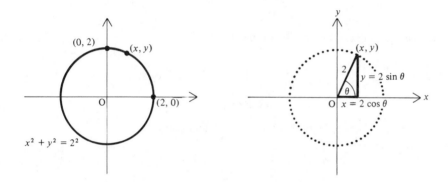

Frequently, however, a parameter cannot be represented graphically,

e.g. the Cartesian equation of the parabola $y^2 = 4ax$ can be replaced by the parametric equations $x = at^2$, $y = 2at$, where t is simply a real number.

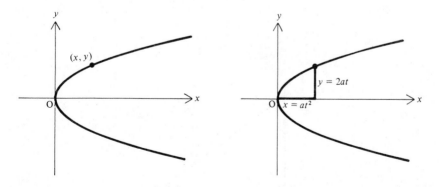

In those cases where the parameter has a geometric meaning, its *geometric* properties can be used in the solution of problems.

EXAMPLE

Show that the curve whose parametric equations are $x = a(1 + \cos\theta)$ and $y = a\sin\theta$, represents a circle. Indicate on a diagram the significance of the parameter θ.

Hence find, in terms of θ, the intercepts on the x and y axes made by the tangent to the circle at the point $P[a(1 + \cos\theta), a\sin\theta]$.

To convert the parametric equations to Cartesian form we eliminate θ as follows

$$x = a(1 + \cos\theta) \quad \Rightarrow \quad \cos\theta = \left(\frac{x}{a} - 1\right) \qquad [1]$$

$$y = a\sin\theta \qquad \Rightarrow \quad \sin\theta = \frac{y}{a} \qquad [2]$$

But $\cos^2\theta + \sin^2\theta \equiv 1$

Hence squaring and adding equations [1] and [2] gives

$$\left(\frac{x}{a} - 1\right)^2 + \left(\frac{y}{a}\right)^2 = 1$$

i.e. $$(x - a)^2 + y^2 = a^2$$

This represents a circle with centre $C(a, 0)$ and radius a.

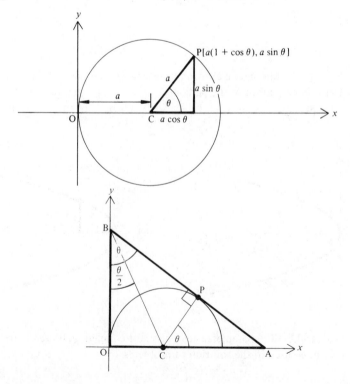

The tangent at P is perpendicular to CP and so is inclined at an angle θ to the y-axis.

Also BC bisects the angle OBP so angle $OBC = \dfrac{\theta}{2}$

Then
$$OB = OC \cot \frac{\theta}{2} = a \cot \frac{\theta}{2}$$

and
$$OA = OB \tan \theta = a \cot \frac{\theta}{2} \tan \theta$$

Thus (without finding the equation of the tangent at P) the intercepts made by the tangent on the x and y axes are

$$a \cot \frac{\theta}{2} \tan \theta \quad \text{and} \quad a \cot \frac{\theta}{2}$$

respectively.

INTERSECTION

Consider the line $y = mx + c$ and the parabola with parametric equations

$$x = at^2, \quad y = 2at$$

Any point which is common to the line and the parabola must have coordinates $(at^2, 2at)$ and these coordinates must also satisfy the equation $y = mx + c$.

Hence
$$2at = mat^2 + c$$

The roots of this quadratic equation in t give the values of the parameter at points of intersection. If the roots are equal, the two points of intersection coincide and the line is a tangent to the parabola.

EXAMPLE 11h

1) Find the value of k if the line $y = k - 2x$ touches the curve with parametric equations $x = t, \quad y = \dfrac{1}{t}$.

Points on the curve have coordinates $\left(t, \dfrac{1}{t}\right)$ so, where the line meets the curve, these coordinates also satisfy the equation of the line

$$y = k - 2x$$

i.e.
$$\frac{1}{t} = k - 2t$$

$\Rightarrow$
$$2t^2 - kt + 1 = 0 \qquad\qquad [1]$$

If the line *touches* the curve there is *only one* common point, so equation [1] has equal roots

i.e. $$(-k)^2 - 4(2)(1) = 0$$

$$\Rightarrow \qquad k = \pm 2\sqrt{2}$$

Alternatively, since the gradient of the curve at any point $\left(t, \dfrac{1}{t} \right)$ is $-\dfrac{1}{t^2}$,

we can deduce that a tangent with gradient -2 touches the curve where

$$-\frac{1}{t^2} = -2, \quad \text{i.e. where} \quad t = \pm \frac{1}{\sqrt{2}}.$$

The point where $t = \dfrac{1}{\sqrt{2}}$ is $\left(\dfrac{1}{\sqrt{2}}, \sqrt{2} \right)$ and the tangent at this point has equation

$$y - \sqrt{2} = -2 \left(x - \frac{1}{\sqrt{2}} \right) \quad \Rightarrow \quad y = 2\sqrt{2} - 2x$$

So $t = \dfrac{1}{\sqrt{2}} \ \Rightarrow \ k = 2\sqrt{2}.$

Similarly $t = -\dfrac{1}{\sqrt{2}} \ \Rightarrow \ k = -2\sqrt{2}.$

Now consider the points of intersection of two curves whose equations are both expressed in parametric form, for instance

a curve (C_1) with equations $\qquad \begin{cases} x = t^2 + 1 \\ y = 2t \end{cases}$

and a curve (C_2) with equations $\qquad \begin{cases} x = 2s \\ y = \dfrac{2}{s} \end{cases}$

If C_1 and C_2 meet at a point then, at that point,

$$x = t^2 + 1 \quad and \quad x = 2s \ \Rightarrow \ t^2 + 1 = 2s \qquad [1]$$

also $\qquad y = 2t \quad and \quad y = \dfrac{2}{s} \ \Rightarrow \ 2t = \dfrac{2}{s} \qquad$ [2]

Solving [1] and [2] by eliminating s gives

$$t^3 + t - 2 = 0$$

$$\Rightarrow \qquad (t - 1)(t^2 + t + 2) = 0$$

So either $t = 1$ or $t^2 + t + 2 = 0$, which gives no real values of t. Hence the given curves have only one point of intersection.

At this point $t = 1$ so $x = 2$ and $y = 2$.

$$\left[\text{As a check we can use the corresponding value of } s\left(=\frac{1}{t}\right) \text{ in } C_2. \right]$$

A sketch to illustrate this result can be drawn by recognising that:

(a) Since $\begin{cases} x = t^2 \\ y = 2t \end{cases} \Rightarrow y^2 = 4x$

which is a standard parabola,

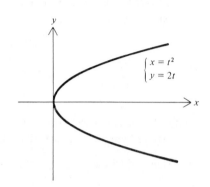

then $\begin{cases} x = t^2 + 1 \\ y = 2t \end{cases}$ represent

a similar parabola where each x
coordinate is increased by
1 unit.

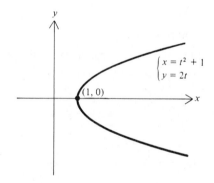

(b) $\begin{cases} x = 2s \\ y = \dfrac{2}{s} \end{cases} \Rightarrow xy = 4$

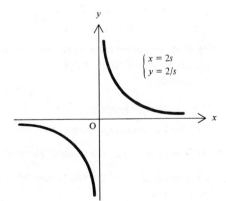

which is a rectangular hyperbola.
(See page 389.)

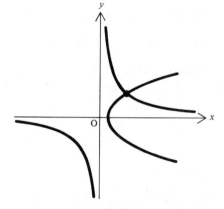

Thus these curves meet only once, as shown.

Note. This example shows that it is useful to recognise a standard parabola or rectangular hyperbola in parametric as well as Cartesian form.

The standard ellipse $\dfrac{x^2}{a^2} + \dfrac{y^2}{b^2} = 1$ should also be recognised from its

parametric equations which are usually given as $\begin{cases} x = a\cos\theta \\ y = b\sin\theta \end{cases}$

EXERCISE 11h

1) Eliminate the parameter from the following pairs of parametric equations:

(a) $x = t^2 - 1$; $y = 3 + t$

(b) $x = t^3$; $y = t^2$

(c) $x = 3t$; $y = \dfrac{3}{t}$

(d) $x = \dfrac{1+t}{t}$; $y = \dfrac{1-t}{t^2}$

(e) $x = 2\cos\theta$; $y = 3\sin\theta$

(f) $x = \sec\theta$; $y = 2\tan\theta$

(g) $x = a\cos 2\theta$; $y = a\cos\theta$

2) Plot the curves represented by the equations given in Question 1, parts (a) to (d) taking values in the range $-3 \leqslant t \leqslant 3$ and parts (e) and (f) taking $0 \leqslant \theta \leqslant 2\pi$.

A curve has parametric equations $x = 2t^2$, $y = 4t$. Find:

3) The Cartesian equation of the curve.

4) The equation of the tangent at the point where $y = 8$.

5) The equation of the chord joining the points on the curve where $t = p$ and $t = q$.

6) The coordinates of the points where the line $y = x - 6$ meets the curve.

7) The value(s) of k if $y = x + k$ is a tangent to the curve.

8) The coordinates of the point(s) of intersection of the curve and the circle $x^2 + y^2 - 2x = 16$.

9) The coordinates of the point(s) of intersection of the curve and the curve whose parametric equations are $x = 8s, \quad y = \dfrac{8}{s}$.

Find the equation of the chord AB in Questions 10–12.

10) At A, $(t = t_1)$ and at B, $(t = t_2)$ on the curve with parametric equations $x = ct, \quad y = \dfrac{c}{t}$

11) A$(ap^2, 2ap)$ and B$(aq^2, 2aq)$ are on the parabola $x = at^2, \quad y = 2at$.

12) The curve has equations $x = t^3, \quad y = t^2$. At A and B, $x = T^3$ and 8 respectively.

13) Find the length of OP where P is a point on the curve given by:

(a) $x = at^2, \quad y = 2at$ (b) $x = ct, \quad y = \dfrac{c}{t}$

(c) $x = t - 1, \quad y = t^2 + 1$ (d) $x = a \cos \theta, \quad y = b \sin \theta$.

Find the equation of the normal to the given curve at the given point and find the coordinates of the point where this normal meets the curve again.

14) $x = at^2, \quad y = 2at; \quad (at_1{}^2, 2at_1)$.

15) $x = ct, \quad y = \dfrac{c}{t}; \quad \left(cp, \dfrac{c}{p}\right)$

16) $x = b(1 - t^2), \quad y = 2b(t - 2); \quad [b(1 - p^2), 2b(p - 2)]$.

Find the coordinates of the midpoint of the chord cut off on the line $2x + y = 7$ by the following curves:

17) $x = 4t^2, \quad y = 8t$ 18) $x = 2t, \quad y = \dfrac{2}{t}$.

19) $x = t^2 - 2, \quad y = 2t + 1$

FURTHER LOCI

When a point Q is constrained to move in a particular way, its possible positions may be defined in terms of another point P which itself satisfies a specified condition.

The examples which follow show how the Cartesian equation of Q can be found in such cases.

EXAMPLES 11i

1) A circle has centre $(2, 0)$ and radius 2 and P is any point on this circle. OP is produced to Q so that OQ = 3OP. Find the equation of the locus of Q.

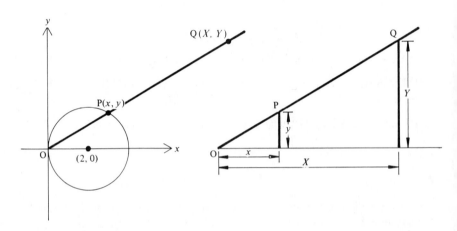

The equation of the given circle is

$$(x-2)^2 + y^2 = 2^2$$

i.e. $$x^2 + y^2 - 4x = 0 \qquad [1]$$

$P(x, y)$ is any point on the circle.
$Q(X, Y)$ is any point on the required locus.

The given condition is   OQ = 3OP

Hence $\begin{cases} X = 3x \\ Y = 3y \end{cases} \Rightarrow \begin{cases} x = \frac{1}{3}X \\ y = \frac{1}{3}Y \end{cases}$

But P is on the given circle so its coordinates satisfy equation [1].

Thus $$\left(\frac{X}{3}\right)^2 + \left(\frac{Y}{3}\right)^2 - 4\left(\frac{X}{3}\right) = 0$$

$\Rightarrow$ $$X^2 + Y^2 - 12X = 0 \qquad [2]$$

Equation [1] contains the coordinates of every point P on the circle, so equation [2] contains the coordinates of every possible point Q and is therefore the equation of the locus of Q.

Note: Because $Q(X, Y)$ is defined in terms of $P(x, y)$ it is necessary to find separate relationships between x and X, y and Y. In this way the given equation linking x and y leads to another equation linking X and Y.

2) $P(ap^2, 2ap)$ is any point on the parabola whose parametric equations are $x = at^2$, $y = 2at$ and S is the point $(a, 0)$. PS is produced to cut the parabola again at Q. Find the Cartesian equation of the locus of M, the midpoint of PQ.

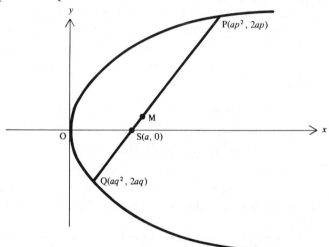

As Q is a point on the parabola, its parametric coordinates are $(aq^2, 2aq)$. Also, since Q is on PS produced, the gradients of SQ and SP are equal,

i.e.
$$\frac{2aq}{aq^2 - a} = \frac{2ap}{ap^2 - a}$$

$\Rightarrow \qquad q(p^2 - 1) = p(q^2 - 1)$

$\Rightarrow \qquad (pq + 1)(p - q) = 0$

But $p \neq q$ therefore $pq = -1$.   [1]

(**Note.** When the chord joining $P(ap^2, 2ap)$ and $Q(aq^2, 2aq)$ passes through the focus $S(a, 0)$, PQ is a focal chord and the property $pq = -1$, derived above, can be quoted in other problems.)

At M
$$\begin{cases} x = \tfrac{1}{2}(ap^2 + aq^2) \quad \Rightarrow \quad \dfrac{2x}{a} = p^2 + q^2 & [2] \\[2em] y = \tfrac{1}{2}(2ap + 2aq) \quad \Rightarrow \quad \dfrac{y}{a} = p + q & [3] \end{cases}$$

Equations [2] and [3] are parametric equations for the locus of M. To find the Cartesian equation of this locus we must eliminate p and q.

[3] $\Rightarrow$ $$\left(\frac{y}{a}\right)^2 = p^2 + 2pq + q^2$$

But $p^2 + q^2 = \dfrac{2x}{a}$ and $pq = -1$ (from [2] and [1])

Hence $\dfrac{y^2}{a^2} = \dfrac{2x}{a} - 2$

Thus the Cartesian equation of the locus of M is $y^2 = 2ax - 2a^2$

(**Note.** In this problem the position of M depends upon the positions of *two* moving points P and Q. *Three* equations are needed in these circumstances to derive the Cartesian equation of M.)

3) Show that the equation of the normal at $P(a \cos\theta, b \sin\theta)$, to the ellipse $b^2 x^2 + a^2 y^2 = a^2 b^2$ is:

$$ax \sin\theta - by \cos\theta = (a^2 - b^2) \sin\theta \cos\theta$$

The normal at P meets the x-axis at Q and the y-axis at R. Find:

(a) the greatest value of the area of the triangle OQR, where O is the origin,

(b) the equation of the locus of the centroid of triangle OQR.

(**Note.** It is not necessary to know any geometry of the ellipse, or even its shape, in order to solve this problem, since the parametric analysis of all curves utilises the same methods.)

Using the parametric equations (based on the given parametric coordinates of P)

we have $\begin{cases} x = a \cos\theta \\ y = b \sin\theta \end{cases}$

Hence $\dfrac{dy}{dx} = \dfrac{dy}{d\theta} \Big/ \dfrac{dx}{d\theta} = -\dfrac{b \cos\theta}{a \sin\theta}$

The gradient of the normal at P is therefore $\dfrac{a \sin\theta}{b \cos\theta}$

and the equation of this normal is

$$y - b \sin\theta = \frac{a \sin\theta}{b \cos\theta}(x - a \cos\theta)$$

$\Rightarrow$ $by \cos\theta - b^2 \sin\theta \cos\theta = ax \sin\theta - a^2 \sin\theta \cos\theta$

$\Rightarrow$ $ax \sin\theta - by \cos\theta = (a^2 - b^2) \sin\theta \cos\theta$

At Q, where the normal meets Ox, $y = 0$

Hence $x = \left(\dfrac{a^2 - b^2}{a}\right) \cos\theta$

Similarly at R, $x = 0$

Hence
$$y = \left(\frac{a^2 - b^2}{-b}\right) \sin \theta$$

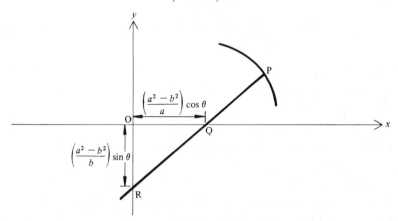

(i) The area, Δ, of triangle OQR is

$$\frac{1}{2}(OQ)(OR) = \frac{1}{2}\left(\frac{a^2 - b^2}{a}\right)\cos\theta \left(\frac{a^2 - b^2}{b}\right)\sin\theta$$

$$= \frac{(a^2 - b^2)^2}{2ab}\sin\theta\cos\theta$$

$$= \frac{(a^2 - b^2)^2}{4ab}\sin 2\theta$$

But the greatest value of $\sin 2\theta$ is 1.
Hence the greatest value of Δ is $(a^2 - b^2)^2/4ab$.

(ii) The coordinates of O, Q and R are:

$$(0, 0), \left(\frac{a^2 - b^2}{a}\cos\theta, 0\right), \left(0, \frac{a^2 - b^2}{-b}\sin\theta\right)$$

Hence at G, the centroid of triangle OQR,

$$\begin{cases} x = \frac{1}{3}\left(0 + \frac{a^2 - b^2}{a}\cos\theta + 0\right) = \left(\frac{a^2 - b^2}{3a}\right)\cos\theta \\ y = \frac{1}{3}\left(0 + 0 + \frac{a^2 - b^2}{-b}\sin\theta\right) = \left(\frac{a^2 - b^2}{-3b}\right)\sin\theta \end{cases}$$

These equations represent the coordinates of G for *any* point P on the given ellipse, so they are the parametric equations of the locus of G. The corresponding Cartesian equation is given by eliminating θ from these two equations as follows:

$$\cos \theta = \frac{3ax}{a^2 - b^2} \quad \text{and} \quad \sin \theta = \frac{-3by}{a^2 - b^2}$$

But
$$\cos^2 \theta + \sin^2 \theta \equiv 1$$

Hence
$$\left(\frac{3ax}{a^2 - b^2}\right)^2 + \left(\frac{-3by}{a^2 - b^2}\right)^2 = 1$$

for all possible positions of G.
Thus the Cartesian equation of the locus of G is

$$9a^2 x^2 + 9b^2 y^2 = (a^2 - b^2)^2$$

(which can be recognised as the equation of another ellipse).

EXERCISE 11i

1) P is any point on the curve $xy = 4$ and O is the origin. Find the equation of the locus of the midpoint of OP.

2) P is any point on the parabola $y^2 = 4x$ and A is the point $(4, 0)$. Q divides AP in the ratio $1:2$. Find the equation of the locus of Q.

3) A parabola has parametric equations $x = 3t^2$, $y = 6t$ and P is any point on the parabola. Find the Cartesian equation of the locus of the midpoint of OP.

4) $P\left(ct, \dfrac{c}{t}\right)$ is any point on the rectangular hyperbola $xy = c^2$. The tangent at P cuts the y-axis at T and the normal at P cuts the x-axis at N. Find the Cartesian equation of the locus of the midpoint of TN.

5) A circle touches the x-axis and also touches the circle $x^2 + y^2 = 4$. Find the equation of the locus of its centre.

6) P is any point on the ellipse whose parametric equations are $x = 3 \cos \theta$, $y = 2 \sin \theta$. The line joining the origin, O, to P is produced to Q where $OQ = 2OP$. Find the Cartesian equation of the locus of Q.

7) A line parallel to the x-axis cuts the curve $y^2 = 4x$ at P and cuts the line $x = -2$ at Q. Find the equation of the locus of the midpoint of PQ.

8) $P(ap^2, 2ap)$ and $Q(aq^2, 2aq)$ are two points on the parabola $y^2 = 4ax$ and PQ passes through the focus $(a, 0)$. Show that, if the tangents at P and Q intersect at T, the locus of T is a straight line and identify this line.

9) A point $P(x, y)$ moves so that its distance from the point $(2, 2)$ is $\sqrt{2}$ times its distance from the line $x + y = 2$. Find and sketch the locus of P.

SUMMARY

1) If $P(X, Y)$ divides the line joining $A(x_1, y_1)$ and $B(x_2, y_2)$ in the ratio $\lambda : \mu$ then

$$X = \frac{\mu x_1 + \lambda x_2}{\mu + \lambda}, \quad Y = \frac{\mu y_1 + \lambda y_2}{\mu + \lambda}$$

2) The centroid of a triangle with vertices (x_1, y_1), (x_2, y_2), (x_3, y_3) has coordinates

$$[\tfrac{1}{3}(x_1 + x_2 + x_3), \tfrac{1}{3}(y_1 + y_2 + y_3)]$$

3) If θ is the acute angle between $y = m_1 x + c_1$ and $y = m_2 x + c_2$ then

$$\tan \theta = \left| \frac{m_1 - m_2}{1 + m_1 m_2} \right|$$

4) If d is the distance from (p, q) to the line $ax + by + c = 0$, then

$$d = \left| \frac{ap + bq + c}{\sqrt{(a^2 + b^2)}} \right|$$

5) $x^2 + y^2 + 2gx + 2fy + c = 0$ is the equation of a circle with centre $(-g, -f)$ and radius $\sqrt{g^2 + f^2 - c}$.

6) Two circles with centres A, B and radii r_1, r_2 respectively:
(a) are orthogonal if $r_1{}^2 + r_2{}^2 = (AB)^2$,
(b) touch if $r_1 + r_2 = AB$ or $|r_1 - r_2| = AB$.

7) $y^2 = 4ax$ is the equation of a parabola with focus $(a, 0)$ and directrix $x = -a$.

8) $\dfrac{x^2}{a^2} + \dfrac{y^2}{b^2} = 1$ is the equation of an ellipse with major and minor axes of lengths $2a$ and $2b$ $(a > b)$. Its eccentricity is e where $b^2 = a^2(1 - e^2)$.

9) Two curves or lines $y = f(x)$, $y = g(x)$ meet where $f(x) = g(x)$. If this equation has equal roots, the curves touch.

MULTIPLE CHOICE EXERCISE 11

(Instructions for answering these questions are given on p. xii.)

(All questions relate to figures in one plane.)

TYPE I

1) A point P moves so that it is equidistant from A and B. The locus of the set of points P is:
(a) a circle on AB as diameter,

(b) a line parallel to AB,

(c) the perpendicular bisector of AB,

(d) a parabola with focus A and directrix B,

(e) none of these.

2) The point dividing $A(1, 2)$ and $B(7, -4)$ in the ratio $1:2$ has coordinates:

(a) $(3, -2)$ (b) $(5, -2)$ (c) $(\frac{8}{3}, -\frac{4}{3})$ (d) $(3, 0)$

(e) $(13, -10)$.

3) The gradient of the normal to the parabola $x = 3t^2$, $y = 6t$ at the point where $t = -2$ is:

(a) $-\frac{1}{2}$ (b) -2 (c) 2 (d) $\frac{1}{2}$ (e) 1.

4) The radius of the circle $2x^2 + 2y^2 - 4x + 12y + 11 = 0$ is:

(a) $\sqrt{29}$ (b) $\sqrt{51}$ (c) $\sqrt{4.5}$ (d) 29 (e) $\sqrt{15.5}$.

5) The parametric equations of a curve are $x = \sec \theta + 1$, $y = \tan \theta - 1$. Its Cartesian equation is:

(a) $y^2 + 3 = x^2$ (b) $x^2 - y^2 - 2x - 2y = 1$

(c) $x^2 - y^2 + 2x + 2y + 1 = 0$ (d) $x^2 - y^2 = 1$ (e) $(x - 1)^2 = y^2$.

6) S is the focus of a parabola $\begin{cases} x = at^2 \\ y = 2at \end{cases}$ and P is any point on the parabola.

The equation of the locus of the midpoint of SP is:

(a) $y^2 = a(2x - a)$ (b) $2y^2 = ax$ (c) $y^2 = 2ax - a$

(d) $(x - a)^2 = 4y^2$.

7)

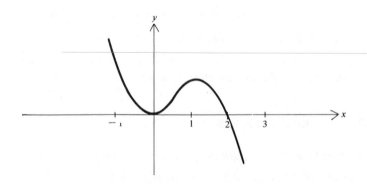

The equation of this curve could be:

(a) $y = x(x - 2)^2$ (b) $y = 2 - x + 2x^2 - x^3$ (c) $y = 2x - x^3$

(d) $y = x^2(x + 2)$ (e) none of these.

8) In an attempt to find the condition that the line $y = mx + c$ should touch the curve $y^2 + 4x = 0$, the following solution was given. State the line where an error first appears.

(a) The line meets the curve where $y^2 + 4\left(\dfrac{y-c}{m}\right) = 0.$

(b) i.e. where $my^2 + 4y - 4c = 0.$

(c) If the line is a tangent this equation has real roots.

(d) The required condition is therefore $mc = 1.$

9) $P(ap^2, ap^3)$ and $Q(aq^2, aq^3)$ are two points on a curve.

(a) The curve is a parabola.

(b) The equation of PQ is $y(p + q) = x(p^2 + pq + q^2) - ap^2q^2.$

(c) The Cartesian equation of the curve is $ax^3 = y^2.$

(d) The curve cuts the x-axis three times.

10) A line with gradient 2 which passes through the point $(k, 0)$ will touch the parabola $y^2 = 4ax$ if and only if:
(a) $k > 0$ (b) $k < 0$ (c) $k = -a$
(d) $k < -a$ (e) $4k = -a$

TYPE II

11) If the equation $ax^2 + by^2 + 2gx + 2fy + c = 0$ represents a circle through the origin,
(a) $g = 0$ and $f = 0.$ (b) $c = 0.$
(c) $a = b.$ (d) $a = -b.$

12) The equations of two curves, C_1 and C_2, are $xy = 1$ and $x^2 + y^2 = 2.$

(a) C_1 is a rectangular hyperbola.

(b) C_1 and C_2 intersect at four distinct points.

(c) C_1 and C_2 have two common tangents.

(d) C_1 is a continuous curve.

13) The parabola $y^2 + 8x = 0$:
(a) has a focal length of two units,

(b) can be represented by $\begin{cases} x = 2t^2 \\ y = 4t, \end{cases}$

(c) is the locus of a point equidistant from the point $(2, 0)$ and the line $x = -2,$

(d) is symmetrical about the x-axis.

14) The parametric coordinates of a point P are $(3 - \cos\theta, 2 + \sin\theta).$
(a) The locus of P is an ellipse.

(b) The parameter θ can be marked on a diagram.

(c) The point $(3, 2)$ is on the locus of P.

(d) The gradient function of the locus of P is $\dfrac{3-x}{y-2}$.

TYPE III

15) (a) A point P is equidistant from a fixed straight line and a fixed point.

(b) The locus of a point P is a parabola.

16) (a) A certain cubic equation is satisfied by $x = 1$ and $x = 2$.

(b) A certain cubic equation has three real roots.

17) (a) The point $(1, 2)$ is on a certain parabola.

(b) The equation of a certain parabola is $y^2 = 4x$.

18) (a) The circles $x^2 + y^2 = r^2$ and $(x-3)^2 + (y+4)^2 = 16$ touch each other.

(b) $r = 1$.

19) A, B and P are points on a circle.

(a) Angle APB is $90°$.

(b) AB is a diameter of the circle.

20) (a) $f(x) = 0$ has two equal roots.

(b) $y = f(x)$ touches the x-axis.

21) (a) $f(x, y) = 0$ is the equation of a circle.

(b) $f(x, y) \equiv x^2 + 2y^2 - 3x - y + 2$.

TYPE IV

22) Find the equation of the tangent at a point A on the curve $ax^2 + by^2 + 2gx + 2fy + c = 0$ if:

(a) the coordinates of A are given,

(b) the values of g and f are given,

(c) the values of a and b are equal,

(d) the curve passes through the origin.

23) Find the equation of a circle whose centre is in the first quadrant given that:

(a) it touches the x-axis,

(b) its centre is on the line $y = x$,

(c) it touches the y-axis,

(d) it touches the line $x + y = 6$.

24) Determine whether the line $y = mx + c$ touches the parabola $y^2 = 4x$ if:

(a) the line is parallel to $3x + 2y = 7$,

(b) the line passes through the focus of the parabola,

(c) the focus of the parabola is on the x-axis,
(d) the line has a negative gradient.

25) $y = x^3 + ax^2 + bx + c$.   Determine the values of a, b and c if:
(a) the curve touches the x-axis at the point $(3, 0)$,
(b) the curve crosses the y-axis at the point $(0, 9)$,
(c) the gradient of the curve when $x = 0$ is 3.

26) Find the equation of the bisector of the angle A in triangle ABC, given:
(a) the equation of BC,
(b) the coordinates of A,
(c) the distance of A from BC,
(d) the coordinates of the centroid of triangle ABC.

TYPE V

27) $x^2 + y^2 - 2x - 4y + 6 = 0$ is the equation of a circle.

28) If the line $y = mx + 2$ is a tangent to the parabola $y^2 = 4x$ there are two possible values of m.

29) No cubic equation has exactly two real roots.

30) The gradient of every tangent to the rectangular hyperbola $xy = c^2$ is negative.

31) The parametric equations $x = 2 \cos \theta$, $y = 3 \sin \theta$, represent a circle.

MISCELLANEOUS EXERCISE 11

1) A triangle has its vertices at A(4, 4), B(−4, 0), C(6, 0).
(a) Find the equation of the circle through the points A, B, C.
(b) Find the coordinates of the point where the internal bisector of the angle BAC meets the x-axis.
(c) Find the equation of the circle which passes through B and touches AC at C.   (U of L)

2) The vertex A of a square ABCD, lettered in the anticlockwise sense, has coordinates $(-1, -3)$. The diagonal BD lies along the line $x - 2y + 5 = 0$.
(a) Prove, by calculation, that the coordinates of C are $(-5, 5)$, and find the coordinates of B and D.
(b) Find the equation of the circle which touches all four sides of the square, confirming that this circle passes through the origin.
(c) Calculate the area of that portion of the square which lies in the first quadrant $(x > 0, y > 0)$.   (C)

3) Show that the points $(-1, 0)$ and $(1, 0)$ are on the same side of the line $y = x - 3$.
Find the equations of the two circles each passing through the points $(-1, 0)$, $(1, 0)$ and touching the line $y = x - 3$.   (U of L)

4) If P, Q are the points (x_1, y_1), (x_2, y_2) respectively, show that the equation of the circle on PQ as diameter is

$$(x - x_1)(x - x_2) + (y - y_1)(y - y_2) = 0.$$

A triangle has vertices at the origin O and the points $A(4, 3)$, $B(\frac{6}{5}, \frac{17}{5})$. The line through O perpendicular to AB and the line through A perpendicular to OB meet at H. Show that the coordinates of H are $(\frac{3}{5}, \frac{21}{5})$. If BH meets OA at D, find the coordinates of D.
Write down the equation of the circle on OB as diameter and verify that this circle passes through D. (U of L)

5) Prove that the point $B(1, 0)$ is the mirror-image of the point $A(5, 6)$ in the line $2x + 3y = 15$.
Find the equation of:
(a) the circle on AB as diameter,
(b) the circle which passes through A and B and touches the x-axis. (U of L)

6) A straight line of gradient m passes through the point $(1, 1)$ and cuts the x and y axes at A and B respectively. The point P lies on AB and is such that $AP:PB = 1:2$. Show that, as m varies, P moves on the curve whose equation is $3xy - x - 2y = 0$.
Find the perpendicular distance of the point $(1, 1)$ from the tangent to this curve at the origin. (JMB)p

7) Find the equation of the circle S which passes through $A(0, 4)$ and $B(8, 0)$ and has its centre on the x-axis. If the point C lies on the circumference of S, find the greatest possible area of the triangle ABC. (U of L)

8) A rhombus ABCD is such that the coordinates of A and C are $(-3, -4)$ and $(5, 4)$ respectively. Find the equation of the diagonal BD of the rhombus. If the side BC has gradient 2, obtain the coordinates of B and of D. Prove that the rhombus has area $21\frac{1}{3}$. (C)

9) A circle with centre P and radius r touches externally both the circles $x^2 + y^2 = 4$ and $x^2 + y^2 - 6x + 8 = 0$. Prove that the x-coordinate of P is $\frac{1}{3}r + 2$, and that P lies on the curve $y^2 = 8(x - 1)(x - 2)$. (U of L)

10) The circle S_1 with centre $C_1(a_1, b_1)$ and radius r_1 touches externally the circle S_2 with centre $C_2(a_2, b_2)$ and radius r_2. The tangent at their common point passes through the origin. Show that

$$(a_1^2 - a_2^2) + (b_1^2 - b_2^2) = (r_1^2 - r_2^2).$$

If, also, the other two tangents from the origin to S_1 and S_2 are perpendicular, prove that $|a_2 b_1 - a_1 b_2| = |a_1 a_2 + b_1 b_2|$.

Hence show that, if C_1 remains fixed but S_1 and S_2 vary, then C_2 lies on the curve

$$(a_1{}^2 - b_1{}^2)(x^2 - y^2) + 4a_1 b_1 xy = 0. \qquad \text{(U of L)}$$

11) Given the points A(2, 14), B(−6, 2), C(12, −10) verify that the triangle ABC is right angled. Calculate the coordinates of:
(a) the point D on AB produced such that   AC = CD,
(b) the point of intersection of the perpendiculars from A, C, D to the
    opposite sides of the triangle ACD. (JMB)

12) Find the equation of the two circles which each satisfy the following conditions:
(a) the axis of x is a tangent to the circle,
(b) the centre of the circle lies on the line $2y = x$,
(c) the point (14, 2) lies on the circle.
Prove that the line $3y = 4x$ is a common tangent to these circles. (C)

13) A circle passes through the points A, B and C which have coordinates $(0, 3)$, $(\sqrt{3}, 0)$ and $(-\sqrt{3}, 0)$ respectively.
Find:
(a) the equation of the circle,
(b) the length of the minor arc BC,
(c) the equation of the circle on AB as diameter.
A line $y = mx - 3$,   of variable gradient m, cuts the circle ABC in two points L and M. Find in Cartesian form the equation of the locus of the midpoint of LM. (AEB)'76

14) Find the equation of the circle which touches the line $3y - 4x - 24 = 0$ at the point $(0, 8)$ and also passes through the point $(7, 9)$. Prove that this circle also touches the axis of x. Find the equations of the tangents to this circle which are perpendicular to the line $3y - 4x - 24 = 0$. (C)

15) (a) A circle with centre $(3, 2)$ touches the line $4x - 3y + 4 = 0$.   Find the equation of the circle and show that it touches the x-axis.
(b) In each of the following pairs of equations, t is a parameter. Sketch the locus given by each pair of equations:
(i) $x = 3 + 5 \cos t$, $y = 4 + 5 \sin t$ $(0 \leqslant t \leqslant \pi)$
(ii) $x = 3 \cos t$, $y = 4 \cos t$ $(0 \leqslant t \leqslant \pi)$
(iii) $x = 3 + t \cos \dfrac{\pi}{3}$, $y = 4 + t \sin \dfrac{\pi}{3}$ $(-\infty < t < \infty)$. (JMB)

16) Find the coordinates of the foot of the perpendicular from the point $(2, -6)$ to the line $3y - x + 2 = 0$. (U of L)

17) Given that $a^2 + b^2 = c^2$,   show that the two circles
$x^2 + y^2 + ax + by = 0$ and $x^2 + y^2 = c^2$ touch each other and find the coordinates of the point of contact.

Two circles, which pass through the origin and the point $(1, 0)$, touch the circle $x^2 + y^2 = 4$. Find the coordinates of the points of contact. Find also the equation of the circle which has these points of contact as the ends of a diameter.   (U of L)

18) If the normal at $P(ap^2, 2ap)$ to the parabola $y^2 = 4ax$ meets the curve again at $Q(aq^2, 2aq)$ prove that $p^2 + pq + 2 = 0$. Prove that the equation of the locus of the point of intersection of the tangents to the parabola at P and Q is

$$y^2(x + 2a) + 4a^3 = 0.$$ (U of L)

19) Obtain the equation of the tangent to the parabola $y^2 = 4x$ at the point $(t^2, 2t)$. The tangents to the parabola at the points $P(p^2, 2p)$ and $Q(q^2, 2q)$ meet on the line $y = 3$. Find the equation of the locus of the midpoint of PQ.

If PQ intersects the x-axis and the y-axis at R and S respectively, find also the equation of the locus of the midpoint of RS.   (AEB)'74

20) Show that the equation of the normal to the parabola $y^2 = 4ax$ at the point $(at^2, 2at)$ is $y + tx = 2at + at^3$.

If this normal meets the parabola again at the point $(aT^2, 2aT)$ show that

$$t^2 + tT + 2 = 0,$$

and deduce that T^2 cannot be less that 8.

The line $3y = 2x + 4a$ meets the parabola at the points P and Q. Show that the normals at P and Q meet on the parabola.   (U of L)

21) The points $P(ap^2, 2ap)$ and $Q(aq^2, 2aq)$ move on the parabola $y^2 = 4ax$ and $p + q = 2$. Show that the chord PQ makes a constant angle with the x-axis and that the locus of the midpoint M of PQ is part of a line which is parallel to the x-axis.

If also the point $R(ar^2, 2ar)$ moves so that $p - r = 2$, find in its simplest form the Cartesian equation of the locus of the midpoint N of PR.   (JMB)

22) The parabolas $x^2 = 4ay$ and $y^2 = 4ax$ meet at the origin and at the point P. The tangent to $x^2 = 4ay$ at P meets $y^2 = 4ax$ again at A, and the tangent to $y^2 = 4ax$ at P meets $x^2 = 4ay$ again at B. Prove that the angle APB is $\arctan\left(\frac{3}{4}\right)$ and that AB is a common tangent to the two parabolas.   (U of L)

23) Show that the equation of the normal to the parabola $y^2 = 4ax$ at the point $P(at^2, 2at)$ is $y + tx = 2at + at^3$.

The normal at P meets the x-axis at G and the midpoint of PG is N.

(a) Find the equation of the locus of N as P moves on the parabola.

(b) The focus of the parabola is S. Prove that SN is perpendicular to PG.

(c) If the triangle SPG is equilateral, find the coordinates of P.   (AEB)'75

24) Prove that the line $y = mx + \dfrac{15}{4m}$ is a tangent to the parabola $y^2 = 15x$

for all non-zero values of m. Using this result, or otherwise, find the equations of the common tangents to this parabola and the circle $x^2 + y^2 = 16$. (U of L)

25) A variable line $y = mx + c$ cuts the fixed parabola $y^2 = 4ax$ in two points P and Q. Show that the coordinates of M, the midpoint of PQ are

$$\left(\frac{2a - mc}{m^2}, \ \frac{2a}{m} \right)$$

Find one equation satisfied by the coordinates of M in each of the following cases:
(a) if the line has fixed gradient m,
(b) if the line passes through the fixed point $(0, -a)$.   (U of L)

26) Sketch the curve given by the parametric equations $x = \sin\theta$, $y = \sin 2\theta$.
(AEB)'66

27) Show that the tangent at the point P, with parameter t, on the curve $x = 3t^2$, $y = 2t^3$ has equation $y = tx - t^3$. Prove that this tangent will cut the curve again at the point Q with coordinates $(3t^2/4, -t^3/4)$
Find the coordinates of the possible positions of P if the tangent to the curve at P is the normal to the curve at Q.   (U of L)

28) Show that the straight line $3y + x = 0$ is tangential to the curve $3y(x^2 - 9) = 2(3x + 1)$.
(AEB)'72

29) Show that the point with coordinates $(2 + 2\cos\theta, 2\sin\theta)$ lies on the circle $x^2 + y^2 = 4x$, and obtain the equation of the tangent to the circle at this point.
The tangents at the points P and Q on this circle touch the circle $x^2 + y^2 = 1$ at the points R and S. Find the coordinates of the point of intersection of these tangents, and obtain the equation of the circle through the points P, Q, R and S.   (U of L)

30) The parametric equations of a curve are $x = \cos 2t$, $y = 4\sin t$. Sketch the curve for $0 \leqslant t \leqslant \frac{1}{2}\pi$.

Show that $\dfrac{dy}{dx} = -\operatorname{cosec} t$ and find the equation of the tangent to the curve at the point $A(\cos 2T, 4\sin T)$.
The tangent at A crosses the x-axis at the point M and the normal at A crosses the x-axis at the point N. If the area of the triangle AMN is $12\sin T$, find the value of T between 0 and $\frac{1}{2}\pi$.   (AEB)'76

31) The foot of the perpendicular from a point P to the straight line $x + y = \sqrt{2}$ is the point R and Q is the point with coordinates $(\sqrt{2}, \sqrt{2})$. If P varies in such a way that $PQ^2 = 2PR^2$, show that its locus is the rectangular hyperbola $xy = 1$.

Find the equation of the tangent to this hyperbola at the point $\left(t, \dfrac{1}{t}\right)$.

This tangent cuts the x-axis at A and the y-axis at B, and C is the point on AB such that $AC:CB = a:b$. Show that the locus of C as t varies is the

rectangular hyperbola $\quad xy = \dfrac{4ab}{(a+b)^2}$.

Determine the two possible values of the ratio $a:b$ such that the straight line $x + y = \sqrt{2}$ is a tangent to the locus of C.   (JMB)

32) A curve is given parametrically by $\quad x = a(5\cos\theta + \cos 5\theta)$,
$$y = a(5\sin\theta - \sin 5\theta).$$
Find the equation of the normal to the curve at the point with parameter θ.
Find those points of the curve in the first quadrant at which the normal is also normal to the curve at another point.   (O)

33) Find the equation of the tangent to the curve $\quad y = \tfrac{5}{12}x^3 - \tfrac{13}{9}x \quad$ at the point P at which $\quad x = x_0$.
Show that the x coordinate of the point Q where this tangent meets the curve again is $-2x_0$, and find the values of x_0 for which the tangent at P is the normal at Q.   (O)

34) Find the equation of the tangent to the circle $\quad x^2 + y^2 = a^2 \quad$ at the point $T(a\cos\theta, a\sin\theta)$. This tangent meets the line $\quad x + a = 0 \quad$ at R. If RT is produced to P so that $\quad RT = TP$, find the coordinates of P in terms of θ and find the coordinates of the points in which the locus of P meets the y-axis.
(U of L)

35) A curve is given by the parametric equations $\quad x = t^2 - 3, \quad y = t(t^2 - 3)$.
(a) Find its Cartesian equation, in a form clear of surds and fractions.
(b) Prove that it is symmetrical about the x-axis.
(c) Show that there are no points on the curve for which $\quad x < -3$.   (C)

36) The tangent and the normal at a point $P(t, e^{-t^2/2})$ on the curve $y = e^{-x^2/2}$ meet the x-axis in T and G respectively, and N is the foot of the ordinate from P. Show that $\quad G = [t(1 - e^{-t^2}), 0] \quad$ and if $NT.GN = e^{-1}$, find the length of PN.   (AEB)'75

37) The parametric equations of a curve are
$$x = 3(2\theta - \sin 2\theta)$$
$$y = 3(1 - \cos 2\theta)$$

The tangent and the normal to the curve at the point P where $\quad \theta = \dfrac{\pi}{4} \quad$ meet the y-axis at L and M respectively.
Show that the area of triangle PLM is $\tfrac{9}{4}(\pi - 2)^2$.   (AEB)'74

38) Show that the equation of the normal to the ellipse $\dfrac{x^2}{a^2} + \dfrac{y^2}{b^2} = 1$ at the

point $P(a \cos \theta, b \sin \theta)$ is $ax \sin \theta - by \cos \theta = (a^2 - b^2) \sin \theta \cos \theta$.

If the normal to the ellipse $\dfrac{x^2}{25} + \dfrac{y^2}{9} = 1$ at a point Q meets the coordinate

axes at A and B respectively, show that, as Q varies, the locus of the
midpoint of AB is another ellipse, and give the coordinates of the foci of this
second ellipse.   (U of L)

39) Show that, if $y = mx + c$ is a tangent to the ellipse
$$b^2 x^2 + a^2 y^2 = a^2 b^2,$$

then $c^2 = a^2 m^2 + b^2$.
Find the equations of the tangents to the ellipse $9x^2 + 16y^2 = 144$ from the
point $P(4, 9)$. These tangents touch the ellipse at the points A and B, and C
is the midpoint of AB. Find the coordinates of A and B, and show that the
straight line PC goes through the origin.   (AEB)'76

40) Sketch the curve $y^2 = (x - a)(x - 2a)^2$.
Show that the normal to either branch of the curve at the point $(2a, 0)$ meets

the curve again at a point where $x = a + \dfrac{1}{a}$.   (U of L)p

41) Show that the circle on the line joining the points (x_1, y_1), (x_2, y_2) as
diameter has the equation
$$(x - x_1)(x - x_2) + (y - y_1)(y - y_2) = 0.$$

Prove that the normal to the rectangular hyperbola $xy = c^2$ at the point

$A\left(ct, \dfrac{c}{t}\right)$ meets the hyperbola again at $B\left(-\dfrac{c}{t^3}, -ct^3\right)$.

Find the point C at which the circle on AB as diameter meets the rectangular
hyperbola again and prove that the origin is the midpoint of AC.   (U of L)

CHAPTER 12

CURVE SKETCHING

The frequent use made of curve sketching throughout this text serves to emphasize the importance of this aspect of mathematical understanding. When the equation of a curve is unfamiliar, determining the shape of the curve can require varied and sometimes extensive consideration. It is impossible to study every type of curve but if we discuss some of those which occur quite frequently the techniques used in these cases can be applied to other curves.

Certain features of a curve can be observed, or easily calculated, from the equation of the curve, for example:

(a) the points where the curve crosses the coordinate axes,
(b) the stationary points,
(c) the linear asymptotes,
(d) regions of the xy plane where the graph does not exist,
(e) the behaviour of y as $x \rightarrow \pm \infty$.

It is not always necessary to consider *all* of these features when examining a particular curve, different equations requiring different approaches. But as each of the above observations will be used in some problems, let us consider the application of each in turn.

(a) *Axes intercepts*

A curve meets (i.e. cuts or touches) the x-axis if the equation $y = 0$ has real solutions,

e.g. for $y = 2 - \dfrac{1}{x}$,

$$y = 0 \implies 2 - \frac{1}{x} = 0 \implies x = \frac{1}{2}$$

Thus the curve *cuts* the *x*-axis at $(\frac{1}{2}, 0)$.
Similarly the intercepts on the *y*-axis are given by the real solution(s) of the
equation $x = 0$,

e.g. for $y(y - 2) = x - 1$, $x = 0$ $\Rightarrow$ $y^2 - 2y + 1 = 0$ $\Rightarrow$ $(y - 1)^2 = 0$

i.e.   when $x = 0$, $y = 1, 1$.
Thus the curve *touches* the *y*-axis at $(0, 1)$.

(b) *Stationary points*
The analysis of stationary points is dealt with in Chapter 5.
But in curve sketching the *absence* of stationary points is just as important as
their presence,

e.g. for $y = 2 - \dfrac{1}{x}$,   we note that $\dfrac{dy}{dx} = \dfrac{1}{x^2}$.   Hence there is no finite value

of *x* for which $\dfrac{dy}{dx} = 0$,   and we deduce that there are no finite stationary

points.

(c) *Linear asymptotes*
If, as either *x* or *y* becomes very large the equation of a curve approximates
to the equation of a straight line, that straight line is an asymptote. For instance

(i)   if $y = 2 - \dfrac{1}{x}$ then, as

    $x \to \pm \infty$ the equation
    approximates to $y = 2$.
    So the line $y = 2$ is an
    asymptote,
    i.e. the curve approaches this
    line as $x \to \pm \infty$.

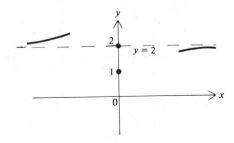

(ii)   if $y = \dfrac{1}{x - 2}$,

    $y \to \pm \infty$ when $x \to 2$.
    So the line $x = 2$ is an
    asymptote.

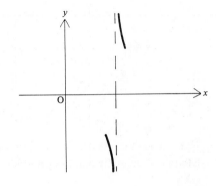

(iii) if $y = \dfrac{x^2 - 3x + 1}{x - 1}$,

by carrying out the division on the R.H.S. the equation can be written

$$y = x - 2 - \frac{1}{x - 1}$$

Then as $x \to \pm\infty$ the equation approximates to $y = x - 2$ and this line is an asymptote.

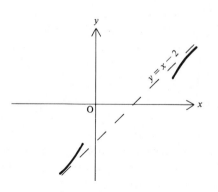

(**Note.** The side from which a curve approaches an asymptote is usually made clear by the positions of axis intercepts or turning points.)

(d) *Empty sections*, i.e. regions where no part of the curve lies, can be located using the method adopted in the following example.

If $y = \dfrac{x - 2}{(x + 2)(x - 1)}$ then the values $x = -2, 1, 2$ are significant. So we complete the table

	$x < -2$	$-2 < x < 1$	$1 < x < 2$	$2 < x$
$x + 2$	$-$	$+$	$+$	$+$
$x - 1$	$-$	$-$	$+$	$+$
$x - 2$	$-$	$-$	$-$	$+$
y	$-$	$+$	$-$	$+$

From these results we can shade areas of the xy plane where no curve lies.

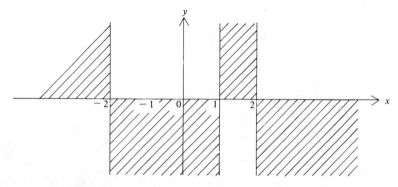

(**Note.** The series of signs of y found by using the above table can be determined mentally and indicated directly on a number line.)

e.g. $y = \dfrac{x-2}{(x+2)(x-1)}$

$\Rightarrow$

| $y<0$ | $y>0$ | $y<0$ | $y>0$ |

$-3 \quad -2 \quad -1 \quad 0 \quad 1 \quad 2 \quad 3 \quad \to x$

Alternatively, if we find the set of possible values of the function $\dfrac{x-2}{(x+2)(x-1)}$, all other values of y correspond to empty sections in the xy plane. This method, which can be used only when a function is quadratic in x, can be applied as follows.

If $y = \dfrac{x-2}{(x+2)(x-1)}$, then

$$yx^2 + (y-1)x + (2-2y) = 0 \qquad [1]$$

The solution set of this equation comprises the coordinates of all points on the corresponding curve. So the values of x that satisfy it must be real.
Using the condition $b^2 - 4ac \geq 0$ (since [1] is quadratic in x) we have

$f(y) \equiv 9y^2 - 10y + 1$

$(y-1)^2 - 4y(2-2y) \geq 0$

or $\quad 9y^2 - 10y + 1 \geq 0$

or $\quad (9y-1)(y-1) \geq 0$

Thus $\quad y \leq \frac{1}{9}$ or $y \geq 1$

Conversely, x is not real if $\frac{1}{9} < y < 1$ so no part of the required curve lies in this region.

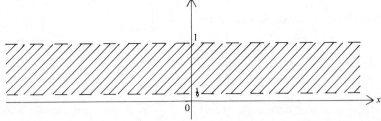

The points where the curve *touches* the boundary lines $y = \frac{1}{9}$ and $y = 1$ correspond to the values of x for which equation [1] has *equal* roots.

i.e. when $\quad x = \dfrac{-(y-1)}{2y} \quad \left(\text{equal roots have value } -\dfrac{b}{2a}\right).$

Thus when $\quad y = \frac{1}{9}, \quad x = 4,$

and when $\quad y = 1, \quad x = 0.$

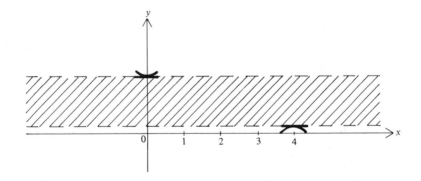

(**Note** that the points $(4, \frac{1}{9})$ and $(0, 1)$ are maximum and minimum points respectively. They can be determined by calculus methods *instead* of the method shown above.)

(e) *The behaviour of y as* $x \to \pm \infty$. If a curve has no obvious asymptotes, the behaviour of y as $x \to \pm \infty$ is often useful,

e.g. if $y = xe^{x^2}$, e^{x^2} is always positive

thus $y \to \infty$ when $x \to \infty$

and $y \to -\infty$ when $x \to -\infty$.

The information gained from the above investigations is usually sufficient for a *sketch* of the corresponding curve to be drawn. So many variations are possible, however, that no standard procedure can be formulated to cover all cases. The examples that follow indicate some of the approaches that can be used.

EXAMPLES 12a

1) Sketch the curve whose equation is $y = \dfrac{x-2}{x-1}$

(a) The curve crosses the axes where $\begin{cases} x = 0 \quad y = 2 \\ x = 2 \quad y = 0 \end{cases}$

(b) There is no real value of x for which $\dfrac{dy}{dx} = 0$ so there are no stationary

points.

(c) Rewriting the equation as $y = 1 - \dfrac{1}{x-1}$

we see that $\begin{cases} y \to 1 \quad \text{when} \quad x \to \pm \infty \\ y \to \pm \infty \quad \text{when} \quad x \to 1. \end{cases}$

So $y = 1$ and $x = 1$ are asymptotes.

(d)

	$x<1$	$1<x<2$	$2<x$
$x-2$	$-$	$-$	$+$
$x-1$	$-$	$+$	$+$
y	$+$	$-$	$+$

We now have sufficient information to sketch the curve.

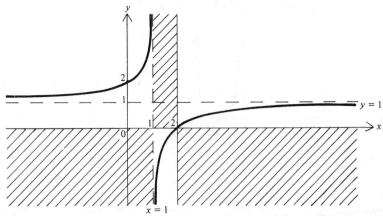

Note that y is undefined when $x=1$ and x is undefined when $y=1$.

2) Sketch the curve $\quad 2y = \dfrac{3x-4}{x(x-1)}$

(a) The axes intercepts are given by
$y=0 \;\Rightarrow\; x=\frac{4}{3}$ i.e. $(\frac{4}{3},0)$
$x=0 \;\Rightarrow\;$ no finite value of y.

(b) As $\;x\to\infty,\;\; y\to 0,$
i.e. $\;y=0\;$ is an asymptote.
Also $\;y\to\infty\;$ as $\;x\to 0\;$ or $1,$
i.e. $\;x=0\;$ and $\;x=1\;$ are asymptotes.

(d) Rearranging the equation (which is quadratic in x) gives
$$2yx^2 - (2y+3)x + 4 = 0 \qquad [1]$$

If x is to have real values, y
must satisfy the condition

$(2y+3)^2 - 4(2y)(4) \geqslant 0$
$\Rightarrow \qquad 4y^2 - 20y + 9 \geqslant 0$
$\Rightarrow \qquad (2y-9)(2y-1) \geqslant 0$
$\Rightarrow \qquad y \leqslant \frac{1}{2}, \;\; y \geqslant \frac{9}{2}$

Hence there are no real points $\;(x,y)\;$ for $\;\frac{1}{2}<y<\frac{9}{2}$
and the graph does not exist for this range.

When $y = \frac{1}{2}$ or $\frac{9}{2}$ the roots of equation [1] above are equal,

so
$$x = \frac{2y + 3}{4y} \qquad \left(-\frac{b}{2a}\right)$$

Thus, when $y = \frac{1}{2}$, $x = 2$,
and, when $y = \frac{9}{2}$, $x = \frac{2}{3}$

(e) When $\begin{cases} x \to +\infty, & y \to 0 \quad \text{from above and} \\ x \to -\infty, & y \to 0 \quad \text{from below.} \end{cases}$

The curve can now be sketched as follows.

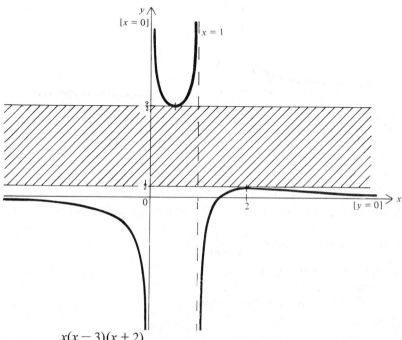

3) If $y = \dfrac{x(x-3)(x+2)}{(x-1)^2}$ find the linear asymptotes and sketch the curve.

(a) Axes intercepts are where $\begin{cases} y = 0 & \Rightarrow \quad x = 0, 3, -2 \\ x = 0 & \Rightarrow \quad y = 0 \quad \text{(already known).} \end{cases}$

(b) When the process of finding $\dfrac{dy}{dx}$ looks formidable, it is usual to omit the precise location of turning points unless they are specifically asked for.

(c) Converting to the form $y = x + 1 - \dfrac{(5x + 1)}{(x - 1)^2}$ (by division) we see that

when $\begin{cases} (x-1)^2 \to 0, & y \to \infty, \\ x \to \infty, & y \to x + 1. \end{cases}$

Therefore $y = x + 1$ is an asymptote,
and $x = 1$ is a *double* asymptote.

(d) Critical values of x are $-2, 0, 1, 3$

Thus

$$\frac{x(x-3)(x+2)}{(x-1)^2} \Rightarrow$$

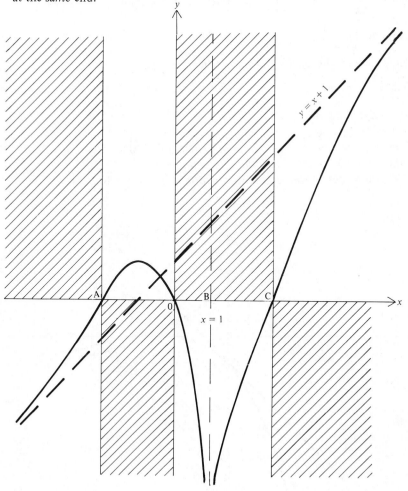

Note that y does not change sign at $x = 1$ because $x - 1$ is a squared factor.

The curve can now be sketched, noting that, because $y = 0$ at both $x = -2$ and at $x = 0$, and because there is no asymptote between these values, the curve must have a maximum turning point between **A** and **O**.

Note also that, although there is no change in the sign of y at $x = 1$, the line $x = 1$ is a *double* asymptote, i.e. the curve approaches it *twice* at the *same* end.

4) Find the turning point(s) on the curve with equation $y = xe^{-x}$ and hence sketch the curve.

If $y = xe^{-x}$, $\dfrac{dy}{dx} = e^{-x} + x(-e^{-x})$

At turning points

$$\frac{dy}{dx} = e^{-x}(1-x) = 0 \ \Rightarrow \ x = 1$$

Further,

$$\frac{d^2y}{dx^2} = -e^{-x}(1-x) + e^{-x}(-1)$$

$$= -\frac{1}{e} \quad \text{when} \quad x = 1$$

Since $\dfrac{dy}{dx} = 0$ and $\dfrac{d^2y}{dx^2} < 0$ when $x = 1$, there is a maximum turning point at $\left(1, \dfrac{1}{e}\right)$.

Now consider the behaviour of y when $x \to \pm\infty$

When $x \to \infty$, $y = \dfrac{x}{e^x} \to 0$ (since e^n is very much greater than n when n is large and positive).

When $x \to -\infty$, $y \to -\infty$

Note also, since e^{-x} is positive for all values of x, that when $x < 0$, $y < 0$, when $x > 0$, $y > 0$ and when $x = 0$, $y = 0$.

Thus the sketch of the curve with equation $y = xe^{-x}$ is as shown.

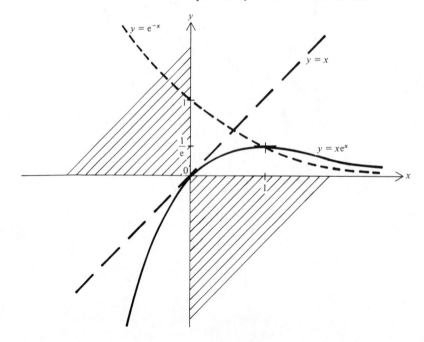

5) Sketch the graph of $y = \dfrac{1}{x}\ln x$, showing clearly any turning points.

There are no values of $\ln x$ for negative values of x, so the curve $y = \dfrac{1}{x}\ln x$ does not exist when $x < 0$.

To determine any turning points we use

$$\frac{dy}{dx} = -\frac{1}{x^2}\ln x + \left(\frac{1}{x}\right)\left(\frac{1}{x}\right) = 0 \;\Rightarrow\; \frac{1}{x^2}(1 - \ln x) = 0$$

i.e. $\ln x = 1 \;\Rightarrow\; x = e.$

$\left(\textbf{Note} \text{ also that } \dfrac{dy}{dx} \to 0 \text{ when } x \to \infty.\right)$

$$\frac{d^2 y}{dx^2} = -\frac{2}{x^3}(1 - \ln x) + \frac{1}{x^2}\left(-\frac{1}{x}\right) < 0 \quad \text{when} \quad x = e$$

Thus $\left(e, \dfrac{1}{e}\right)$ is a maximum turning point.

When $x \to \infty$, $y \to 0$ (since n is much greater than $\ln n$ when n is large).
When $x = 1$, $\ln x = 0 \;\Rightarrow\; y = 0.$
When $x < 1$, $\ln x < 0 \;\Rightarrow\; y < 0.$
When $x > 1$, $\ln x > 0 \;\Rightarrow\; y > 0.$

When $x \to 0$, $\dfrac{1}{x} \to \infty$, $\ln x \to -\infty \;\Rightarrow\; y \to -\infty.$

The curve can now be sketched.

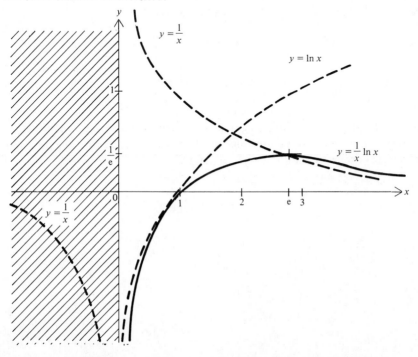

EXERCISE 12a

1) Write down the equations of the linear asymptotes of the curves whose equations are:

(a) $y = \dfrac{1}{1-x}$　　(b) $y = \dfrac{x}{1-x}$　　(c) $y = \dfrac{x^2}{1-x}$

(d) $y = \dfrac{x-2}{(x-4)(x+3)}$　　(e) $y = \dfrac{(x-4)(x+3)}{x-2}$

2) Determine the range(s) of values of y which correspond to real values of x if:

(a) $y = \dfrac{1}{1-x}$　　(b) $y = \dfrac{x}{1+x^2}$　　(c) $y = \dfrac{1+x^2}{x}$

(d) $y = \dfrac{x-2}{(x+1)(x-1)}$　　(e) $y = \dfrac{x}{(1-x)^2}$

(f) $y^2 - x^2 = 4$　　(g) $x^2 + y^2 = 4$　　(h) $y = \dfrac{x-2}{(x+2)(x+3)}$

3) Determine the behaviour of y as $x \to \infty$ and $x \to -\infty$ if:

(a) $y = \dfrac{x}{1+x^2}$　　(b) $y^2 = 4ax$　　(c) $y = x^2 e^{-x}$

(d) $y = x \ln x$　　(e) $y = x^3 + x - 1$　　(f) $y = x^3 + 4x^2 + 2x - 3$

4) For what range(s) of values of x is y positive, when:

(a) $y = x^2 + 5x - 6$　　(b) $y = x^3 - 6x^2 + 11x - 6$

(c) $y = (x-1)e^x$　　(d) $y = \dfrac{x}{1-x}$　　(e) $y = \dfrac{x}{(1-x)^2}$

(f) $y = \dfrac{x-2}{(x+1)(x-1)}$　　(g) $y = x^2 + 4x + 5$

Sketch the graphs of the following functions showing clearly any asymptotes, turning points, points of intersection with the coordinate axes and the behaviour of the curve when x and/or y are very large.

5) $y = \dfrac{1}{1-x}$　　　　　　[Use (1a) and (2a)]

6) $y = \dfrac{x}{1-x}$　　　　　　[Use (1b) and (4d)]

7) $y = \dfrac{x-2}{(x+1)(x-1)}$　　　　[Use (2d) and (4f)]

8) $y = x^2\,e^x$

9) $y = \dfrac{\ln x}{x-1}$

10) $y^2 - x^2 = 4$ $\qquad\qquad$ [Use (2f)]

11) $y = x^3 + 2x^2 - x - 2$

12) $y = \dfrac{x}{(1-x)^2}$ $\qquad\qquad$ [Use (2e) and (4e)]

13) $y = \dfrac{x^2}{(1-x)}$ $\qquad\qquad$ [Use (1c)]

14) $y = \dfrac{x^3 - 2x^2 - x + 3}{(x-1)(x-2)}$

15) $y = \dfrac{e^x}{x}$

16) $y = \dfrac{2x}{x^2+1}$

17) $y = \dfrac{2(x-2)}{(x-1)(x-3)}$

18) $y = \dfrac{(x-3)(x-1)}{(x-2)^2}$

THE EQUATION $y^2 = f(x)$

The curve C_1 with equation $\quad y^2 = f(x), \quad$ can be deduced from the curve C_2 with equation $\qquad y = f(x), \quad$ by noting that,

(a) because $y^2 \geqslant 0$,
$\qquad C_1$ is limited to those values of x for which $\quad f(x) \geqslant 0$.

(b) $y = \pm\sqrt{f(x)} \quad$ so $\;C_1$ is symmetrical about the x-axis.

(c) When $\;f(x) > 1, \qquad \sqrt{f(x)} < f(x) \quad$ but
$\qquad$ when $\;\; 0 < f(x) < 1, \;\; \sqrt{f(x)} > f(x)$.

Consider, for example, $\quad f(x) \equiv (x+1)(x-1)(x-2)$

so that $\qquad\qquad\quad C_1$ is $\;\; y^2 = (x+1)(x-1)(x-2)$

and $\qquad\qquad\qquad\; C_2$ is $\quad\; y = (x+1)(x-1)(x-2)$

The graph of C_2 is shown in diagram (i)

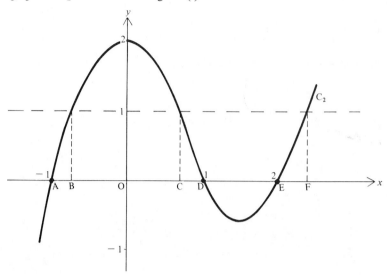

Now $f(x) \geqslant 0$ for $-1 \leqslant x \leqslant 1$ and $x \geqslant 2$ so C_2 exists only within these ranges.

Between A and B, C and D, E and F, $f(x) < 1$ so $\sqrt{f(x)} > f(x)$

Between B and C and right of F, $f(x) > 1$ so $\sqrt{f(x)} < f(x)$

Thus the graph of C_1 is as shown in diagram (ii).

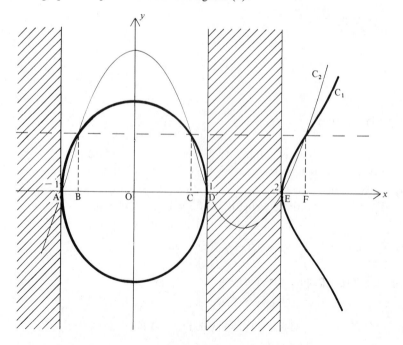

Note. The graph of the curve whose equation is $y = \sqrt{(x+1)(x-1)(x-2)}$ is that part of the curve C_1 *which lies above the x-axis*.

THE EQUATION $y = \dfrac{1}{f(x)}$

The curve C_1 with equation $y = \dfrac{1}{f(x)}$ can be deduced from the curve C_2 with equation $y = f(x)$ by noting that:

(a) for a given value of x, $f(x)$ and $\dfrac{1}{f(x)}$ have the same sign.

(b) If the numerical value of $f(x)$ increases, the numerical value of $\dfrac{1}{f(x)}$ decreases (and conversely).

(c) If $f(x_1) = 1$ then $\dfrac{1}{f(x_1)} = 1$ also, and

if $f(x_2) = -1$ then $\dfrac{1}{f(x_2)} = -1$ also.

(d) If there is a maximum turning point on C_2, i.e. $f(x)$ stops increasing and begins to decrease, then at the *same value of* x, $\dfrac{1}{f(x)}$ stops decreasing and begins to increase i.e. C_1 has a minimum turning point (and conversely).

(e) If $f(x) \to \infty$ for a finite value of x, $(x = x_3)$, then $x = x_3$ is an asymptote to C_2.

But $\dfrac{1}{f(x_3)} = \dfrac{1}{\infty} = 0$ i.e. when $x = x_3$, C_1 meets the x-axis.

Thus, considering again $f(x) \equiv (x+1)(x-1)(x-2)$, the graph of C_1 can be deduced from the known graph of C_2 as shown overleaf.

EXERCISE 12b

Sketch the graph of $f(x)$ and hence the graphs of (a) $y^2 = f(x)$ and (b) $y = \dfrac{1}{f(x)}$ when $f(x)$ is:

1) $x^2 - 3x + 2$ 2) $x^3 - 3x^2 + 2x$ 3) $\dfrac{x}{x-1}$ 4) $\dfrac{x}{x^2-4}$

5) xe^x 6) $\sin x$ 7) $\ln x$ 8) $\tan x$

9) e^{x^2} 10) $e^x \sin x$ (for $-2\pi \leqslant x \leqslant 2\pi$)

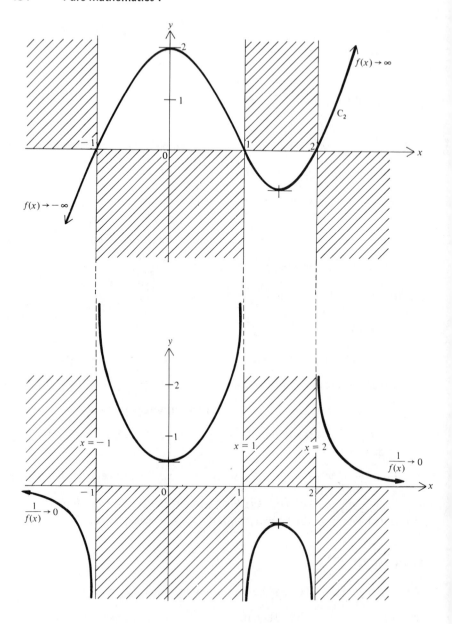

THE MODULUS OF A FUNCTION

If $y = x$, y is negative when x is negative. But if $y = |x|$, y takes the numerical value and *not the sign* of x, i.e. y is *always positive*.
$|x|$ is called the *modulus of* x.
Taking values of x from -3 to $+3$ we have

x	-3	-2	-1	0	1	2	3
$|x|$	3	2	1	0	1	2	3

Thus the graphs of $y = x$ and $y = |x|$ are

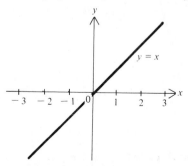
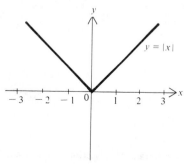

It can be seen that, for negative values of x, $y = |x|$ is the reflection of $y = x$ in the x-axis.

In general, the graph C_1 with equation $y = |f(x)|$ can be obtained from the graph C_2 of $y = f(x)$ by reflecting in the x-axis those points on C_2 for which $f(x) < 0$, the remaining sections of C_2 being retained without change.

For instance, to sketch $y = |x^2 - 3x + 2|$ the graph of $y = x^2 - 3x + 2$ is adapted as follows

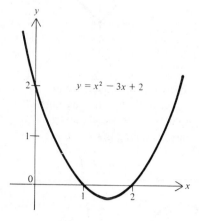
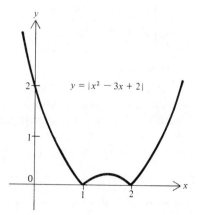

INEQUALITIES

An *equation* involving a modulus describes a line or curve in the xy plane. An *inequality* involving a modulus describes an area in this plane,

e.g. $y > |x^2 - 3x + 2|$ defines the area shaded in the following diagram while the unshaded area corresponds to $y < |x^2 - 3x + 2|$.

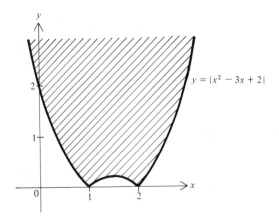

$y = |x^2 - 3x + 2|$

The Effect of a Modulus Sign on a Cartesian Equation

When a section of a graph $y = f(x)$ is reflected in the x-axis, the y-coordinate of every point on that section of the graph changes sign.

Thus the equation of the *reflected section* becomes

$$y = -f(x)$$

e.g. when $y = |x|$ the Cartesian equations are

$$\begin{cases} y = x & \text{for} \quad x > 0 \\ y = -x & \text{for} \quad x < 0 \end{cases}$$

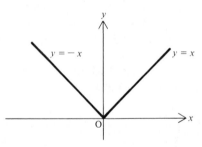

$y = -x$ $y = x$

Intersection

If intersection occurs between two lines or curves whose equations involve a modulus, care must be taken to use the correct pair of Cartesian equations when locating the point(s) of intersection. For example, the points common to $y = |x - 1|$ and $y = |x^2 - 3|$ are seen by sketching these two graphs on the same axes as shown.

- - - $y = |x - 1|$

——— $y = |x^2 - 3|$

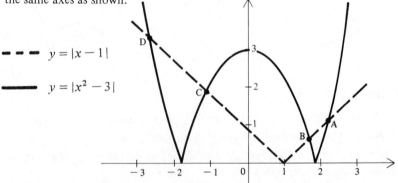

The Cartesian equations of the various sections of the line and curve show that

at A	$y = x - 1$	cuts	$y = x^2 - 3$	[1]
at B	$y = x - 1$	cuts	$y = -(x^2 - 3)$	[2]
at C	$y = -(x - 1)$	cuts	$y = -(x^2 - 3)$	[3]
at D	$y = -(x - 1)$	cuts	$y = x^2 - 3$	[4]

Thus the x coordinates of both A and C are given by the equation

$$x^2 - x - 2 = 0 \qquad \text{(from [1] and [3])}$$

i.e. $$x = -1, 2$$

Hence at A, $x = 2$ and $y = x - 1 \Rightarrow y = 1,$
while at C, $x = -1$ and $y = -(x - 1) \Rightarrow y = 2$
Similarly the x coordinates of both B and D are given by the equation

$$x^2 + x - 4 = 0 \qquad \text{(from [2] and [4])}$$

i.e. $$x = -\tfrac{1}{2}(1 - \sqrt{17}), \quad -\tfrac{1}{2}(1 + \sqrt{17})$$

Hence at B, $$x = -\tfrac{1}{2}(1 - \sqrt{17})$$

and $$y = x - 1 \Rightarrow y = \tfrac{1}{2}(-3 + \sqrt{17})$$

while at D, $$x = -\tfrac{1}{2}(1 + \sqrt{17})$$

and $$y = -(x - 1) \Rightarrow y = \tfrac{1}{2}(3 + \sqrt{17})$$

Alternatively, the equations of both sections of the graph $y = |x - 1|$ are included in the equation $y^2 = (x - 1)^2$.   Similarly both sections of the graph $y = |x^2 - 3|$ are included in the equation $y^2 = (x^2 - 3)^2$.
Thus the x coordinates of all four points of intersection are given by solving

$$(x - 1)^2 = (x^2 - 3)^2 \qquad [5]$$

This method is attractively concise but is not always suitable. For instance equation [5] is quartic and can be solved only if at least two of its roots can be found by using the factor theorem. Otherwise equation [5] cannot be solved exactly and the first method must be used to locate any common points.

EXAMPLES 12c

1) Sketch on the same axes the graphs $y = f(x)$ and $y = g(x)$ where $f(x) \equiv 2 - |x - 1|$ and $g(x) \equiv |x + 2| - 3$.   Indicate on your graph the range(s) of values of x for which $g(x) < f(x)$.

In order to sketch $y = f(x)$ we first draw the graph of the function $|x - 1|$ diagram (i)

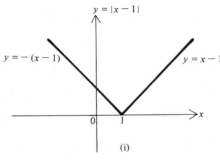

(i)

Then the function $-|x - 1|$ can be drawn as shown in diagram (ii)

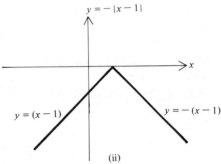

(ii)

When this graph is raised vertically by two units we have the graph of $f(x) \equiv 2 - |x - 1|$, as shown in diagram (iii)

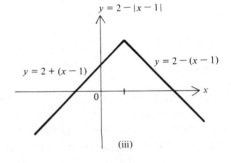

(iii)

Similarly for $g(x)$ we consider the graph of $|x + 2|$ and then lower this graph vertically by three units as in diagram (iv).

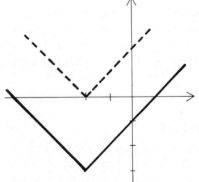

Sketching the graphs of $y = f(x)$ and $y = g(x)$ on the same axes we have

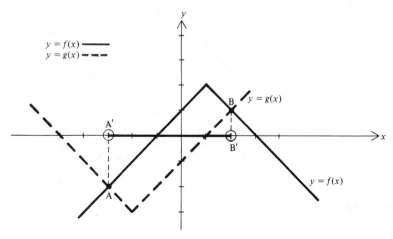

Between A and B, $f(x) > g(x)$.
Thus the range of values of x for which $g(x) < f(x)$ is from A′ to B′ as shown.
Note. The actual coordinates of A′ and B′ can be found, if required, by solving the equations

$$2 - (x - 1) = (x + 2) - 3 \qquad \text{for B′}$$

$$2 + (x - 1) = -(x + 2) - 3 \qquad \text{for A′}$$

2) Shade the area in which a point P can lie if the coordinates of P satisfy the following conditions

$$\begin{cases} y \geqslant |2 - x| \\ y \leqslant 4 - |x| \\ 0 \leqslant y < \sqrt{4x} \end{cases}$$

Considering the conditions separately and shading the area in which P can lie in each case, we have

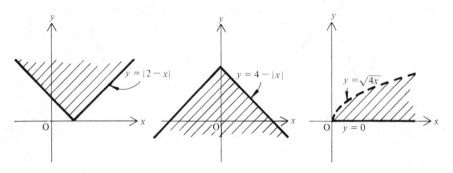

Combining all three conditions shows that P can lie only in the area shaded in the diagram below.

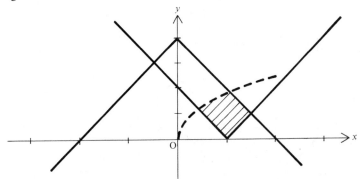

3) Solve the equation $|2x - 3| - |x^2 - 3x - 4| = 1$.

Rearranging the equation gives

$$|2x - 3| - 1 = |x^2 - 3x - 4|$$

or $$f(x) = g(x)$$

where $$f(x) \equiv |2x - 3| - 1$$

and $$g(x) \equiv |x^2 - 3x - 4| \equiv |(x - 4)(x + 1)|$$

Drawing the graphs $y = f(x)$ and $y = g(x)$ on the same axes, we see that they meet at four points.

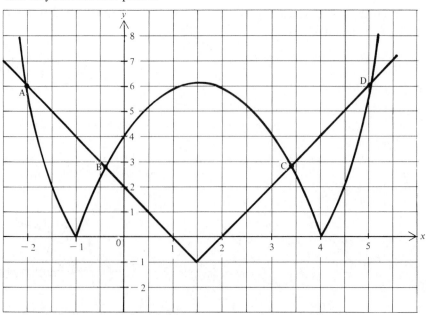

At A $\qquad$ $y = x^2 - 3x - 4$ $\qquad$ meets $\quad y = -(2x-3)-1$

At B $\qquad$ $y = -(x^2 - 3x - 4)$ $\quad$ meets $\quad y = -(2x-3)-1$

At C $\qquad$ $y = -(x^2 - 3x - 4)$ $\quad$ meets $\quad y = (2x-3)-1$

At D $\qquad$ $y = x^2 - 3x - 4$ $\qquad$ meets $\quad y = (2x-3)-1$

Using these equations, and referring to the positions of the points on the diagram, we see that:

At A $\qquad$ $x^2 - x - 6 = 0 \;\Rightarrow\; x = -2$ $\qquad$ (not 3)

at B $\qquad$ $x^2 - 5x - 2 = 0 \;\Rightarrow\; x = \frac{1}{2}(5 - \sqrt{33})$ $\quad$ $\{\text{not } \frac{1}{2}(5 + \sqrt{33})\}$

at C $\qquad$ $x^2 - x - 8 = 0 \;\Rightarrow\; x = \frac{1}{2}(1 + \sqrt{33})$ $\quad$ $\{\text{not } \frac{1}{2}(1 - \sqrt{33})\}$

at D $\qquad$ $x^2 - 5x = 0 \;\Rightarrow\; x = 5$ $\qquad$ (not 0)

Thus the solution of the equation is

$$x = -2,\; \tfrac{1}{2}(5 - \sqrt{33}),\; \tfrac{1}{2}(1 + \sqrt{33}),\; 5$$

EXERCISE 12c

Sketch the following graphs:

1) $y = |2x - 1|$ $\qquad\qquad$ 2) $y = |x(x-1)(x-2)|$

3) $y = |x^2 - 1|$ $\qquad\qquad$ 4) $y = |x^2 + 1|$

5) $y = |\sin x|$ $\qquad\qquad$ 6) $y = |\ln x|$

7) $y = -|\cos x|$ $\qquad\qquad$ 8) $y = 3 + |x + 1|$

9) $y = |2x + 5| - 4$ $\qquad$ 10) $y = |x^2 - x - 20|$

Indicate on a diagram the range(s) of values of x that satisfy the following inequalities. (Do not calculate the boundary values of x.)

11) $|x + 1| > |3x - 5|$ $\qquad$ 12) $|x| < |x^2 - 7x + 6|$

13) $|(x - 1)(x - 2)(x - 3)| > e^{-x}$

14) $|\sin x| < |\cos x|$ $\qquad$ [for $0 \leqslant x \leqslant 2\pi$]

Solve the following equations

15) $|x| = |x^2 - 2|$ $\qquad$ 16) $2 - |x + 1| = |4x - 3|$

17) $|x^2 - 1| - 1 = |3x - 2|$

18) $|x^3 - 3x^2 + 2x| = |x^2 - 2x|$

Shade the areas indicated by the following sets of conditions

19) $0 < x < 2\pi,$ $\qquad$ $0 < y < |\sin x|$

20) $y > |x|,$ $\qquad\qquad$ $y < 4 - |x|$

21) $y < 4 - x^2$, $y > |x - 1|$

22) Each of the four diagrams corresponds to one of the following seven equations.

Arrange the correct equations in the correct order.

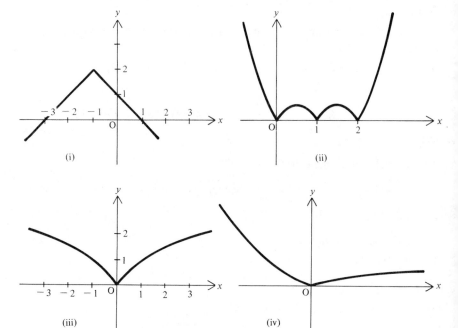

(i)

(ii)

(iii)

(iv)

(a) $y = |x - 1|$ (b) $y = |2x - 3x^2 + x^3|$ (c) $y = |e^{-x}|$
(d) $y = 2 - |x + 1|$ (e) $y = \sqrt{|x|}$
(f) $y = |e^{-x} - 1|$ (g) $y = |x^3|$

POLAR EQUATIONS

The polar frame of reference has already been used to locate a point P by means of its distance r from a fixed point O (*the pole*) and its angular displacement θ from a fixed line Ox (*the initial line*).

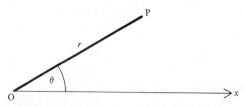

The *polar coordinates* of P are (r, θ).

OP is called the *radius vector* and θ is sometimes called the *vectorial angle*.

For θ, the anticlockwise sense of
rotation is taken to be positive and
clockwise rotation is negative.

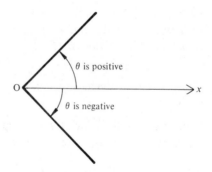

For r, the positive direction is
along OP where P is (r, θ) and
r is negative for points on PO
produced.

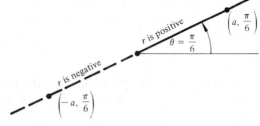

Thus the points with polar coordinates $\left(2, \dfrac{\pi}{4}\right)$, $\left(2, -\dfrac{\pi}{4}\right)$, $\left(-2, \dfrac{\pi}{4}\right)$,

$\left(-2, -\dfrac{\pi}{4}\right)$ are the points A, B, C and D respectively in the diagram below.

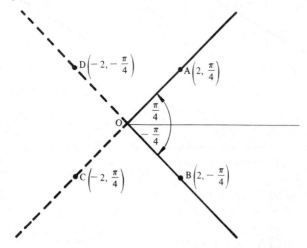

If the point P is unrestricted, then r and θ can independently take any value.
But if P is constrained to a particular path the values of r and/or θ are
limited.

For example, if P lies on a circle
with centre O and radius a then,
for all possible positions of P,

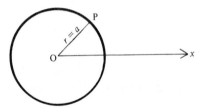

$$r = a.$$

Thus $r = a$ is the *polar equation*
of the locus of P.

In this example the value of r is independent of the value of θ so θ does not
appear in the polar equation of the circle.

Now consider the case when P lies
on a straight line through the pole,
inclined at an angle α to the initial
line Ox.

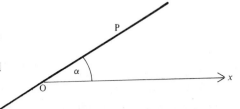

For every position of P, $\theta = \alpha$ and this is the polar equation of the line.
This time θ is independent of r so r does not appear in the equation of the
line.
The pole divides this line into two sections, one made up of points for which r
is positive, the other section made up of points for which r is negative.
The section corresponding to positive values of r is called the part-line
(or half-line) with equation $\theta = \alpha$.

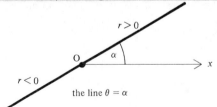

the line $\theta = \alpha$

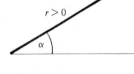

the part-line $\theta = \alpha$

Now consider the set of points P
which lie on a circle with centre C on
the initial line, with diameter $2a$, and
passing through the pole.
If OA is a diameter and P is any
point on the circle then angle OPA is
a right angle.

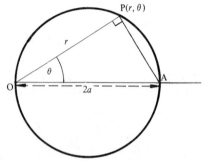

Thus, in triangle OPA, $r = 2a \cos \theta$.
This relationship between r and θ is valid for all possible positions of P and
for no other points. Hence $r = 2a \cos \theta$ is the polar equation of the
specified circle.

Relationship between Polar and Cartesian Coordinates

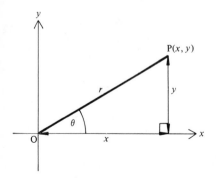

If a point P has Cartesian coordinates (x, y) relative to an origin O and polar coordinates (r, θ) relative to a pole O and initial line Ox, then

$$\begin{cases} x = r \cos \theta \\ y = r \sin \theta \end{cases} \quad \text{or} \quad \begin{cases} r^2 = x^2 + y^2 \\ \tan \theta = \dfrac{y}{x} \end{cases}$$

These relationships can be used to convert a Cartesian equation into polar form and conversely, e.g.
to find the Cartesian equation of the circle with polar equation $r = 2a \cos \theta$

we use $$\cos \theta = \frac{x}{r} \quad \text{and} \quad r^2 = x^2 + y^2$$

Thus $$r = 2a \left(\frac{x}{r} \right)$$

$\Rightarrow$ $$r^2 = 2ax$$

i.e. $$x^2 + y^2 - 2ax = 0$$

Conversely, to find the polar equation of the parabola $y^2 = 4x$ we use

$$y = r \sin \theta \quad \text{and} \quad x = r \cos \theta$$

Thus $$r^2 \sin^2 \theta = 4r \cos \theta$$

$\Rightarrow$ $$r = \frac{4 \cos \theta}{\sin^2 \theta}$$

Note that the Cartesian x-axis corresponds in polar form to the part-lines

$$\theta = 0 \quad \text{and} \quad \theta = \pi$$

and that the Cartesian y-axis corresponds to the part-lines

$$\theta = \frac{\pi}{2} \quad \text{and} \quad \theta = -\frac{\pi}{2} \quad \left(\text{or} \ \frac{3\pi}{2} \right)$$

i.e.

EXERCISE 12d

1) Mark the following points on a diagram:

$$\left(3, \frac{\pi}{4}\right); \quad \left(1, -\frac{\pi}{3}\right); \quad \left(-2, \frac{\pi}{2}\right); \quad \left(-1, -\frac{2\pi}{3}\right); \quad \left(2, \frac{3\pi}{2}\right); \quad \left(-4, \frac{\pi}{4}\right).$$

2) Given the polar coordinates of two points A and B find $(AB)^2$ if A and B are:

(a) $\left(2, \frac{\pi}{4}\right); \left(3, \frac{\pi}{2}\right)$

(b) $\left(-3, -\frac{\pi}{2}\right); \left(4, \frac{5\pi}{6}\right)$

(c) $(4, 0); \left(2, -\frac{\pi}{2}\right)$

(d) $\left(2, \frac{\pi}{3}\right); \left(-2, -\frac{\pi}{3}\right)$

(e) $\left(3, -\frac{\pi}{4}\right); \left(3, \frac{3\pi}{4}\right)$

3) In a polar frame of reference, where O is the pole, given $A(2, 0)$, $B\left(3, \frac{\pi}{3}\right)$ $C\left(-2, -\frac{\pi}{6}\right)$, $D\left(4, -\frac{\pi}{2}\right)$, $E\left(-3, -\frac{\pi}{3}\right)$, state whether each of the following triangles is isosceles, right angled or neither of these.

(a) AOD   (b) BDE   (c) COE   (d) COB
(e) AOB   (f) DOE   (g) BOE

Find the Cartesian equations of the lines or curves whose polar equations are:

4) $r = a$ 5) $r = a \cos 2\theta$ 6) $r^2 = a^2 \sin 2\theta$

7) $r = a(1 + \cos 2\theta)$ 8) $r = d \sec (\theta - \alpha)$

Find the polar equations of the lines or curves whose Cartesian equations are:

9) $(x - a)^2 + (y - a)^2 = a^2$ 10) $\dfrac{x^2}{a^2} + \dfrac{y^2}{b^2} = 1$

11) $y = 2x$ 12) $y = x^2$ 13) $xy = 4$

POLAR CURVE TRACING

The shape of a curve can be determined from its polar equation by listing corresponding values of θ and r and plotting these coordinates. (Values of θ are usually limited to the range $0 \leqslant \theta \leqslant 2\pi$.) Certain observations can reduce the amount of tabulated work however.

(1) If r is a function of $\cos \theta$, the curve is symmetrical about the line $\theta = 0$ [since $\cos(-\theta) = \cos \theta$].
Such a curve can be traced by calculating the values of r for angles in the range $0 \leqslant \theta \leqslant \pi$, plotting these points and then reflecting the curve so drawn, in the initial line.

(2) If r is a function of $\sin \theta$, the curve is symmetrical about the line

$\theta = \dfrac{\pi}{2}$ [since $\sin(\pi - \theta) = \sin \theta$].

Such a curve can be traced by plotting points calculated in the range

$-\dfrac{\pi}{2} \leqslant \theta \leqslant \dfrac{\pi}{2}$ and reflecting this curve in the line $\theta = \dfrac{\pi}{2}$.

(3) If $r = 0$ when $\theta = \alpha$, the line $\theta = \alpha$ is a tangent to the curve at the pole.

EXAMPLES 12e

1) Trace the curve with polar equation $r = a(1 + \cos \theta)$. Deduce the shapes of the curves with equations (a) $r = a(1 - \cos \theta)$ (b) $r = a(1 + \sin \theta)$.

(i) There is symmetry about the initial line (because r depends on $\cos \theta$).

(ii) $r = 0$ when $\cos \theta = -1$, i.e. when $\theta = \pi$.
So the line $\theta = \pi$ is a tangent to the curve at O.

(iii) Values of r are now calculated for convenient angles in the range $0 \leqslant \theta \leqslant \pi$.

θ	0	$\dfrac{\pi}{6}$	$\dfrac{\pi}{4}$	$\dfrac{\pi}{3}$	$\dfrac{\pi}{2}$	$\dfrac{2\pi}{3}$	$\dfrac{3\pi}{4}$	$\dfrac{5\pi}{6}$	π
$\cos \theta$ (to 1 d.p.)	1	0.9	0.7	0.5	0	-0.5	-0.7	-0.9	-1
r		$2a$	$1.9a$	$1.7a$	$1.5a$	a	$0.5a$	$0.3a$	$0.1a$	0

These points are easily located if, first, a framework is prepared comprising lines at each of the values of θ used above, and a series of circles of radii $\dfrac{a}{2}$, a, $\dfrac{3a}{2}$ and $2a$.

Thus

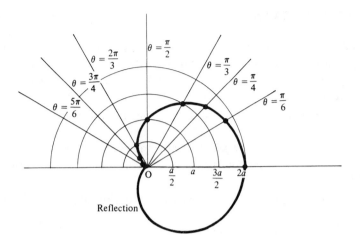

Note. The part-line $\theta = \pi$ is a tangent to the curve at O but the part-line $\theta = 0$ is not.
The shape of the curve at O is called a *cusp*.

(a) When $r = a(1 - \cos \theta)$, the values of r obtained in the table above are reversed in order giving the following curve trace.

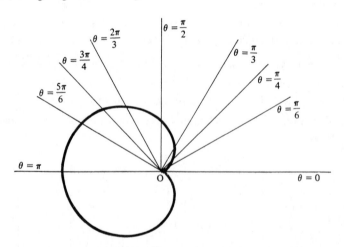

(b) The series of values of $\sin \theta$ as θ goes from $\dfrac{\pi}{2}$ to $\dfrac{3\pi}{2}$ is the same as the series of values of $\cos \theta$ as θ goes from 0 to π.

Thus the curve $r = a(1 + \sin \theta)$ is given by rotating $r = a(1 + \cos \theta)$ through an angle $\dfrac{\pi}{2}$,

i.e.

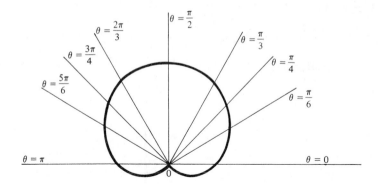

Note. Each of the above curves is called a *cardioid*.

2) N is the foot of the perpendicular from the pole, O, to a given straight line (not passing through O). ON is inclined to the initial line at an angle α and is of length d. Show that the polar equation of the given line is $r = d \sec (\theta - \alpha)$.   Hence sketch the lines   (a) $r \cos \theta = 4$ (b) $r \sin \theta = 3$.

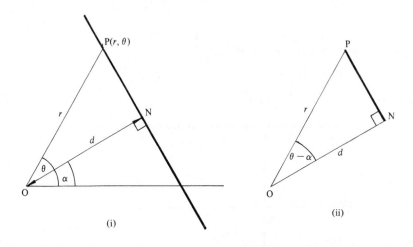

Diagram (i) shows the given data, together with a general point $P(r, \theta)$ on the given line. Diagram (ii) shows the right angle triangle ONP from which we see that

$$ON = OP \cos PON$$

i.e.

$$d = r \cos (\theta - \alpha)$$

As this condition applies to *any* point P on the given line, the polar equation of the given line is $r = d \sec(\theta - \alpha)$.

(a) If $r \cos \theta = 4$, $r = 4 \sec \theta$

This is the equation of a line for which $\begin{cases} d = 4 \\ \alpha = 0 \end{cases}$

i.e.

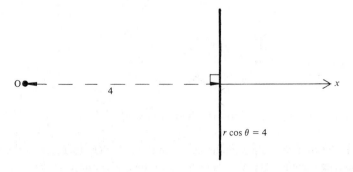

$r \cos \theta = 4$

(b) If $r \sin \theta = 3$, $r \cos \left(\theta - \dfrac{\pi}{2}\right) = 3$

$\Rightarrow$ $r = 3 \sec \left(\theta - \dfrac{\pi}{2}\right)$

This is the equation of a line for which $\begin{cases} d = 3 \\ \alpha = \dfrac{\pi}{2} \end{cases}$

i.e.

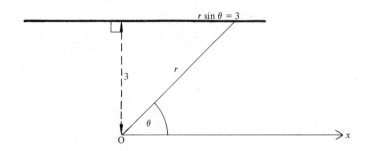

$r \sin \theta = 3$

3) Sketch the curve with equation $r = 2a \cos 2\theta$.

(i) The equation can be written as $r = 2a(2 \cos^2 \theta - 1)$

or as $r = 2a(1 - 2 \sin^2 \theta)$

Thus there is symmetry about the line $\theta = 0$ (r depends on $\cos \theta$)

and there is also symmetry about the line $\theta = \dfrac{\pi}{2}$ (r depends on $\sin \theta$)

Hence we will tabulate values of θ in the range $0 \leqslant \theta \leqslant \dfrac{\pi}{2}$ only, reflecting

the curve so obtained in *both* axes of symmetry.

(ii) When $r = 0$, $\cos 2\theta = 0 \Rightarrow 2\theta = \dfrac{\pi}{2}, \dfrac{3\pi}{2}, \dfrac{5\pi}{2}, \dfrac{7\pi}{2}$.

Thus the part-lines $\theta = \dfrac{\pi}{4}$, $\theta = \dfrac{3\pi}{4}$, $\theta = \dfrac{5\pi}{4}$, $\theta = \dfrac{7\pi}{4}$ are all tangents

to the curve at the pole.

(iii)

θ	0	$\dfrac{\pi}{16}$	$\dfrac{\pi}{8}$	$\dfrac{3\pi}{16}$	$\dfrac{\pi}{4}$	$\dfrac{5\pi}{16}$	$\dfrac{3\pi}{8}$	$\dfrac{7\pi}{16}$	$\dfrac{\pi}{2}$
$\cos 2\theta$	1	0.9	0.7	0.4	0	-0.4	-0.7	-0.9	-1
r	$2a$	$1.8a$	$1.4a$	$0.8a$	0	$-0.8a$	$-1.4a$	$-1.8a$	$-2a$

The polar coordinates in the above table give

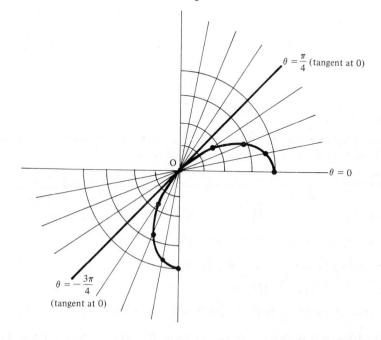

$\theta = \dfrac{\pi}{4}$ (tangent at 0)

$\theta = 0$

$\theta = -\dfrac{3\pi}{4}$
(tangent at 0)

$$\left(\textbf{Note} \text{ that when } \quad \theta = \frac{5\pi}{16}, \frac{3\pi}{8}, \frac{7\pi}{16}, \frac{\pi}{2}, r \text{ is negative.}\right)$$

Reflecting this curve in both axes of symmetry gives the complete curve with equation $r = 2a \cos 2\theta$.

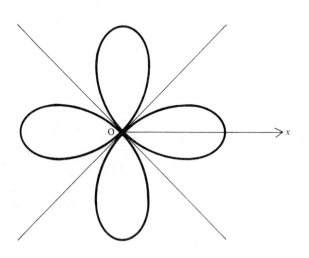

EXERCISE 12e

Trace the curves or lines whose polar equations are given below stating, where appropriate, the equations of the tangents at the pole.

1) $r = 3$

2) $\theta = -\dfrac{2\pi}{3}$

3) $r = 4 \sin \theta$

4) $r = 2 \sin 3\theta$

5) $r = 2 \cos 3\theta$

6) $r = a \cos 2\theta$

7) $r = a \cos \theta$

8) $r^2 = a^2 \cos^2 \theta$

9) $r = k\theta$ when   (a) $k > 0$,   (b) $k < 0$

10) $r = 2 - 3 \cos \theta$ 11) $r = 3 - 2 \cos \theta$

12) $r^2 = a^2 \cos 2\theta$ (*Hint.* If $\cos 2\theta < 0$, r is not real.)

Note. Because it is useful to be familiar with the sketches of the common polar curves (most of which are included in the exercise above), these curves are given on pages 457 and 458.

AREA BOUNDED BY A POLAR CURVE

Consider two points P and Q on a curve whose polar equation is $r = f(\theta)$. P is the point (r, θ) and Q has coordinates $(r + \delta r, \theta + \delta\theta)$ if Q is close to P.

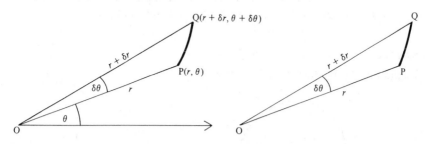

The area, δA, of sector POQ is approximately equal to the area of triangle POQ

i.e.
$$\delta A \simeq \tfrac{1}{2}r(r + \delta r)\sin \delta\theta$$

Hence
$$\frac{\delta A}{\delta\theta} \simeq \tfrac{1}{2}r(r + \delta r)\frac{\sin \delta\theta}{\delta\theta}$$

As $\delta\theta \to 0$: $\delta r \to 0$, $\dfrac{\sin \delta\theta}{\delta\theta} \to 1$ and $\dfrac{\delta A}{\delta\theta} \to \dfrac{dA}{d\theta}$

Therefore
$$\frac{dA}{d\theta} = \tfrac{1}{2}r^2$$

Now consider the area bounded by the two part-lines $\theta = \alpha$, $\theta = \beta$ and the part of the curve between these lines.

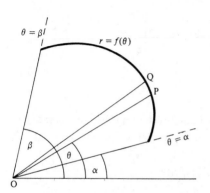

This area comprises an infinite set of elements of the type POQ.

Hence, throughout the whole of the area being considered

$$\frac{dA}{d\theta} = \tfrac{1}{2}r^2$$

so

$$A = \tfrac{1}{2}\int_\alpha^\beta r^2 \, d\theta$$

In deriving this expression it has been assumed that every angle between α and β corresponds to a real value of r. If this condition is not satisfied, α and β are not a suitable pair of boundary values (see Example 12f 3).

EXAMPLES 12f

1) Find the area of the cardioid $r = a(1 + \cos \theta)$.

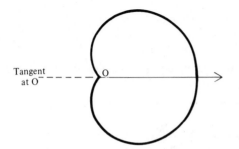

Tangent
at O

The complete area is traced out by taking values of θ from 0 to 2π.

Thus the area A of the cardioid is given by

$$A = \tfrac{1}{2}\int_0^{2\pi} r^2 \, d\theta = \tfrac{1}{2}\int_0^{2\pi} a^2(1 + \cos \theta)^2 \, d\theta$$

$$A = \tfrac{1}{2}a^2 \int_0^{2\pi} \{1 + 2\cos\theta + \tfrac{1}{2}(1 + \cos 2\theta)\} \, d\theta$$

$$= \frac{a^2}{2}\left[\theta + 2\sin\theta + \frac{\theta}{2} + \frac{1}{4}\sin 2\theta\right]_0^{2\pi}$$

$$= \frac{3\pi a^2}{2}$$

Note. The area of the cardioid can also be found by finding the area of the upper half and doubling it (because the curve is symmetrical about $\theta = 0$),

i.e.

$$A = 2 \times \tfrac{1}{2}\int_0^\pi a^2(1 + \cos\theta)^2 \, d\theta$$

2) Find the area of one loop of the curve $r = a \cos 3\theta$.

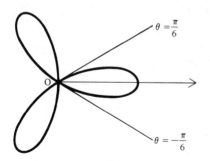

Any loop of this curve lies between values of θ for which $r = 0$

i.e. for which $\cos 3\theta = 0$

$$\Rightarrow \qquad 3\theta = \pm \frac{\pi}{2}, \ \pm \frac{3\pi}{2}, \dots$$

or $\qquad \theta = \pm \frac{\pi}{6}, \ \pm \frac{\pi}{2}, \dots$

One loop lies between $\theta = -\dfrac{\pi}{6}$ and $\theta = +\dfrac{\pi}{6}$

so its area is
$$\frac{1}{2} \int_{-\frac{\pi}{6}}^{\frac{\pi}{6}} a^2 \cos^2 3\theta \ d\theta = \frac{1}{4}a^2 \int_{-\frac{\pi}{6}}^{\frac{\pi}{6}} (1 + \cos 6\theta) \ d\theta$$

$$= \frac{a^2}{4} \left[\theta + \frac{1}{6} \sin 6\theta \right]_{-\frac{\pi}{6}}^{\frac{\pi}{6}}$$

$$= \frac{\pi a^2}{12}$$

3) Find the area enclosed by the curve $r^2 = a^2 \sin 2\theta$.

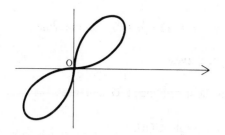

r is real only if $\sin 2\theta \geqslant 0$

i.e.

$0 \leqslant 2\theta \leqslant \pi$ and $2\pi \leqslant 2\theta \leqslant 3\pi$

or

$0 \leqslant \theta \leqslant \dfrac{\pi}{2}$ and $\pi \leqslant \theta \leqslant \dfrac{3\pi}{2}$

Using only those values of θ for which r is real, we see that the enclosed area is given by

$$\int_0^{\frac{\pi}{2}} \tfrac{1}{2} r^2 \ d\theta + \int_\pi^{\frac{3\pi}{2}} \tfrac{1}{2} r^2 \ d\theta$$

or, since the two loops are identical

$$\text{Area} = 2 \int_0^{\frac{\pi}{2}} \tfrac{1}{2} r^2 \, d\theta$$

$$= a^2 \int_0^{\frac{\pi}{2}} \sin 2\theta \, d\theta$$

$$= a^2$$

EXERCISE 12f

In Questions 1–5 find the areas bounded by the given curve and the given radius vectors.
In each case sketch the curve, indicating the required area.

1) $r = a$ between $\theta = 0$ and $\theta = \pi$

2) $r = 2a \sin \theta$ between $\theta = 0$ and $\theta = \dfrac{\pi}{2}$

3) $r = a(1 - \sin \theta)$ between $\theta = 0$ and $\theta = 2\pi$

4) $r = 5\theta$ between $\theta = 0$ and $\theta = \dfrac{\pi}{2}$

5) $r(1 + \cos \theta) = a$ between $\theta = 0$ and $\theta = \dfrac{\pi}{2}$

6) Find the area of one loop of the curve $r = 2a \sin 2\theta$.

7) Sketch the curve $r = 3 - 5 \cos \theta$.   Find the area of the inner loop of this curve.

8) Sketch the curves $r = 2a \cos \theta$ and $r = a(1 + \cos \theta)$ on the same diagram.
Hence find the area enclosed *between* these curves.

9) The line $r = 4 \sec \left(\theta - \dfrac{\pi}{6} \right)$ meets the initial line at   G   and meets the part-line $\theta = \dfrac{\pi}{2}$ at   H. Find the area of triangle   GOH.

10) Find the area enclosed by the curve $r = a(4 + 3 \cos \theta)$.

COMMON POLAR LINES AND CURVES

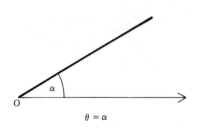

$\theta = \alpha$

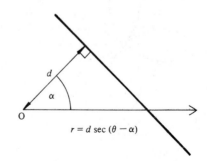

$r = d \sec (\theta - \alpha)$

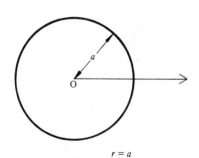

$r = a$

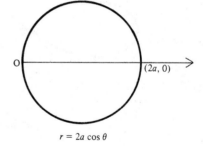

$(2a, 0)$

$r = 2a \cos \theta$

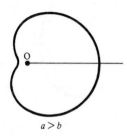

$a > b$

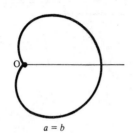

$a = b$

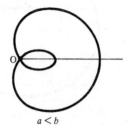

$a < b$

$r = a + b \cos \theta$

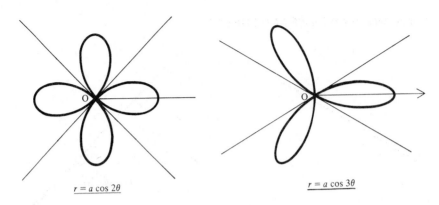

$r = a \cos 2\theta$

$r = a \cos 3\theta$

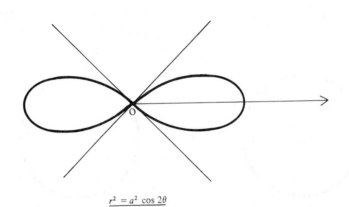

$r^2 = a^2 \cos 2\theta$

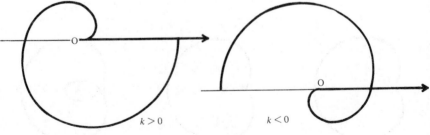

$k > 0$

$k < 0$

$r = k\theta$

MULTIPLE CHOICE EXERCISE 12

(Instructions for answering these questions are given on p. xii.)

TYPE I

1) The number of linear asymptotes of the curve $y = \dfrac{x}{x-1}$ is:

(a) 1   (b) 2   (c) 0   (d) ∞.

2) The polar equation of the circle
in the diagram is:

(a) $x^2 + y^2 - 2ay = 0$

(b) $r = 2a \sin \theta$

(c) $x^2 + y^2 - 2ax = 0$

(d) $r = 2a \cos \theta$

(e) none of these.

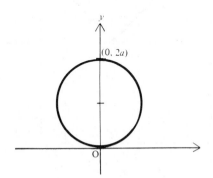

3) The area of one loop of the
curve $r^2 = a^2 \cos 2\theta$ is given by:

(a) $\frac{1}{2}a^2 \displaystyle\int_{-\frac{\pi}{2}}^{\frac{\pi}{2}} \cos 2\theta \ d\theta$

(b) $a^2 \displaystyle\int_{-\frac{\pi}{2}}^{\frac{\pi}{2}} \cos 2\theta \ d\theta$

(c) $a^2 \displaystyle\int_{-\frac{\pi}{4}}^{\frac{\pi}{4}} \cos 2\theta \ d\theta$

(d) $\frac{1}{2}a^2 \displaystyle\int_{-\frac{\pi}{4}}^{\frac{\pi}{4}} \cos 2\theta \ d\theta$

(e) none of these.

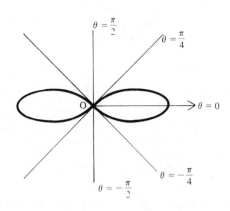

4) The graph with equation $y = |x - 2|$ is:

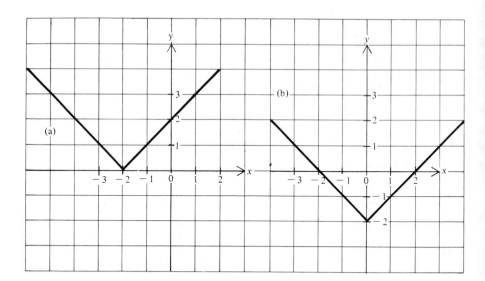

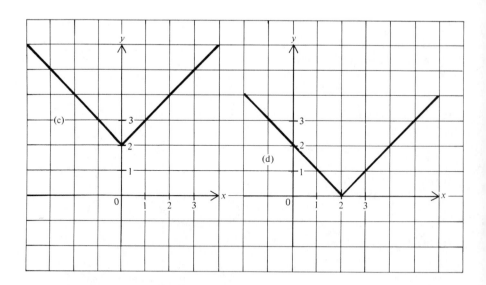

5) The graph of $y = f(x)$ is reflected in the x-axis. The equation of this reflected curve is:

(a) $y = f(-x)$ (b) $y = 1/f(x)$ (c) $y = |f(x)|$
(d) $y = -f(x)$ (e) $y^2 = f(x)$.

6) Which of the following graphs is the sketch of $y = xe^x$?

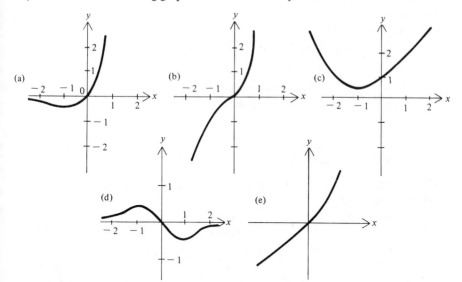

7) The shaded area can be defined
by:
(a) $r \leqslant 2$
(b) $y^2 \leqslant x^2 - 4$
(c) $|x| + |y| \leqslant 2$
(d) $|r| \leqslant 2$.

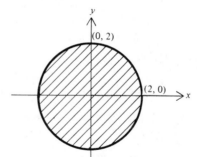

8) The sketch of the equation $r = a + b \cos \theta$ is as shown in the diagram if:
(a) $a = b$
(b) $a > b$
(c) $a < b$
(d) $a = 0$.

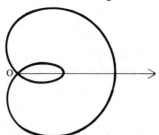

TYPE II

9) Which of the following equations represents a circle?
(a) $x^2 + y^2 - 2x - 4y + 9 = 0$ (b) $r = a$
(c) $(x - 1)^2 + (y - 3)^2 = 0$ (d) $r = a \sin \theta$.

10) On the graph $y = e^{x^2}$:
(a) $y \to \infty$ as $x \to \infty$
(b) $y \to 0$ as $x \to -\infty$
(c) there is no finite turning point
(d) there is one linear asymptote.

11) Which of the following equations represents a tangent to the circle $r = 2a \cos \theta$?

(a) $a = r \sin \theta$ (b) $\theta = -\dfrac{\pi}{2}$

(c) $r = 2a \sec \theta$ (d) $\theta = \pi$.

12) The equation $|x - 1| = |3 - 4x|$:
(a) has only one root
(b) has two roots
(c) can be written as $(x - 1)^2 = (3 - 4x)^2$
(d) can be written as $x - 1 = \pm (3 - 4x)$.

TYPE III

13) (a) $y = 1 - |x + 4|$.
      (b) $y \geqslant -3$.

14) (a) The graph of $y = f(x)$ has exactly two distinct vertical asymptotes.

      (b) The graph of $y = \dfrac{1}{f(x)}$ has exactly two distinct points of intersection
      with the x-axis.

15) (a) $f(x) \equiv x \ln x$.
      (b) $x \not< 0$.

16) (a) The line $\theta = 0$ is a tangent to a circle.

      (b) A circle passes through the pole and has its centre on the line $\theta = \dfrac{\pi}{2}$.

17) (a) $f(x) \equiv (x^2 - 1)(x - 3)$.
      (b) $y^2 = f(x)$ is limited to the range $x \leqslant -1$, $1 \leqslant x \leqslant 3$.

18) (a) $y = |x|$.
      (b) $x = |y|$.

19) (a) The graph of $y^2 = f(x)$ is continuous.
      (b) $f(x) \geqslant 0$.

TYPE IV

20) Find the polar equation of a circle.
(a) The circle passes through the pole.

(b) The radius of the circle is a.

(c) The circle is symmetrical about a diameter.

(d) The point $\left(a\sqrt{2}, \dfrac{\pi}{4}\right)$ is on the circle.

21) Sketch the curve with equation $y = \dfrac{ax + b}{(x - c)(x - d)}$

(a) $x = 2$ is an asymptote.   (b) $x = 3$ is an asymptote.

(c) The point $(4, 0)$ is on the curve.   (d) The point $(0, 2)$ is on the curve.

22) Shade the closed area in which $P(x, y)$ can lie.

(a) $y < |x - 1|$.   (b) $y > |x^2 - 2x|$.   (c) $y > 0$.

TYPE V

23) When using polar coordinates, r is taken always to be positive.

24) The curve $r = a \cos k\theta$ always has k loops.

25) If $y = f(x)$ and $y \to \infty$ as $x \to \infty$, $y = x$ is a linear asymptote to the graph of $y = f(x)$.

26) The gradient of $y = |x|$ at the origin is zero.

27) The graph of $|y| = x$ is symmetrical about the x-axis.

MISCELLANEOUS EXERCISE 12

1) Given that $y = \dfrac{4x^2 + 2x + 1}{x^2 - x + 1}$, determine the range of values taken by

y for real values of x.   (JMB)

2) Sketch the curve whose equation is $y = 1 - \dfrac{1}{x + 2}$.   (JMB)p

3) Draw a sketch-graph of the curve whose equation is
$$y = x^2(2 - x)$$
Hence, or otherwise, draw a sketch-graph of the curve whose equation is
$$y^2 = x^2(2 - x)$$
indicating briefly how the form of the curve has been derived.   (C)

4) Find the stationary points on the graphs of

(a) $y = \dfrac{x - 1}{x^2}$ (b) $y = \dfrac{1}{x} - 1 + \ln x$

Sketch the graphs of these functions.

5) Show that $y = x - 6$ is an asymptote of the curve

$$y = \frac{(x-1)(x-3)}{x+2}$$

Sketch the curve showing clearly the other asymptote and the points of intersection of the curve with the coordinate axes.   (U of L)

6) Find the range of values of k for which the function

$$\frac{x^2 - 1}{(x-2)(x+k)}$$

where x is real, takes all real values.   (JMB)

7) The graph of $y = \dfrac{ax^2 + bx + c}{x^2 + qx + r}$ has the lines $y = 2$, $x = 1$ and
$x = 3$ as asymptotes and a turning point at $(0, 1)$. Find the constants a, b, c, q
and r, and show that the graph has a second turning point.
Sketch the graph showing clearly its turning points and its behaviour as it
approaches the asymptotes.   (U of L)

8) Sketch the graph of $y = \dfrac{x^2 - 4x + 1}{x - 4}$ showing clearly the coordinates of
its turning points and its behaviour as it approaches its asymptotes. Find the
range of values of y for which there are no real values of x.
Sketch also the graph of $y = \dfrac{x - 4}{x^2 - 4x + 1}$.   (U of L)

9) Sketch the graph of $y = \dfrac{2(x+6)(x-4)}{(x-6)(x+4)}$ and give the equations of the
asymptotes.   (AEB)'74

10) Given that $f(x) \equiv \dfrac{2x - x^2}{x^2 - 2x - 3}$
(a) state the values of x for which $f(x) = 0$,   and the values of x for which
    $f(x) > 0$,
(b) show that $f(x) \to -1$ as $x \to \pm\infty$,
(c) sketch the graph $y = f(x)$,   showing particularly where the curve crosses
    the x-axis and how it approaches it asymptotes.   (U of L)

11) Show that for real values of x the expression $\dfrac{2x}{x^2 + 4x + 3}$ cannot lie
between certain values. Find these values. Sketch the curve

$$y = \frac{2x}{x^2 + 4x + 3}$$

determining the stationary values of y and the asymptotes of the curve.

(JMB)p

12) Sketch the graphs of the functions $f(x) \equiv \dfrac{4}{x^2(x + 2)}$ and

$g(x) \equiv \dfrac{4}{x(x + 2)}$.

Determine carefully the values of K or the ranges of real values of K for which the number of distinct, finite roots of the equation $f(x) = K$ is (a) three, (b) two, (c) one, (d) zero.

Find the values of x for which $g(x) < f(x)$. (U of L)

13) Show that the straight line $3y + x = 0$ is tangential to the curve $3y(x^2 - 9) = 2(3x + 1)$ shown in the diagram.

Find the area in the fourth quadrant bounded by the straight line, the y-axis and the curve.

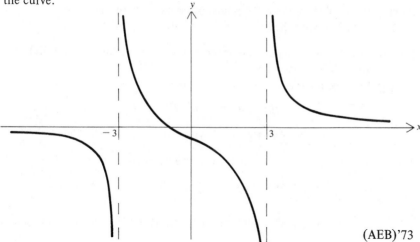

(AEB)'73

14) Find the set of values of y for which x is real when

$$y = \frac{15 + 10x}{4 + x^2}$$

Find also the gradient of this curve when $x = 0$. *Sketch* the curve. (U of L)

15) Show that, if $y = \dfrac{(x + 1)(x - 3)}{x(x - 2)}$ and x is real, y cannot lie between 1 and 4.

Sketch the graph of y against x.

Hence sketch the graph of $y = \left| \dfrac{(x + 1)(x - 3)}{x(x - 2)} \right|$.

16) (a) Sketch the curve $y^2 = (x + 1)^2(3 - x)$ and show that the area enclosed by the loop of the curve is $\frac{256}{15}$.

   (b) Find the area of the sector enclosed by the curve whose equation in polar coordinates is $r = a \sec^2 \dfrac{\theta}{2}$ and the radii $\theta = 0$, $\theta = \alpha$

   where $\alpha < \pi$.   (AEB)'72

17) Sketch the curve $r = a(1 + \sin \theta)$.   Express that part of the area above the initial line as a fraction of the total area bounded by the curve.   (AEB)'67

18) Sketch the curve whose polar equation is

$$r \cos^2 \theta = 1$$

Find the area contained between the curve and the straight lines $\theta = \pm \dfrac{\pi}{4}$.

(AEB)'72

19) Sketch the curve whose polar equation is $r = 1 + 2 \cos 2\theta$ for the range $0 \leqslant \theta \leqslant \pi$.
Prove that the values of θ at the points of intersection of the curve
$r = 1 + 2 \cos 2\theta$ with the line $r = \frac{1}{2} \csc \theta$ are given by $2 \sin 3\theta = 1$.

20) (a) Sketch the curve whose polar equation is $r = a \sin 3\theta$ and find the area of one of its loops.

   (b) Find the polar equation corresponding to

$$(x^2 + y^2)^2 = a^2(x^2 - y^2)$$

   Hence, or otherwise, sketch the curve and calculate the enclosed area.

21) Show by means of a sketch that the curve whose polar equation is

$r = 2 \cos \theta \sin \left(\theta - \dfrac{\pi}{4} \right)$ describes a loop for values of θ in the range

$\dfrac{\pi}{4} \leqslant \theta \leqslant \dfrac{\pi}{2}$.   Show that the point on the loop farthest from the origin is at a

distance from the origin of $1 - \dfrac{1}{\sqrt{2}}$.   (AEB)'74

22) If a curve C has the equation $5x^2 + 4xy + 2y^2 = 1$,   find, by putting $x = r \cos \theta$, $y = r \sin \theta$ or otherwise, the greatest and least distances of points of C from the origin 0.   (O)

CHAPTER 13

VECTORS

VECTORS

Readers will by now appreciate that two and two do not always make four.

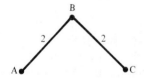

For example, consider three points A, B and C. If B is 2 cm from A and C is 2 cm from B, then C is not, in general, 4 cm from A.

AB, BC and AC are displacements. Each has a magnitude (e.g. the magnitude of AB is the distance from A to B) and is related to a definite direction in space. The displacement from A to B $(\overrightarrow{AB})$ followed by the displacement from B to C $(\overrightarrow{BC})$ is equivalent to the displacement from A to C $(\overrightarrow{AC})$, i.e. $\overrightarrow{AB}$ together with $\overrightarrow{BC}$ is equivalent to $\overrightarrow{AC}$, which is written

$$\overrightarrow{AB} + \overrightarrow{BC} \equiv \overrightarrow{AC}.$$

There are many other quantities which behave in the same way as these displacements and they can all be represented by vectors.
A vector quantity is one that has a magnitude and is also related to a definite direction in space.

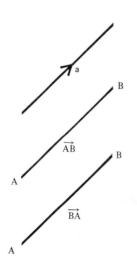

It follows that we may represent a vector by a section of a straight line where the length of the line represents the magnitude of the vector, and the direction of the line (together with the arrow) represents the direction of the vector.

Such vectors are denoted by bold type e.g. **a** (when hand written, by a̲).

Alternatively we may represent a vector by the straight line joining two points A and B, and denote the vector in the direction A to B by $\overrightarrow{AB}$ (or **AB**) and the vector in the opposite direction, B to A, by $\overrightarrow{BA}$ (or **BA**).

Scalars

A scalar quantity is one that is fully defined by magnitude alone and can therefore be represented completely by a real number. For example, length is a scalar quantity as a real number specifies the length of an object completely. Scalar quantities can be compounded using the familiar laws of algebra, and in this case two and two *do* make four.

MODULUS OF A VECTOR

The modulus of a vector **a** is the magnitude of **a**, i.e. the length of the line representing **a**. The modulus of **a** is written $|\mathbf{a}|$ or a.

EQUAL VECTORS

Two vectors are equal if they have the same magnitude and the same direction

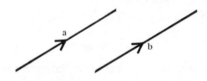

i.e. $\mathbf{a} = \mathbf{b} \iff \begin{cases} |\mathbf{a}| = |\mathbf{b}| \\ \text{the direction of } \mathbf{a} \text{ is the same as the direction of } \mathbf{b} \end{cases}$

It follows from this that a vector may be represented by any line of the right length and direction, i.e. the location of the line in space does not matter.

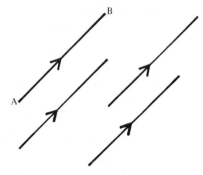

Thus any of the lines in the diagram may represent the vector $\overrightarrow{AB}$.

NEGATIVE VECTORS

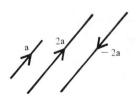

If two vectors **a** and **b** have the same magnitude but opposite directions then we say that

$$\mathbf{a} = -\mathbf{b}$$

i.e. $-\mathbf{a}$ is a vector of magnitude $|\mathbf{a}|$ and direction opposite to that of **a**.

MULTIPLICATION OF A VECTOR BY A SCALAR

If λ is a positive real number then

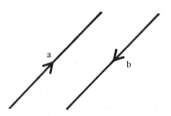

$\lambda\mathbf{a}$ is a vector of magnitude $\lambda|\mathbf{a}|$ and in the same direction as **a**. It also follows that $-\lambda\mathbf{a}$ is a vector of magnitude $\lambda|\mathbf{a}|$ and with direction opposite to that of **a**.

ADDITION OF VECTORS

All vectors are compounded in the same way as the displacements with which we began this chapter. The addition of vectors is defined as follows.

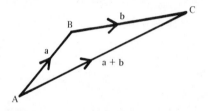

If the sides **AB** and **BC** of triangle ABC represent the vectors **a** and **b**, the third side **AC** represents the vector sum, or resultant, of **a** and **b** and is denoted by $\mathbf{a} + \mathbf{b}$.

Note that **a** and **b** have directions that go in one sense round the triangle (clockwise in the diagram) and that their resultant, **a** + **b**, has a direction in the opposite sense (anticlockwise in the diagram). This is known as the triangle law for addition of vectors. It can be extended to cover the addition of more than two vectors.

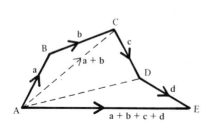

Consider a polygon whose sides AB, BC, CD, DE represent the vectors **a, b, c, d** respectively.
Using the triangle law gives

$$\overrightarrow{AC} = a + b$$

so

$$\overrightarrow{AD} = \overrightarrow{AC} + \overrightarrow{CD} = (a + b) + c$$

and

$$\overrightarrow{AE} = \overrightarrow{AD} + \overrightarrow{DE} = (a + b + c) + d$$

i.e. if vectors **a, b, c, d**, . . . are represented by the sides of an open polygon taken in order (i.e. the direction of the vectors all follow the same sense round the polygon) then the line that closes the polygon, in the opposite sense, represents the vector **a** + **b** + **c** + **d** + ...
The diagram below shows that the operation of addition on vectors is commutative,
i.e. the order in which the addition is performed does not matter.

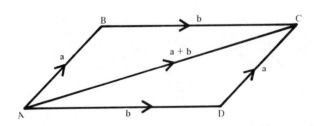

ABCD is a parallelogram, therefore $\overrightarrow{AB}$ and $\overrightarrow{DC}$ both represent **a**
similarly $\overrightarrow{BC}$ and $\overrightarrow{AD}$ both represent **b**

From △ABC, $\overrightarrow{AC} = a + b$

From △ADC, $\overrightarrow{AC} = b + a$

i.e. $a + b \equiv b + a$

EXAMPLES 13a

1) In the triangle ABC, D is the midpoint of AB. $\overrightarrow{AB}$ represents **a** and $\overrightarrow{BC}$ represents **b**.

Express in terms of **a** and **b** the vectors $\overrightarrow{CA}$ and $\overrightarrow{DC}$.

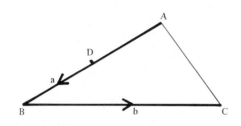

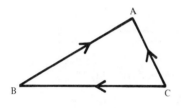

Now
$$\overrightarrow{CA} = \overrightarrow{CB} + \overrightarrow{BA}$$
$$\overrightarrow{BA} = -\overrightarrow{AB} = -\mathbf{a}$$
and
$$\overrightarrow{CB} = -\overrightarrow{BC} = -\mathbf{b}$$

Therefore
$$\overrightarrow{CA} = -\mathbf{a} - \mathbf{b} = -(\mathbf{a} + \mathbf{b})$$
$$\overrightarrow{DC} = \overrightarrow{DB} + \overrightarrow{BC}$$
$$= \tfrac{1}{2}\overrightarrow{AB} + \overrightarrow{BC}$$
$$= \tfrac{1}{2}\mathbf{a} + \mathbf{b}$$

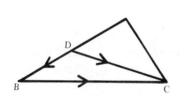

2) Show that $\overrightarrow{AB} + \overrightarrow{AC} = 2\overrightarrow{AD}$ where ABC is a triangle and D is the midpoint of BC.

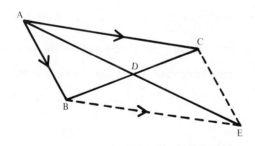

Completing the parallelogram ABEC we see that $\overrightarrow{BE} = \overrightarrow{AC}$.

Therefore $\overrightarrow{AB} + \overrightarrow{AC} = \overrightarrow{AB} + \overrightarrow{BE} = \overrightarrow{AE}$

The diagonals of a parallelogram bisect each other.

Therefore $\qquad\qquad\qquad \overrightarrow{AE} = 2\overrightarrow{AD}$

$\Rightarrow \qquad\qquad\qquad \overrightarrow{AB} + \overrightarrow{AC} = 2\overrightarrow{AD}$

3) If O, A, B, C are four points such that

$$\overrightarrow{OA} = 10\mathbf{a}, \quad \overrightarrow{OB} = 5\mathbf{b}, \quad \overrightarrow{OC} = 4\mathbf{a} + 3\mathbf{b}$$

show that A, B and C are collinear.

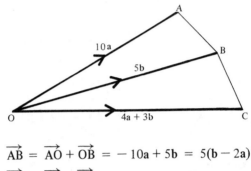

$$\overrightarrow{AB} = \overrightarrow{AO} + \overrightarrow{OB} = -10\mathbf{a} + 5\mathbf{b} = 5(\mathbf{b} - 2\mathbf{a})$$
$$\overrightarrow{BC} = \overrightarrow{BO} + \overrightarrow{OC} = -5\mathbf{b} + 4\mathbf{a} + 3\mathbf{b}$$
$$= 4\mathbf{a} - 2\mathbf{b}$$
$$= -2(\mathbf{b} - 2\mathbf{a})$$

i.e. $\overrightarrow{BC} = -\frac{2}{5}\overrightarrow{AB}$, so $\overrightarrow{AB}$ and $\overrightarrow{BC}$ are in opposite directions, i.e. AB and CB are parallel. Hence since B is a common point A, B and C are collinear.

EXERCISE 13a

1) ABCD is a quadrilateral. Find the single vector which is equivalent to:

(a) $\overrightarrow{AB} + \overrightarrow{BC}$ (b) $\overrightarrow{BC} + \overrightarrow{CD}$ (c) $\overrightarrow{AB} + \overrightarrow{BC} + \overrightarrow{CD}$ (d) $\overrightarrow{AB} + \overrightarrow{DA}$.

2) ABCDEF is a regular hexagon in which $\overrightarrow{BC}$ represents $\mathbf{b}$ and $\overrightarrow{FC}$ represents $2\mathbf{a}$.
Express the vectors $\overrightarrow{AB}$, $\overrightarrow{CD}$ and $\overrightarrow{BE}$ in terms of $\mathbf{a}$ and $\mathbf{b}$.

3) Draw diagrams representing the following vector equations.

(a) $\overrightarrow{AB} - \overrightarrow{CB} = \overrightarrow{AC}$ (b) $\overrightarrow{AB} = 2\overrightarrow{PQ}$ (c) $\overrightarrow{AB} + \overrightarrow{BC} = 3\overrightarrow{AD}$

(d) $\mathbf{a} + \mathbf{b} = -\mathbf{c}$

4) If A, B, C, D are four points such that

$$\overrightarrow{AB} = \overrightarrow{DC} \quad \text{and} \quad \overrightarrow{BC} + \overrightarrow{DA} = 0$$

prove that ABCD is a parallelogram.

5) O, A, B, C, D are five points such that $\overrightarrow{OA} = $ **a**, $\overrightarrow{OB} = $ **b**, $\overrightarrow{OC} = $ **a** + 2**b**, $\overrightarrow{OD} = $ 2**a** − **b**.

Express $\overrightarrow{AB}, \overrightarrow{BC}, \overrightarrow{CD}, \overrightarrow{AC}, \overrightarrow{BD}$ in terms of **a** and **b**.

6) If **a** and **b** are represented by two adjacent sides of a regular hexagon, find, in terms of **a** and **b**, the vectors represented by the remaining sides taken in order.

7)

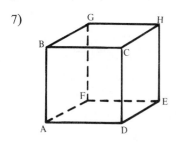

If **a, b, c** are represented by the edges $\overrightarrow{AB}, \overrightarrow{AD}, \overrightarrow{AF}$ of the cube in the diagram, find, in terms of **a**, **b** and **c** the vectors represented by the remaining edges.

8) If O, A, B, C are four points such that $\overrightarrow{OA} = $ **a**, $\overrightarrow{OB} = $ 2**a** − **b**, $\overrightarrow{OC} = $ **b**. Show that A, B and C are collinear.

9)

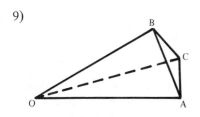

If OABC is a tetrahedron and $\overrightarrow{OA} = $ **a**, $\overrightarrow{OB} = $ **b**, $\overrightarrow{OC} = $ **c**, find $\overrightarrow{AC}, \overrightarrow{AB}, \overrightarrow{CB}$ in terms of **a**, **b**, **c**.

10) For the cube defined in Question 7, find the vectors $\overrightarrow{BE}, \overrightarrow{GD}, \overrightarrow{AH}, \overrightarrow{FC}$ in terms of **a**, **b**, and **c**.

11) If D is the midpoint of the edge OB of the tetrahedron in Question 9, find $\overrightarrow{CD}$ in terms of **b** and **c**. Also, if E is the midpoint of CB, find $\overrightarrow{AE}$ in tefms of **a**, **b** and **c**.

POSITION VECTORS

In general a vector has no specific location in space. There are some quantities that compound in the same way as vectors but which are constrained to a specific position.

Consider, for example, the position of a point A relative to a fixed origin O.

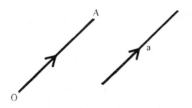

$\overrightarrow{OA}$ is called the position vector of A relative to O. This displacement is unique and *cannot* be represented by the line in the diagram denoting **a** even though **a** = $\overrightarrow{OA}$.

Vectors such as $\overrightarrow{OA}$, representing quantities that have a specific location, are called *tied vectors* or *position vectors*.

Vectors such as **a**, representing quantities not related to a fixed position, are known as *free vectors*.

LOCATION OF A POINT IN A PLANE

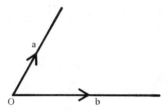

If **a** and **b** are *non-parallel* free vectors and O is a fixed point, then there is one and only one plane containing O and the vectors **a** and **b**. i.e. the vectors **a**, **b** and the fixed point O define a particular plane.

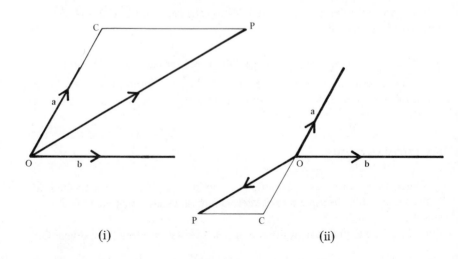

(i)   (ii)

If P is any other point in the plane,

$\overrightarrow{OP}$ is the position vector of P.

If C is the point such that $\overrightarrow{OC}$ is parallel to **a**,

i.e. $$\overrightarrow{OC} = \lambda\mathbf{a}$$

and $\overrightarrow{CP}$ is parallel to **b**,

i.e. $$\overrightarrow{CP} = \mu\mathbf{b}$$

where λ and μ are scalars (both having positive values in diagram (i) and both having negative values in diagram (ii)),

then
$$\overrightarrow{OP} = \overrightarrow{OC} + \overrightarrow{CP}$$
$$= \lambda\mathbf{a} + \mu\mathbf{b}$$

i.e. the position vector of any point P can be expressed in terms of **a** and **b**. Thus **a**, **b** and O form a 'frame of reference' for the position vector of any point in the plane.

The vectors **a** and **b** are known as base vectors.

It also follows that *any vector* equal to $\overrightarrow{OP}$ can be expressed in terms of **a** and **b**,

i.e. any vector (tied or free) that is parallel to the plane can be expressed in the form $\lambda\mathbf{a} + \mu\mathbf{b}$, where λ and μ are scalars.

ANGLE BETWEEN TWO VECTORS

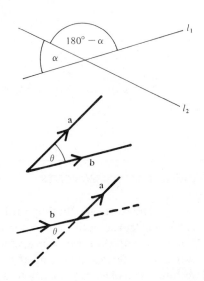

The angle between two lines l_1 and l_2 is ambiguous as it may be α or $180° - \alpha$.

But the angle between two vectors **a** and **b** is unique. It is the angle between their directions when those directions *both converge* or *both diverge* from a point, as illustrated in the diagrams.

EXAMPLES 13b

1) **a** and **b** are base vectors, where $|\mathbf{a}| = 3$, $|\mathbf{b}| = 5$ and $\dfrac{\pi}{3}$ is the angle

between **a** and **b**. The position vector of a point P is $3\mathbf{a} - 2\mathbf{b}$ relative to an origin O. Find $|\overrightarrow{OP}|$ and the angle between $\overrightarrow{OP}$ and **a**.

Drawing OQ parallel to **a**
such that $\quad \overrightarrow{OQ} = 3\mathbf{a}$
then QP parallel to **b**
such that $\quad \overrightarrow{QP} = -2\mathbf{b} \quad$ gives

$$\overrightarrow{OP} = 3\mathbf{a} - 2\mathbf{b}.$$

Now $\quad |\overrightarrow{OQ}| = 3|\mathbf{a}| = 9,$

and $\quad |\overrightarrow{QP}| = 2|\mathbf{b}| = 10.$

Using the cosine formula in $\triangle OPQ$ gives

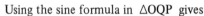

$$OP^2 = 81 + 100 - 2(9)(10) \cos \frac{\pi}{3}$$

$$= 91$$

Therefore $\quad OP = \sqrt{91}$

The angle between $\overrightarrow{OP}$ and **a** is α.

Using the sine formula in $\triangle OQP$ gives

$$\frac{\sin \alpha}{10} = \frac{\sin \dfrac{\pi}{3}}{\sqrt{91}}$$

$\Rightarrow \qquad\qquad\qquad \sin \alpha = 0.9078$

$\Rightarrow \qquad\qquad\qquad \alpha = 65.20°$

So $\quad |\overrightarrow{OP}| = \sqrt{91} \quad$ and $\overrightarrow{OP}$ is inclined at an angle of $65.20°$ to **a**.

CARTESIAN COMPONENTS

Calculations are greatly simplified when the base vectors used are perpendicular and both have a magnitude of unity.
Note that *any* vector whose magnitude is one unit is called a *unit vector*.
A frame of reference consisting of an origin O and a pair of perpendicular vectors has an obvious similarity to the Cartesian xy plane. We formally relate the two by taking as our base vectors the unit vector **i** in the positive direction of the x-axis and the unit vector **j** in the positive direction of the y-axis.

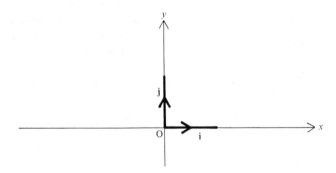

Thus if **A** is the point with coordinates $(3, 2)$

$$ON = 3, \quad so \quad \overrightarrow{ON} = 3i$$

$$NA = 2, \quad so \quad \overrightarrow{NA} = 2j$$

Therefore $\qquad \overrightarrow{OA} = \overrightarrow{ON} + \overrightarrow{NA} = 3i + 2j$

i.e. the position vector of $A(3, 2)$ is $\quad r = 3i + 2j$.

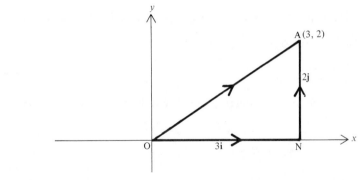

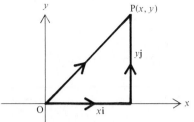

Similarly we see that $\quad \overrightarrow{OP} = xi + yj$
where P is the point (x, y)

so the position vector of any point $P(x, y)$ is $xi + yj$ and the converse of this
statement is also true, i.e.

the position vector $\quad \overrightarrow{OP} = xi + yj \Longleftrightarrow$ P is the point (x, y).

Note that, unless we are told that a vector is a position vector or we are using
vectors to represent a quantity that has a specific position, we may assume that
a given vector is free,

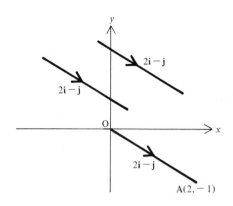

i.e. if a vector is given simply as
$a = 2i - j$ then any line section
in the xy plane of the same
magnitude and direction can represent
$2i - j$.

But if we are told that a is a position
vector $2i - j$ then only the line
$\overrightarrow{OA}$ can represent a.

Now $|\overrightarrow{OA}| = $ length of OA

$$= \sqrt{[(2)^2 + (-1)^2]}$$

i.e. $|2i - j| = \sqrt{[(2)^2 + (-1)^2]}$

$$= \sqrt{5}$$

and in general,

$$|ai + bj| = \sqrt{(a^2 + b^2)}$$

Alternative notation: the vector $ai + bj$ can be denoted by $\begin{pmatrix} a \\ b \end{pmatrix}$ which is known as a column vector or column matrix.

EXAMPLES 13b (continued)

2) Find $\overrightarrow{AB}$ where A is the point $(2, -1)$ and B is the point $(-1, 3)$

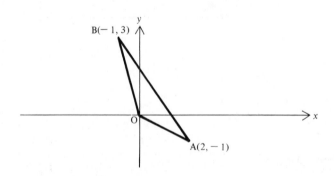

$$\overrightarrow{AB} = \overrightarrow{OB} + \overrightarrow{AO} = \overrightarrow{OB} - \overrightarrow{OA}$$

$$\overrightarrow{OA} = 2i - j \quad \text{and} \quad \overrightarrow{OB} = -i + 3j$$

Therefore $\overrightarrow{AB} = (-i + 3j) - (2i - j)$

$$= -3i + 4j$$

Note that for any two points $A(x_1, y_1)$ and $B(x_2, y_2)$,

$$\overrightarrow{AB} = \overrightarrow{OB} - \overrightarrow{OA}$$

$$= (x_2 - x_1)i + (y_2 - y_1)j$$

and $\quad |\overrightarrow{AB}| = \text{length of AB} = \sqrt{[(x_2 - x_1)^2 + (y_2 - y_1)^2]}$

3) Find in terms of **a** and **b** the position vector of the point dividing the line AB in the ratio $m:n$, where **a** is the position vector of A and **b** is the position vector of B with respect to an origin O.

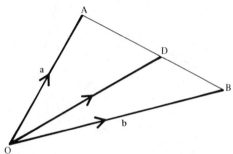

If D is the point on AB such that $\dfrac{AD}{DB} = \dfrac{m}{n}$

then $\qquad AD = \left(\dfrac{m}{m+n}\right) AB \quad$ so $\quad \overrightarrow{AD} = \left(\dfrac{m}{m+n}\right) \overrightarrow{AB}$

Now $\qquad \overrightarrow{AB} = b - a, \quad$ so $\quad \overrightarrow{AD} = \left(\dfrac{m}{m+n}\right)(b - a)$

$$\overrightarrow{OD} = \overrightarrow{OA} + \overrightarrow{AD}$$

$$= a + \left(\dfrac{m}{m+n}\right)(b - a)$$

$$= \dfrac{na + mb}{m+n}$$

This example gives an important result which may be quoted.
To express this result in Cartesian components, let A be the point (x_1, y_1) and B the point (x_2, y_2) so that
$a = x_1 i + y_1 j \quad$ and $\quad b = x_2 i + y_2 j$.

$\Rightarrow \qquad \overrightarrow{OD} = \dfrac{n(x_1 i + y_1 j) + m(x_2 i + y_2 j)}{m+n}$

$$= \left(\dfrac{nx_1 + mx_2}{m+n}\right) i + \left(\dfrac{ny_1 + my_2}{m+n}\right) j$$

i.e.   D is the point $\left(\dfrac{nx_1 + mx_2}{m + n} , \dfrac{ny_1 + my_2}{m + n} \right)$,

confirming, by vector methods, a result obtained on p. 355.

The Position Vector of the Centroid of a Triangle

Consider a triangle ABC where a, b, c are the position vectors of A, B, C respectively.

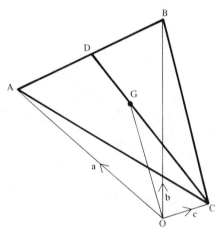

The centroid, G, of this triangle is the point of intersection of its medians, i.e. G is the point which divides CD in the ratio $2:1$, D being the midpoint of AB.

Hence $\overrightarrow{OG} = \dfrac{2\overrightarrow{OD} + \overrightarrow{OC}}{3}$

and $\overrightarrow{OD} = \dfrac{a + b}{2}$

Therefore $\overrightarrow{OG} = \tfrac{1}{3}(a + b + c)$

EXERCISE 13b

1) If $a = 2i + j$, $b = i - 2j$, express, in terms of i and j:
(a) $a + b$ (b) $a - b$ (c) $2a + b$ (d) $-3a + 4b$

2) If the position vector of A is $i - 3j$ and the position vector of B is $2i + 5j$, find:

(a) $|\overrightarrow{AB}|$,

(b) the position vector of the midpoint of AB,

(c) the position vector of the point dividing AB in the ratio $1:3$.

3) If $a = i - 3j$, $b = -2i + 5j$, $c = 3i - j$ find:
(a) $a + b + c$ in terms of i and j,
(b) $|a - b|$,
(c) the position vector of the point dividing the line AC in the ratio $-2:3$, where a is position vector of A and c is the position vector of C.

4) Find a vector which is parallel to the line $y = 2x - 1$.

5) Show that $i + 2j$ is the position vector of a point on the line $x - 2y + 3 = 0$.

6) Find the angle that the vector $3i - 2j$ makes with the x-axis.

7) Show that $\begin{pmatrix} a \\ b \end{pmatrix}$ and $\begin{pmatrix} -b \\ a \end{pmatrix}$ are perpendicular.

8) Show that, for all values of t, the point whose position vector is $r = ti + (2t - 1)j$ lies on the line whose equation is $y = 2x - 1$.

9) Show that
the position vector of P is $t^2 i + 2tj \iff P$ is a point on $y^2 = 4x$.

10) a and b are base vectors where $|a| = 2$, $|b| = 5$ and the angle between a and b is $45°$. Draw diagrams showing the points whose position vectors, relative to an origin O, are:
(a) $a - b$ (b) $a + b$ (c) $2a - b$ (d) $-a + 2b$ (e) $4a + 3b$

SCALAR PRODUCT (OR DOT PRODUCT)

The scalar product of two vectors a and b is the operation which is written as $a.b$ and which is defined as $|a||b| \cos \theta$ where θ is the angle between a and b.
(There is another product operation on vectors, called the vector product which is written as $a \times b$, so the 'cross' must not be used for the scalar product.)

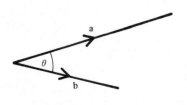

So for any two vectors a and b

$$a.b = ab \cos \theta$$

As $\quad ab \cos \theta \equiv ba \cos \theta$

it follows that

$$a.b \equiv b.a$$

For *parallel vectors*

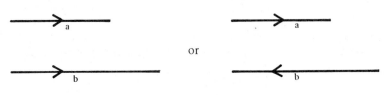

or

$$\mathbf{a.b} = ab \cos 0 \quad \text{or} \quad ab \cos \pi$$

i.e. $\quad\quad\quad \mathbf{a.b} = ab \quad$ for vectors in the same direction

$$\mathbf{a.b} = -ab \quad \text{for vectors in opposite directions}$$

For *perpendicular vectors*

$$\mathbf{a.b} = ab \cos \frac{\pi}{2} = 0$$

In the special case of the Cartesian base vectors in the xy plane we have

$$\mathbf{i.i} = \mathbf{j.j} = 1$$

and $\quad\quad\quad \mathbf{i.j} = \mathbf{j.i} = 0$

The scalar product is distributive across addition

i.e. $\quad\quad\quad \mathbf{a.(b+c)} \equiv \mathbf{a.b} + \mathbf{a.c}$

This result is proved in *Pure Mathematics II*, so for the present we will accept its validity and use it to calculate the scalar product of vectors expressed in terms of $\mathbf{i}$ and $\mathbf{j}$.

If $\quad\quad\quad \mathbf{a} = 2\mathbf{i} - 5\mathbf{j} \quad \text{and} \quad \mathbf{b} = 3\mathbf{i} + 2\mathbf{j}$

then $\quad\quad\quad \mathbf{a.b} = (2\mathbf{i} - 5\mathbf{j}).(3\mathbf{i} + 2\mathbf{j})$

$$= 6\mathbf{i.i} - 15\mathbf{j.i} + 4\mathbf{i.j} - 10\mathbf{j.j}$$

But $\quad\quad\quad \mathbf{i.i} = \mathbf{j.j} = 1 \quad \text{and} \quad \mathbf{i.j} = 0$

so $\quad\quad\quad \mathbf{a.b} = 6 - 10 = -4$

In general if $\quad\quad\quad \mathbf{a} = x_1\mathbf{i} + y_1\mathbf{j} \quad \text{and} \quad \mathbf{b} = x_2\mathbf{i} + y_2\mathbf{j}$

then $$\mathbf{a.b} = (x_1\mathbf{i}+y_1\mathbf{j}).(x_2\mathbf{i}+y_2\mathbf{j})$$
$$= x_1x_2\mathbf{i.i}+y_1x_2\mathbf{j.i}+x_1y_2\mathbf{i.j}+y_1y_2\mathbf{j.j}$$

i.e. $$\mathbf{a.b} = x_1x_2 +y_1y_2$$

so $$(2\mathbf{i}-\mathbf{j}).(5\mathbf{i}+3\mathbf{j}) = (2)(5)+(-1)(3) = 7$$

EXAMPLES 13c

1) Use vector methods to find the angle between

$$3\mathbf{i}-\mathbf{j} \quad \text{and} \quad -4\mathbf{i}+6\mathbf{j}$$

If $\mathbf{a} = 3\mathbf{i}-\mathbf{j}$ and $\mathbf{b} = -4\mathbf{i}+6\mathbf{j}$

then

$$\mathbf{a.b} = (3\mathbf{i}-\mathbf{j}).(-4\mathbf{i}+6\mathbf{j}) = -18$$

but

$$\mathbf{a.b} = ab\cos\theta \quad \text{where} \quad a = \sqrt{10}$$
$$\text{and} \quad b = \sqrt{52}$$

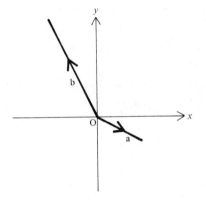

Therefore

$$-18 = \sqrt{10}\sqrt{52}\cos\theta$$

$$\Rightarrow \cos\theta = \frac{-9}{\sqrt{130}}$$

$$\theta = \arccos\left(\frac{-9}{\sqrt{130}}\right)$$

2) Simplify (a) $(\mathbf{a}-\mathbf{b}).(\mathbf{a}+\mathbf{b})$
     (b) $(\mathbf{a}+\mathbf{b}).\mathbf{c}-(\mathbf{a}+\mathbf{c}).\mathbf{b}$

(a) $$(\mathbf{a}-\mathbf{b}).(\mathbf{a}+\mathbf{b}) = \mathbf{a.a}-\mathbf{b.a}+\mathbf{a.b}-\mathbf{b.b}$$

but $$\mathbf{a.b} = \mathbf{b.a} \quad \text{hence} \quad \mathbf{a.b}-\mathbf{b.a} = 0$$

also $$\mathbf{a.a} = a^2 \quad \text{and} \quad \mathbf{b.b} = b^2$$

Therefore $$(\mathbf{a}-\mathbf{b}).(\mathbf{a}+\mathbf{b}) = a^2-b^2$$

(b) $$(\mathbf{a}+\mathbf{b}).\mathbf{c}-(\mathbf{a}+\mathbf{c}).\mathbf{b}$$

$$= \mathbf{a.c}+\mathbf{b.c}-\mathbf{a.b}-\mathbf{c.b}$$

$$= \mathbf{a.c}-\mathbf{a.b}$$

$$= \mathbf{a.(c}-\mathbf{b})$$

EXERCISE 13c

1) Evaluate:
(a) $(i-j).(2i+5j)$ (b) $i.(2i-7j)$ (c) $(-3i+j).(3i+2j)$

2) Find the angle between the following pairs of vectors:
(a) $-2i+5j, i+j$ (b) $-i-3j, i+2j$
(c) $4i+j, i$ (d) $-7i+5j, i-3j$

3) Simplify:
(a) $(a-b).a$ (b) $(a-b).(a-b)$
(c) $(a+b).b-(a+b).a$ (d) $(a+b).c-(a-b).c$

4) Given that a and b are perpendicular, simplify:
(a) $a.b$ (b) $(a-b).b$ (c) $(a+b).a$ (d) $(a-b).(2a+b)$

5) Given that $a=2b$, simplify:
(a) $a.b$ (b) $(a-3b).b$

6) Which of the following statements are true?
(a) $a=b \iff |a|=|b|$
(b) $a.b=ab \iff a=\lambda^2 b$
(c) A is the point $(3,2) \iff \overrightarrow{OA}=3i+2j$
(d) $\overrightarrow{OP}=ti+2tj \iff y=2x$

CHAPTER 14

COMPLEX NUMBERS

IMAGINARY NUMBERS

In all previous work it has been assumed that when any number, k say, is squared the result is either positive or zero, i.e. $k^2 \geqslant 0$. Such numbers are called *real* numbers.

We have already met equations such as $x^2 = -1$ whose roots are clearly not real since when squared they give -1 as the result. To work with equations of this type we need another category of numbers, namely the set of numbers whose squares are negative real numbers. Members of this set are called *imaginary* numbers, examples being $\sqrt{-1}$, $\sqrt{-7}$, $\sqrt{-20}$.

A general member of this set is $\sqrt{-n^2}$ where n is real.

But
$$\sqrt{-n^2} \equiv \sqrt{n^2 \times -1}$$
$$\equiv \sqrt{n^2} \times \sqrt{-1}$$
$$\equiv ni \quad \text{where} \quad i = \sqrt{-1}$$

So we see that every imaginary number can be written in the form ni where n is real and $i = \sqrt{-1}$ (or sometimes nj where $j = \sqrt{-1}$)

e.g.
$$\sqrt{-16} = 4i, \qquad \sqrt{-3} = i\sqrt{3}$$

Imaginary numbers can be added to and subtracted from other imaginary numbers,

e.g.
$$2i + 5i = 7i$$
$$\sqrt{7}i - i = (\sqrt{7} - 1)i$$

The product or quotient of two imaginary numbers is real,

e.g. $\qquad\qquad 2i \times 5i = 10i^2$

But $\quad i^2 = -1 \quad$ since $\quad i = \sqrt{-1}$

Hence $\qquad\qquad 2i \times 5i = -10$

Also $\qquad\qquad 6i \div 3i = 2$

Powers of i can be simplified,

e.g.
$$i^3 = (i^2)(i) = -i$$
$$i^4 = (i^2)^2 = (-1)^2 = 1$$
$$i^5 = (i^4)i = i$$
$$i^{-1} = \frac{i}{i^2} = \frac{i}{-1} = -i$$

COMPLEX NUMBERS

When a real number and an imaginary number are added or subtracted, the expression so formed, which cannot be simplified, is called a complex number, e.g. $2 + 3i, 4 - 7i, -1 + 4i$. A general complex number can be written in the form $a + bi$ where a and b can have any real value including zero.
If $\quad a = 0 \quad$ we have numbers of the form bi, i.e. imaginary numbers.
If $\quad b = 0 \quad$ we have numbers of the form a, i.e. real numbers.
Therefore *the field of complex numbers includes the real number set and the imaginary number set.*

OPERATIONS ON COMPLEX NUMBERS

Addition and Subtraction

Real terms and imaginary terms are compounded in two separate groups,

e.g. $\qquad (2 + 3i) + (4 - i) = (2 + 4) + (3i - i)$
$$= 6 + 2i$$
and $\qquad (4 - 2i) - (3 + 5i) = (4 - 3) - (2i + 5i)$
$$= 1 - 7i$$

Multiplication

The distributive law of multiplication applied to two complex numbers gives their product,

e.g.
$$(2 + 3i)(4 - i) = 8 - 2i + 12i - 3i^2$$
$$= 8 + 10i - 3(-1)$$
$$= 11 + 10i$$

and
$$(2 + 3i)(2 - 3i) = 4 - 6i + 6i - 9i^2$$
$$= 4 + 9$$
$$= 13$$

Note that the product here is a real number. This is because of the special form of the given complex numbers, $2 \pm 3i$.

Any pair of complex numbers of the form $a \pm bi$ have a product which is real, since

$$(a + bi)(a - bi) \equiv a^2 - abi + abi - b^2i^2$$
$$\equiv a^2 + b^2$$

Such complex numbers are said to be *conjugate* and each is the conjugate of the other.

Thus $4 + 5i$ and $4 - 5i$ are conjugate complex numbers and $4 + 5i$ is the conjugate of $4 - 5i$.

Note. If $a + bi$ is denoted by z then its conjugate, $a - bi$, is denoted by $\bar{z}$.

Division

Direct division by a complex number cannot be carried out because the denominator is made up of two independent terms. This difficulty can be overcome by making the denominator real, a process called 'realising the denominator'.

This can be done by using the product property of conjugate complex numbers. Division can therefore be carried out as follows:

$$\frac{2 + 9i}{5 - 2i} = \frac{(2 + 9i)(5 + 2i)}{(5 - 2i)(5 + 2i)}$$

(multiplying numerator and denominator by the conjugate of the denominator)

$$= \frac{10 + 49i + 18i^2}{25 - 4i^2}$$

$$= \frac{-8 + 49i}{29}$$

$$= -\frac{8}{29} + \frac{49}{29}i$$

(**Note.** The real term is given first even when it is negative.)

THE ZERO COMPLEX NUMBER

A complex number is zero if and only if the real term and the imaginary term are each zero,

i.e. $\qquad X + Yi = 0 \iff X = 0 \quad and \quad Y = 0$

EQUAL COMPLEX NUMBERS

Now consider the case

$$a + bi = c + di \qquad\qquad [1]$$

or $\qquad\qquad (a + bi) - (c + di) = 0 \qquad\qquad [2]$

[2] gives $\qquad\qquad (a - c) + (b - d)i = 0$

Hence $\qquad\qquad a - c = 0 \quad and \quad b - d = 0$

i.e. $\qquad\qquad a = c \quad and \quad b = d$

Thus two complex numbers are equal if and only if the real terms and the imaginary terms are separately equal.

Writing $\text{Re}\,(a + bi)$ to indicate the real part of $a + bi$ (i.e. a) and $\text{Im}\,(a + bi)$ to indicate the imaginary part (i.e. bi) we have:

$$a + bi = c + di \iff \begin{cases} \text{Re}(a + bi) = \text{Re}(c + di) \\ \text{Im}(a + bi) = \text{Im}(c + di) \end{cases}$$

A complex number equation is therefore equivalent to two separate equations. This property provides an alternative method for division by a complex number, e.g. to divide $3 - 2i$ by $5 + i$ we say,

let $\qquad\qquad \dfrac{3 - 2i}{5 + i} = p + qi$

hence $\qquad\qquad 3 - 2i = (p + qi)(5 + i)$

$$= 5p + pi + 5qi + qi^2$$

$$= (5p - q) + (p + 5q)i$$

Equating real and imaginary parts gives:

$$\begin{cases} 3 = 5p - q \\ -2 = p + 5q \end{cases}$$

Solving these equations simultaneously gives $p = \frac{1}{2}, \quad q = -\frac{1}{2}.$

Thus $\qquad\qquad (3 - 2i) \div (5 + i) = \frac{1}{2} - \frac{1}{2}i$

The method of equating real and imaginary parts of a complex equation also provides one way of determining the square root of a complex number,

e.g. to find $\sqrt{15 + 8i}$ we say,

let $\qquad\qquad\qquad\qquad \sqrt{15 + 8i} = a + bi \qquad\qquad$ where a and b are real

$\Rightarrow \qquad\qquad\qquad\qquad 15 + 8i = (a + bi)^2$

$$= a^2 - b^2 + 2abi$$

Equating real and imaginary parts gives:

$$a^2 - b^2 = 15 \qquad\qquad\qquad\qquad [1]$$

and $\qquad\qquad\qquad\qquad 2ab = 8 \qquad\qquad\qquad\qquad\qquad [2]$

Using $\quad b = \dfrac{4}{a} \quad$ in [1] gives

$$a^2 - \frac{16}{a^2} = 15$$

$\Rightarrow \qquad\qquad\qquad a^4 - 15a^2 - 16 = 0$

$\Rightarrow \qquad\qquad\qquad (a^2 - 16)(a^2 + 1) = 0$

Thus $\qquad\qquad\qquad a^2 - 16 = 0 \quad$ or $\quad a^2 + 1 = 0$

But a is real so $\quad a^2 + 1 = 0 \quad$ gives no suitable values

and $\qquad\qquad\qquad\qquad a = \pm 4$

Referring to equation [2] we have

a	b
4	1
-4	-1

[**Note.** It is not correct to say $\quad a = \pm 4 \quad$ therefore $\quad b = \pm 1 \quad$ as this offers four different pairs of values for a and b (i.e. $4, 1;\ 4, -1;\ -4, 1;\ -4, -1$) two of which are invalid.]

Hence $\qquad\qquad\qquad \sqrt{15 + 8i} = 4 + i \quad$ or $\quad -4 - i$

$$= \pm(4 + i)$$

This result justifies our original assumption that the square root of a complex number is another complex number.

It is sometimes possible, however, to determine the square root of a complex number simply by observation. In the above example, equation [2] shows that the product of a and b is always half the coefficient of i in the original complex number.

Suitable integral values for a and b can then be checked quite quickly, e.g. to find $\sqrt{8-6i}$ we note that $ab = -3$ so possible values for a and b are: $1, -3;\ 3, -1$.

Checking:
$$(1-3i)^2 = -8-6i$$
$$(3-i)^2 = 8-6i$$

Hence one square root of $8-6i$ is $3-i$ and the other is $-(3-i)$

i.e.
$$\sqrt{8-6i} = \pm(3-i)$$

(**Note.** Unless a and b are integers, this method is unlikely to be useful.)

EXERCISE 14a

1) Simplify: $i^7,\ i^{-3},\ i^9,\ i^{-5},\ i^{4n},\ i^{4n+1}$.

2) Add the following pairs of complex numbers:
(a) $3+5i$ and $7-i$ (b) $4-i$ and $3+3i$
(c) $2+7i$ and $4-9i$ (d) $a+bi$ and $c+di$

3) Subtract the second number from the first in each part of Question 2.

4) Simplify:
(a) $(2+i)(3-4i)$ (b) $(5+4i)(7-i)$ (c) $(3-i)(4-i)$
(d) $(3+4i)(3-4i)$ (e) $(2-i)^2$ (f) $(1+i)^3$
(g) $i(3+4i)$ (h) $(x+yi)(x-yi)$ (i) $i(1+i)(2+i)$
(j) $(a+bi)^2$

5) Realise the denominator of each of the following fractions and hence express each in the form $a+bi$.

(a) $\dfrac{2}{1-i}$ (b) $\dfrac{3+i}{4-3i}$ (c) $\dfrac{4i}{4+i}$ (d) $\dfrac{1+i}{1-i}$

(e) $\dfrac{7-i}{1+7i}$ (f) $\dfrac{x+yi}{x-yi}$ (g) $\dfrac{3+i}{i}$ (h) $\dfrac{-2+3i}{-i}$

6) Solve the following equations for x and y:

(a) $x+yi = (3+i)(2-3i)$ (b) $\dfrac{2+5i}{1-i} = x+yi$

(c) $3+4i = (x+yi)(1+i)$ (d) $x+yi = 2$

(e) $x+yi = (3+2i)(3-2i)$ (f) $x+yi = (4+i)^2$

(g) $\dfrac{x+yi}{2+i} = 5-i$ (h) $(x+yi)^2 = 3+4i$

7) Find the real and imaginary parts of:
(a) $(2-i)(3+i)$ (b) $(1-i)^3$

(c) $\dfrac{3 + 2i}{4 - i}$

(d) $\dfrac{2}{3 + i} + \dfrac{3}{2 + i}$

(e) $\dfrac{1}{x + yi} - \dfrac{1}{x - yi}$

(f) $\left(\cos\dfrac{\pi}{3} + i\sin\dfrac{\pi}{3}\right)^3$

(g) $\left(\cos\dfrac{\pi}{6} + i\sin\dfrac{\pi}{6}\right)^2$

8) Find the square roots of:
(a) $3 - 4i$ (b) $21 - 20i$ (c) $2i$
(d) $15 + 8i$ (e) $-24 + 10i$

COMPLEX ROOTS OF QUADRATIC EQUATIONS

Consider the quadratic equation $x^2 + 2x + 2 = 0$.

Using the formula $x = -\dfrac{b \pm \sqrt{b^2 - 4ac}}{2a}$ we find that $x = \dfrac{-2 \pm \sqrt{-4}}{2}$.

Previously we dismissed solutions of this type, in which $b^2 - 4ac < 0$, as not being real. But now, because $\sqrt{-4} = 2i$, we see that the roots of the given equation are the complex numbers $-1 + i$ and $-1 - i$. Further, the roots are conjugate complex numbers.
An examination of the general quadratic equation $ax^2 + bx + c = 0$, where a, b and c are real numbers, shows that, if $b^2 - 4ac < 0$, the roots of the equation are always conjugate complex numbers.

If $\qquad\qquad\qquad ax^2 + bx + c = 0 \qquad\qquad\qquad$ [1]

and $\qquad\qquad\qquad b^2 - 4ac < 0$

then $\qquad\qquad\qquad x = \dfrac{-b \pm \sqrt{b^2 - 4ac}}{2a}$

or $\qquad\qquad\qquad x = \dfrac{-b}{2a} \pm i\dfrac{\sqrt{4ac - b^2}}{2a}$

Now using $\qquad p = \dfrac{-b}{2a} \quad$ and $\quad q = \dfrac{\sqrt{4ac - b^2}}{2a}$

the roots of equation [1] are:

$$p + qi \quad \text{and} \quad p - qi$$

and these are conjugate complex numbers.
So if one root of a quadratic equation with real coefficients is known to be complex, the other must also be complex and the conjugate of the first. Similarly a quadratic expression whose discriminant is negative, has conjugate complex factors,

e.g. $\qquad\qquad x^2 + 2x + 2 \equiv (x + 1 - i)(x + 1 + i)$

(the factors are found by solving the equation $x^2 + 2x + 2 = 0$)

CUBIC EQUATIONS WITH COMPLEX ROOTS

Consider a cubic equation $f(x) = 0$.

We have seen that a cubic equation with real coefficients has at least one real root so we can always find a real value a such that $f(a) = 0$.

Then $(x - a)$ is a factor of $f(x)$ and the remaining factor must be quadratic, i.e. $(px^2 + qx + r)$

$\Rightarrow$ $$(x - a)(px^2 + qx + r) = 0$$

so that $$x = a \quad \text{or} \quad px^2 + qx + r = 0$$

But the roots of $px^2 + qx + r = 0$ are either both real or are a pair of conjugate complex numbers.

Thus a cubic equation with real coefficients can have either three real roots or one real root and two conjugate complex roots.

In general, if any polynomial equation with real coefficients has complex roots, they occur in conjugate pairs.

Cube Roots of Unity

Consider the equation $x^3 - 1 = 0$

An obvious root is $x = 1$, so $x - 1$ is a factor, giving

$$x^3 - 1 \equiv (x - 1)(x^2 + x + 1) = 0$$

Thus $$x = 1 \quad \text{or} \quad x^2 + x + 1 = 0$$

i.e. $$x = 1 \quad \text{or} \quad \frac{-1 \pm \sqrt{1 - 4}}{2}$$

The roots of $x^3 - 1 = 0$ are therefore

$$1, \quad -\frac{1}{2} + \frac{\sqrt{3}}{2}i, \quad -\frac{1}{2} - \frac{\sqrt{3}}{2}i$$

But if the equation is rewritten in the form

$$x^3 = 1 \quad \text{or} \quad x = \sqrt[3]{1}$$

we see that the three values of x that we have found are the *three cube roots of unity*.

The following properties should be noted:

(1) $$\left(-\frac{1}{2} + \frac{\sqrt{3}}{2}i\right)^2 = \frac{1}{4} - \frac{\sqrt{3}}{2}i + \frac{3}{4}i^2 = -\frac{1}{2} - \frac{\sqrt{3}}{2}i$$

also $$\left(-\frac{1}{2} - \frac{\sqrt{3}}{2}i\right)^2 = \frac{1}{4} + \frac{\sqrt{3}}{2}i + \frac{3}{4}i^2 = -\frac{1}{2} + \frac{\sqrt{3}}{2}i$$

i.e. either complex cube root of 1, when squared, gives the other complex root. For this reason the three cube roots of 1 are usually denoted by 1, ω, ω^2.

(2) The sum of the three cube roots is zero, since

$$1 + \left(-\frac{1}{2} + \frac{\sqrt{3}}{2}i\right) + \left(-\frac{1}{2} - \frac{\sqrt{3}}{2}i\right) = 0$$

i.e.
$$1 + \omega + \omega^2 = 0$$

EXAMPLES 14b

1) Find the complex roots of the equation $2x^2 + 3x + 5 = 0$. If these roots are α and β, confirm the relationships $\alpha + \beta = -\dfrac{b}{a}$ and $\alpha\beta = \dfrac{c}{a}$.

If
$$2x^2 + 3x + 5 = 0$$

then
$$x = \frac{-3 \pm \sqrt{9 - 40}}{4}$$

$\Rightarrow$
$$\alpha = -\frac{3}{4} + \frac{\sqrt{31}}{4}i \qquad \beta = -\frac{3}{4} - \frac{\sqrt{31}}{4}i$$

Hence
$$\alpha + \beta = \left(-\frac{3}{4} + \frac{\sqrt{31}}{4}i\right) + \left(-\frac{3}{4} - \frac{\sqrt{31}}{4}i\right)$$

$$= -\frac{3}{2} = -\frac{b}{a}$$

and
$$\alpha\beta = \left(-\frac{3}{4} + \frac{\sqrt{31}}{4}i\right)\left(-\frac{3}{4} - \frac{\sqrt{31}}{4}i\right)$$

$$= \frac{9}{16} - \frac{31}{16}i^2$$

$$= \frac{40}{16} = \frac{5}{2} = \frac{c}{a}$$

2) One root of the equation $x^2 + px + q = 0$ is $2 - 3i$. Find the values of p and q.

If one root is $2 - 3i$ the other must be $2 + 3i$.

Then
$$\alpha + \beta = 4$$

and
$$\alpha\beta = (2 - 3i)(2 + 3i)$$

$$= 13$$

Now any quadratic equation can be written in the form

$$x^2 - (\text{sum of roots})x + (\text{product of roots}) = 0$$

So the equation with roots $2 \pm 3i$ is

$$x^2 - 4x + 13 = 0$$

i.e. $p = 4$ and $q = 13$

3) Find the complex factors of $x^2 - 2x + 10$ and hence express $\dfrac{6}{x^2 - 2x + 10}$ in partial fractions with complex linear denominators.

Let $x^2 - 2x + 10 = 0 \Rightarrow x = 1 \pm 3i$.
Thus the factors of $x^2 - 2x + 10$ are $(x - 1 - 3i)$, $(x - 1 + 3i)$

i.e. $x^2 - 2x + 10 \equiv (x - 1 - 3i)(x - 1 + 3i)$

Since the required partial fractions have complex denominators, we must prepare for unknown numerators that are complex.

Let $\dfrac{6}{x^2 - 2x + 10} \equiv \dfrac{A + Bi}{x - 1 - 3i} + \dfrac{C + Di}{x - 1 + 3i}$

$$\equiv \frac{(A + Bi)(x - 1 + 3i) + (C + Di)(x - 1 - 3i)}{x^2 - 2x + 10}$$

i.e. $6 \equiv (A + Bi)(x - 1 + 3i) + (C + Di)(x - 1 - 3i)$

When $x = 1 - 3i$

$$6 = (C + Di)(-6i) = 6D - 6Ci$$

Equating real and imaginary parts gives:

$$D = 1 \quad \text{and} \quad C = 0$$

When $x = 1 + 3i$

$$6 = (A + Bi)(6i) = 6Ai - 6B$$

giving $A = 0$ and $B = -1$

Hence $\dfrac{6}{x^2 - 2x + 10} \equiv \dfrac{-i}{x - 1 - 3i} + \dfrac{i}{x - 1 + 3i}$

EXERCISE 14b

1) Solve the following equations:
(a) $x^2 + x + 1 = 0$ (b) $2x^2 + 7x + 1 = 0$
(c) $x^2 + 9 = 0$ (d) $x^2 + x + 3 = 0$
(e) $x^4 - 1 = 0$

2) Form the equation whose roots are:
(a) $i, -i$ (b) $2 + i, 2 - i$ (c) $1 - 3i, 1 + 3i$ (d) $1 + i, 1 - i, 2$

3) Without calculating a, b and c, evaluate $-\dfrac{b}{a}$ and $\dfrac{c}{a}$ if one root of the equation $ax^2 + bx + c = 0$ is:

(a) $2 + i$ (b) $3 - 4i$ (c) i (d) $5i - 12$ (e) $-1 - i$

Explain why this question cannot be answered if the given root is 2.

4) Find the complex factors of:

(a) $x^2 + 4x + 5$ (b) $x^2 - 2x + 17$ (c) $x^2 + x + 1$ (d) $x^3 - 8$

5) Express as partial fractions with complex linear denominators:

(a) $\dfrac{1}{x^2 + 1}$ (b) $\dfrac{4}{x^2 - 4x + 5}$

(c) $\dfrac{16}{x^2 + 4x + 8}$ (d) $\dfrac{2}{x^2 + 4}$ (e) $\dfrac{x + 8}{x^2 + 4x + 13}$

6) If $1, \omega, \omega^2$ are the three cube roots of unity, find the value of:

(a) $\dfrac{(1 + \omega)^2}{\omega}$ (b) $(1 + 2\omega + 3\omega^2)(3 + 2\omega + \omega^2)$ (c) $\omega^7 + \omega^8 + \omega^9$

7) By solving the equation $x^3 + 1 = 0$, find the three cube roots of -1. If one of the complex cube roots is λ express the other in terms of λ. Prove that $1 + \lambda^2 = \lambda$.

THE ARGAND DIAGRAM

For a complex number $a + bi$, a and b are both real numbers. Thus $a + bi$ can be represented by the ordered pair $\begin{pmatrix} a \\ b \end{pmatrix}$. This suggests using a and b as the Cartesian coordinates of a point, A say, and using the vector $\overrightarrow{OA}$ as a visual representation of the complex number $a + bi$.

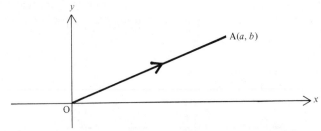

As this idea was introduced by the French mathematician Argand, his name is given to the diagram which demonstrates a complex number in this way. On an Argand diagram real numbers are represented on the x axis and imaginary numbers on the y axis. (For this reason the x and y axes are often called the real and imaginary axes.)

A general complex number $x + yi$ is represented by the line $\overrightarrow{OP}$ where P is the point (x, y).

In an Argand diagram, the magnitude and direction of a line are used to represent a complex number in the same way that a section of a line can be used to represent a vector quantity. Thus the techniques and operations used in vector problems can be applied equally well to complex number analysis.

A COMPLEX NUMBER AS A VECTOR

On an Argand diagram, the complex number $5 + 3i$ can be represented by the line $\overrightarrow{OA}$ where A is the point $(5, 3)$. But any other line with the same length and direction (e.g. $\overrightarrow{BC}$ or $\overrightarrow{DE}$ as shown) is equally representative.

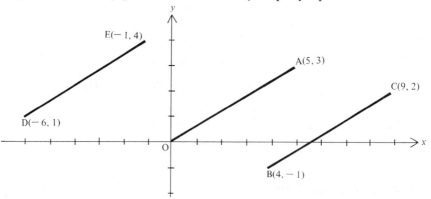

Treated in this way, a complex number behaves as a *free vector*.
If, however, $5 + 3i$ is regarded as a *position vector*, then *only* the line $\overrightarrow{OA}$ represents $5 + 3i$. In this case the *point* A(5, 3) is sometimes taken to represent $5 + 3i$.
The symbol z is often used to denote a complex number, e.g. $z = x + yi$, $z_1 = 5 + 3i$. Used on an Argand diagram, z must be accompanied by an arrow to indicate the direction of the line representing the complex number, e.g.

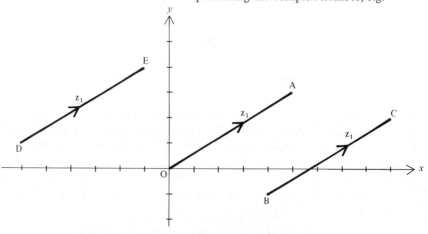

GRAPHICAL ADDITION AND SUBTRACTION

Consider two complex numbers z_1 and z_2 represented on an Argand diagram by $\overrightarrow{OA}$ and $\overrightarrow{OB}$.

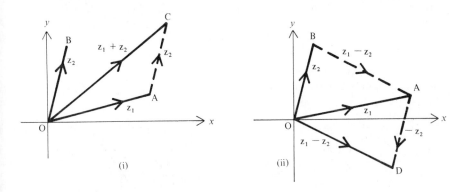

(i)   (ii)

If AC is drawn equal and parallel to OB, then $\overrightarrow{AC}$ also represents z_2.
But the sum of the vectors $\overrightarrow{OA}$ and $\overrightarrow{AC}$ is $\overrightarrow{OC}$.
Therefore, if z_1 and z_2 are represented by $\overrightarrow{OA}$ and $\overrightarrow{OB}$, their sum is represented by the diagonal $\overrightarrow{OC}$ of the parallelogram OACB (see diagram (i)).
If $\overrightarrow{OB}$ represents z_2, $\overrightarrow{AD}$ (equal and parallel to $\overrightarrow{BO}$) represents $-z_2$. Then $\overrightarrow{OD}$ represents $z_1 - z_2$. But $\overrightarrow{OD}$ is equal and parallel to $\overrightarrow{BA}$.
Therefore the line joining B to A $(\overrightarrow{BA})$ represents $z_1 - z_2$.
Similarly the line joining A to B $(\overrightarrow{AB})$ represents $z_2 - z_1$ (see diagram (ii)).
Thus the two diagonals of the parallelogram OACB represent the sum and difference of z_1 and z_2 (see diagram (iii)).

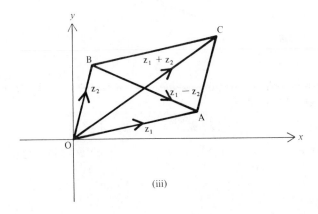

(iii)

Taking $z_1 = x_1 + y_1 i$ and
$z_2 = x_2 + y_2 i$ we see that the
coordinates of C obtained in this
way are $[(x_1 + x_2), (y_1 + y_2)]$.
Hence $\overrightarrow{OC}$ represents the complex
number

$$(x_1 + x_2) + (y_1 + y_2)i$$

the same result as is obtained by
adding separately the real and
imaginary parts of z_1 and z_2,

e.g. if $z_1 = 5 + 2i$

and $z_2 = 3 + 4i$

then:

$$z_1 + z_2 = 5 + 2i + 3 + 4i$$

$$= 8 + 6i$$

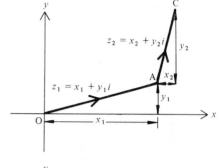

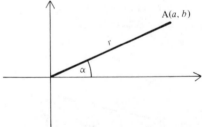

MODULUS AND ARGUMENT

The point $A(a, b)$ can equally well be located using polar coordinates (r, α)
where r is the length of $\overrightarrow{OA}$ and α is the angle between the positive x axis
and $\overrightarrow{OA}$.

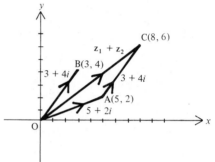

The length of OA is called the *modulus* of the complex number $a + bi$ and is
written $|a + bi|$ so that

$$|a + bi| = r = \sqrt{a^2 + b^2}$$

The angle α is called the argument of $a + bi$ and is written $\text{Arg}\,(a + bi)$,

thus $$\text{Arg}\,(a + bi) = \alpha = \text{Arctan}\,\frac{b}{a}$$

There is an infinite set of angles whose tangent is $\dfrac{b}{a}$ so there is also an infinite set of arguments for $a + bi$. But the position of OA is unique and corresponds to only one value of α in the range $-\pi < \alpha \leqslant \pi$. This value is the *principal argument*, and is written $\arg(a + bi)$ whereas $\text{Arg}(a + bi)$ represents the infinite set of arguments.

(**Note.** An argument is sometimes given the alternative name *amplitude*.)

When working with arguments it is always wise to draw an Argand diagram. Consider, for example, the complex numbers $4 + 3i$, $-4 + 3i$, $-4 - 3i$, $4 - 3i$.

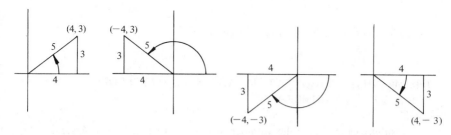

The modulus of each is 5.

The line representing $4 + 3i$ is in the first quadrant, so the principal argument of $4 + 3i$ is positive and acute, and its value is $\arctan \frac{3}{4}$,

i.e. $\qquad\qquad\qquad\qquad \arg(4 + 3i) = 0.644^{c}$

The line representing $-4 + 3i$ is in the second quadrant, so the principal argument of $-4 + 3i$ is positive and obtuse and its value is $\text{Arctan}\left(-\frac{3}{4}\right)$,

i.e. $\qquad\qquad\qquad\qquad \arg(-4 + 3i) = 2.498^{c}$

The line representing $-4 - 3i$ is in the third quadrant so the principal argument of $-4 - 3i$ is negative and obtuse and of value $\text{Arctan}\dfrac{3}{4}$

i.e. $\quad \arg(-4 - 3i) = -2.498^{c}$

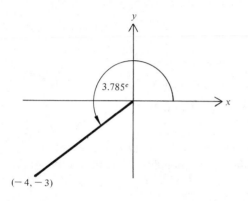

$+ 3.785^{c}$ is *an* argument of $-4 - 3i$ but not the principal argument.

The line representing $4 - 3i$ is in the fourth quadrant so the principal argument of $4 - 3i$ is negative and acute and of value $\arctan(-\tfrac{3}{4})$,

i.e. $$\arg(4 - 3i) = -0.644^c$$

Note. In this example we see why it is not sufficient to say that

$$\arg(a + bi) = \arctan\frac{b}{a}.$$ Both $4 + 3i$ and $-4 - 3i$ have an argument of

value $\arctan\tfrac{3}{4}$ but their arguments are different because they are in different quadrants. Similarly the principal arguments of $4 - 3i$ and $-4 + 3i$ are both $\arctan(-\tfrac{3}{4})$ but are different angles. So, in finding the argument of $a + bi$ we use

$\arctan\dfrac{b}{a}$ *together with a quadrant diagram*.

THE MODULUS, ARGUMENT (POLAR COORDINATE) FORM FOR A COMPLEX NUMBER

Consider a general complex number $x + yi$ represented on an Argand diagram by $\overrightarrow{OP}$ where P has Cartesian coordinates (x, y) and polar coordinates (r, θ).

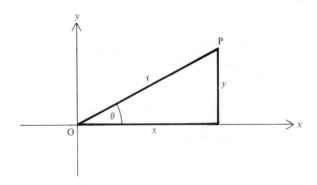

From the diagram we see that $\quad x \equiv r \cos \theta$

$$y \equiv r \sin \theta$$

Hence $\qquad x + yi \equiv r \cos \theta + ir \sin \theta$

i.e. $\qquad x + yi \equiv r(\cos \theta + i \sin \theta)$

(**Note.** For clarity we write $i \sin \theta$ and not $\sin \theta i$.)
If a complex number is given in the form $x + yi$ it can be converted into the form $r(\cos \theta + i \sin \theta)$ simply by finding the modulus r and the argument θ, e.g. for the complex number $1 - i$, we find that

$$r = \sqrt{1^2 + (-1)^2} = \sqrt{2}$$

$$\theta = \arg(1-i) = -\frac{\pi}{4}$$

Hence

$$1 - i = \sqrt{2}\left[\cos\left(-\frac{\pi}{4}\right) + i\sin\left(-\frac{\pi}{4}\right)\right]$$

Conversely a complex number given in polar form can be expressed directly in Cartesian form,

e.g.

$$4\left(\cos\frac{2\pi}{3} + i\sin\frac{2\pi}{3}\right) = 4\left(-\frac{1}{2}\right) + 4\left(\frac{\sqrt{3}}{2}\right)i$$

$$= -2 + 2i\sqrt{3}$$

EXAMPLES 14c

1) Given that $z_1 = 3 - i$ and $z_2 = -2 + 5i$, represent on an Argand diagram the complex numbers z_1, z_2, $z_1 + z_2$, $z_1 - z_2$. Find the modulus and principal argument of $z_1 + z_2$ and $z_1 - z_2$.

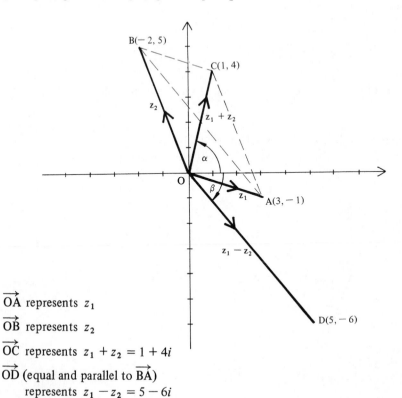

$\overrightarrow{OA}$ represents z_1

$\overrightarrow{OB}$ represents z_2

$\overrightarrow{OC}$ represents $z_1 + z_2 = 1 + 4i$

$\overrightarrow{OD}$ (equal and parallel to $\overrightarrow{BA}$)
    represents $z_1 - z_2 = 5 - 6i$

$$|z_1 + z_2| = \sqrt{1^2 + 4^2} \quad\quad = \sqrt{17} \text{ (length of OC)}$$
$$|z_1 - z_2| = \sqrt{5^2 + (-6)^2} = \sqrt{61} \text{ (length of OD)}$$

$$\arg(z_1 + z_2) = \alpha \quad \text{where} \quad \tan \alpha = \tfrac{4}{1}$$

so $$\arg(z_1 + z_2) = 1.33^c$$

$$\arg(z_1 - z_2) = \beta \quad \text{where} \quad \tan \beta = -\tfrac{6}{5}$$

so $$\arg(z_1 - z_2) = -0.88^c$$

2) Find the modulus and argument of $\dfrac{7-i}{3-4i}$.

First we must express $\dfrac{7-i}{3-4i}$ in the form $a + bi$

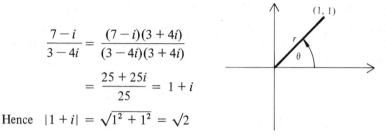

$$\frac{7-i}{3-4i} = \frac{(7-i)(3+4i)}{(3-4i)(3+4i)}$$

$$= \frac{25 + 25i}{25} = 1 + i$$

Hence $|1 + i| = \sqrt{1^2 + 1^2} = \sqrt{2}$

and $\arg(1 + i)$ is a positive acute angle of value $\arctan 1$, i.e. $\dfrac{\pi}{4}$

3) Express in the form $r(\cos \theta + i \sin \theta)$:
(a) $1 - i\sqrt{3}$ (b) 2   (c) $-5i$ (d) $-2 + 2i$.

(a)

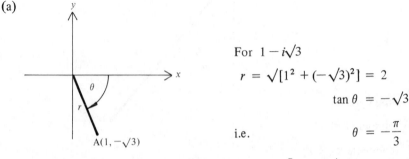

For $1 - i\sqrt{3}$

$$r = \sqrt{[1^2 + (-\sqrt{3})^2]} = 2$$

$$\tan \theta = -\sqrt{3}$$

i.e. $$\theta = -\frac{\pi}{3}$$

Thus $1 - i\sqrt{3} = 2\left[\cos\left(-\dfrac{\pi}{3}\right) + i \sin\left(-\dfrac{\pi}{3}\right)\right]$

(b)

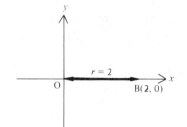

For 2,

$$r = 2 \quad \text{and}$$

$$\theta = 0$$

Thus $2 = 2(\cos 0 + i \sin 0)$

(c)

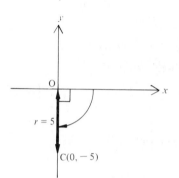

For $-5i$

$$r = 5 \quad \text{and}$$

$$\theta = -\frac{\pi}{2}$$

Thus

$$-5i = 5\left\{\cos\left(-\frac{\pi}{2}\right) + i \sin\left(-\frac{\pi}{2}\right)\right\}$$

(d)

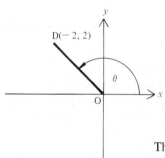

For $-2 + 2i$

$$r = \sqrt{(-2)^2 + 2^2} = 2\sqrt{2}$$

$$\tan \theta = -1$$

i.e. $\theta = \dfrac{3\pi}{4}$

Thus $-2 + 2i = 2\sqrt{2}\left(\cos\dfrac{3\pi}{4} + i \sin\dfrac{3\pi}{4}\right)$

EXERCISE 14c

1) Represent the following complex numbers by lines on Argand diagrams. Determine the modulus and argument of each complex number.

(a) $3 - 2i$ (b) $-4 + i$ (c) $-3 - 4i$ (d) $5 + 12i$

(e) $1 - i$ (f) $-1 + i$ (g) 4 (h) $-2i$

(i) $a + bi$ (j) $1 + i$ (k) $i(1 + i)$ (l) $i^2(1 + i)$

(m) $i^3(1 + i)$ (n) $(3 + i)(4 + i)$ (o) $2\left(\cos\dfrac{\pi}{3} + i \sin\dfrac{\pi}{3}\right)$

(p) $\cos\dfrac{3\pi}{4} + i \sin\dfrac{3\pi}{4}$ (q) $3\left[\cos\left(-\dfrac{5\pi}{6}\right) + i \sin\left(-\dfrac{5\pi}{6}\right)\right]$

2) If $z_1 = 3 - i$, $z_2 = 1 + 4i$, $z_3 = -4 + i$, $z_4 = -2 - 5i$, represent the following by lines on Argand diagrams, showing the direction of each line by an arrow.

(a) $z_1 + z_2$ (b) $z_2 - z_3$ (c) $z_1 - z_3$
(d) $z_2 + z_4$ (e) $z_4 - z_1$ (f) $z_3 - z_4$
(g) z_1 (h) z_4 (i) $z_2 - z_1$ (j) $z_1 + z_3$

3) Express in the form $r(\cos\theta + i\sin\theta)$:
(a) $1 + i$ (b) $\sqrt{3} - i$ (c) $-3 - 4i$ (d) $-5 + 12i$
(e) $2 - i$ (f) 6 (g) -3 (h) $4i$
(i) $-3 - i\sqrt{3}$ (j) $24 + 7i$

4) Express in the form $x + yi$ a complex number represented on an Argand diagram by $\overrightarrow{OP}$ where the polar coordinates of P are:

(a) $\left(2, \dfrac{\pi}{6}\right)$ (b) $\left(3, -\dfrac{\pi}{4}\right)$ (c) $\left(1, \dfrac{2\pi}{3}\right)$ (d) $\left(1, -\dfrac{3\pi}{4}\right)$

(e) $(3, 0)$ (f) $(2, \pi)$ (g) $\left(4, -\dfrac{\pi}{6}\right)$

(h) $\left(1, \dfrac{\pi}{2}\right)$ (i) $\left(3, -\dfrac{\pi}{2}\right)$ (j) $\left(1, -\dfrac{2\pi}{3}\right)$

5) By using $z_1 = x_1 + y_1 i$, $z_2 = x_2 + y_2 i$ show on an Argand diagram the position of the point representing:
(a) $\frac{1}{2}(z_1 + z_2)$ (b) $\frac{1}{3}(2z_1 + z_2)$

6) By solving the equation $z^3 = 1$, find the three cube roots of 1. If $\overrightarrow{OA}$, $\overrightarrow{OB}$ and $\overrightarrow{OC}$ represent the three cube roots on an Argand diagram, show that A, B and C lie equally spaced, on a circle of radius 1 and centre O. Write down each cube root in the form $r(\cos\theta + i\sin\theta)$.

PRODUCTS AND QUOTIENTS

Taking two complex numbers $z_1 = r_1(\cos\theta_1 + i\sin\theta_1)$ and $z_2 = r_2(\cos\theta_2 + i\sin\theta_2)$ we find that:

$$z_1 z_2 = r_1 r_2 (\cos\theta_1 + i\sin\theta_1)(\cos\theta_2 + i\sin\theta_2)$$

$$= r_1 r_2 [\cos\theta_1 \cos\theta_2 + i\sin\theta_1 \cos\theta_2 + i\sin\theta_2 \cos\theta_1 + i^2 \sin\theta_1 \sin\theta_2]$$

$$= r_1 r_2 [\cos\theta_1 \cos\theta_2 - \sin\theta_1 \sin\theta_2 + i(\sin\theta_1 \cos\theta_2 + \cos\theta_1 \sin\theta_2)]$$

$$= r_1 r_2 [\cos(\theta_1 + \theta_2) + i\sin(\theta_1 + \theta_2)]$$

i.e. $z_1 z_2$ gives a complex number with modulus $r_1 r_2$ and argument $\theta_1 + \theta_2$.

We also find that:

$$\frac{z_1}{z_2} = \frac{r_1(\cos\theta_1 + i\sin\theta_1)}{r_2(\cos\theta_2 + i\sin\theta_2)} = \frac{r_1}{r_2}\left[\frac{(\cos\theta_1 + i\sin\theta_1)(\cos\theta_2 - i\sin\theta_2)}{(\cos\theta_2 + i\sin\theta_2)(\cos\theta_2 - i\sin\theta_2)}\right]$$

which simplifies to $\dfrac{r_1}{r_2}[\cos(\theta_1 - \theta_2) + i\sin(\theta_1 - \theta_2)]$,

i.e. $\dfrac{z_1}{z_2}$ gives a complex number with modulus $\dfrac{r_1}{r_2}$ and argument $\theta_1 - \theta_2$.

These results can be expressed as follows:

$$|z_1 z_2| = |z_1||z_2|$$

$$\left|\frac{z_1}{z_2}\right| = \frac{|z_1|}{|z_2|}$$

$$\text{Arg}(z_1 z_2) = \arg z_1 + \arg z_2$$

$$\text{Arg}\left(\frac{z_1}{z_2}\right) = \arg z_1 - \arg z_2$$

Thus
$$|(3 + 4i)(5 - 12i)| = |3 + 4i||5 - 12i|$$
$$= 5 \times 13$$

Similarly
$$\left|\frac{3 + 4i}{5 - 12i}\right| = \frac{5}{13}$$

Further, a complex equation of the type $\left|\dfrac{z - 3}{z + 5i}\right| = 2$ can be transformed

into $|z - 3| = 2|z + 5i|$.

GREATEST AND LEAST VALUES OF $|z_1 + z_2|$

We know that for any two complex numbers z_1 and z_2 represented by $\overrightarrow{OA}$ and $\overrightarrow{OB}$, $z_1 + z_2$ is represented by $\overrightarrow{OC}$ where OACB is a parallelogram.

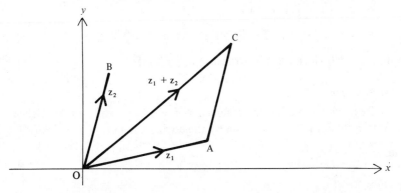

Now $\quad |z_1| = OA, \quad |z_2| = OB = AC \quad$ and $\quad |z_1 + z_2| = OC.$

In any triangle the sum of the lengths of any two sides is greater than the length of the third side.

Thus in triangle OAC

$$OA + AC > OC$$

i.e.

$$|z_1| + |z_2| > |z_1 + z_2|$$

In the special case when OB and OA are themselves parallel (i.e. $\arg z_2 = \arg z_1$) the parallelogram OACB becomes a straight line

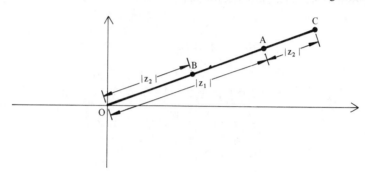

In this case $\qquad\qquad\qquad OA + AC = OC$

i.e. $\qquad\qquad\qquad\qquad |z_1| + |z_2| = |z_1 + z_2|$

Thus $\qquad\qquad\qquad\qquad |z_1 + z_2| \leqslant |z_1| + |z_2| \qquad\qquad\qquad$ [1]

i.e. the greatest possible value of $|z_1 + z_2|$ is $|z_1| + |z_2|$.

Returning to triangle OAC, two further inequalities can be used:

i.e. $\quad OC + CA > OA \quad$ and $\quad CO + OA > AC.$ These, together with the extreme case when O, A, B and C are collinear, give the result

$$|z_1 + z_2| \geqslant |z_1| \sim |z_2| \qquad\qquad\qquad [2]$$

where $\sim$ means 'the positive difference between'.

Thus the least possible value of $|z_1 + z_2|$ is $|z_1| \sim |z_2|$.

Combining [1] and [2] we have

$$|z_1| \sim |z_2| \leqslant |z_1 + z_2| \leqslant |z_1| + |z_2|$$

Note. $|z_1| \sim |z_2|$ can also be written $\Big| |z_1| - |z_2| \Big|.$

EXAMPLES 14d

1) Write down the moduli and arguments of $-\sqrt{3} + i$ and $4 + 4i$. Hence express in the form $r(\cos\theta + i\sin\theta)$ the complex numbers $(-\sqrt{3} + i)(4 + 4i)$ and $\dfrac{(-\sqrt{3} + i)}{(4 + 4i)}$.

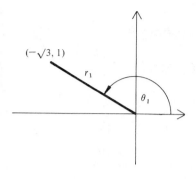

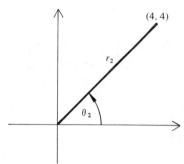

$$r_1 = |-\sqrt{3} + i| = 2$$

$$\theta_1 = \arg(-\sqrt{3} + i) = \frac{5\pi}{6}$$

$$r_2 = |4 + 4i| = 4\sqrt{2}$$

$$\theta_2 = \arg(4 + 4i) = \frac{\pi}{4}$$

Let

$$(-\sqrt{3} + i)(4 + 4i) = r_3(\cos\theta_3 + i\sin\theta_3)$$

then

$$r_3 = r_1 r_2 = 8\sqrt{2}$$

and

$$\theta_3 = \theta_1 + \theta_2 = \frac{5\pi}{6} + \frac{\pi}{4} = \frac{13\pi}{12}$$

Thus

$$(-\sqrt{3} + i)(4 + 4i) = 8\sqrt{2}\left(\cos\frac{13\pi}{12} + i\sin\frac{13\pi}{12}\right)$$

Note. Although this result is quite correct, the angle $\frac{13\pi}{12}$ is not the principal argument since it exceeds π.

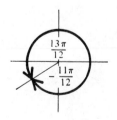

Using the principal argument, which is $-\frac{11\pi}{12}$, we can also express

$(-\sqrt{3} + i)(4 + 4i)$ in the form $8\sqrt{2}\left[\cos\left(-\frac{11\pi}{12}\right) + i\sin\left(-\frac{11\pi}{12}\right)\right]$.

Let
$$\frac{(-\sqrt{3}+i)}{(4+4i)} = r_4(\cos\theta_4 + i\sin\theta_4)$$

then
$$r_4 = \frac{r_1}{r_2} = \frac{2}{4\sqrt{2}} = \frac{\sqrt{2}}{4}$$

and
$$\theta_4 = \theta_1 - \theta_2 = \frac{5\pi}{6} - \frac{\pi}{4} = \frac{7\pi}{12}$$

(this time the value we have found for θ_4 is the principal argument).

Thus
$$\frac{(-\sqrt{3}+i)}{(4+4i)} = \frac{\sqrt{2}}{4}\left(\cos\frac{7\pi}{12} + i\sin\frac{7\pi}{12}\right)$$

2) If $z_1 = 3 + 4i$ and $|z_2| = 13$, find the greatest value of $|z_1 + z_2|$.

If $|z_1 + z_2|$ has its greatest value and also $0 < \arg z_2 < \frac{\pi}{2}$, express z_2 in the form $a + bi$.

Using
$$|z_1 + z_2| \leqslant |z_1| + |z_2|$$

we have
$$|z_1 + z_2| \leqslant \sqrt{3^2 + 4^2} + 13$$

Thus the greatest value of $|z_1 + z_2|$ is 18.

If $\overrightarrow{OA}$ and $\overrightarrow{OB}$ represent z_1 and z_2 where A is the point $(3, 4)$ and B is the point (a, b), both A and B are in the first quadrant $\left(0 < \arg z_2 < \frac{\pi}{2}\right)$.
Now if $\overrightarrow{OC}$ represents $z_1 + z_2$ and $|z_1 + z_2|$ has its greatest value, O, A, B and C are collinear.

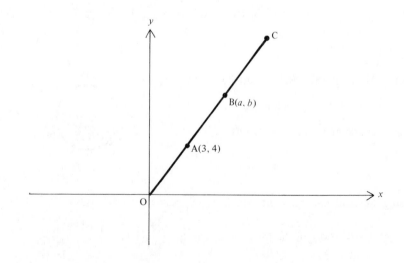

Hence   gradient OB = gradient OA,

i.e.
$$\frac{b}{a} = \frac{4}{3} \qquad [1]$$

also
$$|z_2| = \sqrt{a^2 + b^2} = 13$$

i.e.
$$a^2 + b^2 = 169 \qquad [2]$$

Solving [1] and [2] simultaneously gives

$$a^2 + \left(\frac{4a}{3}\right)^2 = 169$$

i.e.
$$\frac{25a^2}{9} = 169$$

$\Rightarrow$
$$a = \pm\frac{39}{5}$$

But a is positive so $a = \frac{39}{5}$ and $b = \frac{52}{5}$

$\Rightarrow$
$$z_2 = \tfrac{39}{5} + \tfrac{52}{5}i$$

EXERCISE 14d

1) Show that
$$|z_1 - z_2| \leqslant |z_1| + |z_2|$$
and
$$|z_1 - z_2| \geqslant |z_1| \sim |z_2|$$
Draw a clear diagram showing the case when
$$|z_1 - z_2| = |z_1| + |z_2|$$

2) $z_1 = 24 + 7i$ and $|z_2| = 6$.   Find the greatest and least values of $|z_1 + z_2|$.

3) Without first expressing them in the form $a + bi$, determine the modulus and argument of the following:

(a) $2(1 + i)$ (b) $(3 - \sqrt{3}i)(1 - i)$ (c) $\dfrac{(-2 - \sqrt{3}i)}{(\sqrt{3}i - 2)}$

4) If $z = \cos\theta + i\sin\theta$ (i.e. $r = 1$),   show that
$z^2 = \cos 2\theta + i\sin 2\theta$ and that $z^3 = \cos 3\theta + i\sin 3\theta$.
(Do not square or cube $\cos\theta + i\sin\theta$.)

5) Using the fact that if $z = r(\cos\theta + i\sin\theta)$ then
$z^2 = r^2(\cos 2\theta + i\sin 2\theta)$,   find the square roots of $2\sqrt{3} - 2i$.
(*Hint.* Let $z^2 = 2\sqrt{3} - 2i$.)

LOCI

When $z = x + yi$, the point $P(x, y)$ can be anywhere on the Argand diagram. But if a special condition is imposed on z, this affects the possible positions of P.

For instance, if $|z| = 4$ the line OP is of constant length 4 units.

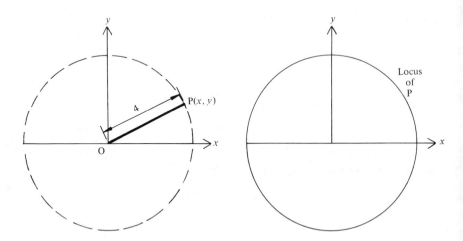

Thus P is restricted to any point on the circumference of a circle with centre O and radius 4, i.e. $|z| = 4$ defines the circle, centre O and radius 4.

This circle, which is the locus of P, has an equation

$$x^2 + y^2 = 4^2 \quad \text{in Cartesian form}$$

$$r = 4 \quad \text{in polar form}$$

and $$|z| = 4 \quad \text{in complex form}$$

Alternatively the Cartesian equation of the locus of P can be found directly from the complex equation, without reference to a diagram, as follows:

$$z = x + yi \quad \text{so} \quad |x + yi| = 4$$

But $$|x + yi| = \sqrt{(x^2 + y^2)}$$

Hence $$\sqrt{(x^2 + y^2)} = 4$$

or $$x^2 + y^2 = 16$$

which is the Cartesian equation of a circle, centre O and radius 4.

(Each of these methods has advantages in different problems. It is unwise to use one of them exclusively.)

Now let us consider the case when $|z - z_1| = 4$ where z_1 is a fixed point representing $x_1 + y_1 i$.

If $P(x, y)$ represents $x + yi$ and $A(x_1, y_1)$ represents $x_1 + y_1 i$ on an Argand diagram, then $z - z_1$ is represented by the line joining A to P. So $|z - z_1| = 4$ means that the length of AP is always 4 units.

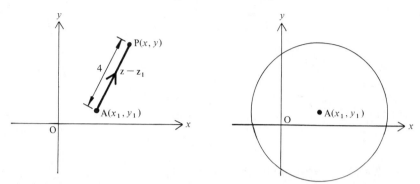

P must therefore be on a circle with centre A and radius 4 units. The Cartesian equation of this circle, which is the locus of P, is

$$(x - x_1)^2 + (y - y_1)^2 = 4^2$$

Note. When $z = x + yi$ is represented by $\overrightarrow{OP}$ where P is the point (x, y), then $\overrightarrow{OP}$ is a position vector. Thus z is a position vector in all problems involving the locus of P.

We have already used the fact that $z - z_1$ is represented on an Argand diagram by *a line joining the two points A and P* (where $z = \overrightarrow{OP}$ and $z_1 = \overrightarrow{OA}$).

Thus $|z - z_1|$ means 'the length of the line AP'.

Before dealing with a locus problem involving an argument, a careful appraisal of the meaning of $\arg(z - z_1)$ is necessary.

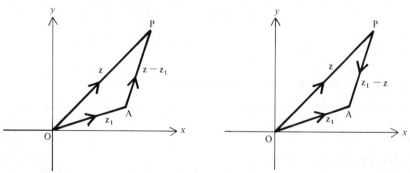

$z - z_1$ is represented by the line joining A to P.
$z_1 - z$ is represented by the line joining P to A.
These lines have opposite directions so

$$\arg(z - z_1) \neq \arg(z_1 - z)$$

One method for determining the argument of a line that does not pass through O is given below.

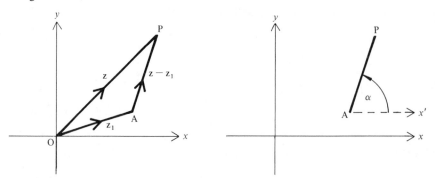

Draw a line Ax', parallel to the *positive* x axis.

(**Note** that this line is drawn through the point representing the *subtracted* complex number.)

The angle $x'AP$, α, is then an argument of $z - z_1$ and if $-\pi < \alpha \leqslant \pi$ it is the principal argument.

Now if $\arg(z_1 - z)$ is required, a line parallel to Ox is drawn through P as shown; then $\arg(z_1 - z) = \beta$

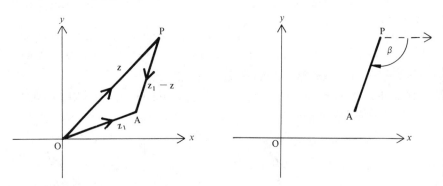

Consider now the locus defined by the condition $\arg(z - z_1) = \dfrac{\pi}{4}$.

If $P(x, y)$ represents $x + yi$
and $A(x_1, y_1)$ represents $x_1 + y_1 i$
then one possible position of P is
shown in the diagram

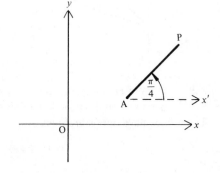

But the length of AP is unspecified so the line AP can be produced

indefinitely and every point on it satisfies the condition $\arg(z - z_1) = \dfrac{\pi}{4}$

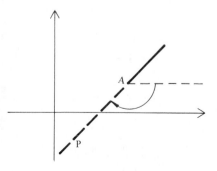

The line PA must *not* be produced

however, since $\arg(z - z_1) \neq \dfrac{\pi}{4}$

when P is below A as shown.

The locus of P in a case like this
is described as a *part-line*.

EXAMPLES 14e

1) If $|z - 3| = 2$,   sketch the locus of $P(x, y)$ which represents z on an Argand diagram. Write down the Cartesian equation of this locus.

$$z - 3 = z - z_1 \quad \text{where} \quad z_1 = 3 + 0i$$

Let $A(3, 0)$ represent z_1
The line AP represents $z - z_1$
But $|z - z_1| = 2$ so the length of AP is always 2 units.
Since A is a fixed point, P lies anywhere on a circle with centre A and radius 2.
This circle is thus the locus of P.

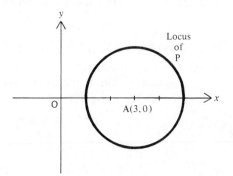

The Cartesian equation of the circle
is

$$(x - 3)^2 + (y - 0)^2 = 2^2$$
$$\Rightarrow \quad x^2 + y^2 - 6x + 5 = 0$$

2) If $z_1 = 3 + i$, $z_2 = -3 - i$ and $z = x + yi$ determine the locus of the set of points $P(x, y)$ on the Argand diagram given that

$$|z - z_1| = |z - z_2|$$

Using $A(3, 1)$ and $B(-3, -1)$, the given condition becomes

$$AP = BP$$

i.e. P is always equidistant from the two fixed points A and B,

i.e. P always lies on the perpendicular bisector of the line AB.

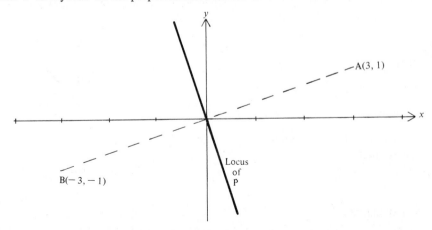

B(−3, −1)

Locus
of
P

A(3, 1)

The locus of P is therefore the perpendicular bisector of AB and its Cartesian equation is $y = -3x$.

3) $P(x, y)$ is the point on an Argand diagram representing $z = x + yi$. If $|z + 2| = 3|z - 2 - 4i|$ show that the Cartesian equation of the locus of P is $x^2 + y^2 - 5x - 9y + 22 = 0$.

To use the geometric approach in this problem we would first write $z + 2$ in the form $z - z_1$ so that it corresponds to a line joining two points,

i.e. $|z - (-2 + 0i)| = 3|z - (2 + 4i)|$

Then, using A(−2, 0), B(2, 4) and P(x, y) we find that AP = 3BP. It is not obvious from this property however, where P can lie, so the method based on the algebraic definition of a modulus is used instead.

Writing $x + yi$ for z we have

$$z + 2 = x + yi + 2 = (x + 2) + yi$$

Then $|z + 2| = \sqrt{(x + 2)^2 + (y)^2}$

So $z - 2 - 4i = x + yi - 2 - 4i = (x - 2) + (y - 4)i$

and $|z - 2 - 4i| = \sqrt{(x - 2)^2 + (y - 4)^2}$

Thus the given condition can be written

$$\sqrt{[(x + 2)^2 + y^2]} = 3\sqrt{[(x - 2)^2 + (y - 4)^2]}$$

i.e. $x^2 + 4x + 4 + y^2 = 9(x^2 - 4x + 4 + y^2 - 8y + 16)$

$\Rightarrow$ $8x^2 + 8y^2 - 40x - 72y + 176 = 0$

$\Rightarrow$ $$x^2 + y^2 - 5x - 9y + 22 = 0$$

Because this is the condition which the coordinates of P must satisfy, it is the equation of the locus of P, which is seen to be a circle.

4) Sketch on an Argand diagram the locus of the set of points $P(x, y)$ where $z = x + yi$, for which $\arg(z - 2 + 3i) = \dfrac{2\pi}{3}$

$$z - 2 + 3i = z - (2 - 3i)$$

So $z - 2 + 3i$ is represented by the line joining $A(2, -3)$ to $P(x, y)$.

Since A is fixed, the line Ax' can be drawn and, taking an angle $\dfrac{2\pi}{3}$ in the anticlockwise sense from Ax', a line AP can be drawn. Any point on this part-line represents P and the line is therefore the locus of P.

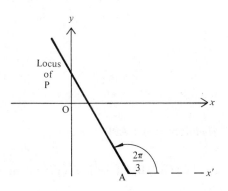

5) If $\arg(z - 1) - \arg(z + 1) = \dfrac{\pi}{4}$, show that the point $P(x, y)$ representing z on an Argand diagram lies on an arc of a circle. Give the coordinates of the centre of this circle.

If A is $(1, 0)$ and B is $(-1, 0)$, $\arg(z - 1) = \alpha$ and $\arg(z + 1) = \beta$ we have:

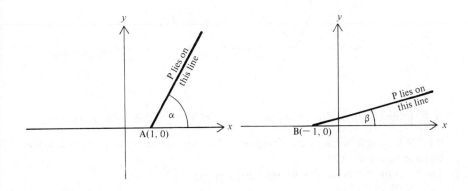

Thus

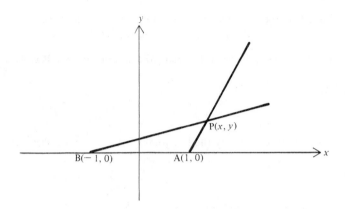

But $\alpha - \beta = \dfrac{\pi}{4}$ and angle $APB = \alpha - \beta$.

Therefore angle $APB = \dfrac{\pi}{4}$,

i.e. the line AB subtends a constant angle $\dfrac{\pi}{4}$ at P, so P must lie on an arc of a circle cut off by the chord AB.

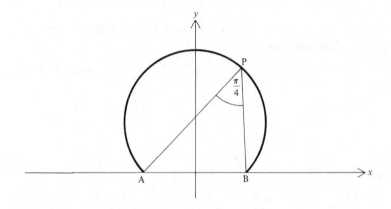

If C is the centre of this circle, then angle $ACB = \dfrac{\pi}{2}$ (the angle at the centre is twice the angle at the circumference) and C lies on the y axis (perpendicular bisector of chord AB).

The locus of P can now be drawn more clearly:

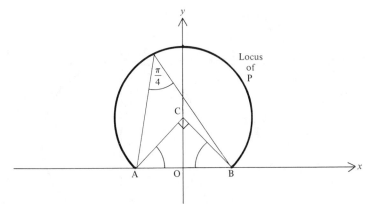

Since OB $=$ OC $\left(\angle CAO = \angle CBO = \dfrac{\pi}{4}\right)$ the coordinates of C are $(0, 1)$.

Note. The Cartesian equation of a circle with centre $(0, 1)$ and radius $\sqrt{2}$
(i.e. AC) is $x^2 + y^2 - 2y = 1$.   This equation represents a *complete* circle
however and not just the major arc which is the locus of P.
Thus the Cartesian equation of the locus of P is

$$x^2 + y^2 - 2y = 1 \quad and \quad y > 0$$

6) If $\mathrm{Re}\left(z - \dfrac{1}{z}\right)$ is zero, find the polar equation of the locus of $P(r, \theta)$ which
represents z on an Argand diagram. Sketch this locus.

If

$$z = r(\cos\theta + i\sin\theta)$$

then

$$\frac{1}{z} = \frac{1}{r(\cos\theta + i\sin\theta)} \times \frac{\cos\theta - i\sin\theta}{\cos\theta - i\sin\theta}$$

$$= \frac{\cos\theta - i\sin\theta}{r}$$

Hence

$$\mathrm{Re}\left(z - \frac{1}{z}\right) = r\cos\theta - \frac{\cos\theta}{r}$$

$$= \frac{1}{r}\cos\theta\,(r^2 - 1)$$

So

$$\mathrm{Re}\left(z - \frac{1}{z}\right) = 0 \;\Rightarrow\; (r^2 - 1)\cos\theta = 0$$

and this is the polar equation of the locus of z when $\mathrm{Re}\left(z - \dfrac{1}{z}\right) = 0$

As $(r^2 - 1)\cos\theta = 0 \;\Rightarrow\; \cos\theta = 0 \quad or \quad r = \pm 1$

the locus is seen to be in two parts:

the line $\qquad\qquad\qquad \theta = \dfrac{\pi}{2}$

and the circle, centre O, of radius 1.   i.e.

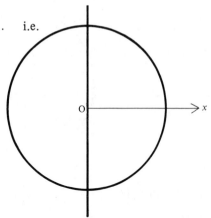

EXERCISE 14e

Sketch the locus of the point $P(x, y)$, where P represents the complex number $z = x + yi$ on an Argand diagram, if:

1) $|z| = 1$

2) $|z - 1| = 3$

3) $|z - 2i| = 3$

4) $|z + 2| = 2$

5) $|z - 1 + i| = 4$

6) $\arg z = \dfrac{\pi}{3}$

7) $\arg (z - 1) = -\dfrac{3\pi}{4}$

8) $\arg (z + 3 - 4i) = \dfrac{\pi}{4}$

9) $|z - 2 - 3i| = |z + 4 - 5i|$

10) $|z| = |z + 4i|$

11) $\arg (z - 4) - \arg z = \dfrac{\pi}{3}$

12) $\arg (z + 2) = \arg (z - 2) - \dfrac{\pi}{6}$

13) $\arg (z - i) + \dfrac{\pi}{4} = \arg (z + i)$

14) $\arg (z - 3) = \dfrac{2\pi}{3} + \arg z$

Express, in complex form, the equations of the following loci:

15) A circle with centre (h, k) and radius r when
(a) $(h, k) = (3, 4);$ $r = 5$ (b) $(h, k) = (5, 0);$ $r = 2$
(c) $(h, k) = (0, 4);$ $r = 4$

16) Write down the polar equation of the locus of $P(r, \theta)$ where $z = r(\cos \theta + i \sin \theta)$ if:

(a) $|z| = 1$ (b) $|z| = 4$ (c) $\arg z = \dfrac{\pi}{4}$ (d) $\arg z = -\dfrac{2\pi}{3}$

17) Write down the Cartesian equation of each locus given in Question 16.

18) Find the equation (i) in complex form, (ii) in Cartesian form, of the perpendicular bisector of the line joining the points A and B where the coordinates of A and B are:

(a) $(4, 0), (-8, 0)$ (b) $(1, 2), (7, -4)$ (c) $(0, 6), (6, 0)$

19) Write down the equation, in argument form, of the following part-lines:

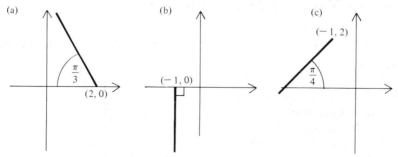

Find the Cartesian equations equivalent to the following equations, in which $z = x + yi$. Sketch the locus of $P(x, y)$ on an Argand diagram.

20) $2|z - 2| = |z + 4i|$ 21) $\text{Im}\,(z^2) = 0$

Find the polar equations of the loci in Questions 22 to 24 using $z = r(\cos\theta + i\sin\theta)$. Sketch the locus of z on an Argand diagram.

22) $\text{Im}\left(z - 1 + \dfrac{4}{z}\right) = 0$ 23) $\text{Re}\left(\dfrac{z - 2}{z}\right) = 0$

24) $\text{Re}\left(z - \dfrac{1}{\bar{z}}\right) = 0$ where $\bar{z}$ denotes the conjugate of z.

25) If $|z_1 - z_2| = |z_1 + z_2|$ show that the arguments of z_1 and z_2 differ by $\dfrac{\pi}{2}$.

26) Sketch the loci defined by the following conditions, determining first the equivalent Cartesian equation where necessary.

(a) $\left|\dfrac{z - 2}{z + 6}\right| = 1$ (b) $\left|\dfrac{z - 3i}{z - 1 + i}\right| = 3$

(c) $\arg\left(\dfrac{z}{z + 5}\right) = \dfrac{\pi}{2}$ (d) $\arg\left(\dfrac{z - 3 - i}{z + 5 - 3i}\right) = \dfrac{\pi}{3}$

FURTHER USE OF THE ARGAND DIAGRAM

Many interesting and varied problems involving complex numbers can be solved very simply by using an Argand diagram. This approach is illustrated by the following examples, in all of which $z = x + yi$, P is a point (x, y) and $\overrightarrow{OP}$ represents z on the Argand diagram.

EXAMPLES 14f

1) If $\arg(z + 3) = \dfrac{\pi}{3}$ find the least value of $|z|$.

Since $\arg(z + 3) = \arg[z - (-3)] = \dfrac{\pi}{3}$, P lies anywhere on the line shown in the diagram.

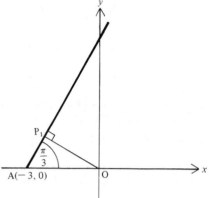

Now $|z|$ means 'the length of OP'.
So we require the point P on the line, which is nearest to O.
This is the point P_1 where OP_1 is perpendicular to AP_1.

Then $OP_1 = OA \sin \dfrac{\pi}{3} = \dfrac{3\sqrt{3}}{2}$

Thus the least value of $|z|$ is $\dfrac{3\sqrt{3}}{2}$

2) Shade the area represented on an Argand diagram by $|z + i| < 4$.

As $z + i = z - (-i)$ the equation can be expressed in the form:

'the length of $AP < 4$ units' where A is $(0, -1)$.

When $AP = 4$ units, P lies on a circle with centre A and radius 4.
When $AP < 4$, P lies inside this circle.
Therefore the area is as shown below.

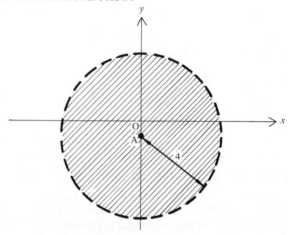

3) If $|z - 3 + 3i| = 2$, find the greatest and least values of $|z + 1|$.

Given $|z - (3 - 3i)| = 2$, we see that $P(x, y)$ lies on a circle of radius 2 and centre $(3, -3)$.
Also $|z + 1| = |z - (-1)|$ means 'the length of AP' where A is the point $(-1, 0)$.

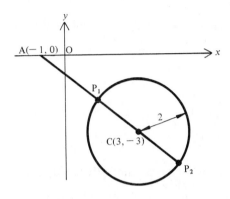

Therefore we require the least and greatest distances from A to any point on the circle.
These are the distances AP_1 and AP_2 where $AP_1 P_2$ passes through the centre C of the circle.

The length of AC $= \sqrt{[3 - (-1)]^2 + [-3]^2} = 5$

Therefore $AP_1 = AC - CP_1 = 5 - 2 = 3$

and $AP_2 = AC + CP_2 = 5 + 2 = 7$

Thus the greatest value of $|z + 1|$ is 7 and the least value of $|z + 1|$ is 3.
(**Note.** If the coordinates of P_1 and P_2 are required, the Cartesian equations of the circle and the line AC can be solved simultaneously.)

4) Indicate on an Argand diagram what is meant by $0 < \arg(z + 2 + 3i) \leqslant \dfrac{\pi}{6}$

$z + 2 + 3i = z - (-2 - 3i)$ which is represented by a line AP where A is the point $(-2, -3)$.
The inclination of this line to the positive x direction must be between

0 and $\dfrac{\pi}{6}$.

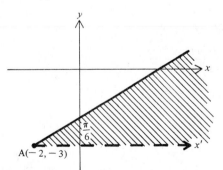

Thus P can lie anywhere in the area shaded in the diagram, so this area is indicated by the given inequality.

EXERCISE 14f

1) Shade on an Argand diagram the areas represented by the following inequalities.

(a) $|z - 1| < 4$

(b) $|z + 3i| > 2$

(c) $|z + 1 - i| < 1$

(d) $\dfrac{\pi}{3} < \arg z < \dfrac{2\pi}{3}$

2) Shade on an Argand diagram the region occupied by the set of points $P(x, y)$ for which $|z| < 5$ _and_ $-\dfrac{\pi}{6} < \arg z < \dfrac{\pi}{6}$, where $z = x + yi$.

3) Find the least value of $|z - 1|$ if:

(a) $\arg(z + 1) = -\dfrac{\pi}{4}$

(b) $|z + 3 - i| = 2$

4) Find the value(s) of z for which:

(a) $|z| = 4$ and $\arg z = \dfrac{\pi}{4}$

(b) $|z + 2 + i| = 5$ and $\operatorname{Re}(z - 1) = 0$

5) Find the points of intersection of the loci on an Argand diagram defined by:

(a) $|z - 1 + 2i| = |z + 1|$ and $|z - 1| = \sqrt{2}$

(b) $\arg z = -\dfrac{\pi}{4}$ and $|z| = 2$

6) Indicate on an Argand diagram, the set of points $P(x, y)$ for which:

(a) $0 \leqslant \arg(z + 1) \leqslant \dfrac{\pi}{3}$ and $|z + i| = 3$

(b) $|z + 3 - 2i| < 4$ and $\arg(z + 1) = \dfrac{5\pi}{6}$

(c) $|z| > 1$, $|z| < 4$ and $\arg z = -\dfrac{3\pi}{4}$

SUMMARY

1) $(a + bi) \pm (c + di) \equiv (a \pm c) + (b \pm d) i$

$(a + bi)(c + di) \equiv (ac - bd) + (ad + bc) i$

$\dfrac{(a + bi)}{(c + di)} \equiv \dfrac{(a + bi)(c - di)}{(c + di)(c - di)} \equiv \dfrac{(a + bi)(c - di)}{c^2 + d^2}$

2) $x + yi = a + bi \iff x = a$ and $y = b$

3) $|a + bi| = \sqrt{a^2 + b^2}$

$\arg(a + bi) = \alpha$ where $\tan \alpha = \dfrac{b}{a}$ and $-\pi < \alpha \leqslant \pi$.

4) $x + yi = r(\cos \theta + i \sin \theta)$ where $r = |x + yi|$ and $\theta = \arg(x + yi)$.

5) If $z_1 = x_1 + y_1 i$ and $z_2 = x_2 + y_2 i$ then:
$$|z_1 + z_2| \leqslant |z_1| + |z_2|$$
$$|z_1 + z_2| \geqslant |z_1| \sim |z_2|$$

6) If $z = a + bi$, $\bar{z} = a - bi$, where $\bar{z}$ and z are conjugate.

7) If a quadratic or cubic equation has any complex roots, they occur in conjugate pairs.

8) If $z_1 = r_1(\cos\theta_1 + i\sin\theta_1)$ and $z_2 = r_2(\cos\theta_2 + i\sin\theta_2)$
then
$$|z_1 z_2| = r_1 r_2$$
$$\text{Arg } z_1 z_2 = \theta_1 + \theta_2$$
$$\left|\frac{z_1}{z_2}\right| = \frac{r_1}{r_2}$$
$$\text{Arg } \frac{z_1}{z_2} = \theta_1 - \theta_2$$

MULTIPLE CHOICE EXERCISE 14

(Instructions for answering these questions are given on p. xii.)

TYPE I

1) The modulus of $12 - 5i$ is:
(a) 119   (b) 7   (c) 13   (d) $\sqrt{119}$ (e) $\sqrt{7}$.

2) On an Argand diagram OP represents a complex number z. The conjugate of z is $\bar{z}$.
If P and Q are the points $(3, 5)$ and $(5, -3)$ then OQ represents:
(a) $-\bar{z}$ (b) $i\bar{z}$ (c) $-z$ (d) iz (e) $-iz$.

3) $\dfrac{3 + 2i}{3 - 2i}$ is equal to:

(a) $\dfrac{5 + 12i}{13}$ (b) $\dfrac{13 + 12i}{13}$ (c) $\dfrac{5 + 6i}{13}$ (d) $\dfrac{5 + 6i}{5}$ (e) $\dfrac{13 + 12i}{5}$.

4) When $\sqrt{3} - i$ is divided by $-1 - i$ the modulus and argument of the quotient are respectively:

(a) $2\sqrt{2}, \dfrac{7\pi}{12}$ (b) $\sqrt{2}, -\dfrac{11\pi}{12}$ (c) $\sqrt{2}, \dfrac{7\pi}{12}$

(d) $2\sqrt{2}, -\dfrac{11\pi}{12}$ (e) $\sqrt{2}, \dfrac{11\pi}{12}$.

5) The set of points on an Argand diagram which satisfy both $|z| \leqslant 4$ and $\arg z = \dfrac{\pi}{3}$, is:

(a) a circle and a line   (b) an infinite part-line

(c) a diameter of a circle   (d) a radius of a circle
(e) Two points only, where a line cuts a circle.

6) The equation $x^2 + 3x + 1 = 0$ has:
(a) no roots   (b) one real and one complex root
(c) two imaginary roots   (d) two real roots
(e) two complex roots.

7) Expressed in the form $r(\cos\theta + i\sin\theta)$, $-2 + 2i$ becomes:

(a) $2\left[\cos\left(-\dfrac{\pi}{4}\right) + i\sin\left(-\dfrac{\pi}{4}\right)\right]$ (b) $2\left(\cos\dfrac{3\pi}{4} + i\sin\dfrac{3\pi}{4}\right)$

(c) $2\sqrt{2}\left[\cos\left(-\dfrac{3\pi}{4}\right) + i\sin\left(-\dfrac{3\pi}{4}\right)\right]$ (d) $2\sqrt{2}\left[\cos\left(-\dfrac{\pi}{4}\right) + i\sin\left(-\dfrac{\pi}{4}\right)\right]$

(e) none of these.

8) On an Argand diagram P is the point (x, y) and A is the point (x_1, y_1).
If $z = x + yi$ and $z_1 = x_1 + y_1 i$, $|z - z_1|$ represents:
(a) a circle with centre z_1 (b) the length of OA
(c) the direction of AP   (d) the diagonal OC of the parallelogram
(e) the length of AP.   OACP

9) $|z_1 - z_2| \leqslant$
(a) $|z_1| - |z_2|$ (b) $|z_2| - |z_1|$ (c) $|z_1| \sim |z_2|$
(d) $|z_1| + |z_2|$ (e) None of these.

TYPE II

10) $-\dfrac{5\pi}{6}$ is an argument of:

(a) $\cos\dfrac{5\pi}{6} - i\sin\dfrac{5\pi}{6}$ (b) $\cos\dfrac{11\pi}{6} + i\sin\dfrac{11\pi}{6}$

(c) $-\sqrt{3} - i$ (d) $\sqrt{3} - i$ (e) $-\sqrt{3} + i$.

11) Which of the following lines can represent z if $z = -2 + 3i$?

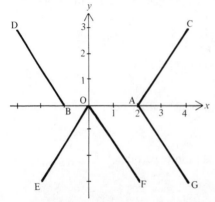

(a) $\overrightarrow{CA}$ (b) $\overrightarrow{BD}$
(c) $\overrightarrow{EO}$ (d) $\overrightarrow{OF}$
(e) $\overrightarrow{GA}$

12) $\bar{z}$ is the conjugate of z:
(a) $|\bar{z}| = |z|$ (b) $\arg z = \arg \bar{z}$ (c) $z\bar{z}$ is real   (d) $z/\bar{z}$ is real
(e) $\bar{z}$ is the mirror image of z in the y axis.

13) In an Argand diagram, $\overrightarrow{OA}$ represents z, $\overrightarrow{OB}$ represents iz, $\overrightarrow{OC}$
represents $i^2 z$, $\overrightarrow{OD}$ represents $-iz$:
(a) ABCD is a straight line,

(b) $\overrightarrow{OC}$ represents $\bar{z}$,

(c) ABCD is a square,

(d) A, B, C and D lie on a circle,

(e) $\overrightarrow{BD}$ represents a real number.

14) $z = x + yi$ and P is the point (x, y) on an Argand diagram. If $|z| = 9$
(a) an equation of the locus of P is $x^2 + y^2 = 9$,
(b) the conjugate of z is $x - yi$,
(c) an equation of the locus of P is $r = 9$,
(d) z is a free vector,
(e) the argument of z is the positive acute angle $\arctan \dfrac{y}{x}$.

15) If z is any cube root of unity, the value of $1 + z + z^2$ can be:

(a) 0   (b) 1   (c) 2   (d) 3   (e) -1.

TYPE III

16) (a) $z = 1 + i$

    (b) $\arg z = \dfrac{\pi}{4}$

17) (a) The sum of the roots of the equation $ax^2 + bx + c = 0$, is $-\dfrac{b}{a}$

    (b) The equation $ax^2 + bx + c = 0$ is such that $b^2 < 4ac$

18) $z = x + yi$ and P is a point (x, y) on an Argand diagram.
(a) $|z + 1 - i| = 3$.
(b) P lies on a circle with centre $(1, -1)$ and radius 3.

19) (a) $|z_1| = |z_2|$
    (b) $\arg z_1 = \arg z_2$

20) (a) $z = 9 - 16i$
    (b) $\sqrt{z} = 3 - 4i$

TYPE V

21) A complex number has only one argument.

22) $|z_1 - z_2| \geqslant |z_1| \sim |z_2|$.

23) Any complex number whose modulus is unity can be expressed as $\cos \theta + i \sin \theta$.

24) If any cube root of 1 is squared, the result is another of the cube roots of 1.

25) A complex number $a + bi$ is zero if $a = -b$.

MISCELLANEOUS EXERCISE 14

1) The two complex numbers z_1, z_2 are represented on an Argand diagram. Show that $|z_1 + z_2| \leqslant |z_1| + |z_2|$.
If $|z_1| = 6$ and $z_2 = 4 + 3i$, show that the greatest value of $|z_1 + z_2|$ is 11 and find its least value. (U of L)p

2) If $z_1 = \dfrac{2-i}{2+i}$, $z_2 = \dfrac{2i-1}{1-i}$, express z_1 and z_2 in the form $a + ib$.
Sketch an Argand diagram showing points P and Q representing the complex numbers $5z_1 + 2z_2$ and $5z_1 - 2z_2$ respectively. (U of L)p

3) If $(1 + 3i)z_1 = 5(1 + i)$, express z_1 and z_1^2 in the form $x + yi$, where x and y are real.
Sketch in an Argand diagram the circle $|z - z_1| = |z_1|$ giving the coordinates of its centre. (U of L)p

4) (a) If $z = 4 - 3i$ express $z + \dfrac{1}{z}$ in the form $a + ib$.

   (b) Find the two square roots of $4i$ in the form $a + ib$.

   (c) If $z_1 = 5 - 5i$ and $z_2 = -1 + 7i$ prove that:

$$|z_1 + z_2| < |z_1 - z_2| < |z_1| + |z_2| \tag{C}$$

5) Express the complex number $\dfrac{5 + 12i}{3 + 4i}$ in the form $a + ib$ and in the form $r(\cos \theta + i \sin \theta)$ giving the values of a, b, r, $\cos \theta$, $\sin \theta$. (C)p

6) The complex numbers $z_1 = \dfrac{a}{1+i}$, $z_2 = \dfrac{b}{1+2i}$ where a and b are real, are such that $z_1 + z_2 = 1$. Find a and b.
With these values of a and b, find the distance between the points which represent z_1 and z_2 in the Argand diagram. (JMB)

7) Find the modulus and argument of. $z_1 = \sqrt{3} + i$. If $z_2 = \sqrt{3} - i$ express $q = z_1/z_2$ in the form $a + bi$ where a and b are real.

Plot z_1, z_2 and q on an Argand diagram and sketch the curve given by the equation $|z_1 - z_2| = |q - z_1|$. (U of L)

8) Given that z is one of the three cube roots of unity, find the two possible values of the expression $z^2 + z + 1$.

Given that ω is a complex cube root of unity, simplify each of the expressions: $(1 + 3\omega + \omega^2)^2$ and $(1 + \omega + 3\omega^2)^2$, and show that their product is equal to 16 and that their sum is -4. (JMB)

9) (a) If $z_1 = 1 - i$ and $z_2 = 7 + i$,   find the modulus of:

   (i) $z_1 - z_2$ (ii) $z_1 z_2$ (iii) $\dfrac{z_1 - z_2}{z_1 z_2}$

   (b) Sketch on an Argand diagram the locus of a point P representing the complex number z, where

   $$|z - 1| = |z - 3i|$$

   and find z when $|z|$ has its least value on this locus. (U of L)

10) (a) If $z = 3 + 4i$,   express in the form $a + bi$:

   (i) $\dfrac{1}{z^2}$ (ii) $\sqrt{z}$

   (b) Sketch on an Argand diagram the curve described by the equation $|z - 3 + 6i| = 2|z|$ and express the equation of this curve in Cartesian form. (U of L)

11) (a) If $a = 3 - i$ and $b = 1 + 2i$,   find the moduli of:

   (i) $2a + 3b$ (ii) $\dfrac{a}{2b}$

   (b) Sketch the locus defined by $|z| = 3$.   If $c = 5 + i$ and $|z| = 3$ find the greatest and least values of $|z + c|$. (C)

12) The points A and B represent the complex numbers z_1 and z_2 respectively on an Argand diagram, where $0 < \arg z_2 < \arg z_1 < \dfrac{\pi}{2}$.

Give geometrical constructions to find the *points* C and D representing $z_1 + z_2$ and $z_1 - z_2$ respectively.

Given that $\arg(z_1 - z_2) - \arg(z_1 + z_2) = \dfrac{\pi}{2}$,   prove that $|z_1| = |z_2|$. (JMB)

13) If $z = \cos\theta + i\sin\theta$ where θ is real, show that:

$$\frac{1}{1 + z} = \frac{1}{2}\left(1 - i\tan\frac{\theta}{2}\right)$$

Express: (a) $\dfrac{2z}{1 + z^2}$ (b) $\dfrac{1 - z^2}{1 + z^2}$ in the form $a + ib$ where a and b are real functions of θ. (C)

14) (a) Given that the complex number z and its conjugate $\bar{z}$ satisfy the equation

$$z\bar{z} + 2iz = 12 + 6i$$

find the possible values of z.

(b) Mark in an Argand diagram the points representing:

(i) the complex numbers $4 + 3i$, $4 - 3i$, $\dfrac{4 + 3i}{4 - 3i}$,

(ii) the three cube roots of unity.   (JMB)p

15) (a) If $z = 3 + 4i$, express $z + \dfrac{25}{z}$ in its simplest form.

(b) If $z = x + yi$, find the real part and the imaginary part of $z + \dfrac{1}{z}$.

Find the locus of points in the Argand diagram for which the imaginary part of $z + \dfrac{1}{z}$ is zero.   (U of L)p

16) (a) Find the square roots of $(5 + 12i)$.

(b) Find the modulus and amplitude of each of the numbers
(i) $(1 - i)$ (ii) $(4 + 3i)$ (iii) $(1 - i)(4 + 3i)$.
If these numbers are represented in an Argand diagram by the points A, B, C calculate the area of the triangle ABC.

(c) Find the ratio of the greatest value of $|z + 1|$ to its least value when $|z - i| = 1$.   (U of L)

17) (a) Find the modulus and one value for the argument of $\dfrac{(i + 1)^2}{(i - 1)^4}$.

Find the two square roots of $5 - 12i$ in the form $a + bi$ where a and b are real. Show the points P and Q representing the square roots in an Argand diagram. Find the complex numbers represented by points R_1, R_2 such that the triangles PQR_1, PQR_2 are equilateral. (U of L)

18) Express each of the complex numbers

$$z_1 = (1 - i)(1 + 2i), \quad z_2 = \frac{2 + 6i}{3 - i}, \quad z_3 = \frac{-4i}{1 - i}$$

in the form $a + bi$, where a and b are real.
Show that $|z_2 - z_1| = |z_1 - z_3|$ and that, for principal values of the

arguments, $\arg(z_2 - z_1) - \arg(z_1 - z_3) = \dfrac{\pi}{2}$.

If z_1, z_2, z_3 are represented by points P_1, P_2, P_3 respectively in an Argand diagram, prove that P_1 lies on the circle with P_2P_3 as diameter.   (U of L)p

19) (a) Express:
(i) $(3 + 2i)^2$ (ii) $\dfrac{1}{(3 + 2i)^2}$ in the form $x + iy$.

(b) If P is the point on an Argand diagram representing the complex number z and $|z - 1| = 3|z + i|$,   sketch the locus of P and express the equation of this locus in Cartesian form.
Find the points on the locus which satisfy the equation
$$|z| = |z - 1 + i|$$
(U of L)

20) (a) Show on an Argand diagram the locus of z when:
(i) $|z - 1 - i| = 2$
(ii) Re $z = 1$ and $-\dfrac{\pi}{3} \leqslant \arg z \leqslant \dfrac{\pi}{4}$.

In each case find the greatest value of $|z|$.

(b) The coordinates of a point P in one Argand diagram are (x, y) and are expressed in complex form $z = x + iy$. If the coordinates of a point Q in a second Argand diagram are (u, v) express the coordinates of Q in complex form w.
If $z = w^2$, find x and y in terms of u and v and show that, if P lies on the circle $x^2 + y^2 = 16$, then Q lies on the circle $u^2 + v^2 = 4$.
(AEB)'73

21) (a) Find the modulus and the argument of each root of the equation
$$z^2 + 4z + 8 = 0$$

If the roots are denoted by α and β, simplify the expression
$$(\alpha + \beta + 4i)/(\alpha\beta + 8i)$$

(b) If $z_1 = -1 + i\sqrt{3}$ and $z_2 = \sqrt{3} + i$, show in an Argand diagram points representing the complex numbers z_1, z_2, $(z_1 + z_2)$ and (z_1/z_2).
(U of L)

22) (a) If $z = x + jy$ and $z_1 = x_1 + jy_1$, where x, y, x_1 and y_1 are real, prove that when $|z + z_1| = |z - z_1|$ then $\dfrac{jz}{z_1}$ is real.
Find the locus of z if $|z + 1| + |z - 1| = 4$.

(b) Prove that the modulus of $2 + \cos\theta + j\sin\theta$ is $(5 + 4\cos\theta)^{\frac{1}{2}}$
Hence show that the modulus of
$$\frac{2 + \cos\theta + j\sin\theta}{2 + \cos\theta - j\sin\theta}$$
is unity.
(AEB)'67

23) (a) Express the following complex numbers in the form $a + ib$, where a and b are real:
(i) $\dfrac{1 - i}{(3 - i)^2}$ (ii) $(c + i)^4$, where c is real.

(b) If $z = x + iy$ and $z^2 = a + ib$ where x, y, a, b are real, prove that $2x^2 = \sqrt{(a^2 + b^2)} + a$.

By solving the equation $z^4 + 6z^2 + 25 = 0$ for z^2, or otherwise, express each of the four roots of the equation in the form $x + iy$.

(JMB)

24) (a) Find the complex number z which satisfies the equation $(2 - 3i)z = 4 + i$.

(b) In an Argand diagram shade the region in which the point representing the complex number z can lie if $|z + 2i - 1| < |z - i|$.   (U of L)

25) (a) If in an electric circuit the two branches are in parallel and have impedances z_1 and z_2 ohms, the total impedance of the circuit, z ohms, is given by $\dfrac{1}{z} = \dfrac{1}{z_1} + \dfrac{1}{z_2}$. If $z_1 = 3 + 4j$ and $z_2 = \dfrac{5}{2} + \dfrac{5}{2}j$ calculate z in the form $x + jy$ and in the form $r \angle \theta$ (i.e. in modulus, argument form).

(b) If z is the point $x + jy$ in the Argand z plane and $\left| \dfrac{z}{z - 3} \right| = \dfrac{1}{2}$, find the locus of z.   (AEB)'66

26) In each part of this question the argument of a complex number z is to satisfy the inequalities $-\pi < \arg z < \pi$.

(a) The complex number z has modulus r and argument θ, where $0 < \theta < \dfrac{1}{2}\pi$.

Find, in terms of r and θ, the modulus and argument of:

(i) z^2 (ii) $\dfrac{1}{z}$ (iii) iz.

(b) If z is any complex number such that $|z| = 1$ prove, using an Argand diagram or otherwise, that $1 \leqslant |2 + z| \leqslant 3$ and that $-\dfrac{1}{6}\pi \leqslant \arg(2 + z) \leqslant \dfrac{1}{6}\pi$.   (C)

CHAPTER 15

PERMUTATIONS
AND COMBINATIONS

PERMUTATIONS AND COMBINATIONS

Three pictures are to be hung in line on a wall. Indicating the different pictures by A, B and C, one order in which they can be hung is A, B, C and another is A, C, B.

Each of these arrangements is called a *permutation* of the three pictures (and there are further possible permutations),

i.e. a permutation is an ordered arrangement of a number of items.

Suppose, however, that seven pictures are available for hanging and only three of them can be displayed. This time a choice has first to be made. Representing the seven pictures by A, B, C, D, E, F and G, one possible choice of the three pictures for display is A, B, and C. Regardless of the order in which they are then hung this group of three is just one choice and is called a combination.

thus
$$\left.\begin{array}{l} A, B, C \\ A, C, B \\ B, A, C \\ B, C, A \\ C, A, B \\ C, B, A \end{array}\right\}$$ are *six* different *permutations* but only *one combination*

i.e. a combination is an unordered selection of a number of items from a given set.

In this chapter we will investigate methods for finding the total number of ways of arranging items or choosing groups of items from a given set. But before we do so it is important to be able to distinguish between permutations and combinations.

Consider the following situation.

A street news vendor stocks ten weekly periodicals and has a display stand with five racks in a vertical column. He clearly cannot display all ten of his magazines so first he must *choose a group of five*. The order in which he picks up his chosen five periodicals is irrelevant; the set of five is only *one combination*. Once he has made his choice he is then able to place the five periodicals in various different orders on the display stand. He is now *arranging* them and each arrangement is a *permutation*,

i.e. a particular set of five periodicals is one combination but can be arranged to give several different permutations.

EXERCISE 15a

In each of the following problems determine, *without* working out the answer, whether you are asked to find a number of permutations, or a number of combinations.

1) How many arrangements of the letters A, B, C are there?

2) A team of six members is chosen from a group of eight. How many different teams can be selected?

3) A person can take eight records to a desert island, chosen from his own collection of one hundred records. How many different sets of records could he choose?

4) The first, second and third prizes for a raffle are awarded by drawing tickets from a box of five hundred. In how many ways can the prizes be won?

5) A London telephone number is a seven digit number. How many London telephone lines are available?

6) One red die and one green die are rolled (each numbered one to six). In how many ways can a total score of six be obtained?

PERMUTATIONS

We will now consider methods for finding a number of permutations in different types of problems.

1) How many arrangements of the letters A, B, C are there?

The first (i.e. L.H.) letter can be

$$\left. \begin{array}{c} A \\ \text{or} \\ B \\ \text{or} \\ C \end{array} \right\} \quad \text{i.e. there are three ways of choosing the first letter.}$$

When the first letter has been chosen there are two letters from which to choose the second; and the possible ways of choosing the first two letters are:

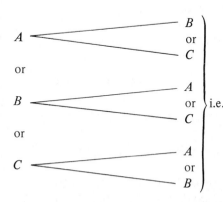

for *each* of the three ways of choosing the first letter there are two ways of choosing the second letter. Hence there are 3×2 ways of choosing the first two letters.

Having chosen the first two letters there is only one choice for the third letter, i.e. for *each* of the 3×2 ways of choosing the first two letters there is only one possibility for the third letter. Hence there are $3 \times 2 \times 1$ ways of arranging the three letters A, B, C.

2) How many three digit numbers can be made from the integers $2, 3, 4, 5, 6$ if:
(a) each integer is used only once,
(b) there is no restriction on the number of times each integer can be used?

(a) the first digit can be any of the integers $2, 3, 4, 5, 6$, i.e. there are five ways of writing down the first digit. Having written down the first digit there are four integers from which the second digit can be taken. So the possible ways of writing down the first two digits are:

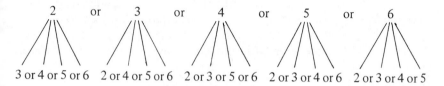

i.e. for *each* of the five ways of writing down the first digit there are four ways of writing down the second digit, i.e. there are 5×4 ways of writing down the first two digits.

The next digit can be taken from the three remaining integers, so for *each* of 5×4 ways of writing down the first two digits there are three ways of writing down the third digit,

i.e. there are $5 \times 4 \times 3$ ways of making the three digit number, i.e. there are 60 such numbers.

(b) In this problem each integer from the set may be selected up to three times. Thus there are 5 ways of taking the first integer and as the integer used for the first place can be used again for the second place there are 5 ways of writing down the second digit. Similarly there are 5 ways of taking the third digit,

i.e. there are $5 \times 5 \times 5$ ways of making the three digit number, so there are 125 such numbers.

We will now look at some examples where the possible arrangements are restricted in some way.

3) How many arrangements of the letters of the word BEGIN are there which start with a vowel?

Starting with the first letter we see that there are 2 possibilities, i.e. E or I.

Having taken the first letter there are 4 possibilities for the second letter, so for *each* of the 2 ways of taking the first letter there are 4 ways of taking the second letter, so there are 2×4 ways of taking the first two letters.

Having removed the first two letters there are 3 possibilities left for the third letter, and having selected the third letter there are 2 possibilities for the fourth letter and only 1 possibility for the fifth letter.

Hence there are $2 \times 4 \times 3 \times 2 \times 1$ ways of arranging the letters B, E, G, I, N in which a vowel comes first, i.e. there are 48 such arrangements.

4) How many even numbers greater than 2000 can be made from the integers 1, 2, 3, 4, if each integer is used only once?

There are two restrictions on the permutations here:

(a) the number is even so the last digit must be either 2 or 4,

(b) the number exceeds 2000 so the first digit must be either 2, 3 or 4.

Beginning with the last digit, we see from (a) that there are 2 possibilities. Having used one of these this leaves only 2 possibilities for the first digit. So for each of the 2 ways of writing down the last digit there are 2 ways of writing down the first digit, i.e. there are 2×2 ways of writing down the first and last digits. There are now only 2 integers from which the second digit can be taken. Having written down the second digit this leaves only 1 integer for third place. Therefore there are $2 \times 2 \times 2 \times 1 = 8$ such numbers.

This argument can be seen clearly in the table below.

Number of	1st digit	2nd digit	3rd digit	4th digit
possibilities	2	2	1	2

Note that we start with the item in the arrangement whose choice from the given set is most restricted.

4) The representatives of five countries attend a conference. In how many ways can they be seated at a round table?

As they are seated at a round table, there is no first or last place to consider. What matters in this arrangement is where each one sits relative to the others, as no one seat is special.

Numbering the chairs 1 to 5, we can say that there are 5 ways of selecting the occupant of the first seat, 4 ways of selecting the occupant of the second seat, ... and so on. Thus there are $5 \times 4 \times 3 \times 2 \times 1$ ways of arranging the representatives in the five seats. But this number includes the arrangements shown below:

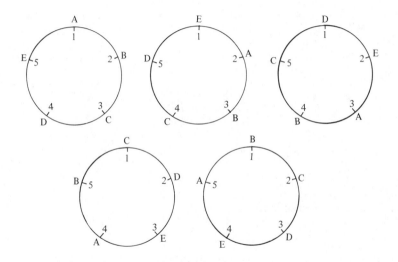

i.e. for any *one* arrangement in the five seats, the representatives can be moved clockwise *five* times and still have the *same* people to the left and to the right of them.

Therefore the $5 \times 4 \times 3 \times 2 \times 1$ ways of arranging the five representatives in numbered seats is *five times* the number of ways of arranging them round a circular table, so there are $\dfrac{5 \times 4 \times 3 \times 2 \times 1}{5}$ ways, or $4 \times 3 \times 2 \times 1$ ways in which the representatives can be seated at the round table.

[This problem does assume that all the chairs are identical and that there are no distinguishing features of the room which might affect the choice of position of a particular delegate (such as not wishing to sit with his back to the window). If considerations such as these do affect a particular arrangement they will be stated.]

5) In how many ways can five beads, chosen from eight different beads be threaded on to a ring?

The number of ways of arranging five beads, taken from eight different beads, in five numbered places is $8 \times 7 \times 6 \times 5 \times 4$.

Thus the number of ways of arranging five (from eight) beads in a circle is $\dfrac{8 \times 7 \times 6 \times 5 \times 4}{5}$ as:

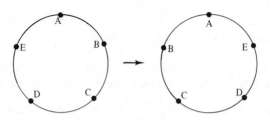

But a ring can be turned over,

i.e.

and these have been counted as two separate arrangements. So the number of circular arrangements is *twice* the number of arrangements on a ring.

Therefore the number of ways of threading five beads, from eight different beads, on a ring is $\dfrac{8 \times 7 \times 6 \times 5 \times 4}{5 \times 2} = 672.$

EXERCISE 15b

1) In how many ways can five different books be arranged on a shelf?

2) How many two digit numbers can be made from the set $\{2, 3, 4, 5, 6, 7, 8, 9\}$, each number containing two different digits?

3) In how many ways can six different shrubs be planted in a row?

4) How many four digit odd numbers can be made from the set $\{5, 7, 8, 9\}$, no integer being used more than once?

5) In how many ways can eight people be seated at a round table?

6) How many numbers greater than 4000 can be made from the set $\{1, 3, 5, 7\}$, if each integer can be used only once?

7) How many arrangements can be made of three letters chosen from PEAT if the first letter is a vowel and each arrangement contains three different letters?

8) A telephone dial is numbered 0 to 9. If the 0 is dialled first, the caller is connected to the STD system. How many local calls (i.e. calls not going through the STD system) can be rung, if a local number has five digits?

9) How many three digit numbers can be made from the set of integers $\{1, 2, 3, 4, 5, 6, 7, 8, 9\}$ if:
(a) the three digits are all different,
(b) the three digits are all the same,
(c) the number is greater than 600,
(d) all three digits are the same and the number is odd?

10) Three boxes each contain three identical balls. The first box has red balls in it, the second blue balls and the third green balls. In how many ways can three balls be arranged in a row if:
(a) the balls are of different colours,
(b) all three balls are of the same colour?

11) In how many ways can eight cows be placed in a circular milking parlour?

12) In how many ways can six different coloured beads be arranged on a ring?

COMBINATIONS

Consider the number of arrangements of four different books on a shelf. We have seen that the number of permutations of the four books is

$$4 \times 3 \times 2 \times 1 = 24.$$

But if the order does not matter, there is only one combination of books. If there are five different books available and only four of them can be placed

on the shelf, the total number of ways in which this can be done is
$$5 \times 4 \times 3 \times 2 = 120.$$
But, for each particular set of four books,
$$4 \times 3 \times 2 \times 1 \text{ permutations} \equiv \text{ one combination}$$
i.e. for any one of the 120 permutations there are 23 other permutations of the same combination of four books.

So the *number of different sets* of four books taken from the five available

books is given by $\dfrac{\text{total number of permutations}}{\text{number of permutations of each set}}$

Generalising this argument we see that if we have n objects from which we select groups of r objects, the total number of possible groups (combinations)

is given by $\dfrac{\text{number of permutations of } r \text{ objects from } n \text{ objects}}{\text{number of permutations of } r \text{ objects among themselves}}$

EXAMPLES 15c

1) How many different hands of four cards can be dealt from a pack of fifty two playing cards?

The order in which the cards are dealt is irrelevant, it is the particular set of four cards that matters.

The number of ways of arranging four cards from fifty two is
$$52 \times 51 \times 50 \times 49.$$
The number of ways of arranging any one set of four cards among themselves is $\qquad 4 \times 3 \times 2 \times 1.$

Therefore the number of combinations of four cards from fifty two cards is

$$\frac{52 \times 51 \times 50 \times 49}{4 \times 3 \times 2 \times 1} = 270\,725$$

In the following examples we investigate problems when the selection is restricted.

2) In how many ways can five boys be chosen from a class of twenty boys if the class captain has to be included?

As one particular boy has to be chosen this leaves four boys to be chosen from nineteen boys.

The number of ways of arranging four boys from nineteen is $19 \times 18 \times 17 \times 16$.
The number of ways of arranging the four among themselves is $4 \times 3 \times 2 \times 1$.
Therefore the number of ways of choosing the four boys to join the captain is

$$\frac{19 \times 18 \times 17 \times 16}{4 \times 3 \times 2 \times 1} = 3876$$

3) A bowl of fruit contains a large number of apples, pears, oranges and bananas. In how many ways can three fruits be chosen if two of them are of the same variety?

The single fruit can be chosen in 4 ways.
The two fruits of the same variety can then be chosen in 3 ways.
(These two must be a variety different from the first, otherwise all three would be the same.)
Hence there are 4 × 3 ways of choosing three fruits, two of which are the same, i.e. the number of combinations is 12.

4) In how many ways can a party of ten children be divided into two groups of five children?

The number of combinations of five children chosen from ten is
$$\frac{10 \times 9 \times 8 \times 7 \times 6}{5 \times 4 \times 3 \times 2 \times 1} = 252$$
Whenever one group of five children is selected, the remaining five children automatically form the other group.
Using the letters A to J to identify the children, two possible selections for the first group are ABCDE and FGHIJ.
The children in the corresponding second group are FGHIJ and ABCDE respectively.
But the division of the ten children into the groups ABCDE/FGHIJ and FGHIJ/ABCDE is the *same* division,
i.e. the 252 combinations of five selected from ten is *twice* the number of divisions into two *equal* groups of five.
Hence there are 126 ways in which the children can be divided into two equal groups.

EXERCISE 15c

1) How many different combinations of six letters can be chosen from the letters A, B, C, D, E, F, G, H, if each letter is chosen only once?

2) In how many ways can the eight letters in Question 1 be divided into two groups of six and two letters?

3) A team of four children is to be selected from a class of twenty children, to compete in a quiz game. In how many ways can the team be chosen if:
(a) any four can be chosen,
(b) the four chosen must include the oldest in the class?

4) A shop stocks ten different varieties of packet soup. In how many ways can a shopper buy three packets of soup if:
(a) each packet is a different variety,
(b) two packets are the same variety?

5) In how many ways can ten different books be divided into two groups of six books and four books?

6) How many different hands of five cards can be dealt from a suit of thirteen cards?

7) In Question 6, if one of the cards dealt is the ace, how many different hands of five cards are there?

8) A large box of biscuits contains nine different varieties. In how many ways can four biscuits be chosen if:
(a) all four are different,
(b) two are the same and the others different,
(c) two each of two varieties are selected,
(d) three are the same and the fourth is different,
(e) all four are the same?

THE FACTORIAL NOTATION

Consider the number of ways of arranging a pack of fifty two playing cards in a row.
This is $52 \times 51 \times 50 \times 49 \times \ldots \times 3 \times 2 \times 1$, which is a *very* large number, is difficult to multiply out, and is cumbersome to write even when left in factor form as above. So we denote this clumsy product by $52!$

In general, $n!$ represents the number $n \times (n-1) \times (n-2) \times \ldots \times 2 \times 1$, i.e. $n!$ means the product of all the integers from 1 to n inclusive and is called 'n factorial'.

EXAMPLES 15d

1) Evaluate $\dfrac{10!}{7!}$

$$\frac{10!}{7!} = \frac{10 \times 9 \times 8 \times 7 \times 6 \times 5 \times 4 \times 3 \times 2 \times 1}{7 \times 6 \times 5 \times 4 \times 3 \times 2 \times 1} = 10 \times 9 \times 8 = 720$$

Note that we have cancelled by $7!$

2) Write $52 \times 51 \times 50 \times 49 \times 48 \times 47$ in factorial form.

$52 \times 51 \times \ldots \times 47$ is $52!$ with the factors $46 \times 45 \times \ldots \times 3 \times 2 \times 1$
(i.e. $46!$) missing.
Multiplying and dividing by $46!$ we have

$$52 \times 51 \times 50 \times 49 \times 48 \times 47$$

$$= \frac{(52 \times 51 \times 50 \times \ldots \times 47) \times (46 \times 45 \times \ldots \times 2 \times 1)}{46!} = \frac{52!}{46!}$$

3) Evaluate $\dfrac{20!}{15!\,5!}$

Cancelling by the larger factorial in the denominator gives

$$\frac{20!}{15!\,5!} = \frac{20 \times 19 \times 18 \times 17 \times 16}{5 \times 4 \times 3 \times 2 \times 1} = 15\,504$$

4) Write $\dfrac{9 \times 8 \times 7}{4 \times 3 \times 2}$ in factorial notation.

$$9 \times 8 \times 7 = \frac{9!}{6!}$$

Therefore

$$\frac{9 \times 8 \times 7}{4 \times 3 \times 2} = \frac{9!}{6!\,4!}$$

5) Factorize $8! - 4(7!)$

As $8! = 8 \times 7!$

$8! - 4(7!)$ has a common factor of $7!$

i.e. $\qquad 8! - 4(7!) = 7!(8 - 4) = 4(7!)$

6) Factorize $(n + 2)! + n^2(n - 1)!$

$$n^2(n - 1)! = [n \times n] \times [(n - 1) \times (n - 2) \times \ldots \times 3 \times 2 \times 1]$$

$$= n \times [n \times (n - 1) \times (n - 2) \times \ldots \times 3 \times 2 \times 1]$$

$$= n \times n!$$

Also $\qquad (n + 2)! = (n + 2) \times (n + 1) \times n!$

Hence $\qquad (n + 2)! + n^2(n - 1)! = (n + 2)(n + 1)n! + n(n!)$

$$= n![(n + 2)(n + 1) + n]$$

$$= n![n^2 + 4n + 2]$$

7) Express as a single fraction $\dfrac{n!}{(n - r)!\,r!} + \dfrac{2(n - 1)!}{(r - 1)!(n - r + 1)!}$

The numerators of both fractions have a common factor $(n - 1)!$
The denominators of both fractions have a common factor $(n - r)!(r - 1)!$

[**Note** $\quad (n - r + 1)! = (n - r + 1)(n - r)(n - r - 1) \times \ldots \times 3 \times 2 \times 1$

$$= (n - r + 1)(n - r)!$$

and $\qquad r! = r(r - 1)!]$

Hence

$$\frac{n!}{(n-r)!(r!)} + \frac{2(n-1)!}{(r-1)!(n-r+1)!} = \frac{(n-1)!}{(n-r)!(r-1)!}\left(\frac{n}{r} + \frac{2}{n-r+1}\right)$$

$$= \frac{(n-1)!}{(n-r)!(r-1)!}\left(\frac{n(n-r+1)+2r}{r(n-r+1)}\right)$$

$$= \frac{(n-1)!}{(n-r)!(r-1)!}\left(\frac{n^2-nr+n+2r}{r(n-r+1)}\right)$$

$$= \frac{(n-1)!}{(n-r+1)!\,r!}(n^2-nr+n+2r)$$

EXERCISE 15d

Evaluate:

1) $3!$ 2) $4!$ 3) $5!$ 4) $6!$ 5) $\dfrac{6!}{4!}$ 6) $\dfrac{12!}{10!}$

7) $\dfrac{15!}{12!}$ 8) $\dfrac{7!}{3!}$ 9) $\dfrac{8!}{2!\,6!}$ 10) $\dfrac{20!}{17!\,3!}$ 11) $\dfrac{15!}{6!\,7!}$ 12) $\dfrac{9!}{2!\,3!\,4!}$

13) $\dfrac{8!}{(4!)^2}$ 14) $\dfrac{(3!)^2}{2!\,4!}$

Write in factorial form:

15) $5 \times 4 \times 3$ 16) 11×10

17) $39 \times 38 \times 37 \times 36 \times 35$ 18) $n(n-1)(n-2)(n-3)$

19) $(n+1)(n)(n-1)$ 20) $(n+5)(n+4)(n+3)(n+2)(n+1)$

21) $(n+r)(n+r-1)(n+r-2)$ 22) $\dfrac{20 \times 19 \times 18}{3 \times 2 \times 1}$

23) $\dfrac{14 \times 13}{3 \times 2 \times 1}$ 24) $\dfrac{8 \times 7 \times 6}{6 \times 5 \times 4}$

25) $\dfrac{n(n-1)(n-2)}{3 \times 2 \times 1}$ 26) $\dfrac{(n-2)(n-3)(n-4)}{4 \times 3 \times 2 \times 1}$

Factorize:

27) $8! + 9!$ 28) $7! - 2(5!)$ 29) $3(10!) + 4(8!)$

30) $n! + (n-1)!$ 31) $(n+1)! - (n-1)!$ 32) $n^2(n-1)! + 2n(n-2)!$

33) $n! + (n-1)! + (n-2)!$ 34) $\dfrac{7!}{3!\,4!} + \dfrac{7!}{2!\,5!}$

35) $\dfrac{7!}{3!} + \dfrac{6!}{4!}$

36) $\dfrac{10!}{8!\,2!} + \dfrac{9!}{7!\,2!} + \dfrac{8!}{6!\,2!}$

37) $\dfrac{n!}{r!} + \dfrac{(n-1)!}{(r+1)!}$

38) $\dfrac{2(n+1)!}{r!} - \dfrac{3n!}{(r+1)!}$

39) $\dfrac{n!}{r!(n-r)!} + \dfrac{2(n+1)!}{r!(n-r+1)!}$

40) $\dfrac{n!r^2}{(n-r)!r!} - \dfrac{2(n-1)!}{(r-1)!(n-r)!}$

FURTHER WORK ON PERMUTATIONS AND COMBINATIONS

We have seen that the number of different ways of arranging five objects chosen from eight different objects is $8 \times 7 \times 6 \times 5 \times 4$ or, using factorial notation, $\dfrac{8!}{3!}$.

From this we can generalise to say that:
the number of permutations of r objects chosen from a set of n different objects is $\dfrac{n!}{(n-r)!}$.

Using ^nP_r as a symbol for 'the number of permutations of r objects taken from n different objects' we may write

$$^nP_r = \dfrac{n!}{(n-r)!}$$

We have also seen that the number of ways of choosing (irrespective of order) five objects from eight is $\dfrac{8 \times 7 \times 6 \times 5 \times 4}{5 \times 4 \times 3 \times 2 \times 1}$ or $\dfrac{8!}{5!3!}$.

Generalising again we can say that:
the number of combinations of r objects selected from n different objects is $\dfrac{n!}{(n-r)!\,r!}$.

Using ^nC_r as a symbol for 'the number of combinations of r objects taken from n different objects' we may write

$$^nC_r = \dfrac{n!}{(n-r)!\,r!}$$

Thus $^6P_4 = \dfrac{6!}{2!} = 6 \times 5 \times 4 \times 3 = 360$

and $^6C_4 = \dfrac{6!}{2!\,4!} = \dfrac{6 \times 5}{2 \times 1} = 15$

Also $^nC_{n-r} = \dfrac{n!}{[n-(n-r)]!(n-r)!} = \dfrac{n!}{r!(n-r)!} = {}^nC_r$

i.e.
$$^nC_{n-r} = {}^nC_r$$

Now ^nC_n means the number of ways of choosing n objects from n objects, and as there is obviously only one way of choosing n objects from n objects, $^nC_n = 1$.

But using the definition

$$^nC_r = \frac{n!}{r!(n-r)!}$$

in the case when $r = n$ we have

$$^nC_n = \frac{n!}{n!0!}$$

and this is equal to unity only if we define $0!$ as having the value 1, i.e.
$$0! = 1$$
We will now consider some more varied problems involving permutations and combinations in which we will use the factorial notation.

Permutations Involving some Identical Objects

Consider the number of possible arrangements of the letters of the word DIGIT.

This word contains two I's which are identical, but which can be distinguished by adding suffixes, i.e. $D\,I_1\,G\,I_2\,T$.

Then the number of permutations of the letters $D\,I_1\,G\,I_2\,T$ is $5!$

But this number includes separately the two permutations

$$D\,I_1\,G\,I_2\,T \quad \text{and} \quad D\,I_2\,G\,I_1\,T$$

so that the arrangement DIGIT is counted twice in the $5!$ permutations. Because I_1 and I_2 can be arranged in $2!$ ways, every distinct arrangement of the letters DIGIT is included $2!$ times in the permutations of $D\,I_1\,G\,I_2\,T$.

Hence the number of arrangements of the letters $D\,I\,G\,I\,T$ is $\dfrac{5!}{2!} = 60$.

Now consider the number of permutations of the letters

$$D\,E\,F\,E\,A\,T\,E\,D$$

There are three E's and two D's.

The number of permutations of $D_1\,E_1\,F\,E_2\,A\,T\,E_3\,D_2$ is $8!$

But E_1, E_2, E_3 can be arranged in $3!$ ways, and D_1, D_2 can be arranged in $2!$ ways, so

the number of arrangements of $D_1\,E_1\,F\,E_2\,A\,T\,E_3\,D_2$ is $3! \times 2!$ times the number of arrangements of $D\,E\,F\,E\,A\,T\,E\,D$.

Therefore the number of permutations of $D\,E\,F\,E\,A\,T\,E\,D$ is $\dfrac{8!}{3!\,2!} = 3360$.

Generalising this argument,

the number of permutations of n objects, r of which are identical is $\dfrac{n!}{r!}$.

Further, the number of permutations of n objects, p of which are alike and q of which are alike (but different from the set of p objects) is $\dfrac{n!}{p!\,q!}$.

EXAMPLES 15e

1) In how many of the possible permutations of the letters of the word ADDING are the two D's:
(a) together, (b) separate?

(a) The number of permutations in which the D's are together can be found easily by bracketing the D's and treating them as one item in the arrangements of A, (DD), I, N, G. There are now five different items which can be arranged in 5! ways.

(b) As the D's are either together or separate,

(number of permutations without restriction)

$-$ (number of arrangements with D's together)

$=$ (number of arrangements with D's separate)

Now the number of arrangements without restriction is $\dfrac{6!}{2!}$.

Hence the number of arrangements in which the D's are separated is

$\dfrac{6!}{2!} - 5! = 240.$

Note. The number of permutations in which D and A are next to each other is found in a similar way to (a) above, but these two letters can be written (DA) or (AD). Therefore there are twice as many arrangements of A, D, D, I, N, G in which A and D are adjacent than when the two D's are adjacent.

Independent Permutations or Combinations

Consider the number of ways in which three bottles of wine and ten cans of beer can be selected from a cellar containing thirty different bottles of wine and fifteen different cans of beer.
The choice of the bottles of wine in no way affects the choice of the cans of beer,
i.e. these two combinations are independent of each other.
The three bottles of wine can be chosen in $^{30}C_3$ ways. The ten cans of beer can be chosen in $^{15}C_{10}$ ways. For *each* of the $^{30}C_3$ ways of choosing the bottles of wine there are $^{15}C_{10}$ ways of choosing the cans of beer so there are

$$^{30}C_3 \times {}^{15}C_{10}$$

ways of choosing the bottles of wine and the cans of beer.

Consider the number of ways in which a code number of two letters followed by three digits can be made if no letter or digit is repeated in any one code. In this problem we are considering the number of permutations of two letters followed by the number of permutations of three digits. These two sets of permutations are independent of each other as the arrangement of the letters has no affect on the arrangement of the digits. There are $^{26}P_2$ ways of arranging the letters and $^{10}P_3$ ways of arranging the digits in the code.
For *each* of the $^{26}P_2$ permutations of the letters there are $^{10}P_3$ permutations of the digits.
Therefore the number of permutations of two letters followed by three digits is $^{26}P_2 \times {}^{10}P_3$.
Generalising from these examples we can say that:
the number of permutations, P_1, of objects from one set *followed by* the number of permutations, P_2, of objects from an *independent* set is $P_1 \times P_2$.
This is also true for combinations of objects from two independent sets and can be extended to cover more than two sets.

Mutually Exclusive Permutations or Combinations

Care must be taken to distinguish between problems involving:

(a) *both* one set of objects *and* another set of objects, which involves a product of permutations, or

(b) *either* one set of objects *or* another set of objects.

Consider, for example, the number of ways in which a number greater than 20 can be made from the integers 2, 3, 4, no integer being repeated.
The number may contain *either* two digits *or* it may contain three digits. It *cannot* contain *both* two digits *and* three digits. Such sets of permutations are said to be mutually exclusive.
The number of ways of making a two digit number is $^3P_2$.
The number of ways of making a three digit number is $^3P_3$.
These two cases cover all possible permutations,
so there are $^3P_2 + {}^3P_3$ numbers greater than 20.
Now consider the number of ways in which a class of twenty children can be divided into two groups of six and fourteen respectively, if the class contains one set of twins who are not to be separated.
If we consider the number of ways of choosing six children from twenty, the choice of the six will determine the members of the other group.
However in choosing six children we have to consider two cases, viz,
either six children including the twins
or   six children excluding the twins.

As a combination of six children cannot both include and exclude the twins, the two separate cases are mutually exclusive.

If the twins are included, four more children must be chosen from the remaining eighteen, and this can be done in $^{18}C_4$ ways.

If the twins are excluded, the group can be chosen in $^{18}C_6$ ways.

These two cases cover all possible combinations, so there are

$$^{18}C_4 + {}^{18}C_6$$

ways of dividing the children into two groups as stated.

Generalising from these two examples we can say that if a choice (ordered or not) divides into two cases (A and B) which are mutually exclusive then

(number of choices of *either A or B*)

= (number of choices of A) + (number of choices of B)

This can be extended to cover more than two mutually exclusive cases.

To summarize: If P_1 and P_2 are two permutations (or combinations) whose values are p_1 and p_2 then

when P_1 and P_2 are independent

$$P_1 \text{ followed by } P_2 \text{ (or } P_1 \text{ and } P_2) = p_1 \times p_2$$

when P_1 and P_2 are mutually exclusive

$$(\text{either } P_1 \text{ or } P_2) = p_1 + p_2.$$

EXAMPLES 15e (continued)

2) How many permutations of the letters of the word DEFEATED are there in which the E's are separated from each other?

The most direct way of answering this problem is first to remove the E's and consider the number of permutations of the letters D F A T D, of which there are $\dfrac{5!}{2!} = 60$.

If for any one of these permutations (e.g. D F A T D) the E's are reinserted so as to be separated from each other, the three E's can be placed in three of the six positions indicated.

$$\uparrow D \uparrow F \uparrow A \uparrow T \uparrow D \uparrow$$

The number of ways of choosing three positions from six is $^6C_3 = 20$.

Hence there are twenty ways of inserting the E's into *each* of the sixty permutations of DFATD, so there are 20×60, or 1200, permutations of the letters of DEFEATED in which no two E's are adjacent.

(**Note** that the permutations of DFATD and the choice of the possible positions for the E's are independent.)

3) A box contains seven billiard balls, three of which are red, two black, one white and one green. In how many ways can three balls be chosen?

This problem divides itself into three mutually exclusive cases, viz.

(a) Three balls of the same colour;
there is only one way of choosing these (the three red balls).

(b) Two balls of the same colour, the other being a different colour.
As there are two ways of choosing the balls of the same colour (two red or two black) and three other colours in each case from which the third ball can be chosen, there are 2×3, i.e. 6 ways of choosing three balls, when two of them are of the same colour.

(c) All three balls of different colour.
There are four colours to choose from, so the three balls can be chosen in $^4C_3$ ways, i.e. 4 ways.
Thus there are $1 + 6 + 4$ ways of choosing the three balls, i.e. 11 ways.

4) Four books are taken from a shelf of eighteen books, of which six are paperback, and twelve are hardback. In how many of the possible combinations of four books is at least one a paperback?

If at least one paperback is to be included then the combinations could
include   one paperback,
or   two paperbacks,
or   three paperbacks,
or   four paperbacks.
As all of these cases are mutually exclusive we could work out the number of combinations in each case and sum them to find the number of combinations which include at least one paperback.
However a more direct approach is to say that the only combination we do *not* want is the one made up entirely of hardbacks. So we need consider only two cases,

i.e.   any four books (there are $^{18}C_4$ such combinations)

and   four hardbacked books (there are $^{12}C_4$ such combinations)

Hence $^{18}C_4 = {}^{12}C_4 +$ (number of combinations with at least one paperback).
Therefore the number of combinations including at least one paperback is

$$^{18}C_4 - {}^{12}C_4 = \frac{18!}{4!\,14!} - \frac{12!}{4!\,8!}$$

$$= 2565$$

EXERCISE 15e

1) Find how many numbers between 10 and 300 can be made from the digits $1, 2, 3$, if:
(a) each digit may be used only once,
(b) each digit may be used more than once?

2) How many combinations of three letters taken from the letters
A, A, B, B, C, C, D are there?

3) A mixed team of ten players is chosen from a class of thirty, eighteen of whom are boys and twelve of whom are girls. In how many ways can this be done if the team has five boys and five girls?

4) Find the number of permutations of the letters of the word MATHEMATICS.

5) Find the number of permutations of four letters from the word MATHEMATICS.

6) How many of the permutations in Question 5 contain two pairs of letters that are the same?

7) Find the number of ways in which twelve children can be divided into two groups of six if two particular boys must be in different groups.

8) In how many of the permutations in Question 4 do all the consonants come together?

9) A team of two pairs, each consisting of a man and a woman, is chosen to represent a club at a tennis match. If these pairs are chosen from five men and four women, in how many ways can the team be selected?

10) A bridge team of four is chosen from six married couples to represent a club at a match. If a husband and wife cannot both be in the team, in how many ways can the team be formed?

11) Two sets of books contain five novels and three reference books respectively. In how many ways can the books be arranged on a shelf if the novels and reference books are not mixed up?

12) A box contains ten bricks, identical except for colour. Three bricks are red, two are white, two are yellow, two are blue and one is black. In how many ways can three distinguishable bricks be:
(a) taken from the box   (b) arranged in a row?

13) In how many of the arrangements in a row of all ten bricks in Question 12 are:
(a) the three red bricks separated from each other,
(b) just two of the red bricks next to each other?

14) In a multiple choice question there is one correct answer and four wrong answers to each question. For two such questions, in how many ways is it possible to select the wrong answer to both questions?

15) In Question 14, if a correct answer scores one mark and a wrong answer scores zero, in answering three such questions in how many ways is it possible to score:
(a) 0,   (b) 1,   (c) 2?

16) How many even numbers less than 500 can be made from the integers 1, 2, 3, 4, 5, each integer being used only once?

17) Four boxes each contain a large number of identical balls, those in one box are red, those in a second box are blue, those in a third box are yellow and those in the remaining box are green. In how many ways can five balls be chosen if:
(a) there is no restriction   (b) at least one ball is red?

18) How many different hands of four cards can be dealt from a pack of fifty-two playing cards if at least one of the cards is an ace?

19) In how many ways can four tins of fruit be chosen from a supermarket offering ten varieties if at least two of the tins are of the same variety?

20) In how many ways can three letters from the word GREEN be arranged in a row if at least one of the letters is E?

SUMMARY

$$n! = n(n-1)(n-2)(n-3)\ldots\ldots(3)(2)(1)$$

$$0! = 1$$

$$^nP_r = \frac{n!}{(n-r)!} \qquad ^nC_r = \frac{n!}{(n-r)!\,r!} \qquad ^nC_r = {}^nC_{n-r}$$

The number of permutations of r objects chosen from n *different* objects is
$$^nP_r = \frac{n!}{(n-r)!}$$

The number of permutations of n objects, r of which are identical, is $\dfrac{n!}{r!}$

The number of circular arrangements of r objects chosen from n *different* objects is $\dfrac{n!}{(n-r)!r}$

The number of combinations of r objects chosen from n *different* objects is
$$\frac{n!}{(n-r)!\,r!}$$

If P_1 and P_2 are two permutations (or combinations) whose values are p_1 and p_2 then:
when P_1 and P_2 are independent

$$P_1 \text{ followed by } P_2 \text{ (or } P_1 \text{ and } P_2) = p_1 \times p_2$$

when P_1 and P_2 are mutually exclusive

$$(\text{either } P_1 \text{ or } P_2) = p_1 + p_2.$$

MISCELLANEOUS EXERCISE 15

1) Find the number of ways in which a committee of 4 can be chosen from 6 boys and 6 girls
(a) if it must contain 2 boys and 2 girls,
(b) if it must contain at least 1 boy and 1 girl,
(c) if either the oldest boy or the oldest girl must be included but not both.
(U of L)p

2) n boxes are arranged in a straight line and numbered 1 to n. Find:
(a) in how many ways n different articles can be arranged in the boxes, one in each box, so that a particular article A is in box 2;
(b) in how many ways the n articles can be arranged in the boxes so that the article A is in neither box 1 nor box 2 and a given article B is not in box 2.
Deduce the number of ways in which the articles can be arranged so that A is not in box 1 and B is not in box 2. (JMB)

3) A forecast is to be made of the results of five football matches, each of which can be a win, a draw or a loss for the home team. Find the number of different possible forecasts, and show how this number is divided into forecasts containing 0, 1, 2, 3, 4, 5 errors respectively. (U of L)p

4) Find in factor form the number of ways in which 20 boys can be arranged in a line from right to left so that no two of three particular boys will be standing next to each other. (U of L)p

5) Find how many distinct numbers greater than 5000 and divisible by 3 can be formed from the digits 3, 4, 5, 6 and 0, each digit being used at most once in any number. (JMB)

6) A certain test consists of seven questions, to each of which a candidate *must* give one of three possible answers. According to the answer that he chooses, the candidate *must* score 1, 2, or 3 marks for each of the seven questions. In how many different ways can a candidate score exactly 18 marks in the test? (U of L)p

7) A tennis club is to select a team of three pairs, each pair consisting of a man and a woman, for a match. The team is to be chosen from 7 men and 5 women. In how many different ways can the three pairs be selected? (U of L)p

8) n red counters and m green counters are to be placed in a straight line. Find the number of different arrangements of the colours.
A town has n streets running from south to north and m streets running from west to east. A man wishes to go from the extreme south-west intersection to the extreme north-east intersection, always moving either north or east along one of the streets. Find the number of different routes he can take. (JMB)

9) Show that there are 126 ways in which 10 children can be divided into two groups of 5. Find the number of ways in which this can be done
(a) if the two youngest children must be in the same group,
(b) if they must not be in the same group.   (U of L)p

10) A committee of three people is to be chosen from four married couples. Find in how many ways this committee can be chosen
(a) if all are equally eligible,
(b) if the committee must consist of one woman and two men,
(c) if all are equally eligible except that a husband and wife cannot *both* serve on the committee.   (U of L)p

11) Find the number of integers between 1000 and 4000 which can be formed by using the digits 1, 2, 3, 4
(a) if each digit may be used only once,
(b) if each digit may be used more than once.   (U of L)p

12) In how many different ways can the letters of the word MATHEMATICS be arranged? In how many of these arrangements will two A's be adjacent? Find the number of arrangements in which all the vowels come together.
(U of L)p

13) Code numbers, each containing three digits, are to be formed from the nine digits 1, 2, 3, . . . , 9. In any number no particular digit may occur more than once.
(a) How many different code numbers may be formed, and in how many of these will 9 be one of the three digits selected?
(b) In how many numbers will the three digits occur in their natural order (i.e. the digits being in ascending order of magnitude reading from left to right, e.g. 359)?   (C)p

CHAPTER 16

SERIES

SEQUENCES

Consider the following sets of numbers

(a) $2, 4, 6, 8, 10, \ldots$

(b) $1, 2, 4, 8, 16, \ldots$

(c) $4, 9, 16, 25, 36, \ldots$

In each set the numbers are in a given order and there is an obvious rule for obtaining the next number and as many subsequent numbers as we wish to find. For example (b) can be continued as follows, $32, 64, 128, 256, \ldots$.

Such sets are called sequences and each member of the set is called a term of the sequence.

Thus *a sequence is a set of terms in a defined order with a rule for obtaining the terms.*

SERIES

When the terms of a sequence are added, e.g. $1 + 2 + 4 + 8 + 16 + \ldots$ a series is formed.

If the series stops after a finite number of terms it is called a finite series.

Thus $1 + 2 + 4 + 8 + 16 + 32 + 64$ is a finite series of seven terms.

If the series does not stop but continues indefinitely it is called an infinite series.

Thus $1 + \frac{1}{2} + \frac{1}{4} + \frac{1}{8} + \frac{1}{16} + \frac{1}{32} + \ldots + \frac{1}{1024} + \ldots$ is an infinite series.

Consider again the series $1 + 2 + 4 + 8 + 16 + 32 + 64$.

As each term is a power of 2 we can write this series in the form

$$2^0 + 2^1 + 2^2 + 2^3 + 2^4 + 2^5 + 2^6.$$

All terms of this series are of the form 2^r, so that 2^r is a general term and we can define the series as follows:
the sum of terms of the form 2^r where r takes all integral values in order from 0 to 6 inclusive.

THE SIGMA NOTATION

Using Σ as a symbol for 'the sum of terms such as' we can redefine our series more concisely as $\Sigma\, 2^r$, r taking all integral values from 0 to 6 inclusive, or even more briefly

$$\sum_{r=0}^{6} 2^r$$

Placing the lowest and highest value that r takes, below and above the sigma symbol respectively, indicates that r also takes all integral values between these extreme values.

Thus $\displaystyle\sum_{r=2}^{10} r^3$ means the sum of all terms of the form r^3 where r takes all integral values from 2 to 10 inclusive,

i.e. $$\sum_{r=2}^{10} r^3 = 2^3 + 3^3 + 4^3 + 5^3 + 6^3 + 7^3 + 8^3 + 9^3 + 10^3$$

Note that a finite series, when written out, should always end with the last term even if several intermediate terms are omitted, e.g. $3 + 6 + 9 + \ldots + 99$.

The series $$1 + \tfrac{1}{2} + \tfrac{1}{4} + \tfrac{1}{8} + \tfrac{1}{16} + \ldots$$

may also be written in the sigma notation. The continuing dots after the last term indicate that the series is infinite (i.e. there is *no* last term).
Each term of series above is a power of $\tfrac{1}{2}$.
Thus a general term of the series can be written as $(\tfrac{1}{2})^r$.
The first term is 1 or $(\tfrac{1}{2})^0$, so the first value that r takes is zero. There is no last term of this series, so there is no upper limit for the value of r.
Therefore $1 + \tfrac{1}{2} + \tfrac{1}{4} + \tfrac{1}{8} + \tfrac{1}{16} + \ldots$ may be written as

$$\sum_{r=0}^{\infty} (\tfrac{1}{2})^r$$

Note that when a given series is rewritten in the sigma notation it is as well to check that the first few values of r give the correct first few terms of the series.

Writing a series in the sigma notation, apart from the obvious advantage of brevity, allows us to select a particular term of a series without having to write down all the earlier terms.

For example, consider the series $\sum\limits_{r=3}^{10} (2r+5)$.

The first term is the value of $2r+5$ when $r=3$, i.e. $2 \times 3 + 5 = 11$.

The last term is the value of $2r+5$ when $r=10$, i.e. 25.

The fourth term is the value of $2r+5$ when r takes its fourth value in order from $r=3$, i.e. when $r=6$.

Thus the fourth term of $\sum\limits_{r=3}^{10} (2r+5)$ is $2 \times 6 + 5 = 17$.

EXAMPLE 16a

Write the following series in the sigma notation:

(a) $1 - x + x^2 - x^3 + \ldots$

(b) $2 - 4 + 8 - 16 + \ldots + 128$.

(a) A general term of this series is $\pm x^r$, having a positive sign when r is even and a negative sign when r is odd.

Because $(-1)^r$ is positive when r is even and negative when r is odd, the general term can be written $(-1)^r x^r$.

The first term of this series is 1, or x^0

Hence $1 - x + x^2 - x^3 + \ldots = \sum\limits_{r=0}^{\infty} (-1)^r x^r$.

(b) $2 - 4 + 8 - 16 + \ldots + 128 = 2 - (2)^2 + (2)^3 - (2)^4 + \ldots + (2)^7$.

So a general term is of the form $\pm 2^r$, being positive when r is odd and negative when r is even,

i.e. the general term is $(-1)^{r+1} 2^r$.

Hence $2 - 4 + 8 - 16 + \ldots + 128 = \sum\limits_{r=1}^{7} (-1)^{r+1} 2^r$.

EXERCISE 16a

1) Write the following series in the sigma notation:

(a) $1 + 8 + 27 + 64 + 125$

(b) $2 + 4 + 6 + 8 + 10 + \ldots + 20$

(c) $3 + 6 + 9 + 12 + 15 + \ldots + 99$

(d) $\frac{1}{2} + \frac{1}{3} + \frac{1}{4} + \frac{1}{5} + \ldots + \frac{1}{50}$

(e) $1 + \frac{1}{3} + \frac{1}{9} + \frac{1}{27} + \ldots$

(f) $-4 - 1 + 2 + 5 + 8 + \ldots + 17$

(g) $8 + 4 + 2 + 1 + \frac{1}{2} + \ldots$

2) Write down the first three terms and, where possible, the last term of the following series:

(a) $\displaystyle\sum_{r=1}^{\infty} \frac{1}{r}$ (b) $\displaystyle\sum_{r=0}^{5} r(r+1)$ (c) $\displaystyle\sum_{r=0}^{20} r!$

(d) $\displaystyle\sum_{r=0}^{\infty} \frac{1}{(r^2+1)}$ (e) $\displaystyle\sum_{r=-1}^{8} r(r+1)(r+2)$ (f) $\displaystyle\sum_{r=0}^{\infty} a^r(-1)^{r+1}$

3) For the following series, write down the term indicated, and the number of terms in the series.

(a) $\displaystyle\sum_{r=1}^{9} 2^r$,   3rd term   (b) $\displaystyle\sum_{r=-1}^{8} (2r+3)$,   5th term

(c) $\displaystyle\sum_{r=-6}^{-1} \frac{1}{r(2r+1)}$,   last term   (d) $\displaystyle\sum_{r=0}^{\infty} \frac{1}{(r+1)(r+2)}$,   20th term

(e) $\displaystyle\sum_{r=2}^{\infty} (-1)^r \frac{2^r}{r!}$,   4th term   (f) $8+4+0-4-8-12\ldots-80.$
                                                                     15th term

(g) $\displaystyle\sum_{r=1}^{\infty} \left(\frac{1}{2}\right)^r$, nth term

ARITHMETIC PROGRESSIONS

Consider the sequence $5, 8, 11, 14, 17, \ldots, 29$.
Each term of this sequence exceeds the previous term by 3, so the sequence can be written in the form

$$5, (5+3), (5+2\times3), (5+3\times3), (5+4\times3), \ldots, (5+8\times3)$$

This sequence is an example of an arithmetic progression, where an arithmetic progression (A.P.) is a sequence in which any term differs from the proceeding term by a constant, called the common difference. The common difference may be positive or negative,
e.g. the first 6 terms of an A.P., whose first term is 8 and whose common difference is -3, are $8, 5, 2, -1, -4, -7$,

or $8, [8+1(-3)], [8+2(-3)], [8+3(-3)], [8+4(-3)], [8+5(-3)]$

In general, if an A.P. has a first term a, and a common difference d, the
first four terms are $a, (a+d), (a+2d), (a+3d)$,
the general term is $a+rd$,
and the nth term, u_n, is $a+(n-1)d$.

Thus an A.P. with n terms can be written as

$$a, (a + d), (a + 2d), \ldots, [a + (n - 1)d].$$

EXAMPLES 16b

1) The 8th term of an A.P. is 11 and the 15th term is 21. Find the common difference, the first term of the series, and the nth term.

If the first term of the series is a and the common difference is d, then the 8th term is $a + 7d$,

i.e. $\qquad\qquad\qquad a + 7d = 11 \qquad\qquad\qquad$ [1]

and the 15th term is $a + 14d$,

i.e. $\qquad\qquad\qquad a + 14d = 21 \qquad\qquad\qquad$ [2]

[2] $-$ [1] gives $\qquad 7d = 10 \;\Rightarrow\; d = \frac{10}{7}$

and $\qquad\qquad\qquad\qquad a = 1$

i.e. the first term is 1 and the common difference is $\frac{10}{7}$.
Hence the nth term is $\quad a + (n - 1)d = 1 + (n - 1)\frac{10}{7} = \frac{1}{7}(10n - 3)$.

2) The nth term of an A.P. is $12 - 4n$. Find the first term and the common difference.

If the nth term is $12 - 4n$, the first term $(n = 1)$ is 8.
The second term $(n = 2)$ is 4.
Therefore the common difference is -4.

The Sum of an Arithmetic Progression

Consider the sum of the first ten even numbers.
This series is an A.P., and writing it in normal and in reverse order we have

$$S = \;\; 2 + \;\; 4 + \;\; 6 + \;\; 8 + \ldots + 18 + 20$$

$$S = 20 + 18 + 16 + 14 + \ldots + \;\; 4 + \;\; 2$$

Adding gives $\qquad 2S = 22 + 22 + 22 + 22 + \ldots + 22 + 22$

As there are ten terms in this series we have

$$2S = 10 \times 22$$

$\Rightarrow \qquad\qquad\qquad S = 110$

This process is known as finding the sum from first principles. Applying this process to a general A.P., gives formulae for the sum which may be quoted and used, unless a proof from first principles is specifically asked for.

Consider the sum of the first n terms (S_n) of an A.P. whose last term is l,

i.e. $S_n = a + (a+d) + (a+2d) + \ldots + (l-d) + l$

reversing $S_n = l + (l-d) + (l-2d) + \ldots + (a+d) + a$

adding $2S_n = (a+l) + (a+l) + (a+l) + \ldots + (a+l) + (a+l)$

as there are n terms we have

$$2S_n = n(a+l)$$

$\Rightarrow$ $$S_n = \frac{n}{2}(a+l) \qquad \text{i.e. (number of terms)} \times \text{(average term)}$$

Also, writing the nth term, l, as $a + (n-1)d$ we have

$$S_n = \frac{n}{2}[a + a + (n-1)d]$$

or $$S_n = \frac{n}{2}[2a + (n-1)d]$$

Either of these formulae can now be used to find the sum of the first n terms of an A.P.

EXAMPLES 16b (continued)

3) Find the sum of the following series:

(a) An A.P. of eleven terms whose first term is 1 and whose last term is 6

(b) $\displaystyle\sum_{r=1}^{8}\left(2 - \frac{2r}{3}\right)$

(a) We know the first and last terms, and the number of terms so, using
$S_n = \dfrac{n}{2}(a+l)$,   we have

$$S_{11} = \tfrac{11}{2}(1+6) = \tfrac{77}{2}$$

(b) $\displaystyle\sum_{r=1}^{8}\left(2 - \frac{2r}{3}\right) = \tfrac{4}{3} + \tfrac{2}{3} + 0 - \tfrac{2}{3} - \ldots - \tfrac{10}{3}.$

This is an A.P. with 8 terms where $a = \tfrac{4}{3}$, $d = -\tfrac{2}{3}$.

Using $S_n = \dfrac{n}{2}[2a + (n-1)d]$ gives

$$S_8 = 4[\tfrac{8}{3} + 7(-\tfrac{2}{3})] = -8$$

4) In an A.P. the sum of the first ten terms is 50 and the 5th term is three times the 2nd term. Find the first term and the sum of the first 20 terms.

If a is the first term and d is the common difference, and there are n terms

using $S_n = \dfrac{n}{2}[2a + (n-1)d]$ gives

$$S_{10} = 50 = 5(2a + 9d) \qquad [1]$$

Now using $u_n = a + (n-1)d$ gives

$$u_5 = a + 4d \quad \text{and} \quad u_2 = a + d$$

Therefore $\qquad a + 4d = 3(a + d) \qquad [2]$

$[1] \Rightarrow \qquad \left.\begin{array}{r} 2a + 9d = 10 \\ 2a - d = 0 \end{array}\right\} \Rightarrow \quad d = 1 \quad \text{and} \quad a = \tfrac{1}{2}$

$[2] \Rightarrow$

so the first term is $\tfrac{1}{2}$.

The sum of the first twenty terms is given by

$$S_{20} = 10(1 + 19 \times 1) = 200$$

5) Show that the terms of $\displaystyle\sum_{r=1}^{n} \ln 2^r$ are in arithmetic progression.

Find the sum of the first 10 terms of this series and the least value of n for which the sum of the first $2n$ terms exceeds 1000.

By putting $r = 1, 2, 3 \ldots$ we have

$$\sum_{r=1}^{n} \ln 2^r = \ln 2 + \ln 2^2 + \ln 2^3 + \ldots + \ln 2^n$$

$$= \ln 2 + 2\ln 2 + 3\ln 2 + \ldots + n\ln 2$$

from which we see that there is a common difference of $\ln 2$ between successive terms, so the terms of this series are in arithmetic progression.

$$\sum_{r=1}^{10} \ln 2^r = \ln 2 + 2\ln 2 + 3\ln 2 + \ldots + 10\ln 2$$

$$= (1 + 2 + 3 + \ldots + 10)\ln 2$$

$$= \tfrac{10}{2}(1 + 10)\ln 2 = 55\ln 2$$

$$\sum_{r=1}^{2n} \ln 2^r = (1 + 2 + 3 + \ldots + 2n)\ln 2$$

$$= \dfrac{2n}{2}[1 + 2n]\ln 2$$

So we require the least value of n for which $n(1 + 2n)\ln 2 > 1000$

As $\ln 2$ is positive $\quad 2n^2 + n > 1443$

$\Rightarrow \qquad 2n^2 + n - 1443 > 0$

$\Rightarrow \qquad n > 26.6$

(26.6 is the positive root of $2n^2 + n - 1443 = 0$)

But n must be an integer so
the least value of n for which $n(2n + 1)\ln 2 > 1000$ is 27.
Note that the sum of the first n natural numbers,
i.e. $1 + 2 + 3 + \ldots + n$ is an A.P. in which $a = 1$ and $d = 1$ so

$$\sum_{r=1}^{n} r = \frac{n(n + 1)}{2}$$

This is a result that may be quoted, unless a proof is specifically asked for.

6) The sum of the first n terms of a series, S_n, is given by $S_n = n(n + 3)$.
Find the fourth term of the series and show that the terms are in arithmetic
progression.

If the terms of the series are $a_1, a_2, a_3 \ldots a_n$
then $\qquad\qquad\qquad\qquad a_1 + a_2 \ldots + a_n = n(n + 3)$

therefore when $\begin{cases} n = 4 \\ n = 3 \end{cases}$ we have $\begin{cases} a_1 + a_2 + a_3 + a_4 = 28 \\ a_1 + a_2 + a_3 = 18 \end{cases}$

hence a_4, the fourth term of the series, is 10.

Now if $\qquad S_n$ (i.e. $a_1 + a_2 + \ldots + a_n) = n(n + 3) = n^2 + 3n$
then $\qquad S_{n-1}$ (i.e. $a_1 + a_2 + \ldots + a_{n-1}) = (n - 1)[(n - 1) + 3] = n^2 + n - 2$
Hence $\qquad\qquad\qquad S_n - S_{n-1} = a_n = 2n + 2,$
i.e. the nth term of the series is $2n + 2$.
Further $\quad S_{n-2}$ (i.e. $a_1 + a_2 + \ldots + a_{n-2}) = (n - 2)[(n - 2) + 3] = n^2 - n - 2$
so that $\qquad\qquad\qquad S_{n-1} - S_{n-2} = a_{n-1} = 2n,$
i.e. the $(n - 1)$th term of the series is $2n$.
Therefore $\qquad\qquad a_n - a_{n-1} = (2n + 2) - 2n = 2,$
i.e. there is a common difference of 2 between successive terms of the series
and hence it is an A.P.

EXERCISE 16b

1) Write down the fifth term and the nth term of the following A.P.s:
(a) $1, 5, \ldots$ $\qquad$ (b) $2, 1\frac{1}{2}, \ldots$ $\qquad$ (c) first term 5, common difference 3

(d) $\sum_{r=1}^{n} (2r - 1)$ $\quad$ (e) $\sum_{r=1}^{n} 4(r - 1)$ $\quad$ (f) first term 6, common difference -2

(g) first term p, common difference q

(h) first term 10, last term 20, 6 terms $\qquad$ (i) $\sum_{r=0}^{n} (3r + 3)$

2) Find the sum of the first ten terms of each of the series given in
Question (1), parts (a) to (g) inclusive.

3) The 9th term of an A.P. is 8 and the 4th term is 20. Find the first term
and the common difference.

4) The 6th term of an A.P. is twice the 3rd term and the first term is 3. Find the common difference and the 10th term.

5) The nth term of an A.P. is $\frac{1}{2}(3-n)$. Write down the first three terms and the 20th term.

6) Find the sum, to the number of terms indicated, of the following A.P.s:

(a) $1 + 2\frac{1}{2} + \dots$, 6 terms

(b) $3 + 5 + \dots$, 8 terms

(c) the first twenty odd integers

(d) $a_1 + a_2 + a_3 + \dots + a_8$ where $a_n = 2n + 1$

(e) $4 + 6 + 8 + \dots + 20.$

(f) $\sum\limits_{r=1}^{3n} (3 - 4r)$

(g) $S_n = n^2 - 3n$, 8 terms

(h) $S_n = 2n(n + 3)$, m terms

7) The sum of the first n terms of an A.P. is S_n where $S_n = n^2 - 3n$. Write down the fourth term and the nth term.

8) The sum of the first n terms of a series is given by $S_n = n(3n - 4)$. Show that the terms of the series are in arithmetic progression.

9) In an arithmetic progression, the 8th term is twice the 4th term and the 20th term is 40. Find the common difference and the sum of the terms from the 8th to the 20th inclusive.

10) How many terms of the A.P., $1 + 3 + 5 + \dots$ are required to make a sum of 1521?

11) Find the least number of terms of the A.P., $1 + 3 + 5 + \dots$ that are required to make a sum exceeding 4000.

12) Find the least number of terms required for the sum of the series $\sum\limits_{r=1}^{n} (3 - 2r)$ to be less than -100.

GEOMETRIC PROGRESSIONS

Consider the sequence
$$12, 6, 3, 1.5, 0.75, 0.375, \dots$$

Each term of this sequence is half the preceeding term so the sequence may be written
$$12, 12(\tfrac{1}{2}), 12(\tfrac{1}{2})^2, 12(\tfrac{1}{2})^3, 12(\tfrac{1}{2})^4, 12(\tfrac{1}{2})^5, \dots$$

Such a sequence is called a geometric progression (G.P) which is a sequence where each term is a constant factor times the preceeding term. This constant multiplying factor is called the common ratio, and it may have any real value.

Hence, if a G.P. has a first term of 3 and a common ratio of -2 the first four terms are

$$3, 3(-2), 3(-2)^2, 3(-2)^3$$

or
$$3, -6, 12, -24$$

In general if a G.P. has a first term a, and a common ratio r, the first four terms are

$$a, ar, ar^2, ar^3$$

and the nth term, u_n, is ar^{n-1}.

Thus a G.P. with n terms can be written

$$a, ar, ar^2, \ldots, ar^{n-1}$$

Sum of a Geometric Progression

Consider the sum of the first eight terms, S_8, of the G.P. with first term 1 and common ratio 3

$$S_8 = 1 + 1(3) + 1(3)^2 + 1(3)^3 + \ldots + 1(3)^7$$

then

$$3S_8 = \quad 3 + 3^2 + 3^3 + \ldots + 3^7 + 3^8$$

Hence

$$S_8 - 3S_8 = \quad 1 + 0 \qquad\qquad + \ldots + \quad 0 \quad - 3^8$$

Therefore

$$S_8(1 - 3) = 1 - 3^8$$

$$\Rightarrow \qquad S_8 = \frac{1 - 3^8}{1 - 3} = \frac{3^8 - 1}{2}$$

Applying this process to a general G.P. gives a formula for the sum of the first n terms.

Consider the sum of the first n terms, (S_n), of a G.P. whose first term is a and whose common ratio is r,

i.e.
$$S_n = a + ar + ar^2 + \ldots + ar^{n-1}$$

multiplying by r gives

$$rS_n = \quad ar + ar^2 + \ldots + ar^{n-1} + ar^n$$

Hence
$$S_n - rS_n = a - ar^n$$

$$\Rightarrow \qquad S_n(1 - r) = a(1 - r^n)$$

$$\Rightarrow \qquad S_n = \frac{a(1 - r^n)}{1 - r}$$

This formula can now be used to find the sum of the first n terms of any G.P.

If $r > 1$ the formula may be written $\dfrac{a(r^n - 1)}{r - 1}$

EXAMPLES 16c

1) The 5th term of a G.P. is 8, the third term is 4, and the sum of the first ten terms is positive. Find the first term, the common ratio, and the sum of the first ten terms.

If the first term is a and the common ratio is r then the nth term is ar^{n-1}

Thus	we have	$ar^4 = 8$	$(n = 5)$
and		$ar^2 = 4$	$(n = 3)$
dividing gives		$r^2 = 2$	
$\Rightarrow$		$r = \pm\sqrt{2}$ and $a = 2$	

Using
$$S_n = \frac{a(r^n - 1)}{r - 1}$$

we have either
$$S_{10} = \frac{2[(\sqrt{2})^{10} - 1]}{\sqrt{2} - 1} = \frac{62}{\sqrt{2} - 1} \qquad (r = \sqrt{2}),$$

or
$$S_{10} = \frac{2[(-\sqrt{2})^{10} - 1]}{-\sqrt{2} - 1} = \frac{-62}{\sqrt{2} + 1} \qquad (r = -\sqrt{2}),$$

But we are given $S_{10} > 0$, so we deduce that
$$r = \sqrt{2} \quad \text{and} \quad S_{10} = \frac{62}{\sqrt{2} - 1}$$

2) Find the sum of the first n terms of the G.P. $1 - x + x^2 - x^3 + \ldots$

This series may be written $1 + 1(-x) + 1(-x)^2 + 1(-x)^3 + \ldots$

$$\text{so} \quad a = 1 \quad \text{and} \quad r = -x$$

Therefore
$$S_n = \frac{1 - (-x)^n}{1 + x} = \frac{1 - (-1)^n x^n}{1 + x} = \frac{1 + (-1)^{n+1} x^n}{1 + x}$$

3) The sum of the first n terms of a series is $3^n - 1$. Show that the terms of this series are in geometric progression and find the first term, the common ratio and the sum of the second set of n terms of this series.

If the series is
$$a_1 + a_2 + \ldots + a_n$$

then
$$S_n = a_1 + a_2 + \ldots + a_n = 3^n - 1$$

and
$$S_{n-1} = a_1 + a_2 + \ldots + a_{n-1} = 3^{n-1} - 1$$

therefore the nth term,

$$a_n = S_n - S_{n-1} = (3^n - 1) - (3^{n-1} - 1) = 3^n - 3^{n-1}$$
$$= 3^{n-1}(3 - 1) \qquad = (2)3^{n-1}$$

But the nth term of a G.P., with first term 2 and common ratio 3, is $(2)3^{n-1}$

Hence this series is a G.P. with a first term of 2 and a common ratio of 3. The sum of the second set of n terms is

(the sum of the first $2n$ terms) $-$ (the sum of the first n terms)

$$
\begin{aligned}
&= S_{2n} - S_n \\
&= (3^{2n} - 1) - (3^n - 1) \\
&= 3^{2n} - 3^n \\
&= 3^n(3^n - 1)
\end{aligned}
$$

4) A prize fund is set up with a single investment of £2000 to provide an annual prize of £150. The fund accrues interest at 5% p.a. paid yearly. If the first prize is awarded one year after the investment, find the number of years for which the full prize can be awarded.

After one year the value of the fund is the initial investment of £2000, plus 5% interest, less one £150 prize.
If £ P_n is the value of the fund after n years we have

$$P_1 = 2000 + \frac{5}{100}(2000) - 150 = 2000\left(\frac{105}{100}\right) - 150$$

$$P_2 = P_1 + \frac{5}{100}P_1 - 150 = P_1\left(\frac{105}{100}\right) - 150 = 2000\left(\frac{105}{100}\right)^2 - 150\left(\frac{105}{100}\right) - 150$$

$$P_3 = P_2\left(\frac{105}{100}\right) - 150 = 2000\left(\frac{105}{100}\right)^3 - 150\left(\frac{105}{100}\right)^2 - 150\left(\frac{105}{100}\right) - 150$$

$$P_n = 2000\left(\frac{105}{100}\right)^n - 150\left(\frac{105}{100}\right)^{n-1} - 150\left(\frac{105}{100}\right)^{n-2} - \ldots - 150$$

$$= 2000\left(\frac{105}{100}\right)^n - \left[150 + 150(1.05) + \ldots + 150(1.05)^{n-1}\right]$$

The expression in the square brackets is a G.P. with n terms.

Hence

$$P_n = 2000\left(\frac{105}{100}\right)^n - \left[\frac{150(1 - 1.05^n)}{1 - 1.05}\right]$$

$$= 2000(1.05)^n - 3000[(1.05)^n - 1]$$

$$= 3000 - 1000(1.05)^n$$

The fund can no longer award the full prize while $P_n \geqslant 0$,

i.e. $3000 - 1000(1.05)^n \geqslant 0$

$$\Rightarrow \qquad\qquad (1.05)^n \leqslant 3$$

$$\Rightarrow \qquad\qquad n \leqslant \frac{\log 3}{\log 1.05} = 22.5$$

But the number of years is an integer.

Therefore the full prize can be awarded for 22 years.

EXERCISE 16c

1) Write down the fifth term and the nth term of the following G.P.s:

(a) $2, 4, 8, \ldots$ (b) $2, 1, \frac{1}{2}, \ldots$ (c) $3, -6, 12, \ldots$

(d) first term 8, common ratio $-\frac{1}{2}$ (e) first term 3, last term $\frac{1}{81}$, 6 terms

2) Find the sum, to the number of terms given, of the following G.P.s.

(a) $3 + 6 + \ldots$ 6 terms   (b) $3 - 6 + \ldots$ 8 terms

(c) $1 + \frac{1}{2} + \frac{1}{4} + \ldots$ 20 terms   (d) first term 5, common ratio $\frac{1}{5}$, 5 terms

(e) first term $\frac{1}{2}$, common ratio $-\frac{1}{2}$, 10 terms

(f) first term 1, common ratio -1, 2001 terms.

3) The 6th term of a G.P. is 16 and the 3rd term is 2. Find the first term and the common ratio.

4) Find the common ratio, given that it is negative, of a G.P. whose first term is 8 and whose 5th term is $\frac{1}{2}$.

5) The nth term of a G.P. is $(-\frac{1}{2})^n$. Write down the first term and the 10th term.

6) Evaluate $\displaystyle\sum_{r=1}^{10} (1.05)^r$.

7) Find the sum to n terms of the following series:

(a) $x + x^2 + x^3 + \ldots$ (b) $x + 1 + \dfrac{1}{x} + \ldots$ (c) $1 - y + y^2 - \ldots$

(d) $x + \dfrac{x^2}{2} + \dfrac{x^3}{4} + \dfrac{x^4}{8} + \ldots$ (e) $1 - 2x + 4x^2 - 8x^3 + \ldots$

8) Find the sum of the first n terms of the G.P. $2 + \frac{1}{2} + \frac{1}{8} + \ldots$ and find the least value of n for which this sum exceeds 2.65.

9) The sum of the first 3 terms of a G.P. is 14. If the first term is 2, find the possible values of the sum of the first 5 terms.

10) Evaluate $\displaystyle\sum_{r=1}^{10} 3(3/4)^r$.

11) A mortgage is taken out for £10 000 and is repaid by annual instalments of £2000. Interest is charged on the outstanding debt at 10%, calculated

annually. If the first repayment is made one year after the mortgage is taken out find the number of years it takes for the mortgage to be repaid.

12) A bank loan of £500 is arranged to be repaid in two years by equal monthly instalments. Interest, *calculated monthly*, is charged at 11% p.a. on the remaining debt. Calculate the monthly repayment if the first repayment is one month after the loan is made.

CONVERGENCE OF SERIES

If a piece of string, of length l, is cut up by first cutting it in half and keeping one piece, then cutting the remainder in half and keeping one piece, then cutting the remainder in half and keeping one piece, . . . and so on, the sum of the cut lengths is

$$\frac{l}{2} + \frac{l}{4} + \frac{l}{8} + \frac{l}{16} + \ldots$$

As this process can (in theory) be carried on indefinitely the series formed above is infinite.

After several cuts have been made the remaining part of the string will be very small indeed, so that the sum of the cut lengths will be very nearly equal to the total length, l, of the original piece of string and, the more cuts that are made, the closer to l this sum becomes,

i.e. if after n cuts, the sum of the cut lengths is

$$\frac{l}{2} + \frac{l}{2^2} + \frac{l}{2^3} + \ldots + \frac{l}{2^n}$$

then as $n \to \infty$,

$$\frac{l}{2} + \frac{l}{2^2} + \ldots + \frac{l}{2^n} \to l$$

or

$$\lim_{n \to \infty} \left[\frac{l}{2} + \frac{l}{2^2} + \ldots + \frac{l}{2^n} \right] = l$$

l is called the sum to infinity of this series.

In general, if S_n is the sum of the first n terms of any series and if $\lim_{n \to \infty} \cdot [S_n]$ exists and is finite the series is said to be convergent

and the sum to infinity, S_∞, is given by

$$S_\infty = \lim_{n \to \infty} [S_n]$$

Thus, for example, the series $\frac{l}{2} + \frac{l}{2^2} + \frac{l}{2^3} + \ldots$ is convergent as its sum to infinity is l.

But the series $1 + 2 + 3 + 4 + \ldots + n$ has a sum given by $S_n = \frac{1}{2}n(n + 1)$ (see p. 560) so that, as $n \to \infty$, $S_n \to \infty$ and so this series does not converge. It is said to be divergent.

For any A.P., $S_n = \dfrac{n}{2}(2a + [n-1]d)$, and as $n \to \infty$ it can be seen that $S_n \to \infty$, i.e. any A.P. is divergent.

THE SUM TO INFINITY OF A G.P.

The series discussed above, viz. $\dfrac{l}{2} + \dfrac{l}{2^2} + \dfrac{l}{2^3} + \ldots$

is a G.P. whose first term is $\frac{1}{2}l$ and whose common ratio is $\frac{1}{2}$. Its sum to infinity, which we have seen is l, can also be found using the alternative method below.

The sum of the first n terms, S_n, of this series is given by

$$S_n = \frac{\frac{1}{2}l\{1-(\frac{1}{2})^n\}}{1-\frac{1}{2}} = l\{1-(\tfrac{1}{2})^n\}$$

As n increases, $(\frac{1}{2})^n$ decreases, i.e. as $n \to \infty$, $(\frac{1}{2})^n \to 0$

therefore
$$\lim_{n \to \infty} [l\{1-(\tfrac{1}{2})^n\}] = l$$

Thus the sum to infinity of the series $\dfrac{l}{2} + \dfrac{l}{2^2} + \ldots$ is l.

This approach can also be applied to a general G.P. $a + ar + ar^2 + \ldots$

where
$$S_n = \frac{a(1-r^n)}{1-r}$$

If $|r| < 1$, then $\lim_{n \to \infty} r^n = 0$

and
$$\lim_{n \to \infty} S_n = \lim_{n \to \infty} \left[\frac{a(1-r^n)}{1-r}\right] = \frac{a}{1-r}$$

If $|r| > 1$, $\lim_{n \to \infty} r^n = \infty$ and the series does not converge.

Therefore, provided $|r| < 1$, a G.P. converges to a sum of $\dfrac{a}{1-r}$

i.e. for a G.P. $\quad S_\infty = \dfrac{a}{1-r} \iff |r| < 1$

Arithmetic Mean

If three numbers, p_1, p_2, p_3, are in arithmetic progression then p_2 is called the *arithmetic mean* of p_1 and p_3.

If $p_1 = a$, we may write p_2, p_3 as $a+d, a+2d$ respectively:

hence $\quad p_2 = a+d = \frac{1}{2}(2a+2d) = \frac{1}{2}(p_1+p_3)$

i.e. the arithmetic mean of two numbers x and y is $\frac{1}{2}(x+y)$.

Geometric Mean

If p_1, p_2, p_3 are in geometric progression, p_2 is called the *geometric mean* of p_1 and p_3.

If $p_1 = a$, then we may write $p_2 = ar$, $p_3 = ar^2$

thus $p_1 p_3 = a^2 r^2 = p_2{}^2 \Rightarrow p_2 = \sqrt{p_1 p_3}$,

i.e. the geometric mean of two numbers x and y is $\sqrt{xy}$.

EXAMPLES 16d

1) Determine whether the series given below converge. If they do give their sum to infinity.

(a) $3 + 5 + 7 + \ldots$ (b) $1 - \dfrac{1}{4} + \dfrac{1}{16} - \dfrac{1}{64} + \ldots$ (c) $3 + \dfrac{9}{2} + \dfrac{27}{4} + \ldots$

(a) $3 + 5 + 7 + \ldots$ is an A.P. $(d = 2)$ and so does not converge.

(b) $1 - \dfrac{1}{4} + \dfrac{1}{16} - \dfrac{1}{64} + \ldots = 1 + (-\tfrac{1}{4}) + (-\tfrac{1}{4})^2 + (-\tfrac{1}{4})^3 + \ldots$

which is a G.P. where $r = -\tfrac{1}{4}$, i.e. $|r| < 1$

So this series does converge to $\dfrac{1}{1 - (-\tfrac{1}{4})} = \dfrac{4}{5}$, using $S_\infty = \dfrac{a}{1 - r}$.

(c) $3 + \tfrac{9}{2} + \tfrac{27}{4} + \ldots = 3 + 3(\tfrac{3}{2}) + 3(\tfrac{9}{4}) + \ldots$

$$= 3 + 3(\tfrac{3}{2}) + 3(\tfrac{3}{2})^2 + \ldots$$

This series is a G.P. where $r = \tfrac{3}{2}$.

As $|r| > 1$, the series does not converge.

2) Find the condition on x so that $\displaystyle\sum_{r=0}^{\infty} \dfrac{(x-1)^r}{2^r}$ converges.

Evaluate this expression when $x = 1.5$

$$\sum_{r=0}^{\infty} \frac{(x-1)^r}{2^r} = 1 + \frac{x-1}{2} + \left(\frac{x-1}{2}\right)^2 + \ldots$$

This series is a G.P. with common ratio $\dfrac{x-1}{2}$

so the series converges if $\left| \dfrac{x-1}{2} \right| < 1$,

i.e. if $-1 < \dfrac{x-1}{2} < 1$

$\Rightarrow$ $-2 < x - 1 < 2$

$\Rightarrow$ $-1 < x < 3$

When $x = 1.5$,   the series converges

and
$$\sum_{r=0}^{\infty} \frac{(x-1)^r}{2^r} = \sum_{r=0}^{\infty} (\tfrac{1}{4})^r = 1 + \tfrac{1}{4} + (\tfrac{1}{4})^2 + \ldots$$

$$= \frac{1}{1 - \tfrac{1}{4}} = \frac{4}{3}$$

$\left(\text{using}\ \ S_\infty = \dfrac{a}{1-r}\ \ \text{where}\ \ r = \tfrac{1}{4},\ \ a = 1\right).$

3) Express the recurring decimal $0.1\dot{5}7\dot{6}$ as a fraction in its lowest terms.

$$0.1\dot{5}7\dot{6} = 0.1\overline{576}\,\overline{657}\,\overline{657}\,\overline{6576}\ldots$$

$$= 0.1 + 0.0576 + 0.000\,057\,6 + 0.000\,000\,057\,6 + \ldots$$

$$= \frac{1}{10} + \frac{576}{10^4} + \frac{576}{10^7} + \frac{576}{10^{10}} + \ldots$$

$$= \frac{1}{10} + \frac{576}{10^4}\left[1 + \frac{1}{10^3} + \frac{1}{10^6} + \ldots\right]$$

$$= \frac{1}{10} + \frac{576}{10^4}\left[1 + \frac{1}{10^3} + \left(\frac{1}{10^3}\right)^2 + \ldots\right]$$

Now the series in the square bracket is a G.P. whose first term is 1, and whose common ratio is $\dfrac{1}{10^3}$. Hence it has a sum to infinity equal to $\dfrac{1}{1 - 10^{-3}} = \dfrac{10^3}{999}$

Therefore $0.1\dot{5}7\dot{6} = \dfrac{1}{10} + \dfrac{576}{10^4} \times \dfrac{10^3}{999} = \dfrac{1}{10} + \dfrac{576}{9990} = \dfrac{1575}{9990} = \dfrac{35}{222}$

4) The 3rd term of a convergent G.P. is the arithmetic mean of the 1st and 2nd terms.
Find the common ratio and, if the first term is 1, find the sum to infinity.
If the series is $a + ar + ar^2 + ar^3 + \ldots$

then
$$ar^2 = \tfrac{1}{2}(a + ar)$$

$\Rightarrow$
$$2r^2 - r - 1 = 0$$

$\Rightarrow$
$$(2r + 1)(r - 1) = 0$$

i.e.
$$r = -\tfrac{1}{2} \text{ or } 1$$

As the series is convergent, the common ratio is $-\tfrac{1}{2}$.

When $a = 1$ and $r = -\tfrac{1}{2}$, $S_\infty = \dfrac{1}{1 + \tfrac{1}{2}} = \tfrac{2}{3}$

EXERCISE 16d

1) Determine whether the series given below converge:

(a) $4 + \dfrac{4}{3} + \dfrac{4}{3^2} + \ldots$
(b) $9 + 7 + 5 + 3 + \ldots$

(c) $20 - 10 + 5 - 2.5 + \ldots$

(d) $\dfrac{5}{10} + \dfrac{5}{100} + \dfrac{5}{1000} + \ldots$

(e) $p + 2p + 3p + \ldots$

(f) $3 - 1 + \frac{1}{3} - \frac{1}{9} + \ldots$

2) Find the range of values of x for which the following series converge:

(a) $1 + x + x^2 + x^3 + \ldots$

(b) $x + 1 + \dfrac{1}{x} + \dfrac{1}{x^2} + \ldots$

(c) $1 + 2x + 4x^2 + 8x^3 + \ldots$

(d) $1 - (1 - x) + (1 - x)^2 - (1 - x)^3 + \ldots$

(e) $(a + x) + (a + x)^2 + (a + x)^3 + \ldots$

(f) $(a + x) - 1 + \dfrac{1}{a + x} - \dfrac{1}{(a + x)^2} + \ldots$

3) Find the sum to infinity of the series in Question 1 that are convergent:

4) Express the following recurring decimals as fractions:

(a) $0.1\dot{6}\dot{2}$ (b) $0.\dot{3}\dot{4}$ (c) $0.0\dot{2}\dot{1}$

5) The sum to infinity of a G.P. is twice the first term. Find the common ratio.

6) If $\ln y$ is the arithmetic mean of $\ln x$ and $\ln z$ show that y is the geometric mean of x and z.

THE BINOMIAL THEOREM

When an expression such as $(1 + x)^4$ is expanded, the coefficients of the terms in the expansion can be obtained from Pascal's Triangle (see Chapter 2, p. 36). The same approach could be used to expand $(1 + x)^{20}$, say, but clearly the construction of the triangular array would be tedious in this case. A more general method known as the Binomial Theorem, will now be developed for expanding powers of $(1 + x)$.

Consider $(1 + x)^6 \equiv (1 + x)(1 + x)(1 + x)(1 + x)(1 + x)(1 + x)$.

When the six brackets are expanded, each term in the expansion is obtained by multiplying together either x or 1 from each of the six brackets.

Taking 1 from each bracket we get 1 as a term in the expansion.

Taking x from only one bracket and 1 from the other five, we get $1 \times x$. But this can be done six times because we can choose to take x from each of the six brackets in turn, (i.e. in $^6C_1$ ways).

So the x term in the expansion is $6x$.

Taking x from the first two brackets and 1 from the remaining brackets we get $1 \times x^2$.

Taking x from the first and third brackets and 1 from the remaining brackets we get another $1 \times x^2$. This can be done $^6C_2$ times, as the number of ways in which two brackets can be selected from six brackets is $^6C_2$.

So the x^2 term in the expansion is $^6C_2 x^2$.

It can similarly be shown that the coefficients of $x^3, x^4 \ldots$ are $^6C_3, {}^6C_4 \ldots$

Thus, arranging the expansion of $(1 + x)^6$ as a series of ascending powers of x,

$$(1 + x)^6 \equiv 1 + {}^6C_1 x + {}^6C_2 x^2 + {}^6C_3 x^3 + {}^6C_4 x^4 + {}^6C_5 x^5 + {}^6C_6 x^6$$

$$\equiv 1 + 6x + 15x^2 + 20x^3 + 15x^4 + 6x^5 + x^6$$

The R.H.S. of this identity is called the *series expansion* of $(1 + x)^6$.

Note that the expansion has 7 (i.e. $6 + 1$) terms.

This argument can be generalised as follows.

If n is any positive integer, $(1 + x)^n$ can be expanded to give a series of terms in ascending powers of x where the term containing x^r is obtained by multiplying together x's from r brackets and 1's from the remaining brackets. There are ${}^nC_r$ ways in which r brackets can be chosen from n brackets, so the term containing x^r is ${}^nC_r x^r$.

Hence

$$(1 + x)^n \equiv {}^nC_0 + {}^nC_1 x + {}^nC_2 x^2 + \ldots + {}^nC_r x^r + \ldots + {}^nC_n x^n \qquad [1]$$

$$\equiv 1 + nx + \frac{n(n-1)}{2!} x^2 + \ldots + \frac{n(n-1)\ldots(n-r+1)}{r!} x^r + \ldots + x^n \qquad [2]$$

This result is known as the Binomial Theorem and may be written more briefly as

$$(1 + x)^n \equiv \sum_{r=0}^{n} {}^nC_r x^r$$

where n is a positive integer.

Note that:

(a) the expansion of $(1 + x)^n$ is a finite series with $n + 1$ terms,

(b) the coefficient of x^r, ${}^nC_r$, when written $\dfrac{n(n-1)(n-2)\ldots(n-r+1)}{r!}$

has r factors in the numerator,

(c) the term containing x^2 is the third term, the term in x^3 is the fourth term, and *the term in x^r is the $(r+1)$th term*, so the rth term is ${}^nC_{r-1} x^{r-1}$,

(d) the form of the expansion given in [1] is useful when the coefficient of a large power of x is required, or when the general term is required. The form of the expansion given in [2] is useful when the first few terms of an expansion are required. Thus *both* forms of the Binomial Theorem should be memorized.

Now consider $(a + x)^n$, where n is a positive integer

$$(a + x)^n \equiv a^n \left(1 + \frac{x}{a}\right)^n$$

replacing x by $\dfrac{x}{a}$ in the binomial series gives

$$(a + x)^n \equiv a^n \left[{}^nC_0 + {}^nC_1 \left(\frac{x}{a}\right) + {}^nC_2 \left(\frac{x}{a}\right)^2 + \ldots + {}^nC_r \left(\frac{x}{a}\right)^r + \ldots + {}^nC_n \left(\frac{x}{a}\right)^n \right]$$

$$\equiv {}^nC_0 a^n + {}^nC_1 a^{n-1} x + {}^nC_2 a^{n-2} x^2 + \ldots + {}^nC_r a^{n-r} x^r + \ldots + {}^nC_n x^n$$

$$\equiv a^n + na^{n-1} x + \frac{n(n-1)}{2} a^{n-2} x^2 + \ldots + x^n$$

i.e. if n is a positive integer

$$(a + x)^n \equiv \sum_{r=0}^{n} {}^nC_r a^{n-r} x^r$$

EXAMPLES 16e

1) Write down the first three terms in the expansion in ascending powers of x of:

(a) $\left(1 - \dfrac{x}{2}\right)^{10}$,   (b) $(3 - 2x)^8$

(a) Using the result [2] above and replacing x by $-\dfrac{x}{2}$ and n by 10 we have

$$\left(1 - \frac{x}{2}\right)^{10} = 1 + (10)\left(-\frac{x}{2}\right) + \frac{10 \times 9}{2!}\left(-\frac{x}{2}\right)^2 + \ldots$$

$$= 1 - 5x + \tfrac{45}{4}x^2 + \ldots$$

(b) Using the general result above and replacing a by 3, x by $-2x$ and n by 8 we have

$$(3 - 2x)^8 \equiv \sum_{r=0}^{8} {}^8C_r (3)^{8-r}(-2x)^r$$

Therefore the first three terms of this series $(r = 0, 1, 2)$ are

$$3^8 + 8 \times (3)^7(-2x) + \frac{8 \times 7}{2}(3)^6(-2x)^2$$

i.e. $3^8 - 16 \times 3^7 x + 112 \times 3^6 x^2$

2) In the expansion of $(2a - b)^{20}$ as a series in ascending powers of b find:

(a) the term containing a^3 (b) the sixth term.

$$(2a - b)^{20} \equiv \sum_{r=0}^{20} {}^{20}C_r (2a)^{20-r}(-b)^r$$

(a) The term containing a^3 is the term for which $20 - r = 3$ i.e. $r = 17$.
Therefore the required term is $^{20}C_{17}(2a)^3(-b)^{17}$

$$= \frac{20!}{17!3!}(8a^3)(-b)^{17}$$

$$= -9120a^3b^{17}$$

(b) As the first term of the series is the term for which $r = 0$,   the sixth term
of the series is that for which $r = 5$,

i.e. $^{20}C_5(2a)^{15}(-b)^5 = -\dfrac{20!}{15!5!}2^{15}a^{15}b^5$

(**Note** that the numerical part of this term is so large that it is better left in factor
form rather than multiplied out.)

3) If n is a positive integer find the coefficient of x^r in the expansion of
$(1 + x)(1 - x)^n$ as a series of ascending powers of x.

$$(1 + x)(1 - x)^n \equiv (1 - x)^n + x(1 - x)^n$$

$$\equiv \sum_{r=0}^{n} {}^nC_r(-x)^r + x \sum_{r=0}^{n} {}^nC_r(-x)^r$$

$$\equiv \sum_{r=0}^{n} {}^nC_r(-1)^r x^r + \sum_{r=0}^{n} {}^nC_r(-1)^r x^{r+1}$$

$$\equiv [1 - {}^nC_1 x + {}^nC_2 x^2 \ldots + {}^nC_{r-1}(-1)^{r-1}x^{r-1} + {}^nC_r(-1)^r x^r + \ldots + (-1)^n x^n]$$
$$+ [x - {}^nC_1 x^2 + \ldots + {}^nC_{r-1}(-1)^{r-1}x^r + {}^nC_r(-1)^r x^{r+1} + \ldots + (-1)^n x^{n+1}]$$

$$\equiv \sum_{r=0}^{n} [{}^nC_r(-1)^r + {}^nC_{r-1}(-1)^{r-1}]x^r$$

Hence the coefficient of x^r is

$${}^nC_r(-1)^r + {}^nC_{r-1}(-1)^{r-1}$$

$$= (-1)^r [{}^nC_r - {}^nC_{r-1}]   \text{which simplifies to}$$

$$= (-1)^r \frac{n!(n - 2r + 1)}{r!(n - r + 1)!}$$

4) Evaluate $\text{Re}(z^6)$ where $z = 2 + 3i$.

$$(2 + 3i)^6 \equiv \sum_{r=0}^{6} {}^6C_r(2)^{6-r}(3i)^r$$

Now $\text{Re}\,(z^6)$ is the sum of the terms containing even powers of $3i$ in the binomial expansion of z^6,

i.e. $\text{Re}\,(z^6) = {}^6C_0(2)^6(3i)^0 + {}^6C_2(2)^4(3i)^2 + {}^6C_4(2)^2(3i)^4 + {}^6C_6(2)^0(3i)^6$

$= 2^6 - 15(2^4)(3^2) + 15(2)^2(3)^4 - (3)^6$

$= 2035$

USE OF SERIES FOR APPROXIMATIONS

Consider $(1 + x)^{20}$ and its binomial expansion

$$(1 + x)^{20} \equiv 1 + 20x + \frac{(20)(19)}{2!}x^2 + \frac{(20)(19)(18)}{3!}x^3 + \ldots + x^{20}$$

This identity is valid for all values of x so if, for example, $x = 0.01$ we have

$$(1.01)^{20} = 1 + 20(0.01) + \frac{(20)(19)}{2!}(0.01)^2 + \frac{(20)(19)(18)}{3!}(0.01)^3$$
$$+ \ldots + (0.01)^{20}$$

i.e.

$$(1.01)^{20} = 1 + 0.2 + 0.019 + 0.001\,14 + 0.000\,048\,45 + \ldots + 10^{-40}$$

Because the value of x (i.e. 0.01) is small, we see that adding successive terms of the series makes progressively smaller contributions to the accuracy of $(1.01)^{20}$.

In fact, taking only the first four terms gives $(1.01)^{20} \simeq 1.220\,14$

This approximation is correct to three decimal places as the fifth and succeeding terms do not add anything to the first four decimal places.

In general, for any positive integer n,

$$(1 + x)^n \equiv 1 + nx + \frac{n(n-1)}{2!}x^2 + \ldots + x^n$$

and, if x is small (i.e. successive powers of x quickly become negligible in value) the sum of the first few terms of the expansion of $(1 + x)^n$ gives an approximate value for $(1 + x)^n$.

The number of terms required to obtain a good approximation depends on two considerations:

(a) the value of x (the smaller x is, the fewer are the terms needed to obtain a good approximation),

(b) the accuracy required (an answer correct to 3 s.f. needs fewer terms than an answer correct to 6 s.f.).

When finding an approximation, the binomial expansion of $(1 + x)^n$ and *not* $(a + x)^n$ should be used.

The following examples illustrate these points.

EXAMPLES 16e (continued)

5) By substituting 0.001 for x in the expansion of $(1-x)^7$ find the value of $(1.998)^7$ correct to six significant figures.

$$(1-x)^7 \equiv 1 - 7x + \frac{(7)(6)}{2!}x^2 + \ldots - x^7$$

and

$$(1.998)^7 = (2-0.002)^7 = 2^7(1-0.001)^7$$

$$= 2^7(1-x)^7 \quad \text{when} \quad x = 0.001$$

Hence

$$(1.998)^7 = 2^7 \left[1 - 7(0.001) + \frac{(7)(6)}{2!}(0.001)^2 - \frac{(7)(6)(5)}{3!}(0.001)^3 \right.$$

$$\left. + \ldots - (0.001)^7 \right]$$

$$\simeq 128\,[1 - 0.007 + 0.000\,021 - 0.000\,000\,035]$$

To give an answer correct to 6 s.f. we must work to 7 s.f. This requires the expression in brackets to be worked to 7 d.p. so only the first three terms need be considered,

i.e. $\qquad (1.998)^7 = 128(1 - 0.007 + 0.000\,021\,0) \quad$ to 7 s.f.

$$= 127.107 \quad \text{to 6 s.f.}$$

Note that $(1.998)^7$ is rational, i.e. it has an exact numerical value. The use of the series expansion enables us to find a value of $(1.998)^7$ correct to as many significant figures as we wish *without* first having to find the exact value.

6) If x is so small that x^2 and higher powers can be neglected show that

$$(1-x)^5 \left(2 + \frac{x}{2}\right)^{10} \simeq 2^9(2-5x)$$

Using the binomial expansion of $(1-x)^5$ and neglecting terms containing x^2 and higher powers of x we have

$$(1-x)^5 \simeq 1 - 5x.$$

Similarly $\qquad \left(2 + \frac{x}{2}\right)^{10} \equiv 2^{10}\left(1 + \frac{x}{4}\right)^{10}$

$$\simeq 2^{10}\left[1 + 10\left(\frac{x}{4}\right)\right]$$

Therefore $\qquad (1-x)^5 \left(2 + \frac{x}{2}\right)^{10} \simeq 2^{10}(1-5x)\left(1 + \frac{5x}{2}\right)$

$$\simeq 2^9(1-5x)(2+5x)$$

$$\simeq 2^9(2-5x)$$

again neglecting the term in x^2.

The graphical significance of this approximation is interesting:

If
$$y = (1-x)^5 \left(2 + \frac{x}{2}\right)^{10}$$

then for values of x close to zero

$$y \simeq 2^9 (2 - 5x)$$

which is the equation of a straight line,

i.e. $y = 2^9(2 - 5x)$ is the tangent to $y = (1-x)^5 \left(2 + \frac{x}{2}\right)^{10}$ at the point
where $x = 0$.

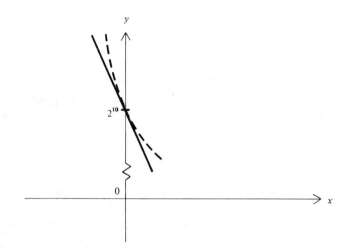

Note that the function $2^9(2 - 5x)$ is called a *linear approximation* for the
function $(1-x)^5 \left(2 + \frac{x}{2}\right)^{10}$ in the region where $x \simeq 0$.

EXERCISE 16e

1) Write down the first four terms in the binomial expansion of:

(a) $(1 + 3x)^{12}$ (b) $(1 - 2x)^9$ (c) $(2 + x)^{10}$

(d) $\left(1 - \frac{x}{3}\right)^{20}$ (e) $\left(2 - \frac{3}{2}x\right)^7$ (f) $\left(\frac{3}{2} + 2x\right)^9$

2) Write down the term indicated in the binomial expansions of the following
functions:

(a) $(1 - 4x)^7$, 3rd term   (b) $\left(1 - \frac{x}{2}\right)^{20}$, 6th term

(c) $(2 - x)^{15}$, 12th term   (d) $(2 - 3x)^{30}$, 9th term

(e) $(p - 2q)^{10}$, 5th term   (f) $(3a + 2b)^8$, 2nd term

(g) $(1-2x)^{12}$, term containing x^4 (h) $\left(2+\dfrac{x}{2}\right)^9$, term containing x^5

(i) $(a+b)^8$, term containing a^3 (j) $(2a-3b)^{10}$, term containing b^8

3) Write down the binomial expansions of the following functions as series of ascending powers of x as far as, and including, the term in x^2 :

(a) $(1+x)(1-x)^9$ (b) $(1-x)(1+2x)^{10}$

(c) $(2+x)\left(1-\dfrac{x}{2}\right)^{20}$ (d) $(1+x)^2(1-5x)^{14}$

4) Find Re (z^8) where $z = 2-i$.

5) Find Im (z^{10}) where $z = 1 + \frac{1}{2}i$.

6) For each of the functions given in Question 3, find the term containing x^r and hence express the expansion in the sigma notation.

7) Find the coefficient of x^r in the expansion of $(1-2x)(1+4x)^8$ as a series of ascending powers of x.

8) Repeat Question 7 for $(1-x)(2+x)^{20}$.

9) By substituting 0.01 for x in the binomial expansion of $(1-2x)^{10}$, find the value of $(0.98)^{10}$ correct to four decimal places.

10) By substituting 0.05 for x in the binomial expansion of $\left(1+\dfrac{x}{5}\right)^6$, find the value of $(1.01)^6$ correct to four significant figures.

11) By using the binomial expansion of $(2+x)^7$, show that $(2.08)^7 = 168.439$ correct to 3 d.p.

12) Show that, if x is small enough for x^2 and higher powers of x to be neglected, the function $(x-2)(1+3x)^8$ has a linear approximation of $-2-47x$.

13) If x is so small that x^3 and higher powers of x are negligible, show that $(2x+3)(1-2x)^{10} \simeq 3 - 58x + 500x^2$.

14) By neglecting x^2 and higher powers of x find linear approximations for the following functions in the immediate neighbourhood of $x = 0$.

(a) $(1-5x)^{10}$ (b) $(2-x)^8$ (c) $(1+x)(1-x)^{20}$

EXTENDING THE BINOMIAL THEOREM

It has been shown that, for a positive integral value of n

$$(1+x)^n \equiv 1 + nx + \frac{n(n-1)}{2!}x^2 + \frac{n(n-1)(n-2)}{3!}x^3 + \ldots + x^n \quad [1]$$

Now consider the infinite series $1 - x + x^2 - x^3 + \ldots$

This is a G.P. whose common ratio is $-x$ and, provided that $|x| < 1$, whose sum to infinity is $\dfrac{1}{1 - (-x)}$ i.e. $\dfrac{1}{1 + x}$

i.e. $(1 + x)^{-1} = 1 - x + x^2 - \ldots$ provided $-1 < x < 1$ [2]

Now considering the identity [1] above and substituting -1 for n in the L.H.S. we get $(1 + x)^{-1}$ which is the L.H.S. of [2].

Substituting -1 for n in the first few terms of the R.H.S. of [1] we get $1 - x + x^2 - x^3 + \ldots$ which is the R.H.S. of [2].

Thus the form of the binomial expansion given in [1] appears to be valid for a value of n that is not a positive integer except that

(a) the series expansion does not terminate but carries on to infinity, and

(b) x must be in the range $-1 < x < 1$.

In fact, although it cannot be proved at this stage, the series

$$1 + nx + \frac{n(n-1)}{2!}x^2 + \frac{n(n-1)(n-2)}{3!}x^3 + \ldots$$

has a sum to infinity of $(1 + x)^n$ when n is not a positive integer provided that $-1 < x < 1$.

So for all values of n we can expand $(1 + x)^n$ as a series of ascending powers of x in the form

$$(1 + x)^n = 1 + nx + \frac{n(n-1)}{2!}x^2 + \frac{n(n-1)(n-2)}{3!}x^3 +$$

$$\ldots + \frac{n(n-1)\ldots(n-r+1)}{r!}x^r + \ldots$$

provided that $|x| < 1$.

When using this expansion it should be noted that:

(a) The expansion is valid for $(1 + x)^n$ and not for $(a + x)^n$. To expand $(a + x)^n$ it must first be written as $a^n\left(1 + \dfrac{x}{a}\right)^n$.

(b) The expansion is valid for a *restricted range of values of the variable* and this range must *always* be stated.

(c) The coefficient of x^r, $\dfrac{n(n-1)(n-2)\ldots(n-r+1)}{r!}$, contains r factors in the numerator.

(d) When $n = -1$,

$$(1 + x)^{-1} = 1 - x + x^2 - x^3 + \ldots + (-1)^r x^r + \ldots, \quad |x| < 1$$

and replacing x by $-x$ gives

$$(1-x)^{-1} = 1+x+x^2+x^3+\ldots+x^r+\ldots, \quad |x|<1$$

Both of these series are G.P's with common ratios $-x$ and x respectively. They both occur frequently and are worth memorizing.

EXAMPLES 16f

1) Expand each of the following functions as a series of ascending powers of x up to and including the term containing x^4

(a) $(1+x)^{\frac{1}{2}}$ (b) $(1-2x)^{-3}$

$$(1+x)^n = 1+nx+\frac{n(n-1)}{2!}x^2+\ldots \quad |x|<1 \qquad [1]$$

(a) Replacing n by $\frac{1}{2}$ in [1] gives

$$(1+x)^{\frac{1}{2}} = 1+\tfrac{1}{2}x+\frac{(\frac{1}{2})(\frac{1}{2}-1)}{2!}x^2+\frac{(\frac{1}{2})(\frac{1}{2}-1)(\frac{1}{2}-2)}{3!}x^3$$
$$+\frac{\frac{1}{2}(\frac{1}{2}-1)(\frac{1}{2}-2)(\frac{1}{2}-3)}{4!}x^4+\ldots$$

$$= 1+\frac{x}{2}+\frac{(\frac{1}{2})(-\frac{1}{2})}{2!}x^2+\frac{(\frac{1}{2})(-\frac{1}{2})(-\frac{3}{2})}{3!}x^3$$
$$+\frac{(\frac{1}{2})(-\frac{1}{2})(-\frac{3}{2})(-\frac{5}{2})}{4!}x^4+\ldots$$

$$= 1+\frac{x}{2}-\frac{x^2}{2^3}+\frac{x^3}{2^4}-\frac{(3)(5)}{(2^4)4!}x^4+\ldots$$

$$= 1+\frac{x}{2}-\frac{x^2}{8}+\frac{x^3}{16}-\frac{5x^4}{128}+\ldots \quad \text{provided} \quad |x|<1$$

(b) Replacing n by -3 and x by $-2x$ in [1] gives

$$(1-2x)^{-3} = 1+(-3)(-2x)+\frac{(-3)(-4)}{2!}(-2x)^2$$

$$+\frac{(-3)(-4)(-5)}{3!}(-2x)^3 +\frac{(-3)(-4)(-5)(-6)}{4!}(-2x)^4+\ldots$$

$$= 1+6x+24x^2+\frac{(4)(5)}{2!}2^3x^3+\frac{(5)(6)}{2!}2^4x^4+\ldots$$

$$= 1+6x+24x^2+80x^3+240x^4+\ldots$$

provided $\quad -1<-2x<1$,

i.e. $\qquad \frac{1}{2}> \quad x>-\frac{1}{2}$.

2) Find the coefficient of x^n in the expansion of $(2-x)^{-2}$ as a series of ascending powers of x, stating the range of values of x for which the expansion is valid.

$(2-x)^{-2}$ must be expressed in the form $(1+x)^n$ (see note (a) p. 578).

i.e. $(2-x)^{-2} \equiv 2^{-2}\left(1-\dfrac{x}{2}\right)^{-2}$.

Consider the expansion of $(1-x)^{-2}$

$$(1-x)^{-2} = 1+(-2)(-x)+\frac{(-2)(-3)}{(1)(2)}(-x)^2$$
$$+\frac{(-2)(-3)(-4)}{(1)(2)(3)}(-x)^3+\ldots$$
$$= 1+2x+3x^2+4x^3+\ldots$$

from which we see that the general term is $(r+1)x^r$

Replacing x by $\dfrac{x}{2}$ gives

$$\left(1-\frac{x}{2}\right)^{-2} = 1+x+\ldots+\frac{r+1}{2^r}x^r+\ldots$$

$\Rightarrow$ $(2-x)^{-2} \equiv \dfrac{1}{4}\left(1-\dfrac{x}{2}\right)^{-2} = \dfrac{1}{4}+\dfrac{x}{4}+\ldots+\dfrac{r+1}{2^{r+2}}x^r+\ldots$

Hence the coefficient of x^n $(r=n)$ is $\dfrac{n+1}{2^{n+2}}$

provided that $-1<-\dfrac{x}{2}<1$,

i.e. $2>$ $x>-2$.

3) Expand $\dfrac{1}{(1+3x)(1-2x)}$ as a series of ascending powers of x giving:
(a) the first four terms   (b) the term containing x^n
(c) the range of values of x for which the expansion is valid.

Expressing $\dfrac{1}{(1+3x)(1-2x)}$ in partial fractions gives

$$\frac{1}{(1+3x)(1-2x)} \equiv \frac{3}{5(1+3x)}+\frac{2}{5(1-2x)} \equiv \tfrac{3}{5}(1+3x)^{-1}+\tfrac{2}{5}(1-2x)^{-1}$$

Using $(1+x)^{-1} = 1-x+x^2-x^3+\ldots,$ $-1<x<1$
and replacing x by $3x$ gives

$$\tfrac{3}{5}(1+3x)^{-1} = \tfrac{3}{5}[1-3x+(3x)^2-(3x)^3+\ldots+(-1)^r(3x)^r+\ldots]$$
$$= \tfrac{3}{5}[1-3x+9x^2-27x^3+\ldots(-1)^r 3^r x^r+\ldots]$$
for $-1<3x<1$.

Using $(1-x)^{-1} = 1 + x + x^2 + \ldots$ and replacing x by $2x$ gives

$$\tfrac{2}{5}(1-2x)^{-1} = \tfrac{2}{5}[1 + (2x) + (2x)^2 + (2x)^3 + \ldots + (2x)^r + \ldots]$$
$$= \tfrac{2}{5}[1 + 2x + 4x^2 + 8x^3 + \ldots + 2^r x^r + \ldots]$$

for $-1 < -2x < 1$.

Hence

$$\frac{1}{(1+3x)(1-2x)} \equiv \tfrac{3}{5}(1+3x)^{-1} + \tfrac{2}{5}(1-2x)^{-1}$$

$$= (\tfrac{3}{5} + \tfrac{2}{5}) + (-\tfrac{9}{5} + \tfrac{4}{5})x + (\tfrac{27}{5} + \tfrac{8}{5})x^2 + (-\tfrac{81}{5} + \tfrac{16}{5})x^3 +$$
$$\ldots + \left[\tfrac{3}{5}(3^r)(-1)^r + \tfrac{2}{5}(2^r)\right] x^r + \ldots$$

provided $-\tfrac{1}{3} < x < \tfrac{1}{3}$ and $-\tfrac{1}{2} < x < \tfrac{1}{2}$.

(a) Therefore the first four terms of the series are $1 - x + 7x^2 - 13x^3$.

(b) The term containing x^r is $\tfrac{1}{5}\{2^{r+1} + (-1)^r 3^{r+1}\}x^r$ so the term containing x^n $(r=n)$ is $\tfrac{1}{5}\{2^{n+1} + (-1)^n 3^{n+1}\}x^n$.

(c) The expansion is valid for the range of values of x satisfying $-\tfrac{1}{3} < x < \tfrac{1}{3}$ and $-\tfrac{1}{2} < x < \tfrac{1}{2}$.

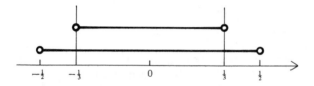

i.e. for $-\tfrac{1}{3} < x < \tfrac{1}{3}$.

4) Expand $\sqrt{\left(\dfrac{1+x}{1-2x}\right)}$ as a series of ascending powers of x up to and including the term containing x^2.

$$\sqrt{\left(\frac{1+x}{1-2x}\right)} \equiv (1+x)^{\frac{1}{2}}(1-2x)^{-\frac{1}{2}}$$

Now
$$(1+x)^{\frac{1}{2}} = \left[1 + \tfrac{1}{2}x + \frac{(\tfrac{1}{2})(-\tfrac{1}{2})}{2!}x^2 + \ldots\right]$$
for $-1 < x < 1$

and
$$(1-2x)^{-\frac{1}{2}} = \left[1 + (-\tfrac{1}{2})(-2x) + \frac{(-\tfrac{1}{2})(-\tfrac{3}{2})}{2!}(-2x)^2 + \ldots\right]$$
for $-1 < 2x < 1$

Hence
$$\sqrt{\left(\frac{1+x}{1-2x}\right)} \equiv (1+x)^{\frac{1}{2}}(1-2x)^{-\frac{1}{2}}$$
$$= (1 + \tfrac{1}{2}x - \tfrac{1}{8}x^2 + \ldots)(1 + x + \tfrac{3}{2}x^2 + \ldots)$$

$$= 1 + (\tfrac{1}{2}x + x) + (\tfrac{1}{2}x^2 - \tfrac{1}{8}x^2 + \tfrac{3}{2}x^2) + \ldots$$
$$= 1 + \tfrac{3}{2}x + \tfrac{15}{8}x^2 + \ldots$$

provided $-1 < x < 1$ *and* $-\tfrac{1}{2} < x < \tfrac{1}{2}$,
i.e. $-\tfrac{1}{2} < x < \tfrac{1}{2}$.

It is interesting to compare the methods used in the last two examples.
In Example 3, the function is expressed as the sum of two binomials and the series is obtained by adding two binomial expansions.
In Example 4 the function is expressed as a product of two binomials and the series is obtained by multiplying two binomial expansions.
The first method has the advantage that it is
(a) very much easier to add series than it is to multiply them, and
(b) the general term of the series can be found by simply adding together the general terms of the component series.
It is almost impossible to find a general term when two infinite series are multiplied together. Therefore *whenever possible a compound function should be expressed as a sum of simpler functions before it is expanded as a series* and when this is not possible, a compound function should be expressed as a product of simpler functions.

Further Approximations

We saw on p. 574 how a series expansion can be used to find an approximate value of $(1.01)^{20}$ without having to calculate the exact value. The following example shows how to find approximate numerical values (to any degree of accuracy required) for irrational quantities.

EXAMPLE 16f (continued)

5) By substituting 0.02 for x in $(1-x)^{\frac{1}{2}}$ and its expansion, find $\sqrt{2}$ correct to five decimal places.

$$(1-x)^{\frac{1}{2}} = 1 - \tfrac{1}{2}x + \frac{(\tfrac{1}{2})(-\tfrac{1}{2})(-x)^2}{2!} + \frac{(\tfrac{1}{2})(-\tfrac{1}{2})(-\tfrac{3}{2})(-x)^3}{3!} + \ldots$$

$$= 1 - \frac{x}{2} - \frac{x^2}{8} - \frac{x^3}{16} + \ldots$$

This expansion is valid for $-1 < x < 1$ and is therefore valid when $x = 0.02$
Substituting 0.02 for x gives

$$(0.98)^{\frac{1}{2}} = 1 - 0.01 - 0.000\,05 - 0.000\,000\,5 + \ldots$$

$$\Rightarrow \qquad \sqrt{\frac{98}{100}} = 0.989\,949\,5 \text{ to seven d.p.}$$

$\Rightarrow$ $\frac{7}{10}\sqrt{2} = 0.989\,949\,5$ to seven d.p.

Hence $\sqrt{2} = \dfrac{9.899\,495}{7}$ to six d.p.

    $= 1.414\,213$ to six d.p.

i.e. $\sqrt{2} = 1.414\,21$ correct to five d.p.

EXERCISE 16f

Expand the following functions as series of ascending powers of x up to and including the term in x^3. In each case give the range of values of x for which the expansion is valid.

1) $(1 - 2x)^{\frac{1}{2}}$ 2) $(3 + x)^{-1}$ 3) $\left(1 + \dfrac{x}{2}\right)^{-\frac{1}{2}}$

4) $\dfrac{1}{(1 - x)^2}$ 5) $\sqrt{\dfrac{1}{1 + x}}$ 6) $(1 + x)\sqrt{(1 - x)}$

7) $\dfrac{x + 2}{x - 1}$ 8) $\dfrac{2 - x}{\sqrt{(1 - 3x)}}$ 9) $\dfrac{1}{(2 - x)(1 + 2x)}$

10) $\sqrt{\left(\dfrac{1 + x}{1 - x}\right)}$ 11) $\left(1 + \dfrac{x^2}{9}\right)^{-1}$

12) $\left(1 + \dfrac{1}{x}\right)^{-1}$ $\left[Hint: \left(1 + \dfrac{1}{x}\right)^{-1} \equiv \left(\dfrac{x + 1}{x}\right)^{-1} \equiv \dfrac{x}{1 + x}\right]$

Find the coefficient of x^n in the expansion of each of the following functions as a series of ascending powers of x.

13) $(1 - x)^{-2}$ 14) $(1 + 2x)^{-3}$ 15) $(1 - 2x)(1 + x)^{-1}$

16) $\dfrac{1}{(1 + 2x)(3 - x)}$ 17) $(1 - x)^{\frac{3}{2}}$ 18) $\dfrac{1 + 2x}{1 - 2x}$

19) Expand $\left(1 + \dfrac{1}{p}\right)^{-3}$ as a series of descending powers of p, as far as and including the term containing p^{-4}. State the range of values of p for which the expansion is valid. $\left(Hint: \text{ replace } x \text{ by } \dfrac{1}{p} \text{ in } (1 + x)^{-3}\right)$.

20) By substituting 0.08 for x in $(1 + x)^{\frac{1}{2}}$ and its expansion find $\sqrt{3}$ correct to four significant figures.

21) By substituting $\dfrac{1}{10}$ for x in $(1 - x)^{-\frac{1}{2}}$ and its expansion find $\sqrt{10}$ correct to six significant figures.

22) Use a suitable binomial expansion to find $\sqrt{1.01}$ correct to five decimal places.

23) Use a suitable binomial expansion to find $\sqrt[3]{8.4}$ correct to seven significant figures.

24) Expand $\sqrt{\dfrac{1 + 2x}{1 - 2x}}$ as a series of ascending powers of x up to and including the term in x^2. By substituting $x = \dfrac{1}{100}$ find an approximation for $\sqrt{51}$, stating the number of significant figures to which your answer is accurate.

25) If x is so small that x^2 and higher powers of x may be neglected show that $\dfrac{1}{(x - 1)(x + 2)} \simeq -\frac{1}{2} - \frac{1}{4}x$.

26) By neglecting x^3 and higher powers of x, find a quadratic function that approximates to the function $\dfrac{1 - 2x}{\sqrt{1 + 2x}}$ in the region close to $x = 0$.

MACLAURIN'S THEOREM

We have shown how binominal functions can be expressed as a series of ascending powers of x, and how the series can be used to find approximate numerical values for certain irrational numbers. There are many other functions (such as e^x, $\cos x$ ... etc) which can be expressed as infinite series. Using these series we can find approximate values for quantities such as e^2, $\cos\dfrac{\pi}{6}$, and in all cases the approximation can be calculated to any degree of accuracy required.

If $f(x)$ is any function of x, and supposing that $f(x)$ can be expanded as a series of ascending powers of x, then

$$f(x) = a_0 + a_1 x + a_2 x^2 + a_3 x^3 + a_4 x^4 + a_5 x^5 + \ldots + a_r x^r + \ldots \quad [1]$$

where $a_0, a_1, a_2 \ldots$ are constants to be evaluated.

Substituting 0 for x in [1] gives $f(0) = a_0$,   i.e. $a_0 = f(0)$.

Differentiating [1] w.r.t. x gives

$$f'(x) = a_1 + 2a_2 x + 3a_3 x^2 + 4a_4 x^3 + 5a_5 x^4 + \ldots \quad [2]$$

Hence

$$f'(0) = a_1 \qquad\qquad\qquad\qquad \text{i.e.} \quad a_1 = f'(0)$$

Differentiating (2) w.r.t. x gives

$$f''(x) = 2a_2 + 3 \times 2a_3 x + 4 \times 3a_4 x^2 + 5 \times 4a_5 x^3 + \ldots \quad [3]$$

Hence

$$f''(0) = 2a_2 \qquad\qquad\qquad\qquad \text{i.e.} \quad a_2 = \dfrac{f''(0)}{2}$$

Similarly

$$f'''(x) = 3 \times 2a_3 + 4 \times 3 \times 2a_4 x + 5 \times 4 \times 3a_5 x^2 + \dots$$

so $\quad f'''(0) = 3!a_3$ $\qquad\qquad$ i.e. $\quad a_3 = \dfrac{f'''(0)}{3!}$

$$f''''(x) = 4 \times 3 \times 2a_4 + 5 \times 4 \times 3 \times 2a_5 x + \dots$$

so $\quad f''''(0) = 4!a_4$ $\qquad\qquad$ i.e. $\quad a_4 = \dfrac{f''''(0)}{4!}$

From which we can deduce that, after differentiating $f(x)$ r times,

$$f^r(x) = r!a_r + (r+1)(r) \dots 2a_{r+1}x + \dots$$

so $\quad f^r(0) = r!a_r$ $\qquad\qquad$ i.e. $\quad a_r = \dfrac{f^r(0)}{r!}$

Substituting these values of $a_0, a_1, a_2 \dots$ in [1] we get

$$f(x) = f(0) + f'(0)x + \frac{f''(0)}{2!}x^2 + \frac{f'''(0)}{3!}x^3 + \dots + \frac{f^r(0)}{r!}x^r + \dots$$

This is known as *Maclaurin's Theorem* and the series obtained is called the Maclaurin expansion of $f(x)$. The series can be found only if:

(a) the nth derivative of $f(x)$ exists for all n. i.e. it is possible to find $f^n(x)$ for $n = 1, 2, 3, \dots$,

(b) $f^n(0)$ is defined for $n = 1, 2, 3, \dots$,

(c) the series obtained converges to $f(x)$. The discussion of the convergence of series is beyond the scope of this book, but it is found that some expansions converge to $f(x)$ for all values of x and others converge to $f(x)$ for a limited range of values of x. In the following applications of Maclaurin's Theorem, the range of values of x for which the expansion is valid will be stated, but not proved.

If, for a given function, any of these conditions are not fulfilled, there is no Maclaurin series of ascending powers of x corresponding to the function. One assumption which was made in deriving the general Maclaurin expansion of $f(x)$ and which cannot be proved at this stage is that the derivative of a series is equal to the derivative of its sum to infinity, i.e. that

$$\frac{d}{dx} \sum_{r=0}^{\infty} a_r x^r = \sum_{r=0}^{\infty} \frac{d}{dx} (a_r x^r)$$

We will now use Maclaurin's Theorem to find the series expansion of some important functions.

EXAMPLES 16g

1) $f(x) \equiv e^x$.

Using $\qquad f(x) = f(0) + f'(0)x + \dfrac{f''(0)}{2!}x^2 + \dots + \dfrac{f^r(0)}{r!}x^r + \dots$

we have

$$f(x) \equiv e^x \qquad\qquad f(0) = e^0 = 1$$

$$f'(x) \equiv e^x \qquad\qquad f'(0) = e^0 = 1$$

$$f''(x) \equiv e^x \qquad\qquad f''(0) = e^0 = 1$$

$$\cdots\cdots\cdots\cdots \qquad\qquad \cdots\cdots\cdots\cdots$$

$$f^r(x) \equiv e^x \qquad\qquad f^r(0) = e^0 = 1$$

hence

$$e^x = 1 + x + \frac{x^2}{2!} + \frac{x^3}{3!} + \frac{x^4}{4!} + \ldots + \frac{x^r}{r!} + \ldots$$

This is valid for all values of x.

2) $f(x) \equiv \ln x$.

Using $f(x) = f(0) + f'(0)x + \ldots$
we have $f(x) \equiv \ln x$, $f(0) = \ln 0$ which is undefined.
Therefore *ln x cannot be expanded as a series of ascending powers of x.*

3) $f(x) \equiv \ln(1 + x)$.

Using

$$f(x) = f(0) + f'(0)x + \frac{f''(0)}{2!}x^2 + \ldots + \frac{f^r(0)}{r!}x^r + \ldots$$

we have

$$f(x) \equiv \ln(1 + x) \qquad\qquad f(0) = \ln 1 = 0$$

$$f'(x) \equiv \frac{1}{1 + x} \qquad\qquad f'(0) = 1$$

$$f''(x) \equiv -\frac{1}{(1 + x)^2} \qquad\qquad f''(0) = -1$$

$$f'''(x) \equiv +\frac{2}{(1 + x)^3} \qquad\qquad f'''(0) = 2$$

$$f''''(x) \equiv -\frac{2 \times 3}{(1 + x)^4} \qquad\qquad f''''(0) = -3!$$

$$\cdots\cdots\cdots\cdots\cdots \qquad\qquad \cdots\cdots\cdots\cdots\cdots$$

$$f^r(x) \equiv (-1)^{r+1}\frac{(r-1)!}{(1 + x)^r} \qquad\qquad f^r(0) = (-1)^{r+1}(r-1)!$$

hence

$$\ln(1 + x) = 0 + x - \frac{x^2}{2!} + \frac{2!x^3}{3!} - \frac{3!x^4}{4!} + \ldots + \frac{(-1)^{r+1}(r-1)!}{r!}x^r \ldots$$

i.e. $\ln(1 + x) = x - \dfrac{x^2}{2} + \dfrac{x^3}{3} - \dfrac{x^4}{4} + \ldots + \dfrac{(-1)^{r+1}x^r}{r} + \ldots$

This series is valid for $-1 < x \leqslant 1$.

4) $f(x) \equiv \cos x$.

Using $\qquad f(x) = f(0) + f'(0)x + \dfrac{f''(0)}{2!}x^2 + \ldots + \dfrac{f^r(0)}{r!}x^r + \ldots$

we have
$$f(x) \equiv \cos x \qquad\qquad f(0) = \cos 0 \quad = 1$$
$$f'(x) \equiv -\sin x \qquad\qquad f'(0) = -\sin 0 = 0$$
$$f''(x) \equiv -\cos x \qquad\qquad f''(0) = -\cos 0 = -1$$
$$f'''(x) \equiv \sin x \qquad\qquad f'''(0) = \sin 0 \quad = 0$$
$$f''''(x) \equiv \cos x \qquad\qquad f''''(0) = \cos 0 \quad = 1$$

Hence $\qquad \cos x = 1 + (0)\,x - \dfrac{1}{2!}x^2 + (0)\dfrac{x^3}{3!} + \dfrac{1}{4!}x^4 \ldots$

$$= 1 - \frac{x^2}{2!} + \frac{x^4}{4!} - \frac{x^6}{6!} + \ldots$$

From this we see that the series consists of even powers of x only, and that the terms containing $x^2, x^6, x^{10}, \ldots$ are negative, but the terms containing x^4, x^8, $x^{12}, \ldots$ are positive. A general term of the series is $\pm \dfrac{x^{2r}}{(2r)!}$ and is positive if r is even, negative if r is odd, so the general term can be written $(-1)^r \dfrac{x^{2r}}{(2r)!}$

Hence $\qquad \cos x = 1 - \dfrac{x^2}{2!} + \dfrac{x^4}{4!} - \dfrac{x^6}{6!} + \ldots + \dfrac{(-1)^r x^{2r}}{(2r)!} + \ldots$

This is valid for all values of x.

EXERCISE 16g

Use Maclaurin's Theorem to expand each of the following as a series of ascending powers of x up to and including the term containing the power of x indicated.

1) e^{-x}, x^4 $\qquad\qquad$ 2) $\sin x, x^7$ $\qquad$ 3) $\ln(1-x), x^4$

4) $(1+x)^n, x^3$ $\qquad\;$ 5) $\tan x, x^3$ $\qquad$ 6) $e^x \ln(1-x), x^2$

7) $(1-x)^{-1}, x^4$ $\qquad$ 8) $e^x \cos x, x^3$ $\quad$ 9) $\tan(e^x), x^2$

STANDARD EXPANSIONS

Summarizing some of the results from the last section we have:

$$e^x = 1 + x + \frac{x^2}{2!} + \frac{x^3}{3!} + \ldots + \frac{x^r}{r!} + \ldots \quad \text{for all values of } x$$

$$\ln(1+x) = x - \frac{x^2}{2} + \frac{x^3}{3} - \frac{x^4}{4} + \ldots + (-1)^{r+1}\frac{x^r}{r} + \ldots \quad \text{for } -1 < x \leqslant 1$$

$$\ln(1-x) = -x - \frac{x^2}{2} - \frac{x^3}{3} - \frac{x^4}{4} - \ldots - \frac{x^r}{r} + \ldots \quad \text{for } -1 \leqslant x < 1$$

$$\cos x = 1 - \frac{x^2}{2!} + \frac{x^4}{4!} - \frac{x^6}{6!} + \ldots + \frac{(-1)^r x^{2r}}{(2r)!} + \ldots \quad \text{for all values of } x.$$

$$\sin x = x - \frac{x^3}{3!} + \frac{x^5}{5!} - \frac{x^7}{7!} + \ldots + \frac{(-1)^r x^{2r+1}}{(2r+1)!} + \ldots \quad \text{for all values of } x.$$

$$(1+x)^n = 1 + nx + \frac{n(n-1)}{2!}x^2 + \ldots \quad \text{for } -1 < x < 1$$

$$(1+x)^{-1} = 1 - x + x^2 + \ldots + (-1)^r x^r + \ldots \quad \text{for } -1 < x < 1$$

$$(1-x)^{-1} = 1 + x + x^2 + \ldots + x^r + \ldots \quad \text{for } -1 < x < 1$$

All the series expansions given above may be quoted *unless* their derivation is asked for.

It is particularly important that the reader should remember the range of values of x for which a particular expansion is valid. The following examples show how these standard expansions can be used to express some compound functions as series, and the reader is advised to note that *whenever possible* a compound function should be expressed as a sum or difference of functions *before* any analysis of it is carried out.

EXAMPLES 16h

1) Expand: (a) $\ln \dfrac{1-2x}{(1+2x)^2}$ (b) $\dfrac{e^{2x}+e^{-2x}}{e^x}$.

as a series of ascending powers of x up to and including the term in x^4. Give the general terms and the range of values of x for which each expansion is valid.

(a) $$\ln\left[\frac{1-2x}{(1+2x)^2}\right] \equiv \ln(1-2x) - \ln(1+2x)^2$$

$$\equiv \ln(1-2x) - 2\ln(1+2x)$$

Using the expansion for $\ln(1-x)$ and replacing x by $2x$ gives

$$\ln(1-2x) = -2x - \frac{(2x)^2}{2} - \frac{(2x)^3}{3} - \frac{(2x)^4}{4} - \ldots - \frac{(2x)^r}{r} - \ldots$$

$$\text{if } -1 \leqslant 2x < 1$$

Using the expansion for ln $(1 + x)$ and replacing x by $2x$ gives

$$2 \ln (1 + 2x) = 2(2x) - \frac{2(2x)^2}{2} + \frac{2(2x)^3}{3} - \frac{2(2x)^4}{4} + \ldots + \frac{2(-1)^{r+1}(2x)^r}{r} + \ldots$$

if $-1 < 2x \le 1$

Hence

$$\ln \left[\frac{1 - 2x}{(1 + 2x)^2} \right] = \left(-2x - 2x^2 - \tfrac{8}{3}x^3 - 4x^4 - \ldots - \frac{2^r x^r}{r} - \ldots \right)$$

$$- \left(4x - 4x^2 + \tfrac{16}{3}x^3 - 8x^4 + \ldots + \frac{2^{r+1}(-1)^{r+1}x^r}{r} + \ldots \right)$$

$$= -6x + 2x^2 - 8x^3 + 4x^4 + \ldots - \frac{2^r x^r}{r} + \frac{2^{r+1}(-1)^r x^r}{r} + \ldots$$

provided $-\tfrac{1}{2} < x \le \tfrac{1}{2}$ and $-\tfrac{1}{2} \le x < \tfrac{1}{2}$.

Therefore the series, up to x^4, is $-6x + 2x^2 - 8x^3 + 4x^4$

The general term is $-\dfrac{2^r}{r} \left[1 - 2(-1)^r \right] x^r$

The expansion is valid for $-\tfrac{1}{2} < x < \tfrac{1}{2}$

(b) $$\frac{e^{2x} + e^{-2x}}{e^x} \equiv \frac{e^{2x}}{e^x} + \frac{e^{-2x}}{e^x} \equiv e^x + e^{-3x}$$

Using the expansion of e^x

$$e^x + e^{-3x} = \left(1 + x + \frac{x^2}{2!} + \frac{x^3}{3!} + \frac{x^4}{4!} + \ldots + \frac{x^r}{r!} + \ldots \right)$$

$$+ \left(1 - 3x + \frac{(-3x)^2}{2!} + \frac{(-3x)^3}{3!} + \frac{(-3x)^4}{4!} + \ldots + \frac{(-3x)^r}{r!} + \ldots \right)$$

$$= 2 - 2x + 5x^2 - \tfrac{13}{3}x^3 + \tfrac{41}{12}x^4 + \ldots + \frac{1 + (-3)^r}{r!} x^r + \ldots$$

As far as the term in x^4, the series is $2 - 2x - 5x^2 - \tfrac{13}{3}x^3 + \tfrac{41}{12}x^4$

The general term is $\dfrac{1 + (-3)^r}{r!} x^r$ and the expansion is valid for all values of x.

2) Show that for small values of x, $\sin^2 x \simeq x^2 - \tfrac{1}{3}x^4$.

Using the identity $\sin^2 x \equiv \tfrac{1}{2}(1 - \cos 2x)$ and the expansion of $\cos x$ we have:

$$\sin^2 x \equiv \tfrac{1}{2} - \tfrac{1}{2}\cos 2x = \tfrac{1}{2} - \tfrac{1}{2}\left(1 - \frac{(2x)^2}{2!} + \frac{(2x)^4}{4!} - \frac{(2x)^6}{6!} + \ldots\right)$$

$$= x^2 - \tfrac{1}{3}x^4 + \tfrac{2}{45}x^6 + \ldots$$

If x is small enough to neglect x^5 and higher powers of x we have

$$\sin^2 x \simeq x^2 - \tfrac{1}{3}x^4$$

3) Expand $\ln \dfrac{x}{x-1}$ as a series of *descending* powers of x as far as the term in x^{-4}. Use your expansion to find $\ln 1.25$ correct to four decimal places.

$$\ln\left(\frac{x}{x-1}\right) = \ln\left\{1 \div \left(\frac{x-1}{x}\right)\right\} = \ln 1 - \ln\left(\frac{x-1}{x}\right)$$

$$= -\ln\left(1 - \frac{1}{x}\right)$$

$$= -\left\{-\left(\frac{1}{x}\right) - \frac{1}{2}\left(\frac{1}{x}\right)^2 - \frac{1}{3}\left(\frac{1}{x}\right)^3 - \frac{1}{4}\left(\frac{1}{x}\right)^4 - \ldots\right\}$$

$$= \frac{1}{x} + \frac{1}{2x^2} + \frac{1}{3x^3} + \frac{1}{4x^4} + \ldots$$

This expansion is valid for $-1 \leqslant \dfrac{1}{x} < 1$

i.e.   for $x \leqslant -1$ and $x > 1$

When $\dfrac{x}{x-1} = 1.25$, $x = 5$ for which the expansion is valid

Therefore $\ln 1.25 = \tfrac{1}{5} + \tfrac{1}{2}(\tfrac{1}{25}) + \tfrac{1}{3}(\tfrac{1}{125}) + \tfrac{1}{4}(\tfrac{1}{625}) + \ldots$

$$= 0.2 + 0.02 + 0.002\,667 + 0.0004 + \ldots$$

$$= 0.223\,067 + \ldots$$

$$= 0.2231 \text{ to 4 d.p.}$$

TAYLOR'S EXPANSION

Maclaurin's expansion cannot be applied to all problems. For instance we have shown that $\ln x$ cannot be expanded as a series of ascending powers of x because $\ln x$ is undefined when $x = 0$. Also the Maclaurin expansion of

$\ln(1+x)$ is valid only for $-1 < x \leqslant 1$ and the series so obtained cannot be used to find $\ln(3)$.

Problems such as these can sometimes be overcome by expanding $f(x)$ as a series of ascending powers of $(x-a)$,

i.e. $f(x) = a_0 + a_1(x-a) + a_2(x-a)^2 + a_3(x-a)^3 + \ldots$

Using a method similar to that for obtaining Maclaurin's expansion gives

$$f(x) = f(a) + f'(a)(x-a) + \frac{f''(a)}{2!}(x-a)^2 + \frac{f'''(a)}{3!}(x-a)^3 + \ldots$$

$$\ldots + \frac{f^r(a)}{r!}(x-a)^r + \ldots$$

This is known as Taylor's Expansion and it is particularly useful for finding an approximation for $f(x)$ when x is close to a.

In this case $(x-a)$ is small, so neglecting $(x-a)^2$ and higher powers of $(x-a)$

$$f(x) \simeq f(a) + f'(a)(x-a)$$

EXAMPLES 16h (continued)

4) Expand $\ln x$ as a series of ascending powers of $(x-a)$ up to the term containing $(x-a)^3$.

Use your expansion to evaluate $\ln 1.01$ correct to five decimal places.

$$f(x) \equiv \ln x \qquad f(a) = \ln a$$

$$f'(x) \equiv \frac{1}{x} \qquad f'(a) = \frac{1}{a}$$

$$f''(x) \equiv -\frac{1}{x^2} \qquad f''(a) = \frac{1}{a^2}$$

$$f'''(x) \equiv \frac{2}{x^3} \qquad f'''(a) = \frac{2}{a^3}$$

Therefore $\ln x = \ln a + \frac{1}{a}(x-a) - \frac{1}{a^2}\frac{(x-a)^2}{2!} + \frac{2}{a^3}\frac{(x-a)^3}{3!} + \ldots$

If $x = 1.01$ and we let $a = 1$, then $(x-a) = 0.01$ which is small.

Hence $\ln 1.01 = \ln 1 + \frac{1}{1}(0.01) - \frac{1}{(1)^2}\frac{(0.01)^2}{2!} + \frac{2}{(1)^3}\frac{(0.01)^3}{3!} + \ldots$

$= 0 + 0.01 - 0.00005 + \ldots$

$= 0.00995$ correct to 5 d.p.

5) Show that if $\left(x - \dfrac{\pi}{4}\right)$ is so small that $\left(x - \dfrac{\pi}{4}\right)^2$ and higher powers may be neglected

$$\tan x \simeq 1 - \frac{\pi}{2} + 2x$$

Using Taylor's Theorem to expand $\tan x$ as a series of ascending powers of $(x - a)$ we have,

$$f(x) \equiv \tan x \qquad f(a) = \tan a$$

$$f'(x) \equiv \sec^2 x \qquad f'(a) = \sec^2 a$$

So $\tan x = \tan a + (\sec^2 a)(x - a) + \dots$

When $a = \dfrac{\pi}{4}$ $\tan x \simeq \tan \dfrac{\pi}{4} + \left(\sec^2 \dfrac{\pi}{4}\right)\left(x - \dfrac{\pi}{4}\right)$

$$\simeq 1 + (2)\left(x - \frac{\pi}{4}\right)$$

Therefore $\tan x \simeq 1 - \dfrac{\pi}{2} + 2x$ when $x \simeq \dfrac{\pi}{4}$

EXERCISE 16h

Expand each of the following functions as a series of ascending powers of x up to and including the term containing x^2. Give the general term and the range of values of x for which the expansion is valid.

1) e^{3x}

2) $\cos \dfrac{x}{2}$

3) $2 \sin x \cos x$

4) $\ln \sqrt{\left(\dfrac{1-x}{1+x}\right)}$

5) $\dfrac{e^x + e^{-x}}{e^x}$

6) $\ln \left[(1 + x)(1 - 2x)^2\right]$

7) $\dfrac{x - e^x}{e^x}$

8) $\cos^2 2x$

9) $e^{3x} + \ln \sqrt{(1 - 2x)}$

10) $\sin^3 x$

Expand each of the following functions as a series of ascending powers of y up to and including the term in y^2.

11) $e^y \ln (1 + y)$ 12) $\tan^2 y$ 13) $e^y \cos y$

14) $\sin (a + y)$

Expand each of the following functions as a series of ascending powers of $\dfrac{1}{x}$ up to and including the term in $\left(\dfrac{1}{x}\right)^3$. Give the range of values of x for which the expansion is valid.

15) $\ln \left(\dfrac{x+1}{x} \right)$ 16) $\ln \left(\dfrac{1+2x}{1+x} \right)$ (*Hint*: divide throughout by x)

17) Use the expansion of e^x to find the value of e correct to seven decimal places.

18) Use the expansion of $\cos x$ to show that when θ is small, $\cos \theta \simeq 1 - \frac{1}{2}\theta^2$.

19) Use the expansion of $\ln \left(\dfrac{1+x}{1-x} \right)$ with $x = \frac{1}{3}$ to find $\ln 2$ correct to three decimal places.

20) Use the expansions of $\ln (1 + x)$ and $\ln (1 - x)$ to find the values of $\ln 1.1$ and $\ln 0.9$ respectively, correct to four decimal places.

21) Find a linear approximation for $f(x) \equiv e^x \cos x$ in the region close to $x = 0$.

22) Use Maclaurin's Theorem to obtain the expansion of $\arcsin x$ as a series of ascending powers of x as far as the term containing x^3. By substituting $x = \frac{1}{2}$ find the value of π to two significant figures.

23) If x is so small that x^3 and higher powers of x may be neglected, show that $\dfrac{e^x}{1+x} \simeq 1 + \frac{1}{2}x^2$.

24) Expand $\cos x \ln x$ as a Taylor series in powers of $(x - 1)$ up to and including the term in $(x - 1)^2$.

25) Expand $\sin x$ as a series of ascending powers of $\left(x - \dfrac{\pi}{6} \right)$ as far as the third term of the series. Use your series to find an approximate value for $\sin 31°$, given that $1° = 0.017$ radians.

26) By using the expansion of $\ln (1 + x)$ as a series of ascending powers of x find a quadratic approximation for $\ln (1 + x)$ when x is small. Use your approximation to find an approximate value for the smallest root of the equation $\ln (1 + x) = x^2$. By sketching the graphs of $f(x) \equiv \ln (1 + x)$ and $f(x) \equiv x^2$ show that the value of your root is exact.

SUMMARY

Arithmetic Progressions
An A.P. with n terms can be written as

$$a + (a + d) + (a + 2d) + \ldots + (a + rd) + \ldots + [a + (n - 1)d]$$

The sum of the first n terms, S_n, is given by

$$S_n = \frac{n}{2} [2a + (n - 1)d]$$

$$= \frac{n}{2}(a + l) \quad \text{where } l \text{ is the last term}$$

The sum of the first n natural numbers is given by

$$\sum_{r=1}^{n} r = \frac{n(n + 1)}{2}$$

Geometric Progressions
A G.P. of n terms can be written as

$$a + ar + ar^2 + \ldots + ar^{(n-1)}$$

$$S_n = \frac{a(1 - r^n)}{1 - r}, \qquad S_\infty = \frac{a}{1 - r}, \quad |r| < 1$$

Arithmetic mean of two numbers x, y is $\frac{1}{2}(x + y)$

Geometric mean of two numbers x, y is $\sqrt{xy}$

Binomial Theorem for a positive integral index

$$(1 + x)^n \equiv \sum_{r=0}^{n} {}^nC_r x^r, \qquad (a + x)^n \equiv \sum_{r=0}^{n} {}^nC_r a^{n-r} x^r$$

Maclaurin's Expansion

$$f(x) = f(0) + f'(0)x + \frac{f''(0)}{2!}x^2 + \frac{f'''(0)}{3!}x^3 + \ldots$$

Taylor's Expansion

$$f(x) = f(a) + f'(a)(x - a) + \frac{f''(a)}{2!}(x - a)^2 + \ldots$$

Series expansions of standard functions

$$(1 + x)^n = 1 + nx + \frac{n(n - 1)}{2!}x^2 + \frac{n(n - 1)(n - 2)}{3!}x^3 + \ldots$$

If n is a positive integer the series is finite and valid for all values of x.
If n is not a positive integer the series is infinite and valid only for
$-1 < x < 1$.

$$e^x = 1 + x + \frac{x^2}{2!} + \frac{x^3}{3!} + \frac{x^4}{4!} + \ldots + \frac{x^r}{r!} + \ldots \quad \text{for all values of } x.$$

$$\ln(1 + x) = x - \frac{x^2}{2} + \frac{x^3}{3} - \frac{x^4}{4} + \ldots + \frac{(-1)^{r+1}x^r}{r} + \ldots$$

$$\text{provided} \quad -1 < x \leqslant 1$$

$$\ln(1-x) = -x - \frac{x^2}{2} - \frac{x^3}{3} - \frac{x^4}{4} - \ldots - \frac{x^r}{r} - \ldots \quad \text{provided} \quad -1 \leqslant x < 1$$

$$\cos x = 1 - \frac{x^2}{2!} + \frac{x^4}{4!} - \frac{x^6}{6!} + \ldots + \frac{(-1)^r x^{2r}}{(2r)!} + \ldots \quad \text{for all values of } x$$

$$\sin x = x - \frac{x^3}{3!} + \frac{x^5}{5!} - \frac{x^7}{7!} + \ldots + \frac{(-1)^r x^{2r+1}}{(2r+1)!} + \ldots \quad \text{for all values of } x$$

MULTIPLE CHOICE EXERCISE 16

(Instructions for answering these questions are given on page xii.)

TYPE I

1) The sum of the series $1 + 5 + 9 + 13 + 17 + 21 + 25 + 29$ is:
(a) 30   (b) 240   (c) 120   (d) 112   (e) 28

2) The sum to infinity of the series $1 + 2x + 4x^2 + 8x^3 + \ldots$, for $-\frac{1}{2} < x < \frac{1}{2}$ is:

(a) $\dfrac{2x}{1-2x}$ (b) $\dfrac{1}{1-2x}$ (c) $\dfrac{1}{1+2x}$ (d) $\dfrac{2}{1-x}$ (e) $1-2x$

3) The first three terms of the series $\sum\limits_{r=0}^{\infty} (-1)^{r+1} 2^r x^{-r}$ are:

(a) $-1 + \dfrac{2}{x} - \dfrac{4}{x^2}$ (b) $1 + \dfrac{2}{x} - \dfrac{4}{x^2}$ (c) $1 + 2x - 4x^2$

(d) $\dfrac{2}{x} - \dfrac{4}{x^2} + \dfrac{8}{x^3}$ (e) none of these.

4) When x is small $(1+x)\ln(1+x)$ is approximately:

(a) $x - \dfrac{x^2}{2}$ (b) $1 + \dfrac{3x^2}{2}$ (c) $1 - \tfrac{1}{2}x^2$ (d) $x + \dfrac{x^2}{2}$

5) 3 is the geometric mean of a and b. Possible values of a and b are:
(a) $5,4$ (b) $0,9$ (c) $3,1$ (d) $4,2$ (e) $9,1$

6) The series $1 - x + 2x^2 - 3x^3 + 4x^4 + \ldots$ may be written more briefly as:

(a) $\sum\limits_{r=0}^{\infty} (-1)^r r x^r$ (b) $1 + \sum\limits_{r=0}^{\infty} (-1)^r r x^r$ (c) $\sum\limits_{r=1}^{\infty} r x^r$

(d) $1 - \sum\limits_{r=0}^{\infty} (-1)^r r x^r$ (e) $\sum\limits_{r=1}^{\infty} (-1)^{r+1} r x^r$

7) The coefficient of x^3 in the binomial expansion of $(2-x)^8$ is:
(a) 1792   (b) 56   (c) -1792 (d) -2000 (e) -448

8) The first non-zero term in the expansion of $(1-x)\ln(1-x)$ is:

(a) 1　　　　(b) x　　　　(c) $-x^2$　　　　(d) $-x$　　　　(e) $-2x$

9) Using the first five terms in the expansion of e^x gives e^2 as approximately:

(a) 7　　　　(b) 5　　　　(c) 8　　　　(d) 9　　　　(e) 3

10) The series $1 - 3x + 9x^2 - 27x^3 + \ldots$ converges to the value:

(a) $\frac{1}{10}$ when $x = 3$　　　　(b) $\frac{1}{2}$ when $x = \frac{1}{3}$　　　　(c) $\frac{2}{3}$ when $x = \frac{1}{6}$
(d) $\frac{3}{2}$ when $x = \frac{1}{9}$　　　　(e) $\frac{1}{4}$ when $x = 1$

TYPE II

11) The sum of the first n terms, S_n, of a given series is given by

$$S_n = \frac{2n^2}{n^2 + 1}.$$

(a) The first two terms of the series are $1, \frac{8}{5}$
(b) The sum of the third and fourth terms is $\frac{24}{85}$
(c) The series converges.

12) $\displaystyle\sum_{r=2}^{12} \frac{2^r}{r}$.

(a) The series has eleven terms.
(b) The series is a G.P.
(c) The third term of the series is $\dfrac{2^3}{3}$.

13) The nth term of a series is $\ln 2^{n+1}$.

(a) The sixth term of the series is $7\ln 2$.
(b) The series is the expansion of $\ln(1 + n)$.
(c) The terms of the series are in arithmetic progression.

14) $(1+x)^3(1-x)^{20}$ is expanded as a series of ascending powers of x.

(a) The series is finite.
(b) The first two terms of the series are $1 - 17x$.
(c) The last term of the series is x^{20}.

15) In the expansion of $\dfrac{1}{(1-x)(1+x)}$ as a series of ascending powers of x:

(a) there are only even powers of x,
(b) the first three terms are $\frac{1}{2} + \frac{1}{2}x^2 + \frac{1}{2}x^4$,
(c) the expansion is valid for $1 \leqslant x \leqslant 1$.

TYPE III

16) (a) $f(x) \equiv \dfrac{1}{1-x}$.

(b) $f(x) = \displaystyle\sum_{r=0}^{\infty} x^r$ for $-1 < x < 1$.

17) (a) $f(x) \equiv \dfrac{1-x^n}{1-x}$.

(b) $f(x) \equiv 1 + x + x^2 + \ldots + x^{n-1}$.

18) (a) The terms of a series are in arithmetic progression.
(b) The sum of the first n terms of a series is $n(3 - 2n)$.

19) (a) $f(\theta) \equiv \cos \theta$.

(b) $f'(\theta) = \theta - \dfrac{\theta^3}{3!} + \dfrac{\theta^5}{5!} + \ldots$

20) (a) $f(x) \equiv \dfrac{e^x}{1 - e^x}$.

(b) The first two terms in the Maclaurin expansion of $f(x)$ as a series of ascending powers of x are $1 - x$.

TYPE V

21) $e = 1 + 1 + \dfrac{1}{2!} + \dfrac{1}{3!} + \dfrac{1}{4!}$ correct to two significant figures.

22) The third term in the binomial expansion of $(2 - 3x)^{10}$ is $(45)(2^7)(3^3)(x^3)$.

23) The coefficient of x^r in the expansion of $\ln(1 - x)^3$ as a series of ascending powers of x is $-\dfrac{3}{r}$.

24) The first term in the expansion of $\ln(2 - x)$ as a series of ascending powers of x is $\ln 2$.

25) If $2^n - 1$ is the sum of the first n terms of a series, $4^n - 1$ is the sum of the first $2n$ terms.

26) 3 is the arithmetic mean of 4 and 1.

27) The fourth term of the series $\displaystyle\sum_{r=0}^{n} (-1)^{(r+1)} 2^r$ is 16.

28) The geometric mean of $3x$ and $12x$ is $6x$.

MISCELLANEOUS EXERCISE 16

1) Find the sum of the first n terms of the arithmetic progression
$2 + 5 + 8 + \ldots$.
Find the value of n for which the sum of the first $2n$ terms will exceed the sum of the first n terms by 224.   (AEB)p '71

2) The sum of n terms of a series is $3n^2$ for $n = 1, 2, 3 \ldots$. Show that the series is an arithmetic progression and find its nth term.   (AEB)p '73

3) The first term of an arithmetic series is $\ln x$ and the rth term is $\ln (xc^{r-1})$.
Show that the sum S_n of the first n terms of the series is $\dfrac{n}{2} \ln (x^2 c^{n-1})$.
      (AEB)p '76

4) The first and last terms of an arithmetic progression are a and l respectively.
If the progression has n terms, prove from first principles that its sum is $\frac{1}{2}n(a + l)$.
A circular disc is cut into twelve sectors whose areas are in arithmetic progression. The area of the largest sector is twice that of the smallest. Find the angle (in degrees) between the straight edges of the smallest sector.   (JMB)p

5) Evaluate the following, giving each answer correct to two significant figures:

(a) $\displaystyle\sum_{n=1}^{20} (1.1)^n$ (b) $\displaystyle\sum_{n=1}^{20} \log_{10}(1.1)^n$.   (JMB)

6) If S_n denotes the sum of the first n terms of the geometric progression $1 + \frac{1}{2} + (\frac{1}{2})^2 + \ldots + (\frac{1}{2})^{n-1} + \ldots$ and if S denotes the sum to infinity, find the least value of n such that $S - S_n < 0.001$.   (O)p

7) A man borrows a sum of money from a building society and agrees to pay the loan (plus interest) over a period of years. If £A is the sum borrowed and $r\%$ the yearly rate of interest charged it can be proved that the amount (£p_n) of each annual instalment which will extinguish the loan in n years is given by the formula

$$P_n = \frac{A(R-1)R^n}{R^n - 1} \quad \text{where} \quad R = 1 + \frac{r}{100}.$$

Assuming this formula calculate P_n (correct to the nearest £) if $A = 1000$, $n = 25$, $r = 6.5$.
Find in its simplest form the ratio $P_{2n} : P_n$. Show that this ratio is always greater than $\frac{1}{2}$ and, if $r = 6.5$, find the least integral number of years for which this ratio is greater than $\frac{3}{5}$.   (C)

8) A rod one metre in length is divided into ten pieces whose lengths are in geometric progression. The length of the longest piece is eight times the length of the shortest piece. Find, to the nearest millimetre, the length of the shortest piece.   (JMB)

9) (a) By using an infinite geometric progression show that $0.43\dot{2}\dot{1}$,

i.e. $0.432\ 121\ 212\ 1\ \ldots$ is equal to $\dfrac{713}{1650}$.

(b) Write down the term containing p^r in the binomial expansion of $(q + p)^n$ where n is a positive integer. If $p = \frac{1}{6}$, $q = \frac{5}{6}$ and $n = 30$, find the value of r for which this term has the numerically greatest value.

(AEB)'74

10) The coefficient of the term in x^n of the series $16 - x - 3x^2 - \ldots$ is $A(\frac{1}{5})^n + B(\frac{2}{3})^n$. Find the values of the constants A and B and hence obtain the next term in the series. Determine the range of values of x for which the series possesses a sum to infinity and find the value of x for which this sum is $\frac{17}{4}$.

(AEB)'72

11) Find the sum $S(x)$ of the series $1 + 2x + 3x^2 + \ldots + (n + 1)x^n$,
(a) by first finding $(1 - x)S(x)$,
(b) by regarding $S(x)$ as the derivative with respect to x of another series.

(O)

12) The first three terms of a geometric progression are also the first, ninth and eleventh terms, respectively, of an arithmetic progression. Given that the terms of the geometric progression are all different, find the common ratio r.
If the sum to infinity of the geometric progression is 8, find the first term and find the common difference of the arithmetic progression.

(JMB)p

13) An infinite geometric progression is such that the sum of all the terms after the nth is equal to twice the nth term. Show that the sum to infinity of the whole progression is three times the first term.

(JMB)p

14) Starting from first principles, prove that the sum of the first n terms of a geometric progression whose first term is a and whose common ratio is r

(where $r \neq 1$) is $\quad S_n = \dfrac{a(1 - r^n)}{1 - r}$

Show that $\qquad \dfrac{S_{3n} - S_{2n}}{S_n} = r^{2n}$

Given that $r = \frac{1}{2}$, find $\displaystyle\sum_{n=1}^{\infty} \dfrac{S_{3n} - S_{2n}}{S_n}$.

15) The binomial expansion of $(1 + x)^n$, n a positive integer, may be written in the form

$$(1 + x)^n = 1 + c_1 x + c_2 x^2 + c_3 x^3 + \ldots + c_r x^r + \ldots$$

Show that, if c_{s-1}, c_s, and c_{s+1} are in arithmetic progression, then $(n - 2s)^2 = n + 2$.
Find possible values of s when $n = 62$.

(AEB) p '71

16) Find the numerical value of the coefficient of x^{11} in the expansion of $\left(x^2 + \dfrac{1}{x}\right)^{10}$ in powers of x. (U of L)

17) Express the function $f(x) \equiv \dfrac{1}{(1+x)(1-3x)}$ in partial fractions.
Find the first four terms in the expansion of $f(x)$ in ascending powers of x, and obtain the coefficient of x^n. (JMB)

18) Expand $z = (1 + ic)^5$ in powers of c, and show that there are five real values of c for which z is real. (JMB)p

19) By putting $x = \dfrac{1}{1000}$ in the expansion of $(1-x)^{\frac{1}{3}}$ in ascending powers of x obtain the cube root of 37 correct to six decimal places. (C)p

20) Prove that when $(1+x)^2(1-x)^{-2}$ is expanded in ascending powers of x, the coefficient of x^n (where $n > 0$) is $4n$. (C)p

21) Show that, for $-1 < x < 1$, the expansion in ascending powers of x of $\left(\dfrac{1+x}{1-x}\right)^{\frac{1}{3}}$ is $1 + \frac{2}{3}x + \frac{2}{9}x^2 + \frac{22}{81}x^3 + \ldots$

By taking $x = 0.02$ obtain the cube root of 357, giving five places of decimals in your answer. (C)p

22) Given that $y = \sqrt{\left(\dfrac{1+x}{2+x}\right)}$:

(a) find the values of a, b, c, in the approximation $y \simeq a + bx + cx^2$, where it is assumed that x is so small that its cube and higher powers may be neglected,

(b) find the constant term and the coefficient of x^n $(n \geqslant 1)$ in the series expansion of $\ln y$ in ascending powers of x. (JMB)p

23) Expand $y = \left(\dfrac{1+2px}{1+2qx}\right)^{\frac{1}{2}}$ and $z = \left(\dfrac{1-px}{1-qx}\right)^{-\frac{1}{2}}$ $(p \neq q)$ in ascending powers of x as far as the terms in x^2.
Given that px and qx are small:

(a) show that if the terms in x^2 and higher powers of x are neglected, the expressions obtained for y and z satisfy the relation $1 + y = 2z$;

(b) find the ratio $p : q$ for which the relation $1 + y = 2z$ still holds if the terms in x^2 are retained but the terms in x^3 and higher powers of x are neglected. (C)

24) Express $f(x) \equiv \dfrac{2x}{(x+1)^2(x^2+1)}$ in partial fractions.

Expand the following functions as series in ascending powers of x up to and including terms indicated in the brackets.

(a) $f(x)$, $[x^3]$ (b) $\log\left[\dfrac{f(x)}{2x}\right]$, $[x^4]$

Assuming the validity of the process, differentiate $\log\left[\dfrac{f(x)}{2x}\right]$ as given in (b)

above to show that as far as the term in x^3, $\dfrac{1}{f(x)}\dfrac{d}{dx}f(x) = \dfrac{1}{x} - 2 - 2x^2 + 4x^3$.

(AEB)'73

25) In the expansion in ascending powers of x of the function $\dfrac{1+x}{1+ax} - e^{bx}$
the coefficients of x and x^2 are both zero. Find the two possible pairs of
values of the constants a and b. Verify that for one of these pairs of values all
coefficients in the expansion are zero.   (JMB)

26) (a) Express $\ln\left(\dfrac{1+x}{1-x}\right)$ as a series of ascending powers of x up to and
including the term in x^5.
Prove that $\frac{2}{3}[1 + \frac{1}{3}(\frac{1}{9}) + \frac{1}{5}(\frac{1}{9})^2 + \ldots] = \ln 2$.

(b) Prove that, if x is so small that x^6 and higher powers of x may be
neglected, then $\dfrac{e^{2x}-1}{e^{2x}+1} \simeq x - \dfrac{x^3}{3} + \dfrac{2x^5}{15}$.   (U of L)

27) If $e^{2y}(1+x) = (1-2x)$ express y as a function of x. Expand y as a
series in ascending powers of x up to and including the term in x^3 and state
the range of values for which the expansion is valid.   (AEB)p'73

28) Write down the first four terms and the rth term u_r in the expansion of
e^{-2x} in ascending powers of x.

Find in its simplest form $\dfrac{u_{r+1}}{u_r}$.

Express $\displaystyle\sum_{r=1}^{\infty} \dfrac{x^{r-1}u_{r+1}}{u_r}$ in the form $k\ln(1+ax)$ and state the values of a

and k.   (AEB)p'76

29) (a) State the formula for the sum, S_n, of the first n terms of the
arithmetic progression $a, a+d, \ldots$. Given that $S_{2m} = 3S_m$ express
a in terms of m and d.

(b) Give the expression for the sum to infinity of the geometric series
$a + ar + ar^2 + \ldots$, stating the range of values of r for which it is
valid.
Express the recurring decimal $0.536\,363\,6\ldots$ as a fraction in its
lowest terms.

(c) Write down the expansion of $\ln (1 + x)$ in powers of x, giving the first three terms and the general term. Calculate $\ln 0.97$ to five significant figures.   (JMB)

30) Write down the expansions of $\ln (1 + x)$ and $\ln (1 - x)$ and in each case state the range of values of x for which the expansions are valid.

Deduce that $\ln \dfrac{y + 1}{y - 1} = \dfrac{2}{y} + \dfrac{2}{3y^3} + \dfrac{2}{5y^5} + \dfrac{2}{7y^7} + \dots$

and state the range of values of y for which the expansion is valid.   (AEB)p '75

31) Find the first non-zero term in the expansion of
$(1 - x)(1 - e^x) + \ln (1 + x)$ in ascending powers of x.   (U of L)

32) If $y = \ln \cos x$,   prove that $\dfrac{d^3 y}{dx^3} + 2 \dfrac{d^2 y}{dx^2} \dfrac{dy}{dx} = 0.$

Hence, or otherwise, obtain the Maclaurin expansion of y in terms of x up to and including the term in x^4. Using $x = \dfrac{\pi}{4}$ show that $\ln 2 \simeq \dfrac{\pi^2}{16} \left(1 + \dfrac{\pi^2}{96}\right).$
     (AEB)'75

33) (a) If $y = \tan \left(x + \dfrac{\pi}{4}\right),$ find the values of the first and second

derivatives of y when $x = 0.$ Hence write down the Maclaurin

series for $\tan \left(x + \dfrac{\pi}{4}\right)$ up to and including the term in x^2.

(b) If $y = x^x$,   show that $\ln y = x \ln x.$ Hence, or otherwise, find the first and second derivatives of y when $x = 1.$ Hence write down the Taylor series for x^x in powers of $(x - 1)$ up to and including the term in $(x - 1)^2$.   (O)

34) By expressing x^x as a power of e and using the exponential series, or otherwise, find (without using tables) the value of x^x when $x = 0.01$, correct to three places of decimals.   ($\ln 10 = 2.30$ to two decimal places).
     (O)

35) If $\sin y = \tfrac{1}{2} \cos x$,   prove that $\dfrac{d^2 y}{dx^2} = \tan y \left\{\left(\dfrac{dy}{dx}\right)^2 - 1\right\}.$

If $-\tfrac{1}{2}\pi < y < \tfrac{1}{2}\pi$,   obtain the expansion of y in ascending powers of x as far as and including the term in x^2, and show that the coefficient of x^3 is zero.
     (C)

36) If $y = \tan (e^x - 1)$,   prove that $\dfrac{d^2 y}{dx^2} = \dfrac{dy}{dx} (1 + 2e^x y).$

Obtain the Maclaurin expansion of y as far as the term in x^4 inclusive.   (C)

37) Use Maclaurin's theorem to expand $\ln \cos x$ in a series of ascending powers of x as far as the term in x^4. Check your result by replacing $\cos x$ by $1 - \frac{1}{2}x^2 + \frac{1}{24}x^4$ and using the logarithmic series.

By substituting $x = \frac{1}{4}\pi$, obtain a value for $\ln 2$ to two decimal places, and explain why you might expect a result less than the true value. (C)

38) If $y = \ln \tan x$, find $\dfrac{dy}{dx}$ and $\dfrac{d^2y}{dx^2}$, and prove that

$$\frac{d^3y}{dx^3} = 4(3 + \cos 4x)/\sin^3 2x.$$

Obtain the expansion of $\ln \tan (x + \frac{1}{4}\pi)$ in powers of x as far as the term in x^3. (C)

39) Obtain the Taylor expansion of the function $x^2 \ln x$ in powers of $(x - 1)$ as far as the term in $(x - 1)^4$. (U of L)p

CHAPTER 17

PROBABILITY

Imagine that you have bought five tickets for a raffle and that 500 tickets altogether have been sold. Assuming that any one of the 500 tickets is as likely as any other to be drawn for first prize, you would say that you had 5 chances in 500, or a chance of 1 in 100, of winning first prize.

In this chapter we develop methods to deal with problems concerned with chance events.

Terminology and notation are introduced to enable us to refer to certain categories of situations precisely and more briefly.

Probability gives us a measure for the likelihood that something will happen. However it must be appreciated that probability can never predict the number of times that an occurrence actually happens. But being able to quantify the likely occurrence of an event is important because most of the decisions that affect our daily lives are based on likelihoods and not on absolute certainties. For example, if it is known that it is likely to rain on two days out of five days at a place where you are taking a holiday, it does *not* mean that it *will* rain on four days out of a ten day holiday but you are likely to decide to take a raincoat with you.

AN EVENT

An event is a defined occurrence or situation. For example:

(a) tossing a coin and the coin landing head up,

(b) scoring a six on the throw of a die,

(c) winning the first prize in a raffle,

(d) being dealt a hand of four cards which are all clubs.

A particular event is denoted by a capital letter, e.g. $A, B, \ldots$ etc.

POSSIBILITY SPACE

In each of the examples given there is an implied set of circumstances from which there are several possible outcomes, including the event(s) described. This set of possible outcomes is called the *possibility space*.
Thus in:

(a) the event is one of the possible ways in which the coin can land, viz. head up, H, or tail up, T, i.e. the possibility space is $\{H, T\}$,

(b) the event is one of the possible ways of scoring on the throw of a die, i.e. the possibility space is $\{1, 2, 3, 4, 5, 6\}$,

(c) the possibility space is $\{$all the tickets in the draw$\}$,

(d) the possibility space is $\{$all the different combinations of four cards that can be obtained from fifty-two cards$\}$.

Now consider the following situation. A bag contains three white balls and two black balls, and one ball is removed from the bag. The possibility space is the set

$$\{\circ, \circ, \circ, \bullet, \bullet\}$$

If the event denoted by A is 'the removal of a white ball' the possibilities for A are the members of the set $\{\circ, \circ, \circ\}$.
Denoting 'the possibilities for the event A' as the set $\{A\}$ we can write
$\{A\} = \{\circ, \circ, \circ\}$
and we note that $\{A\}$ is a subset of $\{\circ, \circ, \circ, \bullet, \bullet\}$.
In general, if E is an event then $\{E\}$ is a subset of $\{$possibility space$\}$.

PROBABILITY THAT AN EVENT OCCURS

The probability that an event, A, occurs is defined as the number of ways in which A can happen expressed as a fraction of the number of ways in which all *equally likely* events, including A, occur. The term 'equally likely' is important, for example if a coin is bent so that when tossed it is more likely to land heads up than tails up, then the likely events (as the result of tossing the coin) are that the coin lands heads up or lands tails up, but they are *not* equally likely.
The *probability of an event, A, occurring is denoted by* $P(A)$

Hence

$$P(A) = \frac{\text{No. of ways in which } A \text{ occurs}}{\text{No. of ways in which all equally likely events, including } A, \text{ occur}}$$

or

$$P(A) = \frac{\text{No. of members of } \{A\}}{\text{No. of members of } \{\text{possibility space}\}}$$

where all members are equally likely.

This is the basic definition of probability. All other developments of probability theory are derived from this definition, and a large number of problems can be answered directly from it.

As $\{A\}$ is a subset of $\{\text{possibility space}\}$ the numerator of this fraction is always less than, or equal to, the denominator, so for any event A

$$0 \leqslant P(A) \leqslant 1$$

If $P(A) = 1$ this means that the event is an absolute certainty.

If $P(A) = 0$ this means that the event is an absolute impossibility.

For example, if one ball is taken from a bag containing only red balls

$$P(\text{ball is red}) = 1 \quad \text{and} \quad P(\text{ball is blue}) = 0.$$

EXAMPLES 17a

1) A pack of felt tipped pens contains five red pens and four blue pens. If one pen is withdrawn at random what is the probability that it is blue?

> The term 'at random' means that all possibilities are *equally likely*.

Thus the possibility space contains 9 equally likely events. If A is the removal of a blue pen, then A can occur in 4 equally likely ways (i.e. $\{A\}$ has 4 members).

Thus $P(A) = 4/9.$

2) If one card is drawn at random from a pack of fifty-two playing cards what is the probability that it is an ace?

There are 52 equally likely events, i.e. the drawing of any one of the fifty-two playing cards.

An ace can be drawn in 4 equally likely ways.

Therefore $P(\text{ace}) = 4/52 = 1/13.$

3) Four cards are drawn at random from a pack of fifty-two playing cards. Find the probability that the four cards are all clubs.

If A is the event 'the withdrawal of four clubs', then as there are thirteen clubs in the pack, there are $^{13}C_4$ different combinations of four clubs.

So A can occur in $^{13}C_4$ ways, i.e. $\{A\}$ contains $^{13}C_4$ members.

But any distinct combination of four cards from the pack is equally likely and there are $^{52}C_4$ of these, i.e. $\{\text{possibility space}\}$ contains $^{52}C_4$ members.

Thus $P(A) = \dfrac{^{13}C_4}{^{52}C_4} = \dfrac{13! \, 4! \, 48!}{4! \, 9! \, 52!} = \dfrac{11}{4165} = 0.003$ to 3 d.p.

4) Four letters are chosen at random from the word DEALING.
Find the probability that:

(a) exactly two vowels are chosen,

(b) at least two vowels are chosen.

DEALING has three vowels and four consonants.
Four letters (without restriction) can be chosen in $^7C_4$ ways.
i.e. the {possibility space} has $^7C_4$ members.

(a) A selection containing two vowels (out of E A I) will also contain two
consonants (out of D L N G). As these are independent combinations, the
number of ways of choosing four letters containing exactly two vowels is
$^3C_2 \times {}^4C_2$.
Therefore the probability of selecting four letters, exactly two of which are
vowels, is

$$\frac{^3C_2 \times {}^4C_2}{^7C_4} = \frac{18}{35}$$

(b) If the selection contains at least two vowels, then either it contains two
vowels or it contains three vowels and these are mutually exclusive
combinations. The number of combinations containing just two vowels is

$$^3C_2 \times {}^4C_2 = 18$$

The number of combinations containing three vowels is

$$^3C_3 \times {}^4C_1 = 4$$

Therefore the number of ways of selecting four letters containing at least
two vowels is $18 + 4 = 22$.
Hence the probability of four letters chosen at random containing at least
two vowels is

$$\frac{22}{^7C_4} = \frac{22}{35}$$

THE PROBABILITY THAT AN EVENT DOES NOT HAPPEN

If, in a possibility space of n equally likely occurences, the event A occurs
r times, there are $n - r$ occasions when A does not happen.
'The event A does not happen' is denoted by $\bar{A}$ (and reads as 'not A').

Thus $\qquad P(A) = \dfrac{r}{n}$ and $P(\bar{A}) = \dfrac{n-r}{n} = 1 - \dfrac{r}{n}$

i.e. $\qquad\qquad\qquad P(\bar{A}) = 1 - P(A)$

or $\qquad\qquad\qquad P(A) + P(\bar{A}) = 1$

This relationship is most useful in the 'at least one' type of problem as illustrated below.

EXAMPLES 17a (continued)

5) If four cards are drawn at random from a pack of fifty-two playing cards find the probability that at least one of them is an ace.

If A is a combination of four cards containing at least one ace (i.e. either one ace, or two aces, or three aces, or four aces)
then $\bar{A}$ is a combination of four cards containing *no* aces.

Now $\quad P(\bar{A}) = \dfrac{\text{No. of combinations of four cards with no aces}}{\text{Total no. of combinations of four cards}}$

$$= {}^{48}C_4/{}^{52}C_4 = 0.72 \text{ to 2 d.p.}$$

Using $P(A) + P(\bar{A}) = 1$ we have

$$P(A) = 1 - P(\bar{A}) = 1 - 0.72 = 0.28 \text{ to 2 d.p.}$$

6) Four balls are taken at random out of a box containing six red and four black balls. What is the probability that at least two red balls are removed?

The number of ways in which any four balls from the box of ten can be removed is ${}^{10}C_4$, i.e. {possibility space} has ${}^{10}C_4$ members.
If A is a selection of four balls, at least two of which are red, then A contains either two, or three or four red balls.
Thus $\bar{A}$ is a selection containing either no red balls or one red ball. As $\bar{A}$ involves two mutually exclusive combinations, and A involves three mutually exclusive combinations, we will consider $P(\bar{A})$.
The number of combinations of four balls containing either one red ball or no red balls is ${}^4C_4 + {}^6C_1 \times {}^4C_3$, i.e. $\{\bar{A}\}$ has ${}^4C_4 + {}^6C_1 \times {}^4C_3$ members.

So $\quad P(\bar{A}) = \dfrac{{}^4C_4 + {}^6C_1 \times {}^4C_3}{{}^{10}C_4} = 0.119 \text{ to 3 d.p.}$

Therefore $\quad P(A) = 1 - P(\bar{A}) = 1 - 0.119 = 0.88 \text{ to 2 d.p.}$

Note that in each of the examples the word *random* is taken to mean that any choice is as likely as any other choice.

EXERCISE 17a

1) One integer is chosen at random from the set $\{1, 2, 3, 4, 5, 6, 7, 8\}$. What is the probability that it is a prime number?

2) Two books are taken at random from a shelf containing five paper backs and four hard backs. What is the probability that both are paper backs?

3) Three cards are drawn at random from a pack of fifty-two playing cards. What is the probability that at least one card is red?

4) Two balls are taken at random from a box containing three black, three red and three yellow balls. Find the probability that:

(a) neither of the balls removed is red,   (b) at least one is red,

(c) both are red.

5) If the letters of the word BOOK are arranged at random, what is the probability that the two O's are separated?

6) A two digit number is made by choosing two integers, at random, from the set $\{1, 2, 3, 4, 5, 6\}$. If each integer may be used more than once, what is the probability that the number is:

(a) divisible by 2,   (b) not divisible by 5.

7) What is the probability that three letters chosen at random from the letters GREEN contain:

(a) the letter N,   (b) two E's,   (c) at least one E?

8) A team of four people is chosen at random from a group of three men and four women. Find the probability that there are:

(a) no men in the team,   (b) at least two men in the team.

9) A box contains four white counters and one red counter. If two counters are removed at random, what is the probability that the red counter is not removed?

10) A number is made by choosing two or three digits at random from the set $\{1, 2, 3, 4\}$. If each digit can be chosen more than once to make a number, what is the probability of a number less that 200 being made?

The solutions of the examples examined so far have used methods developed for permutations and combinations, and where this is possible it is usually the most direct approach. However the likely occurrence of some events cannot be worked out exactly because the factors affecting whether that event occurs or does not occur cannot be measured or counted. For example, the probability that you would hit the bulls eye on a dartboard with one throw of a dart would depend on how much you had practiced, how much natural talent for playing darts you had, how tired you were, how much alcohol you had drunk, how good a dart you were using etc., etc . . . all of which are impossible to quantify.

EMPIRICAL PROBABILITY

A method which can be adopted in the example given above is to throw the dart several times (each throw is a trial) and count the number of times you hit the bulls eye (a success) and the number of times you miss (a failure). Then an empirical value for the probability that you hit the bulls eye with any one throw is

$$\frac{\text{No. of successes}}{\text{No. of successes} + \text{No. of failures}}$$

If the number of throws is small this will not give a particularly good estimate, but for a large number of throws the result will be more reliable.

When the probability of the occurrence of an event A cannot be worked out exactly, an empirical value can be found by adopting the approach described above, that is:

(a) making a *large number* of trials, by setting up the conditions in which the event may or may not occur (i.e. perform an experiment and note the outcome),

(b) counting the number of times the event does occur, i.e. the number of successes,

(c) calculating the value of

$$\frac{\text{No. of successes}}{\text{No. of trials (i.e. successes} + \text{failures)}}$$

The probability of the event A occuring is then defined as

$$P(A) = \lim_{n \to \infty} \left(\frac{r}{n} \right)$$

where r is the number of successes in n trials, and $n \to \infty$ means that the number of trials is large (but what should be taken as 'large' depends on the problem).

For some categories of events a theoretical probability can be found which may or may not prove to be correct in practice. For example, if a coin is tossed and we assume that it is *equally* likely to land head or tail up, the probability of a head on any one toss is $\frac{1}{2}$. For a particular coin the empirical probability of a head on any one toss can be found by experiment (i.e. tossing the coin several times, etc). If, by experiment, the coin if found to be equally likely to land head up or tail up it is said to be *unbiased* or *fair*.

On the other hand if in 100 tosses, say, it is found that the coin lands head up 80 times it is reasonable to assume that the coin is *biased*, i.e. it is *not* equally likely to land head or tail up.

Thus if a coin is known to be unbiased the probability of its landing head up on any one toss is $\frac{1}{2}$.

Similarly if a die (numbered 1 to 6) is thrown, and it is known to be unbiased, the probability of throwing a six, say, is $\frac{1}{6}$.

If a coin is known to be biased, so that it is twice as likely to land head up than tail up, then the number of *equally* likely results of tossing that coin are head up twice and tail up once. So the probability of tossing a head with this coin is $\frac{2}{3}$.

We will now look at problems involving the occurrence of two or more events. They can loosely be divided into two categories:

(a) 'either . . . or' events such as the probability of scoring *either* 5 *or* 6 with one throw of a die,

(b) 'both . . . and' events such as the probability of selecting *both* an orange *and* apple from a bowl of mixed fruit.

Some events fall into both categories, such as the probability of obtaining at least one head when two coins are tossed, as this event involves *either* two heads *or* (*both* one head *and* one tail). At this stage we will investigate the two categories separately.

MUTUALLY EXCLUSIVE EVENTS

Two events are mutually exclusive if the occurrence of either event excludes the possibility of the occurrence of the other event, i.e. *either* one event *or* the other event but *not both* can occur.

For example if a number is chosen at random from the set

$$\{3, 4, 5, 6, 7, 8, 9, 10, 11, 12\}$$

and A is the selection of a prime number,

B is the selection of an odd number,

C is the selection of an even number,

then A and C are mutually exclusive as none of the numbers in this set is both prime and even.

But A and B are not mutually exclusive as some numbers are both prime and odd (viz. 3, 5, 7, 11). The probability of either A or C can be found as follows:

A can be selected in 4 ways out of 10 equally likely selections.
Therefore $P(A) = \frac{4}{10}$.
C can be selected in 5 ways out of 10 equally likely selections.
Therefore $P(C) = \frac{5}{10}$.
The number of selections containing either A or C is $4 + 5$ and as there are 10 equally likely choices,

$$P(\text{either } A \text{ or } C) = \frac{4 + 5}{10} = \frac{4}{10} + \frac{5}{10} = P(A) + P(C)$$

'Either A or C' is denoted by $A \cup C$

So we can write the result above as $P(A \cup C) = \frac{9}{10}$.
Generalising we may say that if E_1 and E_2 are mutually exclusive events and

E_1 occurs in r out of n equally likely occurrences

E_2 occurs in s out of n equally likely occurrences

then $E_1 \cup E_2$ occurs in $r + s$ out of n equally likely occurrences.

Hence
$$P(E_1) = \frac{r}{n}, \quad P(E_2) = \frac{s}{n}, \quad P(E_1 \cup E_2) = \frac{r+s}{n}$$

i.e.
$$P(E_1 \cup E_2) = P(E_1) + P(E_2)$$

EXAMPLES 17b

1) An unbiased die is in the form of an octohedron (eight faces), numbered 1, 2, 3, 3, 4, 5, 5, 6. When the die is thrown the score is taken from the face on which the die lands. What is the probability of obtaining a score of at least five from one throw?

A score of at least 5 means a score of either 5 or 6 and these are mutually exclusive events.
A score of 5 can occur in two ways as the die is equally likely to land on any one of eight faces, two of which show 5.

i.e.
$$P(5) = \tfrac{2}{8}$$

Similarly
$$P(6) = \tfrac{1}{8}$$

Therefore
$$P(5 \cup 6) = \tfrac{2}{8} + \tfrac{1}{8} = \tfrac{3}{8}$$

INDEPENDENT EVENTS

Two events are independent if the occurrence or non occurrence of one event has no influence on the occurrence or non occurrence of the other event, i.e. neither the occurrence of the first event nor the occurrence of the second event has any effect on the other.
For example if a coin is tossed and a die (numbered 1 to 6) is thrown, the way in which the coin lands in no way affects the possible ways in which the die can land (and vice versa) so throwing a head (say) with the coin and a six (say) with the die are independent events. If both are fair, then $P(H) = \tfrac{1}{2}$ and $P(6) = \tfrac{1}{6}$.
Now for each of the two ways in which the coin can land there are six ways in which the die can land,

so
$$P(H \text{ and } 6) = \frac{1 \times 1}{2 \times 6} = \frac{1}{2} \times \frac{1}{6} = P(H) \times P(6)$$

Now consider a bag which contains three white and two red balls. If a ball is removed at random and then replaced and a second ball is then removed, we again have two independent events. If a red ball is taken out first (R_1) and a white ball is taken out second (W_2)

then
$$P(R_1) = \tfrac{2}{5} \quad \text{and} \quad P(W_2) = \tfrac{3}{5}$$

Also $$P(R_1 \text{ followed by } W_2)$$

$$= \frac{2 \text{ ways for } R_1 \times 3 \text{ ways for } W_2}{5 \text{ possibilities for the first } \times 5 \text{ possibilities for the second}}$$

i.e. $$P(R_1 \text{ followed by } W_2) = \frac{2 \times 3}{5 \times 5} = \frac{2}{5} \times \frac{3}{5} = P(R_1) \times P(W_2)$$

Denoting 'E_1 and E_2' by $E_1 \cap E_2$
We can say in general that
if E_1 and E_2 are independent events then

$$P(E_1 \cap E_2) = P(E_1) \times P(E_2)$$

EXAMPLES 17b (continued)

2) Three coins are tossed simultaneously. Two of the coins are fair and one is biased so that a head is twice as likely as a tail. Find the probability that all three coins turn up heads.

There are three coins involved so let them be a, b, c, and let H_a be the event of coin a landing head up and H_b and H_c the events that coins b and c land head up.
If a and b are the fair coins, and c is the biased coin

$$P(H_a) = \tfrac{1}{2}, \quad P(H_b) = \tfrac{1}{2}, \quad P(H_c) = \tfrac{2}{3}$$

The way in which any one of the coins lands has no influence on the way in which either of the others lands, i.e. H_a, H_b, H_c are independent events.

Therefore $$P(H_a \cap H_b \cap H_c) = P(H_a) \times P(H_b) \times P(H_c)$$

$$= \tfrac{1}{2} \times \tfrac{1}{2} \times \tfrac{2}{3} = \tfrac{1}{6}$$

CONDITIONAL EVENTS

Consider again the bag containing three white and two red balls. If the first ball removed is *not* replaced in the bag then the possibility space for the removal of a second ball has been reduced by one member, i.e. the events are no longer independent.
If the second ball taken out is red (R_2) then the number of ways in which this can occur depends on which ball was removed first.
Thus if a white ball was removed, R_2 can happen in two ways out of four equally likely occurrences,

i.e. $$P(R_2 \text{ given that the first ball removed is white}) = \tfrac{2}{4}$$

This is called the conditional probability that R_2 occurs when W_1 has occurred.

It is written as $P(R_2|W_1)$ and read as 'the probability that R_2 occurs given that W_1 has occurred' or more briefly as 'the probability of R_2 given W_1'.

i.e.
$$P(R_2|W_1) = \tfrac{1}{2}$$

If a red ball was removed first, R_2 can happen in only one out of four equally likely occurrences.

i.e. the conditional probability that R_2 occurs when R_1 has already occurred is $\tfrac{1}{4}$

i.e.
$$P(R_2|R_1) = \tfrac{1}{4}$$

COMPOUND PROBABILITY

Now consider the probability of removing a white ball first (W_1) *and* a red ball second (R_2).

There are 3×2 ways in which a white ball first and a red ball second may be removed.

There are 5×4 ways in which any one ball followed by another ball may be removed.

i.e.
$$P(W_1 \cap R_2) = \frac{3 \times 2}{5 \times 4} = \frac{3}{5} \times \frac{2}{4}$$

Now
$$P(W_1) = \tfrac{3}{5} \quad \text{and} \quad P(R_2|W_1) = \tfrac{2}{4}$$

i.e.
$$P(W_1 \cap R_2) = P(W_1) \times P(R_2|W_1)$$

This is called the *compound probability* of W_1 and R_2 occurring.

In general if E_1 and E_2 are two events, the compound probability of both E_1 and E_2 occurring is given by

$$P(E_1 \cap E_2) = P(E_1) \times P(E_2|E_1)$$

Note that if E_1 and E_2 are independent events

then
$$P(E_1 \cap E_2) = P(E_1) \times P(E_2) \Rightarrow P(E_2|E_1) = P(E_2)$$

Summarizing the notation introduced and the results obtained we have:

If A and B are two events

$$A \cap B \quad \text{means both } A \text{ and } B$$

$$A \cup B \quad \text{means either } A \text{ or } B$$

$$A \mid B \quad \text{means } A \text{ given that } B \text{ has already occurred}$$

$$P(A \cup B) = P(A) + P(B) \quad \text{when } A \text{ and } B \text{ are mutually exclusive}$$

$$P(A \cap B) = P(A) \times P(B \mid A) \quad \text{which reduces to}$$

$$\left.\begin{array}{l} P(A \cap B) = P(A) \times P(B) \\[4pt] \Rightarrow \quad P(B \mid A) = P(B) \end{array}\right\} \quad \text{when } A \text{ and } B \text{ are independent}$$

EXAMPLES 17b (continued)

3) An unbiased die, marked 1 to 6, is rolled twice. Find the probability of:

(a) rolling two sixes,

(b) the second throw being a six, given that the first throw is a six,

(c) getting a score of ten from the two throws,

(d) throwing at least one six,

(e) throwing exactly one six.

If the die is unbiased it is equally likely to land on any face, so a score of 6 is just one of six equally likely scores, i.e. $P(6) = \frac{1}{6}$.
Also, each roll of the die is an independent event.

(a) As throwing six on the first roll (6_1) and throwing six on the second roll (6_2) of the die are independent events, the probability of throwing both a six at the first roll and a six at the second roll is

$$P(6_1) \times P(6_2) = \frac{1}{6} \times \frac{1}{6} = \frac{1}{36}$$

i.e. $$P(6_1 \cap 6_2) = \frac{1}{36}$$

(b) The probability of the second throw being six, given that the first throw is six, is $P(6_2 \mid 6_1) = \frac{1}{6}$ $(= P(6_2))$.

(c) A score of ten is obtained either from two fives, or from six followed by four, or from four followed by six, and these are mutually exclusive events. Therefore $P(\text{score of ten})$

$$= P(5_1 \cap 5_2) + P(6_1 \cap 4_2) + P(4_1 \cap 6_2)$$

$$= \frac{1}{36} + \frac{1}{36} + \frac{1}{36} = \frac{1}{12}$$

(d) If A is the throwing of at least one six
then $\bar{A}$ is not throwing a six (i.e. $1, 2, 3, 4,$ or 5) on either of the throws. The probability of not getting a six on one throw is $\frac{5}{6}$.

Hence $$P(\bar{6}_1 \cap \bar{6}_2) = P(\bar{A}) = \frac{5}{6} \times \frac{5}{6} = \frac{25}{36}$$

Using $$P(A) = 1 - P(\bar{A})$$

$$P(A) = 1 - \frac{25}{36} = \frac{11}{36}$$

(e) If exactly one six is thrown then

    either   the first throw is six and second throw is not six

    or   the first throw is not six and the second throw is six

    and these are mutually exclusive events

Now $\qquad\qquad\qquad\qquad P(6_1 \cap \bar{6}_2)$ is $\dfrac{1}{6} \times \dfrac{5}{6} = \dfrac{5}{36}$

and $\qquad\qquad\qquad\qquad P(\bar{6}_1 \cap 6_2)$ is $\dfrac{5}{6} \times \dfrac{1}{6} = \dfrac{5}{36}$

Therefore the probability of obtaining exactly one six,

$$P(6_1 \cap \bar{6}_2) + P(\bar{6}_1 \cap 6_2) = \frac{5}{36} + \frac{5}{36} = \frac{5}{18}$$

4) A bag contains five red and six black counters. The counters are removed one at a time without replacement. If the counters are taken out at random find the probability that:

(a) the first two counters removed are red,

(b) the second counter removed is red.

Let R_1 and R_2 denote respectively the removal of a red counter first and a red counter second, and B_1 denote the removal of a black counter first.

(a) $P(R_1 \cap R_2) = P(R_1) \times P(R_2 | R_1)$

    now R_1 can happen in 5 out of 11 equally likely withdrawals,

    i.e. $P(R_1) = \frac{5}{11}$

    and $R_2 | R_1$ can happen in 4 out of 10 equally likely withdrawals,

    i.e. $P(R_2 | R_1) = \frac{4}{10}$.

    Therefore $P(R_1 \cap R_2) = \dfrac{5}{11} \times \dfrac{4}{10} = \dfrac{2}{11}$

(b) If the second counter is red, then either the first is red and the second is red, or the first is black and the second is red.

    i.e. $\qquad\qquad P(R_2) = P[(R_1 \cap R_2) \cup (B_1 \cap R_2)]$

$$= P(R_1)P(R_2 | R_1) + P(B_1)P(R_2 | B_1)$$

$$= \frac{2}{11} + \frac{6}{11} \times \frac{5}{10}$$

$$= \frac{5}{11}$$

Note that this result can be obtained by disregarding the first counter as no condition is placed on it, i.e. the second counter, if it is red can be any one of five red counters.

As any one of the eleven counters is equally likely to be the second counter removed, $P(R_2) = \frac{5}{11}$.

5) Three people A, B and C, gamble for a prize by rolling a die. The first person to roll a six wins. If the die is unbiased and they roll the die in the order A, then B, then C, find the probability that:

(a) A wins on the first throw,   (b) C wins at his first attempt,

(c) B wins at his third attempt,   (d) A wins.

Each throw of the die is an independent event and, on any one throw, the probability of a six is $\frac{1}{6}$ and the probability of not throwing a six is $\frac{5}{6}$.

(a) The probability that A wins on the first throw, $P(A_1)$, is $\frac{1}{6}$.

(b) A and B roll the die first so they must both fail at their first attempt $(\bar{A}_1$ and $\bar{B}_1)$ if C is to win at his first attempt. These three events are independent, so the probability that C wins at his first attempt is

$$P(\bar{A}_1) \times P(\bar{B}_1) \times P(C_1) \;=\; \frac{5}{6} \times \frac{5}{6} \times \frac{1}{6} \;=\; \frac{25}{216}$$

(c) If B wins at his third attempt then A has had three failures, B two failures, and C two failures. Again these are independent events. Therefore the probability that B wins at his third attempt is

$$P(\bar{A}_1) \times P(\bar{B}_1) \times P(\bar{C}_1) \times P(\bar{A}_2) \times P(\bar{B}_2) \times P(\bar{C}_2) \times P(\bar{A}_3) \times P(B_3) \;=\; (\tfrac{5}{6})^7 \times \tfrac{1}{6}$$

(d) If A wins, then

either	A wins on his first throw
or	A wins on his second throw
or	A wins on his third throw
or	A wins on his fourth throw

and so on.
Therefore $P(A$ wins$)$ on his

first throw is $P(A_1) = \tfrac{1}{6}$

second throw is $P(\bar{A}_1) \times P(\bar{B}_1) \times P(\bar{C}_1) \times P(A_2) \;=\; (\tfrac{5}{6})^3 \times \tfrac{1}{6}$

third throw is $[P(\bar{A}_1 \cap \bar{B}_1 \cap \bar{C}_1) \times P(\bar{A}_2 \cap \bar{B}_2 \cap \bar{C}_2)] \times P(A_3) \;=\; (\tfrac{5}{6})^6 \times \tfrac{1}{6}$

fourth throw is $[(\tfrac{5}{6})^3]^3 \times P(A_4) \;=\; (\tfrac{5}{6})^9 \times \tfrac{1}{6}$

and so on.
These are mutually exclusive events, therefore

$$P(A) \;=\; \tfrac{1}{6} + \tfrac{1}{6}(\tfrac{5}{6})^3 + \tfrac{1}{6}(\tfrac{5}{6})^6 + \tfrac{1}{6}(\tfrac{5}{6})^9 + \ldots$$

This is an infinite G.P. with first term $\frac{1}{6}$ and common ratio $(\frac{5}{6})^3$ and so has a sum to infinity of

$$\frac{(\frac{1}{6})}{1-(\frac{5}{6})^3} = \frac{36}{91}$$

Therefore the probability that A wins is $\frac{36}{91}$.

6) In a game of darts, the probability that a particular player aims at and hits the 'treble twenty' with one dart is 0.4. How many throws are necessary so that the probability of hitting the treble twenty at least once exceeds 0.9?

The probability of hitting the treble twenty, $P(A)$, is 0.4 on one throw, so the probability of not hitting the treble twenty, $P(\bar{A})$, is 0.6 on one throw, and $P(\bar{A})$ in two throws is $(0.6)^2$, etc.
The probability of hitting the treble twenty at least once in n throws

$= 1-$ (probability of not hitting the treble twenty on all n throws)

So in n throws $P(A$ at least once$) = 1 - (0.6)^n$.
So for $P(A$ at least once$)$ to exceed 0.9 in n throws

$$1 - (0.6)^n > 0.9$$

$\Rightarrow \qquad\qquad (0.6)^n < 0.1$

$\Rightarrow \qquad\qquad n \log 0.6 < \log 0.1$

$\Rightarrow \qquad\qquad n > \dfrac{\log 0.1}{\log 0.6} \qquad\qquad$ (as $\log 0.6 < 0$)

$\Rightarrow \qquad\qquad n > 4.5$

Therefore five throws are necessary.

7) In a group of students, 10% are left-handed, 8% are short-sighted and 2% are both left-handed and short-sighted.
(a) Given that a student is short-sighted, find the probability that he is left-handed.
(b) Find the probability that a left-handed student is also short-sighted.

If $P(\text{l.h.})$ represents the probability that a student is left-handed and if $P(\text{s.s.})$ represents the probability that a student is short-sighted then

$$P(\text{l.h.}) = 0.1, \quad P(\text{s.s}) = 0.08 \quad \text{and} \quad P(\text{l.h.} \cap \text{s.s.}) = 0.02$$

a) Now $\qquad\qquad P(\text{l.h.} \cap \text{s.s.}) = P(\text{l.h.|s.s.}) \times P(\text{s.s.})$

$\Rightarrow \qquad\qquad 0.02 = P(\text{l.h.|s.s.}) \times 0.08$

$\Rightarrow \qquad\qquad P(\text{l.h.|s.s.}) = 0.25$

b) $P(\text{l.h.} \cap \text{s.s.}) = P(\text{s.s.}|\text{l.h.}) \times P(\text{l.h.})$

⇒ $0.02 = P(\text{s.s.}|\text{l.h.}) \times 0.1$

⇒ $P(\text{s.s.}|\text{l.h.}) = 0.2$

EXERCISE 17b

1) Two unbiased coins are tossed. Find the probability of:
(a) two heads,   (b) at least one head,   (c) exactly one head.

2) Three unbiased coins are tossed. Find the probability of:
(a) three tails,   (b) at least one tail.

3) The probability of an archer hitting the bulls eye with any one shot is $\frac{1}{5}$.
Find the probability that:
(a) he hits the bulls eye with his second shot,
(b) he hits the bulls eye exactly once in three shots,
(c) he hits the bulls eye at least once in four shots.

4) In a multiple choice examination each question has five possible answers,
only one of which is correct. If a candidate chooses his answers at random find
the probability that, in a test of ten such questions, he gets none right.

5) Two coins are tossed. One coin is fair and the other is biased so that
throwing a head is three times as likely as throwing a tail. Find the probability
that:
(a) on one toss of both coins they both land head up,
(b) on two tosses of both coins, two tails are thrown both times,
(c) on two tosses, at least one head is thrown.

6) Two unbiased normal dice are thrown. On one throw find the probability of:
(a) two 1's,   (b) a score of 3,   (c) a score of at least 4.

7) An unbiased die in the shape of a tetrahedron has its faces numbered
$1, 2, 3, 4$.
The score is taken from the face on which it lands. Find the probability that:
(a) on one throw 4 is scored,
(b) on two throws a total of 2 is scored,
(c) on three throws a total of at least 4 is scored.

8) Two people, A and B, play a game by tossing a fair coin, and the first to
toss a head wins. If A tosses first find the probability that:
(a) A wins on his first toss,   (b) B wins on his first toss,
(c) A wins on his second toss.

9) Two people, A and B, play a game by rolling two fair dice; the first to roll
a double six wins. If A goes first, find the probability that:
(a) B wins on his first throw,   (b) A wins on his second throw.

10) Two people, A and B, play a game by tossing a fair coin and the first to toss a head wins. If A goes first find the probability that A wins.

11) Three people, A, B and C play a game by rolling a fair die and the first to roll a six wins. If they play in the order A then B then C, find the probability that B wins.

12) Two people, A and B, play a game by drawing a card from a pack of fifty-two playing cards. The first to draw an ace wins. The cards drawn are not put back in the pack and they play in the order A, B. Find the probability that:
(a) A wins on his first draw,   (b) A wins on his second draw,
(c) A wins on his third draw.

13) A boy at a rifle range has a probability of $\frac{2}{3}$ of hitting a target with any one shot. Find the probability that he first hits a target with his third shot. How many shots are necessary for the probability of his hitting at least one target to be greater than $\frac{4}{5}$?

14) A box of screws contains 5% defective screws. If a screw is taken at random from the box, what is the probability that it is defective? How many times does this have to be repeated before the probability of removing at least one defective screw is 0.5?

15) In a card game for four players, a pack of fifty-two cards is dealt round so that each player receives thirteen cards. A hand that contains no card greater than nine is called a yarborough. How many hands are necessary for the probability of at least one hand to be a yarborough is greater than $\frac{1}{2}$? (Ace ranks high.)

16) A shelf has fifteen paperbacked and twelve hardbacked books on it. A book is selected at random and removed and then a second and a third book are similarly selected and removed. Find the probability that:
(a) the first three books removed are paperbacks,
(b) the third book removed is a hardback.

17) Three balls are selected at random in order from a box containing 2 red, 3 yellow and 4 black balls. Find the probability that the third ball is yellow, given that the first is red and the second is black if:
(a) the balls are not put back in the box after selection,
(b) the balls are replaced in the box after each selection.

TREE DIAGRAMS

The number of ways in which an event A occurs and the number of occurrences of equally likely events is not always obvious from the statement of a problem. This is particularly true of compound events which can involve three or more separate events.

Consider, for example, tossing three coins. The number of ways in which the coins can land with 2 heads and a tail showing is three, because if the coins are numbered 1, 2, 3 for identification, then they can land

$$H_1 \, H_2 \, T_3 \quad \text{or} \quad H_1 \, T_2 \, H_3 \quad \text{or} \quad T_1 \, H_2 \, H_3$$

Using the methods already developed,

$$P(2H \text{ and } T)$$

$$= P[\text{either } (H_1 \cap H_2 \cap T_3) \text{ or } (H_1 \cap T_2 \cap H_3) \text{ or } (T_1 \cap H_2 \cap H_3)]$$

$$= P(H_1) \times P(H_2) \times P(T_3) + P(H_1) \times P(T_2) \times P(H_3) + P(T_1) \times P(H_2) \times P(H_3)$$

An alternative approach to problems such as this is to construct a diagram which shows all the likely outcomes and the probability of a particular outcome can then be found from the basic definition.

Consider again tossing three coins, numbered 1, 2 and 3. The likely outcomes of tossing coin 1 are a head or a tail. Starting at the left-hand side of the page we draw two branches, each branch being one of the possible events.

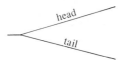

Fig. 1

When coin 2 is tossed there are two possible events (head or tail) for each of the possible outcomes of the toss of coin 1, so we draw two branches at the end of each branch in Fig. 1 to illustrate this.

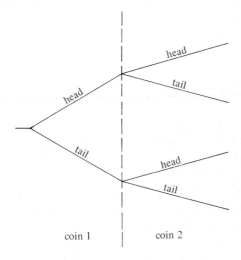

coin 1   coin 2   Fig. 2

Following the branches from the left (trunk) of the diagram through to the right we see that the likely events when tossing two coins are $H_1 H_2$ or $H_1 T_2$ or $T_1 H_2$ or $T_1 T_2$.

For each of these events there are two likely occurrences when coin 3 is tossed. Branching again gives the diagram in Fig. 3.

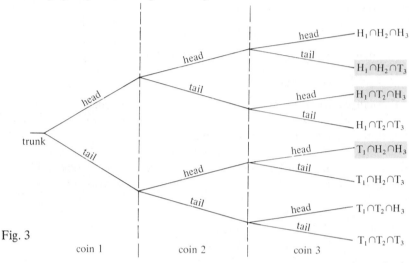

Fig. 3

coin 1   coin 2   coin 3

Starting from the trunk we see that there are eight different routes (branches) which can be followed, i.e. there are eight likely events, three of which give two heads and a tail.

If the coins are all *unbiased* then these eight events are *equally* likely.

Thus $P(2\text{H and T}) = \dfrac{3 \text{ events}}{8 \text{ equally likely events}} = \dfrac{3}{8}$

Note (i)   the eight compound events on the right of Fig. 3 are mutually exclusive,

   (ii) the events along any one route from left to right are independent.

EXERCISE 17c

1) Draw a tree diagram to represent the likely outcomes of tossing a coin and rolling a tetrahedral die numbered 1 to 4.

2) Draw a tree diagram to represent the likely outcomes of tossing two coins and rolling a tetrahedral die numbered 1 to 4. From your diagram find the probability of obtaining a head and a tail and a score of 4, assuming that the coins and the die are unbiased.

3) A die in the form of a cube is numbered 1, 1, 2, 2, 3, 4. Draw a tree diagram to illustrate the likely outcomes of rolling this die and tossing a coin. If both are unbiased, what is the probability of rolling a 2 on the die and tossing a head on the coin?

Tree diagrams can be used to find probabilities when the possible outcomes of an experiment are not equally likely.

Consider, for example, tossing two coins, one of which is fair and one of which is biased so that a tail is twice as likely as a head.

When the fair coin is tossed there are two *equally* likely events, viz. a head or a tail.

When the biased coin is tossed there are two distinct events but they occur in the proportion of one head to two tails, so there are three equally likely events, i.e. 1 head and 2 tails.

Drawing a tree diagram and stating with the fair coin we have

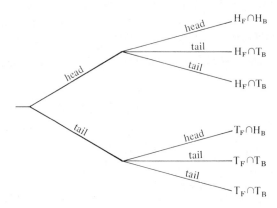

Hence there are six equally likely events

and

$$P(\text{2 heads}) = \tfrac{1}{6}$$

$$P(\text{head and tail}) = \tfrac{3}{6} = \tfrac{1}{2}$$

$$P(\text{2 tails}) = \tfrac{2}{6} = \tfrac{1}{3}$$

As consecutive branches represent independent events we can simplify this diagram by drawing branches for each distinct event (i.e. not for each equally likely event) and writing the probability of that event on the branch.

i.e.

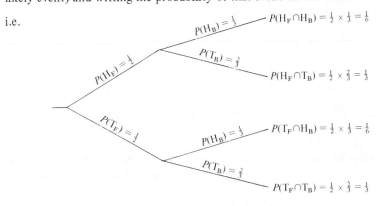

As the events at the right hand end of the figure are mutually exclusive we have

$$P(\text{head and tail}) = \tfrac{1}{3} + \tfrac{1}{6} = \tfrac{1}{2}$$

In most problems we are concerned only with the probability of one event, A, occurring. Tree diagrams may also be used to simplify the work in such problems by showing the probabilities of A and of $\bar{A}$. This is illustrated in the following examples.

EXAMPLES 17d

1) Three dice, each numbered 1 to 6 are rolled. One die is fair and the others are biased so that for each of them a six is twice as likely as any other score. Find the probability of rolling exactly two sixes.

Using a tree diagram to show the probabilities of 6 or $\bar{6}$ and starting with the fair die we have

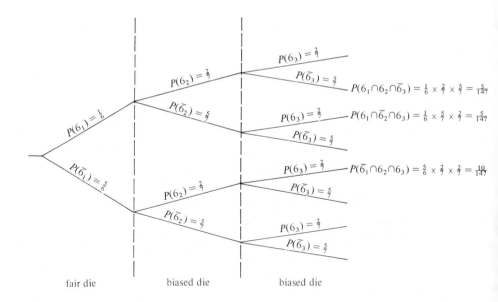

Therefore $P(\text{exactly two sixes})$ is

$$\frac{5}{147} + \frac{5}{147} + \frac{10}{147} = \frac{20}{147}$$

2) One of the dice described in Example 1 is chosen at random and on two throws it shows a six on each occasion. What is the probability that a biased die has been chosen?

There are two independent events to consider in this problem, i.e. choosing a die and rolling it twice. In the second event our only concern is whether a six shows up on both rolls or whether a six does not show up on both rolls. Drawing a tree diagram and starting with a choice of die gives

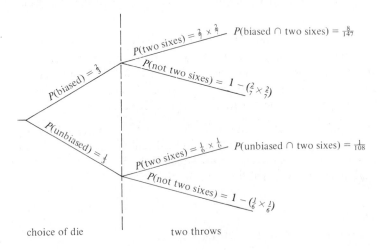

choice of die   two throws

If A is the rolling of two sixes and B is the selection of a biased die, the probability that the die is biased, given that two sixes are rolled, is $P(B \mid A)$. Using $P(A \cap B) = P(A)P(B \mid A)$ and the tree diagram above, we see that the probability of getting two sixes on any of the three dice is

$$\frac{8}{147} + \frac{1}{108} = \frac{337}{5292} = P(A)$$

But the probability of choosing a biased die and getting two sixes is

$$\frac{8}{147} = \frac{288}{5292} = P(A \cap B)$$

So the probability that the die is biased is $\frac{288}{337} = 0.85$ to 2 d.p.

EXERCISE 17d

Use a tree diagram to help in the solution of the following problems.

1) A tetrahedral die, marked $1, 2, 3, 4$ is rolled twice. If the die is unbiased find the probability of getting a total score of three.

2) Two coins and a tetrahedral die numbered 1 to 4 are tossed. If all are unbiased, find the probability that a head, a tail and a 1 are tossed.

3) In a multiple choice test of three questions there are five alternative answers given to the first two questions and four alternative answers given to the last

question. If a candidate guesses answers at random, what is the probability that he will get:

(a) exactly one right,   (b) at least one right?

4) Three coins are tossed. Two of them are fair and one is biased so that a head is three times as likely as a tail. Find the probability of getting two heads and a tail.

5) A local telephone call goes through three independent sets of equipment, the outgoing telephone, the automatic exchange and the receiving telephone. If the probability of failure on the outgoing telephone is 0.05, on the exchange is 0.01 and on the receiving telephone is 0.04, find the probability that if a call is not connected it is at least partly the fault of the exchange.

6) One box of chocolates contains five hard centres and three soft centres. Another box of chocolates contains eight hard centres and seven soft centres. A chocolate is chosen at random from either of the boxes. If it is a soft centred one find the probability that it came from the first box.

POSSIBILITY SPACE AND SAMPLE POINTS

The use of a tree diagram is helpful for dealing with a problem when the number of independent events is small and the outcome under consideration of each independent event is also small, otherwise we end up with a tree with so many branches that it is tedious to draw and difficult to use.

For example, if two dice, numbered 1 to 6, are thrown and *each* likely outcome has to be considered, the tree would have thirty six branches at the end.

An alternative is to tabulate all the *equally* likely outcomes. Thus, if the dice are unbiased, we can list the equally likely outcomes as shown, where the left hand figure in each bracket is the score on one die and the right hand figure is the score on the other die.

(1, 1)	(2, 1)	(3, 1)	(4, 1)	(5, 1)	(6, 1)
(1, 2)	(2, 2)	(3, 2)	(4, 2)	(5, 2)	(6, 2)
(1, 3)	(2, 3)	(3, 3)	(4, 3)	(5, 3)	(6, 3)
(1, 4)	(2, 4)	(3, 4)	(4, 4)	(5, 4)	(6, 4)
(1, 5)	(2, 5)	(3, 5)	(4, 5)	(5, 5)	(6, 5)
(1, 6)	(2, 6)	(3, 6)	(4, 6)	(5, 6)	(6, 6)

This complete set of all the *equally likely* outcomes is called the *possibility space*. Each of the *equally likely* outcomes is called a *sample point*. (In this example there are thirty six sample points in the possibility space.)

From this possibility space we can find, for example, the probability of a score of at least 7.

The number of sample points for which the score is at least 7 (in the shaded area) is twenty one.

Hence
$$P(\text{at least } 7) = \frac{21}{36} = \frac{7}{12}$$

(The use of any other method would require consideration of the mutually exclusive events: either a score of 7 or of 8 or of 9 ... or of 12.)

EXAMPLES 17e

1) Two unbiased tetrahedral dice numbered 1 to 4 are thrown. Set up the possibility space and use it to find:
(a) the probability that at least one 4 is thrown,
(b) a total score of 5 is thrown.

In each sample point the left hand number is the score on one of the dice and the right hand number is the score on the other die.
Hence the possibility space is

$$
\begin{array}{llll}
(1,1) & (1,2) & (1,3) & (1,4) \\
(2,1) & (2,2) & (2,3) & (2,4) \\
(3,1) & (3,2) & (3,3) & (3,4) \\
(4,1) & (4,2) & (4,3) & (4,4)
\end{array}
$$

and we see that there are sixteen sample points.

(a) The subset of points with at least one 4, {A}, contains 7 points.
    Hence $P(A) = \frac{7}{16}$.
(b) The subset of points giving a total score of 5, {B}, contains 4 points.
    Hence $P(B) = \frac{4}{16} = \frac{1}{4}$.

'Either ... Or' Situations Involving Events that are Not Mutually Exclusive

Using the example above, let us now consider the probability that *either* at least one 4 is thrown *or* a total score of 5 is thrown, i.e. $P(A \cup B)$. From the possibility space we see that

the subset of sample points in {A}

and the subset of sample points in {B}

are not mutually exclusive because the sample points (4, 1) and (1, 4) are in both subsets.
The situation becomes clearer if we rearrange the sample points in the possibility space by

placing the points in {A} in one circle

and the points in {B} in a second overlapping circle

so that the points in both $\{A\}$ and $\{B\}$, i.e. in $\{A \cap B\}$ are in the section common to both circles and the remaining points are outside both circles.

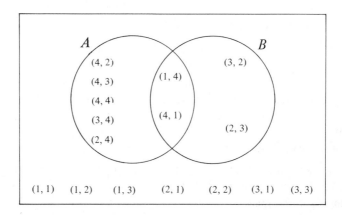

Such a diagram is called a *Venn Diagram*.
From this diagram we see that
the number of points in either $\{A\}$ or $\{B\}$, i.e. in $\{A \cup B\}$,
is *not* [(the number of points in $\{A\}$) + (the number of points in $\{B\}$)]
because this includes the points $(1, 4), (4, 1)$ twice.
Now the points $(1, 4)$ and $(4, 1)$ are in *both* $\{A\}$ *and* $\{B\}$, i.e. in $\{A \cap B\}$.
Hence

$$(\text{points in } \{A \cup B\}) = (\text{points in } \{A\}) + (\text{points in } \{B\}) - (\text{points in } \{A \cap B\})$$

$$= 7 + 4 - 2$$

As there are 16 sample points in total, we have

$$P(A \cup B) = \frac{7 + 4 - 2}{16}$$

$$= \frac{7}{16} + \frac{4}{16} - \frac{2}{16}$$

$$= P(A) + P(B) - P(A \cap B)$$

The result can be generalised as follows:
If a possibility space contains n points,
the subset of possibilities for an event A contains r points,
the subset of possibilities for an event B contains s points
and t points are common to $\{A\}$ and $\{B\}$.

The Venn Diagram illustrating this information is

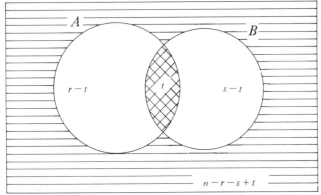

The circle A contains r sample points $\Rightarrow P(A) = \dfrac{r}{n}$

The circle B contains s sample points $\Rightarrow P(B) = \dfrac{s}{n}$

The intersection of the two circles, $A \cap B$, contains t points

$$\Rightarrow P(A \cap B) = \frac{t}{n}.$$

Thus the number of points in either $\{A\}$ or $\{B\}$, i.e. in $\{A \cup B\}$,

is $r + s - t \Rightarrow P(A \cup B) = \dfrac{r + s - t}{n}$

Therefore $\qquad P(A \cup B) = \dfrac{r}{n} + \dfrac{s}{n} - \dfrac{t}{n}$

i.e. $\qquad P(A \cup B) = P(A) + P(B) - P(A \cap B)$

Note that the left hand lune (the unshaded part) contains $r - t$ points which are in $\{A\}$ but not in $\{B\}$, i.e. this lune contains the points in $\{A \cap \bar{B}\}$. Similarly the points in the right hand lune are those in $\{B\}$ but not in $\{A\}$, i.e. in $\{B \cap \bar{A}\}$.

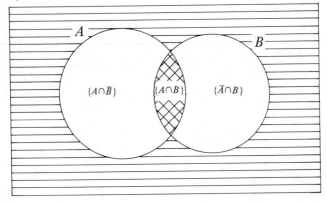

Note that if A and B are mutually exclusive $\{A \cap B\}$ contains no points, i.e.

$$\Rightarrow P(A \cup B) = P(A) + P(B)$$

EXAMPLES 17e (continued)

2) If A and B are two events such that $P(A) = \frac{1}{3}$, $P(B) = \frac{2}{9}$, $P(A \mid B) = \frac{1}{2}$. Find (a) $P(A \cap B)$, (b) $P(A \cup B)$, (c) $P(B \mid \bar{A})$.

Let   the number of points in the possibility space be n

           the number of points in $\{A\}$ be r

           the number of points in $\{B\}$ be s

and   the number of points in $\{A \cap B\}$ be t

Drawing a Venn Diagram we have

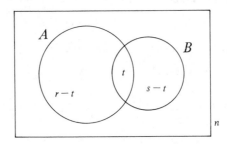

From the diagram

$$P(A) = \frac{r}{n} \qquad\qquad \text{i.e.} \quad \frac{r}{n} = \frac{1}{3} \qquad\qquad [1]$$

$$P(B) = \frac{s}{n} \qquad\qquad \text{i.e.} \quad \frac{s}{n} = \frac{2}{9} \qquad\qquad [2]$$

and $P(A \mid B) = \dfrac{\text{No. of ways } A \text{ occurs, given that } B \text{ has occurred}}{\text{No. of ways } B \text{ occurs (i.e. the equally likely events)}}$

$$= \frac{t}{s} \qquad\qquad \text{i.e.} \quad \frac{t}{s} = \frac{1}{2} \qquad\qquad [3]$$

(a) Now $P(A \cap B) = \dfrac{t}{n}$

$(3) \div (2) \Rightarrow$ $\dfrac{t}{n} = \dfrac{1}{9} \qquad\qquad [4]$

i.e. $P(A \cap B) = \dfrac{1}{9}$

(b) Using $P(A \cup B) = P(A) + P(B) - P(A \cap B)$

we have $P(A \cup B) = \dfrac{1}{3} + \dfrac{2}{9} - \dfrac{1}{9}$

$$= \dfrac{4}{9}$$

(c) $P(B \mid \bar{A}) = \dfrac{\text{No. of ways in which } B, \text{ and 'not } A\text{', occurs}}{\text{No. of ways in which 'not } A\text{' occurs}}$

$$= \dfrac{s-t}{n--r}$$

from [1], [2] and [4] we have $r = \tfrac{1}{3}n$

$$s = \tfrac{2}{9}n$$

$$t = \tfrac{1}{9}n$$

Therefore $\dfrac{s-t}{n-r} = \dfrac{\tfrac{2}{9}n - \tfrac{1}{9}n}{n - \tfrac{1}{3}n} = \dfrac{1}{9} \div \dfrac{2}{3} = \dfrac{1}{6}$

i.e. $P(B \mid \bar{A}) = \dfrac{1}{6}.$

Note that:

(i) if two events E_1 and E_2 are independent

then $P(E_1) = P(E_1 \mid E_2)$

In this problem $P(A) \neq P(A \mid B)$

so A and B are not independent.

(ii) if two events E_1 and E_2 are mutually exclusive

then $P(E_1 \cap E_2) = 0.$

In this problem $P(A \cap B) \neq 0$

so A and B are not mutually exclusive.

EXERCISE 17e

1) Set up a possibility space for the toss of two fair coins and a fair tetrahedral die numbered 1 to 4. From your possibility space find the probability of obtaining:

(a) a head, a tail and a four,   (b) at least one head and a four.

2) Two cubical dice are tossed. Both dice are unbiased and one is numbered 1 to 6, the other is numbered 1, 2, 2, 3, 3, 4. Set up the possibility space and use it to find the probability of obtaining:
(a) a score of 5,
(b) a score of at least 5,
(c) a three on either of the two dice, but not on both.

3) Two tetrahedral dice, both numbered 1 to 4 are tossed. If one die is fair and the other is biased so that a four is twice as likely as any other score, set up the possibility space and use it to find the probability that:
(a) at least one four is thrown,
(b) a total score of four is obtained,
(c) either at least one four is thrown or at least one three is thrown.

4) A football match may be either won, drawn or lost by the home team, so there are three ways of forecasting the result of any one match, one correct and two incorrect. Find the probability of forecasting at least three correct results.

5) In a group of twenty students all of whom are studying Physics or Mathematics or both, ten are studying Physics and fifteen are studying Mathematics. Find the probability that a student chosen at random is:
(a) studying Physics,
(b) studying Physics and Mathematics,
(c) studying Physics but not Mathematics.
Illustrate your results on a Venn Diagram.

6) Three unbiased dice, each numbered 1, 1, 2, 2, 3, 3 are tossed. Find the probability of throwing either at least one 2 or at least one 3.

7) Two normal fair dice, numbered 1 to 6, are tossed simultaneously. What is the probability of obtaining a total score greater than 7 if at least one of the dice scores 5.

8) A and B are two events such that $P(A) = \frac{1}{4}$ and $P(B) = \frac{1}{3}$ and $P(A \cup B) = \frac{1}{2}$. Find $P(A \cap B)$.

9) A and B are two events such that $P(A) = \frac{2}{7}$ and $P(A \cap B) = \frac{1}{5}$. If A and B are independent find $P(B)$ and $P(A \cup B)$.

10) A and B are two events such that $P(A) = \frac{2}{5}$, $P(A|B) = \frac{1}{3}$, $P(B|A) = \frac{1}{2}$.
Find $P(A \cup B)$ and state, with reasons, whether A and B are mutually exclusive.

EXPECTATION

The word 'expectation' or 'expected' is used to mean the 'average' outcome of an experiment or the 'average' result in a series of experiments. The precise meaning of 'expectation' depends on the way it is used. The following examples illustrate its use in different situations.

EXAMPLES 17f

1) Two unbiased dice, numbered 1 to 6, are tossed one hundred and forty four times. Find the number of times that double 6 is expected.

$P(\text{double } 6) = \frac{1}{36}$,
i.e. as the dice are unbiased, a double 6 can occur in one way out·of the thirty six equally likely ways in which the dice may land,
i.e. in thirty six throws we would expect one double 6 so in one hundred and forty four throws we would expect double 6 to appear on $\frac{144}{36}$ occasions,
i.e. the *expected* number of times that double 6 occurs in one hundred and forty four throws is four.

In general, if the probability of an event A is $P(A)$, then in n trials the expected number of times that A occurs is $nP(A)$.

2) The probability that it will snow on any one day in December is 0.05. On how many days is it expected to snow in December?

There are thirty one days in December.
So the numbers of days on which snow is expected is $31 \times 0.05 = 1.55$,
i.e. it is expected to snow on one or two days in December.

3) Three coins are tossed thirty times. Find the expected number of heads if two of the coins are fair and one is biased so that a tail is twice as likely as a head.

Tossing the three coins thirty times is equivalent to

$$\begin{cases} \text{tossing one fair coin sixty times} \\ \text{tossing the biased coin thirty times} \end{cases}$$

$\Rightarrow$ $\begin{cases} \text{the expected number of heads } = \frac{1}{2}(60) = 30 \\ \text{the expected number of heads } = \frac{1}{3}(30) = 10 \end{cases}$

Hence the total number of heads expected is 40.

4) Four counters are drawn at random from a bag containing five red and two black counters. Find the expected number of red counters.

In this context 'expected' means the average number of red counters expected to be drawn in any one trial.
If we find the total number of red counters that are likely in n trials (i.e. in n repeated drawing of four counters from the seven in the bag), then the average number per trial can be found.
The four counters can be

either   (2R and 2B) or (3R and 1B) or (4R)

$$P(2R \cap 2B) = \frac{^5C_2 \times 1}{^7C_4} = \frac{2}{7}$$

$$P(3R \cap 1B) = \frac{^5C_3 \times {}^2C_1}{^7C_4} = \frac{4}{7}$$

$$P(4R) = \frac{^5C_4}{^7C_4} = \frac{1}{7}$$

Therefore in n trials we would expect

two red counters to appear $\frac{2}{7} \times n$ times $\Rightarrow \frac{4n}{7}$ red counters

three red counters to appear $\frac{4}{7} \times n$ times $\Rightarrow \frac{12n}{7}$ red counters

four red counters to appear $\frac{1}{7} \times n$ times $\Rightarrow \frac{4n}{7}$ red counters

so we would expect a total of $\left(\frac{4n}{7} + \frac{12n}{7} + \frac{4n}{7} \right)$ red counters in n trials,

i.e. an average of $\frac{20}{7}$ counters per trial.

Expected Gain or Loss

We often encounter a situation in which the outcome involves either a gain or loss of money. If such a situation is repeated many times the average (expected) gain or loss can be found using the following definition.

If there is a probability p of winning a sum of money £L the expected gain is £Lp.

EXAMPLES 17f (continued)

5) Two people, A and B, roll an unbiased die. The first to toss a 6 wins £10. Find A's expected winnings, if he goes first.

The probability that A wins, $P(A$ wins$)$, is

either $P(A$ wins on his first throw$) = P(A_1)$

$$= \tfrac{1}{6}$$

or $P(A$ wins on his second throw$) = P(\bar{A}_1$ and $\bar{B}_1$ and $A_2)$

$$= (\tfrac{5}{6})^2 \times \tfrac{1}{6}$$

or $P(A$ wins on his third throw$) = P(\bar{A}_1 \cap \bar{B}_1 \cap \bar{A}_2 \cap \bar{B}_2 \cap A_3)$

$$= (\tfrac{5}{6})^4 \times \tfrac{1}{6}$$

and so on,

i.e. $$P(A \text{ wins}) = \tfrac{1}{6} + (\tfrac{5}{6})^2(\tfrac{1}{6}) + (\tfrac{5}{6})^4(\tfrac{1}{6}) + \ldots$$

This is a G.P. with first term $\tfrac{1}{6}$ and common ration $(\tfrac{5}{6})^2$ and hence with a sum to infinity of

$$\frac{\tfrac{1}{6}}{1 - (\tfrac{5}{6})^2} = \frac{6}{11}$$

Hence $P(A \text{ wins}) = \tfrac{6}{11}$
So A's expected winnings are $\quad £10 \times \tfrac{6}{11} = £5.45$
This result is interpreted as meaning that if A and B played several times, with A going first each time, then on average A would expect to win £5.45 per game.

6) The probability of a candidate passing an examination at any one attempt is $\tfrac{3}{5}$. He carries on entering until he passes and each entry costs him £1. Find the expected cost of his passing the examination.

The probability of passing at the first attempt is $\tfrac{3}{5}$ and the cost is £1.
The probability of failing at the first attempt but passing at the second attempt is $(\tfrac{2}{5})(\tfrac{3}{5})$ and the cost is £2.
The probability of failing at the first two attempts but passing at the third attempt is $(\tfrac{2}{5})^2(\tfrac{3}{5})$ and the cost is £3.
And so on.
Therefore the expected cost is

$$£[\tfrac{3}{5}(1) + \tfrac{3}{5}(\tfrac{2}{5})(2) + \tfrac{3}{5}(\tfrac{2}{5})^2(3) + \tfrac{3}{5}(\tfrac{2}{5})^3(4) + \ldots$$
$$= £\tfrac{3}{5}[1 + 2(\tfrac{2}{5}) + 3(\tfrac{2}{5})^2 + 4(\tfrac{2}{5})^3 + \ldots]$$

Now the series $1 + 2x + 3x^2 + 4x^3 + \ldots$

$$= \frac{d}{dx}(x + x^2 + x^3 + x^4 + \ldots$$

$$= \frac{d}{dx}[(1-x)^{-1} - 1] \quad \text{if } |x| < 1$$

$$= \frac{1}{(1-x)^2} \qquad\qquad \text{if } |x| < 1$$

Therefore, replacing x by $\tfrac{2}{5}$,

$$1 + 2(\tfrac{2}{5}) + 3(\tfrac{2}{5})^2 + 4(\tfrac{2}{5})^3 + \ldots = \frac{1}{(1-\tfrac{2}{5})^2} = \frac{25}{9}$$

Hence the expected cost of passing the examination is

$$£\tfrac{3}{5} \times \tfrac{25}{9} = £\tfrac{5}{3} = £1.67$$

This result must be interpreted as being the cost per candidate averaged out for several candidates with the same probability $(\frac{2}{3})$ of passing at any one attempt. Obviously it will cost any one candidate an integral multiple of £1.

7) Three people A, B and C play a game of chance where the probability of A's winning is $\frac{3}{10}$. The winner gets a prize of £100. If a fourth person D wishes to buy A's place in the game, what would be a fair price for D to pay?

As the probability of A's winning is $\frac{3}{10}$, his expected gain is $\frac{3}{10} \times$ £100 = £30. Therefore £30 is a fair price to pay for A's place in the game.

EXERCISE 17f

In this exercise you may assume, where necessary,
that $\quad 1 + x + x^2 + x^3 + \ldots = (1-x)^{-1}$
and that $\quad 1 + 2x + 3x^2 + 4x^3 + \ldots = (1-x)^{-2}$.

1) Two unbiased dice, each numbered 1 to 6, are tossed. If this is repeated fifty times what is the expected number of times that a score of 11 will occur?

2) At a certain seaside resort the probability of rain on any one day in July is 0.02. If I take fourteen days holiday at that resort in July, on how many days should I expect it to rain?

3) In a class of thirty pupils, the probability that any one pupil is absent on any one day is 0.04. How many absence marks are expected in the class register for one week (five days)?

4) Two coins are tossed where one is fair and the other is biased so that a head is three times as likely as a tail. Find the probability that:
(a) two heads are tossed,
(b) one head and a tail are tossed.
In twenty such tosses, find the expected number of heads.

5) A pack of fifty-two playing cards is dealt round to four people so that each player gets thirteen cards. If a yarborough is a hand containing no card greater than a 9 (ace ranks high), find the expected number of yarboroughs in one thousand deals.

6) A 'lucky dip' box contains six tokens which can be exchanged for a prize. Four of the tokens are for books and two of the tokens are for records. A person selects three tokens at random, collects his prizes and then replaces the tokens in the box. If twenty people each have such a 'dip' in the box, what is the expected number of records to be given as prizes?

7) A team of four people is chosen at random from four men and three women. Find the expected number of women in the team.

8) Five books are chosen at random from a shelf of five paperbacks and four hardbacks. Find the expected number of hardbacks chosen.

9) Three people A, B, C gamble for a prize of £100 by rolling an unbiased die, numbered 1 to 6, in the order A, then B, then C and so on. The first to roll a 6 wins the prize. If B wishes to sell his chance of winning to another person, what is a fair price to ask?

10) The probability of a person scoring at least 100 at a pintable is $\frac{1}{5}$. If it costs him 5p for each attempt to score at least 100, and he carries on until he does score at least 100, how much is it expected to cost him?

11) At a shooting gallery, the probability that a boy hits a target with any one shot is $\frac{2}{5}$. If it costs him 5p per shot, and he wins 50p each time he hits a target, find his expected gain if he shoots until he does hit a target.

12) There is a probability of $\frac{1}{21}$ of getting the jackpot on a fruit machine. The jackpot pays out £1 and each go costs 10p. If the machine is played until the jackpot pays out, what is the expected loss?

SUMMARY

For any event A

$$P(A) + P(\overline{A}) = 1$$

For any two events A, B

$$P(A \cap B) = P(A) \times P(B \mid A)$$
$$P(A \cup B) = P(A) + P(B) - P(A \cap B)$$

If A and B are independent (i.e. the occurrence of A has no influence on the likely occurrence of B and vice-versa)

then $\qquad P(B \mid A) = P(B) \;\Rightarrow\; P(A \cap B) = P(A) \times P(B)$

If A and B are mutually exclusive (i.e. A and B cannot both occur)

then $\qquad P(A \cap B) = 0 \;\Rightarrow\; P(A \cup B) = P(A) + P(B)$

MULTIPLE CHOICE EXERCISE 17

(*Instructions for answering these questions are given on page xii.*)

TYPE I

1) A couple decide to have three children. Assuming that the birth of a boy or girl is equally likely, the probability that they have two girls and then a boy is:
(a) $\frac{3}{8}$ (b) $\frac{1}{2}$ (c) $\frac{1}{8}$ (d) $\frac{1}{6}$ (e) $\frac{1}{3}$.

2) A possibility space for the equally likely results of rolling one fair die and one biased die with a 6 twice as likely as any other score, contains the following number of sample points:
(a) 36   (b) 30   (c) 6   (d) 42   (e) 18.

3) Two books are chosen from three paperbacks and three hardbacks. The number of ways in which this can be done, given that at least one is a paperback, is:

(a) 24   (b) 15   (c) 9   (d) 12   (e) 3.

4) If A and B are events such that $P(A) = \frac{1}{2}$, $P(B) = \frac{1}{2}$, $P(A \cap B) = \frac{1}{4}$.

$P(A \cup B) =$

(a) 1   (b) $\frac{3}{4}$ (c) $\frac{1}{4}$ (d) $\frac{1}{16}$ (e) none of these.

5) A biased coin, such that a tail is twice as likely as a head, is tossed thirty times. The expected number of heads is:

(a) 15   (b) 20   (c) 10   (d) 1   (e) 30.

6) A and B are independent events and $P(A) = \frac{1}{4}$, $P(B) = \frac{1}{4}$, $P(\bar{A} \cap \bar{B})$ is:

(a) $\frac{3}{4}$ (b) $\frac{1}{16}$ (c) $\frac{1}{4}$ (d) $\frac{1}{2}$ (e) $\frac{9}{16}$

7) The number of ways in which two letters can be selected from the letters of the word POSSIBLE is:

(a) $^{7}C_2 + 1$ (b) $^{8}C_2$ (c) $^{8}P_2$ (d) $^{7}C_2$ (e) $^{8}P_2 + 1$.

8) The number of permutations of the letters of POSSIBILITY is:

(a) $^{11}P_{11}$ (b) $\dfrac{11!}{3! \, 2!}$ (c) $(^{11}P_3)(^{11}P_2)$ (d) $^{11}P_8$ (e) $^{8}P_7 + ^{8}P_6$.

9) The shaded area in the Venn diagram represents:

(a) A (b) $\bar{B}$ (c) $A \cap \bar{B}$ (d) $A \cup \bar{B}$ (e) $A \,|\, B$.

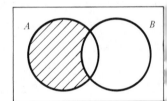

10) Six people are seated at a circular table.
The number of ways in which this can be done is:

(a) $6!$ (b) $5!$ (c) $\dfrac{6!}{2}$ (d) $\dfrac{5!}{2}$ (e) $^{6}P_5$.

TYPE II

11) A and B are two independent events with $P(A) = 0.3$ and
$P(B) = 0.4$.

(a) $P(A \cup B) = 0.7$ (b) $P(A \cap B) = 0.12$

(c) $P(A|B) = 0.3$ (d) $P(A|\bar{B}) = 0.4$.

12) Two different numbers are chosen at random from the set $\{1, 2, 3, 4, 5, 6\}$

(a) If the choice is ordered, it can be made in $^{6}P_2$ ways.

(b) The probability of the second number chosen being 2 is $\frac{1}{6}$.

(c) The probability of choosing the ordered pair $(4, 5)$ is $\frac{1}{30}$.

13) A and B are mutually exclusive events.
(a) $P(A \cup B) = P(A) + P(B)$.
(b) $P(A|B) = P(A)$.
(c) $P(A \cap B) = 0$.

14) Two unbiased tetrahedral dice, each numbered one to four, are rolled.
(a) There are 16 ways in which the two dice can land.
(b) A total score of 4 and the throwing of at least one 1 are mutually exclusive events.
(c) The probability of not throwing a double 4 is the same as the probability of scoring at least 7.

15) Two people, A and B, toss a fair coin. Whoever tosses a head first wins 2p, and they toss in the order A then B.
(a) The probability that A wins on his second toss is $\frac{1}{8}$.
(b) The probability that B wins on his second toss is $\frac{1}{8}$.
(c) If they carry on playing a large number of games, without changing the order of play, they each expect to win an average of 1p per game.
(d) A wins and B wins are mutually exclusive events.

TYPE III

16) (a) A and B are independent events.
(b) $P(A \cap B) = P(A) \times P(B)$.

17) (a) $P(A) = a$, $P(B) = b$.
(b) $P(A \cup B) = a + b - ab$.

18) (a) $P(A \cap B) = 0$.
(b) $P(A|B) = \frac{1}{2}$.

19) (a) $P(E) = r/n$.
(b) $P(\bar{E}) = \dfrac{n-r}{n}$.

TYPE IV

20) Three people A, B and C draw a card at random from a pack of fifty-two playing cards. Find the probability that A is the first to draw an ace.
(a) A draws first.
(b) The cards are not replaced.
(c) B draws second.

21) Find $P(A|\bar{B})$.
(a) A and B are not independent events.
(b) $P(A) = \frac{1}{3}$.
(c) $P(A \cap B) = \frac{1}{5}$.

22) A committee is chosen at random from a group of ten people. Find the expected number of men on the committee.
(a) There are four members of the committee.
(b) There are six men in the group.
(c) The committee of four are seated at a round table.

23) Find the probability that a candidate who guesses at random obtains at least 40% in a multiple choice test.
(a) There are one hundred questions.
(b) Each question has five alternative answers, only one of which is correct.
(c) Each correct answer scores one mark, no marks are given for an incorrect answer.

TYPE V

24) The probability of tossing a head with a biased coin is $\frac{1}{2}$.

25) The number of ways of choosing two cards from a pack of fifty-two playing cards is $^{52}P_2$.

26) If A and B are mutually exclusive, $P(A|B) = 0$.

27) If $P(A) = \frac{1}{3}$, $P(\bar{A}) = \frac{2}{3}$.

28) If $P(A \text{ and } B) = 0$ then either $P(A) = 0$ or $P(B) = 0$.

29) The number of arrangements of the letters of the word EVERY is 24.

30) The number of ways of choosing two different letters from the set $\{A, B, C, D\}$ is 4.

MISCELLANEOUS EXERCISE 17

1) An unbiased die marked $1, 2, 2, 3, 3, 3,$ is rolled three times. Find the probability of getting a total score of 4.   (U of L)

2) (a) A delegation of 4 persons is to be formed from 5 married couples. Find the number of ways in which the delegation can be chosen if it contains:
 (i) 2 married couples,
 (ii) at least one man and one woman,
 (iii) no married couple.
 (b) Three normal unbiased dice are thrown. Find the probability that there will be:
 (i) no sixes,
 (ii) at least one six,
 (iii) exactly one six.

 Find also the probability that the three dice all show the same number when thrown.   (AEB)'76

3) A well-shuffled pack of 52 playing cards is dealt out to four players, each receiving 13 cards. Show that the probability that a particular player receives the four aces is 0.0026.

How many deals are necessary in order that the probability of a particular player receiving all four aces in at least one game exceeds 0.5? (O)

4) If A and B are events and
$$p(A) = \tfrac{8}{15},$$
$$p(A \text{ and } B) = \tfrac{1}{3},$$
$$p(A|B) = \tfrac{4}{7},$$

calculate $p(B)$, $p(B|A)$ and $p(B|\bar{A})$ where $\bar{A}$ is the event 'A does not occur'.

State, with reasons, whether A and B are (i) independent, (ii) mutually exclusive. (C)

5) A committee of four people is chosen at random from a set of seven men and three women. Determine:
(a) the probability that there is at least one woman on the committee,
(b) the probability that there is at least one man on the committee,
(c) the expected number of women on the committee,
(d) the expected number of men on the committee. (C)

6) An event has probability p of success and $q(= 1 - p)$ of failure. Independent trials are carried out until at least one success and at least one failure have occurred. Find the probability that r trials are necessary $(r \geqslant 2)$ and show that this probability equals $(\tfrac{1}{2})^{r-1}$ when $p = \tfrac{1}{2}$.

A couple decide that they will continue to have children until either they have both a boy and a girl in the family or they have four children. Assuming that boys and girls are equally likely to be born, what will be the expected size of their completed family? (O)

7) A mother has found that 20% of the children who accept invitations to her children's birthday parties do not come. For a particular party she invites 12 children but has available only 10 party hats. What is the probability that there is not a hat for every child who comes to the party?

The mother knows that there is a probability of 0.1 that a child who comes to a party will refuse to wear a hat. If this is taken into account, what is the probability that the number of hats will not be adequate? (O)

8) (a) Find the probability that the fourth power of any positive integer n ends in the digit 6.
   (b) In a certain tournament in which games cannot result in a draw A plays B until one of them has won a total of three games. If p is the probability that A wins any individual game he plays against B and if $q = 1 - p$, find in terms of p and q the probabilities that,

    (i) A wins the first three games,

    (ii) a decision is reached in the third game,

    (iii) A wins the match in the fourth game,

    (iv) a decision is reached in the fourth game.

If $p = \frac{2}{3}$, determine the probability of A winning the match before the sixth game.

(AEB)'75

9) (a) Find how many numbers between 3000 and 4000 can be formed using only the digits 1, 2, 3 and 4, no digit being repeated.

    (b) A bag contains 4 red and 6 black balls. One ball is drawn at random; if it is black it is replaced in the bag, but if it is red it is not replaced. A second ball is then drawn. X denotes the event 'The first ball is red' and Y denotes the event 'The second ball is red'. Find the probabilities

    (i) $P(X)$, (ii) $P(Y$ given $X)$, (iii) $P(Y)$, (iv) P(either X or Y but not both).

(C)

10) One of three coins is biased so that the probability of obtaining a head is twice as great as the probability of obtaining a tail. The other two coins are fair. One of the three coins is chosen at random and tossed three times, showing a head on each occasion. Using a tree diagram, or otherwise, find the probability that the chosen coin is biased.

(U of L)

11) (a) In how many ways can a hand of 13 cards be dealt from a normal pack of 52 cards, all of which are different? Assuming that each deal is equally likely, what is the probability of being dealt 13 cards all of the same suit?

    [Answers to both parts should be left in factorial form.]

    (b) If A and B are independent events, the probabilities of which in a certain trial are a and b respectively, what are the probabilities of:

    (i) both A and B occurring,

    (ii) event A occurring but not B,

    (iii) neither A nor B occurring?

If these trials are repeated n times with no change in the values of a and b, what is the probability that neither A nor B will occur? If $a = b = 0.01$, find how many trials are required before this probability becomes less than 0.5.

(AEB)'75

12) Eight trees are planted in a circle in random order. If two of the trees are diseased and later die, what is the probability that the two dead trees are next to each other?

If four of them are diseased find (i) the probability that at least two of them are next to each other, and (ii) the probability that all four are next to each other.

(C)

13) When a boy fires an air-rifle the probability that he hits the target is p.

(a) Find the probability that, firing 5 shots, he scores at least 4 hits.

(b) Find the probability that, firing n shots $(n \geqslant 2)$, he scores at least two hits.

(C)p

14) (a) Four men, two women and a child sit at a round table. Find the number of ways of arranging the seven people if the child is seated (i) between the two women, (ii) between two men.

(b) A die with faces numbered 1 to 6 is biased so that
$P\{\text{score is } r\} = kr, \quad (r = 1, \ldots, 6).$ Find the value of k.

If the die is thrown twice, calculate the probability that the total score exceeds 10.   (C)

15) (a) The results of eleven football matches (as win, lose or draw) are to be forecast. Out of all possible forecasts, find how many will have eight correct and three incorrect results.

(b) An unbiased die in the shape of a regular dodecahedron has twelve faces with the numbers 2, 2, 4, 4, 4, 6, 6, 10, 10, 10, 12, 12, showing separately on the faces. The result of a throw is the number showing on the uppermost face. Each of four players throws the die twice and scores the sum of the two results. What is the probability that all of the four players in succession will each obtain a score greater than six?

(c) An unbiased die in the shape of a cube shows 1, 2, 3, 4, 5, 6 on its six separate faces. It is tossed until it lands the same way up twice running. Find the probability that this requires r tosses.   (AEB)'74

16) (a) A box contains six dice, one of which is unfairly biased. If two dice are chosen at random simultaneously from this box, what is the probability that one of them will be biased?

(b) A uniform unbiased die is constructed in the shape of a regular tetrahedron with faces numbered 2, 2, 3 and 4, and the score is taken from the face on which the die lands. If two such dice are thrown together what total scores are possible at each throw and what is the probability of each score? What is expected to be the average score over a long series of throws?

What is the probability of scoring:
(i) exactly 6 on each of three successive throws,
(ii) more than 4 on at least one of three successive throws?   (AEB)'74

17) (a) Show that it is more probable to get at least one six with a throw of three dice than to get two sixes with any one of fifteen throws of two dice.

(b) There are three identical boxes each containing a sum of money, no two boxes containing the same amount. A man chooses a box as follows: he first takes a box at random (call it A) and sees how much is in it. He then takes one of the other two boxes at random (call it B) and sees how much is in it. If box B contains more than box A, then the man chooses box B; if box B contains less than box A then he chooses the third box (call it C). Find the probability that he will choose:
(i) the box containing the greatest value of money,
(ii) the box containing the smallest value of money.   (AEB)'76

18) Suppose that letters sent by first and second class post have probabilities of being delivered a given number of days after posting according to the following table (weekends are ignored).

Days to delivery	1	2	3
1st class	0.9	0.1	0
2nd class	0.1	0.6	0.3

The secretary of a committee posts a letter to a committee member who replies immediately using the same class of post. What is the probability that four or more days are taken from the secretary posting the letter to receiving the reply if (a) first class, (b) second class post is used?
The secretary sends out four letters and each member replies immediately by the same class of post. Assuming the letters move independently, what is the probability that the secretary receives (a) all the replies within three days using first class post, (b) at least two replies within three days using second class post?
(O)

19) In a game of chance each player throws two unbiased dice and scores the difference between the larger and smaller numbers which arise. Two players compete and one or the other wins if, and only if, he scores at least 4 more than his opponent. Find the probability that neither player wins. (JMB)p

20) Six lines are drawn in a plane and produced to give all their points of intersection. If no three lines are concurrent, and no two parallel, show that there are 15 points of intersection. If three of these points are chosen at random show that the probability that they are all on one of the given lines is $\frac{12}{91}$. Find also the probability that if four points are chosen at random, they are not all on one of the given lines. (C)

21) In a certain examination paper there are 10 questions. Each question has 5 suggested answers, and the candidates have to choose the right one in each question. Suppose that candidate X chooses answers entirely at random, so that he is equally likely to choose any one of the 5 answers in each question. Calculate the probability that he will score at least 3 correct answers out of 10.
(C)p

22) A tennis player A has a probability of $\frac{2}{3}$ of winning a set against a player B. A match is won by the player who first wins three sets. Find the probability that A wins the match.

CHAPTER 18

NUMERICAL APPLICATIONS

In the major part of this book we have concentrated on the general analysis of functions and in this Chapter we examine a variety of numerical applications of this work.

REDUCTION OF A RELATIONSHIP TO LINEAR FORM

If it is thought that a certain relationship exists between two variable quantities, this hypothesis can be tested by experiment, i.e. by giving one variable certain values and measuring the corresponding values of the other variable. The experimental data collected can then be displayed graphically. If the graph shows points that lie approximately on a straight line (allowing for experimental error) then a linear relationship between the variables (i.e. a relationship of the form $Y = mX + c$) is indicated. Further, the gradient of the line (m) and the vertical axis intercept (c) provide the values of the constants.

On the other hand, if the relationship is not of a linear form, the points on the graph will lie on a section of a curve. It is very difficult to identify the equation of a curve from a section of it, so the form of a non-linear relationship cannot be verified in this way. Non-linear relationships, however, can often be reduced to a linear form. The following examples illustrate some of the relationships which can be verified by plotting experimental data in a form which gives a straight line.

Linear Relationships

An elastic string is fixed at one end and a variable weight is hung on the other end. It is believed that the length of the string is related to the weight by a linear law. Use the following experimental data to confirm this belief and find the particular relationship between the length of the string and the weight.

Weight (W) in newtons	1	2	3	4	5	6	7	8
Length (l) in metres	0.33	0.37	0.4	0.45	0.5	0.53	0.56	0.6

If l and W are related by a linear law, then $l = aW + b$ where a and b are constants. Plotting l against W gives

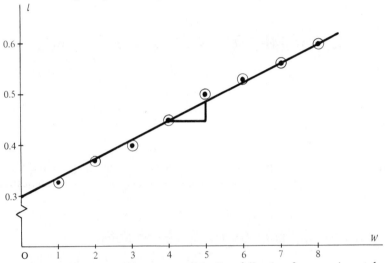

We see that the points do all lie on a straight line (allowing for experimental error) so that l and W are connected by a relationship of the form $l = aW + b$.

Now we draw the line of 'closest fit' (i.e. the line which has the points distributed above and below it as evenly as possible, which is not necessarily the line through the most points).

By measurement from the graph

$$\text{gradient} = 0.04$$

$$\text{intercept on the vertical axis} = 0.3$$

So comparing $l = aW + b$
 with $Y = mX + c$ } we have $a = 0.04, \quad b = 0.3$

i.e. within the limits of experimental accuracy

$$l = 0.04W + 0.3$$

Note that in finding the gradient, the *increase in l is taken from the vertical scale used* and the *increase in W is taken from the horizontal scale used* and these two scales are not necessarily the same.

Relationships of the Form $y = ax^n$

The following data, collected from an experiment, is believed to be related by a law of the form $p = aq^n$ where a and n are constants.

q	1	2	3	4	5	6
p	0.5	0.63	0.72	0.8	0.85	0.9

If the relationship $p = aq^n$ is correct then
$$p = aq^n \iff \log p = n \log q + \log a$$

Comparing $\log p = n \log q + \log a$
$$\left.\begin{array}{l} \log p = n \log q + \log a \\ Y = mX + c \end{array}\right\} \qquad [1]$$
with $Y = mX + c$

we see that $\log p$ and $\log q$ are related by a linear law. So if $\log p$ is plotted against $\log q$ we expect a straight line.

$\log q$	0	0.30	0.48	0.60	0.70	0.78
$\log p$	-0.30	-0.20	-0.14	-0.10	-0.07	-0.05

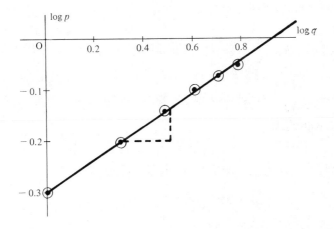

This straight line graph confirms that there is a linear relationship between $\log q$ and $\log p$.

By measurement, the gradient of the line is 0.33.
Reading from the graph, the intercept on the vertical axis is -0.3.

From [1], $n = 0.33$ and $\log a = -0.3 \Rightarrow a = 0.5$

(**Note** that these values of a and n are approximate. Apart from experimental errors in the data, selecting the line of best fit is a personal judgement and so is subject to slight variations which affect the values obtained for a and n.)
The data given confirms that p and q are related by the law $p = aq^n$ where $a \simeq 0.5$ and $n \simeq 0.33$.

Relationships of the Form $y = ab^x$

A relationship of the form $y = ab^x$ where a and b are constant can be reduced to a linear relationship by taking logs, since

$$y = ab^x \iff \log y = x \log b + \log a$$

Comparing $\log y = x \log b + \log a$

with $Y = mX + c$

we see that plotting values of $\log y$ against corresponding values of x gives a straight line whose gradient is $\log b$ and whose intercept on the vertical axis is $\log a$.

Relationships of the Form $\dfrac{1}{y} + \dfrac{1}{x} = \dfrac{1}{a}$

If a is a constant, $\dfrac{1}{y} + \dfrac{1}{x} = \dfrac{1}{a}$ is a linear relationship between $\left(\dfrac{1}{y}\right)$ and $\left(\dfrac{1}{x}\right)$

i.e. if values of $\left(\dfrac{1}{y}\right)$ are plotted against corresponding values of $\left(\dfrac{1}{x}\right)$, a straight line graph will result.

By comparing $\left(\dfrac{1}{y}\right) = -\left(\dfrac{1}{x}\right) + \left(\dfrac{1}{a}\right)$

with $Y = mX + c$

we note that the gradient of the graph should be -1 and the intercept on the $\left(\dfrac{1}{y}\right)$ axis gives the value of $\dfrac{1}{a}$.

EXAMPLES 18a

1) In an experiment, values of a variable y were measured for selected values of a variable x.

The results are shown in the table below. It is believed that x and y are related by a law of the form $2y + 10 = ab^{(x-3)}$. Confirm this graphically and find approximate values for a and b.

x	10	12	15	20	21
y	37.5	90	320	2440	3700

If $2y + 10 = ab^{(x-3)}$, taking logs of both sides gives

$$\log(2y + 10) = (x - 3)\log b + \log a$$

which is of the form $\qquad\qquad Y = mX + c$

where $\qquad \left.\begin{array}{l} Y \equiv \log(2y + 10) \\ X \equiv x - 3 \end{array}\right\}$ and $\left\{\begin{array}{l} m = \log b \\ c = \log a \end{array}\right.$

i.e. $\log(2y + 10)$ and $(x - 3)$ obey a linear law.

So we tabulate corresponding values of $x - 3$ and $\log(2y + 10)$ from the given values of x and y:

$x - 3$	7	9	12	17	18
$\log(2y + 10)$	1.9	2.3	2.8	3.7	3.9

Then plotting $\log(2y + 10)$ against $x - 3$ gives the graph below.

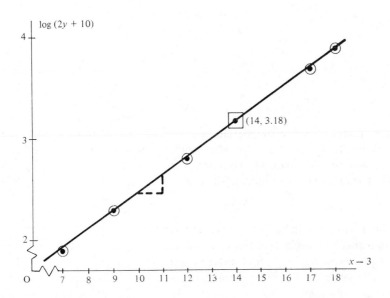

The straight line shows that there is a linear relationship between $\log(2y + 10)$ and $(x - 3)$, confirming that $2y + 10 = ab^{x-3}$.

Measurement from the graph gives gradient $\simeq 0.175$

i.e. $$\log b \simeq 0.175$$

$\Rightarrow$ $$b \simeq 1.49$$

In all graphical work the scales should be chosen to give the greatest possible accuracy, i.e. the range of values given in the table should have as much spread as possible. This sometimes means that the horizontal scale does not include zero and the value of c cannot then be read from the graph. In these circumstances, which arise in this example, we find c by using the equation $Y = mX + c$ together with the measured value of m (i.e. 0.175) and the co-ordinates of any point P on the graph (*not* a pair of values of X and Y from the table).

Thus $$3.18 = (0.175)(14) + c$$

$\Rightarrow$ $$c = 0.73$$

i.e. $$\log a \simeq 0.73$$

$$a \simeq 5.37$$

2) It is known that two variables x and y are related by the law

(a) $ae^y = x^2 - bx$

(b) $y = \dfrac{1}{(x-a)(x-b)}$

In each case state how you would reduce the law to a linear form so that a straight line graph could be drawn from experimental data.

When attempting to reduce a relationship between x and y to a form from which a straight line graph can be drawn, the given equation must be expressed in the form

$$Y = mX + c$$

where X and Y are variable terms, values for which must be calculable from the given data, i.e. X and Y must not contain unknown constants. On the other hand m and c must be constants, but may be unknown.

Now X and Y may be functions of x and/or y, as

$$f(xy) = mg(xy) + c$$

is a linear relationship between $f(xy)$ and $g(xy)$.

So to reduce a non-linear relationship to a linear form we:

(1) try to express it in a form containing three terms,

(2) make one of those terms constant,

(3) remove unknown constants from the coefficient of one of the variable terms.

These objectives will now be applied to the problem in hand.

(a) $ae^y = x^2 - bx$.

This equation has three terms, one of which becomes constant when we divide by x, giving

$$a\frac{e^y}{x} = x - b$$

We also have a variable term (x) whose coefficient is a known constant. This equation may now be written as

$$x = a\frac{e^y}{x} + b$$

Comparing with $\qquad\qquad Y = mX + c$

we see that if values of x are plotted against corresponding values of $\dfrac{e^y}{x}$, a straight line will result (whose gradient is a and whose intercept on the vertical axis is b).

(b) $y = \dfrac{1}{(x-a)(x-b)}$.

Although this form suggests the use of partial fractions, this approach increases the number of times the unknown constants appear.
It is better to invert the equation giving

$$\frac{1}{y} = (x-a)(x-b)$$

$\Rightarrow\qquad\qquad$ $$\frac{1}{y} = x^2 - x(a+b) + ab$$

$\Rightarrow\qquad\qquad$ $$x^2 - \frac{1}{y} = (a+b)x - ab$$

We can now compare this form with $\quad Y = mX + c$.

Thus plotting values of $x^2 - \dfrac{1}{y}$ against corresponding values of x will give a straight line (whose gradient is $a+b$ and whose intercept on the vertical axis is $-ab$).

EXERCISE 18a

1) Reduce each of the given relationships to the form $\quad Y = mX + c$. In each case give the functions equivalent to X and Y and the constants equivalent to m and c.

(a) $\dfrac{1}{y} = ax + b$ (b) $y(y - b) = x - a$ (c) $ae^x = y(y - b)$

In each of the following questions, the table gives sets of values for the related variables and the law which relates the variables. By drawing a straight line graph find approximate values for a and b.

2) $y = ax + ab$.

x	3	5	7	10
y	-2	2	6	12

3) $s = ab^{-t}$.

t	1	2	3	4
s	1.5	0.4	0.1	0.02

4) $r^2 = a\theta - b$.

θ	1	4	10	25	40
r	1.6	2	2.6	3.8	4.7

5) $ay = b^x$.

x	5	6	7	8
y	1.07	2.13	4.27	8.53

6) $\dfrac{a}{V} + \dfrac{b}{L} = 1$.

V	2.5	3	5.5	7	12
L	2.5	1.5	0.79	0.7	0.6

7) $y = (x - a)(x - b)$.

x	1	2	3	4	5	6
y	-6	-4	0	6	14	24

INTEGRATION USED AS A PROCESS OF SUMMATION

We have seen in Chapter 10 that

$$\lim_{\delta x \to 0} \sum_{x=a}^{b} y\, \delta x = \int_a^b y\, dx$$

i.e. integration is a process of summation.

Thus if Q is a quantity made up of elements δQ where $Q = \Sigma\, \delta Q$ and if δQ can be expressed approximately in the form $f(x)\, \delta x$, i.e. $\delta Q \simeq f(x)\, \delta x$ then $Q \simeq \Sigma f(x)\, \delta x$.

Further, if as $\delta x \to 0$ the approximation becomes more accurate,

then $$Q = \lim_{\delta x \to 0} \Sigma f(x)\, \delta x = \int f(x)\, dx$$

AREAS BOUNDED BY CURVES

We have seen in Chapter 10 how this process of summation is used to find areas. The area is divided into elements which are strips. Each element is approximately rectangular. For example, to find the area bounded by the x-axis, and the curve $y = x^2 - 1$, we divide it into vertical strips.

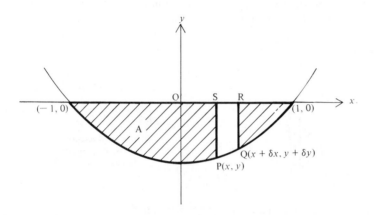

Taking PQRS as a typical element, where $P(x, y)$, $Q(x + \delta x, y + \delta y)$ are points on $y = x^2 - 1$, then the area of PQRS $\simeq y\,\delta x$ (taking the element as approximately rectangular),

i.e.
$$A \simeq \sum_{x=-1}^{1} y\,\delta x$$

Therefore

$$A = \lim_{\delta x \to 0} \sum_{x=-1}^{1} y\,\delta x = \int_{-1}^{1} y\,dx = \int_{-1}^{1} (x^2 - 1)\,dx$$

$$= [\tfrac{1}{3}x^3 - x]_{-1}^{1} = [\tfrac{1}{3} - 1] - [-\tfrac{1}{3} + 1] = -\tfrac{4}{3}$$

We expect $\Sigma y\,\delta x$ to give a negative result for $-1 \leqslant x \leqslant 1$ as y is negative in this range.

Therefore the area is $\tfrac{4}{3}$ square units.

Note that the area A is symmetrical about the y axis, i.e. about $x = 0$.

Hence A can also be found by writing

$$A \simeq 2 \sum_{x=0}^{1} y\,\delta x$$

i.e.
$$A = 2 \int_{0}^{1} (x^2 - 1)\,dx$$

We will now look further at problems in which areas are to be found and will adopt the following procedure:

(a) Draw a diagram clearly showing the area to be found.

(b) Take a vertical or horizontal strip for the element of area. The choice should be made so that:

(i) The element has the same format throughout the area. For example, if the area shown in the diagram is divided into vertical strips, some of the elements are bounded at the top by the curve and others by the line. On the other hand, horizontal strips are all bounded on the left by the y axis and on the right by the curve.

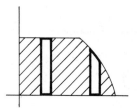   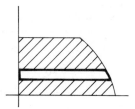

Thus horizontal strips should be chosen in this case.

(ii) The element does *not* have both ends on the *same* curve.

e.g.

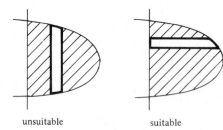

unsuitable   suitable

(c) Taking the element as being approximately rectangular, write down its approximate area.

(d) Sum, by integrating, the areas of the elements over the specified range.

EXAMPLES 18b

1) Find the area of the region of the xy plane which satisfies the following inequalities:

$$y \geqslant x^2, \qquad y \leqslant 4$$

$y \geqslant x^2$　defines the curve　$y = x^2$　and the region above the curve.

$y \leqslant 4$　defines the line　$y = 4$　and the region below the line.

$\left. \begin{array}{l} y = x^2 \\ y = 4 \end{array} \right\}$　intersect at　$(2, 4)$　and　$(-2, 4)$.

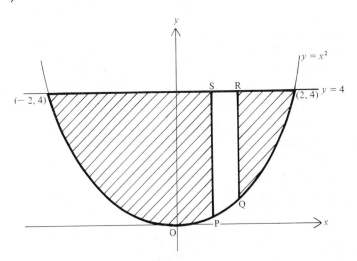

Hence the area to be found is the shaded region in the diagram. Taking PQRS as a typical element of area,

where　　　　　　　　　　　$SP = y_S - y_P$

and as　　　　　　　　　　$y_P = x^2, \quad y_S = 4$

we may write

$$SP = 4 - x^2 \quad \text{and} \quad SR = \delta x$$

So the area of　PQRS　$\simeq (4 - x^2) \, \delta x$.

Therefore the　total area　$\simeq 2 \sum_{x=0}^{2} (4 - x^2) \, \delta x$.

as the area is symmetrical about　$x = 0$,

i.e.　　　total area $= 2 \int_0^2 (4 - x^2) \, dx$

$$= 2 \left[4x - \frac{x^3}{3} \right]_0^2 = 2(8 - \tfrac{8}{3}) = 10\tfrac{2}{3} \text{ square units}$$

2) Find the area of the region of the xy plane defined by the inequalities

$$y^2 \leqslant 1 - x \quad \text{and} \quad y \leqslant x + 1$$

$y^2 = 1 - x$ is the parabola shown in the diagram.

Writing $y^2 \leqslant 1 - x$

as $\qquad x \leqslant 1 - y^2$

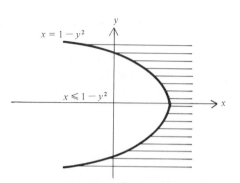

we see that this inequality is satisfied by all points on and to the left of $x = 1 - y^2$.

$y \leqslant x + 1$ defines the line $y = x + 1$ and the region below the line.

$\left.\begin{array}{l} y^2 = 1 - x \\ y = 1 + x \end{array}\right\}$ intersect where $y^2 + y - 2 = 0$,

i.e. at $(-3, -2)$ and $(0, 1)$.

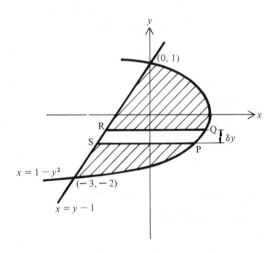

Taking a horizontal strip PQRS as a typical element of area (a vertical strip would change its structure at $x = 0$ and would have both ends on the parabola for $x > 0$) we have

$$\text{SP} = x_\text{P} - x_\text{S}$$

where $\qquad x_\text{P} = 1 - y^2 \quad$ and $\quad x_\text{S} = y - 1$

Hence $\qquad \text{SP} = (1 - y^2) - (y - 1) = 2 - y^2 - y$

Note that although it might appear from the diagram that for the range under consideration, SP should be the sum of the x co-ordinates of S and P, the length of a horizontal line is *always* the difference of x co-ordinates, e.g.

$$x \qquad AB = x_B - x_A = 3 - (-2)$$

$A(-2, 0) \quad O \qquad B(3, 0)$

Area of the element $\qquad \simeq (2 - y^2 - y) \, \delta y$

Therefore the total area $\simeq \sum_{y=-2}^{1} (2 - y^2 - y) \, \delta y$

i.e. $\qquad$ total area $= \int_{-2}^{1} (2 - y^2 - y) \, dy$

$$= \left[2y - \frac{y^3}{3} - \frac{y^2}{2} \right]_{-2}^{1}$$

$$= \left(2 - \frac{1}{3} - \frac{1}{2} \right) - \left(-4 + \frac{8}{3} - \frac{4}{2} \right) = \frac{9}{2}$$

Therefore the area is $4\frac{1}{2}$ square units.

EXERCISE 18b

1) Find the area of the region of the xy plane bounded by the curve $y = x^2 - 4$ and the line:
(a) $y = 0$ (b) $y = -\frac{7}{4}$ (c) $y = \frac{9}{4}$.

2) Find the area of the region of the xy plane bounded by the curve $y^2 = x + 1$ and the line:
(a) $x = 0$ (b) $x = 3$.

3) Find the area of the region of the xy plane defined by the following inequalities:

(a) $y \geqslant e^x$, $x \geqslant 0$, $y \leqslant e$ (b) $y \geqslant \frac{1}{x}, 0 < y \leqslant 2, 0 < x \leqslant 2$.

4) Find the area of the region of the xy plane defined by the following inequalities:
(a) $y \leqslant x(1 - x)$, $y \geqslant x - 1$
(b) $x \leqslant y(1 - y)$, $x \geqslant y - 1$.

5) Find the area of the region of the xy plane defined by the following inequalities:

$$y \geqslant (x + 1)(x - 2), \quad y \leqslant x$$

MEAN VALUES

Consider the part of the curve $y = f(x)$ for values of x in the range $a \leqslant x \leqslant b$.

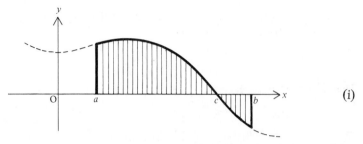

(i)

The mean value of y in this range is the average value of y for that part of the curve.

The sum of the ordinates (i.e. values of y) between $x = a$ and $x = c$ occupies the shaded area above the x axis and is positive.

This area is

$$\int_a^c y \, dx$$

Hence the sum of the ordinates between $x = a$ and $x = c$ is

$$\int_a^c y \, dx$$

Similarly, the sum of the ordinates between $x = c$ and $x = b$ is given by

$$\int_c^b y \, dx$$

(which gives the negative result we expect as the ordinates are negative in this range).

Hence the sum of the ordinates for the complete range is given by

$$\int_a^b y \, dx$$

If y_m is the mean value of y in this range,

i.e.

(ii)

the sum of the mean value ordinates occupies the area of the shaded rectangle and so is equal to

$$(b - a) y_m$$

Now the sum of the ordinates in (i) is equal to the sum of the mean value ordinates in (ii).

Therefore $\qquad (b-a)y_m = \int_a^b y\,dx$

i.e. the mean value of y, y_m, in the range $a \leqslant x \leqslant b$ is given by

$$y_m = \frac{1}{b-a}\int_a^b y\,dx$$

EXAMPLES 18c

1) Find correct to 3 d.p. the mean value of the function $\sin\theta$ for $0 \leqslant \theta \leqslant \pi$.

If $y = \sin\theta$, then for $0 \leqslant \theta \leqslant \pi$

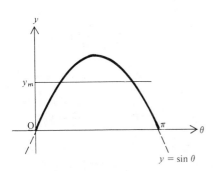

$y = \sin\theta$

$$y_m = \frac{1}{\pi-0}\int_0^\pi \sin\theta\,d\theta$$

$$= \frac{1}{\pi}\left[-\cos\theta\right]_0^\pi$$

$$= \frac{1}{\pi}[1+1] = \frac{2}{\pi}$$

$$= 0.637 \text{ correct to 3 d.p.}$$

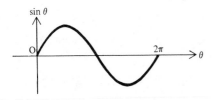

Note that for $0 \leqslant \theta \leqslant 2\pi$, the mean value of $\sin\theta$ is zero as the part of the curve for $\pi \leqslant \theta \leqslant 2\pi$ is the reflection in the x axis of the part of the curve for $0 \leqslant \theta \leqslant \pi$.

2) Find the mean value of the function $f(x) \equiv 4x(x-1)(x-2)$ for the range $-1 \leqslant x \leqslant 1$.

If $y = 4x(x-1)(x-2) = 4x^3 - 12x^2 + 8x$
then for the range $-1 \leqslant x \leqslant 1$

$$y_m = \frac{1}{1-(-1)}\int_{-1}^1 (4x^3 - 12x^2 + 8x)\,dx$$

$$= \frac{1}{2}\left[x^4 - 4x^3 + 4x^2\right]_{-1}^{1}$$

$$= -4$$

i.e. the mean value of $f(x) \equiv 4x(x-1)(x-2)$ for the range $-1 \leqslant x \leqslant 1$
is -4.

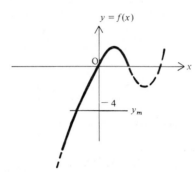

EXERCISE 18c

Find the mean value of the given function for the range of values of the variable indicated.

1) $(x-1)(x-2)$ $1 \leqslant x \leqslant 2$

2) $\dfrac{1}{t}$ $\frac{1}{2} \leqslant t \leqslant 1$

3) e^{-x} $1 \leqslant x \leqslant 5$

4) $\tan \theta$ $0 \leqslant \theta \leqslant \dfrac{\pi}{4}$

5) $\sin^2 \theta$ $0 \leqslant \theta \leqslant 2\pi$

6) $\dfrac{1}{(x-1)(x-2)}$ $-1 \leqslant x \leqslant 0$

7) $\cos^3 \theta$ $0 \leqslant \theta \leqslant \dfrac{\pi}{2}$

8) xe^x $1 \leqslant x \leqslant 2$

9) $\sqrt{1-x^2}$ $0 \leqslant x \leqslant 1$

10) $\dfrac{1}{\sqrt{(1-x^2)}}$ $0 \leqslant x \leqslant \frac{1}{2}$

VOLUMES OF REVOLUTION

If part of a curve is rotated about a straight line, the solid formed is called a solid of revolution.
Note that such a solid is always symmetrical about the axis of rotation.

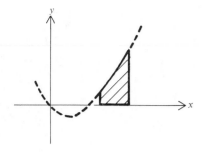

Consider the part of the curve

$$y = x(2x - 1)$$

between $x = 1$ and $x = 2$.

If the area enclosed by this section
of the curve, the x axis and the
ordinates $x = 1$, $x = 2$,   is
rotated about the x axis, a solid
of revolution is formed.

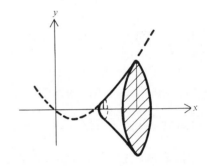

Note that any cross section perpendicular to the axis of rotation is circular.
To calculate the volume of this solid we can divide it up into sections by making
cuts perpendicular to the axis of rotation.

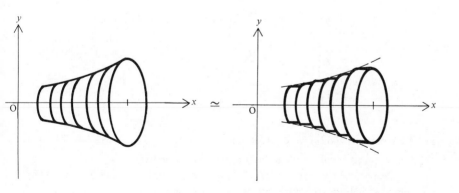

If the cuts are reasonably close, each section obtained is approximately
cylindrical. Hence the volume of the solid is approximately equal to the sum of
the volumes of these cylinders.

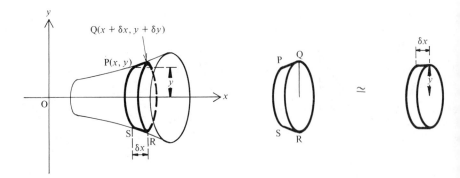

Taking PQRS as a typical element where $P(x, y)$ is on the curve $y = x(2x - 1)$ then the section PQRS is approximately a cylinder whose radius is y and whose 'height' is δx.

Hence the volume of PQRS $\simeq \pi y^2\, \delta x$

Therefore the volume, V, of the solid $\simeq \sum\limits_{x=1}^{2} \pi y^2\, \delta x$

The smaller δx is, the closer this approximation is to V

i.e.
$$V = \lim_{\delta x \to 0} \sum_{x=1}^{2} \pi y^2\, \delta x$$

$$= \int_{1}^{2} \pi\, y^2\, dx$$

$$= \pi \int_{1}^{2} x^2 (2x - 1)^2\, dx = \pi \int_{1}^{2} (4x^4 - 4x^3 + x^2)\, dx$$

$$= \pi \left[\frac{4x^5}{5} - x^4 + \frac{x^3}{3} \right]_{1}^{2}$$

$$= \frac{182\pi}{15}$$

i.e. the volume is $182\pi/15$ cubic units.

Most volumes of revolution can be found in the same way,
i.e. by taking sections perpendicular to the axis of rotation, to give elements of volume, δV, which are approximately cylindrical,

i.e.
$$\delta V \simeq \pi (\text{radius})^2 \times \text{thickness}$$

$$V = \int dV$$

EXAMPLES 18d

1) Find the volume generated when the area between $y = e^x$, the x axis, the y axis and $x = 1$ is rotated through one revolution about the x axis.

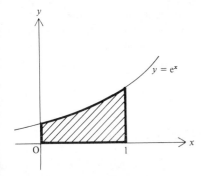

The area described is the shaded region in the diagram on the left.

When this area is rotated about the x axis the solid of revolution formed is shown in the diagram below.

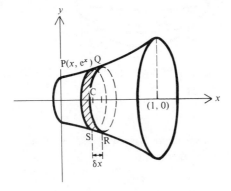

Taking the section PQRS as a typical element of volume then the volume of PQRS, (δV) is approximately that of a cylinder, radius CP and thickness δx,

i.e.
$$V \simeq \sum_{x=0}^{1} \pi(e^x)^2 \, \delta x$$

Therefore
$$V = \pi \int_0^1 e^{2x} \, dx = \left[\frac{\pi}{2} e^{2x} \right]_0^1 = \frac{\pi}{2}(e^2 - 1)$$

Hence the volume generated is $\frac{\pi}{2}(e^2 - 1)$ cubic units.

2) The area defined by the inequalities
$$y \geqslant x^2 + 1, \quad x \geqslant 0, \quad y \leqslant 2$$
is rotated completely about the y axis. Find the volume of the solid generated.

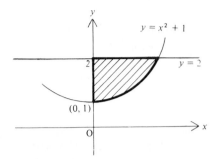

The shaded region in the diagram on the left is that defined by the given inequalities.

Rotating this area about the y axis gives the solid shown below

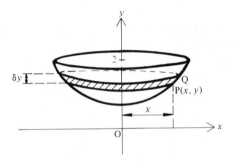

If $P(x, y)$ is any point on the curve $y = x^2 + 1$, sections through P and Q perpendicular to Oy give an element of volume δV, where

$$\delta V \simeq \pi x^2 \, \delta y$$

Therefore

$$V = \int_1^2 \pi x^2 \, dy$$

As $y = x^2 + 1$,

$$x^2 = y - 1$$

Hence

$$V = \pi \int_1^2 (y - 1) \, dy$$

$$= \pi \left[\frac{y^2}{2} - y \right]_1^2$$

$$= \frac{\pi}{2}$$

3) The area enclosed by the curve $y = 4x - x^2$ and the line $y = 3$ is rotated about the line $y = 3$. Find the volume of the solid generated.

The curve $\left.\begin{array}{l} y = 4x - x^2 \\ \text{and the line} \quad y = 3 \end{array}\right\}$ intersect at $(1, 3)$ and $(3, 3)$.

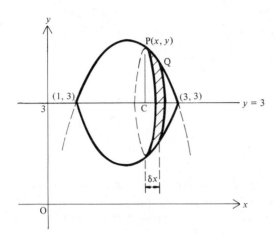

If P and Q are points on $y = 4x - x^2$, sections through P and Q, perpendicular to $y = 3$, give an element of volume δV, where $\delta V \simeq \pi(\text{PC})^2 \, \delta x$.

Now
$$\begin{aligned} \text{PC} &= (y_P) - 3 \\ &= (4x - x^2) - 3 \\ &= -x^2 + 4x - 3 \end{aligned}$$

Hence
$$V = \int_1^3 \pi(-x^2 + 4x - 3)^2 \, dx$$

$$= \pi \int_1^3 (x^4 - 8x^3 + 22x^2 - 24x + 9) \, dx$$

$$= \pi \left[\frac{x^5}{5} - 2x^4 + \frac{22}{3}x^3 - 12x^2 + 9x \right]_1^3$$

$$= \frac{16\pi}{15}$$

4) The area of the region defined by the inequalities

$$y^2 \leqslant x, \quad y \geqslant x$$

is rotated about the x axis.
Find the volume of the solid generated.

The area defined by

$$y^2 \leqslant x, \quad y \geqslant x$$

is the shaded region in the diagram.

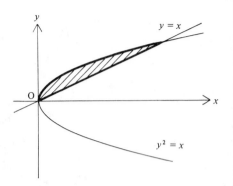

The points of intersection of $\left.\begin{array}{l} y^2 = x \\ y = x \end{array}\right\}$ are $(0,0), \ (1,1)$

When this area is rotated about Ox the solid generated is bowl shaped on the outside with a conical hole inside. The cross section this time is not a simple circle but is an annulus, i.e. the area between two concentric circles.

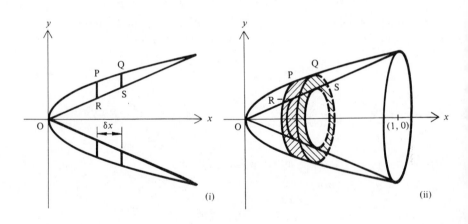

(i)

(ii)

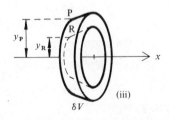

(iii)

If δV is a typical element of this volume as shown in diagram (iii),

where   P is a point on $y^2 = x$

and   R is a point on $y = x$

then the cross section is an annulus.
The area of the cross section through P and R is $\pi y_P^2 - \pi y_R^2$

$$= \pi[(\sqrt{x})^2 - x^2]$$

Therefore $\delta V \simeq \pi(x - x^2)\,\delta x$

Hence $V = \int_0^1 \pi(x - x^2)\,dx$

$$= \pi\left[\frac{x^2}{2} - \frac{x^3}{3}\right]_0^1 = \frac{\pi}{6}$$

i.e. the volume generated is $\dfrac{\pi}{6}$ cubic units.

EXERCISE 18d

In each of the following questions, find the volume generated when the area defined by the following sets of inequalities is rotated completely about the x axis.

1) $0 \leqslant y \leqslant x(4 - x)$.

2) $0 \leqslant y \leqslant e^x$, $0 \leqslant x \leqslant 3$.

3) $0 \leqslant y \leqslant \dfrac{1}{x}$, $1 \leqslant x \leqslant 2$.

4) $0 \leqslant y \leqslant x^2$, $-2 \leqslant x \leqslant 2$.

5) $y^2 \leqslant x$, $x \leqslant 2$.

In each of the following questions, the area bounded by the curve and line(s) given is rotated about the y axis to form a solid. Find the volume generated.

6) $y = x^2$, $y = 4$.

7) $y = 4 - x^2$, $y = 0$.

8) $y = x^3$, $y = 1$, $y = 2$,   for $x \geqslant 0$.

9) $y = \ln x$, $x = 0$, $y = 0$, $y = 1$.

10) Find the volume generated when the area enclosed between $y^2 = x$ and $x = 1$ is rotated about the line $x = 1$.

11) The area defined by the inequalities

$$y \geqslant x^2 - 2x + 4,\quad y \leqslant 4$$

is rotated about the line $y = 4$.   Find the volume generated.

12) The area enclosed by $y = \sin x$ and the x axis for $0 \leqslant x \leqslant \pi$ is rotated about the x axis. Find the volume generated.

13) The area enclosed by $y = x^2$ and $y^2 = x$ is rotated about the x axis. Find the volume generated.

14) The area defined by the inequalities

$$0 \leqslant y \leqslant x^2, \quad 0 \leqslant x \leqslant 2$$

is rotated about the y axis. Find the volume generated.

15) Find the volume generated when the area in the first quadrant enclosed by $y = |x^2 - 1|$, and the line $y = 1$ is rotated about the line $y = 1$.

CENTROID OF AREA

The centroid of an area is the *point* about which the area is evenly distributed,

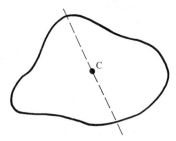

In particular the centroid lies on any line of symmetry.
Hence, the centroid of a rectangle is the point of intersection of its diagonals. The centroid of a circle is the centre. The centroid of a triangle is the point of intersection of the medians.

First Moment of Area

If the point C is the centroid of an area A, the first moment of A about an axis is defined as $A \times$ (distance of C from the axis).

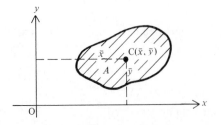

In particular, for an area A in the xy plane, whose centroid is the point $C(\bar{x}, \bar{y})$.

The first moment of A about Ox is $A\bar{y}$.
The first moment of A about Oy is $A\bar{x}$.

To Find the Coordinates of the Centroid of an Area

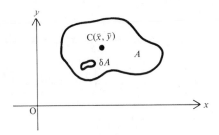

If $C(\bar{x}, \bar{y})$ is the centroid of an area A
and if δA is the area of a small element of A, then first moment of A about any axis

$$= \sum (\text{first moment of } \delta A \text{ about the same axis})$$

Hence $A\bar{x} = \sum (\text{first moment of } \delta A \text{ about } Oy)$

and $A\bar{y} = \sum (\text{first moment of } \delta A \text{ about } Ox)$

For example consider the area bounded by the curve

$$y = x(2 - x) \quad \text{and the } x \text{ axis}$$

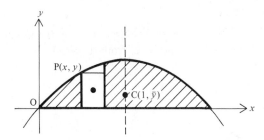

From symmetry the centroid C lies on $x = 1$, i.e. $\bar{x} = 1$.
To find $\bar{y}$ we consider the first moment about Ox.
Taking a vertical strip through P as shown we have $\delta A \simeq y\, \delta x$.
The centroid of the element is approximately a distance of $y/2$ from Ox.
Therefore the first moment of δA about $Ox \simeq (y\,\delta x)\frac{1}{2}y$.

So $$A\bar{y} \simeq \sum_{x=0}^{2} \frac{1}{2}y^2\, \delta x$$

But $$A = \int_{0}^{2} y\, dx$$

Therefore
$$\bar{y}\int_0^2 y\,dx = \int_0^2 \frac{1}{2}y^2\,dx$$

$$\bar{y}\int_0^2 x(2-x)\,dx = \frac{1}{2}\int_0^2 x^2(2-x)^2\,dx$$

$$\bar{y}\left[x^2 - \frac{x^3}{3}\right]_0^2 = \frac{1}{2}\left[\frac{4x^3}{3} - x^4 + \frac{x^5}{5}\right]_0^2$$

$$\frac{4\bar{y}}{3} = \frac{8}{15}$$

$$\bar{y} = \tfrac{2}{5}$$

i.e. the centroid of this area is the point $(1, 2/5)$ and the first moment of this area about Ox is $8/15$.

CENTROID OF VOLUME

The centroid of a volume is the *point* about which the volume is evenly distributed.

Hence *the centroid of a volume of revolution lies on the axis of rotation*.

First Moment of Volume

If the point C is the centroid of a volume V, the first moment of V about an axis is defined as $V \times$ (distance of C from the axis).

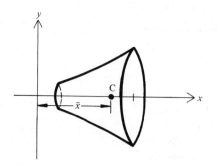

In particular for a volume V, obtained by rotating an area about the x axis, whose centroid is the point $C(\bar{x}, 0)$.

The first moment of V about Ox is zero.
The first moment of V about Oy is $V\bar{x}$.
Further, if δV is the volume of an element of V,

$$V\bar{x} = \sum (1\text{st moment of } \delta V \text{ about } Oy)$$

For example, the area between $y^2 = x$ and $x = 4$ is rotated about the x axis, giving the solid of revolution shown below

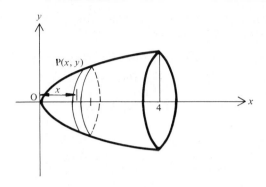

Taking a section of thickness δx as shown, gives

$$\delta V \simeq \pi y^2 \, \delta x$$

Hence $$V = \int_0^4 \pi y^2 \, dx = \pi \int_0^4 x \, dx = \pi \left[\frac{x^2}{2}\right]_0^4 = 8\pi$$

Also the distance from Oy of the centroid of the element is approximately x.
So the first moment of δV about Oy is $\simeq (\pi y^2 \, \delta x)x$.
Using $V\bar{x} = \Sigma$ (first moment of δV about Oy) we have

$$8\pi\bar{x} \simeq \sum_{x=0}^4 (\pi y^2 x \, \delta x)$$

i.e. $$8\pi\bar{x} = \int_0^4 \pi x^2 \, dx = \pi \left[\frac{x^3}{3}\right]_0^4 = \frac{64\pi}{3}$$

$\Rightarrow$ $$\bar{x} = \tfrac{8}{3}$$

i.e. the centroid of this volume is the point $(\tfrac{8}{3}, 0)$

and the first moment of this volume about Oy is $\dfrac{64\pi}{3}$.

EXERCISE 18e

1) A region of the xy plane is defined by the inequalities
$$0 \leqslant y \leqslant \sin x, \quad 0 \leqslant x \leqslant \pi$$
Find: (a) the area of the region,
    (b) the first moment of this area about the x axis,
    (c) the coordinates of the centroid of this area.

Find also:

    (d) the volume obtained when this area is rotated completely about the x axis,

    (e) the first moment of this volume about the y axis,

    (f) the centroid of this volume.

2) Repeat Question 1 for the region of the xy plane defined by the inequalities

$$0 \geqslant y \geqslant x(x-1)$$

3) A region of the xy plane is bounded by the curve $y = x^2$ and the line $y = 4$.

Find: (a) the area of this region,

    (b) the first moment of this area about the x axis,

    (c) the y coordinate of the centroid of the area.

Find also:

    (d) the volume obtained when this area is rotated about the y axis,

    (e) the first moment of the volume about the x axis,

    (f) the centroid of this volume.

4) Repeat Question 3 for the area bounded by $y = \ln x$, $x = 0$, $y = 0$, $y = 1$.

5) Find the x coordinate of the centroid of the area bounded by

$$y = e^x, \ y = 0, \ x = 0, \ x = 2.$$

APPROXIMATE METHODS FOR EVALUATING A DEFINITE INTEGRAL

To determine the indefinite integral $\int f(x) \, dx$ we must find a function whose differential is $f(x)$, and this is not always possible.

But the definite integral $\int_a^b f(x) \, dx$ is a number which represents the area between $y = f(x)$, the x axis and the ordinates $x = a$ and $x = b$.

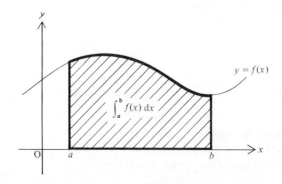

So, even if $\int f(x)\,dx$ cannot be found, an approximate value for $\int_a^b f(x)\,dx$

can be found by evaluating the appropriate area using another method.
We will now look at two methods for finding the approximate value of an area
bounded partly by a curve and the x axis.

1. The Trapezium Rule

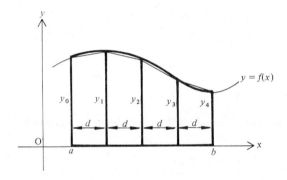

If the area represented by $\int_a^b f(x)\,dx$ is divided into strips, each of width d,

as shown above, then each such strip is approximately a trapezium.
The area of a trapezium

$$= \text{(half sum of } \parallel \text{ sides)} \times \text{(distance between them).}$$

Using the sum of the areas of these strips as an approximation for the actual
value of the area we have

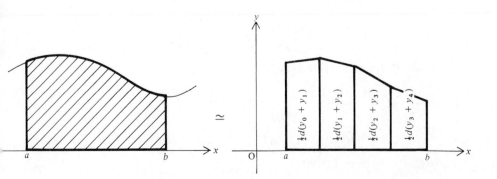

i.e. $\int_a^b f(x)\,dx \simeq \frac{1}{2}d(y_0 + y_1) + \frac{1}{2}d(y_1 + y_2) + \frac{1}{2}d(y_2 + y_3) + \frac{1}{2}d(y_3 + y_4)$

$$\simeq \frac{1}{2}d\left[y_0 + 2y_1 + 2y_2 + 2y_3 + y_4 \right]$$

This method is known as the *trapezium rule* and here we used it with five ordinates (i.e. y_0 to y_4).

Note that the ordinates must be evenly spaced (i.e. the widths of all strips must be the same).

Generalizing, used with n ordinates the trapezium rule is

$$\int_a^b f(x)\,dx \simeq \frac{1}{2}d\left[y_0 + 2y_1 + \ldots + 2y_{n-2} + y_{n-1} \right]$$

EXAMPLES 18f

1) Use the trapezium rule, with five ordinates, to evaluate $\int_0^{0.8} e^{x^2}\,dx$.

For five ordinates, $(y_0, y_1, \ldots, y_4)$ evenly spaced in the range $0 \leqslant x \leqslant 0.8$, we need to divide this range into four equal parts.

i.e.

Thus $\qquad\qquad d = 0.2$

and if $\qquad\qquad y = e^{x^2}$

$$y_0 = 1$$

$$y_1 = e^{0.04} = 1.0408$$

$$y_2 = e^{0.16} = 1.1735$$

$$y_3 = e^{0.36} = 1.4333$$

$$y_4 = e^{0.64} = 1.8965$$

Hence the trapezium rule gives

$$\int_0^{0.8} e^{x^2}\,dx \simeq \frac{1}{2}d\left[y_0 + 2y_1 + 2y_2 + 2y_3 + y_4 \right]$$

$$\simeq \frac{1}{2}(0.2)[1 + 2(1.0408) + 2(1.1735) + 2(1.4333) + 1.8965]$$

$$\simeq 1.0192$$

2. Simpson's Rule

Suppose that the area represented by $\int_a^b f(x)\,dx$ is divided by the ordinates y_0, y_1, y_2 into two strips each of width d as shown below. A particular parabola can be found passing through the three points with the same coordinates.

Simpson's rule uses the area under that parabola as an approximation for the value of the area under the curve $y = f(x)$.

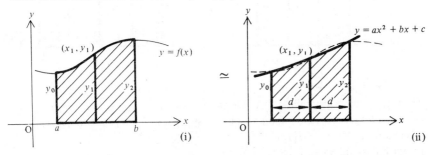

(i)   (ii)

i.e.
$$\int_a^b f(x)\,dx \simeq \int_{x_1-d}^{x_1+d} (ax^2 + bx + c)\,dx \qquad [1]$$

If $y = ax^2 + bx + c$ is the parabola through the ordinates as shown, then $(x_1 - d, y_0), (x_1, y_1), (x_1 + d, y_2)$ are on this parabola,

i.e.
$$y_0 = a(x_1 - d)^2 + b(x_1 - d) + c \qquad [2]$$
$$y_1 = ax_1^2 + bx_1 + c \qquad [3]$$
$$y_2 = a(x_1 + d)^2 + b(x_1 + d) + c \qquad [4]$$

Now the area in diagram (ii) is

$$\int_{x_1-d}^{x_1+d} (ax^2 + bx + c)\,dx$$

$$= \frac{a}{3}\{(x_1 + d)^3 - (x_1 - d)^3\} + \frac{b}{2}\{(x_1 + d)^2 - (x_1 - d)^2\}$$
$$+ c\{(x_1 + d) - (x_1 - d)\}$$

which simplifies to

$$\frac{d}{3}\left[2a\{(x_1 + d)^2 + (x_1 + d)(x_1 - d) + (x_1 - d)^2\} + 3b(2x_1) + 6c\right]$$

Then using [2], [3] and [4] we find that

$$\int_{x_1-d}^{x_1+d} (ax^2 + bx + c)\,dx = \frac{d}{3}(y_0 + 4y_1 + y_2)$$

From [1] $$\int_a^b f(x)\, dx \simeq \tfrac{1}{3}d[y_0 + 4y_1 + y_2]$$

This argument can be extended to another two strips, also of width d.

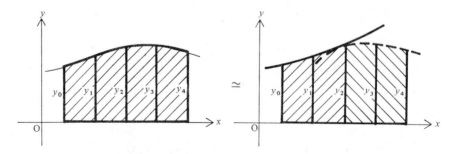

The area under the parabola through y_0, y_1, y_2 is $\tfrac{1}{3}d[y_0 + 4y_1 + y_2]$
and the area under the parabola through y_2, y_3, y_4 is $\tfrac{1}{3}d[y_2 + 4y_3 + y_4]$
giving a total area of $\tfrac{1}{3}d[y_0 + 4y_1 + 2y_2 + 4y_3 + y_4]$
i.e. with five ordinates, Simpson's Rule is

$$\int_a^b f(x)\, dx \simeq \tfrac{1}{3}d[y_0 + 4y_1 + 2y_2 + 4y_3 + y_4]$$

This argument can be extended to cover any even number of strips, i.e. any *odd* number of ordinates.
Hence Simpson's Rule, with $(2n + 1)$ ordinates, is

$$\int_a^b f(x)\, dx \simeq \tfrac{1}{3}d[y_0 + 4y_1 + 2y_2 + 4y_3 + 2y_4 + \ldots + 4y_{2n-1} + y_{2n}]$$

Note that the use of Simpson's Rule requires an *odd* number of ordinates. For ease of computation, the ordinates used can be arranged in the form

$$\frac{d}{3}\left[(1\text{st} + \text{last}) + 4(2\text{nd} + 4\text{th} + \ldots) + 2(3\text{rd} + 5\text{th} + \ldots)\right]$$

or $$\frac{d}{3}\left[y_0 + y_{2n} + 4\sum_{r=1}^{2n} y_{2r-1} + 2\sum_{r=1}^{2n-2} y_{2r}\right]$$

EXAMPLES 18g (continued)

2) Use Simpson's Rule with five ordinates to find an approximate value for

$$\int_0^\pi \sqrt{\sin\theta}\, d\theta$$

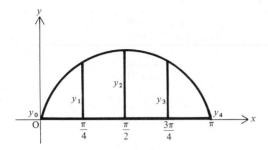

Taking five ordinates from $\theta = 0$ to $\theta = \pi$ gives four strips each of width $\pi/4$

$$y = \sqrt{\sin \theta}$$

$\Rightarrow$

$$y_0 = \sqrt{\sin 0} \quad = 0$$

$$y_1 = \sqrt{\sin \pi/4} \quad = 2^{-\frac{1}{4}} = 0.8409$$

$$y_2 = \sqrt{\sin \pi/2} \quad = 1$$

$$y_3 = \sqrt{\sin (3\pi/4)} = 2^{-\frac{1}{4}} = 0.8409$$

$$y_4 = \sqrt{\sin \pi} \quad = 0$$

Hence, using Simpson's rule,

$$\int_0^\pi \sqrt{\sin \theta} \, d\theta \simeq \tfrac{1}{3}d[y_0 + y_4 + 4(y_1 + y_3) + 2y_2]$$

$$\simeq \left(\frac{1}{3}\right)\left(\frac{\pi}{4}\right)\left[0 + 8(0.8409) + 2\right]$$

$$\simeq 2.2848$$

3) Estimate, to 4 decimal places, $\int_0^1 \frac{1}{1 + x^2} \, dx$ using five ordinates and applying

(a) the trapezium rule   (b) Simpson's rule.

Using the substitution $x = \tan \theta$,   or otherwise, evaluate $\int_0^1 \frac{1}{1 + x^2} \, dx$

and hence determine the accuracy of your estimated values. (Take π as 3.1416.)

Taking five ordinates from $x = 0$ to $x = 1$ gives four strips, each of width 0.25.

So

$$y = \frac{1}{1 + x^2}$$

$\Rightarrow$

$$y_0 = \frac{1}{1 + (0)^2} \quad = 1$$

$$y_1 = \frac{1}{1 + (0.25)^2} = 0.9412$$

$$y_2 = \frac{1}{1 + (0.5)^2} = 0.8$$

$$y_3 = \frac{1}{1 + (0.75)^2} = 0.64$$

$$y_4 = \frac{1}{1 + (1)^2} = 0.5$$

(a) Using the trapezium rule

$$\int_0^1 \frac{1}{1+x^2}\, dx \simeq \frac{0.25}{2}\left[1 + 0.5 + 2(0.9412 + 0.8 + 0.64)\right]$$

$$\simeq 0.7828$$

(b) Using Simpson's rule

$$\int_0^1 \frac{1}{1+x^2}\, dx \simeq \frac{0.25}{3}\left[1 + 0.5 + 4(0.9412 + 0.64) + 2(0.8)\right]$$

$$\simeq 0.7854$$

If $x \equiv \tan\theta, \ldots dx \equiv \ldots \sec^2\theta\, d\theta$

then $\int_0^1 \dfrac{1}{1+x^2}\, dx \simeq \int_0^{\frac{\pi}{4}} \dfrac{1}{\sec^2\theta}\sec^2\theta\, d\theta$

$$= \int_0^{\frac{\pi}{4}} d\theta = \left[\theta\right]_0^{\frac{\pi}{4}} = \frac{\pi}{4}$$

Taking $\pi = 3.1416$, $\dfrac{\pi}{4} = 0.7854$.

Hence, the trapezium rule gave an estimate correct to 2 d.p. and Simpson's rule gave an estimate correct to 4 d.p. (at least).

Approximate Integration using a Series Expansion

Consider $\int_0^{0.1} \dfrac{1}{\sqrt{(1+x^2)}}\, dx$.

Since the range of values of x for which the integral is required is $0 \leqslant x \leqslant 0.1$ x is small for the whole of the range.

Using the binomial theorem,

$$\frac{1}{\sqrt{(1+x^2)}} \equiv (1+x^2)^{-\frac{1}{2}} = 1 - \tfrac{1}{2}x^2 + \tfrac{3}{8}x^4 \ldots$$

Because x is small, terms in x^5 and higher powers do not affect the fourth decimal place and can be neglected.

Hence
$$\int_0^{0.1} \frac{1}{\sqrt{(1+x^2)}}\, dx \simeq \int_0^{0.1} (1 - \tfrac{1}{2}x^2 + \tfrac{3}{8}x^4)\, dx$$

$$\simeq \left[x - \tfrac{1}{6}x^3 + \tfrac{3}{40}x^5 \right]_0^{0.1}$$

$$\simeq 0.0998 \text{ to 4 d.p.}$$

Note that we are assuming that the integral of a function is equal to the sum of the integrals of terms of a series expansion of that function.

EXERCISE 18f

Estimate the values of the following definite integrals, taking the number of ordinates given in each case, using (a) the trapezium rule, (b) Simpson's rule.

1) $\displaystyle\int_0^{\frac{\pi}{2}} \frac{1}{1 + \cos x}\, dx$ 3 ordinates   2) $\displaystyle\int_0^{0.4} \sqrt{(1-x^2)}\, dx$ 5 ordinates

3) $\displaystyle\int_0^{\pi} (\sin^4 \theta)\, d\theta$,   5 ordinates   4) $\displaystyle\int_0^{0.6} xe^x\, dx$,   7 ordinates

5) $\displaystyle\int_1^{e} \ln x\, dx$,   5 ordinates   6) $\displaystyle\int_0^{1} \sqrt{(1+x^2)}\, dx$,   11 ordinates

Use the first three terms of the appropriate series expansion as an approximation for each of the functions to be integrated. Hence estimate the values of the following definite integrals.

7) $\displaystyle\int_0^{0.1} \ln(1+x)\, dx$ 8) $\displaystyle\int_0^{0.02} xe^x\, dx$

9) $\displaystyle\int_{0.1}^{0.2} \sqrt[3]{(1-x^2)}\, dx$ 10) $\displaystyle\int_0^{\frac{\pi}{4}} \sqrt{(1+\cos^2\theta)}\, d\theta$

(*Hint*: use a binomial expansion to give a series of powers of $\cos \theta$.)

SMALL INCREMENTS

If $P(x, y)$ is a general point on the curve $y = f(x)$ and $Q(x + \delta x, y + \delta y)$ is a point on the curve close to P then δx is a small increase in x and $\delta y = f(x + \delta x) - f(x)$ is the corresponding increase in y.

Now

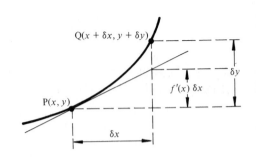

$$\lim_{\delta x \to 0} \left(\frac{\delta y}{\delta x} \right) = \frac{dy}{dx} = f'(x)$$

so, for small values of δx,

$$\frac{\delta y}{\delta x} \simeq f'(x)$$

or
$$\boxed{\delta y \simeq f'(x)\, \delta x}$$

i.e. for a small increment in x, the corresponding increment in y is approximately $f'(x)\, \delta x$. The smaller δx is, the better is the approximation. This approximation may be used to estimate the value of a function which is close to a known value. For example, the value of $\ln 1$ is known so $\ln (1.1)$ can be estimated.

$$y = \ln x \implies \frac{dy}{dx} = f'(x) = \frac{1}{x}$$

Hence
$$\delta y \simeq \frac{1}{x} \delta x$$

When $x = 1$ and $\delta x = 0.1$, $\delta y \simeq \frac{1}{1}(0.1)$

But $\delta y = (y + \delta y) - y = \ln (1.1) - \ln 1$

Hence
$$\ln (1.1) - \ln 1 \simeq 0.1$$
$$\ln 1.1 \simeq \ln 1 + 0.1$$
$$\simeq 0.1$$

Note that the approximation $\delta y \simeq f'(x)\, \delta x$ is equivalent to the linear approximation obtained from a Taylor expansion as follows.

If $P(a, f(a))$ is a point on
$y = f(x)$
and $(x, f(x))$ is a point close to P
then from the diagram

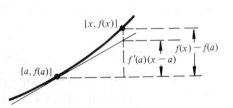

$$f(x) - f(a) \simeq f'(a)(x - a)$$

or
$$f(x) \simeq f(a) + f'(a)(x - a)$$

Using this result to estimate $\ln 1.1$,
we let $f(x) \equiv \ln x$, $a = 1$, $x = 1.1$

$\Rightarrow$ $$\ln 1.1 \simeq \ln 1 + \frac{1}{1}(0.1)$$

i.e. $$\ln 1.1 \simeq 0.1$$

EXAMPLES 18g

1) Using $y = \sqrt{x}$, estimate $\sqrt{101}$.

$$y = \sqrt{x} \Rightarrow \frac{dy}{dx} = \frac{1}{2}x^{-\frac{1}{2}}$$

Using $\delta y \simeq \dfrac{dy}{dx}\delta x$ we have $\delta y \simeq \dfrac{1}{2\sqrt{x}}\delta x$.

when $x = 100$ and $\delta x = 1$, $\delta y = \sqrt{101} - \sqrt{100}$

Hence $$\sqrt{101} - \sqrt{100} \simeq \left(\frac{1}{2\sqrt{100}}\right)(1)$$

$$\sqrt{101} \simeq 0.05 + 10$$

$$\simeq 10.05$$

2) Use $f(\theta) \equiv \sin\theta$ to find an approximate value for
(a) $\sin 31°$ (b) $\sin 29°$. (Take $1° = 0.0175^c$.)

If $f(\theta) \equiv \sin\theta$, $f'(\theta) \equiv \cos\theta$.

Using $\delta y \simeq \dfrac{dy}{dx}\delta x$ in the form $\delta(f(\theta)) \simeq f'(\theta)\,\delta\theta$,

we have $$\delta(\sin\theta) \simeq (\cos\theta)\,\delta\theta$$

(a) When $\theta = \dfrac{\pi}{6}$ and $\delta\theta = 0.0175$, $\delta(\sin\theta) = \sin\left(\dfrac{\pi}{6} + 0.0175\right) - \sin\dfrac{\pi}{6}$.

Hence $$\sin 31° - \sin 30° \simeq \left(\cos\frac{\pi}{6}\right)(0.0175)$$

$$\sin 31° \simeq \frac{\sqrt{3}}{2}(0.0175) + \frac{1}{2}$$

$$\simeq 0.5152$$

(b) Again $\theta = \dfrac{\pi}{6}$ but $\delta\theta = -0.0175$ (as the angle *decreases* from $30°$ to $29°$)

so $$\delta(\sin\theta) = \sin\left(\frac{\pi}{6} - 0.0175\right) - \sin\frac{\pi}{6}$$

Hence $\sin\left(\dfrac{\pi}{6} - 0.0175\right) - \sin\dfrac{\pi}{6} \simeq \left(\cos\dfrac{\pi}{6}\right)(-0.0175)$

$$\sin\left(\frac{\pi}{6} - 0.0175\right) \simeq \frac{\sqrt{3}}{2}(-0.0175) + \frac{1}{2}$$

$$\simeq 0.4848$$

i.e. $\sin 29° \simeq 0.4848$

Note that whenever the derivatives of trig functions are used in calculations, the angle must be measured in radians.

RATE OF CHANGE PROBLEMS

The identity $\dfrac{dy}{dx} \equiv \dfrac{dy}{dt} \times \dfrac{dt}{dx}$ is useful when solving certain rate of change problems.

EXAMPLES 18g (continued)

3) A spherical balloon is blown up so that its volume increases at a constant rate of $2\,\text{cm}^3/\text{s}$.
Find the rate of increase of the radius when the volume of the balloon is $50\,\text{cm}^3$.

If, at time t, the volume of the balloon is V and the radius is r then

$$V = \tfrac{4}{3}\pi r^3 \;\Rightarrow\; \frac{dV}{dr} = 4\pi r^2 \qquad\qquad [1]$$

We wish to find $\dfrac{dr}{dt}$. We are given $\dfrac{dV}{dt}(=2)$ and we know that $\dfrac{dV}{dr} = 4\pi r^2$.

Using $\dfrac{dr}{dt} = \dfrac{dr}{dV} \times \dfrac{dV}{dt} = \dfrac{dV}{dt} \Big/ \dfrac{dV}{dr}$

gives $\dfrac{dr}{dt} = \dfrac{2}{4\pi r^2} = \dfrac{1}{2\pi r^2}$

From [1], when $V = 50$, $r = 2.29$

$\Rightarrow$ $\dfrac{dr}{dt} = \dfrac{1}{2\pi(2.29)^2} = 0.03$

So the radius is increasing at $0.03\,\text{cm/s}$ when the volume is $50\,\text{cm}^3$.

EXERCISE 18g

1) Using $y = \sqrt[3]{x}$, find approximate values for:
(a) $\sqrt[3]{9}$ (b) $\sqrt[3]{28}$ (c) $\sqrt[3]{126}$.

2) Using $f(\theta) \equiv \cos \theta$, find approximate values for:
(a) $\cos 29°$ (b) $\cos 61°$ (c) $\cos 44°$.
(Take $1° = 0.0175^c$.)

3) If $f(x) \equiv x \ln (1 + x)$, find an approximation for the increase in $f(x)$ when x increases by δx.
Hence estimate the value of $\ln (2.1)$, given that $\ln 2 = 0.6931$.

4) If $y = \tan x$, find an approximation for δy when x is increased by δx.
Use your approximation to estimate the value of $\tan \dfrac{9\pi}{32}$.

5) Using $f(x) \equiv x^{\frac{1}{5}}$, find an approximate value for $\sqrt[5]{33}$.

6) Ink is dropped on to blotting paper forming a circular stain which increases in area at the rate of $5 \text{ cm}^2/\text{s}$. Find the rate of change of the radius when the area is 30 cm^2.

7) A container in the shape of a right circular cone of height 20 cm and radius 5 cm is held vertex downward and filled with water which then drips out from the vertex at the rate of $5 \text{ cm}^3/\text{s}$. Find the rate of change of the height of water in the cone when it is half empty (measured by volume).

8) The surface area of a cube is increasing at the rate of $10 \text{ cm}^2/\text{s}$. Find the rate of increase of the volume of the cube when the edge is of length 12 cm.

NUMERICAL TRIGONOMETRY

Triangles are involved in many practical measurements (e.g. in surveying). A triangle has three sides and three angles and is specified by the values of any three members of the set {3 sides, 2 angles}. (The sum of the three angles of a triangle is $180°$, so if any two angles are known the third angle can be calculated directly and is therefore not an independent item.) From any set of data sufficient to define a triangle, the remaining sides and angles can be calculated. This is called *solving* the triangle and requires the use of one of the various formulae that relate the sides and angles of a triangle.
The two relationships that are used most frequently are the sine rule and the cosine rule.

The Sine Rule

In a triangle ABC in which the sides BC, CA and AB are denoted by a, b and c as shown, and A, B, C are used to denote the angles at the vertices A, B, C respectively,

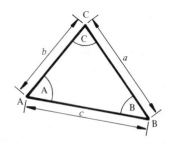

$$\frac{a}{\sin A} = \frac{b}{\sin B} = \frac{c}{\sin C}$$

Proof

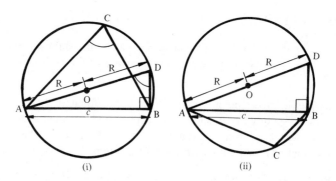

(i)   (ii)

Taking O and R as the centre and radius of the circle through A, B and C, draw the diameter AD and join DB.

Then $\angle ABD = 90°$ (angle in a semicircle).

Thus $c = 2R \sin D$

But, in diagram (i)   C = D   (angles in same segment)

and in diagram (ii) $C = 180° - D$ (opposite angles of a cyclic quadrilateral).

Thus, in both diagrams, $\sin C = \sin D$.

Hence $c = 2R \sin C \Rightarrow \dfrac{c}{\sin C} = 2R$

Similarly $\dfrac{b}{\sin B} = 2R$ and $\dfrac{a}{\sin A} = 2R$

$\Rightarrow$ $\dfrac{a}{\sin A} = \dfrac{b}{\sin B} = \dfrac{c}{\sin C} = 2R$

Any pair of these three equal ratios comprises an equation containing two sides and two angles.

Thus the sine rule can be used to solve a triangle in which we know either two sides and one angle

or   two angles and one side

provided that one given side is opposite to a given angle.

For instance, if in a triangle ABC it is known that $A = 73°$, $B = 49°$ and $a = 12.2$ cm   then, as a, A and B are known, b can be calculated using

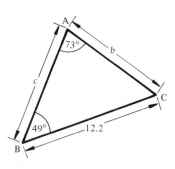

$$\frac{a}{\sin A} = \frac{b}{\sin B} \Rightarrow b = \frac{12.2 \sin 49°}{\sin 73°}$$

$$= 9.63 \text{ cm}$$

Also, since $C = 180° - 73° - 49° = 58°$,

c can be calculated using

$$\frac{a}{\sin A} = \frac{c}{\sin C} \Rightarrow c = \frac{12.2 \sin 58°}{\sin 73°}$$

$$= 10.82 \text{ cm}$$

In this example the given data defines one and only one triangle.

Sometimes, when two sides and one angle are specified, two different triangles can be found from the given data.

The Ambiguous Case

Suppose that we have to solve the triangle in which $A = 24°$, $c = 2.6$ cm and $a = 1.1$ cm.

Knowing a, A and c we can find C using

$$\frac{a}{\sin A} = \frac{c}{\sin C} \Rightarrow \sin C = \frac{2.6}{1.1} \sin 24°$$

i.e. $\sin C = 0.9614$

Now in a triangle, arcsin 0.9614 can be either $74°$

or $106°$

But we must check whether these angles are possible values for C, by considering:

(a) if $C = 74°$ and $A = 24°$ (given)

$74° + 24° = 98°$ which is less than $180°$.

So $74°$ is a possible value for C corresponding to $B = 82°$.

(b) if $C = 106°$ and $A = 24°$

$106° + 24° = 130°$ which is less than $180°$.

So $106°$ is *also* a possible value for C corresponding to $B = 50°$.

Thus we see that there are two possible triangles including the given data.

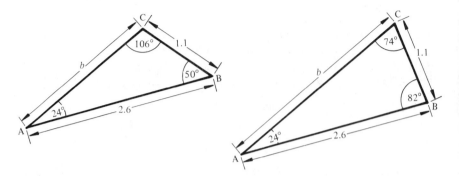

This is known as the *ambiguous case* and it can easily be understood if an attempt is made to *construct* the triangle from the given specification.

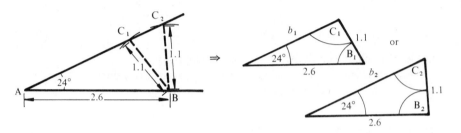

The constructions show that there are two possible positions for C, each corresponding to a different pair of values for B and b.

In each case the solution of the triangle can be completed using $\dfrac{a}{\sin A} = \dfrac{b}{\sin B}$

Note. It must not be thought that there are *always* two possible triangles when one angle and two sides are given. The following example shows that this is not necessarily so.

If $A = 37°$, $a = 4.59 \text{ cm}$ and $c = 2.1 \text{ cm}$ show that there is only one possible triangle ABC and find the remaining angles.

Using $$\frac{a}{\sin A} = \frac{c}{\sin C} \Rightarrow \sin C = \frac{2.1}{4.59}\sin 37°$$

i.e. $$\sin C = 0.2753$$

Now, in a triangle, arcsin $0.2753 = 16°$ or $164°$.

If $C = 16°$ and $A = 37°$, $A + C = 53°$ $(< 180°)$.
So $16°$ is a possible value for C corresponding to $B = 127°$.

If $C = 164°$ and $A = 37°$, $A + C = 201°$ $(> 180°)$.
So $164°$ is *not* a possible value for C.

Hence there is only one triangle defined by the given data and its other angles are $C = 16°$ and $B = 127°$

EXERCISE 18h

In Questions 1–7, the data refers to triangle ABC, using standard notation.

1) $B = 35°$, $a = 2.7$ cm, $b = 5.1$ cm; find A.

2) $C = 52°$, $B = 49°$, $c = 8.62$ cm; find b.

3) $b = 3.8$ cm, $A = 25°$, $a = 1.8$ cm; find C.

4) $A = 87°$, $B = 63°$, $c = 3.2$ cm; find a.

Solve the triangles specified in Questions 5–7.

5) $a = 9.12$ cm, $b = 4.87$ cm, $A = 63°$.

6) $c = 5.73$ cm, $B = 19.3°$, $b = 2.16$ cm.

7) $B = 83.2°$, $C = 43.7°$, $a = 19.86$ cm.

8) In a triangle PQR, angle $PQR = \theta$ and angle $QPR = 2\theta$. Prove that

$$\cos \theta = \frac{p}{2q}.$$

9) Prove the sine rule for a triangle ABC by drawing AD perpendicular to BC, meeting BC at D. Consider two cases (a) B acute, (b) B obtuse.

10) From a point P on the same level as the base of a tower, the angle of elevation of the top of the tower is $24.8°$. From a point Q, 5 m vertically above P, the angle of elevation of the top of the tower is $18.7°$. Find the height of the tower.

The Cosine Rule

We have seen that the sine rule can be applied only when we know an angle opposite to a given side. It cannot be used, for instance, when two sides and the included angle are given. Such cases can usually be solved using the cosine rule

$$a^2 = b^2 + c^2 - 2bc \cos A$$

A reminder of the proof of this formula is given below.

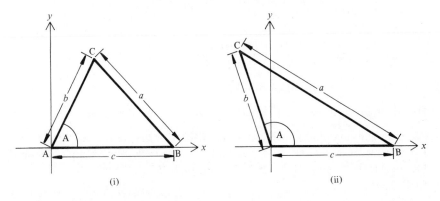

(i)   (ii)

Placing the triangle ABC on Cartesian axes as shown, we see that the coordinates of C are:

$(b \cos A, b \sin A)$ in diagram (i)

$(-b \cos \{\pi - A\}, b \sin A) = (b \cos A, b \sin A)$ in diagram (ii)

and the coordinates of B are $(c, 0)$ in both cases.

Thus
$$a^2 = (b \sin A)^2 + (b \cos A - c)^2$$
$$= b^2 \sin^2 A + b^2 \cos^2 A - 2bc \cos A + c^2$$

i.e.
$$a^2 = b^2 + c^2 - 2bc \cos A$$

Similarly it can be shown that

$$b^2 = c^2 + a^2 - 2ca \cos B$$

and
$$c^2 = a^2 + b^2 - 2ab \cos C$$

The calculation involved in using this formula is less straightforward than that required in using the sine formula. Consequently the cosine rule is used only when the sine rule is inapplicable. For instance, suppose that we have a triangle ABC in which $a = 17.5$ cm, $b = 8.4$ cm and $c = 11.9$ cm and we are asked to find the largest angle.

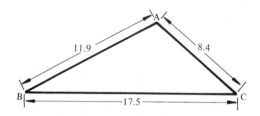

The longest side is opposite to the largest angle so we must find A.
But as a, b, c are given, we must use the cosine rule (rearranged so that A can be found conveniently), i.e.

$$\cos A = \frac{b^2 + c^2 - a^2}{2bc}$$

$$= \frac{(8.4)^2 + (11.9)^2 - (17.5)^2}{2(8.4)(11.9)}$$

$$= -0.4706$$

Hence $A = 118.1°$ (obtuse because $\cos A$ is negative)

When the solution of a triangle requires initial use of the cosine rule, the remaining sides or angles can be found using the sine rule. For example, to solve the triangle PQR given that $p = 12.1$ cm, $q = 7.3$ cm and $R = 37.5°$ we first use

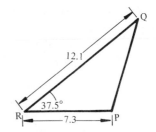

$$r^2 = p^2 + q^2 - 2pq \cos R$$

$$\Rightarrow \quad r = 7.72 \text{ cm}$$

Now the sine rule can be used since
we know r, R, p and q.

Thus $\qquad \dfrac{r}{\sin R} = \dfrac{q}{\sin Q} \Rightarrow \sin Q = \dfrac{7.3 \sin 37.5°}{7.72} = 0.5756$

Hence $\quad Q = 35.1°$.   (Q cannot be obtuse as q is not the longest side of the
triangle)
Finally P can be found using $180° - Q - R$.

EXERCISE 18i

Solve the following triangles.

1) $a = 4$, $b = 7$, $c = 5$.

2) $A = 79°$, $b = 27$ cm, $c = 19.8$ cm.

3) $a = 15.73$ m, $B = 121°$, $c = 23.15$ m.

4) $a = 12.84$ m, $b = 6.58$ m, $c = 8.13$ m.

5) Using the cosine formula $\quad \cos A = \dfrac{b^2 + c^2 - a^2}{2bc}$:

(a) show that A is acute if $a^2 < b^2 + c^2$,

(b) show that A is obtuse if $a^2 > b^2 + c^2$,

(c) verify Pythagoras' Theorem by taking $A = 90°$.
Illustrate each of these cases by a diagram.

6) Find the angles of a triangle whose sides are in the ratio $2:3:4$.

The Cotangent Formula

If D divides the side AB of triangle ABC in the ratio $m:n$ then, with the
notation shown in the diagram below,

$$(m + n) \cot \theta = m \cot \alpha - n \cot \beta$$

Proof

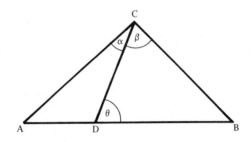

In triangle ACD,

$$\frac{CD}{\sin A} = \frac{AD}{\sin \alpha}$$

$\Rightarrow$

$$CD = \frac{AD \sin (\theta - \alpha)}{\sin \alpha}$$

In triangle BCD,

$$\frac{CD}{\sin B} = \frac{BD}{\sin \beta}$$

$\Rightarrow$

$$CD = \frac{BD \sin (\theta + \beta)}{\sin \beta}$$

Hence

$$\frac{AD \sin (\theta - \alpha)}{\sin \alpha} = \frac{BD \sin (\theta + \beta)}{\sin \beta}$$

But

$$AD:DB = m:n$$

Hence

$$\frac{m \sin (\theta - \alpha)}{\sin \alpha} = \frac{n \sin (\theta + \beta)}{\sin \beta}$$

$\Rightarrow$

$$m\left\{\frac{\sin \theta \cos \alpha - \cos \theta \sin \alpha}{\sin \alpha}\right\} = n\left\{\frac{\sin \theta \cos \beta + \cos \theta \sin \beta}{\sin \beta}\right\}$$

$$m\{\cot \alpha - \cot \theta\} = n\{\cot \beta + \cot \theta\}$$

i.e.

$$(m + n) \cot \theta = m \cot \alpha - n \cot \beta$$

Note. If D is the midpoint of AB we have

$$2 \cot \theta = \cot \alpha - \cot \beta$$

The cotangent formula, particularly in this special case, is valuable in the solution of certain equilibrium problems in mechanics.

Further Properties of Triangles

Although the sine and cosine formulae are the two formulae most frequently used there are a number of other relationships between the sides and angles of a triangle.

A selection of some of the most useful of these alternative formulae is given below.

1) $a \cos B = c - b \cos A$

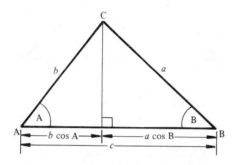

2) $\dfrac{a - b}{a + b} = \tan \dfrac{A - B}{2} \tan \dfrac{C}{2}$.

The proof of this formula is derived by using the sine rule in the form

$$\frac{a}{\sin A} = \frac{b}{\sin B} = k, \text{ say} \quad \Rightarrow \quad \begin{cases} a = k \sin A \\ b = k \sin B \end{cases}$$

Hence

$$\frac{a - b}{a + b} = \frac{k(\sin A - \sin B)}{k(\sin A + \sin B)}$$

$$= \frac{2 \cos \dfrac{A + B}{2} \sin \dfrac{A - B}{2}}{2 \sin \dfrac{A + B}{2} \cos \dfrac{A - B}{2}}$$

$$= \tan \frac{A - B}{2} \cot \frac{A + B}{2}$$

But $A + B + C = \pi \Rightarrow \dfrac{A + B}{2} = \dfrac{\pi}{2} - \dfrac{C}{2}$

i.e.

$$\cot \frac{A + B}{2} = \tan \frac{C}{2}$$

So

$$\frac{a - b}{a + b} = \tan \frac{A - B}{2} \tan \frac{C}{2}$$

3) In triangle ABC the bisector of the angle A divides BC in the ratio $c:b$.

Proof

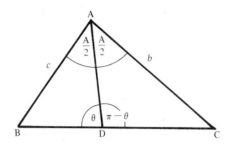

In triangle ABD,

$$\frac{BD}{\sin \frac{1}{2}A} = \frac{c}{\sin \theta}$$

and in triangle ACD

$$\frac{DC}{\sin \frac{1}{2}A} = \frac{b}{\sin (\pi - \theta)} = \frac{b}{\sin \theta}$$

Hence

$$\frac{\sin \frac{1}{2}A}{\sin \theta} = \frac{BD}{c} = \frac{DC}{b}$$

$\Rightarrow$ $\qquad$ $BD:DC = c:b$

The reader should also be familiar with the following geometric properties of a triangle.

The *perpendicular bisectors of the sides* of a triangle ABC are concurrent at a point called the *circumcentre*, which is the centre of the circle through A, B and C (the circumcircle).

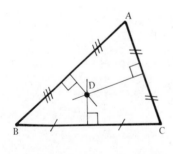

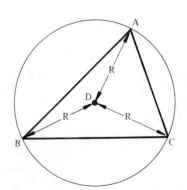

The *bisectors of angles* A, B and C are concurrent at a point called the *incentre* which is the centre of the circle that touches all three sides (the inscribed circle).

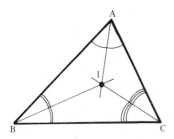

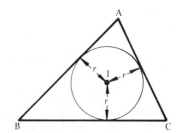

The *altitudes* are concurrent, at a point H called the *orthocentre* and the *medians* are concurrent, at a point G called the *centroid*.

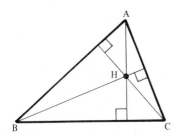

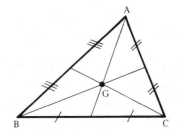

Area of a Triangle

The area of a triangle can be found using:
(a) half base × perpendicular height,
(b) $\frac{1}{2}ab \sin C$ etc.,
(c) $\sqrt{s(s-a)(s-b)(s-c)}$ where $s = \frac{1}{2}(a+b+c)$.

EXAMPLES 18j

1) In a surveying exercise, P and Q are two points on land which is inaccessible. To find the distance PQ, a line AB of length 200 metres is drawn so that P and Q are on opposite sides of AB. The following angles are measured:

$\angle \text{ABP} = 60°$, $\angle \text{ABQ} = 46°$, $\angle \text{BAP} = 30°$ and $\angle \text{BAQ} = 67°$. Find the distance PQ.

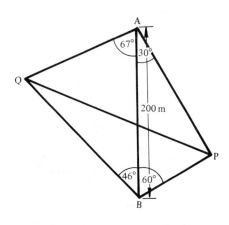

In triangle APB,

$$\angle \text{APB} = 90°$$

Hence $AP = 200 \cos 30°$

$$= 173.2 \text{ m}$$

In triangle ABQ,

$$\angle \text{AQB} = 67°$$

Hence

$$\frac{AQ}{\sin 46°} = \frac{200}{\sin 67°}$$

$\Rightarrow$ $AQ = 156.3 \text{ m}$

Then in triangle APQ, the cosine rule gives

$$PQ^2 = AP^2 + AQ^2 - 2 \cdot AP \cdot AQ \cos 97°$$
$$= (173.2)^2 + (156.3)^2 - 2(173.2)(156.3)(-0.1219)$$

Hence $PQ = 247 \text{ m}$

2) In a triangle ABC, $BC = 7.4 \text{ cm}$, $AC = 4.1 \text{ cm}$ and $\angle ACB = 66°$. Calculate:
(a) the other angles of the triangle,
(b) the area of the triangle,
(c) the distance of I, the incentre of the triangle, from BC.

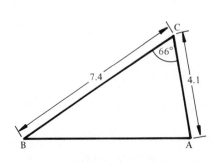

(a) Using the cosine formula,

$$c^2 = (7.4)^2 + (4.1)^2$$
$$- 2(7.4)(4.1) \cos 66°$$

gives $c = 6.85 \text{ cm}$

Then the sine rule gives

$$\frac{\sin A}{7.4} = \frac{\sin B}{4.1} = \frac{\sin 66°}{6.85}$$

Hence

$$A = 80.8°$$

and $B = 33.2°$

(b) The area of the triangle, Δ, is given by

$$\Delta = \tfrac{1}{2}ab \sin C = \tfrac{1}{2}(7.4)(4.1) \sin 66°$$
$$= 13.86 \text{ cm}^2$$

(c) The incentre, I, is the centre of the inscribed circle and is therefore equidistant (r) from all three sides.

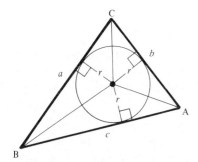

Joining AI, BI and CI we see that

Area of $\triangle$AIB $= \tfrac{1}{2}cr$

Area of $\triangle$BIC $= \tfrac{1}{2}ar$

Area of $\triangle$CIA $= \tfrac{1}{2}br$

Thus the total area of $\triangle$ABC $= \tfrac{1}{2}r(a + b + c) = 9.18r$

But this area is 13.86

Hence the distance of I from BC is $r = 1.5$ cm.

EXERCISE 18j

1) Calculate the area of triangle ABC if:
   (a) $B = 57°$, $a = 3.4$, $c = 14.1$,
   (b) $a = 11$, $b = 10$, $c = 15$,
   (c) $a = 8$, $c = 9$, $A = 52°$.

2) In a triangle PQR, $p = 2.8$ m and $q = 4.5$ m. If the area of the triangle is 5.84 m^2 find the two possible values of r.

3) Prove that the area of a triangle ABC can be found from

$$\frac{1}{2}c^2 \frac{\sin A \sin B}{\sin (A + B)}$$

4) Given a triangle ABC prove that:

   (a) $c \sin \dfrac{A - B}{2} = (a - b) \cos \dfrac{C}{2}$,

   (b) $\cot C = \dfrac{a}{c} \operatorname{cosec} B - \cot B$,

   (c) $abc = 4\Delta R$ where Δ is the area of the triangle and R is the radius of the circumcircle of the triangle.

5) ABC is a triangle and D is a point on BC such that BD = $\frac{1}{3}$DC. Angle BAD is 20° and angle CAD is 30°. Find angle ACB.

6) A quadrilateral ABCD is right-angled at B and D whilst the angle DAB = 132°. If DA = 4 and AB = 7, find the lengths of the diagonals of the quadrilateral and the radius of the inscribed circle of the triangle ABC.

(U of L)

7) Show that for any plane triangle ABC

$$\tan \frac{A}{2} \tan \frac{B-C}{2} = \frac{b-c}{b+c}$$

Three towns A, B and C are all at sea level. The bearings of towns B and C from A are 36° and 247° respectively. If B is 120 km from A and C is 234 km from A calculate the distance and bearing of town B from C.

(AEB)'72

8) ABC is a triangle in which none of the angles is obtuse. The perpendicular AD from A to BC is produced to meet the circumcircle of the triangle at E. If D is equidistant from A and E prove that the triangle must be right-angled. If, alternatively, the incentre of the triangle is equidistant from A and E, prove that cos B + cos C = 1.

(U of L)

THREE-DIMENSIONAL PROBLEMS

One of the difficulties which many people experience in this topic, arises when attempting to illustrate a three-dimensional situation on a two-dimensional diagram. The following hints may help in producing a clear representation of the 3-D problem from which appropriate calculations can be made.

1) Vertical lines should be drawn vertically on the page.

2) Lines in the East–West direction should be drawn horizontally on the page. North–South lines are shown as inclined at an acute angle to the East direction.

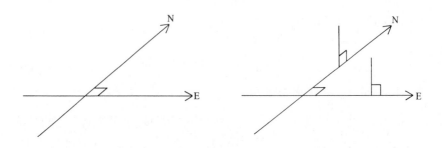

3) All angles that are 90° in three dimensions should be marked as right angles on the diagram, particularly those that do not *appear* to be 90°.

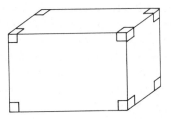

4) Perspective drawing is rarely used, so parallel lines are drawn parallel in the diagram.

5) When viewing a 3-D object, some of its sides are usually not visible. It is helpful to indicate these by broken lines.

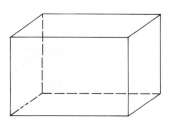
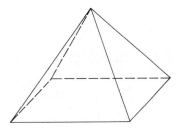

6) In a situation involving two points in the foreground and an object in the background it is usually clearer to draw the object *between* the two points.

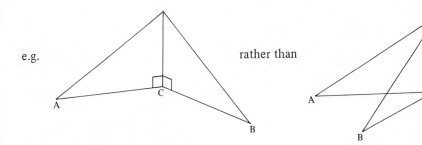

The following facts and definitions should also be known.

1) Two non-parallel planes meet in a line called the common line.

2) (a) A straight line which is perpendicular to a plane is perpendicular to every line in that plane, i.e. if PN is at right angles to the plane Π shown in diagram (i) below, PN is perpendicular to NA, NB, NC, ND . . . *and* to all parallel lines not passing through N (since, in three-dimensions, an angle between two non-intersecting lines is defined as the angle between two intersecting lines parallel to the given pair).

   (b) A line which is perpendicular to any two non-parallel lines in a plane, is perpendicular to the plane.
   Thus in diagram (ii) PN, being perpendicular to PA and PB, is perpendicular to the plane Π containing A, B and N.

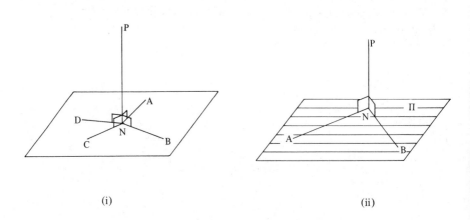

(i)   (ii)

3) The angle between a line and a plane is defined as the angle between the line and its projection on the plane (its projection can be thought of as the shadow cast on the plane by a distant light vertically above the line).

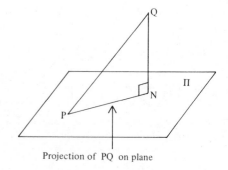

Projection of PQ on plane

4) Consider any point P on the common line of two planes Π_1 and Π_2.
If PA and PB are drawn at right angles to the common line so that PA is in
Π_1 and PB is in Π_2, then angle APB is the angle between Π_1 and Π_2.

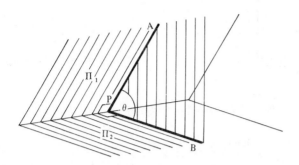

Note. If one of the planes, Π_2 say, is horizontal, then PA is called a *line of
greatest slope* in the plane Π_1.

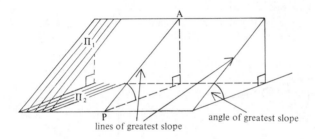

lines of greatest slope   angle of greatest slope

EXAMPLES 18k

1) Given the rectangular solid
(a cuboid) shown in the diagram,
find:

(a) the angle between AC′ and
the plane ABB′A′,

(b) find the angle between the
planes ACD′ and DCD′.

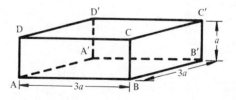

(a) To find the angle between AC′ and plane ABB′A′, we need the projection
of AC′ in the plane. So we drop a perpendicular from C′ to the plane, i.e.
the line C′B′, and join the foot of the perpendicular to A, i.e. B′A.
The required angle lies between C′A and B′A.
Drawing separate diagrams showing the base and the section through A, C′
and B′ we have

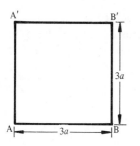
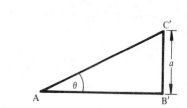

and

Thus
$$\text{AB}' = \sqrt{(3a)^2 + (3a)^2} = 3a\sqrt{2}$$

and
$$\tan\theta = \frac{a}{3a\sqrt{2}}$$

Hence the required angle, θ, is $13.3°$.

(b) The line common to the planes ACD' and DCD' is CD'. If M is the midpoint of this line, then:
   MD is perpendicular to $\text{D}'\text{C}$ in plane DCD' and
   MA is perpendicular to $\text{D}'\text{C}$ in plane $\text{D}'\text{CA}$.
   Thus ϕ is the angle between the planes DCD' and $\text{D}'\text{CA}$.

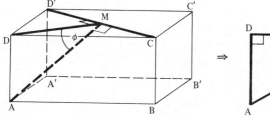
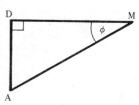

But
$$\text{DM} = \tfrac{1}{2}\text{DC}' = \tfrac{1}{2}(3a\sqrt{2})$$

Hence
$$\tan\phi = a \div \left(\frac{3a\sqrt{2}}{2}\right) = \frac{\sqrt{2}}{3}$$

i.e.   the required angle is $25.2°$

2) Three points A, B and C are on a horizontal line such that   AB $= 70$ m and   BC $= 35$ m.   The angles of elevation of the top of a tower are arctan $1/13$ at A, arctan $1/15$ at B and arctan $1/20$ at C. The foot of the tower is at the same level as A, B and C. Find:
(a) the height of the tower,
(b) the distance from the foot of the tower to the line ABC.

(a)

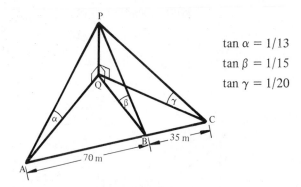

$$\tan \alpha = 1/13$$
$$\tan \beta = 1/15$$
$$\tan \gamma = 1/20$$

Let the height of the tower, PQ, be h.

Then $$h = QA \tan \alpha = QB \tan \beta = QC \tan \gamma$$

$\Rightarrow$ $$QA = 13h, \quad QB = 15h, \quad QC = 20h$$

Now consider the base triangle ABCQ

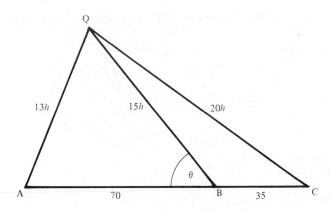

Using the cosine formula in AQB and CQB, we have

$$\cos \theta = \frac{(70)^2 + (15h)^2 - (13h)^2}{2(70)(15h)}$$

and $$\cos (\pi - \theta) = -\cos \theta = \frac{(35)^2 + (15h)^2 - (20h)^2}{2(35)(15h)}$$

Hence $\quad\dfrac{(70)^2 + (15h)^2 - (13h)^2}{2(70)(15h)} = \dfrac{(20h)^2 - (15h)^2 - (35)^2}{2(35)(15h)}$

$\Rightarrow \qquad\qquad 4900 + 56h^2 = 2(175h^2 - 1225)$

$\Rightarrow \qquad\qquad\qquad 7350 = 294h^2$

Hence $\qquad\qquad\qquad\qquad h = 5\,\text{m}.$

(b)

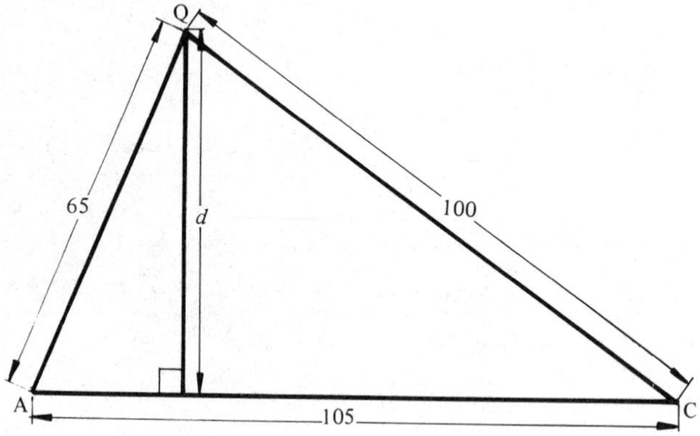

The area of triangle AQC is given by $\frac{1}{2}(105)(d)$ where d is the distance from Q to AC.

The area is also given by $\sqrt{s(s-a)(s-q)(s-c)}$

where $\qquad\qquad s = \frac{1}{2}(100 + 105 + 65) = 135$

Hence $\qquad\dfrac{105d}{2} = \sqrt{(135)(35)(30)(70)}$

$\Rightarrow \qquad\qquad d = 60\,\text{m}$

EXERCISE 18k

1) The angle of elevation of the top of a building from a point P due South of it is 28°. From a point Q due East of the building the angle of elevation is 41°. If PQ is 160 m, find the height of the building.

2) A, B and C are three points on horizontal ground. B is 80 m North of A and C is 50 m West of B. PB and QC are two vertical poles of heights 12 m and 20 m respectively. Find:
(a) the angle between the plane APQ and the horizontal plane,
(b) the angle between AQ and the horizontal plane.

3) The base PQ of an isosceles triangle PQR, is horizontal. The plane containing PQR is inclined to the horizontal at $66°$. If angle PRQ is $42°$ find the angle between PR and the horizontal plane.

4) The elevation of an aircraft is noted simultaneously by three observers A, B and C stationed in a straight horizontal line with AB = BC = 200 m. If the angles of elevation of the aircraft from A, B and C are $35°, 40°$ and $45°$ respectively, find the height of the aircraft.

5) Find the angle between the diagonals of a cube.

6) The diagram shows a roof whose base is a rectangle 16 m long and 10 m wide. Each long face is a trapezium inclined at $\arctan \frac{4}{3}$ to the horizontal and each end face is an isosceles triangle inclined at $\arctan \frac{5}{3}$ to the horizontal.

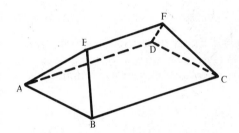

Calculate
(a) the height of the ridge (EF) above the base,
(b) the length of the ridge,
(c) the angle between AE and the horizontal,
(d) the total surface area of the roof.

MISCELLANEOUS EXERCISE 18

1) The table below shows the values of y obtained experimentally for the given values of x. Show graphically that, allowing for small errors of observation, there is a relation of the form $y - 2 = k(1 + x)^n$ and find approximate values of k and n.

x	4	8	15	19	24
y	2.45	2.60	2.80	2.89	3.00

(AEB)p'74

2) The variables x and y are believed to satisfy a relationship of the form $y = ab^x$, where a and b are constants. Show graphically that the values obtained in an experiment and shown in the table below do verify the relationship. From your graph calculate approximate values of a and b.

x	1	2	3	4	5
y	14.1	15.8	17.8	19.9	22.4

(AEB)p'76

3) The variables x and y below are believed to be related by a law of the form $y = \ln(ax^2 + bx)$, where a and b are constants. Plot a suitable graph to show that this is so and determine the probable values of a and b.

x	1	2	3	4	5	6
y	-1.897	0.588	1.599	2.262	2.757	3.153

(AEB)'71

4) In an experiment, sets of values of the related variables (x, y) are obtained. State how you would determine whether x and y were related by a law of the form:

(a) $y = a^{x+b}$,

(b) $ay^2 = (x + b) \ln x$,

where in each case a and b are unknown constants. State briefly how you would be able to determine the values of a and b for each law.   (AEB)p'72

5) The following values of x and y are believed to obey a law of the form

$$y = \frac{a}{bx + c}$$ where a, b and c are constants. Show that they do obey this law

and hence estimate the value of the ratios $a:b:c$.

x	0	1	2	3	4	
y	1.00	0.67	0.50	0.40	0.33	(AEB)p'73

6) It is thought that two variables s and t satisfy a relation of the form $(s/t)^a = be^{-t}$, where the constants a and b are positive integers. Show by drawing a linear graph that this is supported by the following table of measured values, and find the values of a and b.

t	0.2	0.4	0.6	0.8	1.0	
s	1.09	1.96	2.67	3.22	3.64	(U of L)p

7) It is known that two variables x and y are connected by a relation of the form $y = kx^n$, where k and n are constant integers.
From the given table of approximate values of x and y, find graphically the values of k and n.

x	1.34	3.58	7.60	12.1	14.8	
y	208.0	10.9	1.14	0.283	0.155	(U of L)p

8) Sketch the curve $2y = x(x - 4)(2x - 5)$.
The line $y = x$ cuts this curve at the origin O and at A and B where A is between O and B. Find the area bounded by the arc OA of the curve and the line $y = x$.   (U of L)p

9) Find the area of the region bounded by the curve $y = e^x$, the tangent at the point $(1, e)$ and the y-axis.
Find the volume of the solid formed by rotating this region through a complete revolution about (a) the x-axis, (b) the y-axis.   (O)

10) C is the curve given by $y = \sin x$ for values of x between 0 and π.
Find:
(a) the area of the region enclosed between C and the x-axis;
(b) the volume of the solid formed by rotating this region through four right angles about the x-axis.   (C)

11) Find the x-coordinate of the turning point of the curve whose equation is

$$y = \frac{a}{x} + \ln x$$

where $x > 0$ and $a > 0$, and determine whether this turning point is a maximum or a minimum. Deduce the range of values of the constant a for which $y \geqslant 0$ for all $x > 0$.

In the case when $a = 1$, find the area and the x-coordinate of the centroid of the region bounded by the curve, the x-axis and the ordinates $x = 1$ and $x = 2$. Express both answers in terms of $\ln 2$.   (JMB)

12) A curve joining the points $(0, 1)$ and $(0, -1)$ is represented parametrically by the equations

$$x = \sin \theta, \quad y = (1 + \sin \theta) \cos \theta,$$

where $0 \leqslant \theta \leqslant \pi$. Find dy/dx in terms of θ, and determine the x, y coordinates of the points on the curve at which the tangents are parallel to the x-axis and of the point at which the tangent is perpendicular to the x-axis. Sketch the curve.

The region in the quadrant $x \geqslant 0$, $y \geqslant 0$ bounded by the curve and the coordinate axes is rotated about the x-axis through an angle of 2π. Show that the volume swept out is given by

$$V = \pi \int_0^1 (1 + x)^2 (1 - x^2) \, dx$$

Evaluate V, leaving your result in terms of π.   (JMB)

13) Use the substitution $x = a \sin \theta$ to evaluate $\int_0^a \sqrt{a^2 - x^2} \, dx$.

The area A included between the parts of the two curves $x^2 + y^2 = 4$ and $4x^2 + y^2 = 16$ for which $x \geqslant 0$ and $y \geqslant 0$ is rotated completely about the x-axis. Find the volume of the solid thus formed.

Calculate the first moment of the area A about the y-axis and the x-coordinate of the centroid of this area.   (AEB)'74

14) Show that the curves whose equations are

$$y - 1 = x^3 \quad \text{and} \quad y + 3 = 3x^2$$

intersect at a point on the x-axis, and find the coordinates of this point. Show also that the only other point at which the curves meet is $(2, 9)$ and that the curves have a common tangent there. Sketch the two curves on the same diagram.

Show that the area of the finite region bounded by the curves is $27/4$, and find the x-coordinate of the centroid of this region.   (JMB)

15) A region R of the plane is defined by $y^2 - 4ax \leq 0$, $x^2 - 4ay \leq 0$, $x + y - 3a \leq 0$. Find:
(a) the area of R,
(b) the volume obtained by rotating R through 4 right angles about the x-axis. (O)

16) The curves $cy^2 = x^3$ and $y^2 = ax$ (where $a > 0$ and $c > 0$) intersect at the origin O and at a point P in the first quadrant. The areas of the regions enclosed by the arcs OP, the x-axis and the ordinate through P are A_1 and A_2 for the two curves; the volumes of the two solids formed by rotating these regions through four right angles about the x-axis are V_1 and V_2 respectively. Prove that $A_1/A_2 = \frac{3}{5}$ and $V_1/V_2 = \frac{1}{2}$. (O)

17) Sketch the curve whose equation is

$$y = 1 - \frac{1}{x + 2},$$

indicating any asymptotes which the curve possesses.
The region bounded by the curve, the x-axis and the ordinates $x = 0$ and $x = 2$ is denoted by R. Find:

(a) the area of R;
(b) the x-coordinate of the centroid of R;
(c) the volume swept out when R is rotated about the x-axis through an angle of 2π. (JMB)

18) Sketch the curves $y = e^x$ and $y = e^{-x}$ for $-2 \leq x \leq 2$.
The interior of a wine glass is formed by rotating the curve $y = e^x$ from $x = 0$ to $x = 2$ about the y-axis. If the units are centimetres, find, correct to two significant figures, the volume of liquid that the glass contains when full. (C)

19) Find the area of the region in the first quadrant enclosed between the curve $y = x^2 + 4$, the line $y = 8$ and the y-axis.
The region is rotated through four right angles about Oy to form a uniform solid. Find the volume and the coordinates of the centre of mass of the solid. (C)

20) (a) Integrate $\dfrac{1}{x(1 - x^2)}$ with respect to x.

(b) Find the mean value of $\sin^5 x$ over the range $x = 0$ to $x = \pi/2$.

(c) Obtain an approximate value of $\displaystyle\int_{\frac{\pi}{6}}^{\frac{\pi}{2}} \sqrt{\sin \theta}\, d\theta$ by the trapezium rule using five ordinates. Work as accurately as your tables will allow. (AEB)'74

21) Apply Simpson's rule using five ordinates to find an approximate value of

$\int_0^\pi \sin^{\frac{3}{2}} x \, dx.$ (AEB)p'75

22) The table below gives values of $f(x) = \sqrt{1 - x^2/25}$ correct to three decimal places for values of x between 0 and 5, at intervals of 0.5.

x	0	0.5	1.0	1.5	2.0	2.5	3.0	3.5	4.0	4.5	5.0
$f(x)$	1.000	0.995	0.980	0.954	0.917	0.866	0.800	0.714	0.600	0.436	0.000

Use Simpson's rule with eleven ordinates to calculate an approximate value of

$\int_0^5 \sqrt{1 - x^2/25} \, dx.$ Hence find an approximate value, to one decimal place, for

the area of the ellipse $x^2/25 + y^2/4 = 1.$
Using an appropriate substitution, evaluate the above integral and hence find an approximate value for π, to two decimal places.   (AEB)'74

23) R is the region in the first quadrant bounded by the y-axis, the x-axis from 0 to $\frac{1}{2}\pi$, the line $x = \frac{1}{2}\pi$ and part of the curve $y = (1 + \sin x)^{\frac{1}{2}}.$

(a) Show that, when R is rotated about the x-axis through four right angles, the volume of the solid formed is $\frac{1}{2}\pi(\pi + 2).$

(b) Use the trapezium rule with three ordinates to show that the area of R is approximately $0.63\pi.$ (C)p

24) Tabulate values of $f(x) = \sqrt{\{27 + (x - 3)^3\}}$ for integral values of x from 0 to 6 inclusive and sketch the graph of $y = f(x)$ for the interval $0 \leqslant x \leqslant 6.$

Given that $F(t) \equiv \int_0^t f(x) \, dx,$ use Simpson's rule and the calculated values

of $f(x)$ to estimate $F(2)$ and $F(6).$ (U of L)

25) Let $y = \sqrt[3]{x}.$ Write down $\dfrac{dy}{dx}$, and hence find an expression for the

approximate small change in y when x changes by a small amount $\delta x.$ Use your result to estimate the cube root of 1001.   (C)p

26) Tabulate, to three decimal places, the values of the function $f(x) = \sqrt{(1 + x^2)}$ for values of x from 0 to 0.8 at intervals of 0.1.

Use these values to estimate $\int_0^{0.8} f(x) \, dx$:

(a) by the trapezium rule, using all the ordinates,

(b) by Simpson's rule, using only ordinates at intervals of 0.2.   (U of L)

27) Find the coordinates of the centroid of the finite region bounded by the curve $y = x^2$, the lines $x = 2$, $x = 3$ and the x-axis. (U of L)p

28) Estimate $\int_{\frac{\pi}{6}}^{\frac{\pi}{2}} (\sin x)\, dx$ by Simpson's rule, using five ordinates and giving your answer to two decimal places. (U of L)p

29) Given that $y = \sqrt{\left(\dfrac{1+x}{2+x}\right)}$ determine the value of $\dfrac{dy}{dx}$ when $x = 2$,

and deduce the approximate increase in the value of y when x increases in value from 2 to $2 + \epsilon$ (ϵ small). (JMB)p

30) Write down the first three terms in the Maclaurin expansion of $e^x \sin x$. Using these terms as an approximation for $e^x \sin x$, estimate the value of

$$\int_0^{0.2} e^x \sin x \, dx.$$

31) For the function $f(x) \equiv x^2 \ln (1 + x)$, find $f'(x)$ and $f''(x)$.
(a) Find an approximation for $\delta [x^2 \ln (1 + x)]$ in terms of x and δx.
(b) Find the first two non-zero terms in the Maclaurin expansion of $f(x)$ and

use them to estimate the value of $\displaystyle\int_0^{0.1} x^2 \ln (1 + x)\, dx$.

32) If $y = \tan 2\theta$, find $\dfrac{dy}{d\theta}$ and $\dfrac{d^2 y}{d\theta^2}$.
(a) Write down an approximation for δy in terms of θ and $\delta\theta$. Hence

estimate the value of $\tan \dfrac{19\pi}{20}$.

(b) Find the first two terms in the Maclaurin expansion of $\tan 2\theta$ in powers

of θ. Use these terms to find an approximate value for $\displaystyle\int_0^{0.5} \tan 2\theta \, d\theta$.

33) Prove that, for a plane triangle PQR,

$$\cot P = \frac{q}{p} \operatorname{cosec} R - \cot R$$

To an observer standing at the top of a cliff the bearings of two ships P and Q at sea are $5°39'$ and $127°14'$ respectively. If to the observer the angle of depression of ship P is $14°23'$ whilst that of ship Q is $10°7'$, calculate the bearing of ship Q from ship P. (AEB)'71

34) A vertical pole with base at the point O, stands on the west bank of a river with straight parallel banks running from north to south. Two points A and B both lie on the eastern bank, A to the south and B to the north of the pole. The distance AB is $2a\sqrt{7}$ and the angle AOB is $150°$. If the angle of

elevation of the top of the pole, P, is $45°$ from A and $30°$ from B, find in terms of a, both the height of the pole and the width of the river.   (AEB)'71

35) α, β, γ are the angles of elevation of the top of a vertical tower from three distinct points A, B, C on a horizontal straight line that is on the same level as the foot of the tower but does not necessarily pass through it. The distances AB and BC are both equal to d. Prove that the height h of the tower is given by

$$h^2(\cot^2\alpha - 2\cot^2\beta + \cot^2\gamma) = 2d^2.$$

If $2\cot\beta = \cot\alpha + \cot\gamma$, prove that the line ABC passes through the foot of the tower.   (O)

36) The triangle ABC is inscribed in a circle centre O. The line CD is perpendicular to the plane of the circle and of length 10 cm.
If the angle $DAC = 68°12'$, the angle $DBC = 60°19'$ and AB = 6.67 cm, calculate the angle subtended by each of the sides of the triangle ABC at the centre O.   (AEB)'65

37) Show that, in the triangle ABC, the length of the perpendicular from C to AB is given by

$$\frac{ab}{c}\sin C.$$

The triangle ABC, in which $\angle ACB = 150°$, lies in a horizontal plane and D is the point at a distance 2 units vertically above C. Given that $\angle DBC = 30°$ and $\angle DAC = 45°$, find the lengths of the sides of the triangle ABC and the length of the perpendicular from C to AB.   (JMB)

38) Prove that in any plane triangle ABC

$$\cot A = \frac{b}{a}\operatorname{cosec} C - \cot C$$

The bearing of an aeroplane from a fixed observation post is $052°$, the aeroplane being at a horizontal distance of 6 miles from the post. Calculate the flight direction of the aeroplane if, when it passes due east of the observation post, it is at a horizontal distance of 9 miles from the post. If the height of the aeroplane when its bearing is $052°$ is h and its angle of elevation from the observation post is the same in both positions calculate, in terms of h, the height of the aeroplane when it is due east of the observation post.   (AEB)'67

39) P is a point so that PA is 6 m and PA is perpendicular to the horizontal plane of triangle ABC. When BC is 5 m and the angles of elevation of P from B and C are arctan 1 and arctan $\frac{6}{7}$ respectively, find the angle between the planes PBC and ABC.   (AEB)p'77

40) The diagram represents a solid figure in which ABCD is a horizontal rectangle.

   AB = 13 cm,   BC = 8 cm,   AX = DX = 9 cm,   BX = CX = 7 cm

Calculate:

(a) the height of X above the plane ABCD,

(b) the angle of inclination to the horizontal of the edge AX,

(c) the angle of inclination to the horizontal of the face AXB.

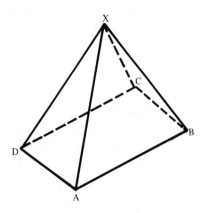

(AEB)p'73

41) A vertical mast stands on the north bank of a river with straight parallel banks running from east to west. The angle of elevation of the top of the mast is α when measured from a point A on the south bank distant $3a$ to the east of the mast and β when measured from another point B on the south bank distant $5a$ to the west of the mast. Prove that the height of the mast is

$$4a/(\cot^2\beta - \cot^2\alpha)^{\frac{1}{2}}$$

and that the angle of elevation θ measured from a point midway between A and B is given by the equation

$$2\cot^2\theta = 3\cot^2\alpha - \cot^2\beta \qquad \text{(U of L)}$$

ANSWERS

Exercise 1a — p. 4

1) a) equation b) identity c) identity
   d) equation e) equation f) identity
   g) equation h) identity

2) a) $x > 5$ b) $x \leqslant -2$
   c) $x \leqslant 4$ d) $x < 1$

3) a) $0, -2, 10, 10$
   b) $-\frac{1}{2}, -1, -\frac{1}{4}, \frac{1}{3}$
   c) $-14, -18, 0, -14$
   d) $1\frac{2}{3}, 2\frac{1}{2}, -\frac{5}{7}, -\frac{5}{42}$

Exercise 1b — p. 9

1) $\dfrac{3}{2(x-1)} - \dfrac{3}{2(x+1)}$

2) $\dfrac{4}{3(x-4)} - \dfrac{1}{3(x-1)}$

3) $\dfrac{3}{4(x+2)} + \dfrac{1}{4(x-2)}$

4) $\dfrac{2}{3(x-2)} - \dfrac{4}{3(2x-1)}$

5) $\dfrac{3}{x} - \dfrac{2}{x+1}$

6) $\dfrac{3}{x+1} - \dfrac{7}{3x+2}$

7) $\dfrac{3}{2(x-1)} - \dfrac{6}{x-2} + \dfrac{9}{2(x-3)}$

8) $\dfrac{7}{24(x-3)} + \dfrac{7}{8(x+1)} - \dfrac{2}{3x}$

9) $\dfrac{5}{8(x-1)} + \dfrac{3}{4(x-3)} + \dfrac{5}{8(x+3)}$

10) $\dfrac{1}{x-1} - \dfrac{x+1}{x^2+1}$

11) $\dfrac{x-2}{2(x^2-2)} - \dfrac{1}{2(x+4)}$

12) $\dfrac{3}{2x} - \dfrac{x}{2(x^2+2)}$

13) $\dfrac{22}{19(x-3)} + \dfrac{1-6x}{19(2x^2+1)}$

14) $\dfrac{3}{5(x+2)} - \dfrac{3}{5(2x+1)} + \dfrac{x-1}{5(x^2+1)}$

15) $\dfrac{1}{x} - \dfrac{6x+3}{2x^2-1} + \dfrac{2}{x-1}$

16) $\dfrac{1}{x-1} - \dfrac{1}{x-2} + \dfrac{2}{(x-2)^2}$

17) $\dfrac{2}{x} - \dfrac{1}{x^2} - \dfrac{3}{2x+1}$

18) $\dfrac{3}{x} - \dfrac{9}{3x-1} + \dfrac{9}{(3x-1)^2}$

19) $\dfrac{3}{4(x-1)} - \dfrac{1}{4(x+1)} - \dfrac{x}{2(x^2+1)}$

20) $1 + \dfrac{1}{2(x-1)} - \dfrac{1}{2(x+1)}$

Exercise 1c — p. 11

1) $-6, 1$ 2) $0, \frac{7}{3}$ 3) $1, 1$
4) $2, \infty$ 5) $1, -\frac{3}{2}$ 6) a, a
7) $1, \infty$ 8) $0, \frac{2}{3}$ 9) $2, \frac{1}{2}$
10) $\sqrt{2}, -\sqrt{2}$

Exercise 1d — p. 13

1) $2, 1$ 2) $1.46, -5.46$
3) $-\frac{1}{2}, -3$ 4) $1 \pm \sqrt{(1-a)}$

5) $a \pm \sqrt{(a^2-b)}$ 6) $\dfrac{-b \pm \sqrt{(b^2-4ac)}}{2a}$

7) equal   8) not real
9) real and distinct   10) not real
11) equal   12) real and distinct
13) ± 24 14) ± 12
15) $6\frac{1}{4}$ 18) $q^2 - 4p = 0$

Exercise 1e – p. 16

1) a) $3, 2$ b) $-\frac{7}{4}, -\frac{3}{4}$
   c) $4, -4$ d) $-1, -8$
   e) k, k^2 f) $\dfrac{a+2}{a}, -1$

2) a) $x^2 - 3x + 4 = 0$
   b) $2x^2 + 4x + 1 = 0$
   c) $15x^2 - 5x - 6 = 0$
   d) $4x^2 + x = 0$
   e) $x^2 - ax + a^2 = 0$
   f) $x^2 + (k+1)x + k^2 - 3 = 0$
   g) $abx^2 - b^2x + ac^2 = 0$

3) a) $\frac{4}{5}$ b) $5\frac{1}{2}$ c) -1 d) 5
   e) -6 f) $-\frac{2}{5}$ g) $\frac{8}{11}$ h) $\frac{4}{7}$ i) $\frac{4}{25}$

4) a) $x^2 - 6x + 11 = 0$
   b) $3x^2 - 2x + 1 = 0$
   c) $x^2 + 2x + 9 = 0$
   d) $3x^2 + 2x + 3 = 0$
   e) $x^2 + 8 = 0$

5) $x^2 - 2x - 3 = 0$
6) $2x^2 - 5x - 4 = 0$
7) $5x^2 + 13x + 7 = 0$
8) $2x^2 + 3x - 1 = 0$
9) $\frac{11}{12}$ 10) -6
11) $cx^2 + bx + a = 0$
12) $a = c$ 14) 1

Multiple Choice Exercise 1 – p. 18

1) e   2) e   3) a   4) b
5) d   6) b   7) d   8) b
9) b, c   10) c   11) a, b, c   12) a
13) E 14) D 15) B 16) C
17) I 18) A 19) a   20) F
21) F 22) T 23) F
24) T 25) F

Miscellaneous Exercise 1 – p. 20

1) $\dfrac{1}{1-x} + \dfrac{2}{2x+1} - \dfrac{3}{(2x+1)^2}$

2) $\dfrac{1}{x-1} - \dfrac{1}{2(x-1)^2} - \dfrac{2x+1}{2(x^2+1)}$

3) $\dfrac{1}{x-3} - \dfrac{1}{x+2} + \dfrac{2}{(x+2)^2}$

4) $acx^2 - b(a+c)x + a^2 + b^2 + c^2 - 2ac = 0$

5) $qy^2 + 2py + 4 = 0$
7) $x^2 + pqx + q^3 = 0$
8) $3y^2 - (p+8)y + 3 = 0$,
   $p = -2$ or -14

Exercise 2a – p. 23

1) 4 2) 2 3) $\frac{1}{3}$ 4) $\frac{4}{7}$
5) $\frac{1}{5}$ 6) 1331 7) $\frac{1}{32}$ 8) 27
9) $\frac{27}{1000}$ 10) $\frac{9}{4}$ 11) 49 12) 0.6
13) 625 14) $\frac{5}{8}$ 15) $\frac{125}{27}$ 16) 1
17) $\frac{4}{3}$ 18) 6 19) 3 20) 4
21) 6 22) 1 23) 16 24) 1
25) $y^{-\frac{3}{4}}$ 26) 1 27) x^5 28) t

29) $\dfrac{x}{x-1}$ 30) $\dfrac{x+4}{(x+2)^{5/2}}$

31) $\dfrac{x^2 + x - 1}{x^2(x+1)^{1/2}}$

32) $m(m-1)^2(m^2+1)^{1/2}$

33) $\dfrac{2t+3}{3(t+1)^{4/3}}$

34) $\dfrac{(p-q)^{1/4} - (p+q)^{1/4}}{p^2 - q^2}$

35) $\dfrac{x^2 - y^2 - (x+y)^{1/2}}{(x-y)^{1/2}}$

36) $-x^3 + x^2 - x$

37) $\dfrac{x^2 - x + 1}{x(x^2+1)}$

Exercise 2b – p. 26

1) a) $2\sqrt{2}$ b) $2\sqrt{3}$ c) $5\sqrt{2}$
   d) $3\sqrt{2}$ e) $10\sqrt{2}$ f) $6\sqrt{2}$
   g) $5\sqrt{5}$ h) $12\sqrt{2}$ i) $15\sqrt{2}$
   j) $20\sqrt{5}$

2) a) $3\sqrt{2} - 2$ b) $3\sqrt{2} - 4$
   c) $9 - \sqrt{3}$ d) 1
   e) $5 - 3\sqrt{3}$ f) $2 - 3\sqrt{2}$
   g) 23 h) $24 + 5\sqrt{5}$
   i) $\sqrt{6} + \sqrt{3} - \sqrt{2} - 1$
   j) $33 - 12\sqrt{6}$
   k) $x - 1$ l) $4x + 1 - 4\sqrt{x}$

3) a) $\dfrac{\sqrt{3}}{3}$ b) $\dfrac{\sqrt{2}}{4}$

   c) $\dfrac{\sqrt{2}}{4}$ d) $\sqrt{2} - 1$

   e) $\dfrac{\sqrt{3}+1}{2}$ f) $3(2 + \sqrt{3})$

   g) $5(\sqrt{5} - 2)$ h) $\frac{2}{3}(2\sqrt{3} + 3)$

i) $\dfrac{2\sqrt{5}-3}{11}$ j) $\sqrt{6}+\sqrt{5}$

k) $\dfrac{2\sqrt{3}-\sqrt{2}}{10}$ l) $2\sqrt{2}$

m) $\dfrac{6-3\sqrt{x}-3x}{(x-1)(x-4)}$ n) $\dfrac{19+6\sqrt{2}}{289}$

Exercise 2c – p. 27

1) a) $\log_5 125 = 3$
   b) $\log_7 49 = 2$
   c) $\log_8 4096 = 4$
   d) $\log_4 32 = \frac{5}{2}$
   e) $\log_{121} 1331 = \frac{3}{2}$
   f) $\log 0.01 = -2$
   g) $\log_5 1 = 0$
   h) $\log_8 \frac{1}{2} = -\frac{1}{3}$
   i) $\log_{1/5} 125 = -3$
   j) $\log_9 \frac{1}{27} = -\frac{3}{2}$
   k) $\log_a 1 = 0$
   l) $\log_\pi 9.8696 = 2$
   m) $\log_q p = 2$
   n) $\log_a b = c$
   o) $\log_x 2 = y$

2) a) $5^4 = 625$ b) $10^3 = 1000$
   c) $3^3 = 27$ d) $10^0 = 1$
   e) $(\frac{1}{2})^{-2} = 4$ f) $25^{1/2} = 5$
   g) $a^0 = 1$ h) $x^2 = y$
   i) $4^q = p$ j) $a^b = 5$
   k) $x^z = y$ l) $q^p = r$

3) a) 3   b) 4   c) $\frac{1}{2}$ d) -2
   e) -1 f) $\frac{2}{3}$ g) $\frac{1}{2}$ h) $\frac{1}{4}$
   i) -1 j) $-\frac{1}{3}$ k) $\frac{1}{2}$ l) 3
   m) 0   n) 0   o) 3   p) b
   q) 10   r) 25

Exercise 2d – p. 31

1) a) $\log a + \log b$
   b) $\log a + \log b + \log c$
   c) $\log a - \log b$
   d) $\log a + \log b - \log c$
   e) $\log a - \log b - \log c$
   f) $-\log a$
   g) $2\log a + \log b$
   h) $\frac{1}{2}\log a - \frac{1}{2}\log b$
   i) $2\log a - \log b$
   j) $3\log a - 1$
   k) $-1 - \log a$
   l) $2 - \frac{1}{2}\log b$

2) a) $\log 12$ b) $\log 3$ c) $\log 3$
   d) $\log 18$ e) $\log \frac{1}{3}$ f) $\log 4$
   g) $\log 30$ h) $\log 20$ i) 2
   j) $-\log(x-1)$ k) 3   l) 0

3) a) 1.6309   b) 0.8614
   c) 1.1620   d) 2.7712
   e) -0.1038 f) 1.2153
4) a) 3.3219   b) 0.8271
   c) 0.2314   d) -1.3368
5) a) $-1, 1$ b) 2   c) 3
6) 2   7) $x = 2, y = 4$
8) $y = 1, x = \frac{1}{2}$ 9) $x = 1, y = 0$
10) 3   12) $x = 1, y = 0$

Exercise 2e – p. 34

1) a) 3   b) 18   c) 47   d) $\frac{35}{16}$ e) $-\frac{16}{27}$
   f) $a^3 - 2a^2 + 6$ g) $c^2 - ac + b$
2) a) yes   b) no   c) no   d) yes
   e) yes   f) yes
3) a) $(x-1)(x+2)(x+1)$
   b) $(x-2)(x^2+x+1)$
   c) $(x-1)(x+1)(x^2+1)$
   d) $(x-2)(x+2)(x^2+x+1)$
   e) $(2x-1)(x^2+1)$
   f) $(3x-1)(9x^2+3x+1)$
   g) $(x+a)(x^2-ax+a^2)$
   h) $(x-y)(x^2+xy+y^2)$
4) -7 5) 5   6) $a = -3$ 7) 1

Exercise 2f – p. 37

1) $1 + 8x + 24x^2 + 32x^3 + 16x^4$
2) $1 - 3x + 3x^2 - x^3$
3) $x^3 - 3x^2 + 3x - 1$
4) $1 - 12y + 54y^2 - 108y^3 + 81y^4$
5) $1 + 7x + 21x^2 + 35x^3 + 35x^4$
   $+ 21x^5 + 7x^6 + x^7$
6) $8x^3 - 12x^2 + 6x - 1$
7) $8x^3 + 12x^2 y + 6xy^2 + y^3$
8) $x^4 + 4x^2 + 6 + \dfrac{4}{x^2} + \dfrac{1}{x^4}$
9) $p^3 - 6p^2 q + 12pq^2 - 8q^3$
10) $x^{10} - 5x^8 y + 10x^6 y^2 - 10x^4 y^3$
    $+ 5x^2 y^4 - y^5$
11) $x^3 - 3x + \dfrac{3}{x} - \dfrac{1}{x^3}$
12) $a^6 - 3a^4 b^2 + 3a^2 b^4 - b^6$
13) $7 + 5\sqrt{2}$ 14) $49 - 20\sqrt{6}$
15) 20   16) 1.030 301
17) 19.4481

Multiple Choice Exercise 2 – p. 38

1) c   2) a   3) a
4) a   5) b   6) e
7) c   8) a, b, c   9) a
10) b, c   11) a, b   12) a, c

13) C 14) A 15) C
16) D 17) T 18) T
19) F 20) T

Miscellaneous Exercise 2 — p. 40

1) $9, \frac{1}{3}$ 2) a) 0.59   b) 1.6
3) a) $\frac{7}{4}$ b) 6   4) $b = a^2$
5) $nm + m - 1$ 6) -1
7) $x = 4, y = 2$ or $x = 9, y = 3$
8) $x = 4, y = 2$
9) $x = 10, y = 4; x = 4, y = 2$
10) $4^8, \frac{1}{4}$
11) $h = 4, g = -1, (x + 2), (x + 3)$
12) $14, x + 1$

14) a) 125 or $\frac{1}{125}$ b) $\dfrac{\sqrt{2}}{2}$

15) $3x^3 - 7x^2 - 2x + 8$
16) $x = 3^6, y = \frac{1}{3}$ or $x = \frac{1}{3}, y = 3^6$

Exercise 3a — p. 47

1) $\frac{11}{4}$ 2) 3   3) 4
4) $\frac{29}{4}$ 5) -2 6) 1
7) least value 4 at $x = 1$
8) least value -12 at $x = -2$
9) least value $-\frac{3}{2}$ at $x = \frac{3}{2}$
10) greatest value $\frac{65}{4}$ at $x = -\frac{7}{2}$
11) least value -10 at $x = 0$
12) greatest value $\frac{49}{12}$ at $x = -\frac{5}{6}$

Exercise 3b — p. 49

1) $x > 1$ 2) $x < -3$
3) $x < \frac{1}{2}$ 4) $x < 3$
5) $x < 1$ 6) all values
7) $x < 1, x > 2$ 8) $x < -1, x > 2$
9) $3 < x < 5$ 10) $-1 < x < \frac{1}{2}$
11) $x < -1, x > 5$ 12) $-\frac{1}{2} < x < \frac{1}{2}$
13) $x < -\frac{2}{5}, x > 1$ 14) $x < -4, x > 2$
15) $-5 < x < \frac{3}{2}$ 16) $1 < x < 2$
17) $-1 < x < \frac{1}{3}$ 18) $-\frac{1}{2} < x < 2$

Exercise 3c — p. 53

1) $x < 2, x > 2\frac{1}{5}$
2) $2 < x < 2\frac{3}{4}$
3) a) $k < 1, k > 9$ b) $0 < k < 1, k > 9$
4) $-4 < k < 0$
5) $c \leqslant -2, c \geqslant 1$
7) $-1 \leqslant \dfrac{4(x - 2)}{4x^2 + 9} \leqslant \dfrac{1}{9}$
8) $k > 1$
9) $-1 \leqslant x < 1$ and $4 < x \leqslant 5$
10) $0 < x < 7$ and $x > 8$

11) $-\dfrac{1}{2} \leqslant \dfrac{2x + 1}{(x^2 + 2)} \leqslant 1$
12) a) $\alpha = 1$ or -3, other root 4 or $\frac{4}{9}$
    b) $\alpha > \frac{5}{2}(\sqrt{2} - 1), \alpha < -\frac{5}{2}(\sqrt{2} + 1)$
13) $-3 < x < 3, x > 5$
14) a) $x < -5.16$ or $x > 1.16$
    b) $-8.24 < x < 0.24$
    c) $-8.24 < x < -5.16$
       or $0.24 < x < 1.16$
15) a) (i) $-4 < x < \frac{3}{2}$
       (ii) $x < 1$ or $2 < x < 3$
    b) real roots, $a = 1$
16) $(m + 1)^2 = ma, a \leqslant 0, a \geqslant 4$
17) $-1\frac{1}{2} < p < \frac{1}{2}$
18) a) $-4 < x < 2, a = 3$
    b) $-2x^2 - 8x + 10$,
       $-50x^2 + 40x + 10$
19) $k < -3\frac{1}{12}, k = -2\frac{1}{12}$
20) $k = -\frac{1}{48}$ a) $k > 1$ b) $k \leqslant \frac{1}{2}$

Exercise 3d — p. 58

1) $\frac{1}{4}, \frac{1}{16}, \frac{1}{64}, 4, 16, 64$ 3) $x = 1$
5) $1, -1$, a) $x > 1, x < -1$
   b) $-1 < x < 1$
6) no meaning

Exercise 4b — p. 67

1) a) 5   b) $\sqrt{2}$ c) $\sqrt{13}$ d) $\sqrt{13}$
   e) $\sqrt{5}$ f) $\sqrt{8}$
2) a) $(\frac{5}{2}, 4)$ b) $(\frac{5}{2}, \frac{1}{2})$ c) $(3, \frac{7}{2})$
   d) $(\frac{1}{2}, 5)$ e) $(-\frac{1}{2}, -1)$
   f) $(-2, -3)$
3) $\sqrt{65}$ 6) $(-\frac{7}{2}, -\frac{1}{2}), \frac{35}{2}$
8) $(0, \frac{9}{2}), 2\frac{1}{2}$ 9) $(-5, -3)$

Exercise 4c — p. 73

1) a) 3   b) $\frac{3}{2}$ c) $\frac{1}{3}$ d) $\frac{3}{4}$
   e) -4 f) 6   g) $-\frac{7}{3}$ h) $-\frac{3}{2}$
   i) k/h
2) a) 1   b) -1 c) 0   d) ∞
3) a) yes   b) no   c) yes
4) a) parallel   b) perpendicular
   c) perpendicular   d) parallel
5) $a = 0, b = 4$ 6) $\frac{45}{2}$
7) 8   8) $b(d - b) = ac$
10) $a^2 + b^2 = 4$ 11) $3k = 2 - h$
12) 5, 35   13) $b^2 = 8a - 16$

Exercise 4d — p. 79

5) no   6) yes   7) no   8) no
9) yes   10) yes

Exercise 4e — p. 85

1) a) $y = 2x$ b) $y = -x$
   c) $3y = x$ d) $4y + x = 0$
   e) $y = 0$ f) $x = 0$
2) a) $2y = x + 2$ b) $2y = x$
   c) $2y = x - 7$
3) a) $2y = x$ b) $2x + y - 6 = 0$
   c) $y + 2x - 4 = 0$ d) $y = 2x + 5$
4) a) $y > x$ b) $y < 4 - 2x$
   c) $y > 2 - x$ d) $y < 2x + 4$
5) a) $3y - 2x = 0$ b) $y + 2x = 0$
6) a) $3y = x + 1$ b) $2y + 3x + 7 = 0$
7) a) $5x - y - 17 = 0$
   b) $x + 7y + 11 = 0$

Exercise 4f — p. 90

1) $2x - y - 8 = 0$ 2) $x - y = 0$
3) a) $3x - 2y + 2 = 0$
   b) $3x - 2y + 7 = 0$
   c) $x = 3$
4) a), c), d) are perpendicular
5) a) $(\frac{2}{5}, \frac{11}{5})$ b) $(-\frac{3}{2}, \frac{5}{2})$
   c) $(-\frac{2}{5}, \frac{4}{5})$ d) $(-\frac{6}{5}, \frac{7}{5})$
6) a) $x + 2y - 5 = 0$
   b) $6y - 16x - 19 = 0$
   c) $16y - 10x = 23$
7) $5y + 4x = 0$ 8) $4y + 5x = 0$
9) $(-5, 0), (\frac{7}{2}, 0), (\frac{16}{7}, \frac{17}{7})$
10) $x + 2y - 11 = 0, \dfrac{11\sqrt{5}}{5}$

Multiple Choice Exercise 4 — p. 92

1) c   2) d   3) b   4) a
5) a   6) c   7) a   8) d
9) b   10) a   11) a, c   12) b
13) a, b   14) c   15) a, c   16) D
17) B   18) C   19) c   20) I
21) A   22) I   23) A   24) T
25) T   26) F   27) T   28) F

Miscellaneous Exercise 4 — p. 95

2) $\frac{1}{4}$
3) $(-5, 0), (6, 0), (-\frac{4}{3}, \frac{11}{3})$
4) $x - y - 4 = 0, 8$
5) $(3, 0), (-2, -5), 5\sqrt{2}$
6) $x + 3y - 11 = 0, (2.6, 2.8)$
7) $y = 2x - 3$
8) $8by = 2ax - 8b^2 - 3a^2$
9) $(x - 2)^2 + y^2 < 16$
10) $(x - 2)^2 + y^2 = 16$
11) $(0, 0), (2, 4)$
12) $A(1 + \sqrt{2}, 1 - \sqrt{2}), B(1 - \sqrt{2}, 1 + \sqrt{2})$

13) $M(1, 1)$, AB is a diameter
14) $(0, 4), (12 + 4\sqrt{2}, 8\sqrt{2}),$
   $(4\sqrt{2}, 4 + 8\sqrt{2})$
15) $(4, 3\frac{1}{2})$ and $(7, 5)$ or $(\frac{2}{5}, -\frac{12}{10})$ and $(2\frac{1}{5}, 9\frac{4}{5})$
16) $(-\frac{2}{11}, \frac{3}{11}), (1, 4), (4\frac{2}{11}, 1\frac{8}{11})$
   $13y - 41x = 11, 7y + 5x = 33,$
   $13y - 41x + 149 = 0$

Exercise 5a — p. 102

1) 6   2) 4   3) -8
4) -5 5) 3   6) -2

Exercise 5b — p. 104

1) $2x$ 2) $3x^2$ 3) $4x^3$
4) 1   5) $6x$ 6) $15x^2$
7) $-\dfrac{2}{x^3}$ 8) $2x + 3$ 9) $2x - 2$

Exercise 5c — p. 107

1) $5x^4$ 2) $10x^9$ 3) $-3x^{-4}$
4) $\frac{1}{3}x^{-2/3}$ 5) $-x^{-2}$ 6) $-\frac{1}{2}x^{-3/2}$
7) $-4x^{-5}$ 8) $\frac{3}{2}x^{1/2}$ 9) 2
10) $-5x^{-6}$ 11) $-\frac{1}{2}x^{-3/2}$ 12) $\frac{1}{4}x^{-3/4}$
13) 0   14) -4 15) $\frac{1}{2}x^{-1/2}$
16) $\frac{2}{3}x^{-1/3}$

Exercise 5d — p. 109

1) $6x - 7$ 2) $4x^3 - 27x^2$
3) $4x - 1$ 4) $2x - 8$
5) $147x^2 - 98x - 1$ 6) $2x - \dfrac{2}{x^2}$
7) $-2x^{-3} - x^{-2}$ 8) $-\dfrac{1}{x^2} + \dfrac{14}{x^3} - \dfrac{12}{x^4}$
9) $\frac{3}{2}\sqrt{x} - \dfrac{1}{2\sqrt{x}}$ 10) $\dfrac{1}{2x^{1/2}} + \dfrac{1}{2x^{3/2}}$
11) $24x^2 + 24x + 6$ 12) $-\dfrac{1}{x^2} - \dfrac{2}{x^3} - \dfrac{3}{x^4}$
13) 2   14) 18   15) $\dfrac{\sqrt{2}}{4}$
16) -1 17) $-\frac{1}{9}$ 18) -1
19) $\frac{1}{9}$ 20) 34   21) 4
22) $-\frac{3}{2}$ 23) -5 24) $\frac{1}{54}$
25) $(1, 3)$ 26) $(\frac{3}{2}, 5)$
27) $(1, 0)$ 28) $(1, -16)$
29) $(-1, 0), (\frac{1}{3}, \frac{4}{27})$
30) $(-4, -\frac{1}{3}), (-2, -3\frac{2}{3})$
31) no points   32) $(\frac{1}{16}, \frac{1}{4})$
33) $(\frac{1}{2}, -1), (-\frac{1}{2}, 3)$ 34) $(-2, \frac{1}{4})$
35) $(1, 1), (-1, -1)$ 36) no points

Exercise 5e — p. 112

1) $y = 2x - 3$ 2) $3x - y - 1 = 0$
3) $x + y + 2 = 0$ 4) $y = 8x + 2$
5) $x - y - 7 = 0$ 6) $y = -2$
7) $x + 2y + 1 = 0, x + 3y + 3 = 0,$
   $x - y = 0, x + 8y + 49 = 0$
   $x + y + 1 = 0, x = 0$
8) $x + 3y + 6 = 0$ 9) $13x - y - 18 = 0$
10) $2y = 6x - 3, 3x + y + 3 = 0,$
    $(-\frac{1}{4}, -\frac{9}{4})$
11) $x + y - 3 = 0, x - y - 2 = 0$
12) $y = 5x - 1, 9x + 3y + 19 = 0$
13) $(1, 1), y = 2x - 1$
14) $(\frac{3}{2}, -\frac{11}{4}), -\frac{29}{4}$
15) $2y = x + 1$ 16) $y = x - 2$
17) $-\frac{87}{32}$ 18) $y = -\frac{121}{8}$

Exercise 5f — p. 114

1) 0 2) $\frac{5}{2}$ 3) ± 2
4) $0, 1, -\frac{1}{4}$ 5) -6 6) $2, -2$
7) $\frac{2}{9}, \frac{34}{9}$ 8) $(-\frac{1}{2}, -\frac{25}{4})$
9) $(\frac{1}{3}, -\frac{41}{27}), (3, -11)$ 10) no points

Exercise 5g — p. 122

1) -2 min, 2 max   2) 4 min, -4 max
3) -3 max   4) 0 min, 4 max
5) $-\frac{49}{8}$ min   6) 0 inflexion
7) 0 min   8) -1 min, 0 max
9) 0 min, $\frac{5}{16}$ max, 2 max
10) $(0, 0), (\frac{2}{3}, \frac{4}{27})$ 11) $(0, 1)$
12) $(1, 2), (-1, -2)$
13) $(0, -4)$ 14) $(\frac{1}{2}, 2\frac{3}{4})$
15) $(-4, -173), (\frac{1}{2}, 9\frac{1}{4})$
16) none   17) $(0, 0)$
18) $(1, -4), (-1, 4)$

Multiple Choice Exercise 5 — p. 123

1) b   2) d   3) a   4) a
5) e   6) c   7) b   8) a, b
9) b, c   10) b, c 11) A 12) C
13) C 14) E 15) I 16) a
17) A 18) I

Miscellaneous Exercise 5 — p. 125

1) 24 2) $-\frac{9}{7}$
3) $2t - 5, 2y - 2, -4v$
4) $1 - \dfrac{1}{x^2}$ 5) 0.289
6) $(\frac{9}{2}, -\frac{33}{4}), x + 2y + 12 = 0$
7) $y = 15x - 16, y = 15x + 16$
8) $4x + 12y + 15 = 0$
9) $(0, 0), (1, -5), (-2, -32)$

10) $2, -2$ 12) $(200\pi)^{-\frac{1}{3}}$
13) $a = 2, b = 2, c = \frac{3}{2}$ 14) $\frac{1}{108}$ m^3
15) $\dfrac{500\pi\sqrt{3}}{9}$ 17) 200 ms^{-1}
18) $\frac{17}{4}$ m   19) $3.5°/$min
20) 1.5 cms^{-1}
21) $y = 8px - 4p^2$
22) $t^2 y + x - 2t = 0; 2$

Exercise 6a — p. 130

1) $\frac{1}{6}\pi; \frac{3}{2}\pi; \frac{2}{3}\pi; \frac{1}{4}\pi; \frac{4}{3}\pi; \frac{5}{6}\pi; \frac{1}{9}\pi; \frac{5}{3}\pi; \frac{1}{8}\pi; \frac{4}{9}\pi$
2) $135°; 150°; 18°; 270°; 330°; 60°;$
   $240°; 15°; 22.5°; 80°$
3) $0.62^c; 1.36^c; 0.96^c; 1.22^c; 0.28^c$
4) $163.9°; 241.2°; 57.3°; 198.8°; 171.9°$

Exercise 6b — p. 133

1) $d = 1.57; r = 4.3; \frac{1}{2}^c; \dfrac{\pi x}{90}; 3.85; 8.26$
2) 3445 km
3) $112° 45'$
4) $4.19; 3.69; 1.78^c; \dfrac{49\pi x}{360}; 2.37; 14.1$
5) $16° 55'$
6) 7.5 cm^2
7) 1.83 cm^2; 32.38 cm^2
8) a) 6.29 cm   b) 9.44 cm^2
   c) 5.54 cm^2
9) a) 18.5 cm   b) 118.9 cm^2
10) 8.94 cm

Exercise 6c — p. 142

1) $87°; \frac{1}{4}\pi; 52°; \frac{1}{3}\pi; 22°; 36°; \frac{1}{6}\pi; \frac{1}{4}\pi;$
   $\frac{1}{4}\pi; 4°$
2) a) $-0.9205, 23°, 157°$
   b) $-0.7071, 45°, 225°$
   c) $0.5, -0.866, -0.5774, \frac{1}{6}\pi$
   d) $-0.866, 0.5, -1.7321, 60°$
   e) $0.1736, -0.9848, -0.1763, 10°$
   f) $-0.5774, 30°, 150°$
   g) $\pm 0.866, \pm 1.7321, 60°, \pm 60°$
   h) $\pm 0.7071, \pm 0.7071, \pm 1, \pm 45°$ or
      $\pm 135°$
3) a) $23.6°, 156.4°, -203.6°, -336.4°$
   b) $\pm 120°, \pm 240°$
   c) $50.2°, 230.2°, -129.8°, -309.8°$
   d) $-41.8°, -138.2°, 221.8°, 318.2°$
   e) $\pm 66.4°, \pm 293.6°$
   f) $-206.6°, -26.6°, 153.4°, 333.4°$
4) a) $36.9°$ b) $-36.9°$ c) $26.6°$

Exercise 6d – p. 155

1) a) $\frac{1}{3}\pi$ b) $\frac{3}{4}\pi$ c) $\frac{1}{3}\pi$
   d) $\frac{1}{6}\pi$ e) $-\frac{1}{2}\pi$ f) 0
   g) $-\frac{1}{4}\pi$ h) $\frac{1}{2}\pi$ i) 0
   j) π k) $-\frac{1}{6}\pi$ l) $\frac{1}{2}\pi$
   m) $39.1°$ n) $63.9°$ p) $70.6°$

2) a) $\frac{1}{4}\pi, \frac{3}{4}\pi$ b) $\frac{5}{6}\pi, -\frac{5}{6}\pi$
   c) $-\frac{1}{3}\pi, \frac{2}{3}\pi$ d) $-32.7°, -147.3°$
   e) $50.9°, -50.9°$
   f) $56.3°, -123.7°$

3) $\sin\theta = 1, \sec\theta = 1.$

Exercise 6e – p. 165

1) a) $\frac{1}{3}\pi, \frac{5}{3}\pi$ b) $\frac{5}{6}\pi, \frac{11}{6}\pi$ c) $\frac{1}{4}\pi, \frac{3}{4}\pi$
   d) $109.5°, 250.5°$ e) $26.6°, 206.6°$
   f) $14.5°, 165.5°$ g) $32.9°, 327.1°$
   h) $203.6°, 336.4°$ i) $36.9°, 216.9°$
   j) $0, \pi, 2\pi, \frac{1}{6}\pi, \frac{11}{6}\pi$
   k) $0, \pi, 2\pi, \frac{2}{3}\pi, \frac{4}{3}\pi$
   l) $14.5°, 165.5°, 90°, 270°$

2) a) $\begin{cases} \frac{1}{3}\pi + 2n\pi \\ \frac{2}{3}\pi + 2n\pi \end{cases}$ b) $\frac{1}{2}(2n+1)\pi$
   c) $-\frac{1}{3}\pi + n\pi$
   d) $\begin{cases} -14.5° + 360n° \\ -165.5° + 360n° \end{cases}$
   e) $\pm 68.2° + 360n°$
   f) $63.4° + 180n°$
   g) $\begin{cases} 19.5° + 360n° \\ 160.5° + 360n° \end{cases}$
   h) $2n\pi$
   i) $\begin{cases} \frac{1}{6}\pi + 2n\pi \\ \frac{5}{6}\pi + 2n\pi \end{cases}$
   and $\begin{cases} -\frac{1}{6}\pi + 2n\pi \\ -\frac{5}{6}\pi + 2n\pi \end{cases} \Rightarrow \pm\frac{1}{6}\pi + n\pi$

Exercise 6f – p. 168

1) a) $22\frac{1}{2}°, 112\frac{1}{2}°, 202\frac{1}{2}°, 292\frac{1}{2}°$
   b) $14.8°, 45.2°, 134.7°, 165.3°,$
      $254.7°, 285.3°$
   c) $63.6°$
   d) $12°, 60°, 84°, 132°, 156°, 204°,$
      $228°, 276°, 300°, 348°$

2) a) $\begin{cases} 1080n° + 125.4° \\ 1080n° + 414.6° \end{cases}$
   b) $4.6° + 45n°$
   c) $\pm 25.5° + 180n°$
   d) $\begin{cases} \frac{1}{2}\pi + 6n\pi \\ \frac{5}{2}\pi + 6n\pi \end{cases}$

Exercise 6g – p. 170

1) $\frac{2}{7}n\pi$ 2) $\frac{1}{5}n\pi$
3) $\frac{1}{2}(4n-1)\pi, \frac{1}{10}(4n+1)\pi$

4) $\frac{1}{18}(2n+1)\pi$ 5) $2n\pi, \frac{1}{7}(2n+1)\pi$
6) $\frac{1}{2}n\pi, \frac{1}{3}n\pi$ 7) $\frac{2}{3}n\pi - \frac{1}{6}\pi, \frac{2}{5}n\pi$
8) $\frac{1}{6}(2n+1)\pi$
9) $0, \frac{1}{5}\pi, \frac{2}{5}\pi, \frac{1}{2}\pi, \frac{3}{5}\pi, \frac{4}{5}\pi, \pi$
10) $\frac{1}{10}\pi, \frac{3}{10}\pi, \frac{1}{2}\pi, \frac{7}{10}\pi, \frac{9}{10}\pi$
11) $0, \frac{1}{9}\pi, \frac{2}{9}\pi, \frac{1}{3}\pi, \frac{5}{9}\pi, \frac{7}{9}\pi, \frac{4}{9}\pi, \pi$
12) $0, \frac{2}{11}\pi, \frac{4}{11}\pi, \frac{6}{11}\pi, \frac{8}{11}\pi, \frac{10}{11}\pi$
13) $-140°, -60°, -20°, 100°$
14) $-115°, -25°, 65°, 155°$
15) $-150°, -110°, 10°, 130°$

Exercise 6h – p. 176

1) $\frac{\pi}{2}, 4$ 2) $\pi, 2$ 3) $4\pi, \frac{1}{4}$

4) $\pi, 2$ 5) $\frac{\pi}{2}, 2$ 6) $2\pi, 1$

7) $\frac{2\pi}{3}, 3$ 8) $\frac{2\pi}{3}, 3$ 9) $8\pi, \frac{1}{4}$

10) $\pi, 2$ 11) $\pi, 1$

12) $\begin{cases} 2n\pi - \frac{7}{12}\pi \\ 2n\pi + \frac{1}{12}\pi \end{cases}$ 13) $\frac{1}{2}n\pi + \frac{1}{24}\pi$

14) $\frac{2}{3}n\pi + \frac{1}{9}\pi \pm \frac{5}{18}\pi$ 15) $\begin{cases} n\pi \\ n\pi + \frac{1}{3}\pi \end{cases}$

Multiple Choice Exercise 6 – p. 177

1) d   2) d   3) c
4) c   5) d   6) c
7) b   8) b, c, d   9) b, d
10) a, c   11) b, c   12) b
13) a, d   14) b, d   15) B
16) D 17) C 18) A
19) C 20) B 21) d
22) d   23) I 24) F
25) T 26) T 27) T
28) F

Miscellaneous Exercise 6 – p. 180

2) $\frac{1}{2}a^2(2\sqrt{3} - \pi)$ 3) $78.79a^2$
4) 12.69 cm
6) a) $2n\pi - \frac{1}{12}\pi, 2n\pi + \frac{7}{12}\pi$
   b) $2n\pi - \frac{1}{12}\pi, 2n\pi + \frac{7}{12}\pi$
7) $\pm\frac{1}{6}\pi, \pm\frac{5}{6}\pi$ 8) $\frac{1}{3}n\pi + \frac{5}{18}\pi$
9) $n\pi - \frac{1}{4}\pi \pm \frac{1}{6}\pi$ 10) $\frac{1}{14}(2n+1)\pi$
12) $\frac{2\pi}{k}; k = 2, \frac{1}{4}(2n+1)\pi, n\pi, \frac{1}{2}(2n+1)\pi$
    $k = \frac{1}{2}, (2n+1)\pi, 4n\pi, (4n+2)\pi$
13) a) $7.5°, 45°, 97.5°$
    b) $17.5°, 107.5°, 145°$
14) $\frac{1}{14}\pi + \frac{2}{7}n\pi, \frac{1}{2}(4n-1)\pi$
15) a) $\frac{2}{5}\pi, \frac{4}{5}\pi, \frac{6}{5}\pi, \frac{8}{5}\pi, 2\pi$

15) b) $k\theta = \begin{cases} 2n\pi + \theta \\ (2n+1)\pi - \theta \end{cases}$; (i) 6   (ii) 2

Exercise 7a − p. 188

1) $\pm 57.7°, \pm 122.3°$
2) $-10.1°, -169.9°$
3) $38.2°, 141.8°$
4) $-135°, 45°$
5) $30°, 150°$
6) $\pm 131.8°, 0°$
7) $\pm 41.4° + 360n°$
8) $-18° + 360n°, -162° + 360n°$
9) $45° + 180n°, -14° + 180n°$
10) $360n° \pm 60°, (2n+1)180°$

20) a) $\dfrac{x^2}{16} - \dfrac{y^2}{25} = 1$

    b) $\dfrac{x^2}{a^2} - \dfrac{y^2}{b^2} = 1$

    c) $y^2(x^2 + 4) = 36$
    d) $(x-1)^2 + (y-1)^2 = 1$

    e) $(x-2)^2 + 1 = \left(\dfrac{2}{y}\right)^2$

    f) $\left(\dfrac{a}{x}\right)^2 + \left(\dfrac{y}{b}\right)^2 = 1$

21) a) $\tan^4 A$ b) 1   c) $\operatorname{cosec}\theta \sec\theta$
    d) $\sec^2\theta$ e) $\tan\theta$
22) a) $-\frac{12}{13}, \frac{12}{5}$ b) $-\frac{4}{5}, -\frac{3}{4}$
    c) $\frac{7}{25}, \frac{24}{25}$ d) 0, 0, straight line

Exercise 7c − p. 193

1) a) $\frac{1}{2}$ b) $\frac{1}{2}$ c) $\frac{1}{4}(\sqrt{6} + \sqrt{2})$
    d) $\frac{1}{4}(\sqrt{6} - \sqrt{2})$ e) $\frac{1}{4}(\sqrt{6} - \sqrt{2})$
    f) $2 + \sqrt{3}$
2) a) $\frac{1}{2}\sqrt{3}, \frac{1}{2}, \sqrt{3}; \frac{1}{2}, -\frac{1}{2}\sqrt{3}, -\frac{1}{3}\sqrt{3};$
     $-\frac{1}{2}, -\frac{1}{2}\sqrt{3}, \pm\infty$
    b) $\frac{1}{2}\sqrt{2}, -\frac{1}{2}\sqrt{2}, -1; -\frac{1}{2}, -\frac{1}{2}\sqrt{3},$
     $\frac{1}{3}\sqrt{3}; -\frac{1}{4}(\sqrt{6} - \sqrt{2}), \frac{1}{4}(\sqrt{6} + \sqrt{2}),$
     $-2 - \sqrt{3}$
    c) $\frac{24}{25}, \frac{7}{24}; -\frac{3}{5}, -\frac{4}{3}; \frac{3}{5}, -\frac{4}{5}, \frac{117}{44}$
    d) $-\frac{4}{5}, -\frac{3}{4} \cdot \frac{12}{13}, -\frac{5}{13}; -\frac{63}{65}, -\frac{16}{65}, \frac{33}{56}$
    e) $\frac{7}{25}, \frac{24}{25}, \frac{7}{25}, \frac{7}{24}; \frac{336}{625}, \frac{527}{625}, 0$
    f) Negative acute, $-\frac{5}{12}$; acute, $\frac{3}{5}, \frac{3}{4}$,
     $\frac{63}{65}, -\frac{56}{63}$
9) $52.5°, 232.5°$ 10) $7.4°, 187.4°$
11) $37.9°, 217.9°$ 12) $15°, 195°$

Exercise 7d − p. 198

1) a) $\sin 28°$ b) $\tan 70°$
    c) $\cos 8\theta$ d) $\tan 6\theta$
    e) $\sqrt{2}\cos 3\theta$ f) $\cos 52°$
    g) $\frac{1}{2}\sin 2\theta$ h) $\cos 68°$

i) $\tan(x + 45°)$
2) a) (i) $\frac{24}{25}, -\frac{7}{25}$ (ii) $-\frac{24}{25}, -\frac{7}{25}$
    b) (i) $\frac{336}{625}, \frac{527}{625}$ (ii) $-\frac{336}{625}, \frac{527}{625}$
    c) (i) $\frac{120}{169}, -\frac{119}{169}$ (ii) $\frac{12\cap}{169}, -\frac{119}{169}$
3) $7, \frac{7}{10}\sqrt{2}, \frac{1}{10}\sqrt{2}; \frac{7}{25}, -\frac{24}{25}$
4) a) $2y = x(1 - y^2)$
    b) $x = 2y^2 - 1$
    c) $y^2(1 - x) = 2$
    d) $y(1 - 2x^2) = 1$
13) $30°, 150°, 270°; 30° + 360n°,$
    $150° + 360n°, 270° + 360n°$
14) $90°, 210°, 270°, 330°; 90° + 180n°,$
    $210° + 360n°, 330° + 360n°$
15) $60°, 300°; \pm 60° + 360n°$
16) $35.26°, 144.74°, 215.26°, 324.74°;$
    $\pm 35.26° + 180n°$
17) $90°, 270°, 45°, 225°; 90° + 180n°,$
    $45° + 180n°$
18) $90°, 270°, 23.58°, 156.42°;$
    $90° + 180n°, 23.58° + 360n°,$
    $156.42° + 360n°$

Exercise 7e − p. 209

1) a) $\frac{24}{25}$ b) $\frac{1}{2}$ c) $-\frac{7}{24}$

2) a) t^2 b) $\dfrac{1}{t}$ c) $\dfrac{1 - t^2}{2t^2}$

    d) $\dfrac{1}{3 + 6t - 5t^2}$ e) $\dfrac{1 - 4t + t^2}{3 - t^2}$

5) a) $0, 67.4°$
    b) $-153°, 130.4°$
    c) $36.9°, 126.9°$
    d) $-90°, 36.8°$
6) a) $2, \frac{1}{6}\pi$ b) $\sqrt{10}, 71.57°$
    c) $5, 36.87°$ d) $\sqrt{2}, \frac{1}{4}\pi$
    e) $\sqrt{29}, 21.80°$
7) a) $2, -2; 300°, 120°$
    b) $28, -22; 286.26°, 106.26°$
    c) $-\dfrac{1}{\sqrt{2}}, \dfrac{1}{\sqrt{2}}; 112.5°, 292.5°, 22.5°,$
     $202.5°$
    d) $-\sqrt{\frac{2}{3}}, \sqrt{\frac{2}{3}}; 125.26°, 305.26°$
    e) $25; 53.13°, 233.13°$
8) a) $2n\pi + \frac{1}{4}\pi$
    b) $360n° + 118.1°$ or $360n° - 36.9°$
    c) $360n°$ or $360n° - 143.2°$
    d) $360n°$ or $360n° - 53.2°$

Exercise 7f − p. 214

1) a) $2\sin 2A \cos A$
    b) $2\cos 4A \cos A$

c) $2 \cos 3A \sin A$

d) $-2 \sin 4A \sin 3A$

e) $-2 \cos 4A \sin A$

f) $2 \sin 3A \sin 2A$

g) $2 \sin 45° \cos 15°$

h) $2 \cos 60° \cos 10°$

i) $2 \sin (A + 45°) \cos (A - 45°)$

j) $2 \cos^2 2A$

2) a) $\sin 3\theta + \sin \theta$

b) $\cos 5\theta + \cos \theta$

c) $\sin 5\theta + \sin 3\theta$

d) $\cos 4\theta - \cos 2\theta$

e) $\cos 2\theta - \cos 6\theta$

f) $\frac{1}{2}(\cos 5\theta + \cos 3\theta)$

g) $1 - \sin 30°$

h) $\frac{1}{2} + \cos 20°$

4) a) $\cos \theta$ b) 0

6) $30°, 90°, 150°, 210°, 270°, 330°$

7) $0°, 45°, 135°, 180°, 225°, 315°, 360°$

8) $0°, 60°, 90°, 120°, 180°, 240°, 270°,$ $300°, 360°$

9) $20°, 90°, 100°, 140°, 220°, 260°,$ $270°, 340°$

10) $22\frac{1}{2}°, 90°, 112\frac{1}{2}°, 202\frac{1}{2}°, 270°, 292\frac{1}{2}°$

11) $0°, 30°, 60°, \ldots 330°, 360°$ and $40°, 80°, 160°, 200°, 280°, 320°$

12) $30°, 45°, 135°, 150°, 225°, 315°$

13) $0°, 60°, 105°, 120°, 165°, 180°,$ $240°, 285°, 300°, 345°, 360°$

14) $10°, 130°, 250°, 330°$

Multiple Choice Exercise 7 — p. 217

1) c   2) b   3) b

4) b   5) c   6) b, c

7) d, e   8) c, d   9) d

10) b, c, d   11) e   12) D

13) B   14) A   15) C

16) D   17) I   18) d

19) F   20) T   21) F

22) F

Miscellaneous Exercise 7 — p. 220

1) a) $\pm \dfrac{8\sqrt{5}}{21}, \pm \dfrac{4\sqrt{5}}{21}$ b) $10.9°, -129°$

2) a) (i) $336.3°, 143.7°$

   (ii) $226.3°, 346.3°$

   b) $\frac{1}{8}$

3) a) $-\frac{63}{65}, \frac{56}{33}$ b) $\frac{1}{6}\pi, \frac{5}{6}\pi$

4) a) $0, 126°52'$ b) $0, 90°, 180°$

5) a) $60°, 109°28'$ b) $113°42', 345°6'$

6) a) $45°, 161°34'. 225°, 341°34'$

   b) $\frac{1}{2}(2n + 1)\pi, \begin{cases} 2n\pi + \frac{1}{6}\pi \\ 2n\pi + \frac{5}{6}\pi \end{cases}$

7) a) $\tan \theta = \left(\dfrac{1 + k}{1 - k}\right) \tan \alpha; 120°, 300°$

   b) $\sin \dfrac{\pi}{10} = \dfrac{\sqrt{5} - 1}{4}$

8) a) $\frac{1}{12}\pi, \frac{5}{12}\pi, \frac{13}{12}\pi, \frac{17}{12}\pi$

9) $\frac{1}{8}\pi, \frac{1}{2}\pi, \frac{5}{8}\pi, \frac{9}{8}\pi, \frac{3}{2}\pi, \frac{13}{8}\pi$

10) $36°52'; -36°52'; 143°8';$ 0 and $-73°44'$

11) b) $72°24', 220°12'$

12) a) $-29°33', 103°17'$

   b) $0, 180°, \pm 75°31'$

13) $\tan\left(\dfrac{\pi}{4} - \dfrac{x}{2}\right); -(2 + \sqrt{3}), 2 - \sqrt{3}$

15) a) $n\pi, \frac{2}{3}n\pi$ b) $26.6°, -153.4°$

16) $2 \sin (\theta - 30°); 70°, 190°, 330°, 310°$

17) a) $32.8°, 147.2°$

   b) $n\pi + \frac{1}{2}\pi, \frac{4}{3}n\pi \pm \frac{1}{3}\pi, \frac{2}{3}n\pi$

19) b) $0, 30°, 60°, 120°, 150°, 180°$

   c) $360n° + 119.6°, 360n° - 13.3°$

20) a) $1 + \sqrt{2}, 2 + \sqrt{3}$

   b) $\cot \frac{1}{2}\theta - \cot 4\theta$

21) $\frac{5}{12}\pi$

22) a) $110°36', 216°52'$ b) $x = y = 60°$

23) a) $n\pi, \frac{1}{4}\pi(4n - 1)$

   b) $\frac{1}{8}\pi(1 + 8n), \frac{1}{24}\pi(8n - 1)$

24) a) $-120°, -90°, -60°, 0°, 60°,$ $90°, 120°$

25) a) $\begin{cases} n\pi + \frac{1}{12}\pi \\ n\pi + \frac{5}{12}\pi \end{cases}$

26) a) $0, 180°, 360°; 54.7°, 125.3°,$ $234.7°, 305.3°$

   b) $257.6°, 349.8°$

27) a) $(6n \pm 1) + \dfrac{3}{\pi} \arctan \frac{5}{12}$

   b) $\frac{1}{2}\pi, \frac{3}{2}\pi, 0.67^c, 2.47^c$

28) a) $n\pi, 2n\pi \pm \frac{1}{6}\pi$

   b) (i) $73.7°, 180°$ (ii) $1, \frac{1}{11}$

29) a) $\sin nt + \sqrt{3} \cos nt,$ $2\sqrt{3} \sin nt + 2 \cos nt;$ $4x^2 + y^2 - 2\sqrt{3}xy = 4$

   b) $199.5°, 340.5°$

30) b) $51.3°, 128.7°$

31) a) $51.7°, 14.8°$

   b) $\pm 90°, \pm 41.4°, 180°$

32) a) (i) $90°, 210°, 330°$

   (ii) $60°, 90°, 180°, 270°, 300°$

   c) $\dfrac{1}{2n + 1}; \dfrac{2n^2 - 1}{2n^2 + 4n + 1}$

33) a) $18°, 30°, 90°, 150°, 162°$

34) a) $199°28', 340°33'$

Exercise 8a — p. 235

2) a) infinite set   b) 2
   c) infinite set
   d) infinite set (all $+ve$)
   e) infinite set   f) 1
3) a) 1.17^c b) 2.55^c c) 0.676^c
4) infinite set
5) 1.28^c
6) 1.16 m

Exercise 8b — p. 239

1) a) $\frac{1}{2}$ b) $\dfrac{\theta^2}{1-2\theta^2}$

  c) $\dfrac{\theta}{2-\theta^2}$ d) 2

  e) $\frac{1}{2}\theta(\sin\alpha + \frac{1}{2}\theta\cos\alpha)$

  f) 1   g) 1

  h) $-\frac{1}{8}(\theta\cos\alpha + 4\sin\alpha)$
3) a) -2.129 b) 0.819
  c) -0.423 d) 0.136
  e) -0.087 f) -0.398

Exercise 8c — p. 245

1) $71.57°; \frac{1}{3}\pi; \frac{1}{4}\pi; 70.53°; 75.52°; \frac{1}{6}\pi;$
   $\frac{1}{6}\pi; -\frac{1}{2}\pi$

7) $\dfrac{2x}{1+x^2}$ 8) $\frac{1}{2}\pi$ 9) $\frac{7}{11}$

10) $\frac{2}{9}$ 11) ± 1 12) 0

Multiple Choice Exercise 8 — p. 246

1) c, d, e   2) b, c   3) d, e
4) a   5) a, c   6) a, b, e
7) d   8) B 9) C
10) D 11) E 12) B
13) B 14) D

Miscellaneous Exercise 8 — p. 248

1) $0.38^c, 2.77^c$
2) a) $1 + \theta - \frac{1}{2}\theta^2$ b) $\frac{1}{2}$ c) $\frac{1}{2}\theta$
  d) $\dfrac{2(\sqrt{2}-\theta)}{2-\theta^2}$

3) a) 0.126   b) 1.664   c) 0.876
4) a) 0.74^c b) $0.67^c, 2.48^c$
  c) $0.93^c, 3.14^c$ d) 0.65^c e) 1.65^c
5) a) (i) 1   (ii) 0
  b) (i) 0   (ii) 0
  c) (i) 1   (ii) 0
  d) (i) ∞ (ii) ∞
  e) (i) ∞ (ii) ∞
  f) (i) 1   (ii) 1

6) 1.001^c
7) $a^2(2\theta - \sin 2\theta); 1.16^c$
8) 2.28^c
9) 0.42^c
10) $0, \frac{1}{2}r^2, \dfrac{\sqrt{3}}{2}r^2$
11) $0, 30°, 70°$
12) $0, \frac{1}{4}\pi$
14) $\frac{1}{10}\sqrt{2}$
15) $\frac{1}{3}\sqrt{3}$

Exercise 9a — p. 252

1) $-\sin x$
2) a) $\cos x - \sin x$ b) $\sin x$
  c) $2\cos\theta$ d) $-4\sin\theta$
  e) $2\cos\theta + 3\sin\theta$ f) $4\cos t$
3) a) $-\sin x$ b) $\dfrac{\sqrt{2}}{2}(\cos x - \sin x)$
  c) $-\frac{1}{2}(\cos x + \sqrt{3}\sin x)$
4) a) -1 b) -1 c) $\dfrac{5\sqrt{2}}{2}$ d) $\dfrac{2\pi}{3} - \dfrac{1}{2}$
5) a) $\dfrac{2\pi}{3}$ b) 2.5^c c) 2.82^c
6) a) $2n\pi, 2n\pi + 1.107^c, 2n\pi - 0.927^c$
  b) $(2n+1)\pi, 2n\pi + 4.25^c$
   $2n\pi + 2.215^c$
7) a) $5x + \sqrt{2}y - 1 - \dfrac{5\pi}{4} = 0$
  b) $18y = (12\pi - 27)x + 9\pi$
   $- 2\pi^2 - 27\sqrt{3}$

Exercise 9b — p. 257

1) $20.1, 0.135, 5.47, 0.819$
2) a) 0.693   b) -1.10
3) a) $2e^x$ b) $2x - e^x$ c) $e^x - \sin x$
4) a) 0   b) -0.69
  c) no stationary values

Exercise 9c — p. 258

1) $2.0568, -0.7277, 3.0680$
2) a) $\ln x - \ln(x+1)$
  b) $\ln 3 + 2\ln x$
  c) $\ln(x+2) + \ln(x-2)$
  d) $\frac{1}{2}\ln(x-1) - \frac{1}{2}\ln(x+1)$
  e) $\ln\cos x - \ln\sin x$
  f) $\ln 4 + 2\ln\cos x$
3) a) $\ln\left[\dfrac{x}{(1-x)^2}\right]$ b) $\ln\dfrac{e}{x}$
  c) $\ln\sin x\cos x$ d) $\ln\left[\dfrac{x^3}{\sqrt{(x-1)}}\right]$

4) a) 2.10   b) 0   c) 1.05   d) 0

Exercise 9d — p. 260

1) a) $\dfrac{2}{x}$ b) $\dfrac{5}{x}$ c) $\dfrac{2}{x}$

d) $-\dfrac{3}{2x}$ e) $-\dfrac{1}{2x}$ f) $\dfrac{1}{2x}$

g) $-\dfrac{5}{2x}$ h) $\dfrac{-1}{4x}$ i) $\dfrac{1}{x \ln 10}$

2) a) $(1, 1)$ b) $(1, -1), (-1, -1)$

Exercise 9e — p. 261

1) $uv, u = e^x, v = x^2 - 1$
2) $e^u, u = 3x^2$
3) $\sin u, u = x^2 - 2$
4) $uv, u = x^3, v = \cos x$
5) $\dfrac{u}{v}, u = \sqrt{(x - 1)}, v = \sqrt{(x + 1)}$
6) $u^4, u = x + 1$
7) $\ln u, u = x^2 - 1$
8) $uv, u = x^2 - 1, v = (x - 2)^5$
9) $\tan u, u = x^2$
10) $\ln u - \ln v, u = x + 1, v = x - 1$
11) $u^2, u = \cos x$
12) $u^{\frac{1}{2}}, u = \ln x$

Exercise 9f — p. 263

1) $3e^{3x}$
2) $\dfrac{2}{x - 1}$
3) $5(x + 1)^4$
4) $-3 \sin \left(3\theta - \dfrac{\pi}{4}\right)$
5) $10x(x^2 + 1)^4$
6) $2xe^{x^2}$
7) $\dfrac{8x}{x^2 + 1}$
8) $3x(3x^2 + 4)^{-\frac{1}{2}}$
9) $10e^{2x}$
10) $2\theta \cos \theta^2$
11) $2 \sin \theta \cos \theta$
12) $3e^{3x}$
13) $-(x + 1)^{-2}$
14) $\dfrac{-2x}{(x^2 + 1)^2}$
15) $-\dfrac{1}{2(x - 1)^{3/2}}$

16) $6(x + 1)^5$
17) $6(2x - 4)^2$
18) $2x\, e^{(x^2 + 2)}$
19) $3 \cos \left(3\theta + \dfrac{\pi}{4}\right)$
20) $-2 \cos \theta \sin \theta$
21) $\dfrac{2x}{x^2 + 2}$
22) $-4(2x + 3)^{-3}$
23) $-(2x - 1)^{-\frac{3}{2}}$
24) $e^x + e^{-x}$
25) $-e^{-x} + 2e^{-2x}$
26) $6 \cos \left(2\theta - \dfrac{\pi}{4}\right)$
27) $-\dfrac{\cos \theta}{\sin^2 \theta}$
28) $\dfrac{\sin \theta}{\cos^2 \theta}$
29) $-6e^{-3x} + 4e^{4x}$
30) $\cot x$
31) $-k\, e^{-kt}$
32) $\dfrac{3}{x - 1}$
33) $\cos x\, e^{\sin x}$
34) $\dfrac{-2 \cos x}{\sin^3 x}$
35) $\dfrac{2 \sin t}{\cos^3 t}$
36) $6 \cos 3\theta\, e^{\sin 3\theta}$
37) $\dfrac{2}{x - 1}$
38) $\dfrac{2\theta}{\sqrt{(\theta^2 + 4)}} \cos (\theta^2 + 4)^{\frac{1}{2}}$
39) $e^x\, e^{e^x}$
40) $-8x \cos (x^2 + 1) \sin (x^2 + 1)$
41) $3e^x \cos e^x$
42) $\dfrac{\sin \theta \cos \theta}{\sqrt{(3 - \cos^2 \theta)}}$
43) $2 \cot 2x$
44) $\dfrac{-e^{\sqrt{1 - x}}}{2\sqrt{1 - x}}$
45) $4x \sin (x^2 + 1) \cos (x^2 + 1)$
46) $-\dfrac{2}{x^2} e^{1/x}$
47) $-\operatorname{cosec} x$
48) $-(e^x + e^{-x})(e^x - e^{-x})^{-2}$

Exercise 9g — p. 267

1) $1 + \ln x$

2) $\dfrac{5x^2 - 4x}{2\sqrt{(x-1)}}$

3) $\dfrac{\ln x - 1}{(\ln x)^2}$

4) $3 \cos x \cos 2x - 6 \sin x \sin 2x$

5) $\sec^2 x$

6) $\dfrac{e^x \,(x-2)}{(x-1)^2}$

7) $1 + \ln (x-1)$

8) $\cos x \ln x + \dfrac{1}{x} \sin x$

9) $\sec x \,(1 + x \tan x)$

10) $-\dfrac{1}{\sin^2 x}$

11) $\dfrac{2}{(x+1)^2}$

12) $e^x \,(\sin x + \cos x)$

13) $\dfrac{e^x \,(\sin x - \cos x)}{\sin^2 x}$

14) $\dfrac{1 + \ln x}{\ln 10}$

15) $\dfrac{-2}{(e^x - e^{-x})^2}$

16) $\dfrac{1 - x^2}{(x^2 + 1)^2}$

17) $\dfrac{x^2 - 1}{x^2}$

18) $\dfrac{2 - x}{2x^2 \sqrt{(x-1)}}$

19) $\dfrac{2x \cos x - \sin x}{2x\sqrt{x}}$

20) $\dfrac{(x-1) \ln (x-1) - x \ln x}{x(x-1) \ln^2 (x-1)}$

21) $- \operatorname{cosec}^2 x$

22) $\dfrac{-2x}{(x-1)^2 (x+1)^2}$

23) $-\dfrac{2}{3(x-2)^2} - \dfrac{1}{3(x+1)^2}$

24) $\dfrac{4}{5(2x-1)^2} - \dfrac{6}{5(x-3)^2}$

25) $-\dfrac{7}{5(x+2)^2} - \dfrac{2}{5(2x-1)^2}$

26) $-\dfrac{5}{2(x-1)^2} + \dfrac{10}{(x-2)^2} - \dfrac{15}{2(x-3)^2}$

27) $\dfrac{10}{3(2x-1)^2} - \dfrac{1}{(x-1)^2} - \dfrac{1}{3(x+1)^2}$

28) $-\dfrac{9}{10(x+3)^2} - \dfrac{x^2 - 6x - 1}{10(x^2 + 1)^2}$

29) $-\dfrac{1}{4(x+1)^2} + \dfrac{1}{4(x-1)^2} - \dfrac{1}{(x-1)^3}$

30) $\dfrac{9}{10(x+3)^2} - \dfrac{9x^2 + 6x - 9}{10(x^2 + 1)^2}$

Exercise 9h — p. 268

1) $3 \cos x \, e^{\sin x}$

2) $- e^{-x} \sin x + e^{-x} \cos x$

3) $\dfrac{-(1 + 2x)}{(1 + x + x^2)^2}$

4) $3 \sin^2 x - 4 \sin^4 x$

5) $\tan x$

6) $7 \cos 7t - \cos t$

7) $2e^{x^2}\left(x \ln 2x^2 + \dfrac{1}{x} \right)$

8) $4 \sec^2 2x \tan 2x$

9) $\dfrac{-1 - \sin x - \cos x}{(1 + \cos x)(1 + \sin x)}$

10) 4

12) $\dfrac{4}{(1-x)^3}$

13) -1

14) $\dfrac{e^x}{2x\sqrt{(x-1)}} \left[(2x^2 - x) \ln x + 2x - 2 \right]$

15) $\dfrac{1}{\sqrt{(1 - x^2)}}$

16) $-\dfrac{1}{2(\frac{1}{2}\pi - 1)^{3/2}}$

17) ± 1 18) $0, 8$

19) $\dfrac{7\pi}{4}$ 20) $-\dfrac{1}{e}$

Exercise 9i — p. 276

1) $2x + 2y \dfrac{dy}{dx} = 0$

2) $2x + y + (x + 2y)\dfrac{dy}{dx} = 0$

3) $2x + x \dfrac{dy}{dx} + y = 2y \dfrac{dy}{dx}$

4) $-\dfrac{1}{x^2} - \dfrac{1}{y^2}\dfrac{dy}{dx} = e^y\dfrac{dy}{dx}$

5) $-\dfrac{2}{x^3} - \dfrac{2}{y^3}\dfrac{dy}{dx} = 0$

6) $\dfrac{x}{2} - \dfrac{2y}{9}\dfrac{dy}{dx} = 0$

7) $\cos x + \cos y\,\dfrac{dy}{dx} = 0$

8) $\cos x \cos y - \sin x \sin y\,\dfrac{dy}{dx} = 0$

9) $e^y + xe^y\,\dfrac{dy}{dx} = 1$

10) $(1 + x)\dfrac{dy}{dx} = 2x - 1 - y$

11) $\dfrac{dy}{dx} = \pm\dfrac{1}{\sqrt{(2x + 1)}}$

14) $\dfrac{x}{(2 - x^2)^{3/2}}$

15) $\pm\dfrac{1}{2\sqrt 2}$

17) $\dfrac{dy}{dx} = -y\ln 2, -\tfrac14 \ln 2$

18) $\tfrac13(\ln 3)^2$

19) $\tfrac12$

22) a) $a^x \ln a + b^x \ln b$

b) $\left(\dfrac{1}{x}\sin x + \cos x \ln x\right)x^{\sin x}$

c) $(\sin x)^x(\ln \sin x + x \cot x)$

d) $(x + x^2)^x\left[\ln (x + x^2) + \dfrac{1 + 2x}{1 + x}\right]$

e) $\dfrac{x}{(x - 1)(x + 3)^2(x^2 - 1)}$
$\times\left(\dfrac{1}{x} - \dfrac{1}{x - 1} - \dfrac{2}{x + 3} - \dfrac{2x}{x^2 - 1}\right)$

f) $\dfrac{6 + 4x - 9x^2}{(x^2 - 1)^{3/2}(3x - 4)^3}$

23) a) $\ln y + \dfrac{x}{y}\dfrac{dy}{dx} = \dfrac{1}{x}$

b) $\ln x\,\dfrac{dy}{dx} + \dfrac{y}{x} = \cot x$

c) $\dfrac{1}{y}\dfrac{dy}{dx} + \dfrac{1}{1 + x} = \tfrac12 \cot x$

d) $\dfrac{\sin x}{y}\dfrac{dy}{dx} + \cos x \ln y = \dfrac{1}{2x}$

24) a) $xx_1 - 3yy_1 = 2(y + y_1)$
b) $x(2x_1 + y_1) + y(2y_1 + x_1) = 6$

26) $3x + 12y - 7 = 0$

Exercise 9j – p. 283

1) a) $y^2 = 4x$ b) $x^2 + y^2 = 1$
c) $xy = 2$

3) a) $\dfrac{1}{t}$ b) $-\cot \theta$ c) $-\dfrac{1}{2t^2}$

4) a) $-\dfrac{1}{2t^3}$ b) $-\csc^3\theta$ c) $\dfrac{1}{2t^3}$

5) $\left(\dfrac{\sqrt 3}{3}, -\dfrac{2\sqrt 3}{9}\right)$ min, $\left(-\dfrac{\sqrt 3}{3}, \dfrac{2\sqrt 3}{9}\right)$ max

6) $\left(2n\pi + \dfrac{\pi}{2}, 1\right)$

7) $e^{-t}\cos t, -e^{-2t}(\cos t + \sin t)$

8) $x + t^2 y - 2t = 0, (2t, 0), \left(0, \dfrac{2}{t}\right)$

9) $tx + 2y - 8t - t^3 = 0,$
$(8 + t^2, 0), (0, \tfrac12 t\{8 + t^2\})$

Multiple Choice Exercise 9 – p. 285

1) b	2) c	3) b	4) a
5) e	6) d	7) a	8) d
9) a	10) b	11) a, b	12) b
13) b, c	14) b, c	15) a	16) C
17) C	18) A	19) D	20) E
21) c	22) c	23) A	24) I
25) F	26) T	27) T	28) F
29) F	30) F		

Miscellaneous Exercise 9 – p. 288

1) a) $A = a^2 - b^2, B = 2ab$
b) $16/(1 - 2x)^3$
c) $\dfrac{x}{y}, (y^2 - x^2)/y^3$

2) $-\tfrac{7}{11}$

3) a) $-\dfrac{(x^3 + 4)}{2x^3(1 + x^3)^{1/2}},$
$\dfrac{1 + 2\cos x - 3\sin x}{(2 + \cos x)(3 - \sin x)}$

4) a) (i) $2\sec 2x$ (ii) $2x/(1 + x^2)^2$
b) $\ln (3\pi/4)$

5) $x^x(1 + \ln x), -1/(1 + x^2)$

6) a) $\left(\dfrac{1}{e}, -\dfrac{1}{e}\right)$ min

6) b) $(0, 2), (0, -2), 3x - 5y + 8 = 0$

9) a) $\dfrac{\sin 2x}{x^3\,e^{3x}} - \dfrac{\cos 2x}{x^2\,e^{3x}} - 3y$ b) 1

10) -2 11) -1.15

13) $1 + \dfrac{1}{3(2x-1)} - \dfrac{2}{3(x+1)}, \dfrac{5}{2}$

14) a) $\tan x + a^x \ln a$ b) $\frac{1}{9}$ max, 1 min

15) b) (iii) $(3t^2 + 3)/2t$

16) a) (i) $1/\sqrt{(x^2+1)}$

   (ii) $4 \sec^2 2x \tan 2x$

   (iii) $(3 \ln 10)\, 10^{3x}$

   b) $\dfrac{dy}{dx} = \dfrac{x}{1-y}$

17) a) $-2e^{-2t}(1-2t)^2$ b) $2/(1+4t^2)$

19) $-\frac{1}{2}$

20) a) (i) $2x \sin 3x + 3x^2 \cos 3x$

   (ii) $\dfrac{2}{x^2}\, e^{-2/x}$ (iii) $\dfrac{2(x-1)}{(2-x)^3}$

   b) max

21) $\begin{cases} (x-1)\ln(1-x) \\ \quad -(x+1)\ln(x+1) - 2x \\ \div (1-x^2)(2 - 2\ln[1-x])^2 \end{cases}$

22) $\left(a\left\{\dfrac{\pi}{2}-1\right\}, a\right), \left(a\left\{\dfrac{3\pi}{2}+1\right\}, a\right),$

   $y = x - a\left(\dfrac{\pi}{2}-2\right),$

   $y = -x + a\left(\dfrac{3\pi}{2}+2\right)$

23) $-2\sqrt{(1-x^2)}$

Note. A constant of integration should be added to every indefinite integral in Chapter 10

Exercise 10a – p. 295

1) $\frac{1}{7}x^7; \frac{3}{4}x^{\frac{4}{3}}; -\frac{1}{3}x^{-3}; -2x^{-\frac{1}{2}}; \frac{4}{5}x^{\frac{5}{4}}; -\frac{1}{6}x^{-6}$

2) $\dfrac{2}{3}x^3 + \dfrac{1}{x} + \dfrac{x^2}{2}$

3) $\frac{2}{3}x^{\frac{3}{2}} + \frac{3}{2}x^{\frac{2}{3}}$

4) $\frac{3}{4}x^4 - \frac{1}{2}x^{-2} + 3x$

5) $-\dfrac{2}{x^2} + \dfrac{1}{x} - \dfrac{x^3}{3}$

6) $\frac{4}{7}x^{\frac{7}{2}} - \frac{5}{3}x^{\frac{3}{5}}$

7) $x^5 - x^3 + 7x$

8) $-2x^{-2} - \frac{1}{3}x^{-3} + x$

9) $6x^{\frac{1}{2}} + 2x^{-\frac{1}{2}}$

10) $\frac{1}{4}x^2 - 4\sqrt{x} - x$

11) $-\frac{1}{3}x^{-3} + \frac{4}{3}x^{\frac{3}{4}} - 4x$

12) $4x^{\frac{3}{2}} - \frac{3}{4}x^4 - x^{-1} + 2x$

Exercise 10b – p. 296

1) $3x; \frac{1}{2}x^6; \frac{81}{2}x^6; \frac{1}{2}(x+1)^6$

2) $x^4 - \frac{5}{2}x^2 + 6x$

3) $\frac{1}{6}(2x-1)^3$

4) $\frac{1}{28}(2+7x)^4$

5) $-\frac{8}{3}(1-x)^{\frac{3}{2}}$

6) $-\frac{2}{5}\sqrt{(2-5x)^3}$

7) $-\dfrac{1}{8(4x+5)^2}$

8) $-\sqrt{1-2x}$

9) $-\frac{3}{14}(3-7x)^{\frac{2}{3}}$

10) $-\dfrac{3}{4(2x+1)^2} - \dfrac{\sqrt{(1-2x)^3}}{3}$

11) $\frac{8}{3}\sqrt{x^3} + \frac{1}{6}\sqrt{(4x+1)^3} + \frac{1}{3}(1-3x)^4$

12) $\dfrac{(px+q)^{r+1}}{p(r+1)}$

13) $\frac{4}{15}(3+5x)^{\frac{3}{4}}$

14) $-\frac{9}{10}(4-5x)^{\frac{2}{3}}$

15) $-\frac{2}{3}\sqrt{(1-x)^3} - 2\sqrt{(1-x)} - \dfrac{1}{1-x}$

Exercise 10c – p. 297

1) $-\frac{1}{2}\cos 2x$

2) $\frac{3}{4}\sin(4x - \frac{1}{2}\pi)$

3) $-\frac{1}{2}\tan(\frac{1}{3}\pi - 2x)$

4) $-\frac{1}{3}\cot(3x + \frac{1}{6}\pi)$

5) $-5\sec(\frac{1}{4}\pi - x)$

6) $-\frac{2}{3}\operatorname{cosec} 3x$

7) $-\frac{2}{3}\cos(3x + \alpha)$

8) $-10\sin(\alpha - \frac{1}{2}x)$

9) $\frac{1}{3}\sin 3x - 3\cos x$

10) $\frac{1}{2}\tan 2x + \frac{1}{4}\cot 4x$

Exercise 10d – p. 299

1) $\frac{1}{2}e^{2x}$ 2) $-3e^{-x}$

3) $\frac{1}{4}e^{4x+1}$ 4) $-\frac{4}{3}e^{5-3x}$

5) $\dfrac{2^x}{\ln 2}$ 6) $\dfrac{3^{2x}}{2\ln 3}$

7) $-\dfrac{3^{1-x}}{\ln 3}$ 8) $2\sqrt{e^x}$

9) $\frac{1}{2}(e^{2x} - e^{-2x})$ 10) $-e^{-3x} - \frac{1}{4}e^{2x}$

Exercise 10e — p. 300

1) $\frac{1}{2}\ln|x|$

2) $2\ln|x|$

3) $\frac{1}{3}\ln|3x+1|$

4) $-\frac{1}{3}\ln|1-3x|$

5) $2\ln|1+2x|$

6) $-\frac{3}{2}\ln|4-2x|$

7) $6\ln|x-2|-3\ln|x-1|$

8) $\dfrac{1}{6(2-3x)^2}$

9) $-\frac{2}{3}\sqrt{2-3x}$

10) $-\frac{1}{3}\ln|2-3x|$

11) $\frac{1}{3}\cos(\frac{1}{2}\pi-3x)$

12) $\frac{1}{12}(4x+1)^3$

13) $\frac{1}{3}x^3$

14) $\dfrac{4^x}{\ln 4}$

15) $-\frac{1}{5}e^{4-5x}$

16) $\frac{1}{4}\tan 4x$

17) $-4\ln|1-x|$

18) $\frac{1}{3}\sqrt{(2x+3)^3}$

19) $\frac{1}{6}e^{6x}$

20) $-\frac{5}{7}\ln|6-7x|$

Exercise 10f — p. 304

1) e^{x^4}

2) $-e^{\cos x}$

3) $e^{\tan x}$

4) e^{x^2+x}

5) $e^{1-\cot x}$

6) $\frac{1}{10}(x^2-3)^5$

7) $-\frac{1}{3}\sqrt{(1-x^2)^3}$

8) $\frac{1}{6}(\sin 2x+3)^3$

9) $-\frac{1}{6}(1-x^3)^2$

10) $\frac{2}{3}\sqrt{(1+e^x)^3}$

11) $\frac{1}{5}\sin^5 x$

12) $\frac{1}{4}\tan^4 x$

13) $\dfrac{1}{3(n+1)}(1+x^{n+1})^3$

14) $-\frac{1}{3}\cot^3 x$

15) $\frac{4}{9}\sqrt{(1+x^{\frac{3}{2}})^3}$

16) $\frac{1}{12}(x^4+4)^3$

17) $-\frac{1}{4}(1-e^x)^4$

18) $\frac{2}{3}\sqrt{(1-\cos\theta)^3}$

19) $\frac{1}{3}\sqrt{(x^2+2x+3)^3}$

20) $\frac{1}{2}e^{x^2+1}$

Exercise 10g — p. 309

1) $x\sin x+\cos x$

2) $e^x(x^2-2x+2)$

3) $\frac{1}{16}x^4(4\ln 3x-1)$

4) $-e^{-x}(x+1)$

5) $3(\sin x-x\cos x)$

6) $\frac{1}{5}e^x(\sin 2x-2\cos 2x)$

7) $\frac{1}{5}e^{2x}(\sin x+2\cos x)$

8) $\frac{1}{32}e^{4x}(8x^2-4x+1)$

9) $-\frac{1}{2}e^{-x}(\cos x+\sin x)$

10) $x(\ln 2x-1)$

11) $x\,e^x$

12) $\frac{1}{72}(8x-1)(x+1)^8$

13) $\sin\left(x+\dfrac{\pi}{6}\right)-x\cos\left(x+\dfrac{\pi}{6}\right)$

14) $\dfrac{1}{n^2}(\cos nx+nx\sin nx)$

15) $\dfrac{x^{n+1}}{(n+1)^2}[(n+1)\ln|x|-1]$

16) $\frac{3}{4}(2x\sin 2x+\cos 2x)$

17) $\frac{1}{5}e^x(\sin 2x-2\cos 2x)$

18) $(2-x^2)\cos x+2x\sin x$

19) $\dfrac{e^{ax}}{a^2+b^2}(a\sin bx-b\cos bx)$

20) $\frac{1}{3}\sin\theta\,(3\cos^2\theta+2\sin^2\theta)$

21) $\frac{1}{2}e^{x^2-2x+4}$

22) $(x^2+1)e^x$

23) $-\frac{1}{4}(4+\cos x)^4$

24) $e^{\sin x}$

25) $\frac{2}{15}\sqrt{(1+x^5)^3}$

26) $\frac{1}{5}(e^x+2)^5$

27) $\cos\left(\dfrac{\pi}{4}-x\right)-x\sin\left(\dfrac{\pi}{4}-x\right)$

28) $\frac{1}{4}e^{2x-1}(2x-1)$

29) $-\frac{1}{20}(1-x^2)^{10}$

30) $\frac{1}{6}\sin^6 x$

Exercise 10h — p. 313

1) $\ln|3+\sin x|$

2) $-\ln|1-e^x|$

3) $\ln(1+x^2)$

4) $-\frac{1}{3}\ln|1-3\tan x|$

5) $-\ln(e^{-x}+1)$

6) $\frac{1}{4}\ln(1+x^4)$

7) $\frac{1}{4}\ln|2\sin 2x+1|$

8) $\ln|\sec x+1|$

9) $\ln|\ln x|$

10) $\ln|x^2+3x+4|$

11) $\ln|\sin x|$

12) $\ln|\tan x+\sec x|$

13) $-\ln|\cot x+\csc x|$

14) $\sqrt{x^2+1}$

15) $-\dfrac{1}{5\sin^5 x}$

16) $2\sqrt{1+\sin x}$

17) $-\dfrac{1}{e^x+4}$

18) $-\dfrac{1}{2\tan^2 x}$

19) $\dfrac{1}{(n-1)\cos^{n-1}x}$

20) $\dfrac{-1}{(n-1)\sin^{n-1}x}$

21) $2\sqrt{1+e^x}$

22) $-\ln|3-\sec x|$

23) $\dfrac{1}{3(2+\cot x)^3}$

24) $\frac{1}{6}\ln|3x^2-6x+1|$

25) $x\arcsin x+\sqrt{1-x^2}$

26) $x\arctan x-\frac{1}{2}\ln(1+x^2)$

27) $x\arccos x-\sqrt{1-x^2}$

28) $x-\ln|x+1|$

29) $x-\frac{1}{2}\ln\left|\dfrac{x-1}{x+1}\right|$

30) $x+\ln\left|\dfrac{x+1}{(x+2)^4}\right|$

31) $x+4\ln|x|$

32) $x+3\ln|x+1|$

33) $\ln|x^2-4|$

34) $\ln\left|\dfrac{u^4}{u+1}\right|$

35) $x+5\ln\left|\dfrac{x}{x+1}\right|$

36) $\ln\left|\dfrac{y-2}{(y-1)^2}\right|$

37) $\ln|z^2-5z+6|$

38) $12\ln\left|\dfrac{(3-x)^3}{(2-x)(4-x)^2}\right|$

39) $\frac{1}{2}\ln\left|\dfrac{(x-1)^7(x+1)}{(2x-1)^7}\right|$

40) $2u-\ln|u+1|-\frac{5}{2}\ln|2u+3|$

Exercise 10i – p. 317

1) a) $\frac{3}{8}\theta-\frac{1}{4}\sin 2\theta+\frac{1}{32}\sin 4\theta$

   b) $\sin\theta-\frac{1}{3}\sin^3\theta$

   c) $\frac{1}{3}\tan^3\theta-\tan\theta+\theta$

   d) $\frac{1}{3}\cos^3\theta-\cos\theta$

e) $\frac{3}{8}\theta+\frac{1}{4}\sin 2\theta+\frac{1}{32}\sin 4\theta$

f) $\frac{1}{4}\tan^4\theta-\frac{1}{2}\tan^2\theta-\ln|\cos\theta|$

2) a) $-(\frac{1}{7}\cos 7\theta+\cos\theta)$

   b) $\frac{1}{21}(3\sin 7\theta+7\sin 3\theta)$

   c) $\frac{1}{16}(2\cos 4\theta-\cos 8\theta)$

   d) $\frac{1}{8}(2\sin 2\theta-\sin 4\theta)$

   e) $-\dfrac{1}{n+m}\cos(n+m)x$
   $-\dfrac{1}{n-m}\cos(n-m)x$

   f) $\dfrac{6}{5}\left(5\sin\dfrac{u}{6}+\sin\dfrac{5u}{6}\right)$

   g) $\dfrac{1}{2(n+m)}\sin(n+m)x$
   $+\dfrac{1}{2(n-m)}\sin(n-m)x$

3) a) $\frac{1}{3}\sin^3 x-\frac{1}{5}\sin^5 x$

   b) $\frac{1}{11}\sin^{11}x-\frac{1}{13}\sin^{13}x$

   c) $\frac{1}{32}(4x-\sin 4x)$

   d) $\frac{1}{5}\tan^5 x+\frac{1}{3}\tan^3 x$

   e) $\frac{1}{10}\cos^5 2x-\frac{1}{6}\cos^3 2x$

   f) $\dfrac{1}{n+1}\sin^{n+1}x-\dfrac{1}{n+3}\sin^{n+3}x$

   g) $\frac{1}{5}\cos^5 x-\frac{1}{3}\cos^3 x$

   h) $\frac{1}{4}\tan^4 x$

Exercise 10j – p. 321

1) $\frac{1}{21}(x+3)^6(3x+2)$

2) $\frac{1}{2}\arctan\dfrac{x}{2}$

3) $-\frac{2}{3}(x+6)\sqrt{3-x}$

4) $\frac{2}{15}(3x-2)(x+1)^{\frac{3}{2}}$

5) $\dfrac{1-5x}{10(x-3)^5}$

6) $\ln|x+\sqrt{1+x^2}|$

7) $\frac{4}{135}(9x+8)(3x-4)^{\frac{3}{2}}$

8) $-\frac{1}{36}(8x+1)(1-x)^8$

9) $\arcsin\dfrac{x}{3}$

10) $\dfrac{5+4x}{12(4-x)^4}$

11) $\frac{1}{2}e^{2x+3}$

12) $\frac{1}{6}(2x^2-5)^{\frac{3}{2}}$

13) $\frac{1}{12}(6x-\sin 6x)$

14) $-\frac{1}{2}e^{-x^2}$

15) $-\frac{1}{8}(\cos 4\theta + 2 \cos 2\theta)$

16) $\frac{1}{110}(10u - 7)(u + 7)^{10}$

17) $-\dfrac{1}{12(x^3 + 9)^4}$

18) $\frac{1}{2} \ln |1 - \cos 2y|$

19) $\frac{1}{2} \ln |2x + 7|$

20) $\arcsin u$

21) $-\frac{2}{9}(1 + \cos 3x)^{\frac{3}{2}}$

22) $\frac{1}{16}(\sin 4x - 4x \cos 4x)$

23) $\frac{1}{2} \ln |x^2 + 4x - 5|$

24) $\frac{1}{3} \ln |(x + 5)^2 (x - 1)|$

25) $-\frac{1}{4}(x^2 + 4x - 5)^{-2}$

26) $-(9 - y^2)^{\frac{3}{2}}$

27) $\frac{1}{13}e^{2x}(2 \cos 3x + 3 \sin 3x)$

28) $x(\ln |5x| - 1)$

29) $\frac{1}{6} \sin 2x(3 - \sin^2 2x)$

30) $-e^{\cot x}$

31) $-2\sqrt{7 + \cos y}$

32) $e^x(x^2 - 2x + 2)$

33) $\frac{1}{2} \ln |x^2 - 4|$

34) $x + \ln \left| \dfrac{x - 2}{x + 2} \right|$

35) $\frac{1}{4} \ln \left| \dfrac{x - 2}{x + 2} \right|$

36) $\frac{1}{30}(3 \sin 5x + 5 \sin 3x)$

37) $-\frac{1}{30} \cos 2\theta (15 - 10 \cos^2 2\theta$
$+ 3 \cos^4 2\theta)$

38) $\frac{1}{15} \cos^3 u(3 \cos^2 u - 5)$

39) $\frac{1}{3} \tan^3 \theta - \tan \theta + \theta$

40) $\frac{1}{4} \tan^4 \theta - \frac{1}{2} \tan^2 \theta - \ln \cos \theta$

41) $\arcsin x + 2\sqrt{1 - x^2}$

42) $\ln |\ln u|$

43) $\frac{1}{27}(9y^2 \sin 3y + 6y \cos 3y - 2 \sin 3y)$

44) $-\ln |1 - \tan x|$

45) $\frac{1}{3}(7 + x^2)^{\frac{3}{2}}$

46) $-\frac{1}{5} \cos \left(5\theta - \dfrac{\pi}{4}\right)$

47) $\sin \theta \{\ln |\sin \theta| - 1\}$

48) $e^{\tan u}$

49) $\dfrac{2x - 1}{10(3 - x)^6}$

50) $\frac{1}{3} \tan^3 x$

Exercise 10k – p. 323

1) $2y = 3x^2 - 8x + 9$

2) $2y = 2x^3 - 5x^2 + 2x + 6$

3) $y = 3e^{2x} - 1$

4) $15(1 - y) = (7 - 5x)^3$

5) $3y = \sin 3x + 4$

6) $y = x(x - 1)$

7) $\left(0, \dfrac{7 - e^3}{3}\right)$

Exercise 10l – p. 326

1) $\frac{3}{2}t^2 + 1; \frac{1}{2}t^3 + t$

2) $1 - \frac{1}{2} \cos 2t; t - \frac{1}{4} \sin 2t + 3$

3) $\dfrac{t}{t + 1}; t + 2 - \ln |t + 1|$

4) a) $\dfrac{1 + \sqrt{7}}{3}$ b) 2 c) $1, \sqrt{2}$

Exercise 10m – p. 330

1) $y^2 = A - 2 \cos x$

2) $x + \dfrac{1}{y} = A$

3) $\ln (x^2 + 1) = 2y + A$

4) $y = A(x - 3)$

5) $x + A = 4 \ln |\sin y|$

6) $u^2 = v^2 + 4v + A$

7) $16y^3 = 12x^4 \ln |x| - 3x^4 + A$

8) $y^2 + 2(x + 1) e^{-x} = A$

9) $\sin x = A - e^{-y}$

10) $2r^2 = 2\theta - \sin 2\theta + A$

11) $u + 2 = A(v + 1)$

12) $y^2 = A + (\ln |x|)^2$

13) $y^2 = Ax(x + 2)$

14) $4v^3 = 3(2 + t)^4 + A$

15) $x^2 = A(1 + y^2)$

16) $e^{-x} = e^{1 - y} + A$

17) $A - \dfrac{1}{y} = 2 \ln |\tan x|$

18) $Ar = e^{\tan \theta}$

19) $y^2 = A - \text{cosec}^2 x$

20) $v^2 + A = 2u - 2 \ln |u|$

21) $2y^2 = x^2 - 1$

22) $e^t(5 - 2\sqrt{s}) = 1$

23) $y^3 = x^3 + 3x - 13$

24) $(6 - y)x = 4$

25) $2(x - 1) = (x + 1)(y + 1)^2$

26) $y = e^x - 2$

27) $s\dfrac{ds}{dt} = k$ 28) $\dfrac{dh}{dt} = k \ln |H - h|$

29) a) $\dfrac{dn}{dt} = 0$ b) $\dfrac{dn}{dt} = k\sqrt{n}$

30) a) $\dfrac{dn}{dt} = K_1 n$ b) $\dfrac{dn}{dt} = \dfrac{K_2}{n}$

c) $\dfrac{dn}{dt} = -K_3$

Exercise 10n — p. 337

1) $10\frac{1}{3}$　　　　　2) 0
3) $\frac{1}{4}(e^4 - 1)$　　4) $\ln\frac{5}{4}$
5) $12\frac{2}{3}$　　　　　6) $-\frac{1}{3}$
7) $\frac{1}{24}(2\pi - 3\sqrt{3})$　8) $\frac{1}{2}e(e^3 - 1)$
9) $\frac{31}{160}$　　　　　10) $\ln 2$
11) 1　　　　　　　12) $\ln\frac{32}{27}$
13) $\ln\sqrt{\frac{5}{2}}$　　　14) $\frac{1}{4}$
15) $\frac{31}{15}$　　　　　16) $\ln 16 - \frac{15}{16}$
17) $\frac{2}{3}$　　　　　　18) 8

19) 5.14　　　　　20) $\frac{1}{2}\left(e^{\frac{\pi}{2}} + 1\right)$

21) 1　　　　　　　22) $\frac{35}{72}$
23) $\frac{1}{3}$　　　　　　24) $\frac{1}{27}(8\sqrt{3} - 9)$
25) $\frac{1}{3}\pi$　　　　　26) $\frac{1}{48}\pi$

27) $-\frac{2}{3}$　　　　　28) $\dfrac{16\sqrt{2}}{3}$

29) $4 - 2\sqrt{2}$　　　30) $\frac{1}{10}\ln\frac{9}{4}$

Exercise 10p — p. 342

1) 12　　　　　　　2) $2\ln 4$
3) $2 - 3\ln 4$　　　4) $\frac{4}{3}$
5) 36　　　　　　　6) 1
7) $\ln\sqrt{2}$　　　　8) $\frac{2}{3}$
9) $e - 1$　　　　　10) 36
11) a) -2　b) 2　c) 0
12) 2

Multiple Choice Exercise 10 — p. 343

1) d　　2) d　　3) b　　4) c
5) b　　6) c　　7) d　　8) d
9) b　　10) a, b　11) b, c　12) a, d
13) b　　14) a, c　15) b　　16) e
17) B　18) D　19) C　20) B
21) A　22) a or c　23) a　24) F
25) F　26) T　27) F

Miscellaneous Exercise 10 — p. 348

1) a) $2 + \ln 2$　b) $\frac{1}{5}$
2) a) $\frac{1}{6}$　　b) $\frac{1}{2}\ln\frac{3}{2}$
3) a) $\frac{1}{12}(1 + x^2)^6$　b) $-\frac{1}{4}e^{-2x}(2x + 1)$
   c) $\frac{1}{2}\ln\frac{32}{27}$　　d) $\frac{2}{15}$
4) a) (i) $\frac{1}{4}e^{4x} - e^{2x} + x$　(ii) $2\sqrt{\sin\theta}$
   b) (i) $\frac{2}{3}$　　　(ii) $\ln\frac{12}{7}$
5) a) $\frac{1}{2}\ln 2$　b) $\frac{4}{15}$
6) a) 78　　b) $\frac{1}{3}\ln 4$　c) $\ln 2$

7) a) 0.13　b) $\frac{1}{12}\ln\left|\dfrac{2 + 3\tan x}{2 - 3\tan x}\right|$

8) a) $\frac{15}{8} - \ln 4$　b) $\frac{7}{3} - \sqrt{2}$　c) $\frac{1}{2}\ln\frac{12}{5}$
9) a) $\frac{3}{4}\ln|x - 3| + \frac{1}{4}\ln|x + 1|$
   b) $\frac{16}{105}$　　c) $y = 1 - e^{-x}$

10) a) 0.275　b) 0.158　c) 0.693

11) a) $\ln 6$　b) $\arccos\dfrac{1}{\sqrt{5}} - \dfrac{1}{4}\pi - \dfrac{1}{10}$

12) a) $2\ln|x + 1| + \dfrac{3}{x + 1}$;
    $\frac{1}{2}(e^{2x} - 4x - e^{-2x})$
    b) $1 - \frac{1}{4}\pi$

13) a) $2\ln\frac{7}{3}$　b) $4(\sqrt{7} - \sqrt{3})$
    c) $2 - \frac{3}{2}\ln\frac{7}{3}$

14) b) (i) $-\ln(1 - \sqrt{x})^2$
       (ii) $\frac{2}{9}(1 - 3\cos^2 x)^{3/2}$

15) a) 0.068　b) $\frac{1}{4}(1 + \sqrt{3})$　c) $\frac{62}{5}, \frac{116}{15}$
16) a) $\frac{1}{2}\cos 2x - \frac{1}{4}\cos 4x$;
    $\frac{2}{3}(x - 1)^{\frac{3}{2}} - 2(x - 1)^{\frac{1}{2}}$
    b) $\frac{1}{18}\pi$

17) a) $\ln\dfrac{x}{\sqrt{(2x + 1)}}$
    b) $x^2 - x + \ln|x + 1|$

18) a) 2　b) $-\dfrac{1}{n}, 0, \dfrac{1}{n}$　c) 0.752

19) $\dfrac{3}{5(x + 1)} - \dfrac{1}{5(2x - 3)}$, $\frac{3}{5}\ln 4 - \frac{7}{10}\ln 3$
20) a) $\frac{3}{2}\ln|2x + 1| - \ln|x - 2|$　b) $\frac{5}{24}$
21) a) $\frac{1}{5}$　b) $\frac{1}{9}(1 - 4e^{-3})$
22) a) $-\frac{2}{5}$　b) $\frac{1}{16}\pi$　c) 1　d) $\ln\frac{32}{27}$
23) a) $\frac{1}{4}$　b) $\frac{1}{3}\sin^3\theta - \frac{1}{5}\sin^5\theta$
    c) $\frac{2}{3}(3 + x)^{\frac{1}{2}}(x - 6)$
24) a) $\ln 6$　b) $\frac{1}{9}\pi$　c) 6
25) $y = x\,e^{1 - x^2}$
26) a) $\ln\frac{2}{3}$　b) $\frac{1}{4}(2x - 1)e^{2x}$
    c) $2\ln\left|\dfrac{y}{2}\right| = x^2$
27) $1 + e^{-y} = 2\sqrt{2}\cos x$
28) $\dfrac{8x}{3x + 5} = 20\arctan 2y + 5\ln(1 + 4y^2)$
29) a) (i) $\frac{1}{2} + \frac{1}{4}\ln 3$　(ii) $\frac{1}{3}$　b) $\frac{4}{3}$
30) $2y = (y + 1)\sqrt{1 + x^2}$
31) $y^2 = x^2 + A$
32) $y^2 = 4x^2 + 5$
33) $(x + 2)(1 - 2y) = Ay$
34) $y = \dfrac{4x}{3 - x}$
35) a) (i) $\frac{1}{2}x^2 - 2x + 4\ln|x + 2|$
       (ii) $-\frac{1}{10}(\cos 5x + 5\cos x)$
    b) $(\ln 3, 3)$; $2\ln 3$

36) $2(\sqrt{3} - \frac{1}{3}\pi)$ 37) $\frac{1}{4}\pi$

38) $1, 4; \frac{45}{4}$

40) a) $y = (2e^{-x^2/2} - 1)^{-1}$

   b) $\dfrac{dy}{dx} = 1 - \dfrac{dz}{dx}; y = e^{-x} + x - 1$

41) $a = x + y; \dfrac{2}{ka} \ln 9$

Exercise 11a — p. 361

1) a) $(1, 5)$ b) $(6, \frac{19}{2})$ c) $(5, 1)$
   d) $(11, -18)$

3) a) $\dfrac{\sqrt{5}}{5}$ b) $\frac{1}{5}$ c) $\left| \dfrac{ma - b + c}{\sqrt{(m^2 + 1)}} \right|$

   d) $\dfrac{3\sqrt{2}}{2}$ e) $\dfrac{ax + by + c}{\sqrt{(a^2 + b^2)}}$

   f) $\left| \dfrac{c}{\sqrt{(a^2 + b^2)}} \right|$

4) a) opposite   b) same   c) same

5) a) $\frac{15}{16}$ b) $\frac{5}{31}$ c) $\left| \dfrac{a_1 b_2 - a_2 b_1}{a_1 a_2 + b_1 b_2} \right|$

6) $\dfrac{3}{14}, \dfrac{9}{83}, \dfrac{3}{29}; \dfrac{9}{\sqrt{170}}, \dfrac{9}{\sqrt{41}}, \dfrac{3}{\sqrt{5}}$

7) a) $(-1, 8)$ b) $(-11, 10)$
   c) $(-3, -14)$

8) $5y = x, y + 5x = 0$

9) $4X^2 + Y^2 - 14X + 12Y - 4XY + 11 = 0$

10) no

11) $14X + 112Y = 93$ and $64X - 8Y = 63$

12) $x + 2y = 7$

Exercise 11b — p. 363

1) $y^2 = 2y + 12x - 13$

2) $x + y = 4$

3) $8x^2 + 9y^2 - 20x = 28$

4) $11y = 3x$ and $99x + 27y + 130 = 0$

5) $x^2 + y^2 - 6x - 10y + 30 = 0$

6) $4x - 3y = 26$ and $4x - 3y + 24 = 0$

7) $x^2 + y^2 = 1$

Exercise 11c — p. 372

1) a) $x^2 + y^2 - 2x - 4y = 4$
   b) $x^2 + y^2 - 8y + 15 = 0$
   c) $x^2 + y^2 + 6x + 14y + 54 = 0$
   d) $x^2 + y^2 - 8x - 10y + 32 = 0$

2) a) $(-4, 1); 5$

   b) $(-\frac{1}{2}, -\frac{3}{2}); \dfrac{3\sqrt{2}}{2}$

   c) $(-3, 0); \sqrt{14}$ d) $(\frac{3}{4}, -\frac{1}{2}); \dfrac{\sqrt{5}}{4}$

   e) $(0, 0); 2$ f) $(2, -3); 3$

   g) $(1, 3); 3$ h) $(-1, \frac{1}{2}); \dfrac{\sqrt{69}}{6}$

3) a) yes   b) yes   c) no   d) yes

4) a) $3y + 4x = 23$ b) $4y = x + 22$
   c) $3y = 2x + 25$

5) a) touch ext.   b) orthog.
   c) touch int.   d) neither
   e) orthog.

6) $9 \pm 6\sqrt{5}$

7) $2mac = a^2 - c^2$

8) $x^2 + y^2 + 4x + 4y + 4 = 0$ or
   $196(x^2 + y^2) + 84(x + y) + 9 = 0$

9) $2x^2 + 2y^2 - 15x - 24y + 77 = 0$

10) $x^2 + y^2 - 8x + 6y = 0$

11) $x^2 + y^2 - 4ax - 6by + 3a^2 + 5b^2 = 0$

12) $x^2 + y^2 - 4x - 6y + 9 = 0$

13) $x^2 + y^2 - 4x - 14y = 116$

14) $x^2 + y^2 - 6y = 0$

15) $y = 0, 8y = 15x$

16) $2y + x = 0$

17) $2x + y = 11, 2x + y + 9 = 0$

18) $(-\frac{4}{3}, \frac{2}{3}); \dfrac{4\sqrt{5}}{3}$

Exercise 11d — p. 378

1) a) $x^2 + y^2 - 10x - 10y - 2xy = 5$
   b) $x^2 - 6x - 8y + 25 = 0$
   c) $y^2 + 4ax = 0$

2) a) $y = x + 1; y + x = 3$
   b) $x = 0; y = 1$
   c) $y = 8x - 2; 16y + 2x = 33$
   d) $2y = x - 4; 2x + y + 12 = 0$

Exercise 11e — p. 383

1) $3, 2$ 2) $4, 3$ 3) $2, 1$

4) $3, 2$ 5) $1, \frac{1}{5}$

6) $4x + 3\sqrt{5}y = 18; 9\sqrt{5}x - 12y = 10\sqrt{5}$

7) $\sqrt{7}x + 4y = 16; 16x - 4\sqrt{7}y = 7\sqrt{7}$

8) $\sqrt{3}x + 2y = 4; 4x - 2\sqrt{3}y = 3\sqrt{3}$

Exercise 11f — p. 391

1) touch at $(1, 2)$

2) $(1, 1)$ and $(-5, 4)$

3) no intersection

4) $(0, 2)$ and $(\frac{8}{5}, \frac{14}{5})$

5) $(1, 0)$ and $(2, 0)$

6) $(0, 0), (-3, 0)$ and $(-2, 0)$

7) touch at $(1, 0)$ and $(2, 0)$

8) touch (inflexion) at $(-3, 0)$;
   meet at $(-2, 0)$
9) touch at $(0, 0)$
10) $k = 8$ 11) $k = \frac{1}{4}$
12) $k = \pm 2\sqrt{19}$ 13) $(-1, 1)$
14) $(0, 0)$ 15) $(-\frac{7}{2}, -\frac{7}{6})$
16) $(\frac{1}{2}, 2)$
17) $(1, 2)$ and $(1, -2)$
18) touch at $(-1, -2)$; cut at $(1, 2)$
    and $(4, \frac{1}{2})$
19) $(0, 0)$ and $(\frac{2}{3}, 2)$
20) a) $\pm 2\sqrt{3}$ b) $-2\sqrt{3} < \lambda < 2\sqrt{3}$
    c) $\lambda < -2\sqrt{3}, \lambda > 2\sqrt{3}$
22) a) (i) $y = \pm (x - 1)$
       (ii) $2y + x = 4$
    b) (i) $y = 0, y = 36(1 - x)$
       (ii) $2y + x = \pm 6\sqrt{2}$
    c) (i) $3y = 2(x - 1), y = 0$
       (ii) $8y + 4x + 3 = 0$

Exercise 11h – p. 402

1) a) $y^2 - 6y + 8 = x$
   b) $x^2 = y^3$
   c) $xy = 9$
   d) $y = x^2 - 3x + 2$
   e) $\dfrac{x^2}{4} + \dfrac{y^2}{9} = 1$
   f) $y^2 = 4(x^2 - 1)$
   g) $2y^2 = a(x + a)$
3) $y^2 = 8x$ 4) $2y = x + 8$
5) $(p + q)y = 2x + 4pq$
6) $(18, 12); (2, -4)$ 7) $k = 2$
8) $(2, 4); (2, -4)$ 9) $(8, 8)$
10) $t_1 t_2 y + x = c(t_1 + t_2)$
11) $(p + q)y = 2x + 2apq$
12) $(T^2 + 2T + 4)y = (T + 2)x + 4T^2$
13) a) $at\sqrt{t^2 + 4}$ b) $\dfrac{c}{t}\sqrt{t^4 + 1}$
    c) $(t^4 + 3t^2 - 2t + 2)^{\frac{1}{2}}$
    d) $(a^2 \cos^2 \theta + b^2 \sin^2 \theta)^{\frac{1}{2}}$
14) $y + t_1 x = 2at_1 + at_1^3$;
    $\left[a\left(\dfrac{2}{t_1} + t_1\right)^2, -2a\left(\dfrac{2}{t_1} + t_1\right) \right]$
15) $py = p^3 x + c(1 - p^4); \left(-\dfrac{c}{p^3}, -cp^3\right)$
16) $y = px + b(p^3 + p - 4)$;
    $\left[-\dfrac{b}{p^2}(p^4 + 3p^2 + 4), \right.$
    $\left. -\dfrac{2b}{p}(p^2 + 2p + 2) \right]$

17) $(\frac{11}{2}, -4)$ 18) $(\frac{7}{4}, \frac{7}{2})$ 19) $(\frac{7}{2}, 0)$

Exercise 11i – p. 408

1) $xy = 1$
2) $9y^2 = 4(3x - 8)$
3) $y^2 = 6x$
4) $y^4 = c^2(c^2 - 2xy)$
5) $x^2 = 4(1 + y)$
6) $4x^2 + 9y^2 = 144$
7) $y^2 = 8(x + 1)$
8) the directrix
9) $xy = 2$

Multiple Choice Exercise 11 – p. 409

1) c	2) d	3) c
4) c	5) b	6) a
7) e	8) c	9) b
10) e	11) b, c	12) a, b
13) a, d	14) b, d	15) C
16) A	17) B	18) A
19) C	20) A	21) D
22) A	23) a or b or c	24) a, c, d
25) b or c	26) I	27) F
28) T	29) T	30) T
31) F		

Miscellaneous Exercise 11 – p. 413

1) a) $x^2 + y^2 - 2x = 24$
   b) $(\frac{8}{3}, 0)$
   c) $x^2 + y^2 - 2x + 5y = 24$
2) a) $(1, 3), (-7, -1)$
   b) $x^2 + y^2 + 6x - 2y = 0$
   c) $1\frac{2}{3}$
3) $x^2 + y^2 + 2y = 1; x^2 + y^2 - 14y = 1$
4) $(\frac{12}{5}, \frac{9}{5}); x(5x - 6) + y(5y - 17) = 0$
5) a) $x^2 + y^2 - 6x - 6y + 5 = 0$
   b) $3x^2 + 3y^2 - 6x - 26y + 3 = 0$
6) $\frac{3}{5}\sqrt{5}$
7) $x^2 + y^2 - 6x = 16$;
   $2(5 + 5\sqrt{5}) = 10(1 + \sqrt{5})$
8) $x + y = 1; (\frac{7}{3}, -\frac{4}{3}), (-\frac{1}{3}, \frac{4}{3})$
11) a) $(-14, -10)$ b) $(2, -\frac{10}{3})$
12) $x^2 + y^2 - 20x - 10y + 100 = 0$,
    $x^2 + y^2 - 40x - 20y + 400 = 0$
13) a) $x^2 + y^2 - 2y = 3$ b) $\frac{4}{3}\pi$
    c) $x^2 + y^2 - \sqrt{3}x - 3y = 0$;
       $(x^2 + y^2 + 2y - 3)(y + 3) = 0$
14) $x^2 + y^2 - 8x - 10y + 16 = 0$;
    $4y + 3x = 7, 4y + 3x = 57$
15) $(x - 3)^2 + (y - 2)^2 = 4$
16) $(\frac{1}{5}, -\frac{3}{5})$
17) $(-a, -b); (1 \pm \sqrt{3}); x^2 + y^2 - 2x = 2$
19) $ty = x + t^2; y = 3; 2x + 3y = 0$

21) $y^2 = 4a(x - a)$

23) a) $y^2 = a(x - a)$ c) $(3a, \pm 2\sqrt{3}a)$

24) $y = \pm (\frac{3}{4}x + 5)$

25) $y = \dfrac{2a}{m}, y(y + a) = 2ax$

27) $(6, \pm 4\sqrt{2})$

29) $x \cos \theta + y \sin \theta = 2 + 2 \cos \theta$;
$(-2, 0); x^2 + y^2 = 2x + 2$

30) $y \sin T + x = 1 + 2 \sin^2 T; \frac{1}{4}\pi$

31) $3 \pm 2\sqrt{2}$

32) $y \sin 2\theta + x \cos 2\theta$
$= a(5 \cos \theta - \cos 3\theta)$;
points where $\theta = \frac{1}{8}\pi, \frac{1}{4}\pi, \frac{3}{8}\pi$

33) $y = (\frac{5}{4}x_0{}^2 - \frac{13}{9})x - \frac{5}{6}x_0{}^3$;
$\pm \frac{1}{3}\sqrt{5}, \pm \frac{2}{3}\sqrt{2}$

34) $x \cos \theta + y \sin \theta = a$;
$$\left[a(2 \cos \theta + 1), \right.$$
$$\left. \dfrac{a}{\sin \theta} (2 \sin^2 \theta - 1 - \cos \theta) \right];$$
$$\left(0, \pm \dfrac{2a}{\sqrt{3}} \right)$$

35) a) $y^2 = x^2(x + 3)$

36) $e^{-\frac{1}{2}}$

38) $(0, \pm \frac{32}{15})$

39) $x = 4, y = x + 5; (4, 0), (-\frac{16}{5}, \frac{9}{5})$

41) $\left(-ct, -\dfrac{c}{t} \right)$

Exercise 12a – p. 430

1) a) $x = 1, y = 0$
   b) $x = 1, y = -1$
   c) $x = 1, x + y + 1 = 0$
   d) $x = -3, x = 4, y = 0$

2) a) all values   b) $|y| \leqslant \frac{1}{2}$
   c) $|y| > 2$
   d) $y \leqslant 1 - \dfrac{\sqrt{3}}{2}, y \geqslant 1 + \dfrac{\sqrt{3}}{2}$
   e) $y \geqslant -\frac{1}{4}$ f) $|y| > 2$
   g) $|y| < 2$
   h) $y \leqslant 9 - 4\sqrt{5}, y \geqslant 9 + 4\sqrt{5}$

3) a) $y \to 0; y \to 0$
   b) $y \to \pm \infty$ as $x \to \infty$; $x \not\to -\infty$
   c) $y \to 0; y \to \infty$
   d) $y \to \infty$ as $x \to \infty; x \not\to -\infty$
   e) $y \to \infty; y \to -\infty$
   f) $y \to \infty; y \to -\infty$

4) a) $x < -6, x > 1$
   b) $1 < x < 2, x > 3$
   c) $x > 1$

d) $0 < x < 1$

e) $x > 0$

f) $-1 < x < 1, x > 2$

g) all values

Exercise 12c – p. 441

15) $-2, -1, 1, 2$ 16) $\frac{2}{3}, \frac{4}{5}$

17) $-4, 3$ 18) $0, 2$

22) d, b, e, f

Exercise 12d – p. 446

2) a) $13 - 6\sqrt{2}$ b) 13
   c) 20 d) 4 e) 36

3) right angled: a), d); isosceles: b), g);
   neither: c), e), f)

4) $x^2 + y^2 = a^2$

5) $(x^2 + y^2)^{3/2} = a(x^2 - y^2)$

6) $(x^2 + y^2)^2 = 2a^2 \tilde{x}y$

7) $(x^2 + y^2)^3 = 4a^2 x^4$

8) $x \cos \alpha + y \sin \alpha = d$

9) $r^2 + a^2 = 2ar(\cos \theta + \sin \theta)$

10) $r^2(a^2 \sin^2 \theta + b^2 \cos^2 \theta) = a^2 b^2$

11) $\theta = \arctan 2$

12) $r = \tan \theta \sec \theta$

13) $r^2 \sin 2\theta = 8$

Exercise 12e – p. 452

3) $\theta = 0, \pi$

4) $\theta = 0, \frac{1}{3}\pi, \frac{2}{3}\pi, \pi, \frac{4}{3}\pi, \frac{5}{3}\pi$

5) $\theta = \frac{1}{6}\pi, \frac{1}{2}\pi, \frac{5}{6}\pi, \frac{7}{6}\pi, \frac{3}{2}\pi, \frac{11}{6}\pi$

6) $\theta = \frac{1}{4}\pi, \frac{3}{4}\pi, \frac{5}{4}\pi, \frac{7}{4}\pi$

7) $\theta = \frac{1}{2}\pi, \frac{3}{2}\pi$

8) $\theta = \frac{1}{2}\pi, \frac{3}{2}\pi$

9) a) $\theta = 0$ b) $\theta = \pi$

10) $\theta = \pm \arccos \frac{2}{3}$

12) $\theta = \frac{1}{4}\pi, \frac{3}{4}\pi, \frac{5}{4}\pi, \frac{7}{4}\pi$

Exercise 12f – p. 456

1) $\frac{1}{2}\pi a^2$ 2) $\frac{1}{2}\pi a^2$

3) $\frac{3}{2}\pi a^2$ 4) $\frac{25}{48}\pi^3$

5) $\frac{1}{3}a^2$ 6) $\frac{1}{2}\pi a^2$

7) $\frac{43}{2} \arccos \frac{3}{5} - 18$ 8) $\frac{1}{2}\pi a^2$

9) $\dfrac{32\sqrt{3}}{3}$ 10) $\frac{41}{2}\pi a^2$

Multiple Choice Exercise 12   p. 459

1) b   2) b   3) d

4) d   5) d   6) a

7) d   8) c   9) b, d

10) a   11) a, b, c   12) b, c, d

13) E 14) C 15) A

16) B 17) D 18) E

19) A 20) c   21) A

22) c   23) F 24) F

25) F 26) F 27) T

Miscellaneous Exercise 12 – p. 463

1) $4 - 2\sqrt{3} \leqslant y \leqslant 4 + 2\sqrt{3}$

4) a) $(2, \frac{1}{4})$ b) $(1, 0)$

6) $|k| \leqslant 1$

7) $2, -4, 3, -4, 3$

8) $2 < y < 6$

10) a) $x = 0, 2$;

 $-1 < x < 0$ and $2 < x < 3$

11) $2 - \sqrt{3}, 2 + \sqrt{3}; (\frac{1}{2}\sqrt{3}, 2 - \sqrt{3})$,

 $(-\frac{1}{2}\sqrt{3}, 2 + \sqrt{3})$;

 $x = -3, x = -1, y = 0$

12) a) $3\frac{3}{8} < K$ b) $K = 3\frac{3}{8}$

 c) $K < 0$ and $0 < K < 3\frac{3}{8}$

 d) $K = 0; -2 < x < 1, x \neq 0$

13) $\ln \frac{9}{8} + \frac{1}{9} \ln 2 - \frac{1}{6}$

14) $-\frac{5}{4} \leqslant y \leqslant 5; \frac{5}{2}$

16) b) $a^2 \tan\frac{\alpha}{2}\left[1 + \frac{1}{3}\tan^2\frac{\alpha}{2}\right]$

17) $\dfrac{3\pi + 8}{6\pi}$ 18) $\frac{4}{3}$

20) a) $\frac{1}{12}\pi a^2$ b) a^2

22) $1, \dfrac{\sqrt{6}}{6}$

Exercise 13a – p. 472

1) a) $\overrightarrow{AC}$ b) $\overrightarrow{BD}$ c) $\overrightarrow{AD}$ d) $\overrightarrow{DB}$

2) $a, b - a, 2b - 2a$

5) $b - a, a + b, a - 3b, 2b, 2a - 2b$

6) $b - a, -a, -b, a - b$

7) $\overrightarrow{DE} = \overrightarrow{CH} = \overrightarrow{BG} = c$,

 $\overrightarrow{DC} = \overrightarrow{EH} = \overrightarrow{FG} = a$,

 $\overrightarrow{FE} = \overrightarrow{GH} = \overrightarrow{BC} = b$

9) $c - a, b - a, b - c$

10) $b + c - a, b - c - a, a + b + c$,

 $a + b - c$

11) $\frac{1}{2}b - c, \frac{1}{2}(c + b) - a$

Exercise 13b – p. 480

1) a) $3i - j$ b) $i + 3j$

 c) $5i$ d) $-2i - 11j$

2) a) $\sqrt{65}$ b) $\frac{1}{2}(3i + 2j)$

 c) $\frac{1}{4}(5i - 4j)$

3) a) $2i + j$ b) $\sqrt{73}$ c) $-3i - 7j$

4) $i + 2j$ 6) $\arctan(-2/3)$

Exercise 13c – p. 484

1) a) -3 b) 2 c) -7

2) a) $66.8°$ b) $171.9°$

 c) $14.03°$ d) $143.97°$

3) a) $a^2 - \mathbf{a.b}$ b) $a^2 + b^2 - 2\mathbf{a.b}$

 c) $b^2 - a^2$ d) $2\mathbf{b.c}$

4) a) 0 b) $-b^2$ c) a^2 d) $2a^2 - b^2$

5) a) $2b^2$ b) $-b^2$

6) (b), (c), (d)

Exercise 14a – p. 490

1) $-i, i, i, -i, 1, i$

2) a) $10 + 4i$ b) $7 + 2i$

 c) $6 - 2i$ d) $(a + c) + (b + d)i$

3) a) $-4 + 6i$ b) $1 - 4i$

 c) $-2 + 16i$ d) $(a - c) + (b - d)i$

4) a) $10 - 5i$ b) $39 + 23i$

 c) $11 - 7i$ d) 25

 e) $3 - 4i$ f) $-2 + 2i$

 g) $-4 + 3i$ h) $x^2 + y^2$

 i) $-3 + i$ j) $(a^2 - b^2) + 2abi$

5) a) $1 + i$ b) $\frac{9}{25} + \frac{13}{25}i$ c) $\frac{4}{17} + \frac{16}{17}i$

 d) i e) $-i$

 f) $\dfrac{x^2 - y^2}{x^2 + y^2} + \dfrac{2xy}{x^2 + y^2}i$

 g) $1 - 3i$ h) $-3 - 2i$

6) a) $x = 9, y = -7$

 b) $x = -\frac{3}{2}, y = \frac{7}{2}$

 c) $x = \frac{7}{2}, y = \frac{1}{2}$

 d) $x = 2, y = 0$

 e) $x = 13, y = 0$

 f) $x = 15, y = 8$

 g) $x = 11, y = 3$

 h) $x = 2, y = 1$ or $x = -2, y = -1$

7) a) $7, -1$ b) $-2, -2$

 c) $\frac{10}{17}, \frac{11}{17}$ d) $\frac{9}{5}, -\frac{4}{5}$

 e) $0, \dfrac{-2y}{x^2 + y^2}$ f) $-1, 0$

 g) $\frac{1}{2}, \frac{1}{2}\sqrt{3}$

8) a) $\pm(2 - i)$ b) $\pm(5 - 2i)$

 c) $\pm(1 + i)$ d) $\pm(4 + i)$

 e) $\pm(1 + 5i)$

Exercise 14b – p. 494

1) a) $-\frac{1}{2} \pm \frac{1}{2}\sqrt{3}i$ b) $-\frac{7}{4} \pm \frac{1}{4}\sqrt{41}$

 c) $\pm 3i$ d) $-\frac{1}{2} \pm \frac{1}{2}\sqrt{11}i$

 e) $\pm 1, \pm i$

2) a) $x^2 + 1 = 0$

 b) $x^2 - 4x + 5 = 0$

 c) $x^2 - 2x + 10 = 0$

 d) $x^3 - 4x^2 + 6x - 4 = 0$

3) a) 4, 5   b) 6, 25   c) 0, 1
   d) $-24, 169$ e) $-2, 2$
4) a) $(x + 2 - i)(x + 2 + i)$
   b) $(x - 1 - 4i)(x - 1 + 4i)$
   c) $(x + \frac{1}{2} - \frac{1}{2}\sqrt{3}i)(x + \frac{1}{2} + \frac{1}{2}\sqrt{3}i)$
   d) $(x + 1 - \sqrt{3}i)(x + 1 + \sqrt{3}i)$
5) a) $\dfrac{i}{2(x + i)} - \dfrac{i}{2(x - i)}$
   b) $\dfrac{-2i}{x - 2 - i} + \dfrac{2i}{x - 2 + i}$
   c) $\dfrac{4i}{x + 2 + 2i} - \dfrac{4i}{x + 2 - 2i}$
   d) $\dfrac{i}{2(x + 2i)} - \dfrac{i}{2(x - 2i)}$
   e) $\dfrac{1 - 2i}{2(x + 2 - 3i)} + \dfrac{1 + 2i}{2(x + 2 + 3i)}$
6) a) 1   b) $3\omega^2$ c) 0
7) $-1, \frac{1}{2} \pm \frac{1}{2}\sqrt{3}i; -\lambda^2$

Exercise 14c – p. 503

1) a) $\sqrt{13}$ b) $\sqrt{17}$
   c) 5   d) 13
   e) $\sqrt{2}, -\frac{1}{4}\pi$ f) $\sqrt{2}, \frac{3}{4}\pi$
   g) 4, 0   h) $2, -\frac{1}{2}\pi$
   i) $\sqrt{a^2 + b^2}, \arctan\dfrac{b}{a}$
   j) $\sqrt{2}, \frac{1}{4}\pi$ k) $\sqrt{2}, \frac{3}{4}\pi$
   l) $\sqrt{2}, -\frac{3}{4}\pi$ m) $\sqrt{2}, -\frac{1}{4}\pi$
   n) $\sqrt{170}$ o) $2, \frac{1}{3}\pi$
   p) $1, \frac{3}{4}\pi$ q) $3, -\frac{5}{6}\pi$
3) a) $\sqrt{2}\left(\cos\dfrac{\pi}{4} + i \sin\dfrac{\pi}{4}\right)$
   b) $2\left\{\cos\left(-\dfrac{\pi}{6}\right) + i \sin\left(-\dfrac{\pi}{6}\right)\right\}$
   c) $5\{\cos(-2.214^c) + i \sin(-2.214^c)\}$
   d) $13 (\cos 1.966^c + i \sin 1.966^c)$
   e) $\sqrt{5}\{\cos(-0.464^c)$
       $+ i \sin(-0.464^c)\}$
   f) $6 (\cos 0 + i \sin 0)$
   g) $3 (\cos \pi + i \sin \pi)$
   h) $4\left(\cos\dfrac{\pi}{2} + i \sin\dfrac{\pi}{2}\right)$
   i) $2\sqrt{3}\left\{\cos\left(-\dfrac{5\pi}{6}\right) + i \sin\left(-\dfrac{5\pi}{6}\right)\right\}$
   j) $25 (\cos 0.284^c + i \sin 0.284^c)$
4) a) $\sqrt{3} + i$ b) $\frac{3}{2}\sqrt{2} - \frac{3}{2}\sqrt{2}i$
   c) $-\frac{1}{2} + \frac{1}{2}\sqrt{3}i$ d) $-\frac{1}{2}\sqrt{2} - \frac{1}{2}\sqrt{2}i$
   e) 3   f) -2

g) $2\sqrt{3} - 2i$ h) i
i) $-3i$ j) $-\frac{1}{2} - \frac{1}{2}\sqrt{3}i$
6) $1, -\frac{1}{2} \pm \frac{1}{2}\sqrt{3}i; 1 (\cos 0 + i \sin 0),$
   $1\left(\cos\dfrac{2\pi}{3} + i \sin\dfrac{2\pi}{3}\right),$
   $1\left\{\cos\left(-\dfrac{2\pi}{3}\right) + i \sin\left(-\dfrac{2\pi}{3}\right)\right\}$

Exercise 14d – p. 509

2) $31; 19$
3) a) $2\sqrt{2}, \frac{1}{4}\pi$ b) $2\sqrt{6}, -\frac{5}{12}\pi$
   c) $1, 2 \arctan \sqrt{3}/2$
5) $\pm 2\left\{\cos\left(-\dfrac{\pi}{12}\right) + i \sin\left(-\dfrac{\pi}{12}\right)\right\}$

Exercise 14e – p. 518

15) a) $|z - 3 - 4i| = 5$ b) $|z - 5| = 2$
    c) $|z - 4i| = 4$
16) a) $r = 1$ b) $r = 4$
    c) $\theta = \frac{1}{4}\pi$ d) $\theta = -\frac{2}{3}\pi$
17) a) $x^2 + y^2 = 1$ b) $x^2 + y^2 = 16$
    c) $y = x \ (x > 0)$ d) $y = \sqrt{3}x \ (x < 0)$
18) a) (i) $|z - 4| = |z + 8|$ (ii) $x = -2$
    b) (i) $|z - 1 - 2i| = |z - 7 + 4i|$
        (iii) $y = x - 5$
    c) (i) $|z - 6i| = |z - 6|$ (iii) $y = x$
19) a) $\arg(z - 2) = \frac{2}{3}\pi$
    b) $\arg(z + 1) = -\frac{1}{2}\pi$
    c) $\arg(z + 1 - 2i) = -\frac{3}{4}\pi$
20) $3x^2 + 3y^2 - 16x - 8y = 0$
21) $x = 0, y = 0$
22) $\theta = 0$ and $r = 2$
23) $r = 2 \cos \theta$
24) $\theta = \dfrac{\pi}{2}$ and $r = 1$

Exercise 14f – p. 522

3) a) $\sqrt{2}$ b) $\sqrt{17} - 2$
4) a) $2\sqrt{2}(1 + i)$ b) $1 + 3i, 1 - 5i$
5) a) $(2, 1), (0, -1)$ b) $(\sqrt{2}, -\sqrt{2})$

Multiple Choice Exercise 14 – p. 523

1) c   2) e   3) a   4) c
5) d   6) d   7) e   8) e
9) d   10) a, c   11) b, e   12) a, c
13) c, d   14) b, c   15) a, d   16) A
17) E   18) D   19) E   20) D
21) F   22) T   23) T   24) F
25) F

Miscellaneous Exercise 14 – p. 526

1) 1   2) $\frac{1}{5}(3 - 4i), \frac{1}{2}(-3 + i)$

3) $2 - i, 3 - 4i, (2, -1)$

4) a) $\frac{104}{25} - \frac{72}{25}i$ b) $\pm\sqrt{2}(1 + i)$

5) $\frac{63}{25}, \frac{16}{25}; \frac{13}{5}, \frac{63}{65}, \frac{16}{65}$

6) $4, -5; 5$

7) $2, \frac{1}{6}\pi; \frac{1}{2}(1 + i\sqrt{3})$

8) $0, 3; 4\omega^2, 4\omega$

9) a) (i) $2\sqrt{10}$ (ii) 10 (iii) $\frac{1}{5}\sqrt{10}$
   b) $\frac{1}{5}(-2 + 6i)$

10) a) (i) $-\frac{7}{625} - \frac{24}{625}i$ (ii) $2 + i$
    b) $x^2 + y^2 + 2x - 4y = 15$

11) a) (i) $\sqrt{97}$ (ii) $\frac{1}{2}\sqrt{2}$
    b) $\pm 3 + \sqrt{26}$

13) a) $\sec\theta$ b) $-i\tan\theta$

14) a) $3 - i, 3 + 3i$

15) a) 6
    b) $\dfrac{x(x^2 + y^2 + 1)}{x^2 + y^2}, \dfrac{y(x^2 + y^2 - 1)}{x^2 + y^2}$;
    $y(x^2 + y^2 - 1) = 0$

16) a) $\pm(3 + 2i)$
    b) (i) $\sqrt{2}, -\frac{1}{4}\pi$ (ii) $5, 0.643^c$
    (iii) $5\sqrt{2}, -0.142^c; 12$
    c) $(\sqrt{2} + 1)/(\sqrt{2} - 1)$

17) a) $\frac{1}{2}, -\frac{1}{2}\pi$
    b) $3 - 2i, -3 + 2i; \pm(2 + 3i)\sqrt{3}$

18) $3 + i, 2i, 2 - 2i$

19) a) (i) $5 + 12i$ (ii) $\frac{5}{169} - \frac{12}{169}i$
    b) $4x^2 + 4y^2 + x + 9y + 4 = 0$;
    $(\frac{1}{4}, -\frac{3}{4}), (-\frac{1}{2}, -\frac{3}{2})$

20) a) (i) $2 + \sqrt{2}$ (ii) 2
    b) $x = u^2 - v^2, y = 2uv$

21) a) $2\sqrt{2}, \frac{3}{4}\pi; 2\sqrt{2}, -\frac{3}{4}\pi; \frac{1}{2}i$

22) $3x^2 + 4y^2 = 12$

23) a) (i) $\frac{7}{50} - \frac{1}{50}i$
    (ii) $(c^4 - 6c^2 + 1) + (4c^3 - 4c)i$
    b) $\pm(1 + 2i); \pm(1 - 2i)$

24) a) $\frac{5}{13} + \frac{14}{13}i$

25) a) $\frac{5}{29}(8 + 9j); 2.08^c\lfloor 48°22'$
    b) $(x + 1)^2 + y^2 = 4$

26) a) (i) $r^2, 2\theta$ (ii) $\dfrac{1}{r}, -\theta$
    (iii) $r, \theta + \frac{1}{2}\pi$

Exercise 15a – p. 532

1) permutation   2) combination
3) combination   4) permutation
5) permutation   6) permutation

Exercise 15b – p. 537

1) 120   2) 56
3) 720   4) 18
5) 5040   6) 12

7) 12   8) 9×10^4
9) a) 504   b) 9   c) 324   d) 5
10) a) 6   b) 3   11) 5040   12) 60

Exercise 15c – p. 539

1) 28   2) 28
3) a) 4845   b) 969
4) a) 120   b) 90
5) 210   6) 1287   7) 495
8) a) 126   b) 252   c) 36   d) 72   e) 9

Exercise 15d – p. 542

1) 6   2) 24   3) 120
4) 720   5) 30   6) 132
7) 2730   8) 840   9) 28
10) 1140   11) 360 360   12) 1260

13) 70   14) $\frac{3}{4}$ 15) $\dfrac{5!}{2!}$

16) $\dfrac{11!}{9!}$ 17) $\dfrac{39!}{34!}$ 18) $\dfrac{n!}{(n-4)!}$

19) $\dfrac{(n+1)!}{(n-2)!}$ 20) $\dfrac{(n+5)!}{n!}$

21) $\dfrac{(n+r)!}{(n+r-3)!}$

22) $\dfrac{20!}{17! \, 3!}$ 23) $\dfrac{14!}{12! \, 3!}$ 24) $\dfrac{8! \, 3!}{5! \, 6!}$

25) $\dfrac{n!}{(n-3)! \, 3!}$ 26) $\dfrac{(n-2)!}{4! \, (n-5)!}$

27) $10(8!)$ 28) $40(5!)$ 29) $274(8!)$

30) $(n + 1)(n - 1)!$

31) $(n^2 + n - 1)(n - 1)!$

32) $n(n^2 - n + 2)(n - 2)!$

33) $n^2(n - 2)!$ 34) $\dfrac{8(7!)}{15(2! \, 4!)}$

35) $\dfrac{29(6!)}{4!}$ 36) $\dfrac{109(8!)}{28(6! \, 2!)}$

37) $\dfrac{(n-1)!}{(r+1)!}(nr + n + 1)$

38) $\dfrac{n!}{(r+1)!}(2nr + 2r + 2n - 1)$

39) $\dfrac{n!}{r! \, (n-r+1)!}(3n - r + 3)$

40) $\dfrac{(nr-2)(n-1)!}{(n-r)! \, (r-1)!}$

Exercise 15e – p. 548

1) a) 10   b) 27   2) 13

3) $\dfrac{18!}{13(5!)^2(7!)}$ 4) $\dfrac{11!}{(2!)^3}$

5) 2454   6) 18

7) 252   8) 75 600

9) 120   10) 240

11) 1440

12) a) 27   b) 109

13) a) $7 \times 7!$ b) $7 \times 7!$

14) 16   15) a) 64   b) 48   c) 12

16) 28   17) a) 56   b) 35

18) 76 145   19) 505

20) 27

Miscellaneous Exercise 15 − p. 551

1) a) 225   b) 465   c) 240

2) a) $(n-1)!$ b) $(n-2)^2(n-2)!$
$(n^2 - 3n + 3)(n-2)!$

3) $1 + 10 + 40 + 80 + 80 + 32 = 243$

4) $16 \times 17 \times 18!$ 5) 126

6) 77   7) 2100

8) $\dfrac{(n+m)!}{n!m!}$, $\dfrac{(n+m-2)}{(n-1)!(m-1)!}$

9) a) 56   b) 70

10) a) 56   b) 24   c) 32

11) a) 18   b) 192

12) $\dfrac{11!}{8}$, $\dfrac{10!}{4}$, $4! \, 7!$

13) a) $504; 168$ b) 84

Exercise 16a − p. 555

1) a) $\displaystyle\sum_{r=1}^{5} r^3$ b) $\displaystyle\sum_{r=1}^{10} 2r$

c) $\displaystyle\sum_{r=1}^{33} 3r$ d) $\displaystyle\sum_{r=2}^{50} \dfrac{1}{r}$

e) $\displaystyle\sum_{r=0}^{\infty} \dfrac{1}{3^r}$ f) $\displaystyle\sum_{r=0}^{7} (-4 + 3r)$

g) $\displaystyle\sum_{r=0}^{\infty} \left(\dfrac{8}{2^r}\right) = \displaystyle\sum_{r=0}^{\infty} 2^{3-r}$

2) a) $1 + \frac{1}{2} + \frac{1}{3} + \ldots$

b) $0 + 2 + 6 + 12 + \ldots + 30$

c) $1 + 1 + 2 + \ldots + 20!$

d) $1 + \frac{1}{2} + \frac{1}{5} + \ldots$

e) $0 + 0 + 6 + 24 + 60 + \ldots + 720$

f) $-1 + a - a^2 + \ldots$

3) a) 8, 9   b) 9, 10

c) 1, 6   d) $\frac{1}{420}, \infty$

e) $-\frac{4}{15}, \infty$ f) $-48, 23$

g) $(\frac{1}{2})^n, \infty$

Exercise 16b − p. 560

1) a) $17, 4n-3$ b) $0, \frac{1}{2}(5-n)$

c) $17, 3n+2$ d) $9, 2n-1$

e) $16, 4(n-1)$ f) $-2, 8-2n$

g) $p + 4q, p + (n-1)q$

h) $18, 8 + 2n$ i) $15, 3n$

2) a) 190   b) $-\frac{5}{2}$ c) 185

d) 100   e) 180   f) -30

g) $5(2p + 9q)$

3) $a = 27\frac{1}{5}, d = -2\frac{2}{5}$ 4) $d = 3; 30$

5) $1, \frac{1}{2}, 0; -8\frac{1}{2}$

6) a) $28\frac{1}{2}$ b) 80   c) 400

d) 80   e) 108   f) $3n(1 - 6n)$

g) 40   h) $2m(m+1)$

7) $4, 2n-4$ 9) $2, 364$

10) 39   11) 64   12) 12

Exercise 16c − p. 565

1) a) $32, 2^n$ b) $\frac{1}{8}, \dfrac{1}{2^{n-2}}$

c) $48, 3(-2)^{n-1}$

d) $\frac{1}{2}, (-1)^{n-1}(\frac{1}{2})^{n-4}$

e) $\frac{1}{27}, (\frac{1}{3})^{n-2}$

2) a) 189   b) -255 c) $2 - (\frac{1}{2})^{19}$

d) $\frac{781}{125}$ e) $\frac{341}{1024}$ f) 1

3) $\frac{1}{2}, 2$ 4) $-\frac{1}{2}$

5) $-\frac{1}{2}, \frac{1}{1024}$ 6) 13.21

7) a) $\dfrac{x - x^{n+1}}{1 - x}$ b) $\dfrac{x^n - 1}{x^{n-2}(x-1)}$

c) $\dfrac{1 + (-1)^{n+1}y^n}{1 + y}$

d) $\dfrac{x(2^n - x^n)}{2^{n-1}(2-x)}$ e) $\dfrac{1 - (-2)^n x^n}{1 + 2x}$

8) $\frac{8}{3}(1 - (\frac{1}{4})^n), 4$ 9) 62 or 122

10) 8.49   11) 8   12) £23.31

Exercise 16d − p. 569

1) a) yes   b) no   c) yes

d) yes   e) no   f) yes

2) a) $-1 < x < 1$

b) $x < -1, x > 1$

c) $-\frac{1}{2} < x < \frac{1}{2}$

d) $0 < x < 2$

e) $-1 - a < x < 1 - a$

f) $x < -1 - a, x > 1 - a$

3) a) 6   c) $13\frac{1}{3}$ d) $\frac{5}{9}$ f) $\frac{9}{4}$

4) a) $\frac{161}{990}$ b) $\frac{34}{99}$ c) $\frac{7}{330}$

5) $\frac{1}{2}$

Exercise 16e – p. 576

1) a) $1 + 36x + 594x^2 + 5940x^3$

   b) $1 - 18x + 144x^2 - 672x^3$

   c) $1024 + 5120x + 11\,520x^2$
   $+ 15\,360x^3$

   d) $1 - \frac{20}{3}x + \frac{190}{9}x^2 - \frac{380}{9}x^3$

   e) $128 - 672x + 1512x^2 - 1890x^3$

   f) $\left(\frac{3}{2}\right)^9 + \frac{3^{10}}{2^7}x + \frac{3^9}{8}x^2 + \frac{7}{2}(3^7)x^3$

2) a) $336x^2$ b) $-\frac{969}{2}x^5$

   c) $-\frac{2^3 15!}{12!}x^{11}$ d) $2^{22}\,3^8\,^{30}C_8 x^8$

   e) $3360p^6 q^4$ f) $16(3a)^7 b$

   g) $7920x^4$ h) $63x^5$

   i) $56a^3 b^5$ j) $20(3^{10})a^2 b^8$

3) a) $1 - 8x + 27x^2$

   b) $1 + 19x + 160x^2$

   c) $2 - 19x + 85x^2$

   d) $1 - 68x + 2136x^2$

4) -527 5) $-\frac{779}{256}i$

6) a) $\displaystyle\sum_{r=0}^{9} (-1)^r \frac{9!\,x^r(10-2r)}{r!\,(10-r)!}$

   b) $\displaystyle\sum_{r=0}^{10} \frac{2^{r-1}(22-3r)10!\,x^r}{r!(11-r)!}$

   c) $\displaystyle\sum_{r=0}^{20} \frac{(-1)^r(21-2r)20!\,x^r}{2^{r-1}r!(21-r)!}$

   d) $\displaystyle\sum_{r=0}^{14} \frac{(-1)^r 14!\,(36r^2 - 936r + 6000)5^{r-2}x^r}{r!(16-r)!}$

7) $\dfrac{4^{r-1}(36 - 6r)8!}{r!\,(9-r)!}$

8) $\dfrac{2^{20-r}20!\,(21-3r)}{r!\,(21-r)!}$

9) 0.8171

10) 1.062

14) a) $1 - 50x$ b) $256 - 1024x$

   c) $1 - 19x$

Exercise 16f – p. 583

1) $1 - x - \dfrac{x^2}{2} - \dfrac{x^3}{2}, -\frac{1}{2} < x < \frac{1}{2}$

2) $\dfrac{1}{3} - \dfrac{x}{9} + \dfrac{x^2}{27} - \dfrac{x^3}{81}, -3 < x < 3$

3) $1 - \dfrac{x}{4} + \dfrac{3x^2}{32} - \dfrac{5}{128}x^3, -2 < x < 2$

4) $1 + 2x + 3x^2 + 4x^3, -1 < x < 1$

5) $1 - \frac{1}{2}x + \frac{3}{8}x^2 - \frac{5}{16}x^3, -1 < x < 1$

6) $1 + \frac{1}{2}x - \frac{5}{8}x^2 - \frac{3}{16}x^3, -1 < x < 1$

7) $-2 - 3x - 3x^2 - 3x^3, -1 < x < 1$

8) $2 + 2x + \frac{21}{4}x^2 + \frac{27}{8}x^3, -\frac{1}{3} < x < \frac{1}{3}$

9) $\frac{1}{2} - \frac{3}{4}x + \frac{13}{8}x^2 - \frac{51}{16}x^3, -\frac{1}{2} < x < \frac{1}{2}$

10) $1 + x + \frac{1}{2}x^2 + \frac{1}{2}x^3, -1 < x < 1$

11) $1 - \dfrac{x^2}{9}, -3 < x < 3$

12) $x - x^2 + x^3, -1 < x < 1$

13) $n + 1$

14) $(-1)^n 2^{n-1}(n+1)(n+2)$

15) $3(-1)^n$

16) $\frac{1}{21}\{3(-1)^n 2^{n+1} + 3^{-n}\}$

17) $\dfrac{3}{2^n n!}\{3 \times 5 \times \ldots \times (2n-5)\}$

18) 2^{n+1}

19) $1 - \dfrac{3}{p} + \dfrac{6}{p^2} - \dfrac{10}{p^3} + \dfrac{15}{p^4}, |p| > 1$

20) 1.732

21) 3.162 28

22) 1.004 99

23) 2.032 793

24) $1 + 2x + 2x^2, \sqrt{51} = 7.1414$

26) $1 - 3x + \frac{7}{2}x^2$

Exercise 16g – p. 587

1) $1 - x + \dfrac{x^2}{2!} - \dfrac{x^3}{3!} + \dfrac{x^4}{4!}$

2) $x - \dfrac{x^3}{3!} + \dfrac{x^5}{5!} - \dfrac{x^7}{7!}$

3) $-x - \dfrac{x^2}{2} - \dfrac{x^3}{3} - \dfrac{x^4}{4}$

4) $1 + nx + \dfrac{n(n-1)}{2!}x^2$
   $+ \dfrac{n(n-1)(n-2)}{3!}x^3$

5) $x + \frac{1}{3}x^3$

6) $-x - \frac{3}{2}x^2$

7) $1 + x + x^2 + x^3 + x^4$

8) $1 + x - \dfrac{x^3}{3}$

9) $\tan 1^c + x \sec^2 1^c$
   $+ \dfrac{x^2}{2!}\sec^2 1^c(1 + 2\tan 1^c)$

Exercise 16h – p. 592

1) $1 + 3x + \frac{9}{2}x^2, \dfrac{3^r x^r}{r!}$, all x

2) $1 - \dfrac{x^2}{8}, \dfrac{(-1)^r x^{2r}}{(2r)! \, 4^r}$, all x

3) $2x, \dfrac{(-1)^r (2)^{2r+1} x^{2r+1}}{(2r+1)!}$, all x

4) $-x, \dfrac{-x^{2r+1}}{2r+1}, -1 < x < 1$

5) $2 - 2x + 2x^2, \dfrac{(-1)^r 2^r x^r}{r!}$, all x

6) $-3x - \frac{9}{2}x^2,$

$\dfrac{x^r}{r}[(-1)^{r+1} - 2^{r+1}], -\frac{1}{2} \leqslant x < \frac{1}{2}$

7) $-1 + x - x^2, \dfrac{(-1)^{r-1}}{(r-1)!}x^r$, all x

8) $1 - 4x^2, \dfrac{(-1)^r (4x)^{2r}}{2(2r)!}$, all x

9) $1 + 2x + \frac{7}{2}x^2, \left(\dfrac{3^r}{r!} - \dfrac{2^{r-1}}{r}\right)x^r,$

$-\frac{1}{2} \leqslant x < \frac{1}{2}$

10) $0, \dfrac{(-1)^r}{4(2r+1)!}(3 - 3^{2r+1})x^{2r+1}$, all x

11) $y + \frac{1}{2}y^2$

12) y^2

13) $1 + y$

14) $\sin a + y \cos a - \dfrac{y^2}{2}\sin a$

15) $\dfrac{1}{x} - \dfrac{1}{2x^2} + \dfrac{1}{3x^3}, x \geqslant 1, x < -1$

16) $\ln 2 - \dfrac{1}{2x} + \dfrac{3}{8x^2} - \dfrac{7}{24x^3}, x \geqslant \frac{1}{2}, x < -\frac{1}{2}$

17) $2.718\,281\,8$

19) 0.693

20) $0.0953, -0.1054$

21) $1 + x$

22) $x + \dfrac{x^3}{3!}, 3.1$

24) $(x-1)\cos 1^c -$
$\dfrac{(x-1)^2}{2}(2\sin 1^c + \cos 1^c)$

25) $\dfrac{1}{2} + \dfrac{\sqrt{3}}{2}\left(x - \dfrac{\pi}{6}\right) - \dfrac{1}{4}\left(x - \dfrac{\pi}{6}\right)^2, 0.515$

26) $x = 0$

Multiple Choice Exercise 16 — p. 595

1) c   2) b   3) a   4) d
5) e   6) b   7) c   8) d

9) a   10) c   11) b, c   12) a
13) a, c   14) a, b   15) a   16) A
17) D 18) B 19) D 20) D
21) T 22) F 23) T 24) T
25) T 26) F 27) F 28) T

Miscellaneous Exercise 16 — p. 597

1) $S_n = \dfrac{n(3n+1)}{2}, n = 7$

2) $6n - 3$ 4) $20°$

5) a) 63 b) 8.7 6) 11

7) £82, 7 years   8) $29\,\text{mm}$

9) b) 5

10) $A = 25, B = -9, -\frac{37}{15}x^3, |x| < \frac{3}{2},$

11) $\dfrac{[1 - (n+2)x^{n+1} + (n+1)x^{n+2}]}{(1-x)^2}$

12) $\frac{1}{4}, 6, -\frac{9}{16}$ 14) $\frac{1}{3}$

15) $35, 27$ 16) 120

17) $1 + 2x + 7x^2 + 20x^3,$
$\frac{1}{4}[(-1)^n + 3^{n+1}]$

18) $z = (1 - 10c^2 + 5c^4)$
$+ i(5c - 10c^3 + c^5)$

19) $3.332\,222$ 21) $7.093\,97$

22) a) $a = \dfrac{1}{\sqrt{2}}, b = \dfrac{1}{4\sqrt{2}}, c = -\dfrac{5}{32\sqrt{2}}$

b) $-\frac{1}{2}\ln 2, \dfrac{(-1)^n}{2n}\left(\dfrac{1}{2^n} - 1\right)$

23) a) $1 + (p-q)x + (\frac{3}{2}q^2 - pq - \frac{1}{2}p^2)x^2,$
$1 + \frac{1}{2}(p-q)x$
$+ (\frac{3}{8}p^2 - \frac{1}{4}pq - \frac{1}{8}q^2)x^2$
b) $-\frac{7}{5}$

24) $\dfrac{1}{x^2+1} - \dfrac{1}{(x+1)^2}$
a) $2x - 4x^2 + 4x^3$
b) $-2x - \frac{2}{3}x^3 + x^4$

25) $a = 1, b = 0$ or $a = -1, b = 2$

26) a) $2x + \dfrac{2x^3}{3} + \dfrac{2x^5}{5}$

27) $-\dfrac{3x}{2} - \dfrac{3x^2}{4} - \dfrac{3x^3}{2}, -\frac{1}{2} \leqslant x < \frac{1}{2}$

28) $1 - 2x + 2x^2 - \frac{4}{3}x^3,$
$(-2x)^{r-1}/(r-1)!,$
$-\dfrac{2x}{r}, k = 2, a = -1$

29) a) $\dfrac{d}{2}(m+1)$ b) $\frac{59}{110}$
c) $-0.030\,459$

30) $|y| > 1$ 31) $\frac{2}{3}x^3$

32) $-\dfrac{x^2}{2} - \dfrac{x^4}{12}$

33) a) $2, 4, 1 + 2x + 2x^2$
    b) $1, 2, 1 + (x - 1) + (x - 1)^2$

34) 0.955 35) $\frac{1}{6}\pi - \frac{1}{6}\sqrt{3}x^2$

36) $x + \frac{1}{2}x^2 + \frac{1}{3}x^3 + \frac{13}{24}x^4$

37) $-\frac{1}{2}x^2 - \frac{1}{12}x^4$, $\ln 2 \simeq 0.68$,
    all terms negative

38) $2x + \frac{4}{3}x^3$

39) $(x - 1) + \frac{3}{2}(x - 1)^2 + \frac{1}{3}(x - 1)^3$
    $- \frac{1}{12}(x - 1)^4$

Exercise 17a — p. 608

1) $\frac{5}{8}$ 2) $\frac{5}{18}$ 3) $\frac{15}{17}$

4) a) $\frac{5}{12}$ b) $\frac{7}{12}$ c) $\frac{1}{12}$

5) $\frac{1}{2}$ 6) a) $\frac{1}{2}$ b) $\frac{5}{6}$

7) a) $\frac{3}{5}$ b) $\frac{3}{10}$ c) $\frac{9}{10}$

8) a) $\frac{1}{35}$ b) $\frac{22}{35}$

9) $\frac{3}{5}$ 10) $\frac{2}{5}$

Exercise 17b — p. 619

1) a) $\frac{1}{4}$ b) $\frac{3}{4}$ c) $\frac{1}{2}$

2) a) $\frac{1}{8}$ b) $\frac{7}{8}$

3) a) $\frac{1}{5}$ b) $\frac{48}{125}$ c) $\frac{369}{625}$

4) $(0.8)^{10} = 0.1074$

5) a) $\frac{3}{8}$ b) $\frac{1}{64}$ c) $\frac{63}{64}$

6) a) $\frac{1}{36}$ b) $\frac{1}{18}$ c) $\frac{11}{12}$

7) a) $\frac{1}{4}$ b) $\frac{1}{16}$ c) $\frac{63}{64}$

8) a) $\frac{1}{2}$ b) $\frac{1}{4}$ c) $\frac{1}{8}$

9) a) $\dfrac{35}{(36)^2} = 0.027$ b) $\dfrac{35^2}{36^3} = 0.026$

10) $\frac{2}{3}$ 11) $\frac{30}{91} = 0.330$

12) a) $\frac{1}{13}$ b) 0.068 c) 0.056

13) $\frac{18}{125}, 4$ 14) $\frac{1}{20}, 14$

15) 1267 16) a) $\frac{7}{45}$ b) $\frac{4}{9}$

17) a) $\frac{3}{7}$ b) $\frac{1}{3}$

Exercise 17c — p. 622

2) $\frac{1}{8}$ 3) $\frac{1}{6}$

Exercise 17d — p. 625

1) $\frac{1}{8}$ 2) $\frac{1}{8}$ 3) a) $\frac{2}{5}$ b) $\frac{13}{25}$

4) $\frac{7}{16}$ 5) 0.1030 6) $\frac{45}{101}$

Exercise 17e — p. 631

1) a) $\frac{1}{8}$ b) $\frac{3}{16}$

2) a) $\frac{1}{6}$ b) $\frac{3}{4}$ c) $\frac{7}{18}$

3) a) $\frac{11}{20}$ b) $\frac{3}{20}$ c) $\frac{4}{5}$

4) 0.3196

5) a) $\frac{1}{2}$ b) $\frac{1}{4}$ c) $\frac{1}{4}$

6) $\frac{26}{27}$ 7) $\frac{7}{11}$

8) $\frac{1}{12}$ 9) $\frac{7}{10}, \frac{11}{14}$

10) $\frac{4}{5}$, no as $P(B|A) \neq 0$

Exercise 17f — p. 636

1) 2.78 2) 0.28 3) 6

4) a) $\frac{3}{8}$ b) $\frac{1}{2}, 25$

5) 2.19 6) 20 7) 1.7

8) 2.2 9) £32.97   10) 25 p

11) 42 p   12) £1.10

Multiple Choice Exercise 17 — p. 637

1) c   2) d   3) d

4) b   5) c   6) e

7) a   8) b   9) c

10) b   11) b, c   12) a, b, c

13) a, c   14) a   15) a, d

16) C   17) E   18) D

19) C   20) c   21) I

22) c   23) A   24) F

25) F   26) T   27) T

28) F   29) F   30) F

Miscellaneous Exercise 17 — p. 640

1) $\frac{1}{36}$

2) a) (i) 10   (ii) 200   (iii) 80
   b) (i) $\frac{125}{216}$ (ii) $\frac{91}{216}$ (iii) $\frac{25}{72}, \frac{1}{36}$

3) 267 (274 from tables)

4) $p(B) = \frac{7}{12}, p(B|A) = \frac{5}{8}, p(B|\bar{A}) = \frac{15}{28}$,
   not independent and not mutually
   exclusive

5) a) $\frac{5}{6}$ b) 1   c) $\frac{6}{5}$ d) $\frac{14}{5}$

6) $pq(p^{r-2} + q^{r-2}), 2\frac{3}{4}$

7) $0.28, 0.11$

8) a) $\frac{2}{5}$
   b) (i) p^3,   (ii) $p^3 + q^3$,   (iii) $3p^3q$
      (iv) $3p^3q + 3pq^3, \frac{64}{81}$

9) a) 6   b) (i) $\frac{2}{5}$ (ii) $\frac{1}{3}$ (iii) $\frac{28}{75}$ (iv) $\frac{38}{75}$

10) $\frac{32}{59}$

11) a) $\dfrac{52!}{39! \, 13!}, \dfrac{4(39! \, 13!)}{52!}$

    b) (i) ab (ii) $a(1 - b)$
       (iii) $(1 - a)(1 - b)$
       (iv) $(1 - a)^n(1 - b)^n, 35$

12) $\frac{2}{7}$, a) $\frac{34}{35}$ b) $\frac{4}{35}$

13) a) $5p^4 - 4p^5$
    b) $1 - np(1 - p)^{n-1} - (1 - p)^n$

14) a) (i) 48   (ii) 288
    b) $\frac{1}{21}, \frac{32}{147}$

15) a) 1320   b) $(\frac{8}{9})^4$ c) $\frac{1}{6}(\frac{5}{6})^{r-2}$

16) a) $\frac{1}{3}$
    b) $\{4, 5, 6, 7, 8\}, \frac{1}{4}, \frac{1}{4}, \frac{5}{16}, \frac{1}{8}, \frac{1}{16}, 5.5$

17) b) (i) $\frac{1}{2}$ (ii) $\frac{1}{6}$

18) $0.01, 0.87, 0.96, 0.085$
19) 0.914
20) $\frac{89}{91}$ 21) 0.322 22) $\frac{64}{81}$

Exercise 18a — p. 651

1) a) $Y = aX + b, Y = \dfrac{1}{y}, X = x, m = a,$
    $c = b$
  b) $Y = y^2 - x, X = y, m = b, c = -a$
  c) $X = e^x/y, Y = y, m = a, c = b$
2) $a = 2, b = -4$ 3) $a = 6, b = 4$
4) $a = 0.5, b = -2$ 5) $a = 30, b = 2$
6) $a = 2, b = \frac{1}{2}$ 7) $a = 3, b = -2$

Exercise 18b — p. 657

1) a) $\frac{32}{3}$ b) $4\frac{1}{2}$ c) $\frac{125}{6}$
2) a) $\frac{4}{3}$ b) $\frac{32}{3}$
3) a) 1 b) $3 - \ln 4$
4) a) $\frac{4}{3}$ b) $\frac{4}{3}$
5) $4\sqrt{3}$

Exercise 18c — p. 660

1) $-\frac{1}{6}$ 2) $2 \ln 2$

3) $\frac{1}{4}(e^{-1} - e^{-5})$ 4) $\dfrac{2}{\pi} \ln 2$

5) $\frac{1}{2}$ 6) $\ln \frac{4}{3}$

7) $\dfrac{4}{3\pi}$ 8) e^2

9) $\dfrac{\pi}{4}$ 10) $\dfrac{\pi}{3}$

Exercise 18d — p. 667

1) $\dfrac{512\pi}{15}$ 2) $\dfrac{\pi}{2}(e^6 - 1)$

3) $\dfrac{\pi}{2}$ 4) $\dfrac{64\pi}{5}$

5) 2π 6) 8π

7) 8π 8) $\dfrac{3\pi}{5}(\sqrt[3]{32} - 1)$

9) $\dfrac{\pi}{2}(e^2 - 1)$ 10) $\dfrac{16\pi}{15}$

11) $\dfrac{16\pi}{15}$ 12) $\dfrac{\pi^2}{2}$

13) $\dfrac{3\pi}{10}$ 14) 8π

15) $\pi(32\sqrt{2} - 40)/15$

Exercise 18e — p. 671

1) a) 2 b) $\dfrac{\pi}{4}$ c) $\left(\dfrac{\pi}{2}, \dfrac{\pi}{8}\right)$

  d) $\dfrac{\pi^2}{2}$ e) $\dfrac{\pi^3}{4}$ f) $\left(\dfrac{\pi}{2}, 0\right)$

2) a) $\frac{1}{6}$ b) $\frac{1}{60}$ c) $(\frac{1}{2}, -\frac{1}{10})$

  d) $\dfrac{\pi}{30}$ e) $\dfrac{\pi}{60}$ f) $(\frac{1}{2}, 0)$

3) a) $\frac{32}{3}$ b) $\frac{128}{5}$ c) $\frac{12}{5}$

  d) 8π e) $\dfrac{64\pi}{3}$ f) $(0, \frac{8}{3})$

4) a) $e - 1$ b) 1 c) $\dfrac{1}{e-1}$

  d) $\dfrac{\pi}{2}(e^2 - 1)$ e) $\dfrac{\pi}{4}(e^2 + 1)$

  f) $\left(0, \dfrac{e^2 + 1}{2(e^2 - 1)}\right)$

5) $\dfrac{e^2 + 1}{e^2 - 1}$

Exercise 18f — p. 679

1) a) 1.049 b) 1.006
2) a) 0.3887 b) 0.3891
3) a) 1.178 b) 1.047
4) a) 0.2727 b) $0.271\,15$
5) a) $0.990\,33$ b) 0.9996
6) a) $1.148\,38$ b) $1.147\,79$
7) $0.004\,842$
8) $0.000\,202\,69$
9) $0.099\,215\,56$
10) 1.0387

Exercise 18g — p. 683

1) a) 2.0833 b) 3.037
  c) 5.0133
2) a) $0.874\,77$ b) 0.4848
  c) $0.719\,48$

3) $\left[\dfrac{x}{1 + x} + \ln(1 + x)\right]\delta x, 0.72$

4) $(\sec^2 x)\,\delta x, 1 + \dfrac{\pi}{16}$

5) 2.0125 6) 0.2575 cm/s
7) -0.1011 cm³/s   8) 30 cm³/s

Exercise 18h — p. 687

1) $17.7°$ 2) 8.26 cm
3) $91.8°$ or $38.2°$ 4) 6.4 cm

5) $B = 28.4°, C = 88.6°, c = 10.23$ cm
6) $A = 99.4°, C = 61.3°, a = 6.45$ cm or
    $A = 42°, B = 118.7°, a = 4.37$ cm
7) $A = 53.1°$ $b = 24.7$ cm, $c = 17.2$ cm
10) 18.7 m

Exercise 18i — p. 689

1) $A = 34°, B = 101.5°, C = 44.5°$
2) $a = 30.3$ cm, $B = 61°, C = 40°$
3) $b = 34$ cm, $A = 23.4°, C = 35.6°$
4) $B = 26°, A = 121.2°, C = 32.8°$
6) $29°, 46.6°, 104.4°$

Exercise 18j — p. 695

1) a) 20.1   b) 55   c) 32.8
2) 4.32 m; 6.13 m
5) 28.5°
6) 13.6, 10.1; 2.5
7) 342.5 km; 056.6°

Exercise 18k — p. 703

1) 72.6 m
2) a) 12.4°   b) 12°
3) 58.5°
4) 934.3 m
5) 70.5°
6) a) $6\frac{2}{3}$ m   b) 8 m   c) 46.2°
    d) 277.7 m²

Miscellaneous Exercise 18 — p. 703

1) $k = 0.2, n = 0.5$
2) $a \doteq 12.6, b = 1.12$
3) $a = 0.75, b = -0.6$
5) $a:b:c = 2:1:2$
6) $a = 2, b = 36$
7) $501, -3$ 8) $4\frac{2}{3}$
9) $\frac{1}{2}e - 1, \pi(\frac{1}{6}e^2 - \frac{1}{2}), 2\pi(1 - \frac{1}{3}e)$
10) $2, \frac{1}{2}\pi^2$
11) a, min, $a \geqslant \dfrac{1}{e}, -1 + 3\ln 2$,

    $(1 + 8\ln 2)/(12\ln 2 - 4)$

12) $(\cos 2\theta - \sin\theta)/\cos\theta, \left(\dfrac{1}{2}, \dfrac{3\sqrt{3}}{4}\right)$,

    $\left(\dfrac{1}{2}, -\dfrac{3\sqrt{3}}{4}\right), (1, 0), \dfrac{13\pi}{10}$

13) $\dfrac{a^2\pi}{4}, 16\pi, \dfrac{8}{3}, \dfrac{8}{3\pi}$

14) $(-1, 0), \frac{1}{5}$
15) $\frac{13}{6}a^2, \frac{59}{15}\pi a^3$
17) a) $2 - \ln 2$ b) $\dfrac{2\ln 2}{2 - \ln 2}$

    c) $\pi(\frac{9}{4} - 2\ln 2)$
18) 40
19) $\frac{16}{3}, 8\pi, (0, \frac{20}{3})$
20) a) $\ln\dfrac{kx}{\sqrt{(1-x^2)}}$ b) $\dfrac{16}{15\pi}$ c) 0.945
21) 1.77
22) $3.91, 31.3, \pi \simeq 3.13$
24) 7.512, 30.005
25) $\dfrac{1}{3x^{2/3}}, \dfrac{\delta x}{3x^{2/3}}, 10.003$

26) a) 0.879,   b) 0.879
27) $(\frac{195}{76}, \frac{633}{190})$ 28) 0.95
29) $\sqrt{3}\,\epsilon/30$ 30) 0.0228

31) a) $\left[\dfrac{x^2}{1+x} + 2x\ln(1+x)\right]\delta x$

    b) 0.000 024 05
32) a) -0.157 b) 0.29
33) 150° 42'
34) $2a; \frac{1}{7}a\sqrt{21}$
36) 170°, 116.6°, 73.4°
37) $2, 2\sqrt{3}, 2\sqrt{7}; \frac{1}{7}\sqrt{21}$
38) 131°, $\frac{3}{2}h$
39) 45.6°
40) a) 2.29   b) 14° 44'   c) 29° 47'

INDEX